S01470

AF598408

Seismic Imaging of Carbonate Reservoirs and Systems

Edited by

Gregor P. Eberli
Jose Luis Masaferro
J. F. "Rick" Sarg

AAPG Memoir 81

Co-Published by

The American Association of Petroleum Geologists
Tulsa, Oklahoma, U.S.A.
and
Shell International Exploration & Production B.V.
The Netherlands

Printed in Canada

ISBN: 0-89181-362-4

AAPG Editor: John C. Lorenz
Geoscience Director: J. B. "Jack" Thomas

This publication is available from:

The AAPG Bookstore
P.O. Box 979
Tulsa, OK U.S.A. 74101-0979
Phone: 1-918-584-2555 or 1-800-364-AAPG (U.S.A. only)
Fax: 1-918-560-2652 or 1-800-898-2274 (U.S.A. only)
E-mail: bookstore@aapg.org
www.aapg.org

Table of Contents

Acknowledgments

The American Association of Petroleum Geologists thanks the following for their generous contribution to *Seismic Imaging of Carbonate Reservoirs and Systems*

Shell International Exploration & Production B. V.

Contributions are applied toward the production costs of publication, thus directly reducing the book's purchase price and making the volume available to a larger readership.

About the Editors

Gregor P. Eberli is Professor at the Division of Marine Geology and Geophysics at the University of Miami. He received his Diploma (M.S. equivalent) and Ph.D. degree (1985) from the Swiss Institute of Technology (ETH) Zürich, Switzerland. He started to work on seismic interpretation during his post-doctoral tenure at the Comparative Sedimentology Laboratory at the University of Miami in 1985–1987. Upon his return to the ETH, he began incorporating seismic sequence stratigraphic concepts in his fieldwork and constructing synthetic seismic models from outcrops. He joined the University of Miami as a faculty in 1991. His analysis of the seismic stratigraphy of Great Bahama Bank sparked two scientific drilling campaigns; the Bahamas Drilling project in 1990 and the Ocean Drilling Program Leg 166 in 1996 on which he served as a co-chief scientist. In his laboratory, he has conducted several petrophysical experiments that now explain the large sonic velocity spectrum in carbonates and help to better understand the log and seismic signature of carbonates. In 2000, he spent a year with the Carbonate Team at Shell Rijswijk, where he became familiar with the newest seismic visualization techniques and realized the power of seismic images to reveal the architecture of carbonate systems. He is currently Chairman of the Division of Marine Geology and Geophysics and Head of the Comparative Sedimentology Laboratory.

Jose Luis Masaferro is a geologist/seismic interpreter since 1998 at Shell Technology Applications and Research Center (Carbonate Development Team) in The Netherlands. He received his Ph.D. from the University of Miami in 1997. The topic of his dissertation was on the interplay of tectonism and carbonate sedimentation in the Bahamas/Cuban foreland. He was a Fulbright Fellow while pursuing his studies in Miami. He also worked as a geologist for Texaco Petrolera Argentina (1989–1991). While in Shell he has been involved in different projects including 3-D seismic volume interpretation, high-resolution sequence stratigraphy and kinematic modeling of compressional structures. Masaferro's research interest is to intergrate 3-D seismic data with borehole data to interpret carbonate depositional systems and translate them into reservoir subsurface models.

J.F. 'Rick' Sarg is currently Stratigraphy Coordinator at the ExxonMobil Exploration Co. He received his Ph.D. in Geology from the University of Wisconsin, Madison, in 1976. MS (1971) and BS (1969) in Geology, University of Pittsburgh, Pittsburgh, Pennsylvania. Sarg has twenty-six years petroleum exploration and production experience in research, supervisory, and operational assignments with Mobil (1976), Exxon (1976–90), Independent Consultant (1990–92), Mobil Technology Company (1992–99) where he attained position of research scientist, and is now with ExxonMobil Exploration (2000–present). He was a member of the exploration research group at Exxon that developed sequence stratigraphy, with an emphasis on carbonate sequence concepts. His involvement in many projects provided him with a worldwide experience in integrated seismic-well-outcrop interpretation of siliciclastic and carbonate sequences. He has authored or co-authored 28 papers on carbonate sedimentology and stratigraphy.

Eberli, G. P., J. L. Masaferro, and J. F. "Rick" Sarg, 2004, Seismic imaging of carbonate reservoirs and systems, *in* Seismic imaging of carbonate reservoirs and systems: AAPG Memoir 81, p. 1–9.

Seismic Imaging of Carbonate Reservoirs and Systems

Gregor P. Eberli
University of Miami, Miami, Florida, U.S.A.

Jose Luis Masaferro
Shell E&P Technology Applications and Research (SEPTAR), Rijswijk, The Netherlands

J. F. "Rick" Sarg
ExxonMobil Exploration Co., Houston, Texas

Recent advances in seismic acquisition, processing, and visualization techniques provide the opportunity to image carbonate reservoir architecture with unprecedented resolution. In particular, the increase in three-dimensional (3-D) seismic data acquisition and the improvements in processing techniques have contributed to these advances and have resulted in higher-resolution imaging of sedimentary bodies. In addition, the analysis of seismic attributes is a developing methodology to quantify the volumes and rock properties of these bodies (Masaferro et al., 2004). These advances in volume visualization that allow imaging of the morphology of ancient carbonate systems rivals images from satellite data in the modern environments (Figure 1). The additional advantage of seismic data is that the imaged deposits can be displayed at various stratigraphic levels thereby documenting the evolution of the depositional environment through time. The paleogeomorphology can now be accurately imaged for carbonate systems constructed by extinct reef builders that have no modern analogs. This capability offers the unique opportunity to exploit 3-D images for questions regarding the growth pattern of different reef communities, their paleoecology, and reservoir heterogeneities in ancient systems. In addition, analyses of seismic attributes are a developing methodology to quantify the volumes and rock properties of sedimentary bodies. When relating the stratigraphic architecture to the forming processes, such information is invaluable to geologists who are often limited to two-dimensional (2-D) sections in outcrops or to the plan view in the modern depositional environment. Deciphering sedimentary processes based on the mutual feedback between seismic data and modern and ancient analogs in outcrops is termed *seismic sedimentology* (Schlager, 2000).

The 16 chapters presented in this volume demonstrate how the combination of geophysical and geological data sets is a powerful method for the understanding of the architecture and heterogeneities in carbonate depositional systems. They provide examples of seismic sedimentology and capture the dynamics of the carbonate system on large exploration scale and on a small reservoir scale. Several chapters show how this knowledge can be directly applied to build improved 3-D reservoir models.

STRENGTH OF SEISMIC DATA

Geometry is probably the most important information that seismic data immediately provides after initial processing. The analysis of geometrical relationships of reflections and reflection terminations resulted in the formulation of the concept of seismic stratigraphy and sequence-stratigraphic facies models. It first improved the prediction in large-scale exploration and is now common practice in reservoir characterization (Vail et al., 1977; Sarg, 1988; Stoudt and Harris, 1995; Kerans and Tinker, 1997; Homewood et al., 2000). Three-dimensional seismic data reveal geometries of subsurface depositional features that are barely visible on

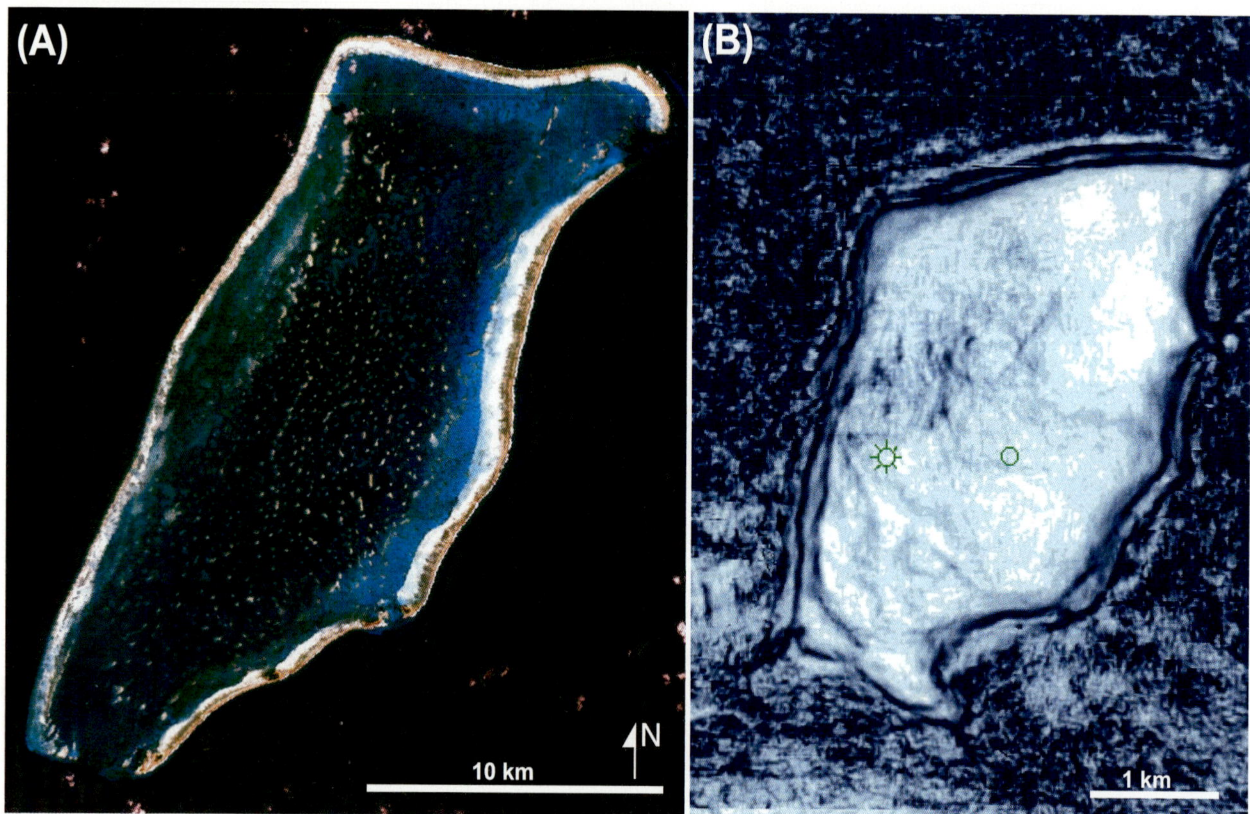

FIGURE 1. Comparison of seismic and satellite data. (A) Satellite image of Glovers Reef offshore Belize displaying the reef rim, white back-reef carbonate sands, and a lagoon with numerous small patch reefs (photo provided by Gene Rankey). (B) The 3-D semblance volume of a Luconian middle Miocene buildup depicts facies belts with high semblance values (dark blue) that correspond with the reef seismic facies, intermediate semblance in the adjacent back-reef deposits, and the low semblance values of the lagoonal seismic facies. Note that the seismic image is capable of recording the breaks in the reef rim. In addition, the seismic image displays three reef facies that document the backstep of the reef margin through time (from Masaferro et al., 2003).

2-D seismic data sets (e.g., Weimer and Davis, 1996). Consequently, subsurface strata that had been interpreted as having a horizontal layering based on core and log data need to be re-evaluated when high-quality seismic data becomes available. For example, Droste and Van Steenwinkel (2004) show that the Cretaceous platforms of Oman have a complex internal architecture rather than a "layer-cake" configuration. Specifically, they show how the Arabian platform margin (the Habshan Formation) in central Oman is composed of a series of clinoforms whose lateral juxtaposition is controlled by relative changes of sea level. They further document that the intrashelf carbonates of the Natih Formation consist of a series of separate platforms that merged by lateral accretion. Melville et al. (2004) demonstrate that, even in the presence of a high density of wells, depositional geometries and patterns from 3-D seismic data provide important new paleogeographic information, in this case for the Bab (Aptian age) and Asab (Late Jurassic) levels. Masaferro et al. (2004) calculated volume dip and azimuth from the seismic cubes to detect previously unnoticed subtle stratigraphic features, such as low-angle progradation units and shoal-type mound seismic facies in the Permian Khuff and Upper Cretaceous Natih E reservoirs in Oman.

Geometries are paramount to delineate facies distribution and internal sequence architecture that, integrated with core, wire-line log suites, and 3-D acoustic impedance cubes, are used to build robust reservoir models. For example, Tinker et al. (2004) show how geometries in combination with facies models derived from cores and outcrops can help to identify the position of the primary reservoir in a sequence-stratigraphic framework. Similarly, 3-D seismic surveys across the upper Paleozoic carbonate buildup on the Horseshoe Atoll, west Texas, display its morphology in detail that would be impossible with well control alone (Saller et al., 2004). Three-dimensional seismic data show that the buildup is composed of stratiform sequences and that porosity and permeability are continuous within these stratiform units. In faulted reservoirs, combining seismic facies, seismic attributes (semblance volumes), and reflection termination is often the key for distinguishing between structural and stratigraphic features (Masaferro et al., 2004).

Karst Features

Repeated exposure and the associated diagenetic changes of carbonate platforms have a major impact on porosity evolution and reservoir quality (Budd et al., 1995; Neuhaus et al., 2004; Vahrenkamp et al., 2004). Tropical karst landscapes exhibit a predominance of positive relief features, for example, tower, cone, polygonal, and pinnacle karst. Although these features are up to 300 m in relief, they are rarely observed on seismic sections (Purdy and Waltham, 1999). Large-scale negative dissolution features, such as sinkholes, caves, and karst collapse, are equally common in carbonate environments. The irregular shape and distribution of these features reflect seismic waves in a chaotic matter, which makes it hard to image them precisely, especially with a 2-D acquisition. On 2-D seismic data, subaerial exposure and negative karst features often need to be inferred from their association with unconformities and erosional truncations or the presence of a chaotic seismic facies (Sarg, 1988; Handford and Loucks, 1993; Moldovanyi et al., 1995). In other cases, concave up sinkhole reflections or irregular hummocky reflection configuration directly indicates negative karst features (Moldovanyi et al., 1995; Macurda, 1997). Three-dimensional data largely overcome the problem of deflected seismic signals in karstified terrains, especially if modern visualization techniques are used in the postprocessing treatment of the data.

Several of the studies in this volume not only identify karst features but display their lateral distribution and arrangement. Isern et al. (2004) describe positive karst features on the Northern Marion Platform. The Miocene–Pliocene carbonate platform systems of the Southeast Asia have undergone multiple Miocene karst events and contain abundant karst features. In the Jintan Platform, offshore Sarawak, several depositional cycles are overprinted by karst (Vahrenkamp et al.,

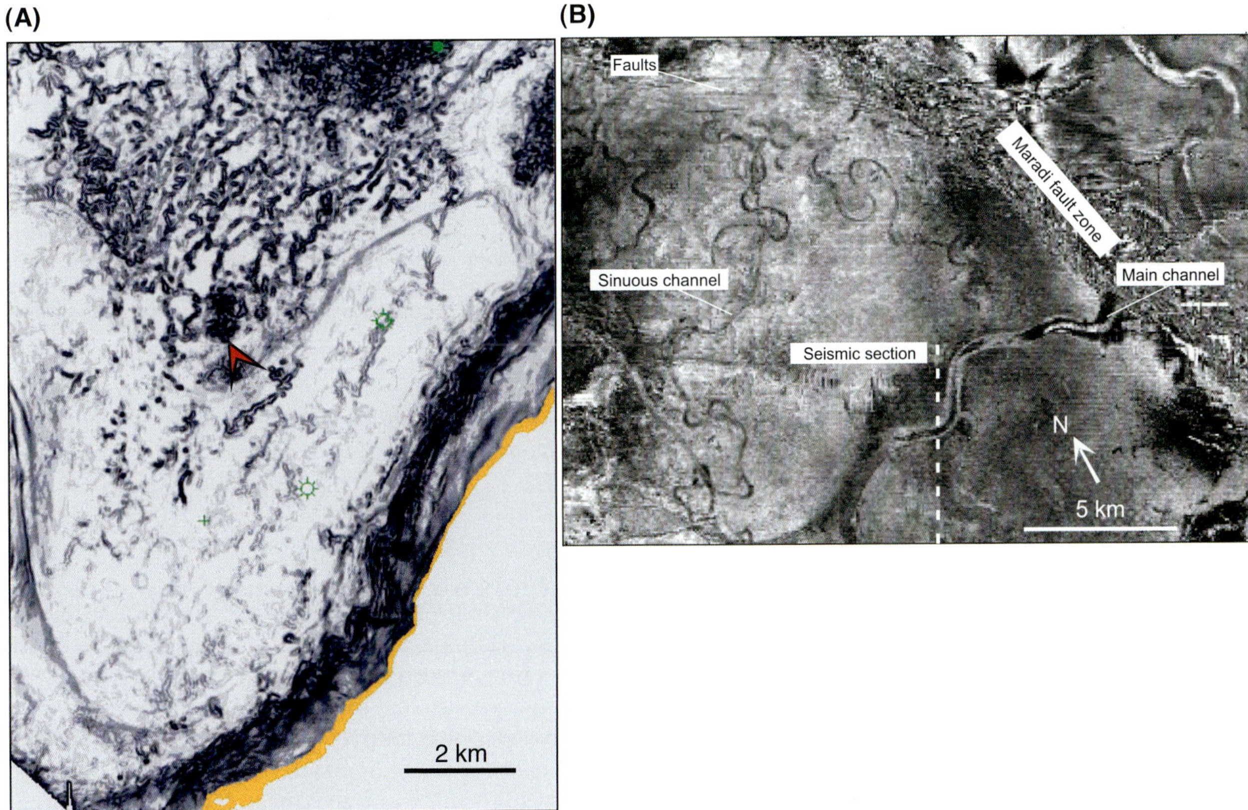

FIGURE 2. Karst features on seismic data. (A) Karstification of Jintan platform Z2 shows a dendritic pattern that resembles in some places a drainage pattern. A dark area (arrow) is interpreted as a collapsed cave in Zone 1A/B at intersecting faults (from Vahrenkamp et al., 2004). (B) Channel complex cut into top of the Natih Formation displayed on an amplitude map along time slice 44 ms below top Natih horizon in north Oman showing a complex of narrow sinuous channels and a wider northeast-southwest–trending main channel (from Droste and Van Steenwinkel, 2004).

2004). The karstification patterns vary widely from pervasive with large karst holes to more localized with a dentritic pattern or even a drainage pattern (Figure 2A; Vahrenkamp et al., 2004). Heubeck et al. (2004) detect on seismic data repeated exposure of a Miocene bank in the Pearl River Mouth Basin by the widespread incipient collapse at or below the scale of seismic resolution and the formation of numerous large sinkholes.

Rivers that meander toward the shelf break during times of lowered sea level often incise attached carbonate ramps and shelves (Esker et al., 1998). Droste and Van Steenwinkel (2004) provide a semblance time slice from the top of the Natih Formation that illustrates such channel incisions on the flat exposed platform (Figure 2B). They recognize several generations of highly sinuous channels that cut in places more than 150 m into the underlying carbonates and reach a width of several hundred meters. The sinuous pattern of the channel is remarkably similar to a meandering stream channel.

Structure, Faults, and Fractures

Reconstructing faults and structure in their accurate 3-D space is a major challenge in subsurface geology. Accurate knowledge of the kinematics and geometry of a structure is, however, essential for volumetric estimations of reservoirs and for the prediction of the orientation, distribution, and density of fractures that formed within this structural framework but are below seismic resolution. Three-dimensional seismic data has become the preferred tool for this task because it allows the interpreter to follow faults and folds throughout the seismic volume. Melville et al. (2004) describe how careful processing improved the seismic quality across an onshore field in Abu Dhabi to precisely image the anticline and associated faulting. Masaferro et al. (2004) point out that modern filtering techniques are needed to take full advantage of the 3-D seismic data for structural analysis. Combined volume dip and azimuth calculated from the seismic cubes and semblance volumes that highlight reflection terminations are such visualization techniques.

Faults and the associated fractures can act as conduits. In carbonates, this quality is especially important, because diagenetic alteration from migrating fluids can change porosity, mineralogy, and reservoir quality. Vahrenkamp et al. (2004) give a good example for focused dissolution (karst) along intersecting faults (Figure 2A). Likewise, faults are the conduits for ascending hydrothermal fluids that can significantly alter the invaded strata. Two studies in this volume document the influence of migrating fluids along faults and their interaction with the stratigraphic architecture on reservoir quality and heterogeneity. Pearson and Hart (2004) used seismic attributes to predict the porosity distribution in the Red River Formation of the Williston Basin. They detected enhanced porosity along faults and flexures that they attributed to fractures that acted locally as preferential pathways for dolomitizing fluids. In the South Dagger Draw field, New Mexico, the primary reservoir is in vuggy algal biostromes and bioherms. The vugs formed by acidic, hydrothermal fluids that migrated upward along joints and were baffled below shales, resulting in stratigraphically defined dissolution zones (Tinker et al., 2004). Although fractures are below seismic resolution, the resulting fracture porosity can potentially be extracted from seismic data. Pranter et al. (2004) characterize fracture and matrix porosity by using 3-D multicomponent seismic data and wire-line logs. They infer fracture density and orientation from the shear-wave anisotropy and use this information to modify permeability models in regions with open fractures.

Geometry of Ancient Reef Builders

Corals are the major reef builder in the Neogene, Pleistocene, and Holocene. The modern coral reefs are the only place where the relationship among water depths, wind and current directions, and the resultant arrangements and dimensions of facies belts can be studied. Consequently, models for ancient reefs rely heavily on these observations. However, the biological communities that form reefs and their associated carbonate environments have changed through time (e.g., James, 1983). Many of the main reef builders are extinct today and their 3-D architecture and lateral distribution are assessed from limited 2-D outcrops. Modern 3-D seismic data provide us with images of these ancient depositional environments that can refine facies models of ancient carbonate communities. For example, algal biostromes and bioherms were dominant during the Upper Carboniferous, and their geometries are vastly different from modern reefs as they form thin (10–20 m thick) laterally extensive buildups with flat lower and knobby upper surfaces (Eberli et al., 2000; Saller et al., 2004; Tinker et al., 2004).

Controls on Depositional Geometries

The growth of a platform is controlled by the interaction of tectonic, eustatic, sedimentation, and climatic processes. Tectonic and eustatic processes combined cause relative changes in sea level, which control the space available for sediments to accumulate on the platform tops (accommodation space). Oceanographic and climatic conditions control the amount and types of sediment that is produced to fill the accommodation space with light, temperature, and nutrients being the most dominant controlling factors (Schlager, 1992). Tropical and subtropical carbonates are mostly produced by carbonate-secreting organisms that live in the photic zone (0–15 m), where light and warm waters are favorable for carbonate production. Carbonate production is highest in this zone and decreases exponentially with depth (Bosscher and Schlager, 1992, 1993). One consequence of the predominance of light-dependent sediment production is that carbonate ramps, shelves, and platforms maintain large areas at shallow water depths (Read, 1985). Although carbonate sediment production is high enough to fill the entire accommodation space up to mean sea level, large parts of carbonate platforms and shelves never fill all the available space. A certain water depth remains because wave, storm, and tidal energy and the resultant water motion suspends sediment and moves sediment within and out of the shallow water areas (Osleger, 1991). Hydraulic energies transport the excess sediment to the adjacent slope, which results in progradation of shelves and platforms (Wilson, 1975; Hine and Neumann, 1977; Bosellini, 1984; Sarg, 1988; Eberli and Ginsburg, 1989; Wilson and Roberts, 1995).

Sea level changes that occur on a variety of frequencies exert a dominant control on carbonate sedimentation (Kendall and Schlager, 1981; Sarg, 1988; Handford and Loucks, 1993). High-frequency sea level changes leave their record in sedimentary cycles on the platform top and in the basinal areas (e.g., Droxler et al., 1983; Goldhammer et al., 1987; Read, 1995; Kerans and Tinker, 1997; Williams et al., 2002). Low-frequency sea level changes are recognized as depositional sequences in outcrop and on seismic data (Sarg, 1988; Handford and Loucks, 1993; Eberli et al., 2001). Generally, carbonate production is highest during sea level highstands because the extent of platform and shelf flooding is greatest during such times. Platform/shelf aggradation with a shoaling trend in the vertical facies succession and maximum progradation occurs during these times. A fall of sea level exposes parts of the platform thereby eliminating the vertical aggradation and subjecting the exposed area to meteoric diagenesis. Depending on the slope profile, the fall decreases the area of sediment production and the amount of shedding to the slope. Reduced sedimentation on the slope during sea level lowstands promotes early diagenesis and the formation of hardgrounds and firmgrounds on the upper slopes (Grammer et al., 1993; Malone et al., 2001). In addition, margin and slope failure triggered by a variety of processes are common during these times (Sarg, 1988; Grammer et al., 1993). Furthermore, the downward shift of the production zone enables the establishment of lowstand terraces, or lowstand platforms on the former slope. For example, a lowstand platform developed during the late middle Miocene sea level fall along the Northern Marion Platform. It can be used to calculate the magnitude of this eustatic sea level fall (86 ± 30 m) in the late middle Miocene (Isern et al., 2002, 2004). During a sea level rise, the facies belts backstep or aggrade depending on the rate of sediment production and rate of accommodation space creation.

The large-scale architecture of carbonate platform and shelves is controlled by the interplay between the rate of relative sea level change and the growth rate (Figure 3; Tucker and Wright, 1990). The rate of relative sea level change is the combined result of the rate of eustatic sea level change and its interference with tectonic subsidence or uplift. The growth rate of the platform is dependent on the rate of carbonate sediment production and the amount of redistribution of the sediment. The rate of vertical growth is also directly related to the rate of relative sea level rise. Short-term carbonate sediment production rates are high and can match or exceed glacio-eustatic sea level rises but sedimentation rates and growth rates of carbonate platforms decrease as time interval increases (Schlager, 1981, 1999). A growth rate of 10^4 μm/yr is sustained during the first 1000 years but decreases to 10^2 μm/yr over 10^7 yrs (Schlager, 1999).

Decreasing growth rates and/or high rates of relative sea level changes cause platforms to backstep and ultimately drown (Figure 3). The respective role of these two parameters is often hard to discern from seismic and core data. In this volume, several authors describe drowned isolated platforms with varying interpretations for the causes of drowning. Bachtel et al. (2004) and Borgomano and Peters (2004) provide strong evidence that increased regional subsidence combined with local faulting caused the Segitiga Platform and Salalah carbonate margin, respectively, to step back and ultimately drown. Bachtel et al. (2004) base this conclusion on the fact that during the drowning, two smaller platforms on a more proximal position of the margin continued to prograde, thus documenting that the ocean environment maintained favorable conditions for carbonate production during this time. Two alternative scenarios are presented to explain the drowning of the middle Miocene Mega Platform in the Luconia Province, offshore Sarawak (Vahrenkamp et al., 2004). The first proposes drowning due to a combination of subsidence and eustatic sea level rise, whereas the second calls for a later drowning after a long period of exposure. In contrast, Bracco Gartner et al. (2004) explain the drowning of the EX Platform in the same province by a decrease of carbonate production. Exposure of platforms prior to drowning and long hiatuses between the shallow-water carbonates and the burying deep-water sediments is observed in the Reinecke Reef (Saller et al., 2004), the Mega Platform (Vahrenkamp et al., 2004), and in the Malampaya buildup (Neuhaus et al., 2004). Obviously, carbonate production on these platforms did not initiate while they subsided through the photic zone, usually

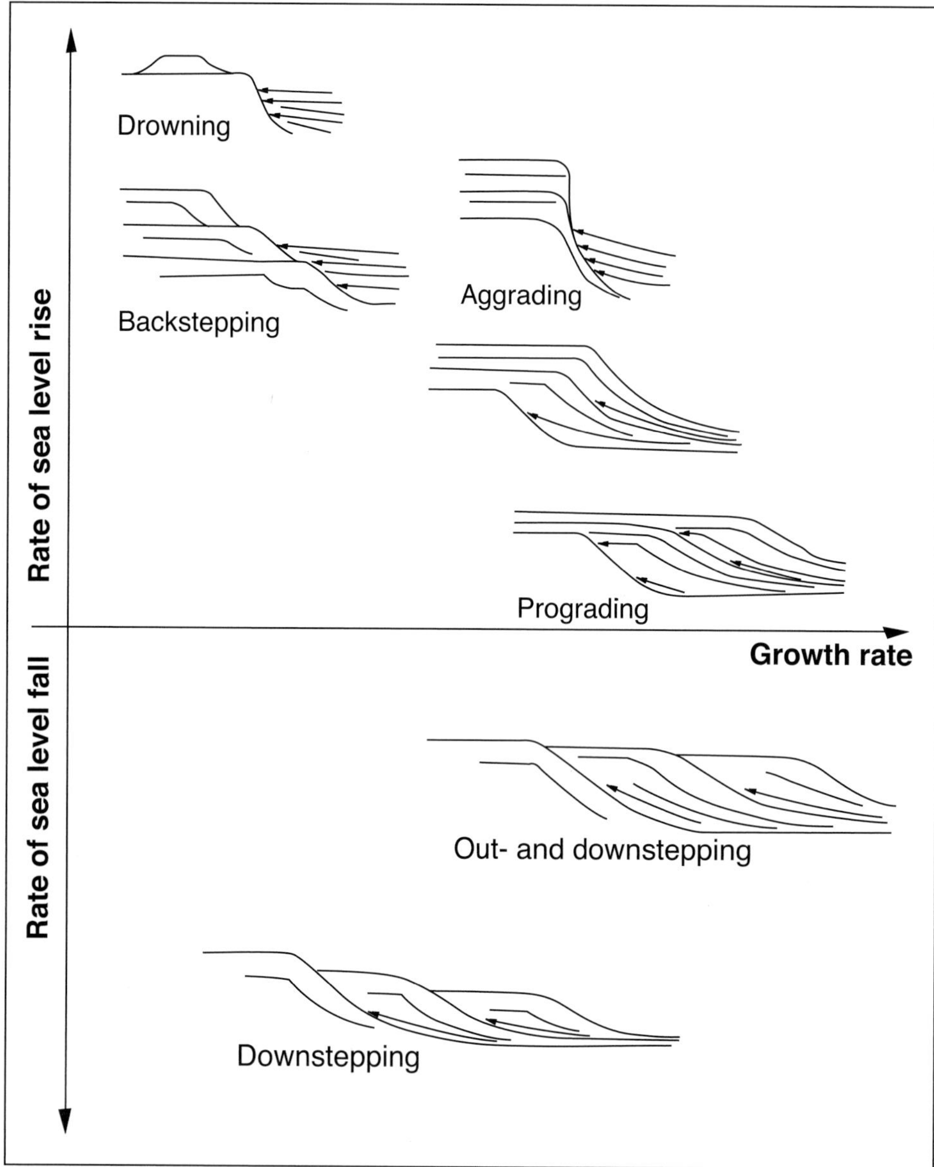

FIGURE 3. Schematic display of platform geometries in relation to rates of relative sea level and sediment production.

the zone of highest sediment production. In addition, currents must have swept all the pelagic sediments after drowning to produce the observed hiatus. This current sweeping still occurs on Southern Marion Platform offshore Australia, where an exposure horizon is covered with a hardground and thin, discontinuous veneer of pelagic sediment (Isern et al., 2002, 2004).

Most carbonate environments produce more sediment than can be accumulated on the platform top and excess sediment is transported offbank. The amount of offbank-transported sediment determines the lateral growth potential of carbonate platforms and shelves. Wind direction plays an important factor in where and how much sediment is deposited in the basinal areas (Hine and Neumann, 1977). Platforms that are subjected to unidirectional trade winds expand along the leeward side whereas the windward part of the platform, where offbank transport is minimal, might aggrade nearly vertical (Eberli and Ginsburg, 1989). In the Maldives, Miocene progradation of several bank margins is bidirectional as a result of seasonal switching of the wind direction after the establishment of the Indian Monsoon (Purdy and Bertram, 1993; Aubert and Droxler, 1996; Belopolsky and Droxler, 2004). Consequently, the lateral growth potential of platforms varies widely even for each side of a platform and asymmetric platform growth is common. In a similar way, ocean currents can cause an asymmetry in platform growth. Along the Southern Marion Platform, for example, south-flowing currents rework the offbank transported sediment and cause the platform to expand into a down-current direction (Isern et al., 2004). During the lifetime of a platform, wind or current regimes and the direction of platform expansion can change, resulting in complicated platform architecture. Bachtel et al. (2004) document how the progradation direction that led to the coalescence of the Segitiga Platform in the East Natuna Sea in Indonesia reversed from north to south in the late Miocene and then changed again during the drowning phase in the Pliocene.

In addition, the degree of progradation and the depositional slope angle are a function of the depth of the basin that the platform is filling. Platforms facing relatively deep basins develop steep slopes of 35° or more, whereas prograding slopes of platforms adjacent to shallow seas generally have gentle slopes of 1–10° (Sarg, 1988). In some cases, the platform margins can be too steep and/or the basin too deep for the platform to prograde and partial infilling of the basin is necessary before the platform is able to prograde (Eberli et al., 2004; Bachtel et al., 2004). Such margins show an evolution from steep, in extreme cases, escarpment-type margins with onlapping basinal sediment, to vertically aggrading and then finally prograding geometry (Figure 3). Decreasing rate of relative sea level change and the concomitant increase of sediment supply to the upper slope can cause the change from aggradation to progradation. Within third- and second-order cycle of relative sea level fall and rise, this turning point can represent the maximum flooding surface (Sarg et al., 1999).

Extensive progradation occurs during long-term lowering of relative sea level by stacking prograding clinoform packages laterally (e.g., Tyrrell and Davis, 1989; Pomar and Ward, 1995; Fitchen, 1997). Each shorter-term sea level cycle produces a prograding sequence but successively in a lower position (Figure 3). Such a downstepping geometry of a prograding complex is described in detail by Belopolsky and Droxler (2004) from the Maldives. The reflection on top of the sequences within the late Middle Miocene complex displays a basinward shift in onlap and/or a downstepping in each successive sequence (Belopolsky and Droxler, 2004).

SUMMARY

The chapters in this volume document the current state of the art in imaging and interpreting carbonate systems on the seismic data. Seismic imaging of carbonate depositional architecture has seen marked improvement over the last 10 years and has allowed interpreters to better delineate the complex histories of carbonate platform sequences. Lithofacies tied to seismic geometries have resulted in a more genetic understanding of the controls on carbonate deposition and early diagenesis. Significant karst events are well expressed at sequence boundaries and allow a more predictive understanding of meteoric alteration. Better imaging of fault and fracture systems has improved mapping of potential fluid migration pathways. The editors of this volume hope that the examples contained here will help carbonate stratigraphers to better interpret the depositional and diagenetic history of their rocks. In addition, we hope it will inspire future research to improve what interpreters "mine" from seismic data as well as improving the tool itself.

REFERENCES CITED

Aubert, O., and A. W. Droxler, 1996, Seismic stratigraphy and depositional signatures of the Maldive carbonate system (Indian Ocean): Marine and Petroleum Geology, v. 13, p. 503–536.

Bachtel, S. L., R. D. Kissling, D. Martono, S. Rahardjanto, P. A. Dunn, and B. A. MacDonald, 2004, Seismic stratigraphic evolution of the Miocene–Pliocene Segitiga Platform, East Natuna Sea, Indonesia: The origin, growth, and demise of an isolated carbonate platform, *in* G. P. Eberli, J. L. Masaferro, and J. F. Sarg, eds., Seismic imaging of carbonate reservoirs and systems: AAPG Memoir 81, p. 309–328.

Belopolsky, A. V., and A. W. Droxler, 2004, Seismic expressions of prograding carbonate bank margins: middle Miocene, Maldives, Indian Ocean, *in* G. P. Eberli, J. L. Masaferro, and J. F. Sarg, eds., Seismic imaging of carbonate reservoirs and systems: AAPG Memoir 81, p. 267–290.

Borgomano, J. R. F., and J. M. Peters, 2004, Outcrop and seismic expressions of coral reefs, carbonate platforms, and adjacent deposits in the Tertiary of the Salalah Basin, South Oman, *in* G. P. Eberli, J. L. Masaferro, and J. F. Sarg, eds., Seismic imaging of carbonate reservoirs and systems: AAPG Memoir 81, p. 251–266.

Bosellini, A., 1984, Progradation geometries of carbonate platforms: Examples from the Triassic of the Dolomites, northern Italy: Sedimentology, v. 31, p. 1–24.

Bosscher, H., and W. Schlager, 1992, Computer simulation of reef growth: Sedimentology, v. 39, p. 503–512.

Bosscher, H., and W. Schlager, 1993, Accumulation rates of carbonate platforms: Journal of Geology, v. 101, p. 345–355.

Bracco Gartner, G. L., W. Schlager, and E. W. Adams, 2004, Seismic expression of the boundaries of a Miocene carbonate platform, Sarawak, Malaysia, *in* G. P. Eberli, J. L. Masaferro, and J. F. Sarg, eds., Seismic imaging of carbonate reservoirs and systems: AAPG Memoir 81, p. 51–365.

Budd, D. A., A. H. Saller, and P. M. Harris, eds., 1995, Unconformities and porosity in carbonate strata: AAPG Memoir 63, 313 p.

Droste, H., and M. Van Steenwinkel, 2004, Stratal geometries and patterns of platform carbonates: The Cretaceous of Oman, *in* G. P. Eberli, J. L. Masaferro, and J. F. Sarg, eds., Seismic imaging of carbonate reservoirs and systems: AAPG Memoir 81, p. 185–206.

Droxler, A. W., W. Schlager, and C. C. Whallon, 1983, Quaternary aragonite cycles and oxygen-isotope record in Bahamian carbonate ooze: Geology, v. 11, p. 235–239.

Eberli, G. P., and R. N. Ginsburg, 1989, Cenozoic progradation of northwestern Great Bahama Bank, a record of lateral platform growth and sea level fluctuations, *in* P. D. Crevello, J. L. Wilson, J. F. Sarg, and J. F. Read, eds., Controls on carbonate platform and basin development: SEPM Special Publication 44, p. 339–351.

Eberli, G. P., A. M., Schwab, and G. M. Grammer, 2000, Anatomy of Ismay algal mound fields— A comparison of outcrop and 3-D seismic data, Paradox Basin, U.S.A., *in* P. W. Homewood and G. P. Eberli, eds., Genetic stratigraphy on the exploration and production scales: Case studies from the Pennsylvanian of the Paradox Basin and the Upper Devonian of Alberta: Bulletin du Centre de Recherches Elf Exploration-Production Mémoire 24, p. 93–107.

Eberli, G. P., F. S. Anselmetti, J. A. M. Kenter, D. F.

McNeill, and L. A. Melim, 2001, Calibration of seismic sequence stratigraphy with cores and logs, *in* R. N. Ginsburg, ed., The Bahamas Drilling Project: SEPM Special Publication 70, p. 241–266.

Eberli, G. P., F. S. Anselmetti, C. Betzler, J.-H. Van Konijnenburg, and D. Bernoulli, 2004, Carbonate platform to basin transitions on seismic data and in outcrops: Great Bahama Bank and the Maiella Platform margin, Italy, *in* G. P. Eberli, J. L. Masaferro, and J. F. Sarg, eds., Seismic imaging of carbonate reservoirs and systems: AAPG Memoir 81, p. 207–250.

Esker, D., G. P. Eberli, and D. F. McNeill, 1998, The structural and sedimentological controls on the reoccupation of quaternary incised valleys, southern lagoon of Belize: AAPG Bulletin, v. 82, p. 2075–2109.

Fitchen, W. M., 1997, Carbonate sequence stratigraphy and its application to hydrocarbon exploration and reservoir development, *in* F. J. Marfurt and A. Palaz, eds., Carbonate seismology: Society of Exploration Geophysicists Geophysical Developments Series 6, p. 121–178.

Goldhammer, R. K., P. A. Dunn, and L. A. Hardie, 1987, High-frequency glacio-eustatic sea level oscillations with Milankovitch characteristics recorded in Middle Triassic platform carbonates in northern Italy: American Journal of Science, v. 287, p. 853–892.

Grammer, G. M., R. N. Ginsburg, and P. M. Harris, 1993, Timing of deposition, diagenesis, and failure of steep carbonate slopes in response to a high-amplitude/high-frequency fluctuation in sea level, Tongue of the Ocean, Bahamas, *in* R. G. Loucks and J. F. Sarg, eds., Carbonate sequence stratigraphy: AAPG Memoir 57, p. 107–131.

Handford, C. R., and R. G. Loucks, 1993, Carbonate depositional sequences and systems tracts— Responses of carbonate platforms to relative sea level, *in* R. G. Loucks and R. G. Handford, eds., Carbonate sequence stratigraphy: AAPG Memoir 57, p. 3–31.

Hine, A. C., and A. C. Neumann, 1977, Shallow carbonate-bank-margin growth and structure, Little Bahama Bank, Bahamas: AAPG Bulletin, v. 61, p. 376–406.

Homewood, P. W., P. Mauriaud, and F. Lafont, 2000, Best practices in sequence stratigraphy for exploration and production: Bulletin du Centre de Recherches Elf Exploration-Production Mémoire 25, 81 p.

Heubeck, C., C. Story, P. Peng, C. Sullivan, and S. Duff, 2004, An integrated reservoir study of the Liuhua 11-1 field using a high-resolution three-dimensional seismic data set, *in* G. P. Eberli, J. L. Masaferro, and J. F. Sarg, eds., Seismic imaging of carbonate reservoirs and systems: AAPG Memoir 81, p. 149–168.

Isern, A. R., F. S. Anselmetti, P. Blum, et al., 2002, Proceedings of the Ocean Drilling Program, Initial Results, v. 194: Ocean Drilling Program, Texas A&M University, College Station, Texas, p. 1–88.

Isern, A. R., F. S. Anselmetti, and P. Blum, 2004, A Neogene carbonate platform, slope, and shelf edifice shaped by sea level and ocean currents, Marion Plateau (northeast Australia), *in* G. P. Eberli, J. L. Masaferro, and J. F. Sarg, eds., Seismic imaging of carbonate reservoirs and systems: AAPG Memoir 81, p. 291–308.

James, N. P., 1983, Reef environment, *in* P. A. Scholle, D. C. Bebout, and C. H. Moore, eds., Carbonate depositional environments: AAPG Memoir 33, p. 344–462.

Kendall, C. G. St. C., and W. Schlager, 1981, Carbonates and relative changes in sea level: Marine Geology, v. 44, p. 181–212.

Kerans, C., and S. Tinker, 1997, Sequence stratigraphy and characterization of carbonate reservoirs: SEPM Short Course Notes No. 40, 130 p.

Macurda Jr., D. B., 1997, Carbonate seismic facies analysis, *in* F. J. Marfurt and A. Palaz, eds., Carbonate seismology: Society of Exploration Geophysicists Geophysical Developments Series 6, p. 95–119.

Malone, M. J., N. C. Slowey, and G. M. Henderson, 2001, Early diagenesis of shallow-water periplatform carbonate sediments, leeward margin, Great Bahama bank (Ocean Drilling Program Leg 166): Geological Society of America Bulletin, v. 113, p. 881–894.

Masaferro, J., R. Bourne, and J. C. Jauffred, 2003, 3-D visualization of carbonate reservoirs: Leading Edge, March, p. 19–25.

Masaferro, J., R. Bourne, and J. C. Jauffred, 2004, Three-dimensional seismic visualization of carbonate reservoirs and structures, *in* G. P. Eberli, J. L. Masaferro, and J. F. Sarg, eds., Seismic imaging of carbonate reservoirs and systems: AAPG Memoir 81, p. 11–42.

Melville, P., O. Al Jeelani, S. Al Menhali, and J. Grötsch, 2004, Three-dimensional seismic analysis in the characterization of a giant carbonate field, onshore Abu Dhabi, United Arab Emirates, *in* G. P. Eberli, J. L. Masaferro, and J. F. Sarg, eds., Seismic imaging of carbonate reservoirs and systems: AAPG Memoir 81, p. 123–148.

Moldovanyi, E. P., F. M. Wall, and Z. J. Yan, 1995, Regional exposure events and platform evolution of Zhujiang Formation carbonates, Pearl River Mouth Basin: Evidence from primary and diagenetic seismic facies, *in* D. A. Budd, A. H. Saller, and P. M. Harris, eds., Unconformities and porosity in carbonate strata: AAPG Memoir 63, p. 125–140.

Neuhaus, D., J. Borgomano, J.-C. Jauffred, C. Mercadier, S. Olotu, and J. Grötsch, 2004, Quantitative seismic reservoir characterization of an Oligocene–Miocene carbonate buildup: Malampaya field, Philippines, *in* G. P. Eberli, J. L. Masaferro, and J. F. Sarg, eds., Seismic imaging of carbonate reservoirs and systems: AAPG Memoir 81, p. 169–184.

Osleger, D. A., 1991, Subtidal carbonate cycles: Implications for allocyclic vs. autocyclic control: Geology, v. 19, p. 917–920.

Pearson, R. A., and B. S. Hart, 2004, Three-dimensional seismic attributes help define controls on reservoir development: case study from the Red River Formation, Williston Basin, *in* G. P. Eberli, J. L. Masaferro, and J. F. Sarg, eds., Seismic imaging of carbonate reservoirs and systems: AAPG Memoir 81, p. 43–58.

Pomar, L., and W. C. Ward, 1995, Sea-level changes, carbonate production and platform architecture: The Llucmajor Platform, Mallorca, Spain, *in* B. U. Haq, ed., Sequence stratigraphy and depositional response to eustatic, tectonic and climatic forcing: Netherlands, Kluwer Academic Publishers, p. 87–112.

Pranter, M. J., N. F. Hurley, and T. L. Davis, 2004, Sequence-stratigraphic, petrophysical, and multicomponent seismic analysis of a shelf-margin reservoir: San Andres Formation (Permian), Vacuum field, New Mexico, United States, *in* G. P. Eberli, J. L. Masaferro, and J. F. Sarg, eds., Seismic imaging of carbonate reservoirs and systems: AAPG Memoir 81, p. 59–90.

Purdy, E. G., and G. T. Bertram, 1993, Carbonate concepts from the Maldives, Indian Ocean: AAPG Studies in Geology No. 34, 56 p.

Purdy, E. G., and D. Waltham, 1999, Reservoir implications of modern karst topography: AAPG Bulletin, v. 83, p. 1774–1794.

Read, J. F., 1985, Carbonate platform facies models: AAPG Bulletin, v. 69, p. 1–21.

Read, J. F., 1995, Overview of carbonate platform sequences, cycle stratigraphy and reservoirs in greenhouse and ice-house worlds, *in* J. F. Read, C. Kerans, and L. J. Weber, eds., Milankovitch sea-level changes, cycles, and reservoirs on carbonate platforms in greenhouse and ice-house worlds: SEPM Short Course No. 35, p. 1–102.

Saller, A. H., S. Walden, S. Robertson, R. Nims, J. Schwab, H. Hagiwara, and S. Mizohata, 2004, Three-dimensional seismic imaging and reservoir modeling of an upper paleozoic "reefal" buildup, Reinecke field, West Texas, United States, *in* G. P. Eberli, J. L. Masaferro, and J. F. Sarg, eds., Seismic imaging of carbonate reservoirs and systems: AAPG Memoir 81, p. 107–122.

Sarg, J. F., 1988, Carbonate Sequence Stratigraphy, *in* C. K. Wilgus, B. S. Hastings, C. G. St. C. Kendall, H. W. Posamentier, C. A. Ross, and J. C. Van Wagoner, eds., Sea-level changes: An integrated approach: SEPM Special Publication 42, p. 155–181.

Sarg, J. F., J. F., Markello, and L. J. Weber, 1999, The second-order cycle, carbonate platform growth, and reservoir, source, and trap prediction, *in* P. M. Harris, A. H. Saller, and J. A. Simo, eds., Advances in carbonate sequence stratigraphy: Application to reservoirs, outcrops, and models: SEPM Special Publication, v. 63, p. 11–34.

Schlager, W., 1981, The paradox of drowned reefs and carbonate platforms: Geological Society of America Bulletin, v. 92, p. 197–211.

Schlager, W., 1992, Sedimentology and sequence stratigraphy of reefs and carbonate platforms: AAPG Continuing Education Course Note Series No. 34, p. 71.

Schlager, W., 1999, Scaling of sedimentation rates and drowning of reefs and carbonate platforms: Geology, v. 27, p. 183–186.

Schlager, W., 2000, The future of applied sedimentology: Journal of Sedimentary Research, v. 70, p. 2–9.

Stoudt, E. L., and P. M. Harris, eds., 1995, Hydrocarbon reservoir characterization: SEPM Short Course No. 34, 357 p.

Tinker, S. W., D. H. Caldwell, D. M. Cox, L. C. Zahm, and L. Brinton, 2004, Integrated reservoir characterization of a carbonate ramp reservoir, South Dagger Draw field, New Mexico: Seismic data are only part of the story, *in* G. P. Eberli, J. L. Masaferro, and J. F. Sarg, eds., Seismic imaging of carbonate reservoirs and systems: AAPG Memoir 81, p. 91–106.

Tucker, M. E., and V. P. Wright, 1990, Carbonate sedimentology: Oxford, Blackwell Scientific Publications, 482 p.

Tyrrell, W. W., and R. G. Davis, 1989, Miocene carbonate shelf margin, Bali-Flores Sea, Indonesia, *in* A. W. Bally, ed., Atlas of seismic stratigraphy, v. 2: AAPG Studies in Geology No. 27, p. 174–179.

Vahrenkamp, V. C., F. David, P. Duijndam, M. Newall, and P. Crevello, 2004, Growth architecture, faulting, and karstification of a middle Miocene carbonate platform, Luconia Province, Offshore Sarawak, Malaysia, *in* G. P. Eberli, J. L. Masaferro, and J. F. Sarg, eds., Seismic imaging of carbonate reservoirs and systems: AAPG Memoir 81, p. 329–350.

Vail, P. R., R. M. Mitchum Jr., and S. Thomson, 1977, Seismic stratigraphy and global changes of sea level, *in* C. E. Payton, ed., Seismic stratigraphy— Applications to hydrocarbon exploration: AAPG Memoir 26, p. 49–212.

Weimer, P., and T. L. Davis, eds., 1996, Applications of 3-D seismic data to exploration and production: AAPG Studies in Geology No. 42/Society of Exploration Geophysicists Geophysical Development Series No. 5, 270 p.

Williams, T., D. Kroon, and S. Spezzaferri, 2002, Middle–upper Miocene cyclostratigraphy of downhole logs and short to long term astronomical cycles in carbonate production of Great Bahama Bank: Marine Geology, v. 185, p. 75–93.

Wilson, J. L., 1975, Carbonate facies in geological history: Berlin, Springer-Verlag, 471 p.

Wilson, P. A., and H. H. Roberts, 1995, Density cascading: Off-shelf sediment transport, evidence and implications, Bahama Banks: Journal of Sedimentary Research, v. 65, p. 45–56.

1

Masaferro, J. L., R. Bourne, and J. C. Jauffred, 2004, Three-dimensional seismic visualization of carbonate reservoirs and structures, *in* Seismic imaging of carbonate reservoirs and systems: AAPG Memoir 81, p. 11–41.

Three-Dimensional Seismic Volume Visualization of Carbonate Reservoirs and Structures

Jose Luis Masaferro

Shell International E&P, Technology Applications & Research (SEPTAR), Rijswijk, The Netherlands

Ruth Bourne

Shell International E&P, Technology Applications & Research (SEPTAR), Rijswijk, The Netherlands

Jean Claude Jauffred

Shell International E&P, Technology Applications & Research (SEPTAR), Rijswijk, The Netherlands

ABSTRACT

In pure carbonate systems, the combined effect of variations in depositional facies and diagenetic alterations plays a key role in controlling variations in sonic velocities and thus in acoustic impedance. As seismic facies are delineated by acoustic impedance contrasts, depositional facies and geometries may be rather poorly defined for various carbonates environments (e.g., shallow-water platform carbonates). Accurate three-dimensional imaging of seismic facies and geometries is critical to construct a realistic, seismically constrained reservoir model. Conventional two- and three-dimensional (2-D and 3-D, respectively) seismic mapping is not an ideal predictive method when attempting to characterize carbonate reservoirs mainly because of the complexity and heterogeneity of carbonate systems.

Three-dimensional image-processing techniques of stacked and migrated data incorporate all three dimensions, which when combined help to identify and highlight events of significance in the data. The result is an attribute cube or volume that can be analyzed and interpreted more objectively by the interpreter than the conventional horizon-based interpretation. We have applied various 3-D image-processing techniques to produce filtered seismic reflectivity data and volume attributes to better visualize and delineate seismic facies, geometries, and the structure of heterogeneous carbonate reservoirs. Structure-oriented filtering was applied to improve signal-to-noise ratios and suppress random noise to obtain a better reflection definition. Combined volume dip and azimuth was calculated from the seismic cubes to detect subtle stratigraphic features, such as low-angle progradation units and shoal-type mounded seismic facies in the Permian Khuff and Upper Cretaceous Natih E reservoirs in Oman. Semblance volumes were used to highlight reflection terminations and distinguish between stratigraphic and structural features. In the Malampaya field (Philippines), neural network classification mapping was applied to the 3-D attribute-generated volumes to extract different seismic facies and properties, which can be related to potential good reservoir zones. Three-dimensional visualization tools were used to image both horizons and faults of a complex inverted structure of a deep Upper Cretaceous restricted marine-lacustrine carbonate reservoir in the Yacoraite Formation, northwest Argentina.

Seismic facies and geometries interpreted from the attribute analyses, combined with interpretation of the original seismic and core-log data, allowed us to construct robust structural and depositional models of carbonate environments that were used as input for static reservoir models.

INTRODUCTION

The acquisition of three-dimensional (3-D) seismic data surveys as an aid to hydrocarbon exploration has become an increasingly standard procedure over the last 10 years (Brown, 1996). The resultant seismic volumes present the interpreter with a less ambiguous image of the subsurface than the data from a two-dimensional (2-D) survey and the increased fidelity and resolution of the data enable detailed interpretations to be made. However, such improved imaging techniques reveal increasingly complex geologic environments for the interpreter to analyze. In particular, for carbonate environments, with their unique depositional systems, it is not always adequate to follow a standard interpretation routine and more sophisticated techniques are demanded (Skirius et al., 1999).

A standard seismic interpretation methodology for reservoir definition in a 3-D seismic survey is commonly based on the identification and mapping of relevant horizons and faults; both on a regional and field scale, progressing to a more detailed interpretation within the reservoir of interest. The resultant surfaces are then combined to produce a consistent geologic model for the region. Such a procedure is often successful and may be especially suitable for areas of relatively simple geology (e.g., "layer-cake" geology) or where the reservoir unit is beyond seismic resolution and therefore no geometries can be identified. This method, however, is highly user intensive, and has the disadvantage of representing 3-D features with 2-D surfaces, which becomes progressively more ambiguous in areas of more complex geology, such as heterogeneous carbonate depositional systems. Furthermore, with all components to the geologic model being hand-picked by an interpreter, the final model has the potential to be unduly (possibly unwittingly) biased by an interpreter's preferred geologic model.

To better image and interpret heterogeneous carbonate reservoirs, we have approached the data analysis in two different ways. The first approach is to improve the signal-to-noise ratio of the seismic data so that the traditional horizon-based interpretation method can be better followed. This may be realized, for example, by applying noise-reduction techniques to improve the quality of the seismic data or by making depositional geometries explicit rather than implicit features.

The second approach is to highlight specific geologic features that have a 3-D extent, and the geometry of which may have little in common with the orientation of the 3-D grid of seismic data. For example, in an environment of hydrocarbon-bearing shoal complexes, there is an immediate focus to the interpretation by initially isolating high-amplitude moundedlike structures within the data set. With both these approaches, combined with well calibration, it is possible to speed up the interpretation process (both in absolute and user time), limit the potential model bias of an interpreter, and improve the quality of the interpretation.

The main purpose of this chapter is to image and extract the 3-D structural and geologic architecture of the reservoir and thus construct reliable subsurface models by integrating the seismic attribute volumes with log, core, and outcrop data (Figure 1). The results show that the use of 3-D visualization and processing methods significantly improved the quality of the seismic data resulting in an essential predictive tool for carbonate reservoir characterization.

THREE-DIMENSIONAL IMAGE-PROCESSING TECHNIQUES

Structure-Oriented Filtering: Filtering for Noise Reduction

These filters are designed to remove background noise from the seismic data by smoothing the data in the direction of the local orientation of the reflections (Hocker and Fehmers, 2002) (Figure 2A). Noise reduction is beneficial in the interpretation process because it makes reflections sharper, thus, improving

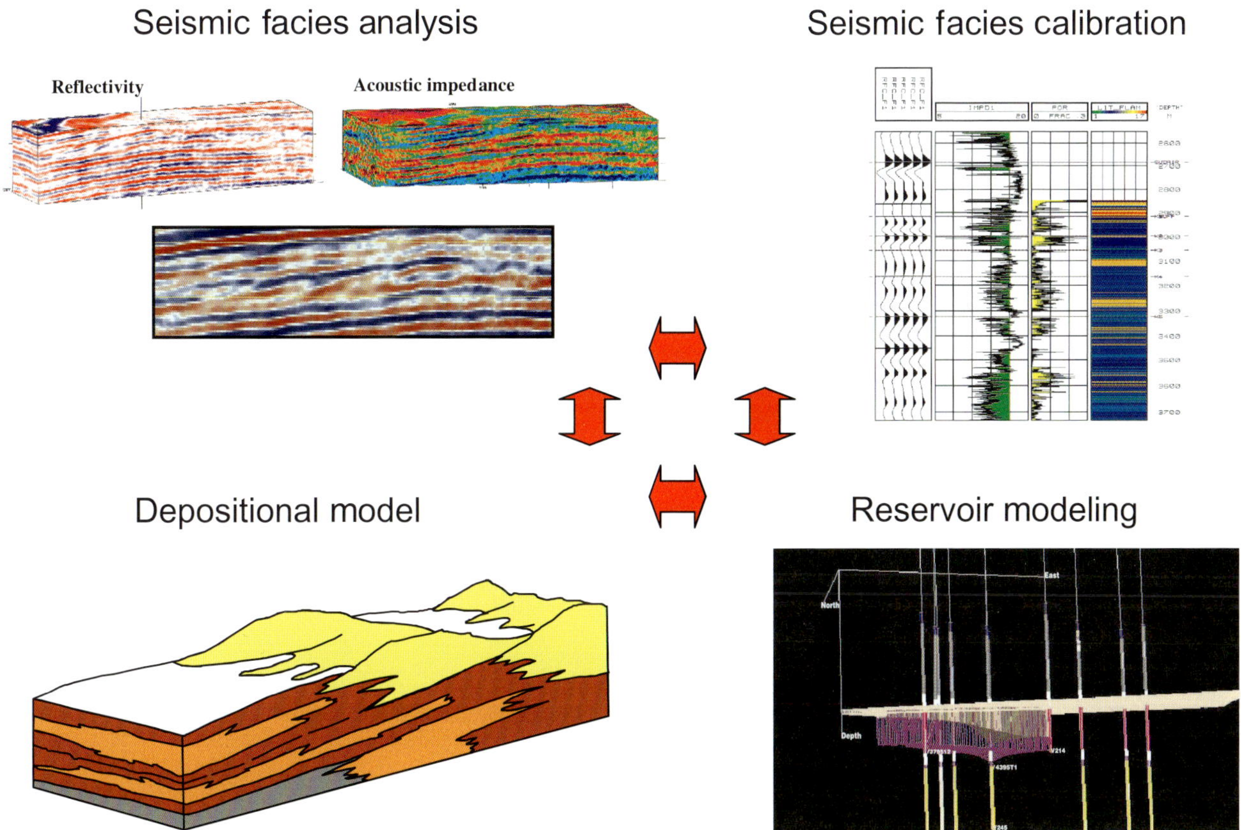

FIGURE 1. Diagram showing the integrated approach followed in this study. See text for discussion.

the resolution of structural and stratigraphic geometries. It also allows horizon autotrackers to track more of the horizon because noisy spikes within the data often cause an autotracker to stop before it has completed horizon tracking. The simplest kind of postprocessing noise-reduction technique is to apply a mean or median filter to a region of the seismic data. Generally, a median filter, which removes only the outlying amplitude values and does not mix the intermediate amplitude values, is the most effective at removing the noise. These filters are, however, commonly applied along the *x-y-z* grid of seismic volumes affecting the seismic signal of the strongly dipping reflections. Application of a mean filter over too great a window size may risk removing some of the pertinent signal, especially at higher frequencies (this can be verified by subtracting the input from the output volume). Such filters can be applied in a more sophisticated manner by orienting the filter along the subsurface structures defined by the seismic data. A more radical approach to reducing the noise so that geometry becomes more apparent is to "skeletonize" the seismic image as shown in Figure 2B. Additionally, in this instance, the continuity of reflections is indicated by color allowing the continuous events to be immediately apparent and clearly displaying their geometry with respect to less continuous reflections.

Semblance

Although the human eye may easily distinguish offsets in reflection sequences on seismic data sets, the observation is implicit rather than explicit. For example, faults are commonly apparent by a lack of reflection. To make them explicit, a filter can be applied to the seismic data to highlight discontinuities rather than continuous regions of data. The resultant image mutes the reflection sequence while highlighting such discontinuities.

Different names have been used to define semblance, such as coherence, continuity, and covariance (Bahorich and Farmer, 1995; Brown, 1996), which describe basically the same concept. Semblance is an expression of similarity between seismic traces (Figure 3A). In the examples shown in this chapter, *dissemblance* is probably a better term because lower values of semblance in the volume output indicate greater similarities between traces. In addition to highlighting faults, semblance volumes can be used to detect subtle stratigraphic discontinuities, in particular when the semblance volume is flattened to a relevant seismic horizon. Volume semblance can also be applied to highlight lateral variations in seismic facies and reflection terminations providing a 3-D spatial distribution of sedimentary features.

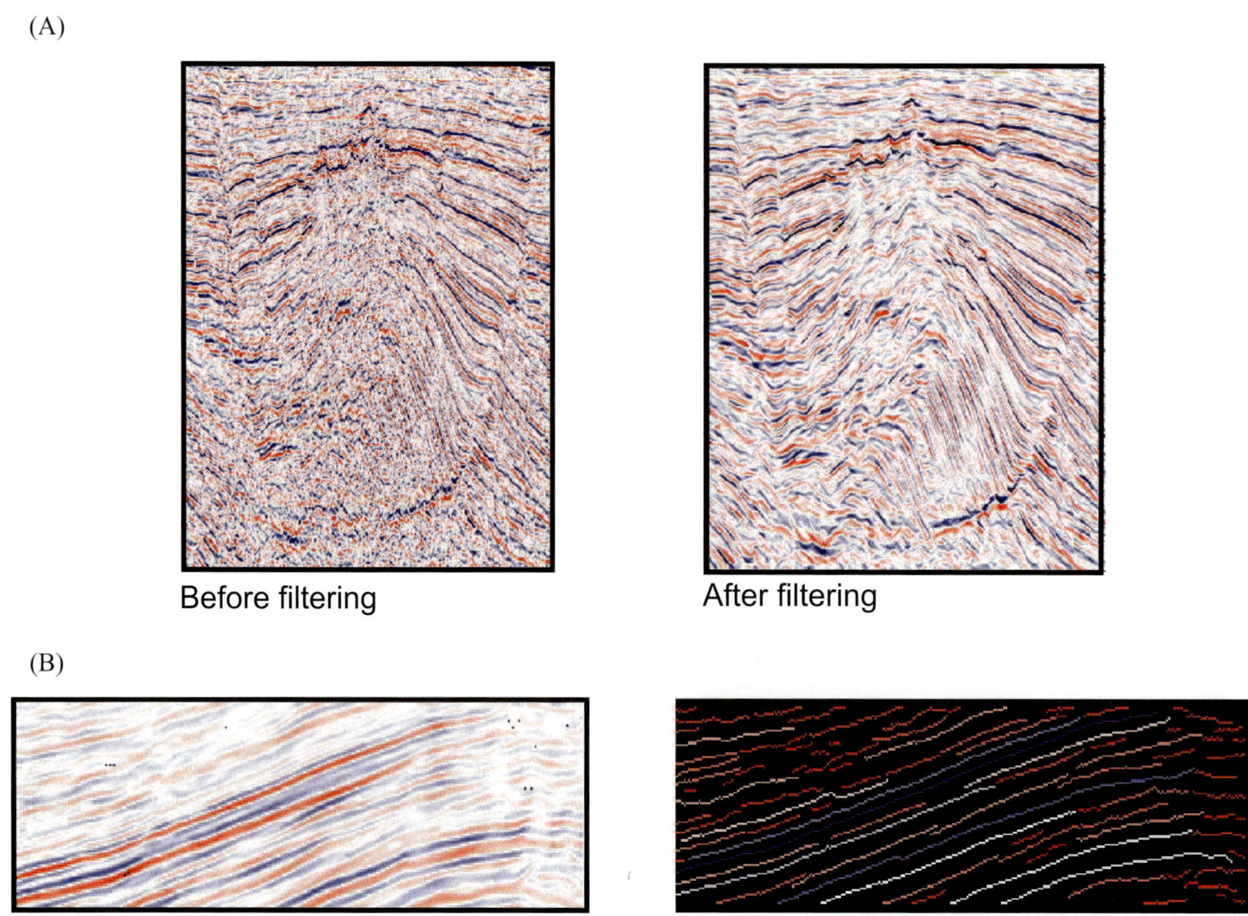

Figure 2. Examples of structure-oriented filtering (SOF) applied to seismic reflectivity volumes. (A) Example showing removal of background noise with a standard footprint filter size. (B) Extreme application of filtering making continuity of reflections more visually obvious.

Structural Vector Attributes

Structural vector attributes are calculated by measuring the direction of the steepest descent within a small seismic reflection volume and is computed by finding the local direction of the best correlation or semblance (Figure 3B). In this way, both reflection dip and azimuth are automatically calculated within the seismic volume. Viewed together as time slices, these images not only indicate the location and dip direction of discontinuities but also give a sense of their orientation.

Texture and Bodychecking

Three-dimensional seismic texture mapping using neural networks allows the grouping of regions of data into different classes based on a combination of 3-D attributes (Figure 4A). For example, in a region consisting predominantly of limestones and shales, it would be possible to define several seismic attributes, which have significantly different values for the two facies. By defining the range of attribute values representing limestones and shale (using a neural network method), it is possible to classify an entire volume into one or other of these facies. To classify a greater number of facies may require a greater number of seismic attributes.

Bodychecking is a technique that searches for connected samples in 3-D volume that occur within a user-defined range of values (Figure 4B). Seismic traces are first converted to voxels, which have an amplitude value and voxel size that correspond to the amplitude and sample interval of the original 3-D volume. A search is then made for adjacent voxels that lie within a specific amplitude range. The resultant 3-D shapes (bodies) may represent significant geologic features. This technique can be useful to locate depositional features for which the seismic grid has no significance, as they can often be difficult to locate on 2-D sections. For example, a 2-D section, which cuts a channel along its central axis, may intersect channel fill at several disconnected locations. The ability to see the channel and its contents in 3-D reduces the ambiguity in interpretation of such structures. To highlight an object from 3-D seismic data, it must be distinguishable from the surroundings by reflection amplitude or some other attribute (such as frequency content). Then, to ensure that only the attributes

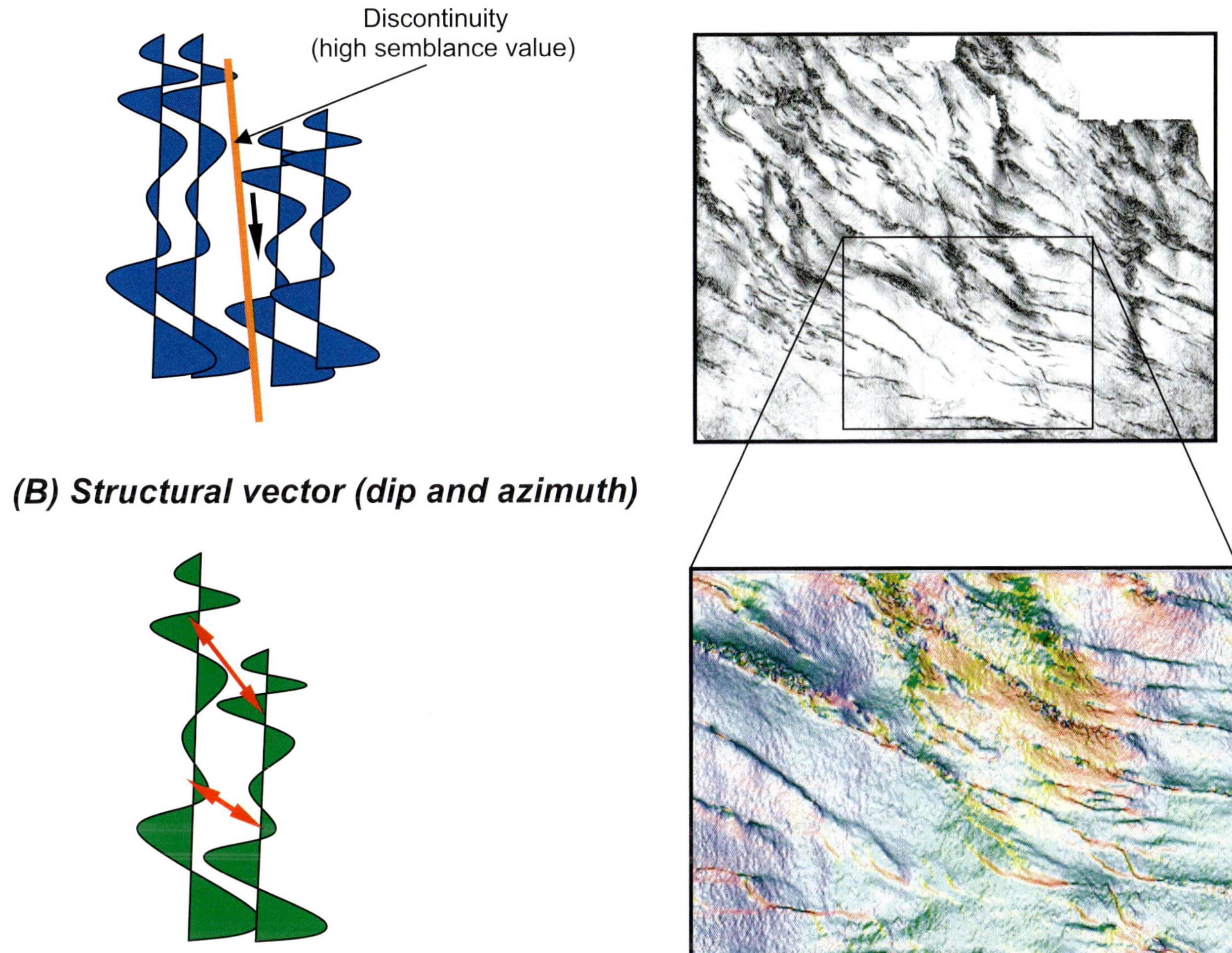

Figure 3. (A) Time slices through semblance volume enhancing discontinuities between traces caused by faulting. Dark colors represent discontinuities in the output data. (B) Time slices through combined dip and azimuth volume illustrating additional features that can be highlighted in addition to the faults imaged in A.

associated with that feature are highlighted, a connectivity criterion must also be applied, e.g., within a particular amplitude-frequency range, only adjacent seismic voxels will be selected. In this way, it is possible to locate many geologic features that may otherwise be lost in the 3-D data cube.

For example, texture mapping was applied to a segmented seismic volume defined by the two horizons corresponding to the top and base of the reservoir in the Malampaya field, Philippines (Nido Formation, Grötsch and Mercadier, 1999; Neuhaus et al., 2004). Two distinctive interpreted seismic facies were identified and then run as training sets through the seismic volume (Figure 5). The first training set represents the seismic character similar to the western part of the buildup (chaotic, steeply dipping discontinuous reflections). The second training set represents seismic facies from the interior of the buildup (high-amplitude, continuous, flat-lying reflections). Several attributes related to amplitude, continuity and similarity between traces, and dip and azimuth were calculated to extract these seismic facies. The results are shown in Figure 6, which represent slices through the calculated texture volume. The blue color represents the calculated texture for the marginal reef-related seismic facies, and the green color represents the more internal seismic lagoonal facies. The two seismic textures were exported to the static reservoir simulator and used to constrain different model scenarios.

Bodychecking was carried out on the porosity volume generated from the acoustic impedance data within the reservoir unit (Neuhaus et al., 2004). The objective was to analyze the porosity distribution and investigate occurrences of sizeable porosity bodies that could be used to constrain the reservoir modeling (Neuhaus et al., this volume). Bodychecking was applied over a wide range of "porosity" thresholds (Figure 7), and

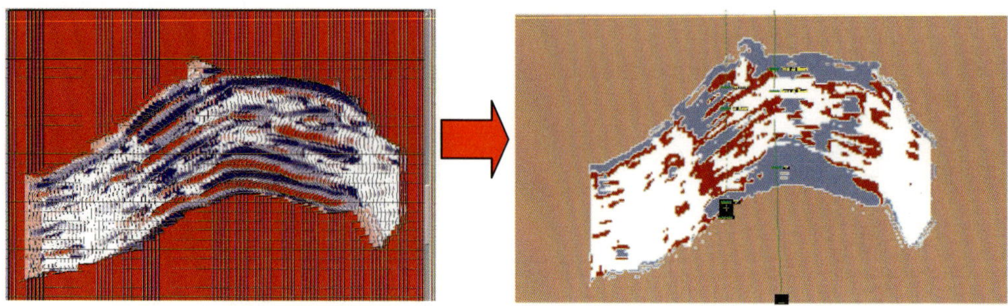

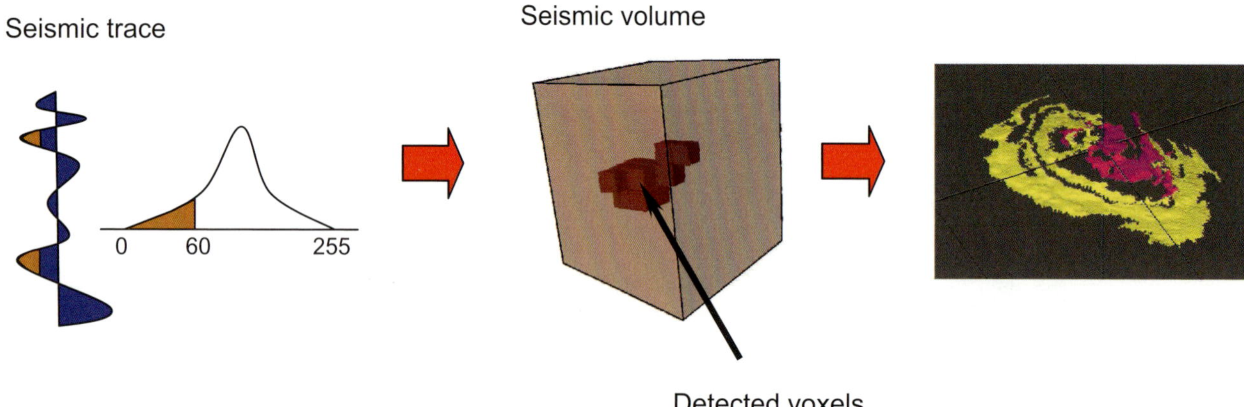

FIGURE 4. (A) Example of application of texture to a seismic reflectivity volume. Result shows a volume output with the extracted seismic facies represented by different colors. (B) Example of bodychecking defined for an amplitude range from 0 to 60. The result shows the extracted bodies contained within that range.

then porosity bodies with 2% threshold were extracted. Figure 7A shows the connected bodies displayed in the same volume with different colors, making it possible to analyze the distribution of the connected porosities and the relationship between porosity ranges. The porosity distribution shows more seismic-derived porosities in the northern part of the Malampaya buildup than in the south.

THREE-DIMENSIONAL IMAGING OF DEPOSITIONAL SYSTEMS AND FACIES IN CARBONATE RESERVOIRS

Prograding Shoal Complex: Khuff Formation, Permian, Oman

The sediments of the Khuff Formation of the study field in northwest Oman were deposited in a shallow, inner-shelf-restricted depositional environment dominated by high-energy oolitic shoals and bars with protected lagoons. Main depositional cycles show the characteristics of trangressive-regressive carbonate-evaporite successions, which consist of several stacked shallowing-upward cycles deposited in subtidal, intertidal, and supratidal (sabkha) environments (Blendinger, 1988; Caline and Droste, 1989; Samiee et al., 1999). The Khuff reservoir can be divided into five major cycles named K5 to K1 (Blendinger, 1988; Caline and Droste, 1989). Only the Top Khuff was cored (K2 mainly) and thus we used this interval to calibrate the 3-D seismic data.

Image Processing, Seismic Facies Description, and Interpretation

The study field is divided by a major northeast-southwest–trending fault into two main blocks (Figure 8). The study 3-D seismic survey is located in the southwestern block in the downthrown side of the fault (Figure 8). Structure-oriented filtering (SOF) was applied to the 3-D seismic reflectivity volume to remove the background noise and thus improve definition-continuity of reflections (Figure 9). A series of seismic sections across the western block of the field shows the improved imaging

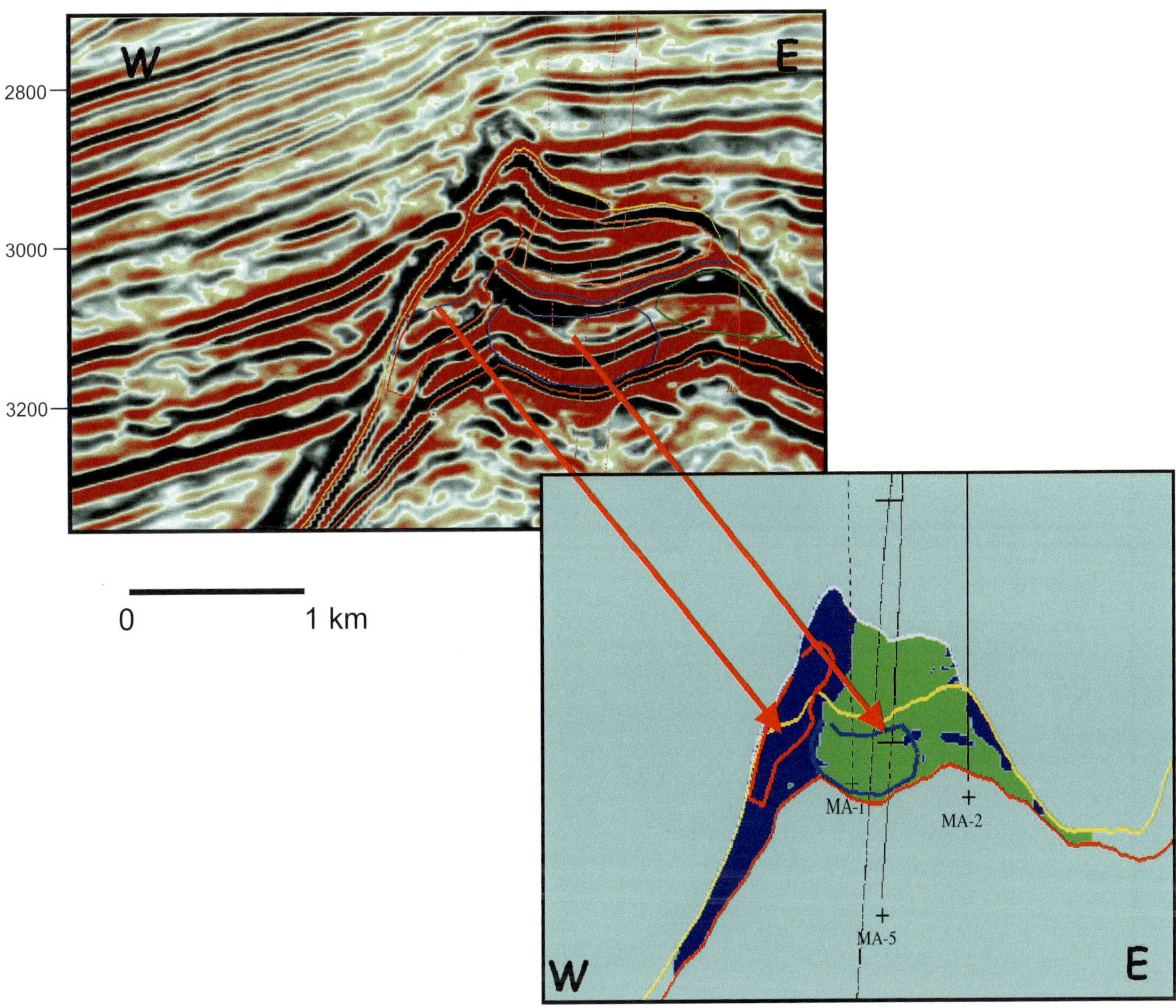

FIGURE 5. Nido Formation, Malampaya field, Philippines. Texture classification applied to prestack depth-migrated data. The texture analysis was based on a combination of amplitude-related attributes. Green colors represent continuous, high-amplitude reflections corresponding to the intrabuildup, lagoonal seismic facies. Blue colors represent chaotic, steeply dipping, discontinuous, marginal seismic facies. Yellow and red interpreted horizons are top and base reservoir, respectively. Red, green, and blue segments are seismic facies defined to calculate texture using a combination of attributes.

of the seismic volume depicting different seismic geometries and facies that change laterally along the reservoir interval (Figures 10–12). A field-scale seismic facies interpretation of some representative lines (with emphasis on the K2), mainly along the western area, was made using the new filtered data.

Khuff seismic facies can be described based on external geometry, internal reflection character, reflectivity, and lateral continuity. Interpreted seismic facies include (1) mounded continuous-discontinuous, (2) progradational shingle, and (3) progradational sigmoid.

The mounded seismic facies is identified by a series of reflections that outline several convex-upward or mound-shaped geometries (Figures 10, 12). The extent of this facies is delineated by a prominent amplitude anomaly (Figure 12). Within the mounded seismic facies, there is clear lateral change from internally discontinuous to more continuous seismic facies. The mounded continuous facies consists of mound structures that show a series of low-angle-dipping internal reflections (Figure 12). Occasionally, the continuous mounded facies varies laterally into more discontinuous to transparent reflections (Figure 13).

The shingled and sigmoid prograding facies occur primarily in the K2 sequence in the western side of the field outside the well area. The shingled prograding facies is identified by a series of shingled, high-amplitude clinoforms (Figures 11, 13A). In map view, these facies are delineated by elongated amplitude anomalies showing the 3-D architecture of the clinoforms.

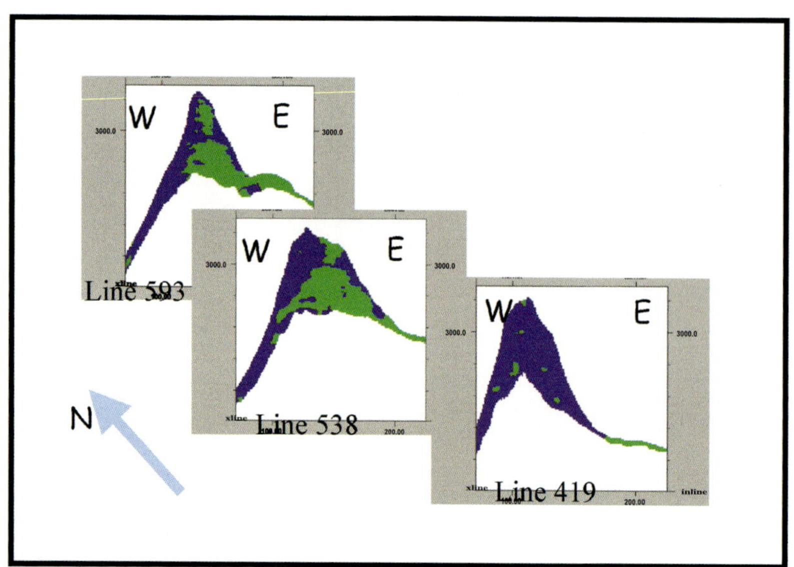

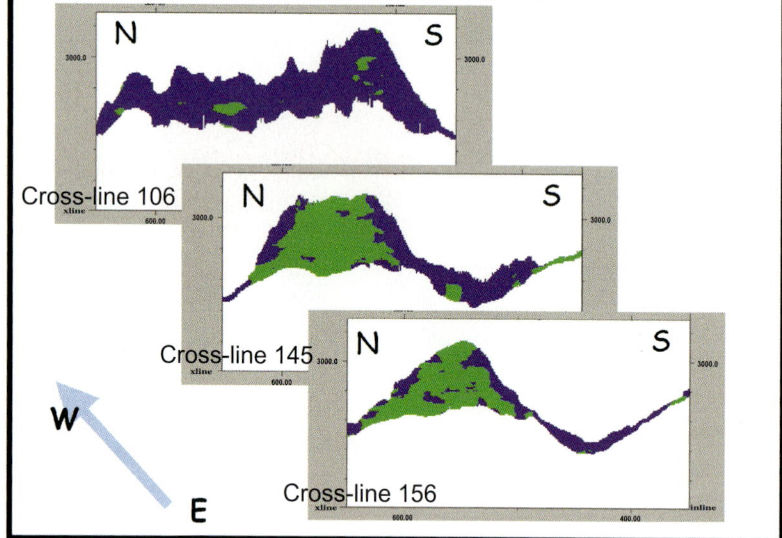

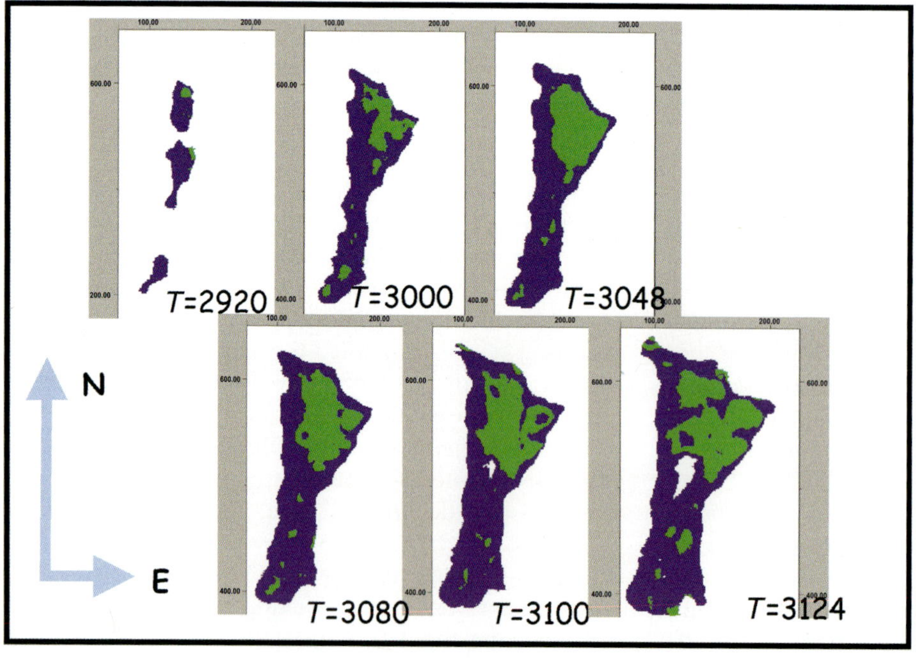

FIGURE 6. Nido Formation, Malampaya field, Philippines. A series of east-west–north-south cross sections and time slices through the texture volume showing seismic facies distribution.

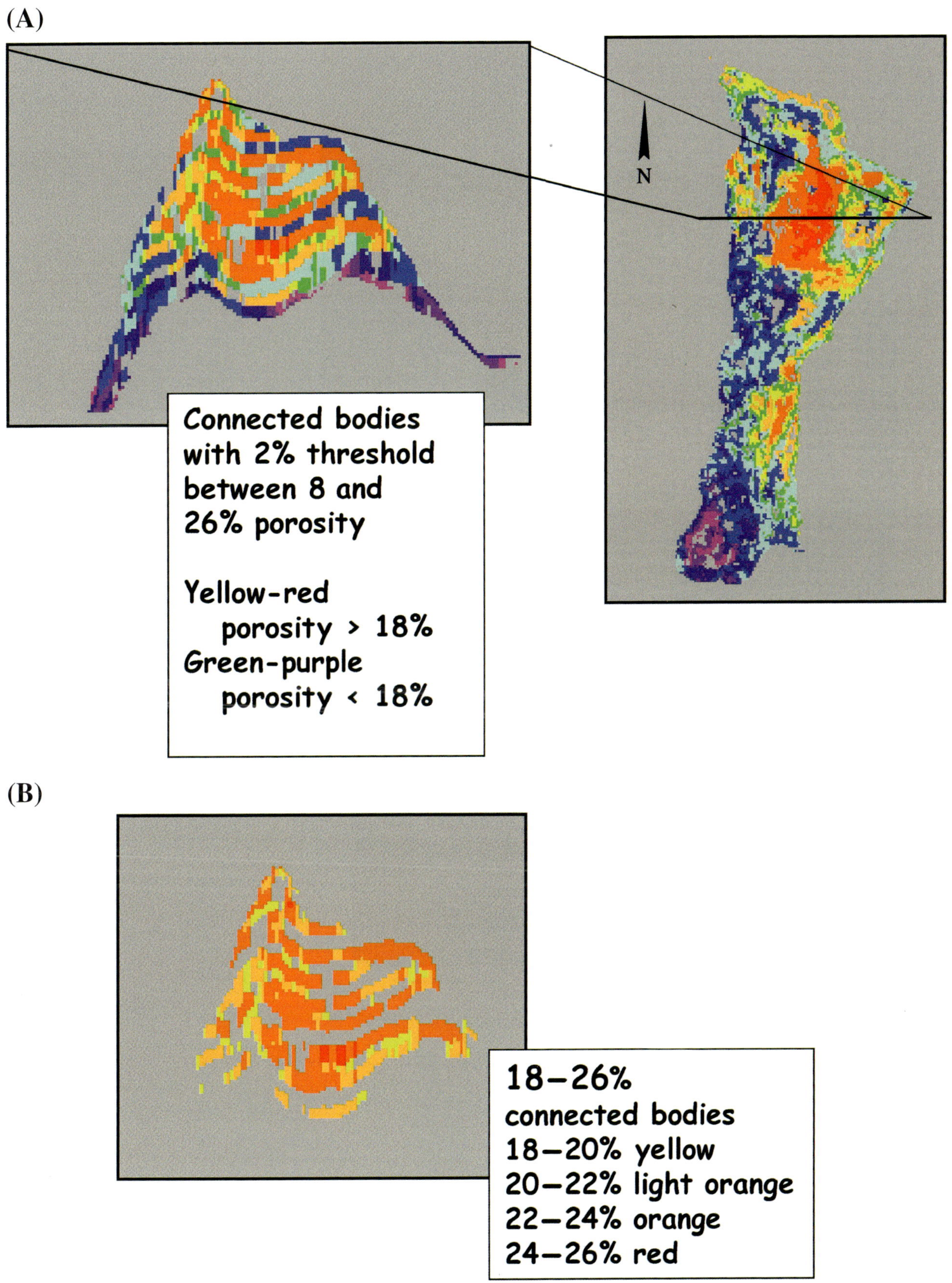

Figure 7. Nido Formation, Malampaya field, Philippines. Bodychecking results for calculated porosity from acoustic impedance data. (A) Cross section and time slices showing porosity distribution. (B) Cross section through porosity volume using a range between 18 and 26%.

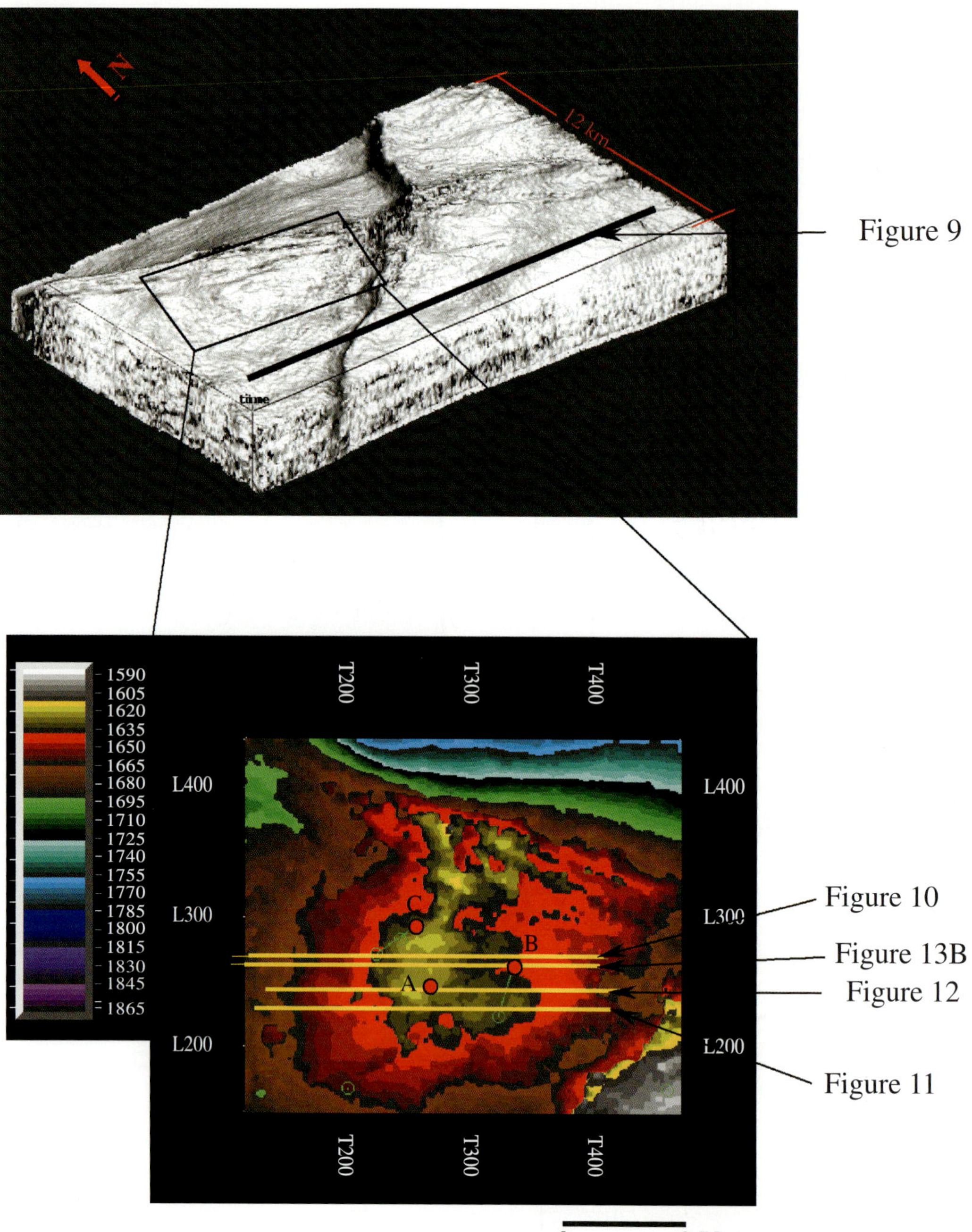

FIGURE 8. Khuff Formation, northern Oman. Dip volume and structure map at the top of K2 reservoir interval. Black square indicates study area. A, B, and C are locations of wells shown in this study. Color bar in two-way traveltime (TWT).

Toward the north, the geometry changes from shingled to low-angle sigmoid facies (Figures 10, 13B, 14). The orientation of the clinoforms indicates a general direction of progradation to the west from the interior of the seismic mound complex.

Calibration of Seismic with Cores and Interpretation

The K2 reservoir interval has an average thickness of about 65 m, which is represented by two seismic reflections created by a change in acoustic impedance given by porosity changes at the base and top of this interval (Figure 10). Core descriptions within K2 show shallowing-upward cycles of cross-bedded grainstones controlled by a decrease in accommodation through time (Figure 14; Caline and Droste, 1989; C. Kerans, personal communication, 2000). Seismic profiles reveal lateral changes in the geometry within K2 interval that can also be related to an overall decrease in accommodation space (Figure 15). Interpretation of geometries from seismic sections depicts an aggradational phase represented mainly by parallel and mounded geometries and a progradational phase represented by shingled and sigmoid geometries.

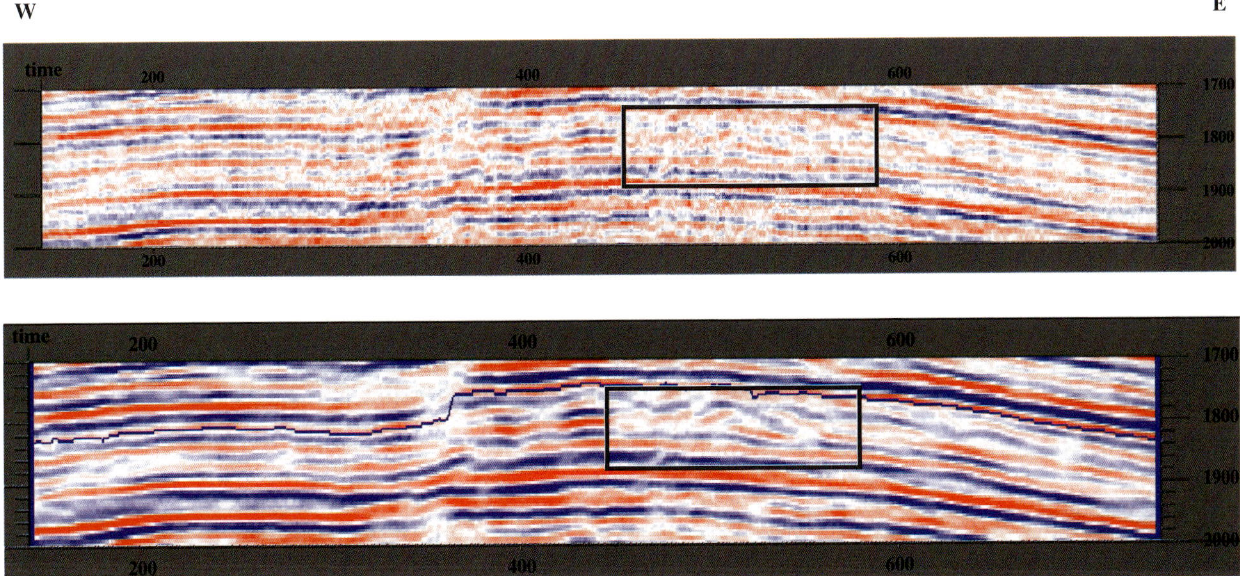

FIGURE 9. Khuff Formation, northern Oman, K2 reservoir. Seismic reflectivity cross sections before and after applying SOF. Note the improvement in resolution of reflection termination and continuity within the indicated black square see Figure 8 for location.

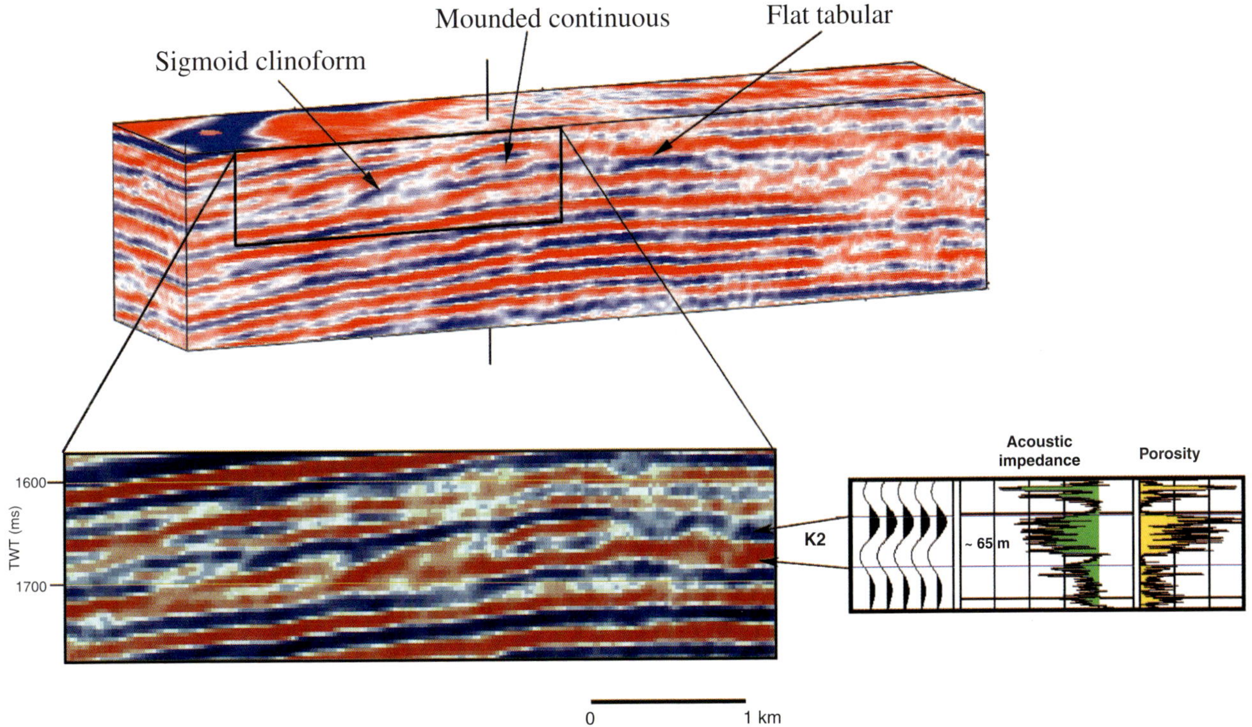

FIGURE 10. Khuff Formation, northern Oman. Structure-oriented filtering volume shows the interpreted seismic facies around the K2 reservoir interval. Synthetic seismogram shows that the reflections were caused mainly by a change in porosity at the top and base of K2. Note change in polarity between the seismic cube and section. See Figure 8 for location. TWT = two-way traveltime.

The seismic section in Figure 15A displays vertical and lateral variations in seismic facies and geometries that ultimately might have an effect on fluid-flow behavior and reservoir performance. The aggradational phase is indicated by parallel, horizontal reflections, which do not show major thickness changes across the field (Figure 13B). However, some inclined reflections (Figure 15B) suggest the existence of prograding

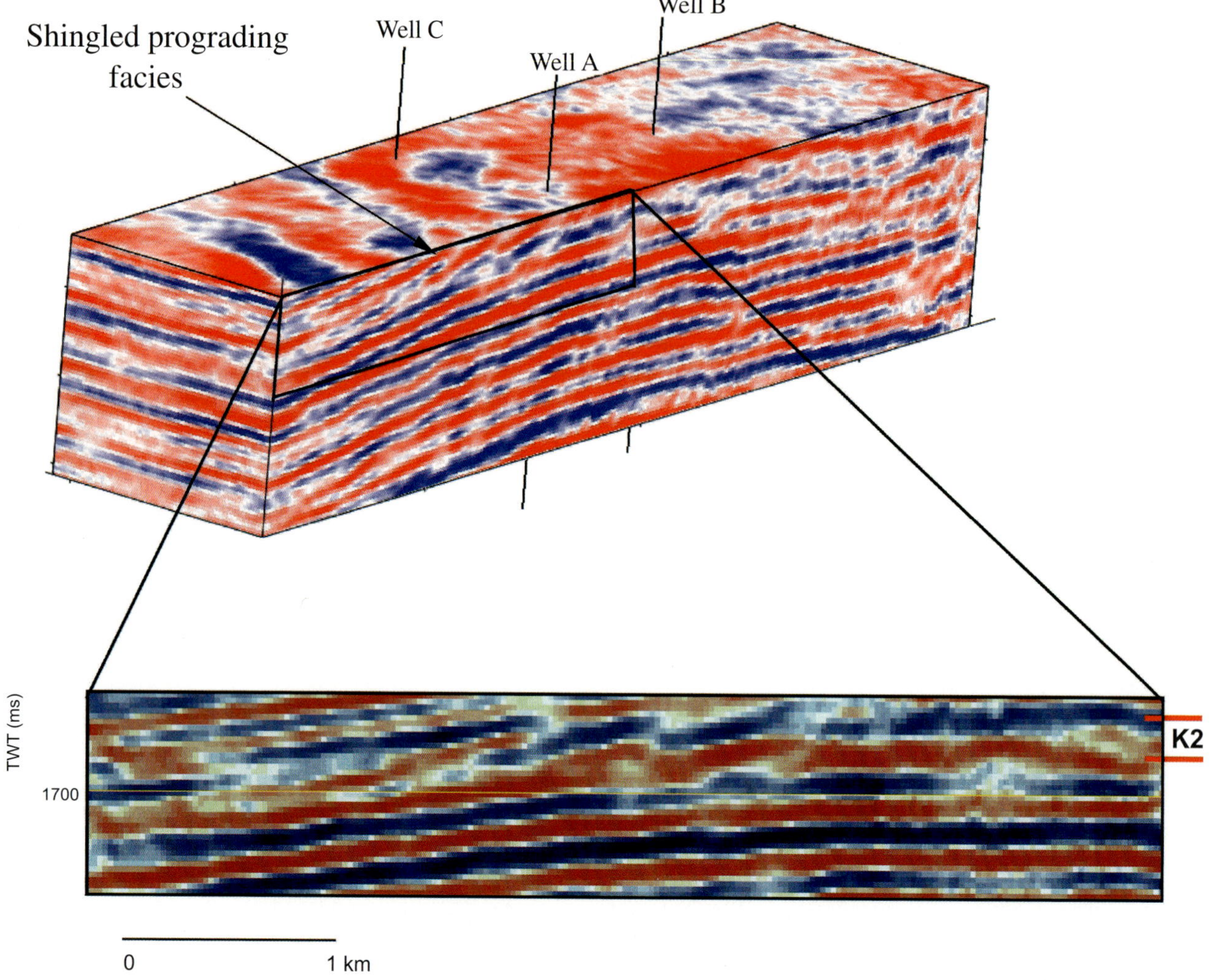

FIGURE 11. Khuff Formation, northern Oman, K2 reservoir. Structure-oriented filtering volume shows the interpreted shingled seismic facies. Top view of the seismic cube shows the 3-D extent and overlapping of prograding units. Note change in polarity between the seismic cube and section. See Figure 8 for location. TWT = two-way traveltime.

geometries that simply might not be resolved by the seismic frequencies. The difference between the highest and lowest point in the mound-shaped seismic facies (Figure 15B) indicates the presence of depositional paleotopography that probably controlled the space available and thus the onset of the westward shingle and sigmoid prograding seismic geometries.

The mounded seismic facies are interpreted to be a result of deposition of a shoal complex, which preserved the local depositional relief. The upper part of K2 was cored in one well (well A in Figures 8, 11, 12) that penetrated the mounded seismic facies. Caline and Droste (1989) interpreted the leaching observed in well A as the result of freshwater zones related to paleogeographic highs. In addition, utilizing Formation MicroImager in well A shows a major cavernous zone that can be interpreted as a leached zone that is the major contributor to flow compared to the rest of the wells. Caline and Droste (1989) also attribute the occurrence of cyclic development of dolomite beds in well A to periods of subaerial exposure.

Progradation is indicated by two geometries: (a) simple sigmoid geometries that prograde out from the mounded complex toward the west (Figures 10, 14, 15A) and (b) shingled geometries that prograde mainly to the west but also to the east (Figures 11, 13). No wells penetrated this prograding complex. Progradation of these units occurs in a downstepping, offlapping manner (Figure 15A), which suggests sea level lowering and a reduction of accommodation space. These forced-regressing geometries imply that the mounded shoal complex was closer to sea level and/or exposed for longer periods during its evolution causing probably significant, more localized leaching events.

The 3-D seismic data enabled us to construct a calibrated depositional model. The resultant reservoir model contains both the stratal patterns and geometries from seismic to core calibration and property distribution from

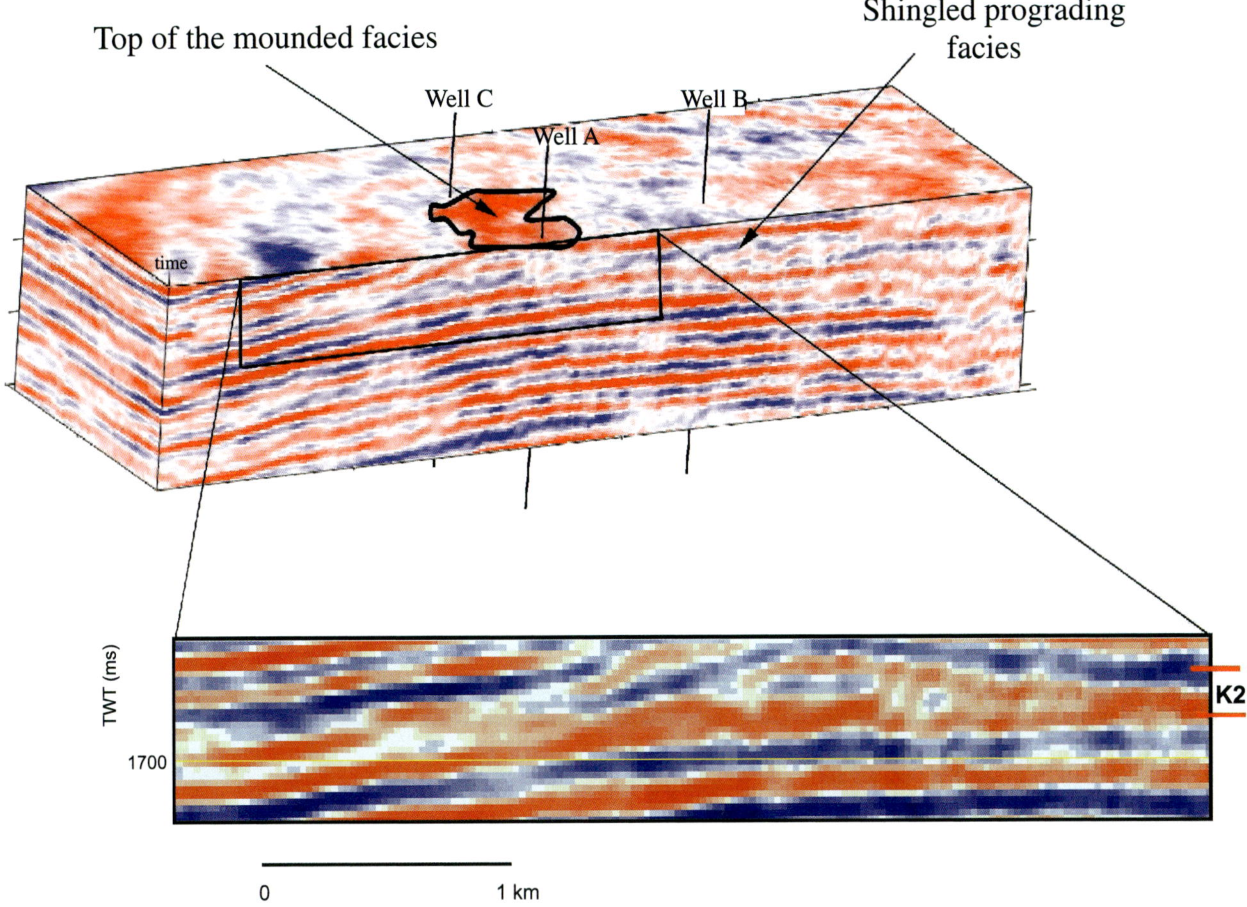

FIGURE 12. Khuff Formation, northern Oman, K2 reservoir. Structure-oriented filtering volume shows the interpreted mounded seismic facies. Top view of seismic cube shows amplitude anomaly (enclosed by the black line) that delineates extent of the mounded facies. Note change in polarity between the seismic cube and section. TWT = two-way traveltime.

the core and log data (Figure 16). This integrated model explained the good production performance of well A, drilled through the mounded shoal complex, when compared to the rest of the nearby wells.

Middle Miocene Isolated Buildup, Luconia Province, Malaysia

Numerous buildups of the late Oligocene–Miocene age form the Luconia carbonate province offshore Sarawak (Epting, 1980; Epting, 1989; Vahrenkamp et al., 2004). Carbonate deposition in Luconia started in the early Miocene on structural highs of faulted upper Eocene to lower Miocene siliciclastics (Ho, 1978; Vahrenkamp et al., 2004). During the middle Miocene, overall growth and demise coincided with a third-order relative sea level cycle.

This study concentrates on a well-defined platform-type buildup located at a depth of approximately 2800 m and with a vertical relief of about 600 m. The entire buildup is tilted slightly (~5–10°) toward the north-northeast. The internal architecture of the buildup was previously interpreted to consist of five reservoir zones based on porosity contrasts (Baumann et al., 1997). The buildup was penetrated by two wells that were drilled around the crestal area of the buildup (Figure 17).

Seismic Facies Description and Interpretation

A series of seismic sections through the study buildup show different depositional geometries and seismic facies that characterize the five reservoir zones (Figures 18–20). The seismic facies can be described based on internal reflection character, lateral continuity, and 3-D geometry. Seismic facies were then calibrated using logs and part of one cored well.

Reservoir zones 1 and 2 exhibit a pronounced asymmetry in terms of stratal seismic geometries (Figures 17, 18). Zone 1 toward the west-southwest end of the platform is characterized by high-amplitude, stacked-reef/mounded seismic facies (Figure 18). The remaining approximately two-thirds of the platform are parallel, onlapping seismic reflections. Zone 2 shows north-northeast–prograding seismic facies, which change to more continuous to transparent reflections (Figures 19,

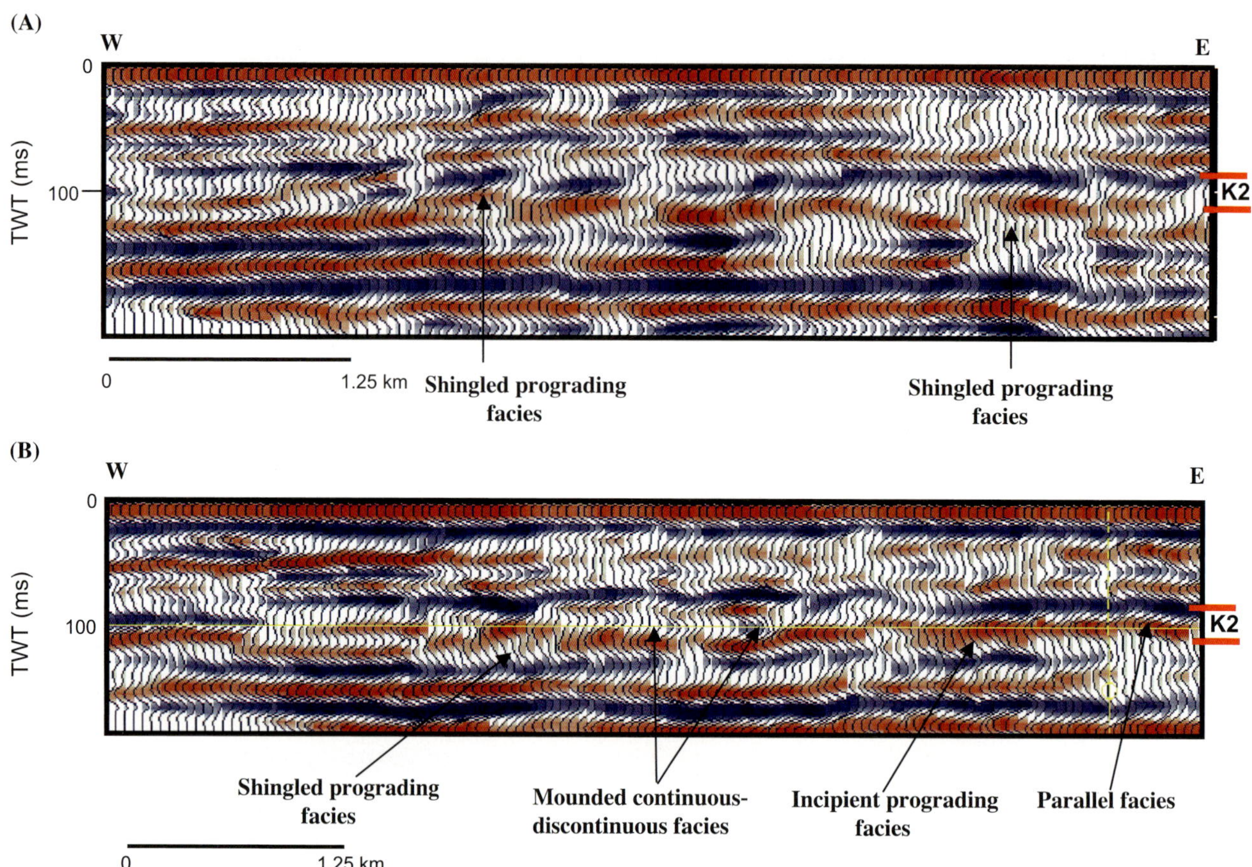

FIGURE 13. Khuff Formation, northern Oman, K2 reservoir. Flattened seismic sections. (A) Same seismic line as in Figure 11 showing main direction of progradation to the west. Some shingled units seem to prograde slightly to the east. (B) Cross section located to the south of line in Figure 10 showing variability of prograding geometries from sigmoids in the north to shingled to the south. Note also incipient prograding reflections that were not probably resolved with seismic resolution. See Figure 8 for location. TWT = two-way traveltime.

20A). Seismic facies in zones 3 and 5 consists of high-amplitude, more or less continuous, flat-lying reflections (Figure 17A). The intermediate reflection (marker reflection in Figure 17C) is continuous throughout the area and was used as a marker horizon to flatten the seismic sections. Seismic facies in zone 4 are characterized by shingled seismic geometries in the southern part of the buildup changing to more continuous, high-amplitude reflections toward the north (Figure 20A). Zone 5 shows more continuous reflections sometimes interrupted by localized mounded facies (Figure 20B).

Seismic images of the reef/back-reef seismic facies were extracted using gate-amplitude extractions, flattened volume semblance, and bodychecking (Figures 18, 19, 21). Root mean square amplitude extractions using an amplitude window of 30 ms were applied to the reflectivity volume to extract the reef/mounded seismic facies in zone 1 (Figure 18). Root mean square amplitude is the square root of the average of the squares of the amplitudes and it is calculated within a defined window (Figure 18B). Because of the extra reflections that are contained within the reef seismic facies, a constant-amplitude window (30 ms) run through the entire seismic volume gives higher-output amplitude values (red and green colors in amplitude map in Figure 18). The result is an amplitude map that shows a narrow distribution of higher amplitudes that correspond to the reef seismic facies confined to the west-southwest part of the buildup. The back-reef/lagoonal seismic facies are also captured by the amplitude extraction as an amplitude anomaly adjacent to the reef seismic facies (Figure 18). Volume semblance was also calculated and then flattened to the continuous reflection interpreted as a flooding surface (Figure 19). A time slice through a flattened semblance volume shows a better-defined, linear seismic reef trend with high output values, whereas lower semblance values represent the onlapping reflections of the back-reef seismic facies, which are distributed more randomly. The lowest semblance values (white in Figure 19) indicate greater similarities between traces that correspond to the more continuous seismic facies in the off-reef, lagoonal setting. Bodychecking was applied to the reflectivity volume to extract the high-amplitude seismic bodies of the reef/back-reef seismic facies by

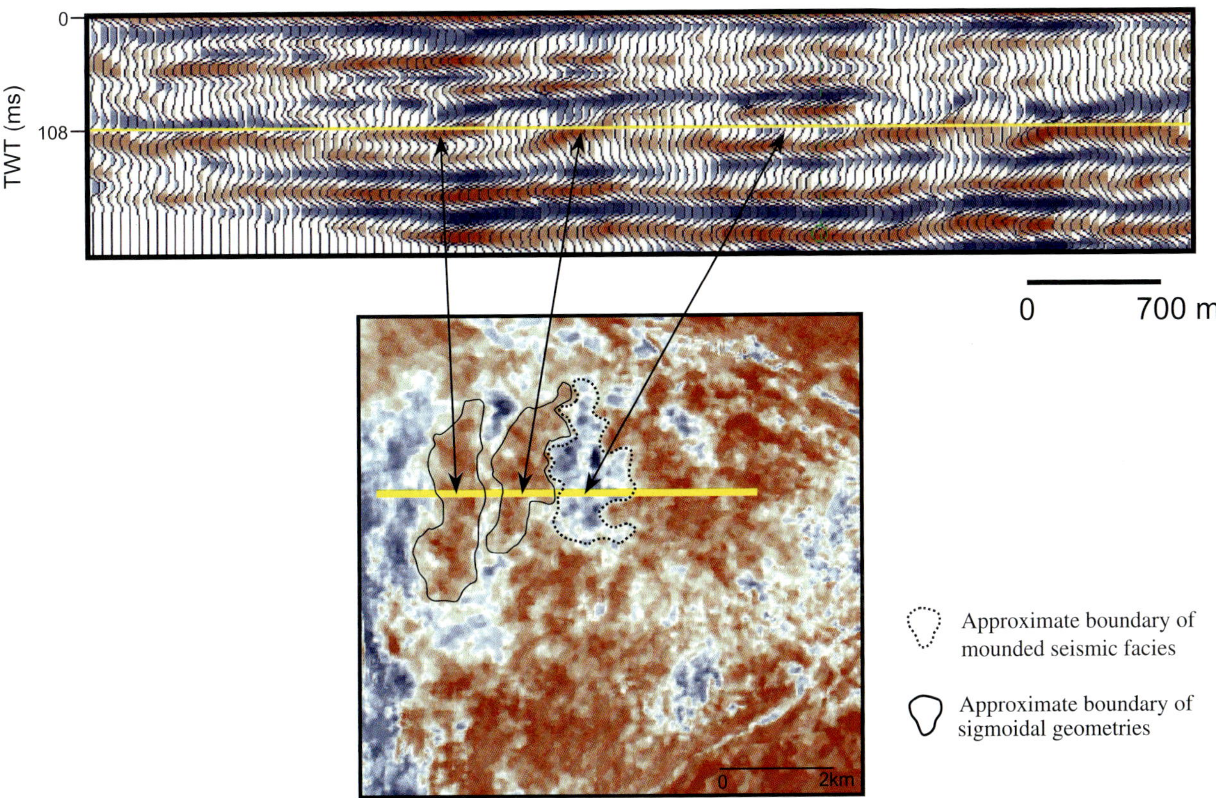

FIGURE 14. Khuff Formation, northern Oman, K2 reservoir. Flattened seismic section and time slice of line shown in Figure 10 showing map extent of progradational sigmoids vs. mounded seismic facies. Time slice was taken at the reservoir level indicated by the yellow line (108 ms flattened two-way traveltime [TWT]).

defining the maximum amplitude range of connected voxels for this particular seismic facies (Figure 21). The result is a 3-D distribution of detected bodies that shows the seismic reef tract and onlapping back-reef facies.

Core description from one of the wells (well 2) and log interpretation performed in both wells were calibrated with seismic facies (Baumann et al., 1997). The tight intervals in the upper part of zones 3 and 5 are interpreted to be slightly argillaceous sediments that were deposited during flooding events (Figure 17; Baumann et al., 1997). Seismic expressions of these events are continuous, horizontal, high-amplitude reflections (Figures 17, 20A) caused by the acoustic impedance contrast between the dense tight flooding events and the porous reservoir intervals in zones 2 and 4. Prograding seismic geometries in zones 2 and 4 indicate a progradation from the platform edge toward the interior with the consequent development of porous intervals deposited in a back-reef lagoonal depositional setting (Baumann et al., 1997). Seismic facies in zone 1 shows stacked seismic reefs indicating a main aggradational phase followed by a widespread regional exposure prior to the final drowning of the buildup (Vahrenkamp, 1996; Vahrenkamp et al., 2004). Obviously, the platform rim aggraded first and then the reefal facies expanded into the back-reef lagoonal area. Reef seismic facies are located at the west-southwest edge of the buildup and are interpreted as a reef complex, which, along with the main north-northeast direction of progradation, indicates the influence of inferred dominant paleowind direction for the middle Miocene in these area (Vahrenkamp, 1996). Simulated wind circulation for the middle Miocene shows two main directions of paleowinds for the summer and winter monsoon winds (Figure 19; Vahrenkamp, 1996; Vahrenkamp et al., 2004). Sea level fluctuations and the effect of the middle Miocene paleowind system were the two main controlling factors that defined the internal architecture of the buildup and the spatial reservoir distribution.

Upper Cretaceous Ramp-Type Carbonate Reservoir, Natih E Formation, Oman

The middle Cretaceous Natih Formation of Oman was deposited on a shallow, west-dipping carbonate platform (Burchette, 1993). The Natih Formation is uppermost Albian, Cenomanian, and lowermost Turonian in age and, in terms of large-scale depositional cycles, it has been described as long-term shoaling-upward sequences (Harris and Frost, 1984; Burchette, 1993; Philip and Al-Maskiry, 1995; van Buchem et al., 1996). The Natih Formation was divided into different units in the subsurface (A–G; Hughes-Clark, 1988) and the Natih E is

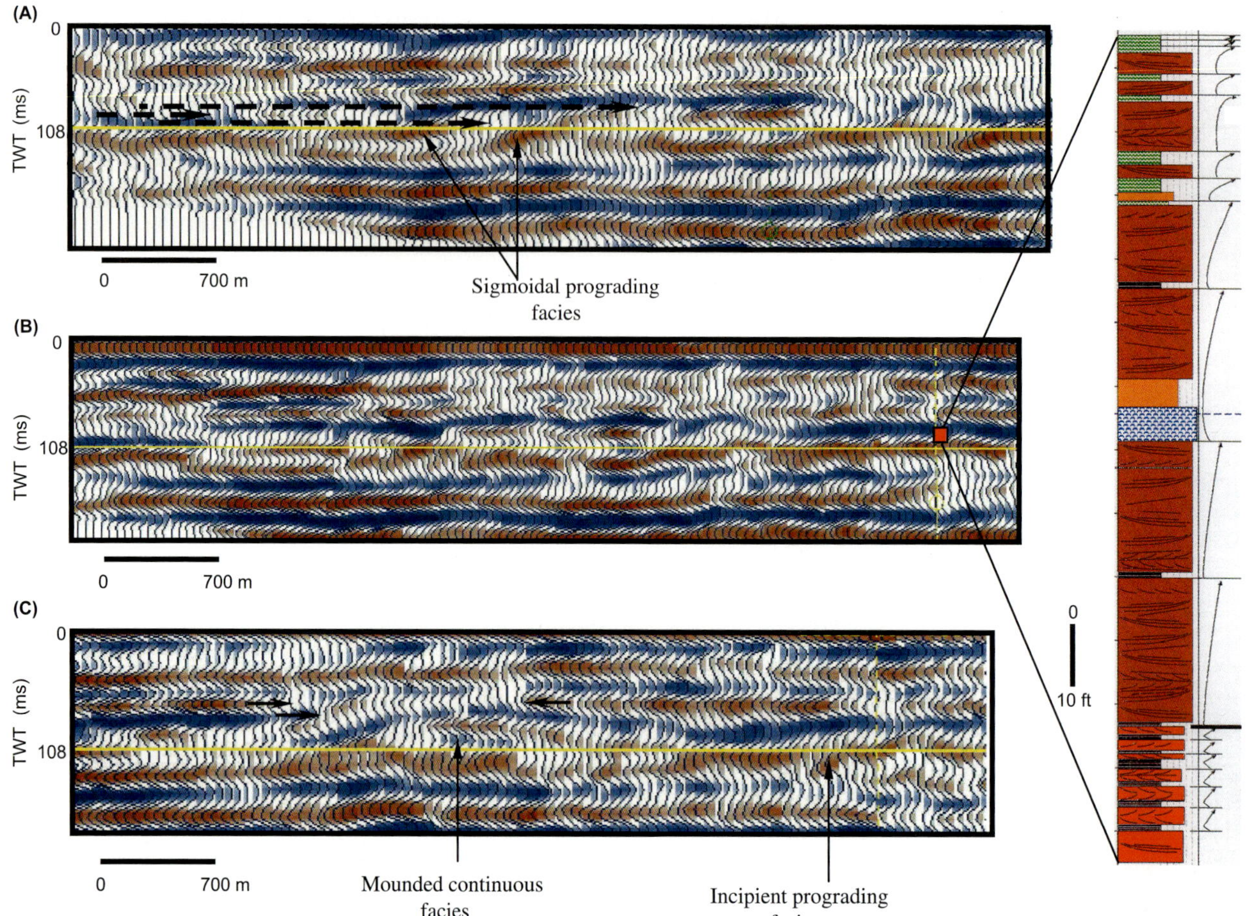

Figure 15. Khuff Formation, northern Oman. Flattened seismic sections showing variability of seismic geometries in a distance of less than 1 km (see Figure 8 for location) at the K2 reservoir level. (A) Seismic section (same line as Figure 14) showing low-angle sigmoids arranged in a downstepping, offlapping manner (dashed arrows) suggesting an overall decrease in accommodation space. (B) Core interpretation showing upward-thinning skeletal-ooid shoal cycles suggesting gradual progradation and decrease of space available represented by the algal laminites capped cycles toward the top of the section. Red colors represent cross-bedded, skeletal-ooid grainstones, orange colors represent ooid-skeletal packstones, dark-gray colors represent skeletal mudstones, blue and white pattern represents thrombolite boundstone, and dashed green pattern represents algal laminites (C. Kerans, personal communication, 2000). Flattened seismic section (same line as shown in Figure 13B) depicts core location and shows that seismic resolution is not enough to resolve prograding geometries. (C) Flattened seismic section showing aggradational topographic relief, which probably created the space needed to develop the seismic-resolvable westward prograding geometries. Black arrows indicate onlapping reflections. TWT = two-way traveltime.

the reservoir interval that is described in this study example.

Image Processing, Seismic Facies Description, and Interpretation

Structure-oriented filtering was applied to the original 3-D seismic data for the study field aiming to (Figure 22):

- reduce or suppress noise, thus, improving lateral reflection continuity and seismic facies especially at the crest top (main reservoir area),
- improve reflection termination and geometries to better define reservoir architecture,
- improve the definition of the Top Natih E horizon by "smoothing out" previous noisy horizon interpretation,
- produce attribute volumes, such as semblance (Figure 23), to constrain structural-stratigraphic interpretation.

The original 3-D seismic data were used as a reference to compare and constrain the image-processed seismic. Three main seismic facies were recognized based on reflection geometry, reflection continuity, and seismic reflectivity:

1) Prograding facies are restricted to the northern part of the field (Figure 24) and consist of low-angle-dipping, high-amplitude reflections. The high-amplitude character of the sigmoids is

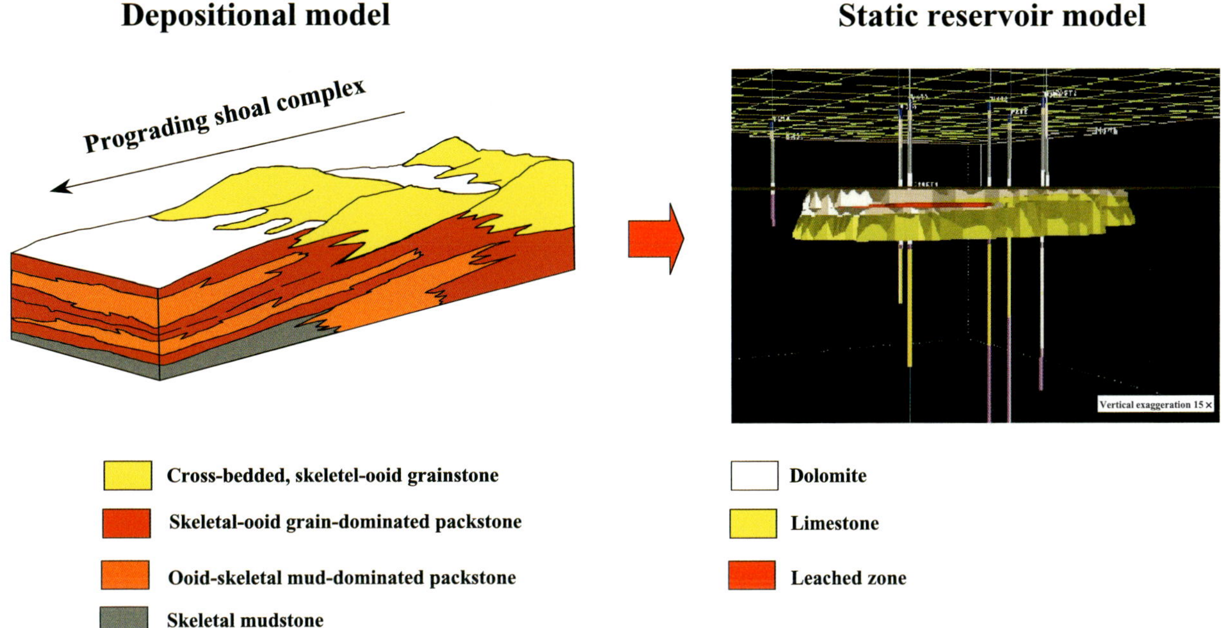

FIGURE 16. Khuff Formation, northern Oman, K2 reservoir. Depositional and static reservoir models interpreted using seismic, core, and log data.

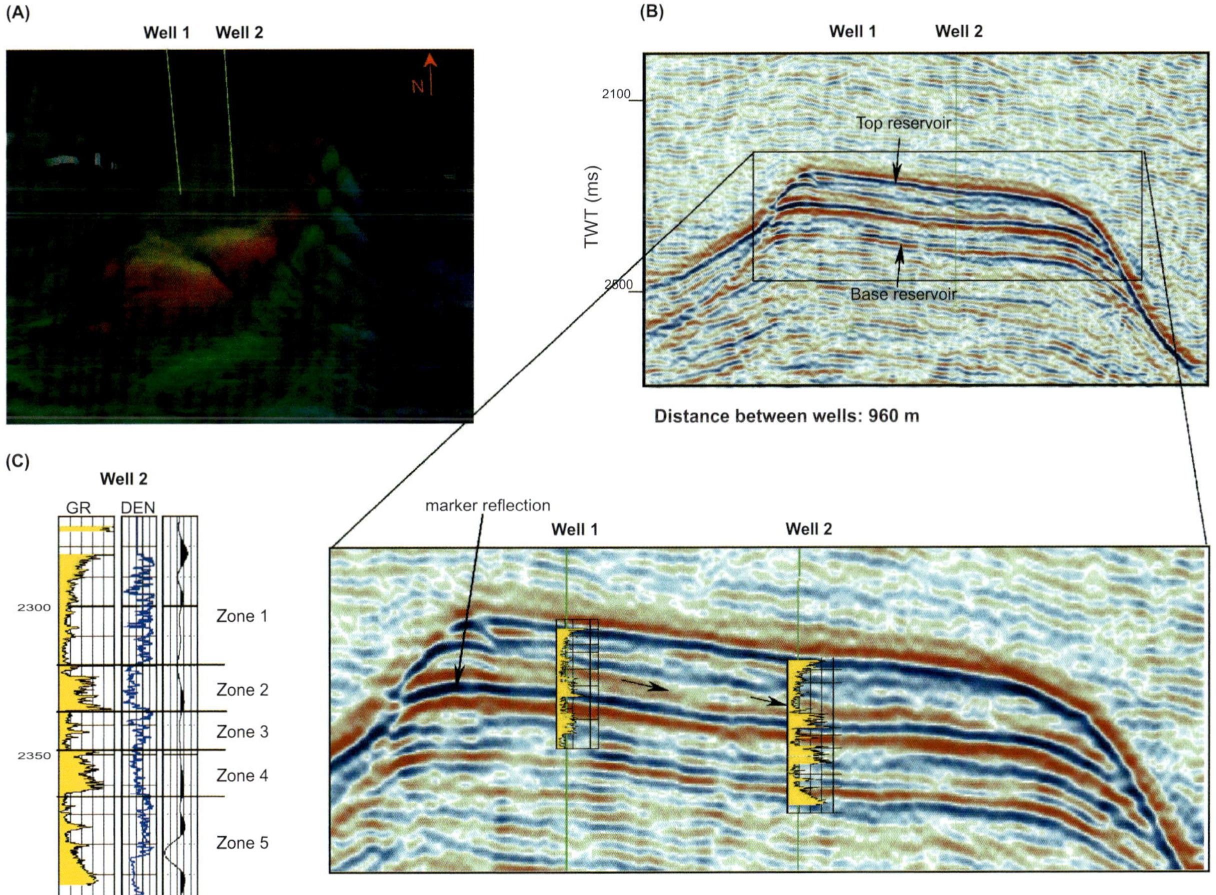

FIGURE 17. Miocene buildup, Luconia Province, Malaysia. (A) Top reservoir 3-D graphs showing distinctive characteristic relief of the Luconian middle Miocene buildup. (B) Seismic section across the two drilled wells with superimposed gamma ray logs. Black arrows are interpreted downlap reflections. TWT = two-way traveltime. (C) Gamma ray (GR), density (DENS), and synthetic seismogram showing the five reservoir zones.

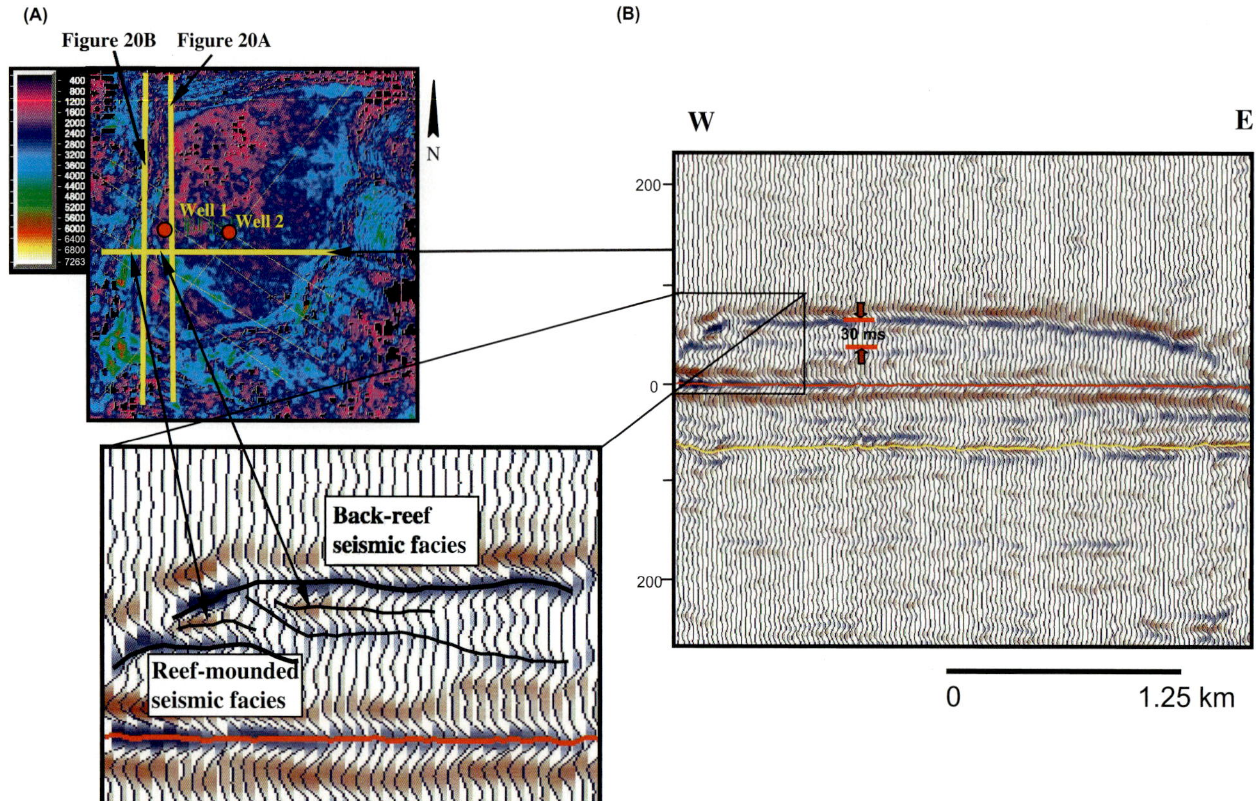

FIGURE 18. Miocene buildup, Luconia Province, Malaysia. (A) Gate amplitude map showing map distribution of the interpreted reef seismic facies. Red and green colors are high-amplitude values and blue and light blue colors are low-amplitude values. (B) Flattened seismic section showing the reef/back-reef seismic facies. Amplitude gate of 30 ms is indicated with red arrows.

interrupted locally by a dimming of amplitudes caused by the overlap between the termination of one sigmoid and the beginning of the next one (Figure 24). Time slices through the flattened reflectivity volume show remarkably well the sigmoid overlapping and thus the east-west trend of the prograding units (Figure 24A). The calculated combined dip and azimuth volume (flattened on Natih E horizon) also shows the trend of progradation (Figure 25).

2) Continuous-semicontinuous facies forms the majority of seismic facies observed within the Natih E interval (Figure 26). It is characterized by high-amplitude, parallel to subparallel seismic reflections. Occasionally, reflection continuity is disrupted, either by faults or by remnant seismic noise.

3) The chaotic to transparent facies is represented by discontinuous reflections with low to moderate reflectivity. This facies is associated with internal faulting and/or nonorganized noise and occurs at the southern crestal part of the field (Figure 24B). Reflection definition in the crestal part of the field is also obscured by possible gas-escape effects. The areal distribution of this facies is shown well in the flattened reflectivity volume at different levels within the reservoir (Figure 24A).

Depositional Seismic Geometries and Core Calibration

A detailed analysis of several seismic sections and volume attributes within the reservoir reveals the presence of seismic geometries that are interpreted as depositional features. Core interpretation from well AG-15 (Smith and Eberli, 1999; Smith et al., 2003) study was used in conjunction with the seismic data to refine the depositional interpretation. Smith and Eberli (1999) interpreted five main facies in the AG-15 60-m core in the Natih E: (1) skeletal pelletal grainstone, (2) rudist algal wackestone-boundstone, (3) burrowed skeletal packstone-wackestone, (4) argillaceous silty dolomite, and (5) brecciated crystalline limestone. The burrowed wackestones located at the base of the core were interpreted as the late transgressive part of a third-order cycle (Figure 26). The uppermost part of the sequence is composed of burrowed skeletal wackestone and rudist boundstones capped by a paleosol unit. This sequence was interpreted to represent the upper regressive part of the third-order

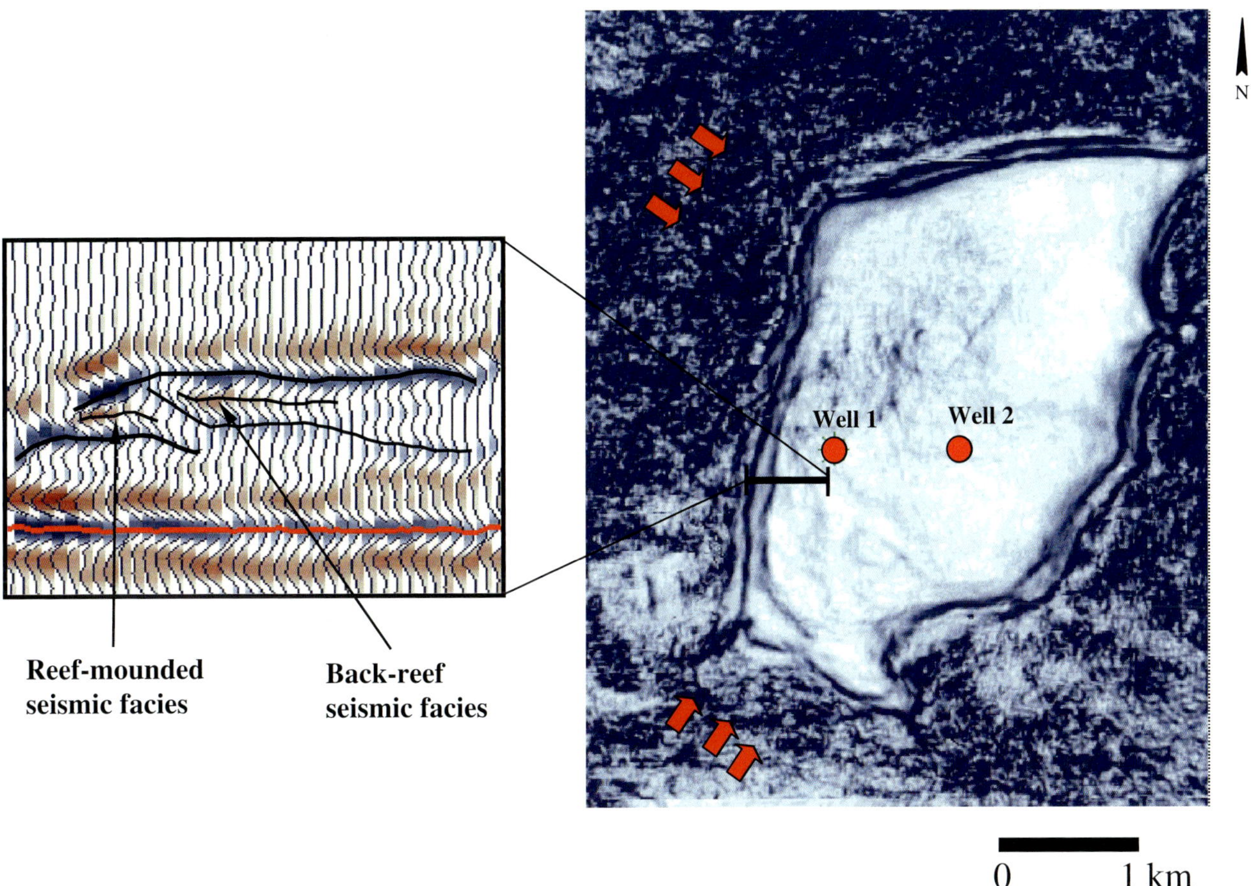

Figure 19. Miocene buildup, Luconia Province, Malaysia. Flattened volume semblance showing high semblance values (dark blue) that correspond to the reef seismic facies. Intermediate semblance values represent the less linear back-reef deposits. Low semblance values indicate flat-lying reflections interpreted as lagoonal seismic facies. Red arrows indicate interpreted paleowind directions for the middle Miocene.

cycle subject to periods of subaerial exposure (Figure 26; van Buchem et al., 1996; Smith and Eberli, 1999). Three medium scale cycles (10–30 m thick) preserved in the part of the Natih E sequence included in the AG-15 core that correlate with the medium-accommodation cycles of the outcrop study (Figure 26; van Buchem et al., 1996). Most of the reservoir grainstones occur within the highstand part of medium-scale cycle 2.

Interpretation of seismic facies and geometries throughout the field shows the basic architecture of the reservoir (Figure 27A). The central part of the field is characterized by the presence of two extra seismic reflections underneath the Top Natih E reflection (negative loop) that cannot be correlated across (Figures 26, 27). Terminations of these two extra reflections to the north indicate an aggradational depositional pattern that probably represents a rudist-shoal complex environment. The two extra reflections that represent the upper grainstone units (cycle 2) seem to continue into the water leg (Figures 26, 27B).

The orientation of the clinoforms indicates progradation of the facies to the north from the interior of interpreted shoal complex geometries (Figures 27A). These clinoform geometries are interpreted to represent a series of shallow-water, prograding carbonates. This interpretation coincides with field observations (van Buchem et al., 1996; Eberli and Smith, personal communication, 1999) wherein strong progradational pulses brought shallow-water carbonates several kilometers out into the basin, but the lack of core control prevents a definite assessment of the lithology that forms these seismic facies.

The Base Natih E is represented by continuous, parallel to subparallel reflections (Figure 27B). The core provides direct information about the lithologic composition of the upper part of these reflections. Based on the core description and regional correlation, we infer that the base of the Natih E reflections represents the transgressive phase of the Natih E sequence (Figure 26). The Top Natih E is a continuous, high-amplitude reflection that can be easily traced across the field. However, in some seismic sections, it appears to change laterally to an erosive surface (Figure 27). The erosive nature of the Top Natih E reflection coincides with a decrease in thickness observed in the Base Natih E/Top Natih E isochore map (Figure 27A). AG-15 core shows the presence of a

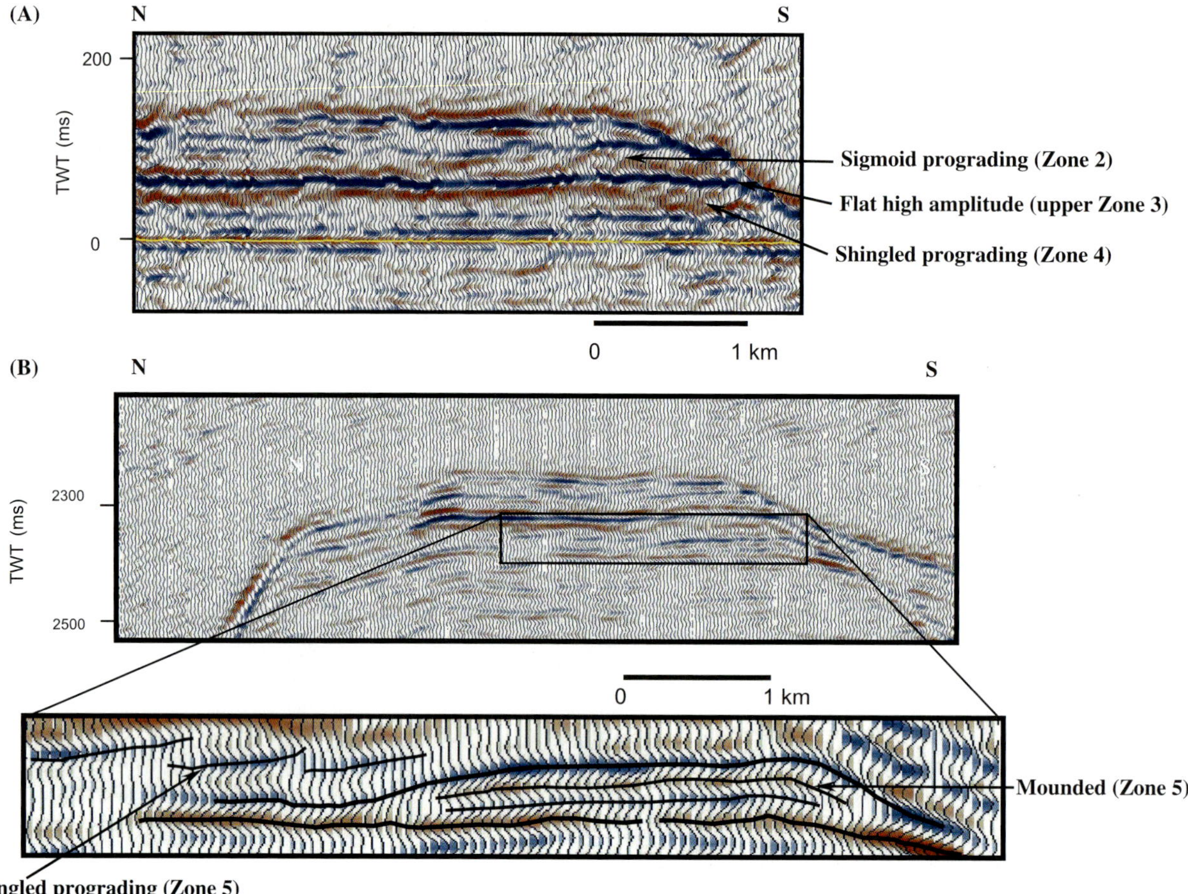

FIGURE 20. Miocene buildup, Luconia Province, Malaysia. Prograding reef system initiated as smaller buildups that then coalesced through time filling in the space available to form a larger platform. (A) Flattened seismic section showing characteristic seismic facies and lateral heterogeneity within the five reservoir zones. See Figure 18 for location. (B) Seismic section showing reef seismic facies and prograding foresets within reservoir zone 5. See Figure 18 for location. TWT = two-way traveltime.

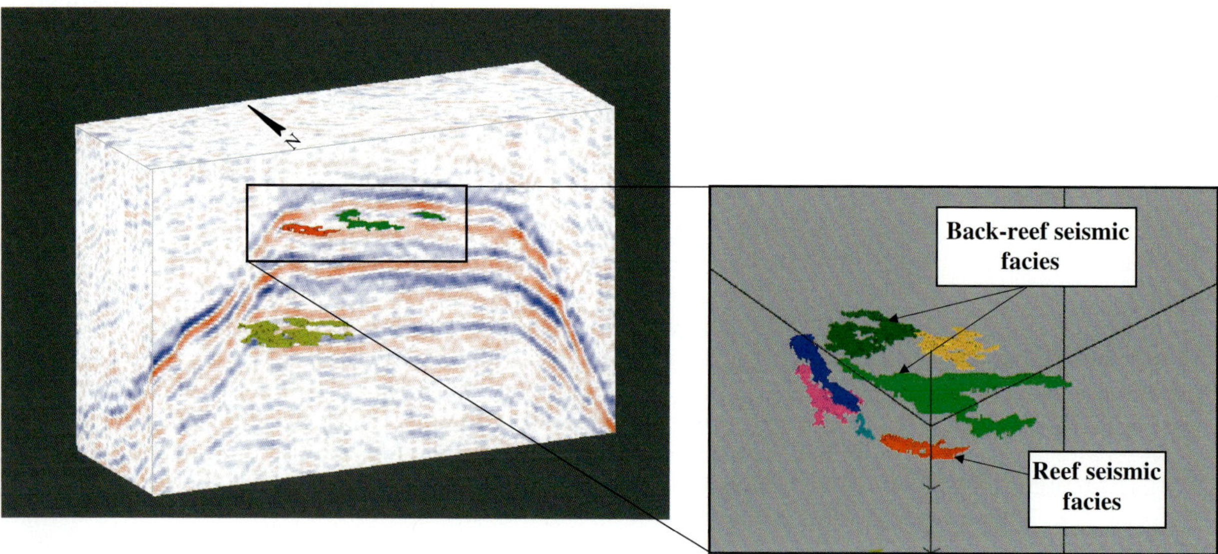

FIGURE 21. Miocene buildup, Luconia Province, Malaysia. Bodychecking on reflectivity volume using calibrated amplitude ranges to extract reef/back-reef seismic bodies. Red, blue, and pink extracted bodies show the linear space distribution of the seismic reef tract. Green and orange extracted bodies represent the back-reef seismic deposits.

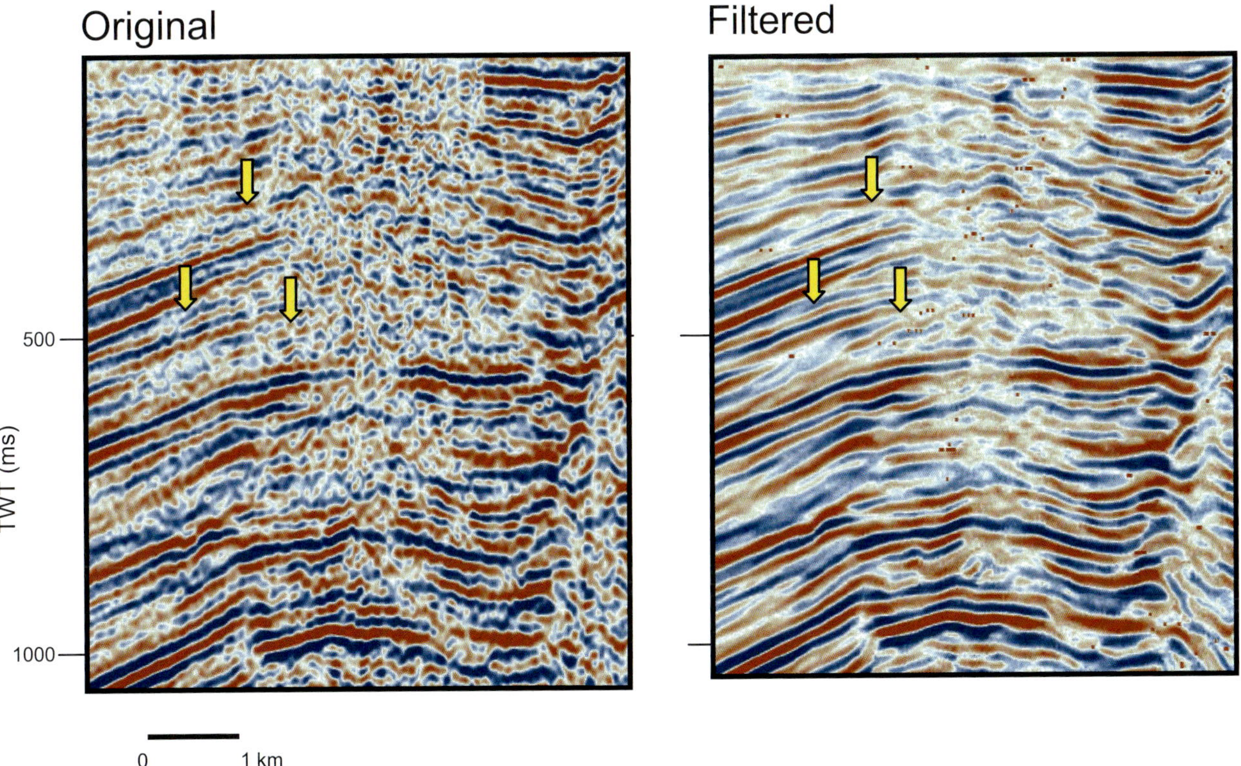

FIGURE 22. Natih E Formation, Oman. Seismic reflectivity cross section before and after applying SOF. Yellows arrows indicate significant improvement in continuity and definition at reflection terminations. Red dots are artifacts from image-filtering process. TWT = two-way traveltime.

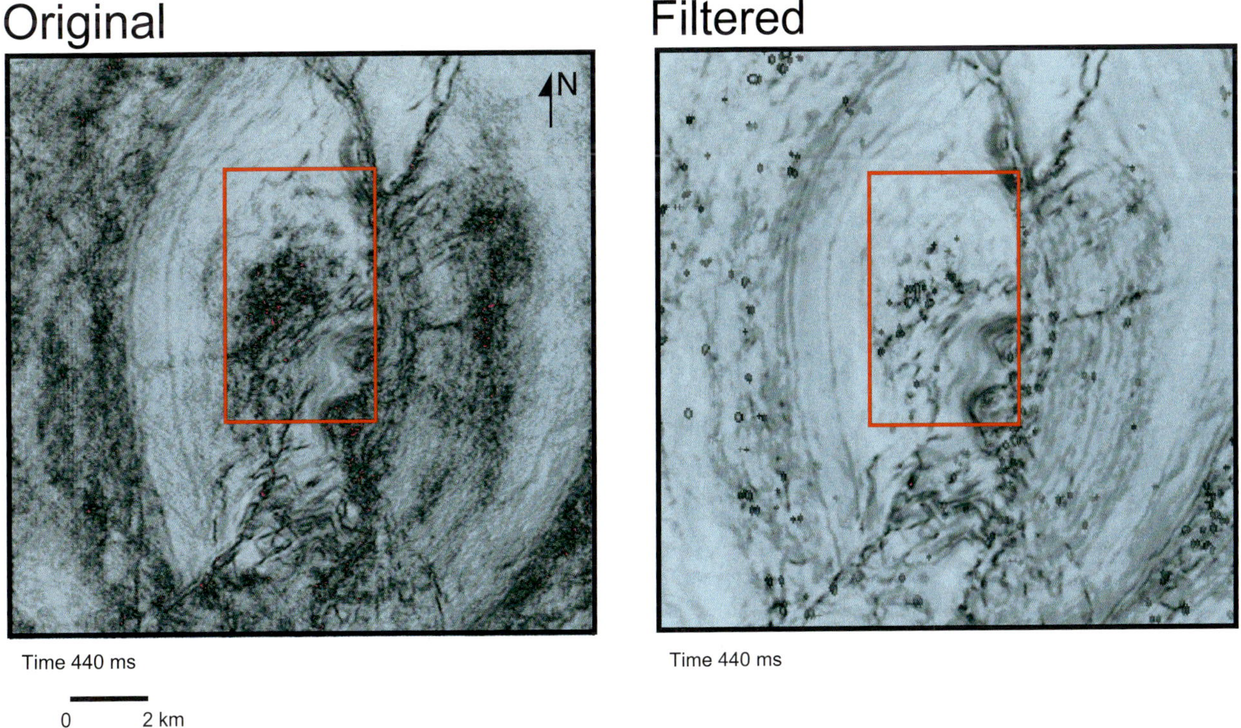

FIGURE 23. Natih E Formation, Oman. Volume semblance calculated from the original reflectivity data and from the image-filtered seismic data. Red square indicates approximately the location of the field.

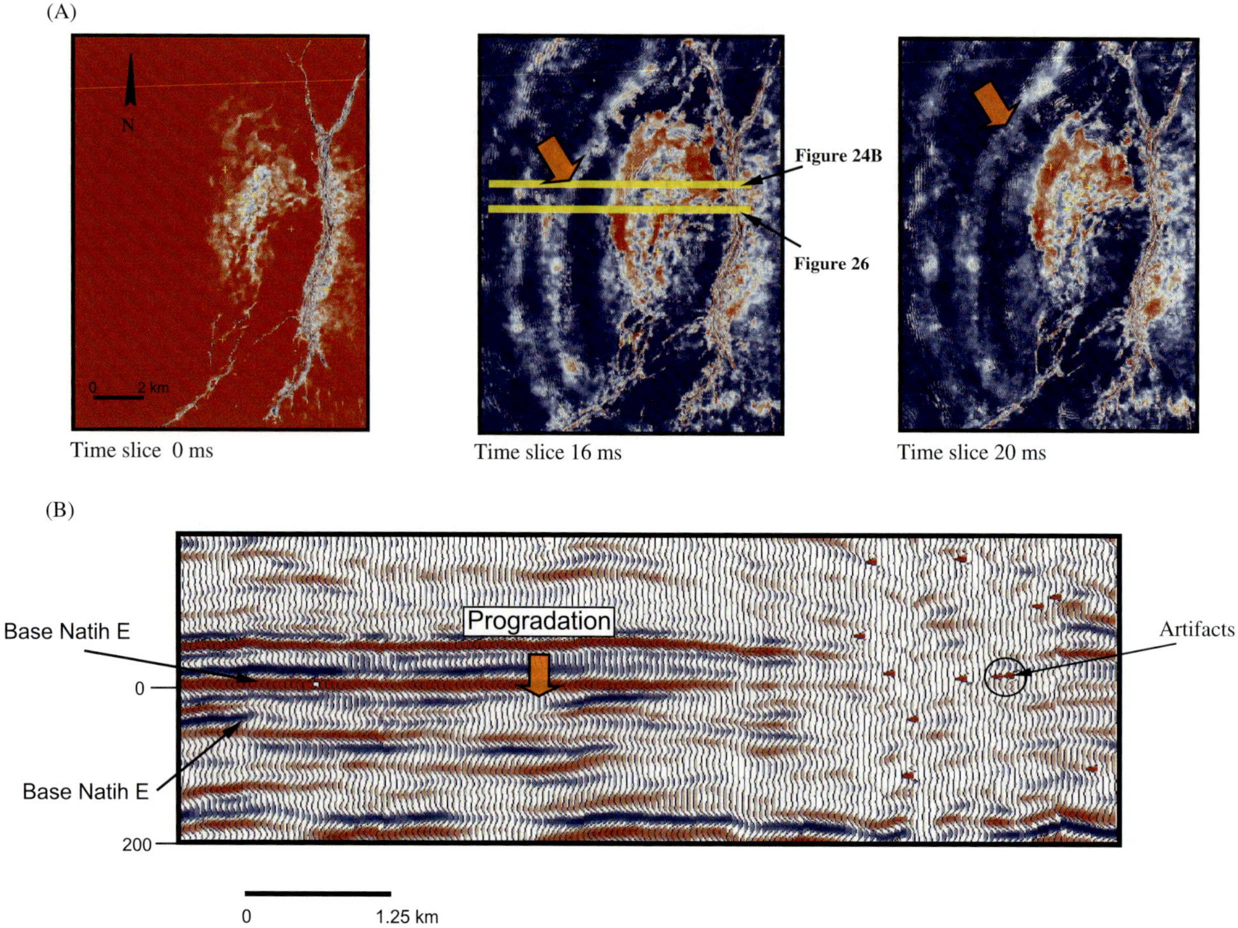

FIGURE 24. Natih E Formation, Oman. (A) Time slices through flattened, filtered reflectivity volume. Time slices at 16 and 20 ms below the flattened reference horizon (Top Natih E) show amplitude dimming caused by the overlapping between the termination of one clinoform and the beginning of the next one (red arrows). (B) Flattened seismic section showing low-angle progradation. Flatten horizon: Top Natih E. Red dots are artifacts from image-filtering process.

well-developed paleosol indicating a period of exposure (Smith and Eberli, 1999). Outcrop observations (van Buchem et al., 1996) indicate the presence of incisions (channels?) at the Top Natih E level.

Analysis of seismic facies and geometries provided an initial framework to constrain the lateral stratigraphic correlation based on core and log data. Permeable units within aggradational geometries observed in the crestal area of the field interpreted as ramp-crest shoal complex seem to continue into the water leg (Figure 27B). This interpretation evolved into one of the subsurface model scenarios that could explain the provenance of high water cuts observed in the field.

THREE-DIMENSIONAL IMAGING OF STRUCTURAL GEOMETRIES

The aim of this section is to show two examples of 3-D imaging of structures in two different carbonate reservoirs applying some of the imaging processing techniques previously described.

The first example is from a giant field in the Middle East and the intention was to image the fault network at different stratigraphic levels (Melville et al., 2004). The structure is a growth-elongated anticline with limbs dipping gently to the southeast and northwest and with the axes plunging to the northeast in the northern part and to the southwest in the southern part (Figures 28, 29). This anticline is offset at the reservoir level by a predominant set of approximately parallel west-northwest–east-southeast–striking faults (Figures 28, 29). Based on the 3-D geometry, the fold consists of a pre-, syn- and post-folding unit. The reservoir occurs within the prefolding units buried at about 1.5 s (two-way traveltime).

Structure-oriented filtering was applied to the reflectivity volume to remove the background noise and to enhance fault and reflection terminations (Figure 28). Volume semblance and volume dip were then calculated from the filtered volume to highlight both stratigraphic and structural discontinuities. Figure 29 shows time slices through semblance and dip volumes at the

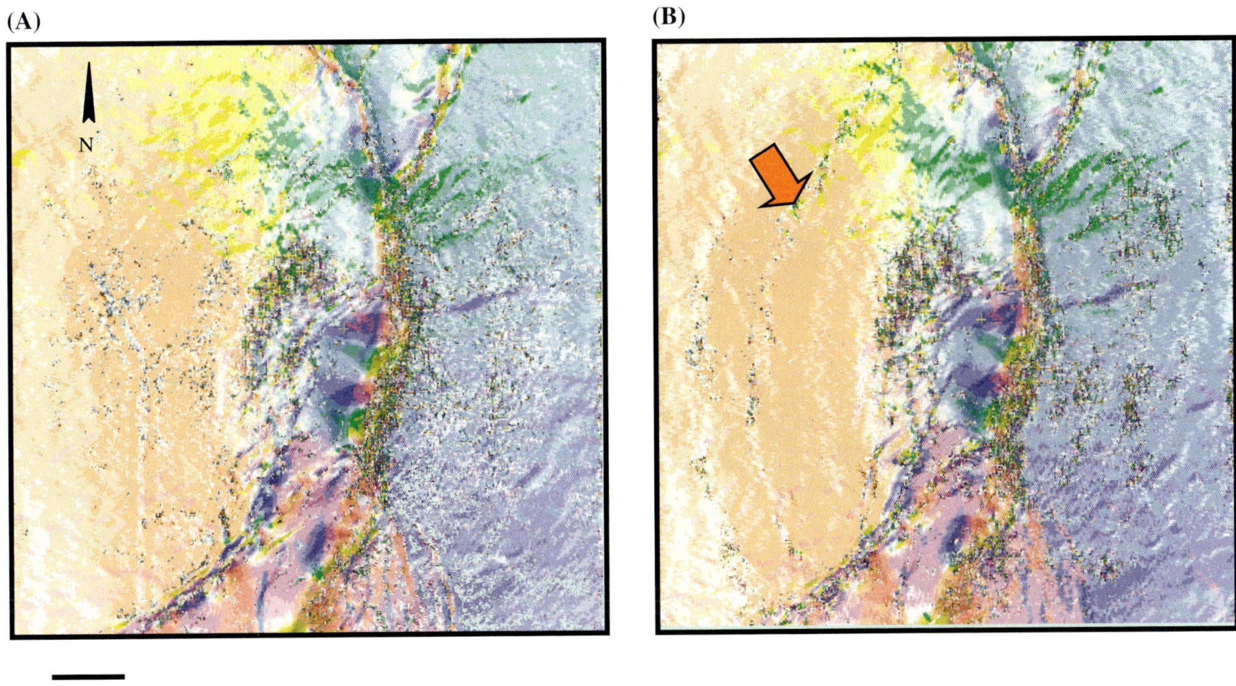

FIGURE 25. Natih E Formation, Oman. Combined dip and azimuth time slices applied to flattened reflectivity data showing orientation of prograding geometries. (A) At 0 time, no preferred orientation is observed. (B) Orientation of clinoforms is obvious at 16 ms below the flattened reference horizon.

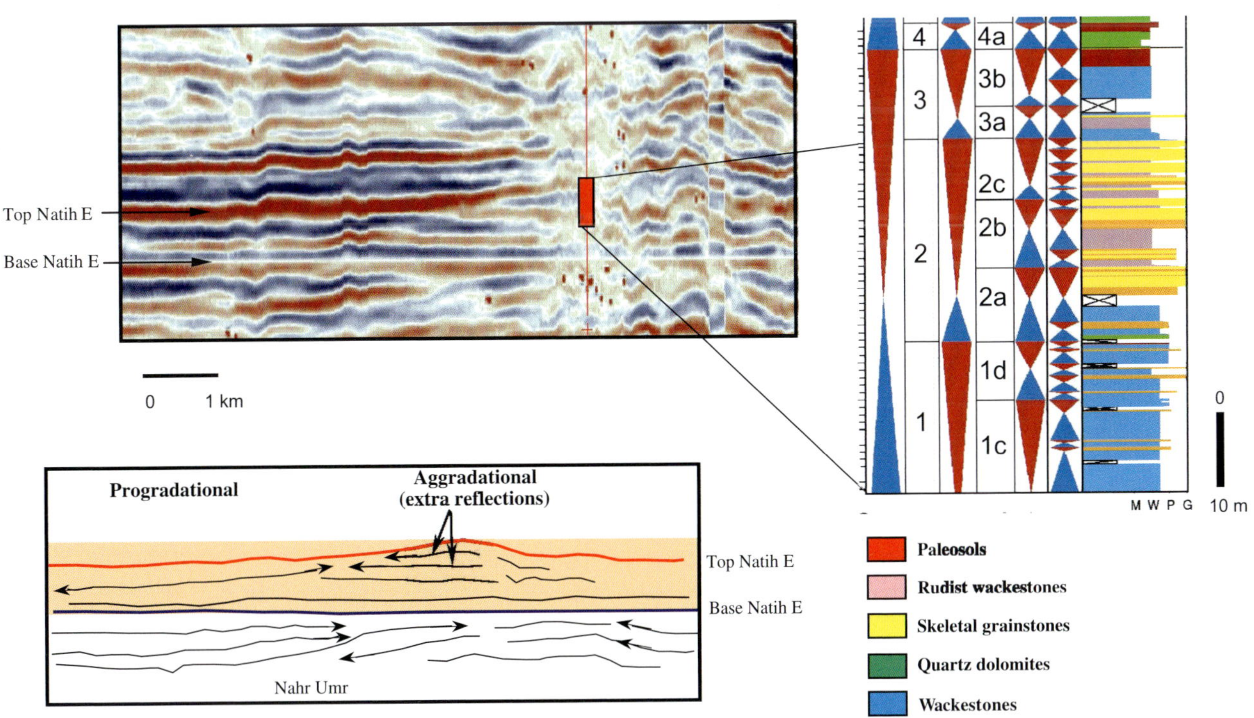

FIGURE 26. Natih E Formation, Oman. Flattened seismic section through AG-15 and cycles interpreted from the AG-15 core. The regressive part of the cycles coincides with the aggradational moundlike geometry observed in the seismic. The skeletal grainstones at the top of the cycle were interpreted as deposited in high-energy shoals in an open platform setting (Smith and Eberli, 1999).

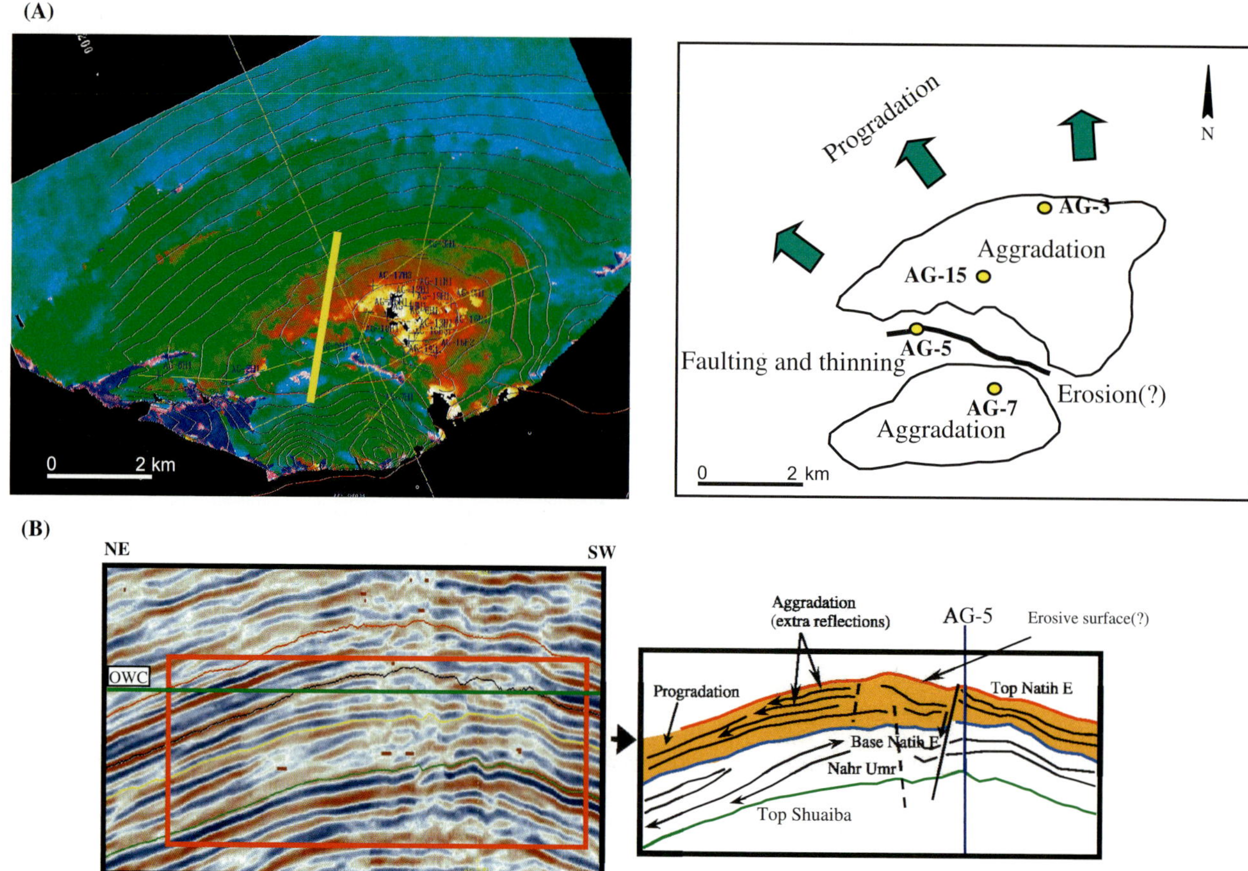

FIGURE 27. Natih E Formation, Oman. (A) Isochore map of the Natih E interval and seismic facies map. (B) Interpreted seismic section across the field. OWC = oil water contact.

reservoir interval (Thamama B). Both volumes show west-northwest–east-southeast–oriented main faults, which cut across the field and secondary smaller faults that form an angle with the main faulting direction. Filtered volume semblance highlights more accurately the offset of the discontinuity on both sides of the fault (inside box in Figure 29) than the dip-filtered volume. Semblance volume indicates clearly the right-lateral component and also the offset distance of the main fault zone.

The second example is from an Upper Cretaceous restricted marine-lacustrine carbonate reservoir in northwest Argentina. The reservoir unit occurs in a deep (~6000 m), complex tectonic setting represented by a tectonically inverted anticline (Figure 30). The study field is located at the western end of the Lomas de Olmedo subbasin approximately 100 km northeast of Santiago de Jujuy (Figure 30). In this area, the Cretaceous–Eocene stratigraphic units were deposited during synrift-postrift phases and were deformed by the Andean orogeny that began in the Miocene (Dewey and Bird, 1970; Jordan et al., 1983). At least three episodes of compression have been recognized in this region, but the last Pleistocene compressional event (Diaguita orogeny) is thought to be responsible for the reactivation of the ancient bounding faults of the Salta Group Basins (Grier et al., 1991; Mon and Salfity, 1995). The stratigraphy of the Salta Group has been discussed by several authors (Figure 30; Moreno, 1970; Salfity, 1979, 1982; Salfity and Marquillas, 1986; Gomez Omil et al., 1989). The maximum thickness of the Salta Group is about 6500 m in the Lomas de Olmedo subbasin. The base of the Salta Group is defined by a regional unconformity, which separates this unit from the underlying Precambrian–Paleozoic units. The top is also defined by a major angular unconformity, which separates the Salta Group form upper Miocene continental, foreland deposits (Oran Group). The Salta Group consists of three separate subgroups: Pirgua (Reyes and Salfity, 1973), Balbuena, and Santa Barbara (Moreno, 1970). Of these subgroups, Balbuena Subgroup is considered the most significant because it includes the Yacoraite Formation, the principal reservoir of the study field and majority of the hydrocarbon occurrences in the basin. The Balbuena Subgroup consists of the Lecho and Yacoraite Formations. Lecho units are continental deposits, whereas Yacoraite units correspond to mixed carbonate siliciclastic sediments deposited in a restricted marine environment.

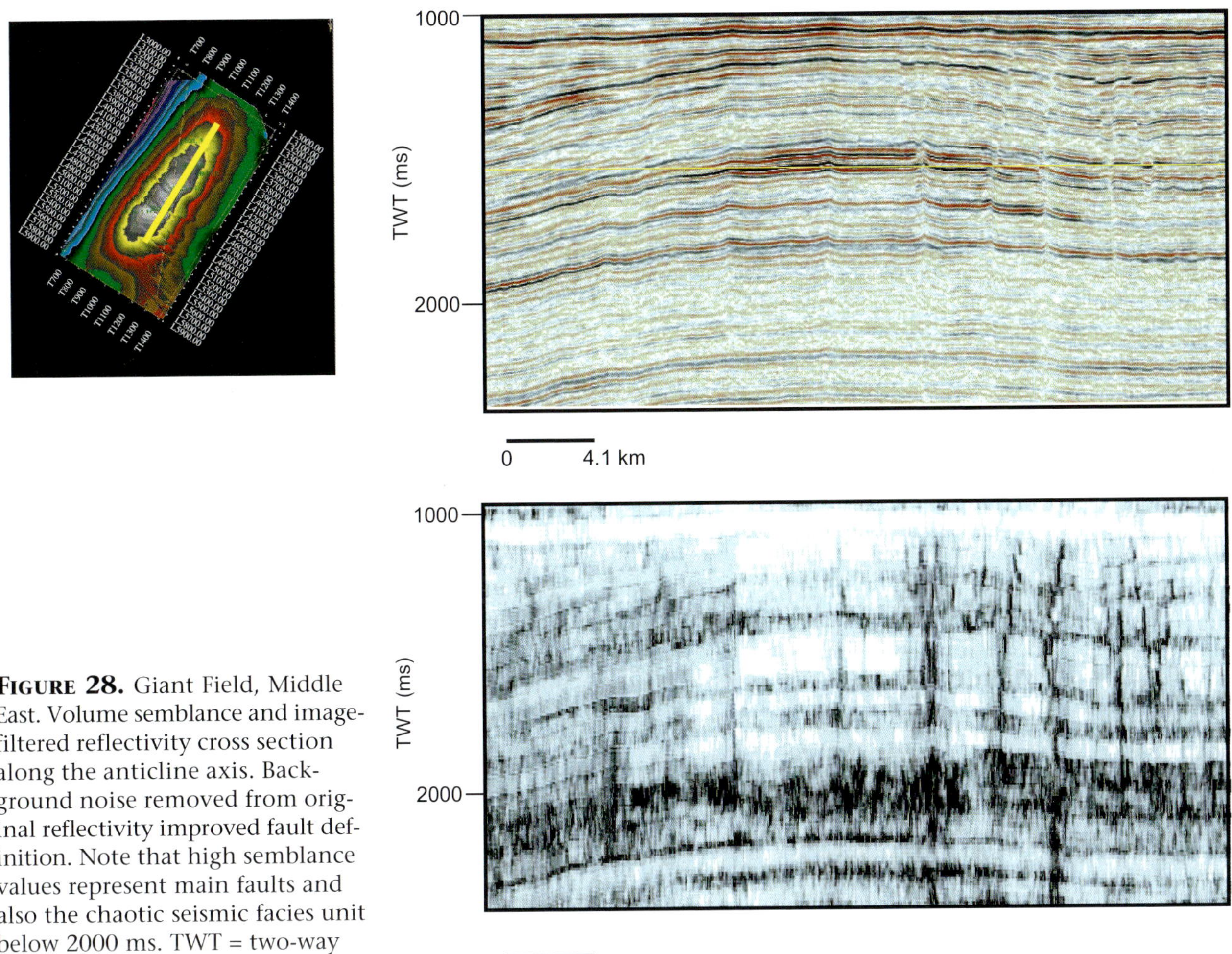

FIGURE 28. Giant Field, Middle East. Volume semblance and image-filtered reflectivity cross section along the anticline axis. Background noise removed from original reflectivity improved fault definition. Note that high semblance values represent main faults and also the chaotic seismic facies unit below 2000 ms. TWT = two-way traveltime.

One of the main goals of the project was to perform a kinematic study of the anticline and then try to construct a predictive model of fracture intensity and orientation. Only one well was drilled, so most of the interpretation relied on how well the base reservoir horizon and the fault system could be seismically imaged. Gaussian curvature analysis was applied to the Base Yacoraite seismic-interpreted horizon (Figure 31; Lisle, 1994). The result was a curvature map of the anticline that was used to calculate potential open fractures related to the kinematic evolution of the structure (Figure 31).

A standard dip-moveout seismic volume was used at the beginning of the project for a preliminary interpretation. The detail structural interpretation was performed using a filtered prestack depth migration seismic volume. Image filtering was applied to remove the high-frequency noise and thus improve the definition of the main 3-D fault planes (Figure 31). In- and cross-line sections from the semblance helped to define the main bounding faults, but semblance time slices were not very accurate to locate minor fault traces because of the incoherent noise within fault zones and also because of the steep dipping of the fold limbs (Figure 31).

The seismic horizon interpretation was performed on the SOF-filtered reflectivity volumes resulting in smoothed surfaces that were used as input for curvature analysis and for prediction of potential open fractures. Image filtering of the reflectivity data helped to improve the 3-D visualization of the fault-related anticline and related seismic horizons (Figure 32).

CONCLUSIONS

Application of 3-D seismic image-processing techniques combined with core and log calibration provided with a robust methodology for the interpretation of carbonate reservoirs and structures. Case studies presented in this chapter demonstrate the use of the different image-processing techniques to highlight key seismic geometries and facies for various types of carbonate systems and structural geometries. These studies showed

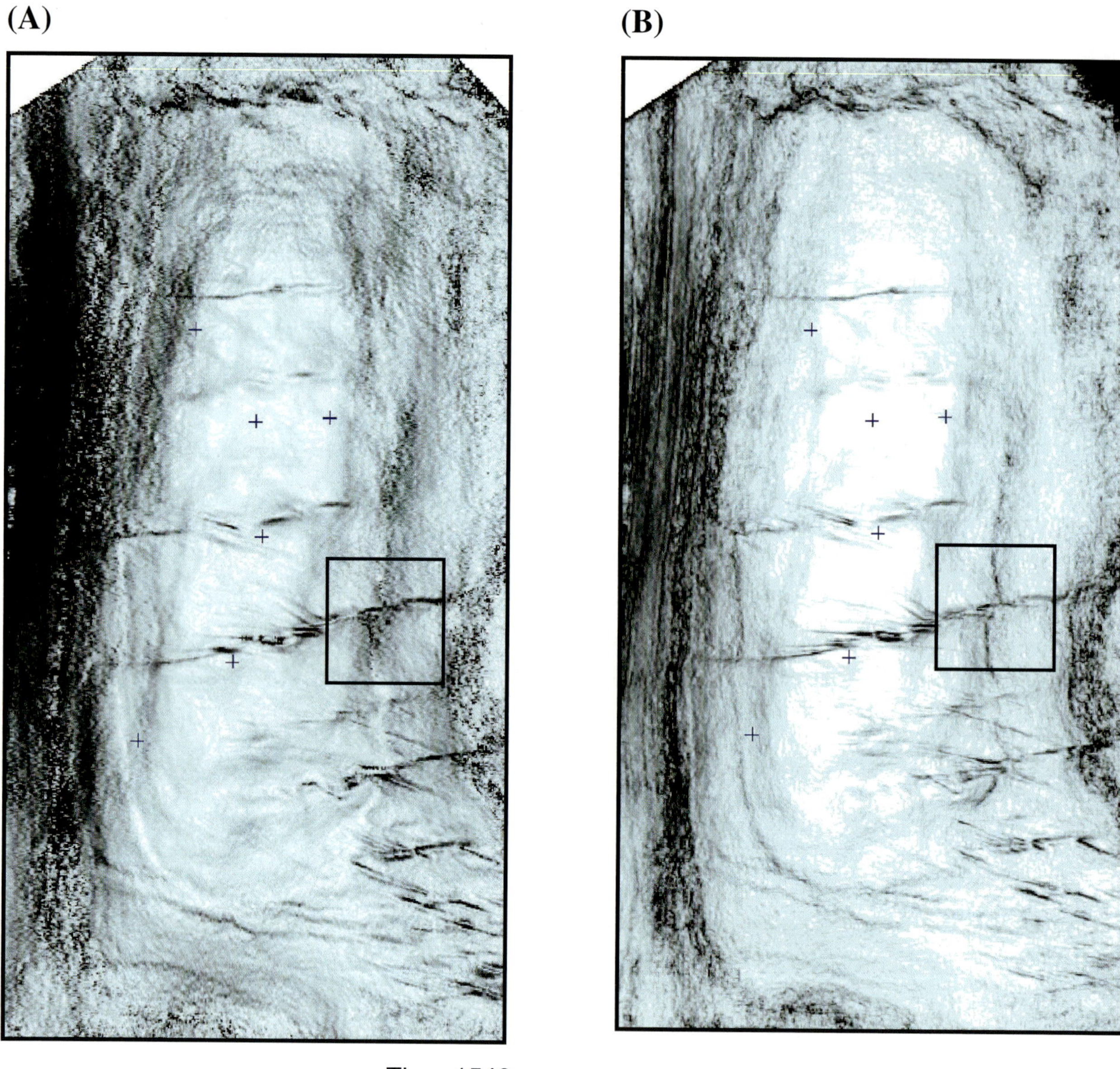

FIGURE 29. Giant Field, Middle East. (A) Volume dip and (B) volume semblance showing offset of main fault zone at the reservoir level.

that accurate seismic imaging of the reservoir architecture becomes an important predictive tool for reservoir characterization because it helped to build a 3-D geologic framework within which depositional facies can be distributed in time and space. Calculation of volume-based attributes produced new volumes of data in addition to reflectivity volumes that helped to extract the 3-D geometries within the reservoir.

In the Permian Khuff example, reservoir SOF considerably improved reflection termination and subsequent delineation of the topographically distinct shoal grainstone complex from the prograding units. Integration of the seismic geometries and facies with core and log data provided a reliable geologic and static model that could explain well performance.

In the middle Miocene Luconian isolated buildup, a combination of volume semblance, gate-amplitude extractions, and bodychecking techniques allowed the identification of depositional geometries within the five reservoir zones. Extracted geometries showed the 3-D spatial distribution of the seismic facies through time resulting in a lateral heterogeneous, dynamic carbonate system. Trangressive parts of the cycles are expressed as flat-lying, high-amplitude reflections, whereas highstand

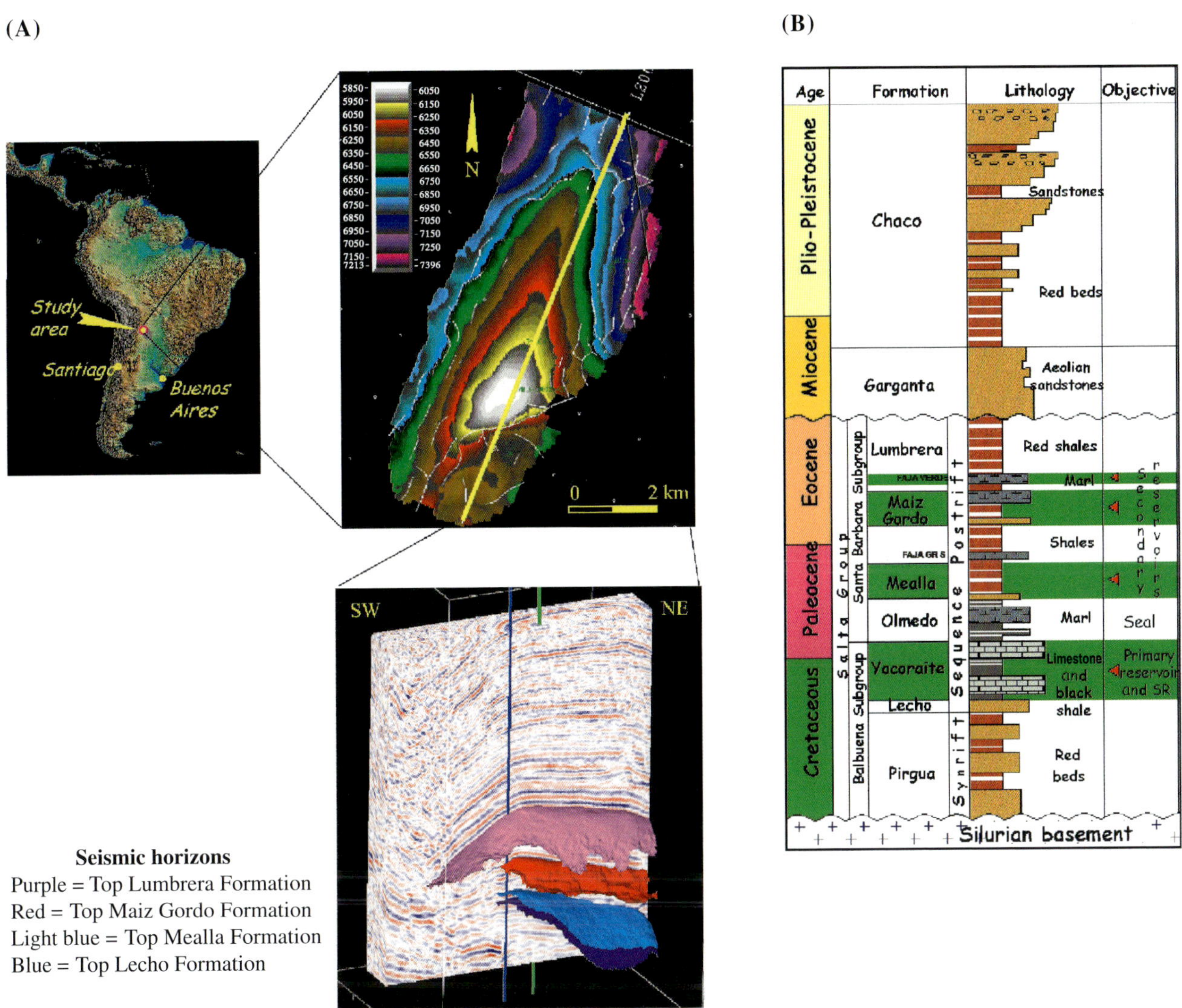

FIGURE 30. Anticlinal oil field, Lomas de Olmedo subbasin, Argentina. (A) Location and structure map at the top of reservoir (Yacoraite Formation) and cross section through imaged-filtered seismic volume. Colored horizons are different interpreted stratigraphic horizons above the reservoir. Dark blue is the interpreted base of the study reservoir. Blue vertical line is the only well drilled in the field. (B) Stratigraphic column for the study area.

conditions provided more accommodation space where aggrading-reef/back-reef and prograding seismic facies developed. Recognition of seismic facies heterogeneities had implications on the final building of the static model in terms of lateral correlation of stratigraphic units between the wells and on the 3-D distribution of petrophysical properties.

Structure-oriented filtering significantly improved definition and continuity of reflections in the Natih E reservoir. Time slices through flattened, image-filtered reflectivity, semblance, and combined volume dip and azimuth helped to delineate the reservoir zone. Integration of seismic interpretation calibrated with the cored well, logs, and outcrop analogs produced a static reservoir model, which was one of the model scenarios that could explain high water cuts in the field.

Texture and bodychecking were successfully applied to the Malampaya seismic data set to quickly identify and classify seismic facies and to extract the seismically detected good porous zones within the reservoir. The results showed the volume distribution of both seismic facies and porosity zones, which were used as input to target potential good wells and as a reference to constrain reservoir-quality distribution.

Semblance and dip-volume attributes combined with image filtering provided an accurate imaging of the fold-to-fault relationship in a giant Middle East carbonate field. The improved definition and continuity of the original seismic reflectivity by image filtering caused a sharp contrast between smooth and fault-disrupted reflections and, as a result, good definition of fault zones reflected in low semblance values. In an anticlinal structure in northwest

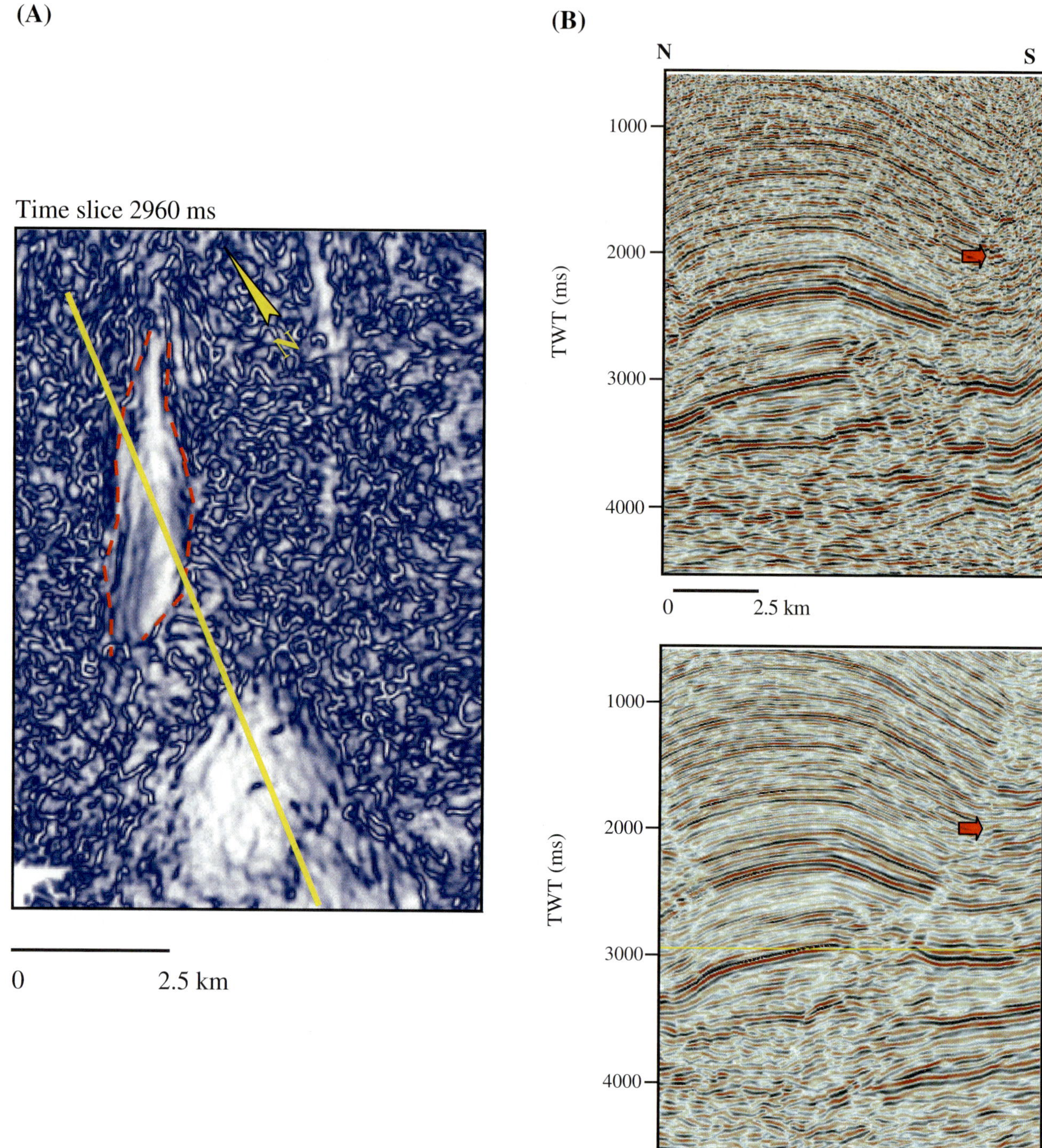

FIGURE 31. Anticlinal oil field, Lomas de Olmedo subbasin, Argentina. (A) Semblance volume slice showing the shape of the anticline at the reservoir level (2960 ms) and the main boundary faults (red dashed lines). Low semblance values show good reflection definition, whereas high semblance values show disrupted fault zones. Note that semblance cannot resolve the smaller faults within the east-west fault zone. (B) Seismic section of the anticline before and after image filtering. Definition and continuity of the reflections and fault planes improved considerably. Yellow line indicates location of semblance time slice. Red arrows indicate better definition of the fault plane in the image-filtered seismic section. TWT = two-way traveltime.

Argentina, image filtering allowed for a better definition of the interpreted horizons below and above the reservoir unit. A good definition of the seismic horizons and faults is critical to (1) construct a geometric kinematic model of the anticline, (2) perform a more accurate curvature analysis for fracture prediction, and (3) map the 3-D geometric fault network that controlled the evolution of the structure.

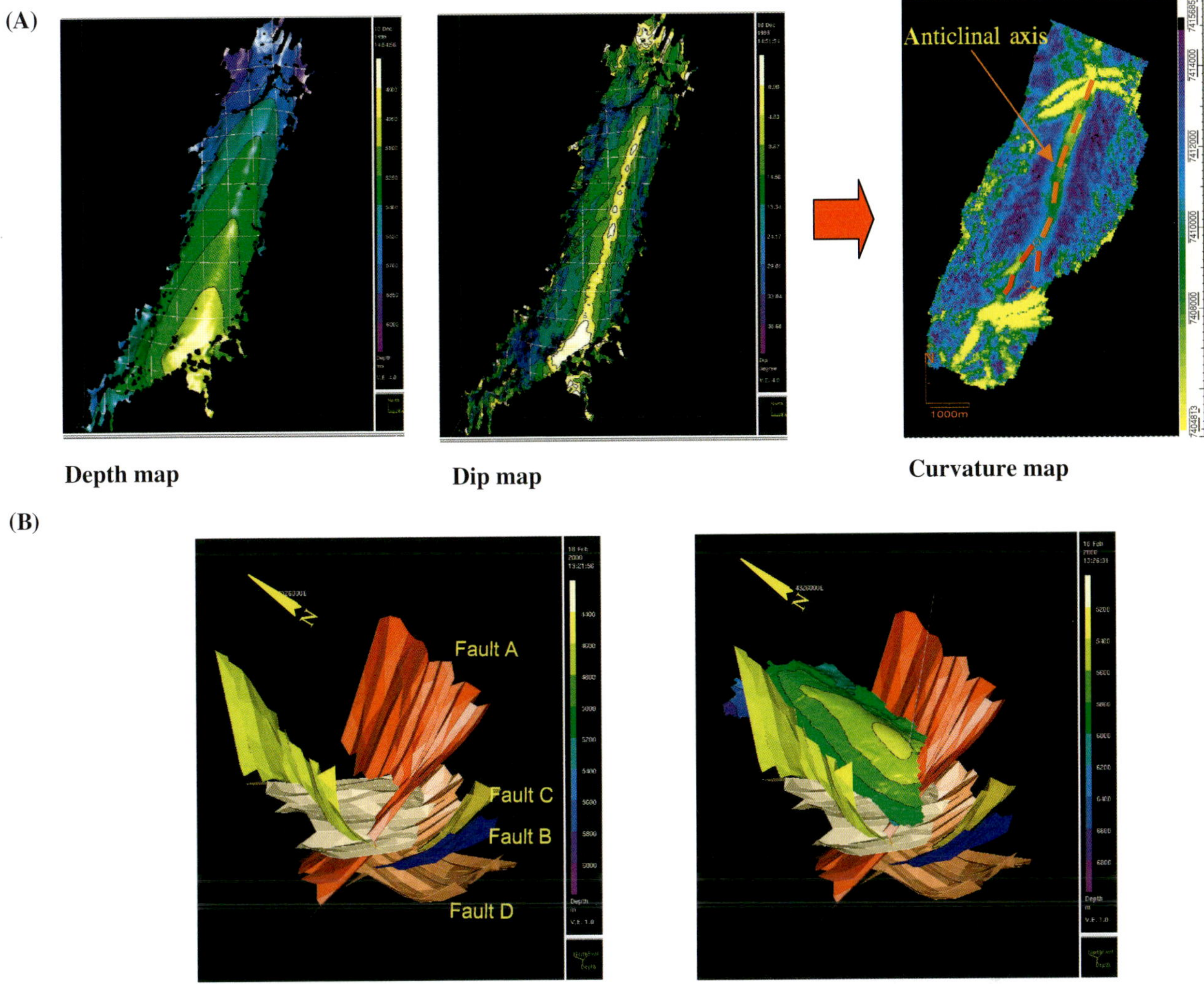

Figure 32. (A) Depth, dip, and calculated curvature map from the base reservoir interpreted horizon. The smoothed interpreted horizon from the imaged-filtered data was critical to calculate curvature and thus predict areas of high fracture intensity. (B) Three-dimensional graphs showing the fault interpretation in relation with the reservoir interpreted horizon. Accurate seismic imaging of the main fault zones was a critical step in the kinematic analysis of the inverted anticline.

ACKNOWLEDGMENTS

The authors thank Shell International Exploration and Production BV for permission to publish this chapter. Petronas, Sarawak Shell Berhad, Abu Dhabi Company for Onshore Oil Operations, Petroleum Development Oman, and Shell Compañia Argentina de Petroleo are gratefully acknowledged for giving permission to publish the data shown in the chapter. We would like to acknowledge the contribution of the Seismic Volume Interpretation Team (VOICE) and the Carbonate Development Team within Shell Technology Applications and Research (SEPTAR).

Our chapter has benefited from numerous discussions with Gregor Eberli, Charles Kerans, Josep Poblet, Mayte Bulnes, Jurriaan Reijs, Updesh Singh, Neil Casson, Jürgen Grötsch, Peter Melville, Jan-Henk van Konijnenburg, Volker Vahrenkamp, and Taury Smith. The manuscript was significantly improved by the reviews of Albert Hine, Mark Grasmueck, and Gregor Eberli. Elena Morettini, Paul Wagner, and Christophe Mercadier provided useful comments to the final version of the chapter.

REFERENCES CITED

Bahorich, M., and S. Farmer, 1995, The coherence cube: Leading Edge, v. 14, p. 1053–1058.

Baumann, A., G. F. Müller, M. K. Gin, and W. S. Heng,

1997, Subsurface indications for porosity upside in B11 field and preliminary evaluation of potential volumetric upside: Sarawak Shell Berhad Note for File EPG-FAD/NFF97/081.

Blendinger, W., 1988, The upper Khuff Formation (K1, K2) of northern Oman: Reservoir development and genetic model: Petroleum Development Oman Exploration Report.

Brown, A., 1996, Interpretation of three-dimensional seismic data: AAPG Memoir 42, 424 p.

Burchette, B., 1993, Mishrif Formation (Cenomanian–Turonian), southern Arabian Gulf: Carbonate platform growth along a cratonic basin margin, *in* J. A. Simo, R. W. Scott, and J. P. Masse, eds., Cretaceous carbonate platforms: AAPG Memoir 56, p. 185–199.

Caline, B., and H. J. Droste, 1989, Diagenetic controls on reservoir development and quality, upper Khuff Formation of the Yibal and Al Huwaisah fields, north Oman: Technical Service Report, Shell International Exploration and Production, 34 p.

Dewey, J., and J. Bird, 1970, Mountain belts and the new global tectonics: Journal of Geophysical Research, v. 75, p. 2625–2647.

Epting, M., 1980, Sedimentology of Miocene carbonate build-ups, Central Luconia, offshore Sarawak: Geological Society of Malaysia Bulletin, v. 12, p. 17–30.

Epting, M., 1989, Miocene carbonate build-ups of central Luconia, offshore Sarawak: *in* A. W. Bally, ed., Atlas of seismic stratigraphy: AAPG Studies in Geology No. 27, p. 168–173.

Gomez Omil, R. J., A. Boll, and R. M. Hernandez, 1989, Cuenca Cretacico— Terciaria del Noroeste Argentino (Grupo Salta), *in* G. Chebli and L. Spalletti, eds., Cuencas sedimentarias Argentinas: Serie Correlacion Geologica No. 6: Universidad de Tucuman, p. 43–64.

Grier, M., J. Slafity, and R. Allmendinger, 1991, Andean reactivation of the Cretaceous Salta rift, northwestern Argentina: Journal of South American Earth Sciences, v. 4, p. 351–372.

Grötsch, J., and C. Mercadier, 1999, Integrated 3-D reservoir modeling based on 3-D seismic: The Tertiary Malampaya and Camago buildups, offshore Palawan, Philippines: AAPG Bulletin, v. 83, p. 1703–1728.

Harris, P. M., and S. H. Frost, 1984, Middle Cretaceous carbonate reservoirs, Fahud field and northwest Oman: AAPG Bulletin, v. 68, p. 649–658.

Ho, K. F., 1978, Stratigraphic framework for oil exploration in Sarawak: Geological Society of Malaysia Bulletin, v. 10, p. 1–13.

Hocker, C., and G. Fehmers, 2002, Fast structural interpretation with structure-oriented filtering: Leading Edge (March), p. 238–243.

Hughes-Clark, M. W., 1988, Stratigraphy and rock unit nomenclature in the oil-producing area of interior Oman: Journal of Petroleum Geology, v. 11, p. 5–60.

Jordan, T., B. Isacks, V. Ramos, and R. Allmendinger, 1983, Mountain building in the Central Andes: Episodes, v. 3, p. 20–26.

Lisle, R. J., 1994, Detection of zones of abnormal strains in structures using Gaussian curvature analysis: AAPG Bulletin, v. 78, p. 1811–1819.

Melville, P., O. Al Jeelani, S. Al Menhali, and J. Grötsch, 2004, Three-dimensional seismic analysis in the characterization of a giant carbonate field, onshore Abu Dhabi, United Arab Emirates, *in* G. P. Eberli, J. L. Masaferro, and J. F. Sarg, eds., Seismic imaging of carbonate reservoirs and systems: AAPG Memoir 81, p. 123–148.

Mon, R., and J. Salfity, 1995, Tectonic evolution of the Andes of northern Argentina, *in* A. Tankard, ed., Petroleum basins of South America: AAPG Memoir 62, p. 269–283.

Moreno, J., 1970, Estratigrafia y paleogeografia del Cretacico superior de la cuenca del Noroeste Argentino, con especial mencion a los Subgrupos Balbuena and Santa Barbara: Asociacion Geologica Argentina Revista, v. XXV, no. 1, p. 9–44.

Neuhaus, D., J. Borgomano, J. C. Jauffred, C. Mercadier, S. Olotu, and J. Grötsch, 2004, Quantitative seismic reservoir characterization of an Oligocene–Miocene carbonate buildup: Malampaya field, Philippines, *in* G. P. Eberli, J. L. Masaferro, and J. F. Sarg, eds., Seismic imaging of carbonate reservoirs and systems: AAPG Memoir 81, p. 169–184.

Philip, J. B., and S. Al-Maskiry, 1995, Cenomanian–early Turonian carbonate platform of northern Oman: Stratigraphy and paleoenvironments: Paleogeography, Paleoclimatology and Paleoecology, v. 119, p. 77–92.

Reyes, F. C., and J. A. Salfity, 1973, Consideraciones sobre la estratigrafia del Cretacico (Subgrupo Pirgua) del Noroeste Argentino: Acta V Congreso Geologico Argentino, v. 3, p. 354–385.

Salfity, J., 1979, Paleogeografia de la Cuenca del Grupo Salta (Cretacico–Eogenico) del Norte de Argentina: VII° Congreso Geologico Argentino, Actas I, p. 505–515.

Salfity, J., 1982, Evolucion paleogeografica del Grupo Salta (Cretacico–Eogenico), Argentina: V° Congreso Latinoamericano de Geologia, Actas I, p. 11–26.

Salfity, J., and R. Marquillas, 1986, Marco tectonico y correlaciones del Grupo Salta (Cretacico–Eoceno), Republica Argentina, *in* Cretacico de America Latina, Primer Simposio, p. 174–188.

Samiee, R., R. Koch, and E. Flügel, 1999, Depositional environment, diagenesis and reservoir characteristics of the Upper Permian–Lower Triassic Khuff Formation (K1 and K2) in the Yibal field, northwest Oman: Internal Report, Shell International Exploration and Production, 32 p.

Skirius, C., S. Nissen, N. Haskell, K. Marfurt, S. Hadley, D. Ternes, K. Michel, I. Reglar, D. D'Amico, F. Deliencourt, T. Romero, R. D'Angelo, and B. Brown, 1999, 3-D seismic attributes applied to carbonates: Leading Edge (March), p. 384–393.

Smith, L. B., and G. P. Eberli, 1999, Reservoir characterization of the Natih E Formation, Al Ghubar field, Oman: Internal Extramural Research Report, Shell International Exploration and Production.

Smith, L. B., G. P. Eberli, J. L. Masaferro, and S. Al-Dhahab, 2003, Discrimination of effective from ineffective porosity in heterogeneous Cretaceous carbonates, Al Ghubar field Oman: AAPG Bulletin, v. 87, p. 1509–1529.

Vahrenkamp, V., 1996, Growth and demise of the Miocene central Luconia carbonate province: Implications for regional Geology and reservoir production behavior: Sarawak Shell Berhad Report No. PELN-96/135.

Vahrenkamp, V., F. David, P. Duijndam, M. Newall, and P. Crevello, 2004, Growth architecture, faulting, and karstification of a middle Miocene carbonate platform, Luconia Province, offshore Sarawak, Malaysia, *in* G. P. Eberli, J. L. Masaferro, and J. F. Sarg, eds., Seismic imaging of carbonate reservoirs and systems: AAPG Memoir 81, p. 329–350.

van Buchem, F. S. P., P. Razin, P. Homewood, J. M. Philip, G. P. Eberli, J. Platel, J. Roger, R. Eschard, G. M. Desaubliaux, T. Boisseau, J. Leduc, R. Labourdette, and S. Cantaloube, 1996, High resolution sequence stratigraphy of the Natih Formation (Cenomian/Turonian) in northern Oman: Distribution of source rocks and reservoir facies: GeoArabia, v. 1, no. 1, p. 65–91.

2

Pearson, R. A., and B. S. Hart, 2004, Three-dimensional seismic attributes help define controls on reservoir development: Case study from the Red River Formation, Williston Basin, *in* Seismic imaging of carbonate reservoirs and systems: AAPG Memoir 81, p. 43–57.

Three-Dimensional Seismic Attributes Help Define Controls on Reservoir Development: Case Study from the Red River Formation, Williston Basin

R. A. Pearson[1]
New Mexico Institute of Mining and Technology, Socorro, New Mexico, U.S.A.

B. S. Hart[2]
New Mexico Bureau of Mines and Mineral Resources, Socorro, New Mexico, U.S.A.

ABSTRACT

The use of three-dimensional (3-D) seismic attributes to predict reservoir properties is becoming widespread in many areas. One of the most underutilized aspects of the methodology is that the property-prediction maps can help geoscientists understand depositional and postdepositional controls on reservoir development. We illustrate this point via a case study that examines partially dolomitized, restricted to open-marine carbonates of the Ordovician Red River Formation in the Williston Basin. We tied log and seismic data, mapped key reflection events in the 3-D seismic volume, calculated the porosity thickness (thickness × sonic porosity) for the porous zone, and then correlated those data with 21 attributes. We derived a relationship between two attributes (the spectral slope from peak to maximum frequency and the ratio of positive to negative samples) and porosity thickness that yielded a 0.88 correlation coefficient between predicted and actual values. This relationship was used to predict the porosity thickness throughout the 3-D seismic area. The resulting porosity distribution shows (1) good porosity development along the flanks of structures that are associated with visible faulting or steep dips at the underlying Winnipeg level, (2) thin (~17–28 ft [~5–8.5 m]) porous zones throughout much of the field, (3) a large, off-structure porosity zone in an area without well control, and (4) small, irregularly distributed porous zones (most likely the result of noise and/or error in the predictive relationship). In areas where faults and flexures are associated with enhanced porosity development, the slope of spectral frequency attribute may be responding to fractures,

[1]*Present address:* Anadarko Petroleum Corporation, The Woodlands, Texas, U.S.A.
[2]*Present address:* Earth and Planetary Sciences, McGill University, Montreal, Québec, Canada.

with more rapid attenuation of high frequencies occurring in these areas. These observations support a diagenetic model where faults and fractures acted locally as preferential pathways for dolomitizing fluids. Away from these zones, the porosity distribution shows some porosity thickness over the entire area that is consistent to drillstem test data that shows depleted pressures in wells drilled in the early 1990s on otherwise isolated structures.

INTRODUCTION

In many reservoirs, understanding and modeling the geologic controls on reservoir quality from sparse well and core data can be problematic, and proposed models may be invalidated as additional data become available. In this chapter, we integrate well and three-dimensional (3-D) seismic data to predict reservoir properties (Figure 1) and use the resulting physical property maps to gain insight into the geologic processes controlling the observed spatial variation in reservoir properties. We illustrate this approach with a case study of the Red River Formation in Brorson field, Williston Basin, Montana.

Zones of porous dolomite within the Red River Formation form the producing interval at Brorson field in Richland County, Montana (Figure 2). As with other Red River fields, the dolomite distribution in this area is complex and difficult to model with sparse well data alone. This complexity is well illustrated by the numerous, sometimes conflicting, models that have been proposed to explain the origin and geometry of Red River reservoirs here and in other areas (see below). As a result, we felt that it would be beneficial to test an

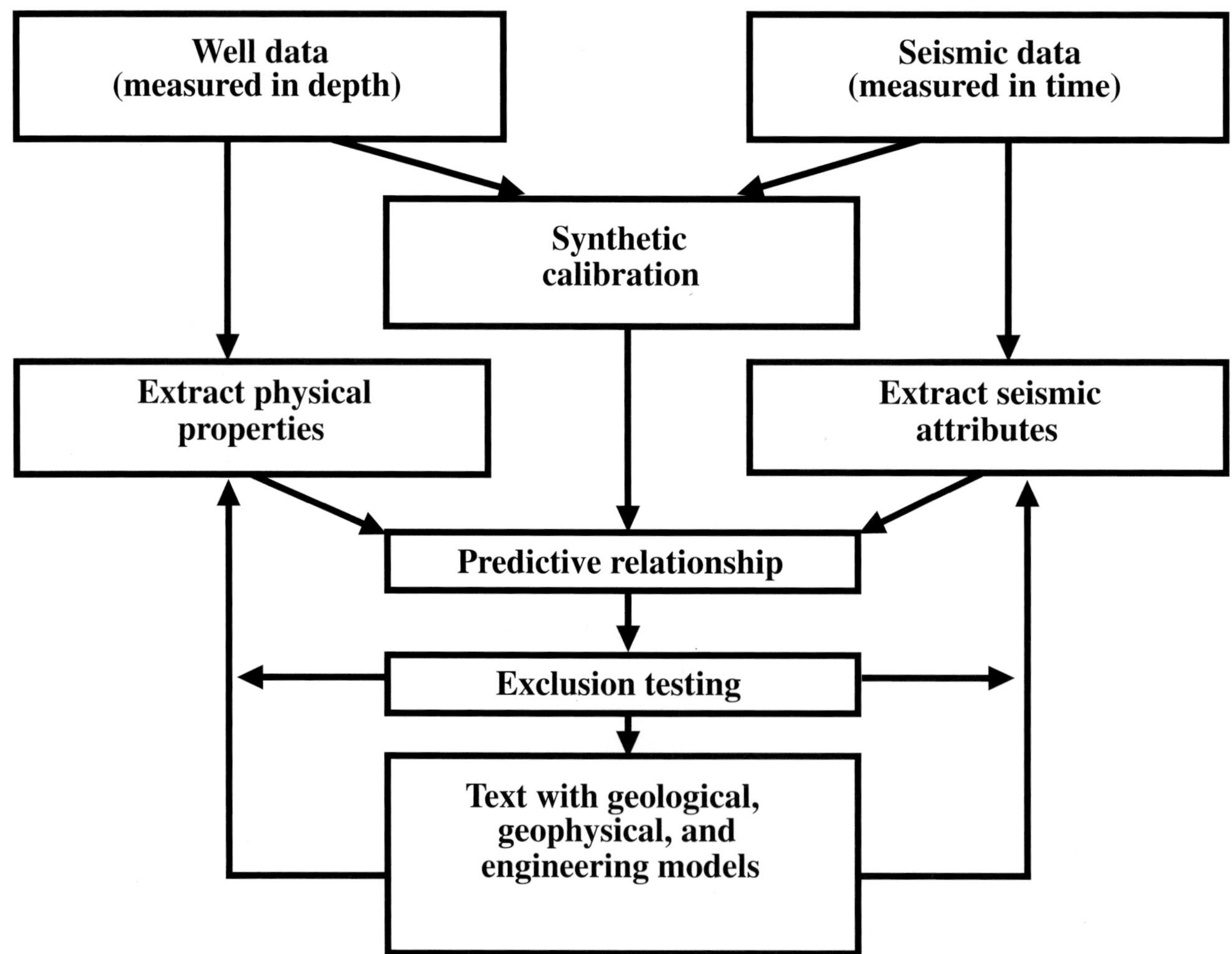

FIGURE 1. The methodology integrates elements of geology, geophysics, and geostatistics and is adapted from methods described by Schultz et al. (1994), Russell et al. (1997), Schuelke and Quirein (1998), and Hart (1999).

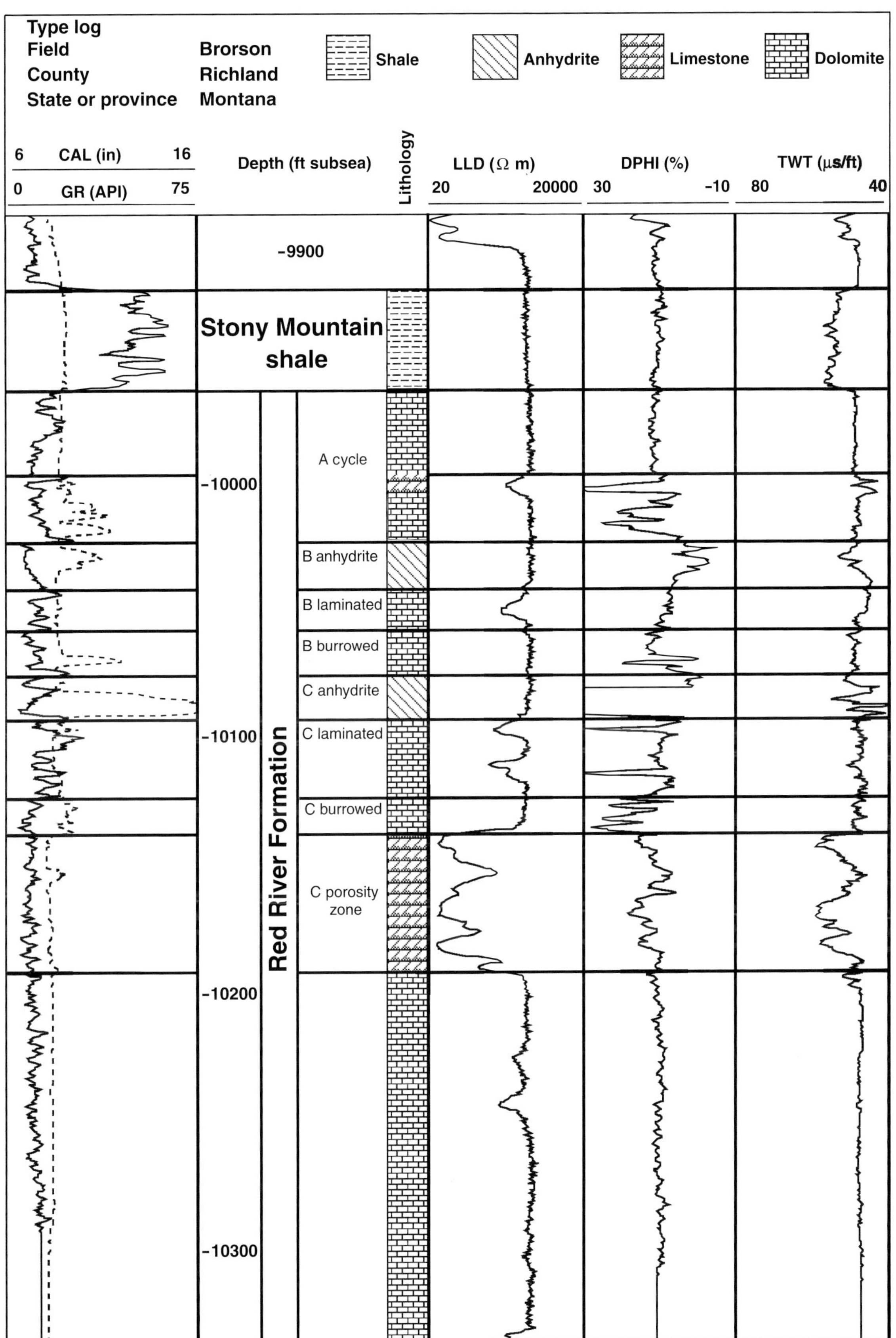

FIGURE 2. Type log for the Red River Formation within the study area. Note the prominent low-velocity interval that is the "C" porous zone. Shown from left to right are caliper (CAL) or gamma-ray (GR), lithology, resistivity (LLD), density porosity (DPHI), and two-way traveltime (TWT) logs.

approach based on the integration of 3-D seismic attributes and log data to characterize the Red River Formation at Brorson field. We sought to image porosity directly, rather than to apply an existing geologic model.

We chose reservoir porosity as the physical property to model in this field because that variable might be seismically detectable (through relationships between porosity and velocity and/or density). Porosity also has reservoir significance in terms of storage capacity and, possibly, relationships to permeability. Furthermore, because porosity is associated with dolomitization in this area, it could be used to track dolomitization and thereby help us to evaluate the controls on diagenesis.

GEOLOGIC FRAMEWORK

Recent publications by Montgomery (1997) and Sippel (1998) summarize much of what is known about the geology and production characteristics of the Red River Formation. The Red River Formation is a sequence of carbonates and evaporites that was deposited in the Williston Basin during the middle–upper Ordovician (Carroll, 1978; Longman et al., 1987). The formation can be divided into three cycles (Figure 2), informally referred to as the "A," "B," and "C" cycles, or zones in stratigraphically descending order (Kohm and Louden, 1978). The lowermost two cycles, the "C" and "B," contain the most complete depositional sequences consisting of variably dolomitized, fossiliferous, burrowed limestone overlain by laminated limestone or dolomite and capped by anhydrite (Kohm and Louden, 1988). The overlying "A" cycle is primarily limestone with a thin interbed of dolomite near the middle that grades laterally into anhydrite toward the basin center (Longman et al., 1987; Figure 2). The individual anhydrite and carbonate members can be correlated on well logs across much of the Williston Basin with only small variations in thickness as the units thin toward the basin edge (Longman et al., 1987). For example, the "C" anhydrite forms a blanket approximately 16–19 ft (~5–6 m) thick across the basin (Longman et al., 1983).

Although there has been some disagreement through the years about the depositional environments represented by these rocks, recent consensus is that the cycles represent shallowing- or "brining-upward" successions. The regional continuity of the individual members implies that the shallowing-upward character of each cycle was in response to processes that acted on a basinal, rather than local, scale.

Porosity development in the Red River Formation is primarily the result of variable dolomitization within the burrowed and laminated members. The patterns of dolomitization within the Red River Formation are complex (rapid lateral and vertical changes in the degree of dolomitization; e.g., figure 3 of Longman et al., 1987), and various models have been proposed to explain how dolomite formed in the Red River Formation. For example, Kohm and Louden (1978, 1988) suggested that magnesium-rich dolomitizing fluids seeped down regionally extensive northeast-trending fracture systems (corresponding to basement lineaments or "zones of slight tectonic adjustment") in Silurian time. They mapped regionally extensive northeast-trending porosity zones (Figure 3A). In contrast, Longman et al. (1983, 1987) suggested that on a small scale, dolomitization occurs in "centers" or lenses with nonporous, cryptocrystalline dolomite grading radially into partially dolomitized, porous zones on the edges (Figure 3B). They propose that dolomitization (at least, in the C cycle, the primary reservoir interval at Brorson field) was the result of downward seepage of magnesium-rich brines that formed during deposition of the cycle-capping anhydrite. In their model, dolomitization was most intense where "holes" (possibly formed by fractures, minor faults, or dewatering features) in the overlying anhydrite provided preferential pathways for dolomitizing fluids. Other authors have recognized that later stages of diagenesis may have had a factor in porosity development as well (see Longman et al., 1992). We note that the maps presented in Figure 3A and B represent different geologic models for dolomitization, but that each map honors existing well data. It is obvious that each has vastly different implications for how operators should explore for, or develop, Red River reservoirs. Developing a methodology for resolving this ambiguity was the primary focus of this study.

STUDY AREA AND DATABASE

Our study area covers a part of the southwest quadrant of T24N, R58E, and extends slightly south into T23N, R58E within the Brorson field, Richland County, Montana (Figure 4). Our database consists of ~6.5 mi^2 (17 km^2) of 3-D seismic data, digital wire-line logs for 28 wells within and around the seismic survey area (nine of which were actually within the 3-D survey area), and production data for seven wells within the 3-D seismic survey area. The well-log suites typically consist of gamma-ray and sonic logs, with the exception of the Dynneson 7-29 well that has a full suite of logs.

METHODS

We correlated all of the available well logs to establish the lateral and vertical distribution of key stratigraphic units as well as the distribution of porosity.

We needed to ensure that the attributes we extracted from the seismic data corresponded (in depth) to the stratigraphic intervals we sought to understand. Accordingly, we used sonic logs (calibrated with velocity surveys) and wavelets extracted from the seismic data to generate synthetic seismograms that could be correlated to the corresponding seismic traces. From the synthetic seismograms, we identified and mapped key horizons within the zone of interest.

We adapted the approach described by Russell et al. (1997; Figure 1) for the seismic attribute analysis. This methodology uses a combination of (a) stepwise linear regression to rank the attribute combinations by their ability to predict the target log property (i.e., the best single attribute, the best pair of attributes, the best combination of three attributes, and so forth), (b) validation testing to determine the optimal number of attributes to use in order to avoid over fitting the data (Schuelke and

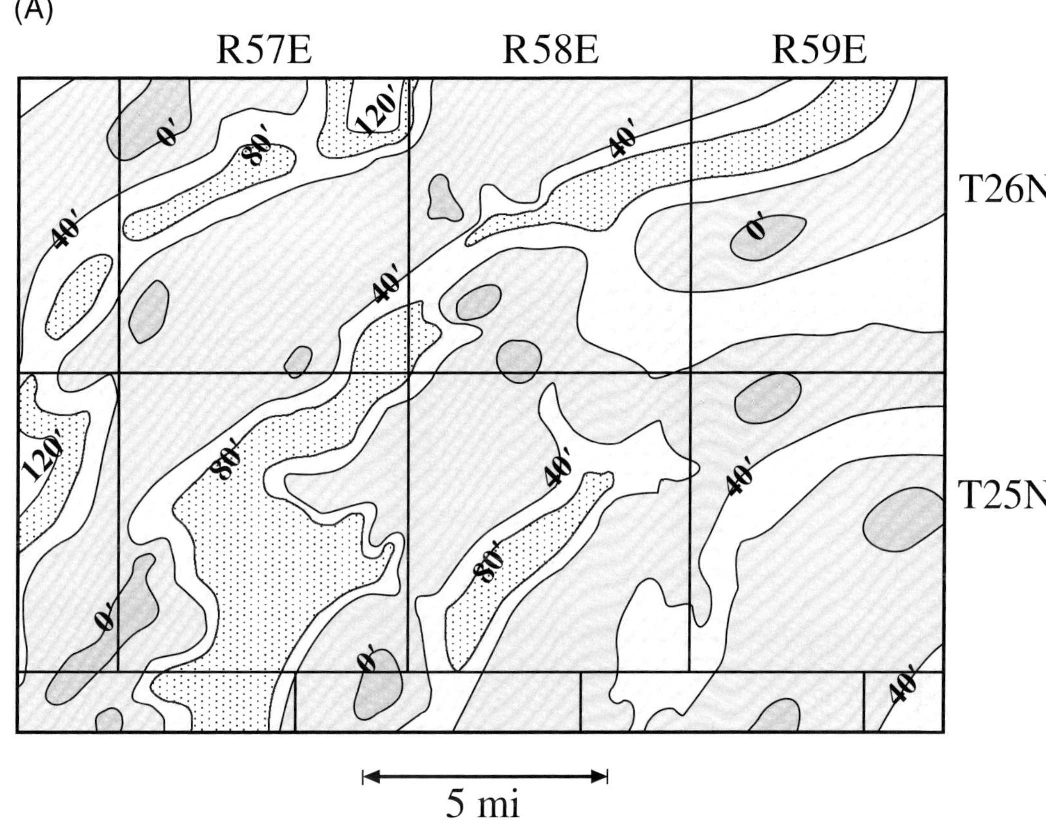

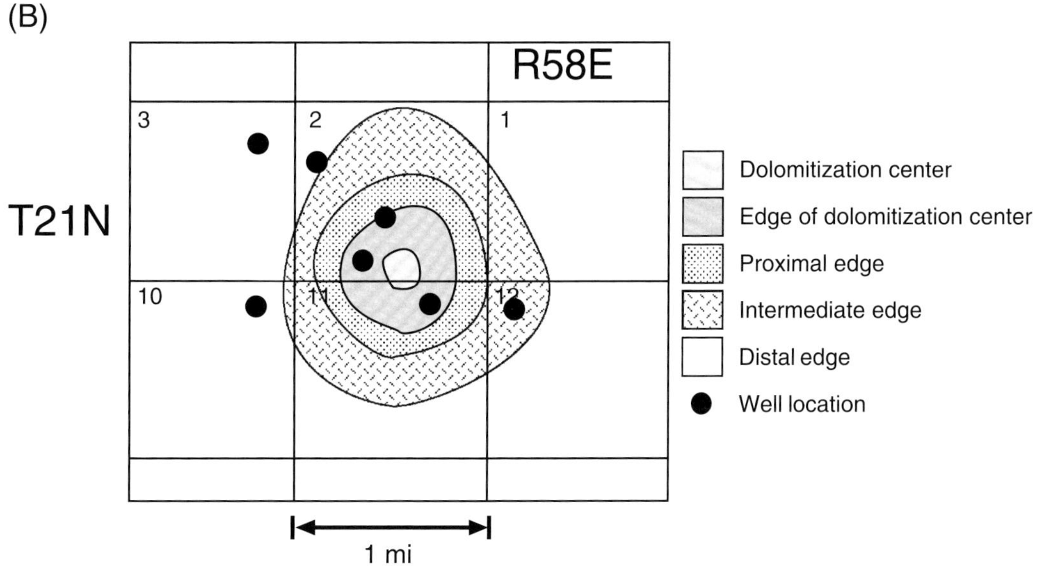

FIGURE 3. (A) Regional porosity distribution as predicted by Kohm and Louden (1978, 1988) for the "C" burrowed member of the Red River Formation in Montana. Contours show the number of feet where porosity exceeds 6% (modified from Kohm and Louden, 1978, 1988). (B) Red River dolomitization model for the Crane field, Montana (Longman et al., 1983; Courtright, 1987). Longman et al. (1983, 1987) argue that linear trends (e.g., panel A) "forced" through the well data are commonly invalidated as additional well data become available.

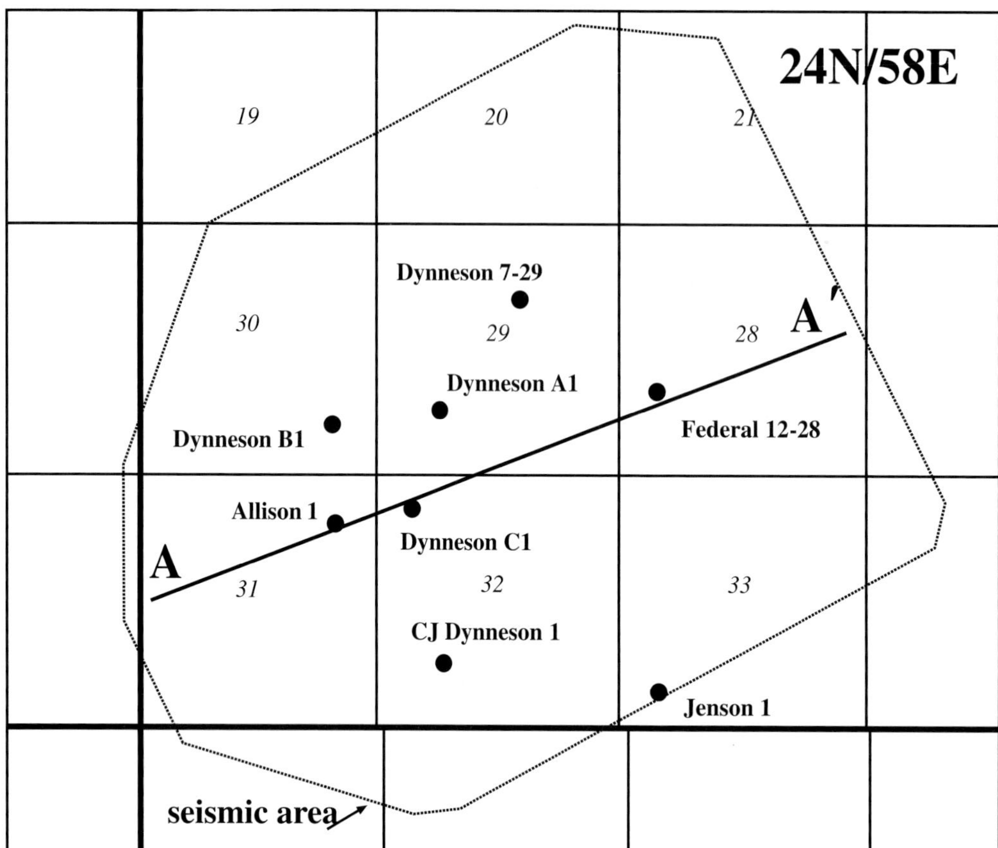

FIGURE 4. Location map showing the seismic survey outline, the locations of wells within the seismic area, and the line of section shown in Figure 6.

Quirein, 1998), and (c) exclusion testing to evaluate the accuracy of the prediction. The ultimate goal of this process is to find a functional relationship of the form:

$$\text{log property} = w_1a_1 + w_2a_2 + w_3a_3 + \ldots + w_na_n$$

where $w_1 \ldots w_n$ is the set of n weights determined by linear regression and $a_1 \ldots a_n$ is the set of n 3-D seismic attributes or nonlinear transformed versions thereof (e.g., $1/x$, x^2, $x^{1/2}$). Russell et al.'s (1997) methodology was designed for a volume-based attribute analysis, but we have adopted it for an interval-based approach. Our interval attributes consist of x, y, z data, where x and y are surface coordinates and z consists of a single numerical value that represents an attribute extracted from the stratigraphic interval of interest (e.g., average amplitude, average frequency). We suggest that this is the most pragmatic approach to follow when (a) the ties between wells and seismic data cannot be adequately established for volume-based analyses, and (b) the interval of interest does not form a single horizon that may be mapped throughout the entire seismic area. As described below, both of these limitations applied in this study.

Although the attribute study enabled us to rapidly find statistically significant relationships, it did not address the geological or geophysical validity of the result. As pointed out by Hirsche et al. (1997), Kalkomey (1997), and Hart (1999) among others, relying on statistics alone has numerous pitfalls. Thus, to ensure that the relationships found are not spurious, we evaluated the resulting predictions to ensure that they are geologically and geophysically plausible as well as statistically significant.

RESULTS

Stratigraphy and Structure

We identified the "A," "B," and "C" cycles of Kohm and Louden (1978; Figure 2) in the Dynneson 7-29 well (which had the most complete suite of logs), then made correlations that allowed us to identify these zones in the other wells. The main productive zone at Brorson field, readily identifiable as a low-velocity and high-porosity zone on logs, is within the burrowed member of the "C" cycle (Figure 2). Unfortunately, these cycles do not all correspond to discrete events in the seismic data and so were not mappable seismically.

From our synthetic seismograms (Figure 5), we were able to obtain a good character match for the principle reflecting horizons. None of our sonic logs extended down to the top of the Winnipeg shale. Accordingly, we based our "Winnipeg" pick on the field operator's experience with data from nearby fields and published examples (e.g., Sippel, 1998).

We identified and mapped (where possible) four key horizons in the seismic data (Figure 6):

- Top Red River: a continuous, relatively high amplitude peak (corresponding to the shale-carbonate contact at the top of the formation) between 2250 and 2300 ms.

- C porosity zone: a discontinuous, variable-amplitude peak best developed on structural flanks and occurring near the middle of the Red River seismic interval (approximately 20 ms below the Top Red River pick). This approximately corresponds to the main producing interval. However, some ambiguity remains because our synthetics did not consistently match the character of the seismic data at this level, possibly because they were missing the side lobes that would be generated from underlying units that were not sampled by the log data. This pick could not be mapped throughout the survey area.
- D marker: a continuous, high-amplitude peak near the base of the formation, except for localized faults and flexures. Note that sonic logs from our area do not extend to this level, and so our synthetics do not reproduce the character of this event (Figure 5). Observation of the seismic and log data together (Figure 6) shows that the D marker is everywhere below the base of the C porosity zone as seen on well logs.
- Winnipeg shale: a moderate-amplitude, continuous trough at 2300–2350 ms that corresponds to the carbonate-shale contact at the base of the Red River Formation.

The top of the Red River has approximately 328 ft (100 m) of relief within the study area (Figure 7), with subsea depths that range from less than 9908 ft (<3020 m) to over 10,236 ft (>3120 m) below sea level. The depth map shows a large star-shaped structural culmination in the central part of the survey area that is flanked by smaller structures. Some of these smaller culminations are drilled, although wells typically do not directly penetrate their crests. Seismic transects across the flanks of these structures suggest that their margins are at least locally defined by faults or high-angle flexures.

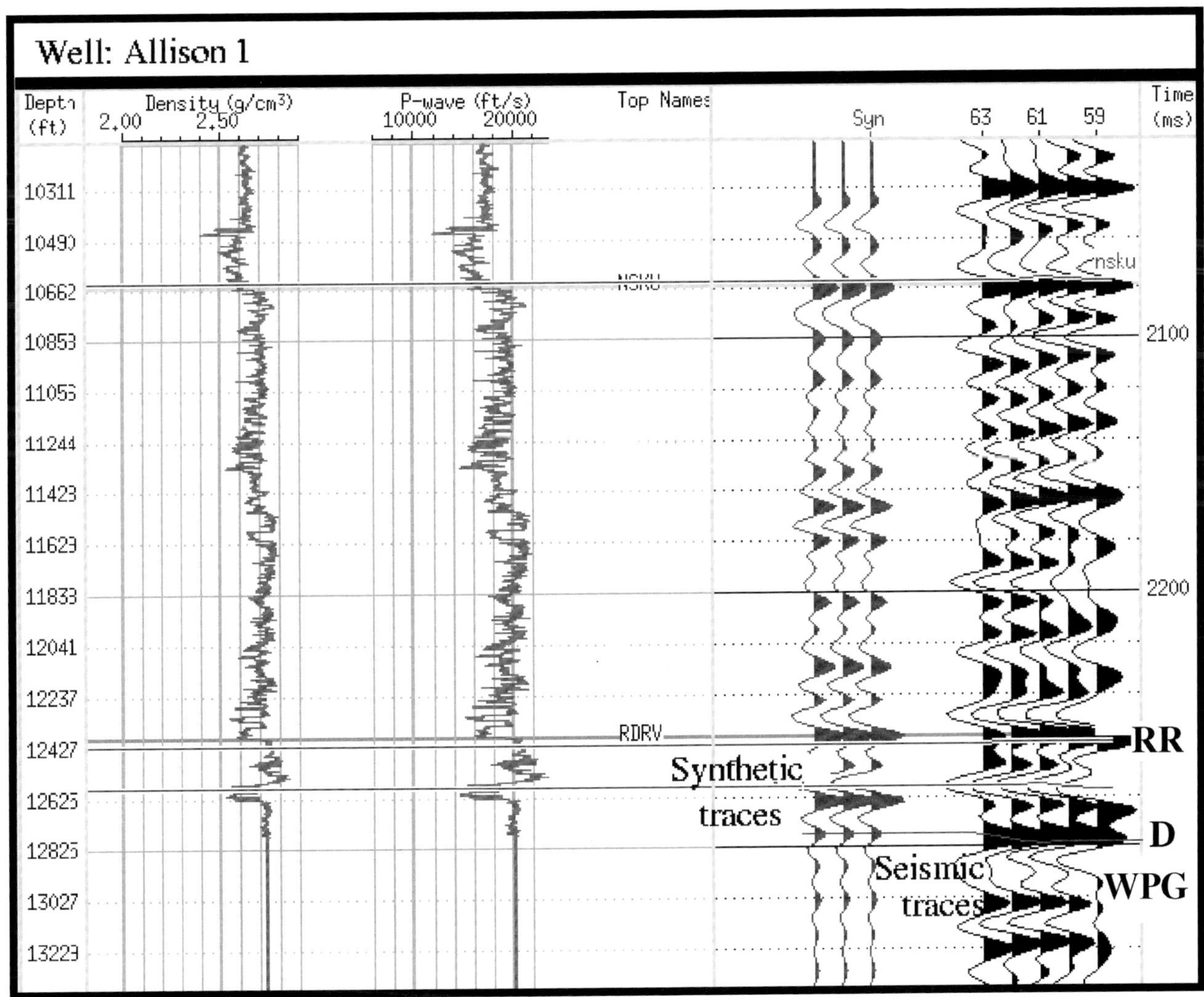

Figure 5. Synthetic seismograms (left) and seismic traces (right) match major reflection events at the Red River level (RR = Top Red River; D = D marker; WPG = Winnipeg shale).

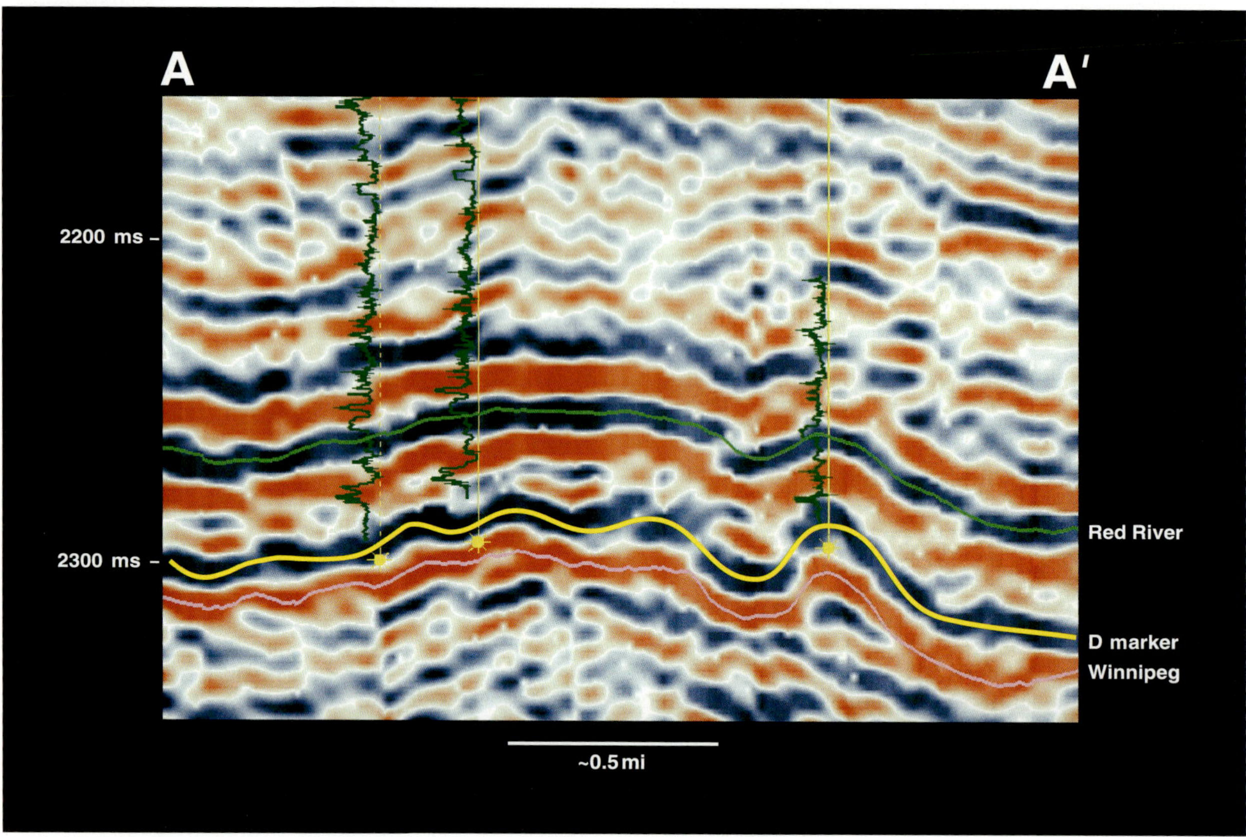

Figure 6. Seismic transect showing the location of the Red River, D marker, and Winnipeg seismic horizons. The overlain logs are sonic porosity (increasing to the left). The cross section location is shown in Figure 4.

An undrilled, structurally low area with a northwest-southeast strike is present along the northeastern margin of the survey area.

Seismic Attribute Analysis

Our attribute analysis was conducted using eight wells within the 3-D seismic survey for which sonic logs were available. Porosity was derived from the sonic logs using a matrix slowness of 43.48 μs/ft and a fluid slowness of 188.68 μs/ft. We chose to model porosity thickness (ϕh), the product of thickness (h, ft) and porosity (ϕ, decimal), at every 0.5 ft log sample and integrated over the C porosity zone.

Because the C porosity zone did not correspond to a mappable seismic reflection, we chose to work with interval attributes. Our extraction window was from the Top Red River to the D marker. Within this window, we extracted 21 amplitude, frequency, and time attributes (Table 1).

The various attributes' ability to predict net pay was assessed both qualitatively and quantitatively using a combination of (1) visual examination of attribute maps, (2) 2- and 3-D crossplots, (3) Spearman rank coefficients (a measure of the strength of both linear and nonlinear correlations), and (4) linear regression models. Several amplitude and frequency attributes were found to correlate well with porosity thickness (Figure 8). The four most strongly correlated of these were ranked using stepwise linear regression to determine the best (in terms of their ability to predict porosity thickness) single attribute, the best pair, and so forth. Validation testing (Figure 9) showed that two is the optimal number of attributes to use. Beyond this, additional attributes only make the predictive relationship too specific to the wells used in the analysis and, thus, less able to predict the excluded wells.

The following linear relationship using two attributes, the slope of spectral frequency and the ratio of positive to negative samples (these attributes will be examined in a subsequent section), was found to give optimal results (Figure 10) for predicting the porosity thickness:

$$\phi h = -5.97 \text{ (slope of spectral frequency)} - 1.99 \times (\text{positive/negative}) + 0.14$$

The correlation coefficient between predicted and actual porosity thickness using this equation is 0.88, and the root mean square error is 0.61.

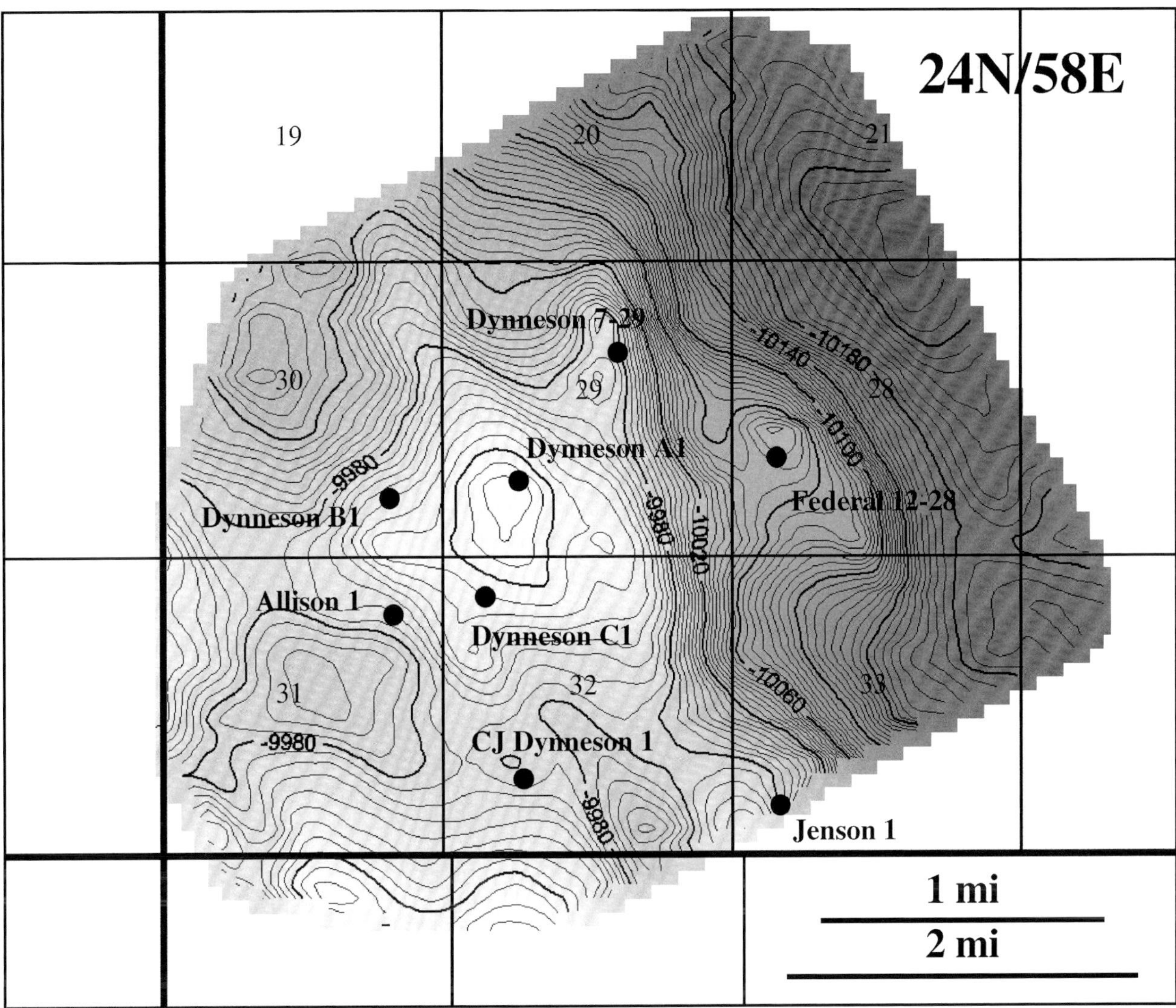

FIGURE 7. Structure map (subsea depth in feet) of the top of the Red River Formation.

Using the predictive relationship found, we created a map of porosity thickness from the seismic data (Figure 11). The values of predicted porosity thickness range from 0 to 6.25 ft (0 to 1.9 m) with the lowest values being concentrated in a few irregularly shaped regions in the northern quarter of the study area. Because of the statistical nature of our method, some negative values of porosity thickness were unavoidable and have been set to zero on the porosity thickness map. The most apparent trend on the map is the tendency for the largest (in area) of the thick-porosity zones to be concentrated on the flanks of structures, although not all of the structural flanks are associated with porosity development. This trend is particularly pronounced in the eastern corner of the study area where the thick-porosity zone southeast of the Federal 12-28 well can be seen wrapping around a structural nose and abruptly terminating where the slope flattens out. These areas, defined arbitrarily by the 4 ft isoline on Figure 11, are generally about 40 ac (161,874 m^2) in size. Smaller thick-porosity zones (less than 1 ac [4047 m^2]) appear to be distributed more randomly. The largest continuous zone, covering approximately 80 ac (323,748 m^2), is in the undrilled northeast part of the survey area, quite low on structure.

DISCUSSION

All attribute-based predictions depend, ultimately, on the strength of the initial well-seismic tie and the quality of the ensuing horizon picks. As noted above, although our synthetic seismograms matched the character of the principle reflecting horizons in the 3-D data, we were unable to consistently match the internal reflection character of the Red River interval. Despite this limitation, our results are statistically sound and, in this section, we will show that our results make sense from geophysical, geological, and engineering perspectives.

Table 1. Spearman Rank Correlation Coefficients for Attributes Extracted in This study.

Spearman rank	*Seismic attribute*
−0.755	Slope spectral frequency
−0.719	Slope of reflection strength
−0.683	Average reflection strength
−0.683	Average peak amplitude
−0.683	Average absolute amplitude
−0.575	Root mean square amplitude
−0.293	Energy half-time
−0.287	Average instantaneous frequency
−0.245	Maximum peak amplitude
−0.240	Slope instantaneous frequency
−0.108	Average trough amplitude
−0.060	Maximum trough amplitude
0.012	Average D zone amplitude
0.084	Peak spectral frequency
0.214	Third dominant frequency
0.263	Average instantaneous phase
0.347	Two-way traveltime to Top Red River
0.407	Red River isochron
0.584	Second dominant frequency
0.608	First dominant frequency
0.700	Ratio of positive to negative samples

The most significant attribute in our regression relationship is the slope of spectral frequency, a measure of how frequencies are absorbed within a given interval (Figure 12). Within the analysis window, the peak spectral frequency is calculated, and the power spectrum is modeled with a multicoefficient polynomial. A linear regression is then performed to determine the slope of the power spectrum (in dB/Hz) between the peak spectral frequency and the maximum usable frequency. Thus, rapid attenuation of high frequencies within the interval is indicated by steep negative slopes. In our case, rapid attenuation is associated with high values of porosity thickness. Attenuation of high frequencies can be because of gas (e.g., low-frequency "shadows" below gas reservoirs), fracturing (frequency-dependent acoustic anisotropy induced by fractures was studied by Pyrak-Nolte, 1999), or perhaps even permeability (e.g., Yamamoto et al., 1995). For reasons described below, we suggest that fractures are probably the main factor contributing to loss of high frequencies in this interval.

The second attribute, the ratio of positive to negative samples, is related to the presence, absence, or degree of development of the peaks and troughs on the seismic traces in the Red River interval (Figure 13). From first principles, it is clear that the development of these events should be related to the internal stratigraphy of the formation (thickness and number of layers, and the acoustic impedance contrasts) and the wavelet (wavelength, phase, type) embedded in the seismic data. Although it would be a seismic thin bed, the presence or absence of a discrete porosity zone at the C horizon (Figure 2) should manifest itself in the seismic character of the Red River Formation. However, based on visual inspection of the seismic and log data and the lack of insights we could obtain from synthetic seismograms, we were unable to identify criteria that can be used to unambiguously separate regions of high and low porosity thickness. Forward modeling of this interval (e.g., Hart and Balch, 2000), which was not undertaken for this project, may shed light on this problem.

It should be noted that although our predictive relationship uses this particular amplitude attribute, several other amplitude attributes correlated almost equally well with porosity thickness (e.g., slope of reflection strength and average reflection strength; Figure 8) and when used in combination with the slope of spectral frequency produce very similar results.

Our ϕh map shows a complex porosity distribution that cannot be described adequately as either dolomitization centers (Longman et al., 1983, 1987) or as large-scale,

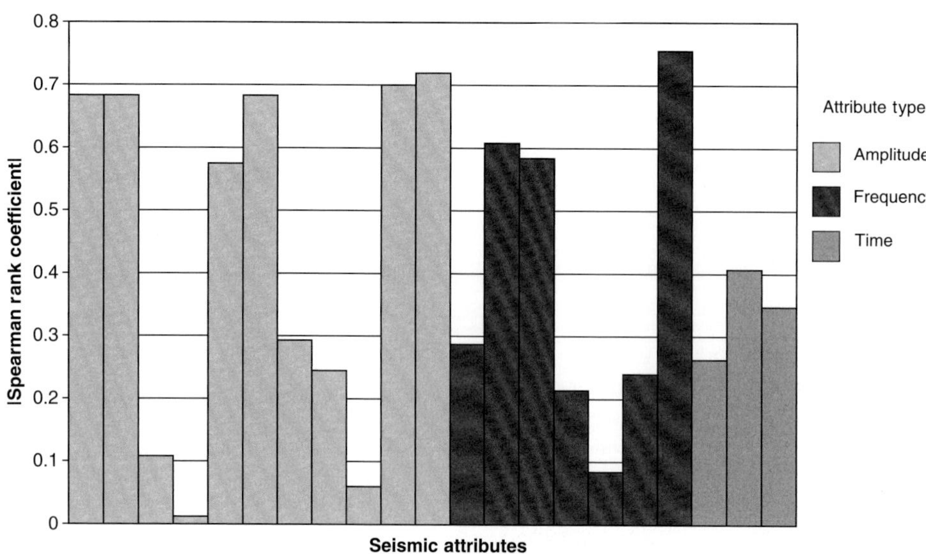

FIGURE 8. Bar graph showing the magnitude of the Spearman rank coefficients for the extracted attributes (1 = perfect correlation; 0 = no correlation).

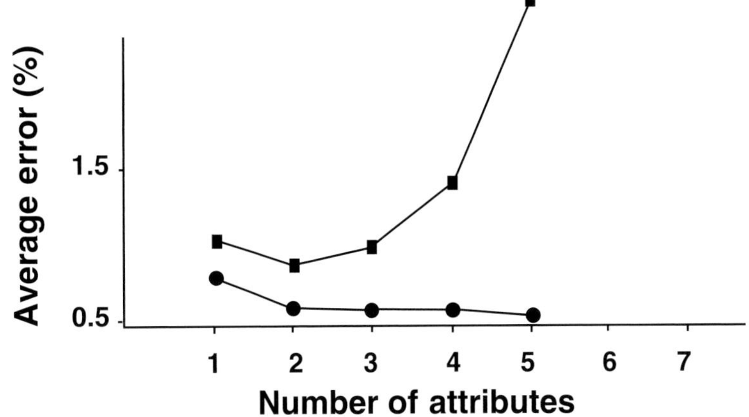

FIGURE 9. Results of validation testing to determine the optimal number of seismic attributes. The average prediction error from exclusion testing increases with the addition of a third attribute indicating two is the optimal number to use.

linear trends (Kohm and Louden, 1978, 1988). Instead, porosity "thicks" are very irregular in shape ranging from elongate (e.g., Section 29, Figure 11) to nearly circular (e.g., Section 30, Figure 11). Three of the larger porosity thicks occur on structural flanks (the large porosity thick in Section 29 south of the Dynneson 7-29 well, a smaller porosity thick along trend with this in Section 32, and a moderate size porosity thick southeast of the Federal 12-28 well crossing Sections 28–33). Clearly, structural position alone is not diagnostic of porosity development as many flanks do not have associated porous zones (e.g., Section 30). Sippel (1998) reported similar results for the Red River Formation in the southern Williston Basin. Using amplitude attributes to make qualitative porosity predictions, he noted that porosity development in the C cycle "tends to be located in low areas and along flanks of structural features." He could not, however, find a "best-fit" isochron or structure map to correlate with the amplitude distribution. In Brorson field, however, seismic transects through the structures show that these porosity zones are best developed where the flanks of the structures are underlain by faults (reverse faults with visible seismic offset) or zones of steep dip (slight flexures or subseismic faults) at the underlying Winnipeg level. This association can be seen by comparing the ϕh map with a Winnipeg shale-dip map (Figure 14). The locations of the Winnipeg dip anomalies roughly coincide with the location of the porosity thicks, with the exception of the apparent dip anomaly near the Jensen 1 well at the southeast margin of the survey (seismic data quality could be an issue at this location). Flanks not underlain by such structures generally do not have thick ϕh zones associated with them. This suggests a possible genetic relationship between these features and porosity development within the Red River C cycle. It is possible that in this area, faults and fractures may have acted as conduits for dolomitizing fluids, resulting in localized

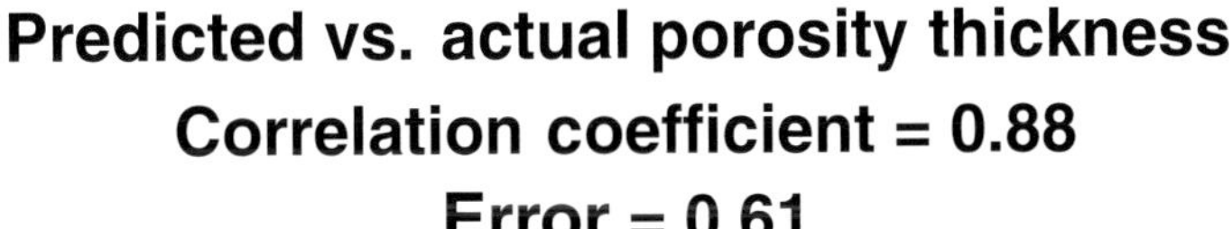

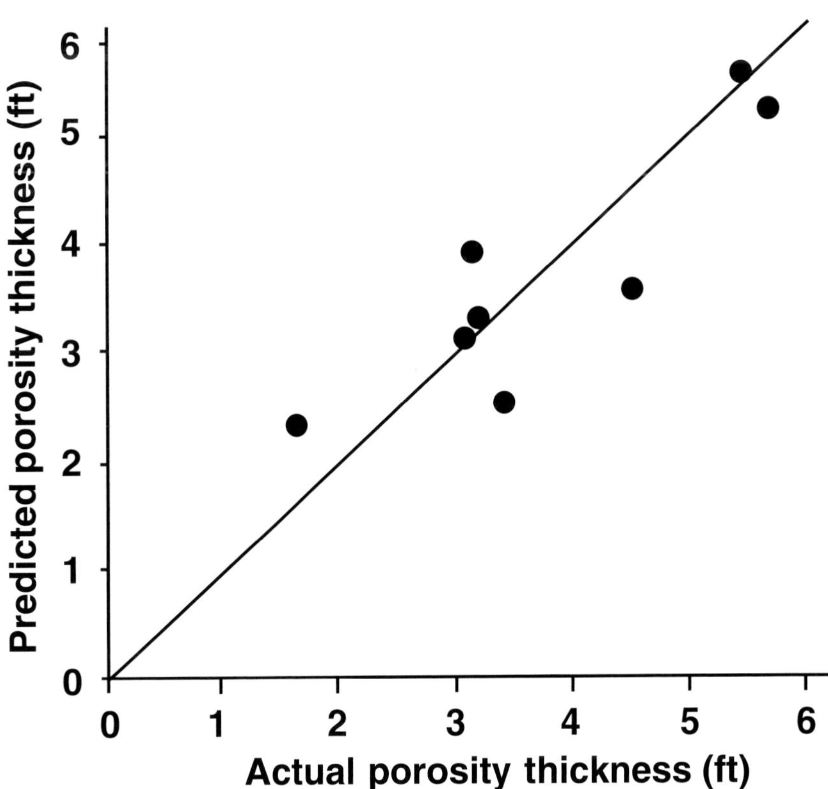

FIGURE 10. Predicted vs. actual porosity thickness. The line marks a perfect 1:1 correlation.

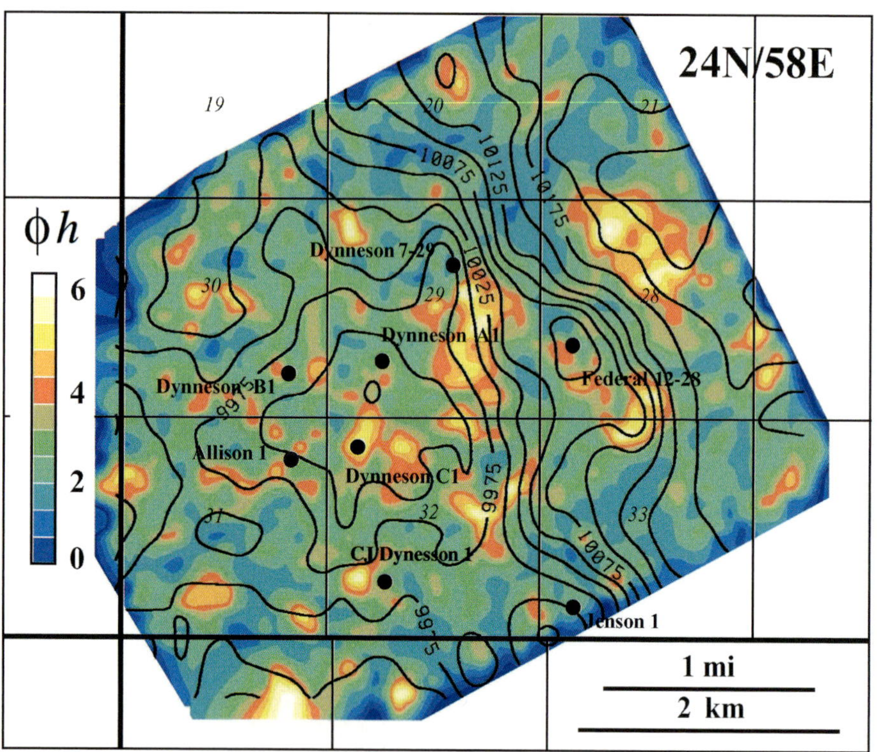

FIGURE 11. Attribute-based prediction of the distribution of porosity thickness within the study area. Colors indicate porosity thickness, and the overlain contour map is the Red River structure (feet below sea level).

zones of enhanced porosity surrounding these features. This is consistent with the observation that the slope of spectral frequency, an attribute that is possibly "seeing" the attenuating effects of fractures, has the most significant correlation with porosity thickness. It should be noted that our map does not address the vertical distribution of porosity. As such, we cannot determine if nonporous dolomite centers occur upsection from these porous zones as would be expected if dolomitization was occurring preferentially along conduits (e.g., as observed by Longman et al., 1983, 1987). A volume-based approach to the attribute analysis (not feasible for this study) may be able to resolve this issue.

Away from the flank-related porosity zones, the ϕh zones are smaller and their distribution is more random. It is possible that some of the aerially restricted porosity thicks are artifacts of nonlinearity in the relationship between seismic attributes and physical properties. It might be that a more accurate (in terms of noise reduction) relationship could have been established using a neural network. For example, Leiphart and Hart (2001) compared results from linear regression (as used here) with those obtained via a probabilistic neural network and determined that the latter gave a better-defined prediction of subsurface physical properties. Based on their results, a neural network might have been able to remove some of the smaller thick-porosity zones (possible artifacts) that appear scattered throughout our map (Figure 11) and give a geologically more reasonable result. Alternatively, although the correlation between porosity zones and underlying structures generally supports the diagenetic model of Longman et al. (1983, 1987), it could be that some of the porosity development in the Red River at Brorson field (e.g., the large zone in the northeast part of the survey area where we have no well control) is not related to localized early-stage, fracture-related dolomitization. Multiple stages of dolomitization and porosity development have been proposed for the Red River Formation (see Longman et al., 1992),

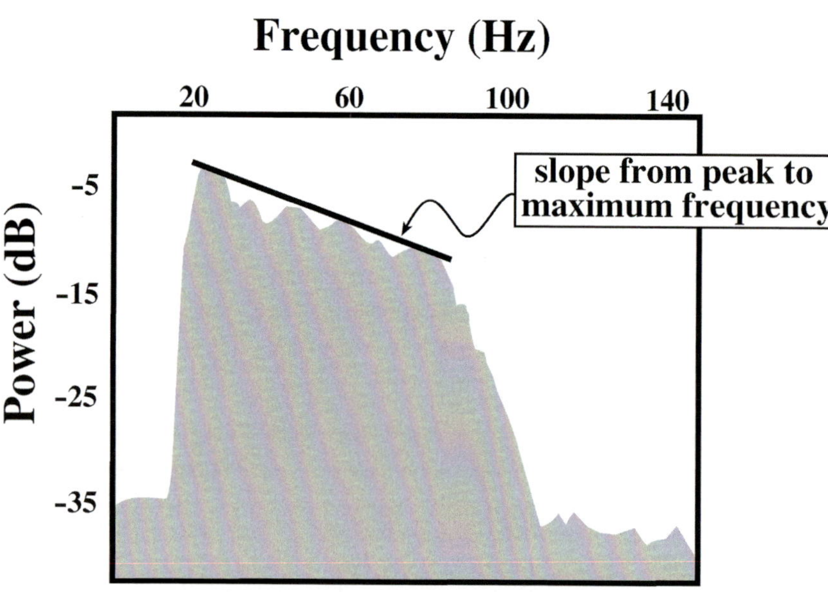

FIGURE 12. The attribute "slope from peak to maximum spectral frequency" reflects how frequencies are absorbed within the analysis window.

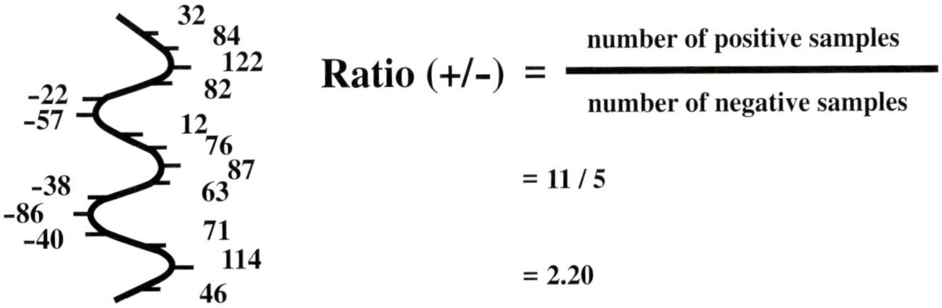

FIGURE 13. The attribute "ratio of positive to negative" counts the number of positive and negative samples within a user-defined window, then expresses those numbers as a ratio.

32 84 122 82 -22 -57 12 76 87 -38 63 -86 -40 71 114 46

$$\text{Ratio (+/-)} = \frac{\text{number of positive samples}}{\text{number of negative samples}}$$

$$= 11 / 5$$

$$= 2.20$$

but to our knowledge, models predicting the spatial distribution of such porosity zones have not been developed. We have no core or other information from these areas at Brorson field that would allow us to further evaluate the origin of porosity.

Our attribute analysis (Figure 11) also suggests that some porosity is developed throughout much of the survey area. If these zones are interconnected in the vertical dimension, it would suggest reservoir continuity over much of the study area. Whereas it is not possible to determine this from an interval-based attribute study such as that employed here, well-test data does locally support this idea. The Federal 12-28 well (Figure 4) was drilled in 1995 to target a previously undrilled small structural culmination on the flank of the main structure. The well found depleted reservoir pressures, and water cut and gas-oil ratios were high before the well was shut in in 1998 (Figure 15B), indicating that this location is in fluid communication with previously drained areas.

Other engineering data from Brorson field are also consistent with our porosity prediction. The two wells with the highest production, the Dynneson C-1 and A-1 wells, are located in or adjacent to large zones of high porosity thickness (Figures 11, 15A). The more marginal wells (e.g., Dynneson B-1) are located away from thick-porosity zones. The Dynneson 7-29 well (Figures 11, 15A), another marginal well, was completed in the Red River Formation in October 1995. This well targeted a small structural culmination that is separate from (northeast of) the main structure, but it was drilled into an area of low porosity according to our map. Production rates were never high in this well, and it was shut-in after a short time to be only a minor producer (Figure 15A).

The relationships among structural location, porosity, and production at Brorson field are complicated, and there is no simple relationship among deep faulting, flank porosity, and high production. At least in part, this is because of the field's development history. The two wells drilled nearest the best-developed flank porosity (Federal 12-28 and Dynneson 7-29, although neither actually hit the best porosity) were drilled after pressures had dropped fieldwide. Ideally, these wells would have been drilled into the large porosity zones early in the development of the field. Other nonflank porosity zones can be productive, especially if they are high on structure and drilled early enough (e.g., Dynneson C1).

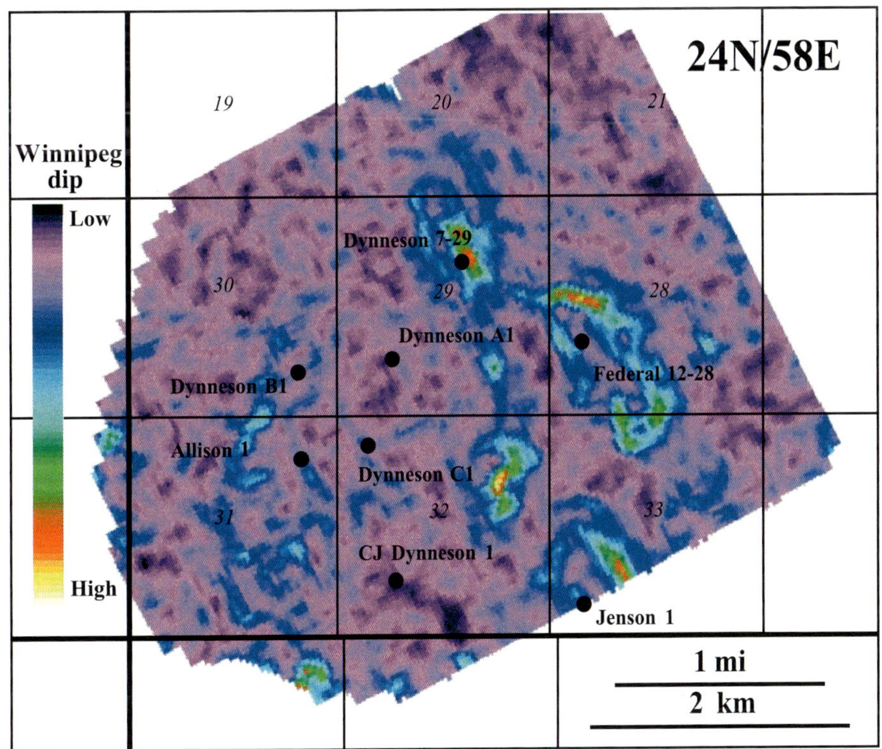

FIGURE 14. Dip map of the Winnipeg horizon (high dips in "hot" colors) (compare to Figure 11). Several of the extensive porosity zones are underlain by steep dips that could be indicative of faulting.

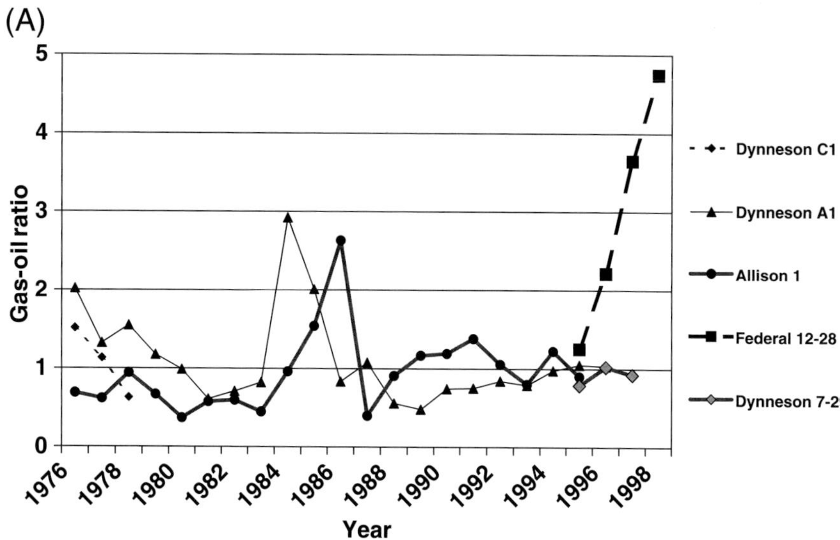

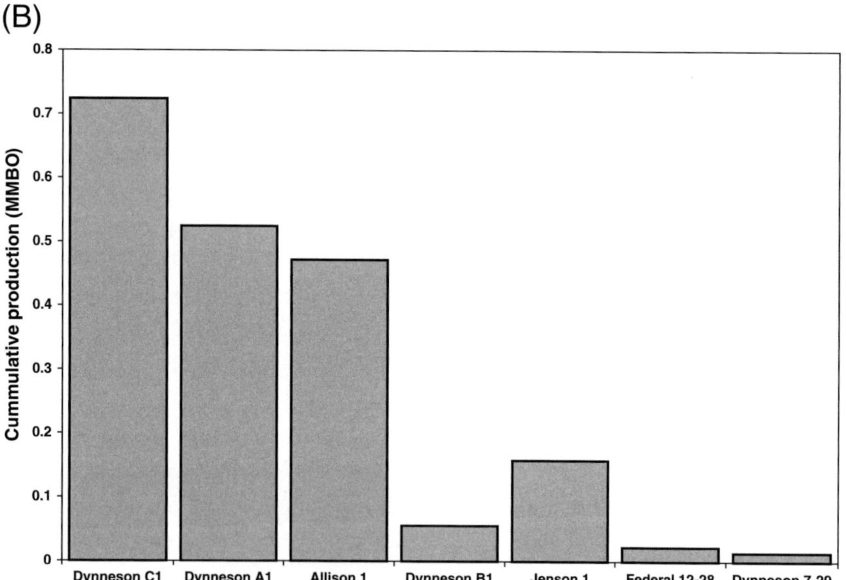

FIGURE 15. (A) Gas-oil ratios through time for Red River producers in the Brorson field. Well locations are shown in Figure 4. (B) Bar graph showing cumulative oil production.

At Brorson field, porosity is best developed on the flanks of small structures that are (commonly) underlain by faults or flexures. However, the distribution of porosity is complex and cannot be predicted from structural location alone. Our results support a diagenetic model whereby dolomitization, and accompanying porosity development, was influenced by the presence of fractures that acted as conduits for downwardly migrating hypersaline brines, although the timing of brine migration cannot be addressed by our results. Other, so far unidentified, diagenetic processes may be responsible for porosity zones that are not the flanks of structures. Our results do not support a previously proposed model (Kohm and Louden, 1978, 1988) wherein porosity development follows linear, large-scale southwest-northeast trends. Instead, our results support a diagenetic model proposed by Longman et al. (1983, 1987) wherein thick zones of porosity development are more localized in nature. However, our results suggest that the shape of the porosity zones is not circular and that there is interconnected porosity over much larger areas than proposed by these latter authors.

This study has provided information that may be used by field operators to assess different field development options. Furthermore, the results provided insights into the controls on reservoir properties that could be used in other areas, perhaps even where 3-D seismic data are unavailable.

CONCLUSIONS

We employed a generally applicable methodology that incorporates elements of geology, geophysics, and geostatistics to image porosity development in the Ordovician Red River Formation of the Williston Basin. Specifically, we integrated porosity derived from sonic logs with two "interval" attributes (slope of spectral frequency and ratio of positive to negative samples) to establish an empirical expression that could be used to predict porosity thickness (ϕh) throughout the 3-D seismic survey area. Our results are statistically significant, geologically plausible, agree with engineering data from the study area, and are based on seismic attributes that have physical meaning. We believe that this combination of analyses should be considered the "best practice" for attribute-based studies of carbonate (or clastic) reservoirs.

ACKNOWLEDGMENTS

Funding for this project was provided by Los Alamos National Laboratory through their Advanced Reservoir Management Project. Data and local knowledge of the Red River Formation were supplied by Flying J Oil

Company. Software was provided by Landmark Graphics Corporation and Hampson-Russell Software Services. We thank these organizations and companies for their support and Guido Bracco Gartner, Gregor Eberli, and an anonymous reviewer for their suggestions that helped improve the focus of this chapter.

REFERENCES CITED

Carroll, W. K., 1978, Depositional and paragenetic controls on porosity development, upper Red River Formation, North Dakota, *in* D. Estelle and R. Miller, eds., The economic geology of the Williston Basin: Williston Basin Symposium, Montana Geological Society 24th Annual Field Conference Guidebook, p. 79–94.

Courtright, T. R., 1987, Richland County Red River dolomite geometry revisited, *in* J. A. Peterson, D. M. Kent, S. B. Anderson, R. H. Pilatzke, and M. W. Longman, eds., Williston Basin, anatomy of a cratonic oil province: Rocky Mountain Association of Geologists, p. 105–108.

Hart, B. S., 1999, Geology plays key role in seismic attribute studies: Oil & Gas Journal, v. 97 (July 12), p. 76–80.

Hart, B. S., and R. S. Balch, 2000, Approaches to defining reservoir physical properties from 3-D seismic attributes with limited well control: An example from the Jurassic Smackover Formation, Alabama: Geophysics, v. 65, p. 368–376.

Hirsche, K., J. Porter-Hirsche, L. Mewhort, and R. Davis, 1997, The use and abuse of geostatistics: Leading Edge, v. 16, p. 253–260.

Kalkomey, C. T., 1997, Potential risks when using seismic attributes as predictors of reservoir properties: Leading Edge, v. 16, no. 3, p. 247–251.

Kohm, J. A., and R. O. Louden, 1978, Ordovician Red River of eastern Montana and western North Dakota: Relationships between lithofacies and production, *in* D. Estelle and R. Miller, eds., The economic geology of the Williston Basin: Williston Basin Symposium, Montana Geological Society 24th Annual Field Conference Guidebook, p. 99–117.

Kohm, J. A., and R. O. Louden, 1988, Red River reservoirs of western North Dakota and eastern Montana, *in* S. M. Goolsby and M. W. Longman, eds., Occurrence and petrophysical properties of carbonate reservoirs in the Rocky Mountain region: Rocky Mountain Association of Geologists, p. 275–290.

Leiphart, D. J., and B. S. Hart, 2001, Comparison of linear regression and a probabilistic neural network to predict porosity from 3-D seismic attributes in lower Brushy Canyon channeled sandstones, southeast New Mexico: Geophysics, v. 66, p. 1349–1358.

Longman, M. W., T. G. Fertal, and J. S. Glennie, 1983, Origin and geometry of Red River dolomite reservoirs, western Williston Basin: AAPG Bulletin, v. 67, p. 744–771.

Longman, M. W., T. G. Fertal, and J. S. Glennie, 1987, Origin and geometry of Red River dolomite reservoirs, western Williston Basin, *in* J. A. Peterson, D. M. Kent, S. B. Anderson, R. H. Pilatzke, and M. W. Longman, eds., Williston Basin, anatomy of a cratonic oil province: Rocky Mountain Association of Geologists, p. 83–104.

Longman, M. W., T. G. Fertal, and J. R. Stell, 1992, Reservoir performance in Ordovician Red River Formation, Horse Creek and South Horse Creek fields, Bowman County, North Dakota: AAPG Bulletin, v. 76, p. 449–467.

Montgomery, S. L., 1997, Ordovician Red River "B": Horizontal oil play in the southern Williston Basin: AAPG Bulletin, v. 81, p. 519–532.

Pyrak-Nolte, L. J., 1999, Imaging seismic wave propagation in fractured media: Proceedings 68th Annual Meeting, Society of Exploration Geophysicists, CD-ROM.

Russell, B., D. Hampson, J. Schuelke, and J. Quirein, 1997, Multiattribute seismic analysis: Leading Edge, v. 16, no. 10, p. 1439–1443.

Schuelke, J. S., and J. A. Quirein, 1998, Validation: A technique for selecting seismic attributes and verifying results: Proceedings 67th Annual Meeting, Society of Exploration Geophysicists, p. 936–939.

Schultz, P. S., S. Ronen, M. Hattori, and C. Corbett, 1994, Seismic guided estimation of log properties, part 1: Leading Edge, v. 13, no. 5, p. 305–315.

Sippel, M. A., 1998, Exploitation of reservoir compartments in the Red River Formation, southern Williston Basin, *in* R. M. Slatt, ed., Compartmentalized reservoirs in Rocky Mountain Basins: Proceedings 1998 Rocky Mountain Association of Geologists Symposium, p. 151–170.

Yamamoto, T., T. Nye, and M. Kuru, 1995, Imaging the permeability structure of a limestone aquifer by crosswell acoustic tomography: Geophysics, v. 60, p. 1634–1645.

3

Pranter, M. J., N. F. Hurley, and T. L. Davis, 2004, Sequence-stratigraphic, petrophysical, and multicomponent seismic analysis of a shelf-margin reservoir: San Andres Formation (Permian), Vacuum field, New Mexico, United States, *in* Seismic imaging of carbonate reservoirs and systems: AAPG Memoir 81, p. 59–89.

Sequence-Stratigraphic, Petrophysical, and Multicomponent Seismic Analysis of a Shelf-Margin Reservoir: San Andres Formation (Permian), Vacuum Field, New Mexico, United States

Matthew J. Pranter[1]
Colorado School of Mines, Golden, Colorado, U.S.A.

Neil F. Hurley
Colorado School of Mines, Golden, Colorado, U.S.A.

Thomas L. Davis
Colorado School of Mines, Golden, Colorado, U.S.A.

ABSTRACT

This chapter describes an integrated approach to reservoir characterization and three-dimensional (3-D) geologic modeling of the San Andres Formation at Vacuum field, New Mexico, United States. We present techniques to identify significant heterogeneities within a carbonate reservoir using stratigraphic, petrophysical, and 3-D multicomponent seismic data. This integrated approach provides a detailed static description of reservoir heterogeneity and improved delineation of the reservoir framework in terms of flow units.

We use a petrophysics-based method to identify hydraulic flow units within a sequence-stratigraphic framework. Flow units are characterized within high-frequency carbonate sequences through analysis of the vertical variation of flow (kh) and storage capacity (ϕh) and pore-throat radius (R35) associated with successions of subtidal, intertidal, and supratidal rocks. Pore-throat radii from cored wells are used to modify the empirically derived Winland equation to estimate values of pore-throat radius in noncored wells. Flow profiles, constructed from log porosities and neural-network permeabilities, are correlated and used to build a 3-D geologic-model framework.

Characterization of both matrix and fracture properties within a reservoir is possible using 3-D multicomponent seismic data and wire-line logs. Compressional- and

[1]*Present address:* University of Colorado, Boulder, Colorado, U.S.A.

shear-wave amplitude attributes together provide more accurate porosity estimates than those determined from compressional-wave data alone. Shear-wave anisotropy measurements provide information about inferred fracture density and orientation that can be used to modify permeability models to account for regions with open fractures.

Because of this study, reservoir-simulation models that incorporate modified permeability distributions more accurately account for unexpected early CO_2-breakthrough times observed in the field. In addition, flow-simulation results indicate that the need to upscale the geologic model was significantly reduced or eliminated by describing flow units using the combined sequence-stratigraphic- and petrophysics-based method.

INTRODUCTION

Vacuum field is located in southeast New Mexico on the northwest shelf of the Permian Basin (Figure 1). Stratigraphic, structural, and diagenetic variability within the shelf-margin carbonates of the Permian San Andres and Grayburg Formations form a very heterogeneous and compartmentalized reservoir. Detailed characterization and modeling of heterogeneities within the reservoir are necessary so that areas of potential bypassed pay can be targeted using supplemental recovery techniques or infill development, including horizontal wells.

The study area (Figure 2) is under waterflood operations and was converted to a partial-field CO_2 flood and monitored to evaluate the effect of gas injection on reservoir performance and recovery. During the CO_2 program, time-lapse, multicomponent (four-dimensional, three-component [4-D, 3-C]) seismic data were acquired to demonstrate the use of such data for reservoir characterization. Another objective involved the ability of repeated surveys to detect and monitor changes within the reservoir. This study provides the baseline static characterization and three-dimensional (3-D) geologic model for the dynamic reservoir characterization project. Static reservoir characterization provides detailed descriptions of reservoir properties that generally do not change with time, namely, porosity, permeability, and flow capacity. Dynamic reservoir characterization that

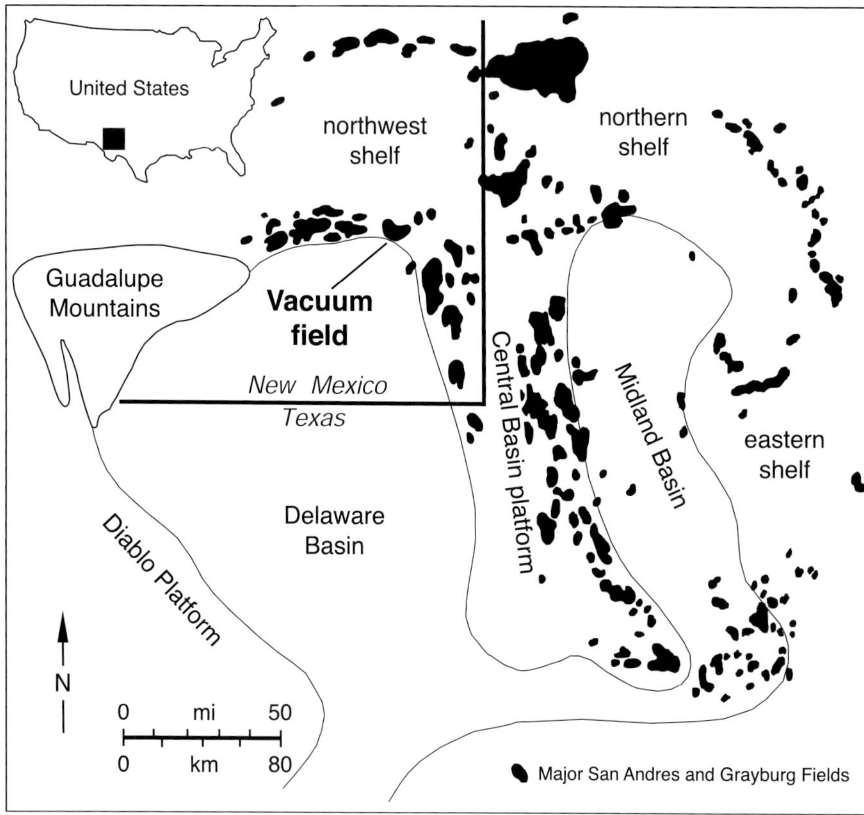

FIGURE 1. Location map showing Vacuum field on the northwest shelf of the Permian Basin. Major San Andres and Grayburg fields are shown as black areas. Modified from Hills (1984).

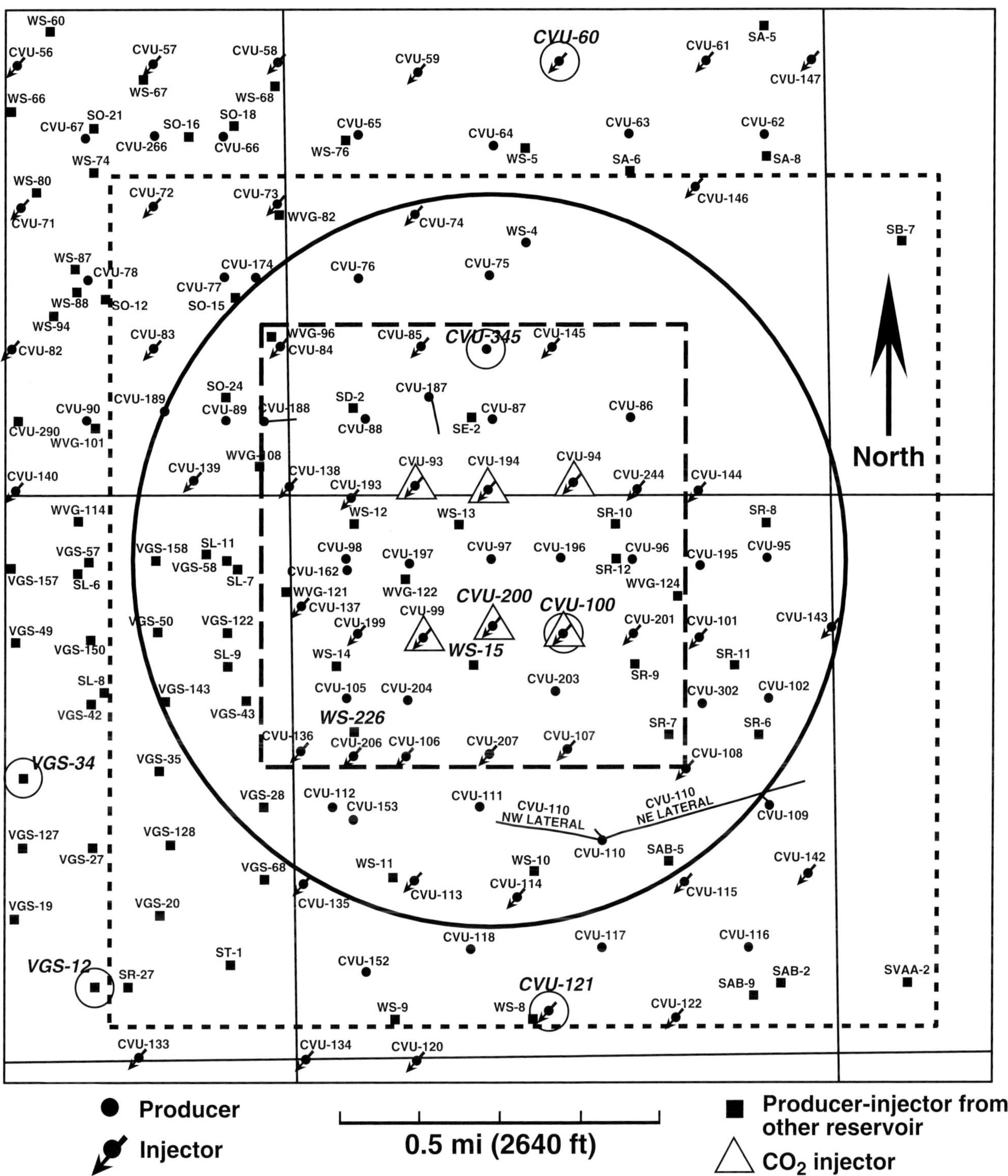

FIGURE 2. Base map of available data in the reservoir characterization area. Cored wells used in this study are indicated by open circles, and CO_2-injection wells are indicated by triangles. VSPs were acquired in wells CVU-200 and WS-15. An FMI log was acquired in well WS-226. The large and small dashed squares indicate the areas where geologic models were constructed. The large solid circle shows the extent of the time-lapse, multicomponent, 3-D seismic surveys. The location of the dual-lateral horizontal well (CVU-110) is shown in the southeast part of the time-lapse area.

incorporates 4-D, 3-C seismology involves monitoring field production to determine reservoir-property variations (e.g., fluid saturation) with time (Benson and Davis, 2000).

This study builds upon other studies and research on the San Andres Formation within Vacuum field. Landes (1970) summarized early work and statistics on Vacuum field. Purves (1990) described lithologic and petrophysical aspects of the San Andres and Grayburg Formations. Several studies have been conducted within the Central Vacuum Unit (CVU) and surrounding area by the Colorado School of Mines Reservoir Characterization Project (RCP). Capello de Passalacqua (1995), Adams (1997), Scuta (1997), and Pranter (1999) established the sequence-stratigraphic framework of the Grayburg and San Andres interval, interpreted the diagenetic history, and addressed the effects of stratigraphy, structure, and diagenesis on reservoir performance. Numerous RCP studies have addressed issues of seismic response, seismic–well-log correlation, attenuation, shear-wave polarization, repeatability of seismic response, regional structural framework, time-lapse logging, fractures, shear-wave amplitude vs. offset, compressional-wave velocity/shear-wave velocity (Vp/Vs) attributes, seismic processing, and shear-wave inversion (Swanson, 1996; Voorhies, 1996; DeVault, 1997; Roche, 1997; Scuta, 1997; Talley, 1997; Blaylock, 1999; Galarraga, 1999; Mattocks, 1998; Mendez-Hernandez, 1999; Lorenzen, 2000). Three-dimensional geologic models for this area of Vacuum field were constructed using well data (Scuta, 1997) and a combination of petrophysical and multicomponent seismic data (Pranter, 1999).

A primary goal of this study was to identify the most significant heterogeneities within the reservoir and incorporate this information into a 3-D geologic model for use in flow simulation. Both sequence-stratigraphic and petrophysics-based methods were used to define reservoir zones. Flow units were characterized within high-frequency carbonate sequences through analysis of the vertical variation of flow (kh) and storage capacity (ϕh) and pore-throat radius (R35) within the reservoir interval. The resulting 3-D geologic model, when combined with multicomponent seismic data, led to improved estimates of reservoir performance using numerical flow models.

Vacuum Field History

Vacuum field is one of the larger oil fields in the Permian Basin that produces from the Grayburg and San Andres Formations. The field is part of a major productive trend along the northwest shelf. Vacuum field was discovered in 1929 and development began in late 1937 after pipeline facilities were built. By 1941, 327 wells were completed on a 40-ac (0.16-km^2) spacing (Purves, 1990; Wehner and Prieditis, 1996). Within the CVU, scattered development drilling associated with primary recovery operations continued until a water-injection program began in 1978. Additional infill development drilling within the CVU continued during 1978 and 1979 on a 20-ac (0.08-km^2) spacing. Wells were drilled on a 10-ac (0.04-km^2) spacing during the late 1980s. During October–December 1995, a one-well pilot CO_2-injection program occurred within the study area. Time-lapse, multicomponent (4-D, 3-C) seismic surveys were acquired during the CO_2 pilot program to investigate the utility of these data for reservoir characterization and to detect and monitor changes in rock-fluid properties associated with the CO_2 injection, soak, and production process (Benson and Davis, 2000). The CO_2 flood was later expanded to six injectors. With the expansion, additional time-lapse, multicomponent seismic surveys were acquired in December 1997 and 1998.

Available Data

Within the study area (Figure 2), 3-D surface seismic volumes consist of four compressional-wave (p-wave) volumes, four "fast" shear-wave (S1-wave) volumes, and four "slow" shear-wave (S2-wave) volumes. These data were acquired through RCP as part of the time-lapse seismic monitoring project associated with one phase of the CO_2-injection program within the CVU. Borehole geophysical data include a nine-component (one 3-C receiver) vertical seismic profile (VSP), walk-away VSP (WAW), and downhole 3-D seismic survey, each acquired in well CVU-200. In addition, two 9-component (12 3-C receivers) VSPs, WAWs, and downhole 3-D seismic surveys were acquired in well WS-15 (pre- and post-CO_2 injection). The downhole 3-D seismic surveys involved data acquisition using the downhole receivers and surface seismic source.

Other data include conventional logs from 120 wells, one FMI (Formation MicroImager) in well WS-226, and log data from a medium-radius, dual-lateral horizontal well (CVU-110). Fifty of the 120 wells are present within the smaller characterization and 3-D model area (Figure 2). Six cores were available (Figure 2), and neural-network-estimated permeability curves were provided by Texaco, the unit operator, for the majority of the wells within the study area.

Injection and production data, primarily consisting of monthly cumulative volumes of fluids injected or produced, were also provided by Texaco. In addition, 17 wells had single or multiple injectivity profiles.

GEOLOGIC SETTING

Within Vacuum field, the Permian San Andres Formation (Guadalupian) consists of approximately 1500 ft (457 m) of dolomites interbedded with a few thin dolomitic siltstones at a depth of approximately 4500 ft (1372 m). However, only the upper 600–800 ft (183–244 m) of the San Andres comprise the main hydrocarbon-bearing interval. The overlying Grayburg Formation (Guadalupian) consists of approximately 250–300 ft (76–91 m) of interbedded dolomite, sandstone, anhydrite, and shale. In general, the Grayburg Formation exhibits much lower reservoir quality than the San Andres Formation in this area.

Vacuum field is associated with an anticlinal feature that developed resulting from a combination of sediment drape, differential compaction, and faulting on the north. Shelf-margin depositional relief and faults bound the reservoir on the south (Figures 3, 4). This feature, combined with the high-frequency cycles that are characteristic of the San Andres Formation, created a stratigraphic-structural trap for hydrocarbons at Vacuum field.

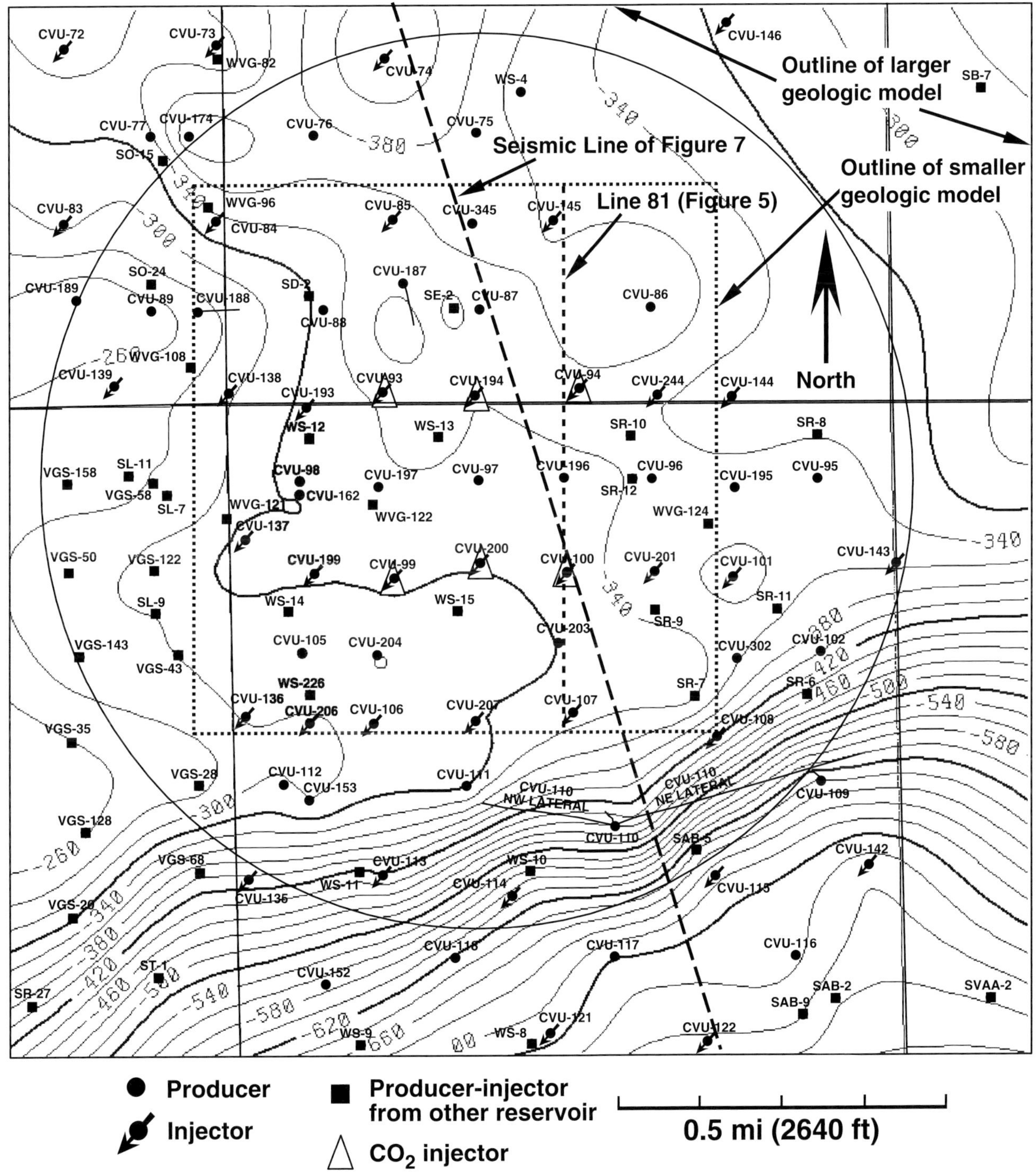

FIGURE 3. San Andres structure contour map. Large circle outlines the time-lapse area, and the smaller dashed square corresponds to the geologic model area. The dashed lines represent the locations of the seismic lines of Figures 5 and 7. The closely spaced contours correspond to the shelf margin. Note that the smaller geologic model area covers an area north of the shelf margin.

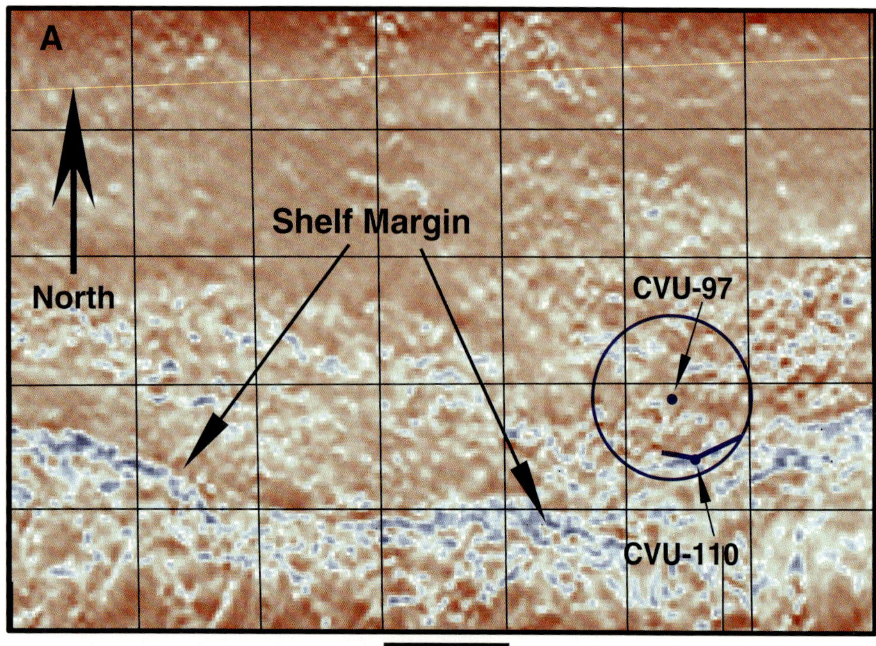

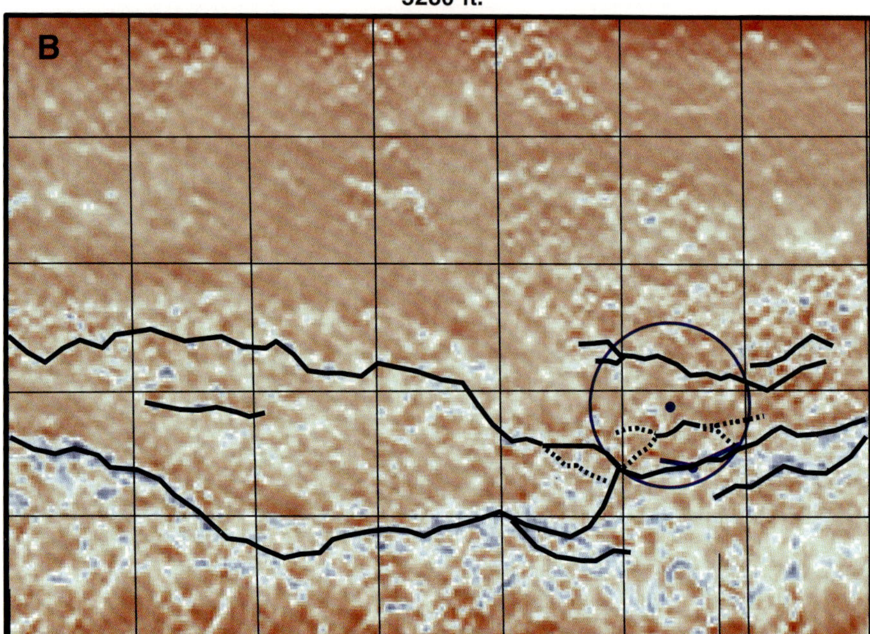

FIGURE 4. Top of Lower San Andres seismic coherency maps: (A) without interpretation and (B) with fault interpretation. Light blue shades indicate areas of low seismic continuity and red shades indicate areas of high seismic continuity. The time-lapse seismic area is denoted by the circle. Modified from Galarraga (1999).

Major discontinuities interpreted within the reservoir interval from regional 3-D seismic-coherency volumes (Talley, 1997; Galarraga, 1999) show a variety of trends, some of which are similar to those fault trends interpreted from detailed structural mapping based only on well data. A regional seismic coherency map of the Lower San Andres horizon (Figure 4) reveals major east-west–trending discontinuities associated with normal faults along the Leonardian-Guadalupian shelf margin within the CVU (Galarraga, 1999). Several east-west trends that exhibit low seismic continuity are apparent within the time-lapse area.

Vertical throw of major faults was estimated from seismic and well data to range from 0 to 70 ft (0 to 21 m) but is generally less than 25 ft (8 m). The major faults are intersected by numerous smaller-scale faults with minor offset (10 ft [3 m] of vertical displacement or less). The estimates of fault displacement are also supported by data from a dual-lateral horizontal well (CVU-110) that was drilled along the shelf margin. Vertical and subvertical fractures within the reservoir interval are observed from core data and borehole images. Within this part of the field, the structure is described as an uplifted and rotated central fault block with downthrown blocks to the north and south.

Vertical displacement along faults at the Upper San Andres horizon is not as pronounced as the displacement at the Lower San Andres level. A north-south seismic

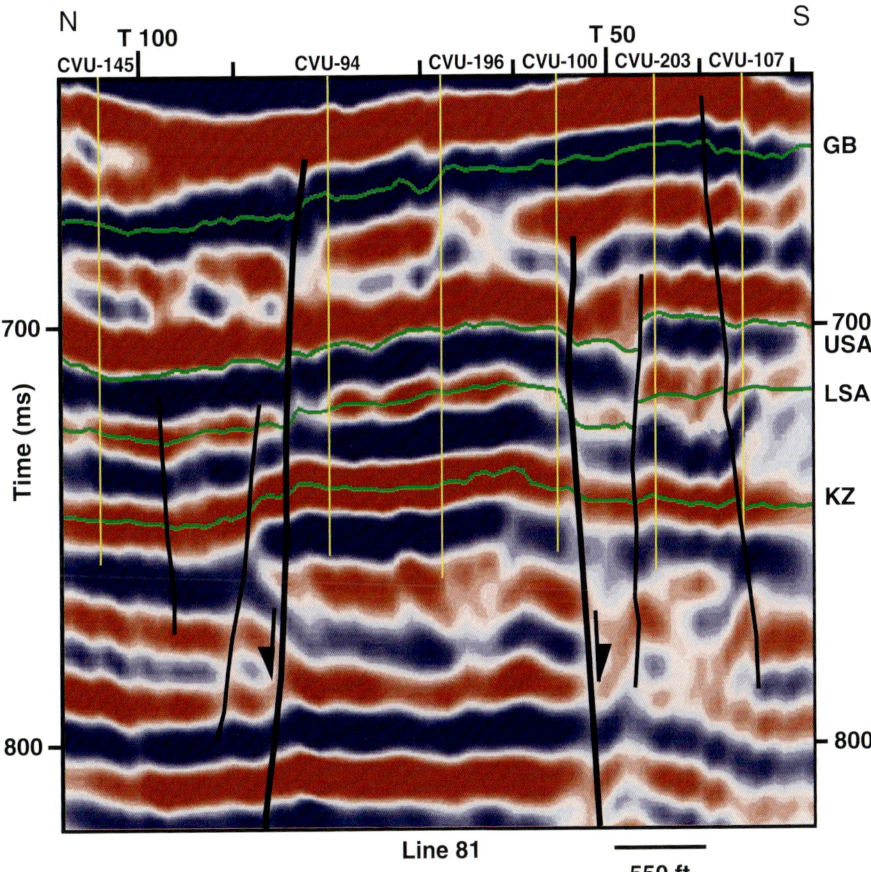

Figure 5. North-south compressional-wave seismic profile (line 81) showing significant faults (black lines). Displayed horizons include the Grayburg (GB), Upper San Andres (USA), Lower San Andres (LSA), and the base of a laterally extensive karst zone within the Lower San Andres (KZ). Red and blue colors correspond to changes in peak and trough amplitudes, respectively (SEG reverse polarity). Well paths are shown as yellow vertical lines. The location of the seismic line is shown in Figure 3.

profile from a 3-D compressional-wave seismic volume shows the key horizons, fault blocks, and locations of the two main bounding faults (Figure 5). Subsidiary faults are also shown.

Four major depositional environments characterize the San Andres Formation within this part of Vacuum field (Figure 6). The primary reservoir rocks consist of peloidal dolopackstones, skeletal dolograinstones, and fusulinid dolopackstones. These rocks alternate with lower-reservoir-quality dolomite intervals that exhibit variable degrees of anhydrite cementation. The Lovington siltstone is characterized by very low matrix permeability and vertically separates the Upper and Lower San Andres in the northwest part of the study area. The Lovington represents eolian silts and sands that were deposited on the platform. Evaporites, supratidal carbonates, and low-permeability siltstones provide the seal for the reservoir. The San Andres in this area represents an overall shallowing-upward interval composed of numerous high-frequency depositional cycles that subdivide the reservoir into alternating zones of high and low reservoir quality. Significant faults, fractures, and features resulting from pervasive diagenesis overprint the primary depositional fabric. The prominent diagenetic processes include dolomitization, karstification, and cementation. These result in additional reservoir complexity (Leary and Vogt, 1990; Adams, 1997).

The San Andres Formation along the northwest shelf can be subdivided into two third-order composite sequences, herein referred to as the Upper and Lower San Andres (Figures 7, 8). Similar to the San Andres in outcrop within the Guadalupe Mountains, the high-frequency sequences in the upper part of the Lower San Andres at Vacuum field record the progradation of subtidal, intertidal, and supratidal facies tracts across a Leonardian platform. The Upper San Andres composite sequence contains approximately nine fourth-order, high-frequency sequences. Only two of the nine high-frequency sequences of the Upper San Andres composite sequence lie directly on top of the Lower San Andres and Lovington on the platform. The San Andres Formation is capped by a regionally extensive subaerial unconformity.

Four San Andres high-frequency sequences comprise the most productive reservoir interval within the study area. These sequences include the upper two high-frequency sequences of the Lower San Andres and the upper two high-frequency sequences of the Upper San Andres. Individual high-frequency sequences are further divided into numerous higher-order (fifth-order) depositional cycles represented by characteristic vertical lithofacies successions (Figure 8).

The sequence-stratigraphic interpretation of the San Andres was guided by key indicator facies that

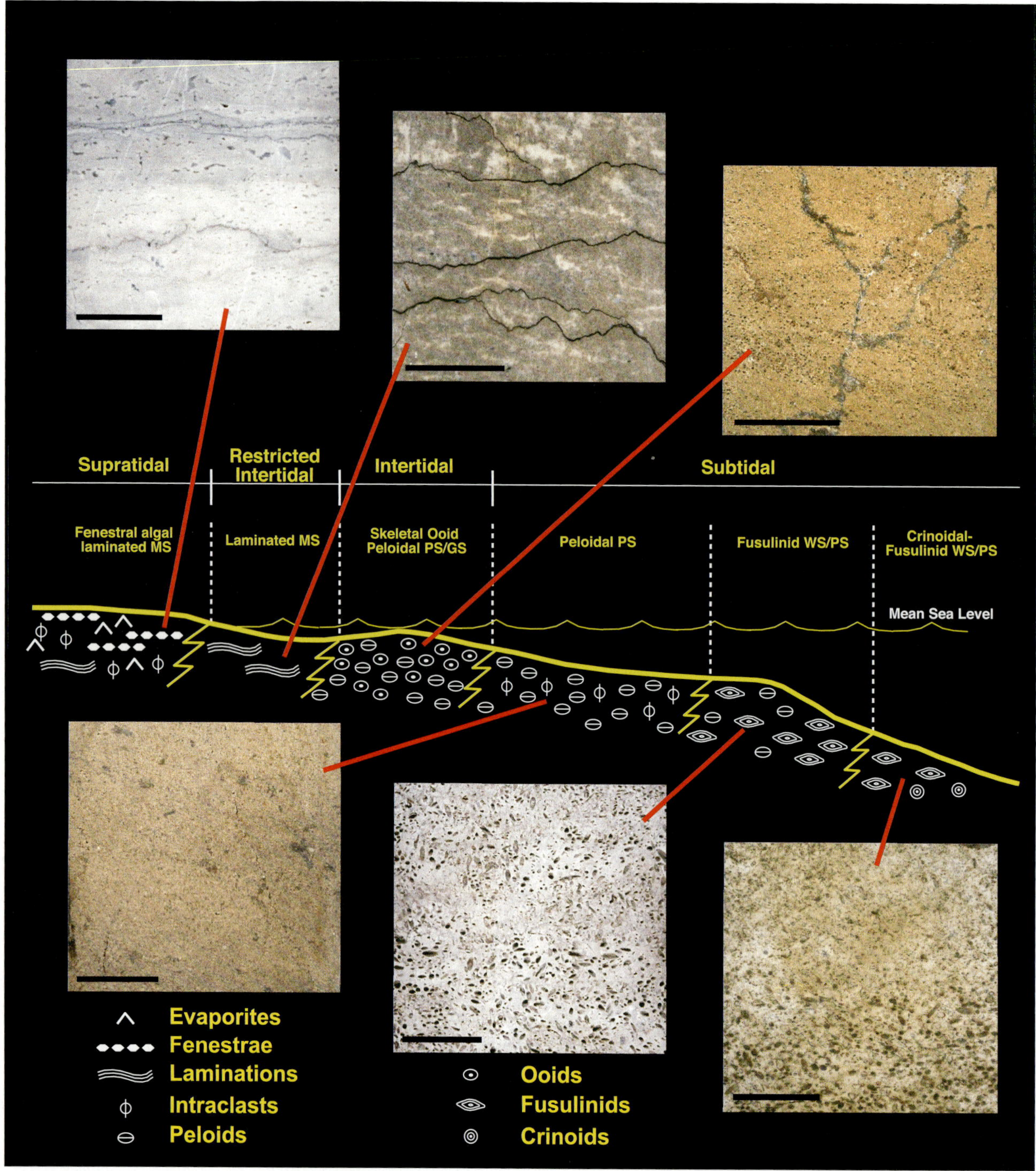

FIGURE 6. Facies model of the San Andres Formation at Vacuum field. The four main facies tracts include subtidal, intertidal, restricted intertidal, and supratidal. Photographs of representative core samples illustrate the carbonate fabric and some sedimentary-diagenetic features associated with key lithofacies. MS = mudstone; WS = wackestone; PS = packstone; GS = grainstone. Black scale bar represents 1 in. (2.54 cm) in all photographs.

represent interpreted depth/energy positions, such as shoreline and fair-weather and storm wave base (Kerans and Tinker, 1997). Three indicator facies were identified based on lithology, allochems, and sedimentary structures. For the San Andres interval, these include fenestral algal laminites, peloidal ooid dolograinstones-dolopackstones, and fusulinid dolowackestones-dolopackstones. Vertical lithofacies successions, in conjunction with key indicator facies and exposure surfaces, were used to define the finer-scale cyclicity of the San Andres Formation within cored wells.

Within each high-frequency sequence (HFS 1–4), individual cycles generally are thickest at the base and become thinner toward the top. Upward cycle thinning, combined with an increase in the ratio of peritidal lithofacies to subtidal lithofacies in successive cycles, reflects decreased accommodation (shallowing-upward trend) associated with carbonate deposition and progradation.

Karst breccias, terrigenous siltstones, increases in the proportion of supratidal lithofacies, and fractures are characteristic features commonly found at or near high-frequency and composite-sequence boundaries.

The analysis of the sequence stratigraphy from core and 3-D seismic data suggests that the San Andres first developed in an overall aggradational (vertically stacked) pattern, followed by a decrease in accommodation

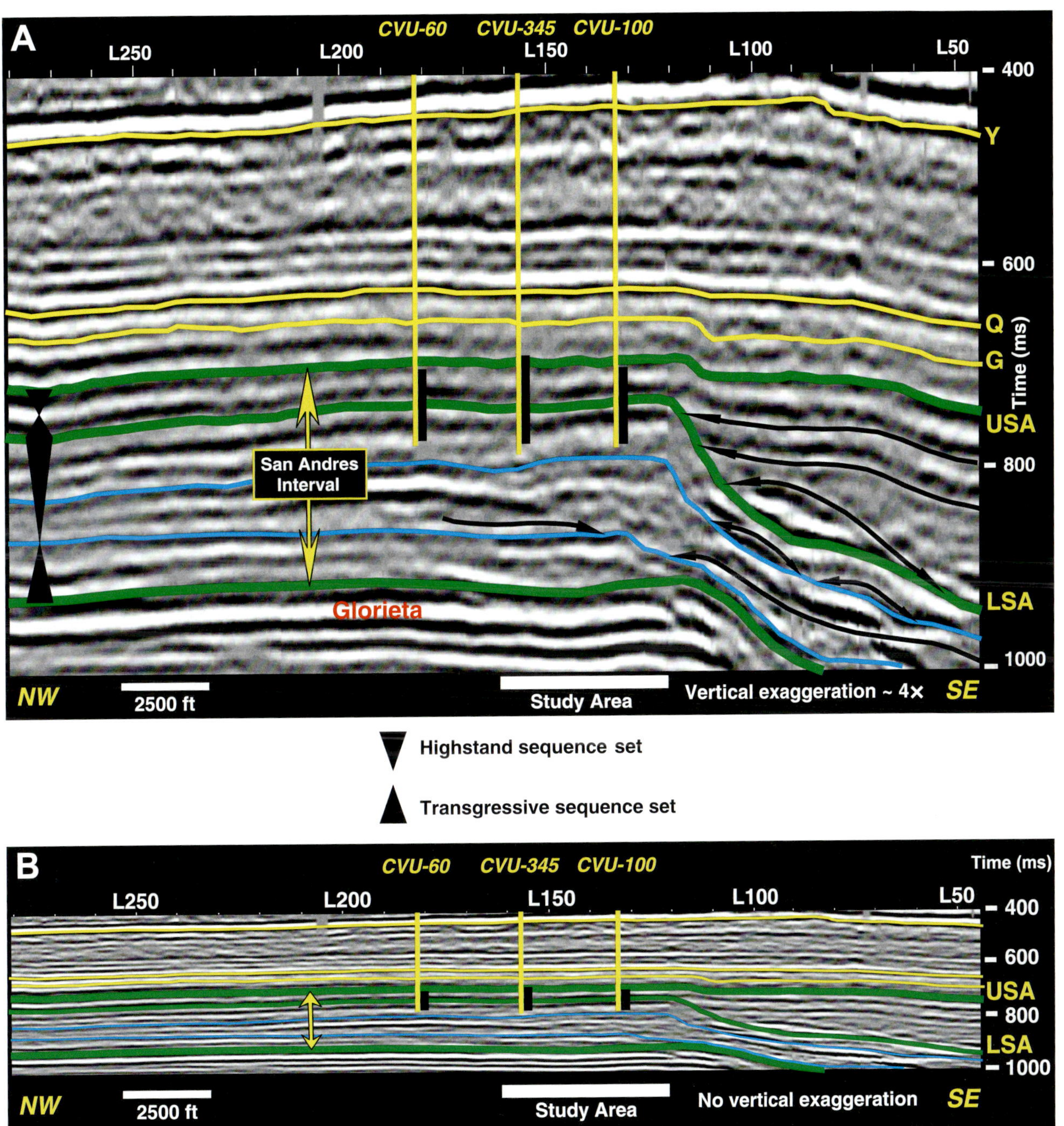

FIGURE 7. Dip-oriented compressional-wave seismic profiles illustrating the San Andres carbonate platform and ramp-shelf margin. The Upper and Lower San Andres composite sequences are shown between the green lines. Profile A shows composite-sequence boundaries (green lines), cored-well locations (yellow vertical lines with black boxes that represent core), stratal geometries, and main stratigraphic horizons: Y = Yates; Q = Queen; G = Grayburg; USA = Upper San Andres; LSA = Lower San Andres. Profile B is the same seismic profile without vertical exaggeration. The location of the study area is denoted by a white bar on both profiles. The location of the seismic line is shown on Figure 3.

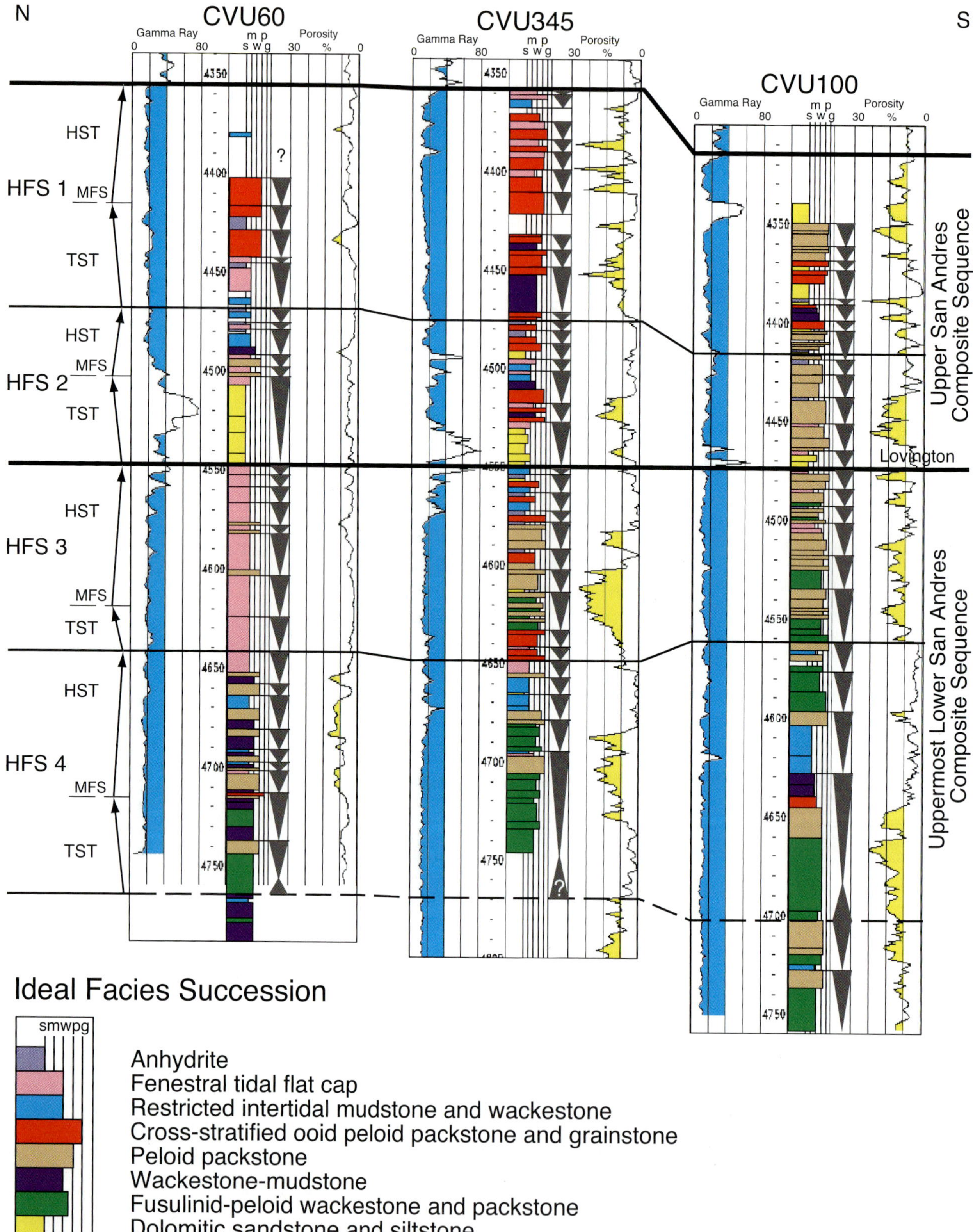

FIGURE 8. North-south profile through CVU-60, CVU-345, and CVU-100 showing gamma ray, neutron porosity, general lithofacies description from core analysis, and interpreted sequence-stratigraphic framework. Profile datum is the top of the Lower San Andres. From core analysis, a simplified description of an ideal vertical lithofacies succession is illustrated. Lithofacies for CVU-345 and -100 were modified from Capello de Passalacqua (1995), Adams (1997), and Scuta (1997). HFS = high-frequency sequence; HST = highstand systems tract; TST = transgressive systems tract; MFS = maximum flooding surface.

space and basinward progradation (seaward-stepping facies tracts). This interpretation is in agreement with several other interpretations of the San Andres Formation from outcrops within the Guadalupe Mountains (Sarg and Lehmann, 1986; Sonnenfeld and Cross, 1993; Kerans, 1995).

PETROPHYSICS-BASED ZONATION

The construction of a representative stratigraphic or structural framework with an appropriate number of layers or flow units is essential in reservoir modeling. Various methods have been proposed to characterize reservoir quality and subdivide carbonate reservoirs into flow units based on descriptions of pore geometry, rock fabric, or rock type (Lucia, 1983, 1995; Ahr, 1991; Lucia et al., 1992; Montgomery et al., 1998; Nevans et al., 1998). These methods commonly involve establishing relationships between petrophysical properties and rock fabric or rock type using core and log information. Rock-fabric units, lithofacies, or depositional facies are then estimated in noncored wells using various techniques. Based on rock-fabric estimates or facies interpretations in wells and interpretations of depositional and diagenetic environments with seismic data, 3-D facies models are built and facies-keyed petrophysical models (e.g., permeability, porosity) are generated. The fine-scale layers associated with these 3-D models are commonly created between key stratigraphic surfaces using interpreted layering schemes (e.g., proportional, onlapping, truncated).

This study used an alternative method in which reservoir zones (flow units) were defined within high-frequency sequences based on vertical variations of flow (kh) and storage capacity (ϕh) and pore-throat radius (R35). These zones were then used to construct the geologic-model framework (Gunter et al., 1997). This method to define flow units can be especially favorable for fields with log data but limited core information, where detailed descriptions of rock fabric are not possible or where finer-scale parasequences (cycles) are difficult to identify. In noncored wells, porosity information is commonly available from porosity logs, thus, only permeability must be estimated. In this study, neural-network-derived values of permeability were used.

Facies Relationships

The San Andres reservoir is composed of facies that were deposited on a relatively low-relief carbonate platform and altered by meteoric, marine, and burial diagenesis. In general, there is an increase in permeability with increasing porosity throughout the reservoir interval associated with both depositional and diagenetic rock fabrics (Figure 9). Extensive dolomitization and karstification have significantly altered the primary carbonate fabric associated with subtidal through supratidal facies. Oolitic dolograinstones, peloidal dolopackstones, and dolomitic mudstones exhibit a wide range of porosity and permeability values within each rock type (Figure 9). Rock-fabric classes exhibit a wide range of porosity and permeability values resulting from effective porosity development associated with

- dolomitization of mudstones,
- porosity and permeability reduction resulting from pervasive anhydrite and dolomite cementation within grainstones and packstones, and
- permeability enhancement related to dissolution of allochems and fracture development.

Montgomery (1998), Nevans et al. (1998), and Tucker et al. (1998) found similar relationships between reservoir quality and rock fabric within other Permian carbonate reservoirs at McElroy (Grayburg) field and North Robertson (Clear Fork) Unit in west Texas. Given the low correlation between depositional texture and reservoir quality, a method based on the direct use of petrophysical data was employed to identify reservoir zones. Within zones exhibiting similar porosity values, connected biomoldic pores and fracture porosity, for example, enhance permeability, whereas separate moldic pores do not increase permeability. Therefore, knowledge of pore types and estimates of pore-throat radius are also important to understand effective porosity vs. total porosity (Lucia, 1999; Martin et al., 1999).

Pore-Throat Radius—R35

Pore-throat radius values from capillary-pressure tests on core plugs and estimates of pore-throat apertures from log data were used to farther support flow-unit interpretations (Appendix). Gunter et al. (1997), Martin et al. (1997), and Hartmann and Beaumont (1999) discussed techniques and advantages of using pore-throat radius values at 35% mercury-injection capillary pressure (R35) to identify reservoir flow units.

Air-brine capillary-pressure tests were conducted on 18 core plugs taken from the CVU-100 core. Core-plug petrophysical data are listed in Table 1. Capillary-pressure and pore-throat radius curves for several core-plug samples illustrate the variability in reservoir quality and flow potential within the reservoir (Figures 10, 11). Flow units with higher reservoir quality and flow potential have relatively flat initial slopes and are shifted to the left on Figure 10 (sample 2 has the highest reservoir quality). The Lovington (sample 7) has very low reservoir quality and limited flow potential and acts as a barrier between the Upper and Lower San Andres.

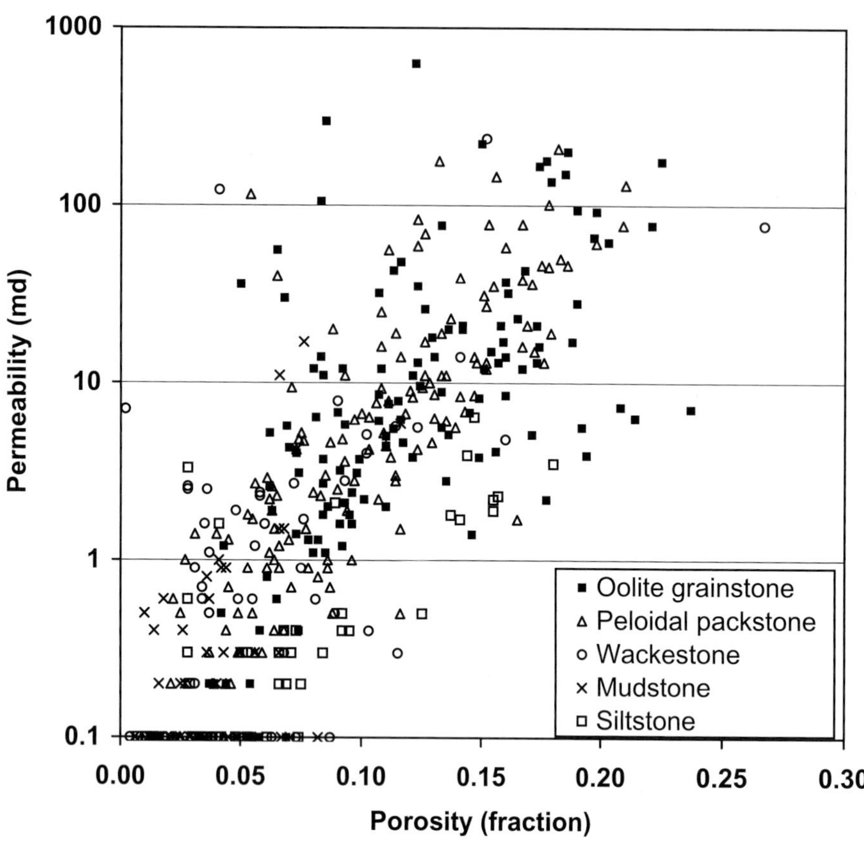

FIGURE 9. Porosity-permeability crossplot for the cored interval in well CVU-100. Data points are labeled according to depositional texture identified from core descriptions in Capello de Passalacqua (1995), Adams (1997), and Scuta (1997). The crossplot shows the range of reservoir quality associated with the observed rock types.

Flow potential is directly related to pore-throat radius or the connectivity between pores. This is clearly illustrated in Figure 11 as sample 2 exhibits the largest pore throats, whereas the Lovington siltstone (sample 7) has extremely small pore throats. For comparison to other petrophysical parameters, pore-throat sorting (PTS) values

Table 1. Core-plug petrophysical data.

Core plug	*Porosity (%)*	*Permeability (md)*	*Pore-throat sorting*	*Pore-throat radius— R35 (μm)*
1	17.1	107.7	1.66	13.0
2	22.4	933.1	1.41	16.8
3	17.0	37.4	2.35	16.1
4	10.4	0.1	3.70	1.5
5	17.3	25.9	1.73	9.5
6	17.8	95.0	1.48	15.4
7	2.9	0.0	–	0.01
8	21.3	146.6	2.29	7.2
9	15.4	78.6	2.38	8.1
10	6.0	0.9	1.39	0.8
11	13.0	35.1	1.86	7.6
12	9.2	6.7	2.67	7.0
13	9.8	23.7	4.21	5.5
14	15.0	7.0	3.70	9.3
15	15.9	2.6	4.33	1.3
16	12.3	8.8	2.41	10.1
17	10.3	10.3	1.80	12.8
18	6.5	1.7	2.96	2.1

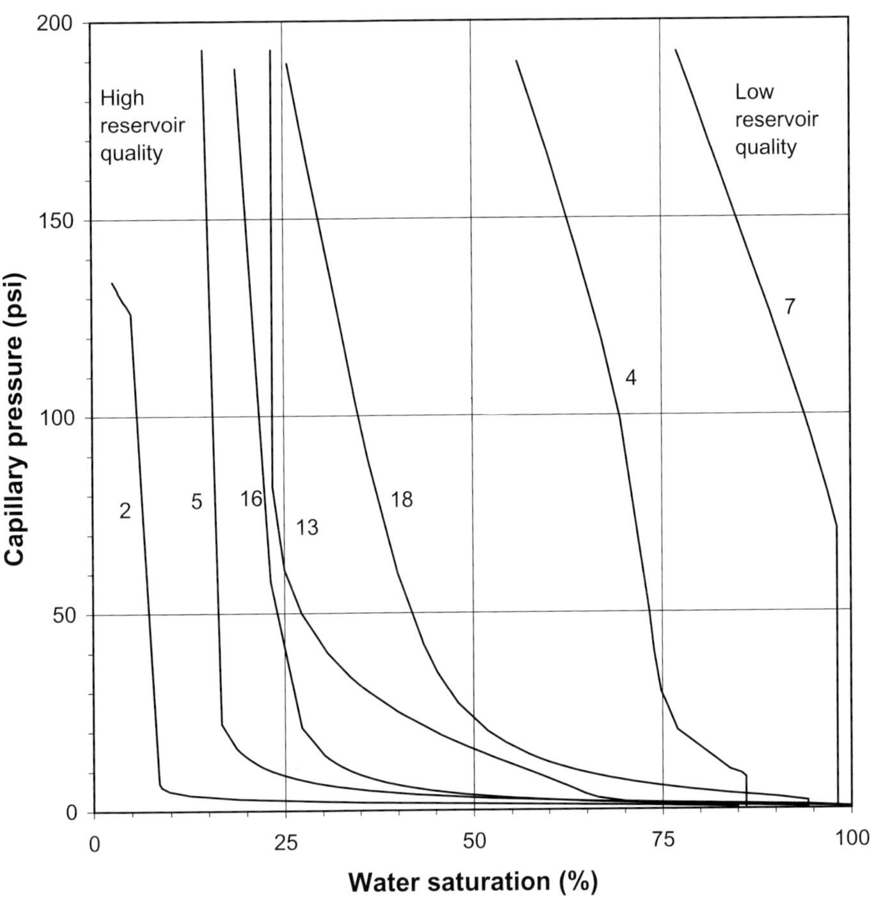

FIGURE 10. Capillary-pressure curves obtained from centrifuge studies of water-saturated core plugs from the San Andres and Lovington siltstone. The capillary-pressure curves show the range of reservoir quality associated with the different rock types. The number next to each curve corresponds to the core-plug sample number from Table 1.

(Appendix) are also listed for each core-plug sample (Table 1).

The geometry and sorting of pores and pore throats were also evaluated by analyzing thin sections with an optical microscope and thin sections and rock surfaces with a scanning electron microscope (SEM) of the same lithology and from the same location as each core plug.

Visual estimates of porosity, pore-throat radius, and pore connectivity (permeability) from thin sections and SEM analysis compare relatively well to estimates obtained from capillary-pressure analysis on core plugs (Pranter, 1999). In general, interparticle and intercrystalline pore types are associated with the most productive lithologies. Intraparticle porosity was significant in some samples, but connectivity between pores was generally low. Therefore, permeability values for these samples were relatively low. A weak relationship exists between depositional texture and reservoir quality. The weak relationship is, in part, a result of pervasive anhydrite cementation and replacement. Porosity within some dolograinstones and dolopackstones is completely filled with anhydrite cement, whereas some dolomudstones and dolowackestones exhibit well-developed intercrystalline porosity. Binary SEM images were created to illustrate the variability in pore geometry and pore-size distribution (Figure 12G–I). The binary images were created using NIH Image (version 1.56), a public-domain image-processing software (Anselmetti et al., 1998). Samples with higher reservoir quality generally have higher values of R35 and exhibit well-sorted pores and pore throats. Some heterogeneous samples with poorly sorted pore throats exhibit greater values of average porosity than samples with well-sorted pore throats. However, permeability of poorly sorted samples is generally lower (Figure 12). Binary images (Figures 12H–I) show the low connectivity (in 2-D) between pores within the poorly sorted sample vs. the well-sorted sample. Estimated R35 values from capillary-pressure data reflect the difference in reservoir quality and PTS observed in these samples (Figure 12).

Information from the capillary-pressure tests on the San Andres samples was used to modify the Winland equation to relate more closely to carbonate pore characteristics of the San Andres reservoir. Pore-throat radius values were related to core-plug porosity and permeability to modify the coefficients of the empirically derived Winland equation using a multivariable linear-regression algorithm (Kuester and Mize, 1973). Wetting-phase (brine) saturations from 10 to 75% were investigated to evaluate the relationship among pore-throat radius, porosity, and permeability at each saturation. The pore-throat radius equation (modified Winland

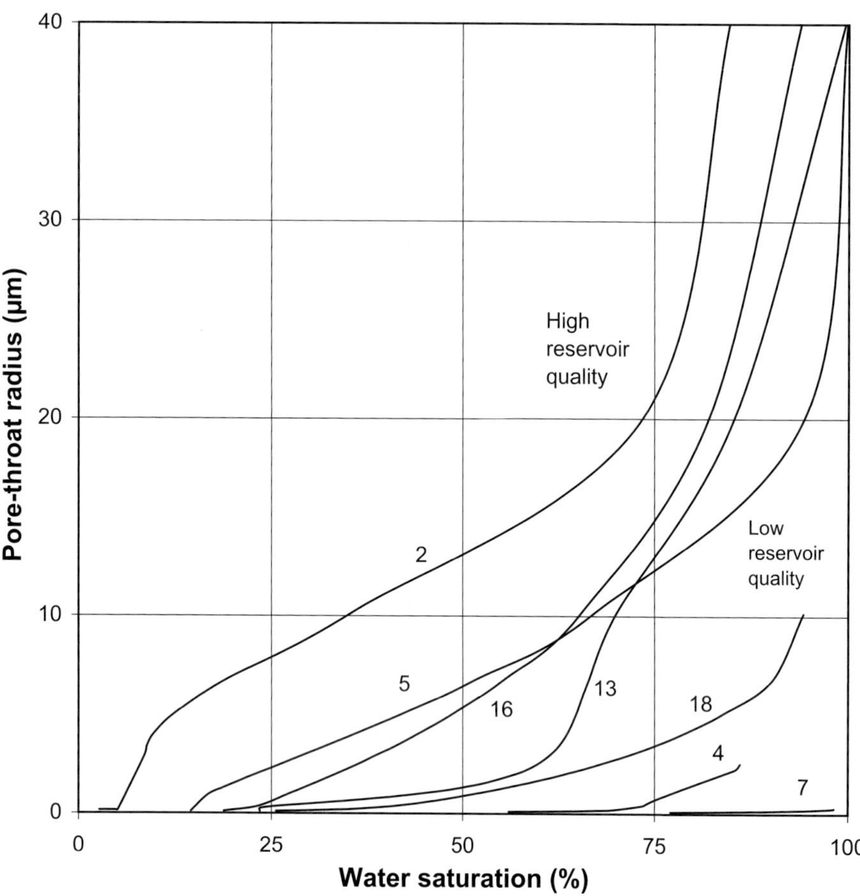

FIGURE 11. Pore-throat radius curves computed from capillary-pressure data on core plugs from the San Andres and Lovington siltstone. The pore-throat radius curves show the range of reservoir quality that is related to the representative pore-throat size of the sample. The number next to each curve corresponds to the core-plug sample number from Table 1.

equation) at 35% saturation resulted in the highest correlation coefficient among the variables. This modified Winland equation was then used to estimate pore-throat radius in noncored wells. R35 curves were generated for each well within the area and used with flow and storage capacity to identify and correlate reservoir flow units.

Use of Flow and Storage Capacity

Reservoir zones within high-frequency sequences were defined based on vertical variations in flow and storage capacity using log data and core. The technique utilizes stratigraphic modified Lorenz (SML) plots (Gunter et al., 1997) or crossplots of cumulative flow (kh) and storage capacity (ϕh) to define flow units within a stratified reservoir. On the SML plot, cumulative flow- and storage-capacity values are plotted in stratigraphic order, starting at the base of the reservoir. This is unlike the original Lorenz plot in which flow-capacity values are arranged from maximum to minimum, despite stratigraphic position (Craig, 1993). Significant inflection points on the SML plot correspond to changes in flow or storage capacity associated with stratification, fractures, porosity, and other factors that affect reservoir quality. These changes are interpreted using straight-line segments that define intervals (flow units) within the reservoir. Segments on the plot that slope greater than 45° are indicative of zones of relatively higher flow potential and lower storage capacity (potential thief zones). Segments that slope less than 45° correspond to zones of greater storage capacity and lower flow capacity (possible barriers). Segments that trend near 45° represent zones characterized by similar flow and storage capacity.

SML plots were generated for the San Andres interval using wells within the study area that had porosity logs and permeability estimates from neural network analysis (Figure 13). Core and log values of cumulative storage (ϕh) and flow capacity (kh) for a given well generally exhibited very similar trends. SML plots were interpreted using straight-line segments to define flow units that adequately described the vertical variability within the reservoir. Figure 14 illustrates an interpreted SML plot for cored-well CVU-345. Within this well, the Lower San Andres is characterized by thicker zones with significant storage capacity, whereas the Upper San Andres is more thinly layered and vertically heterogeneous. The low-permeability Lovington siltstone exhibits limited flow potential and acts as a barrier between the Upper and Lower San Andres, as noted by relatively horizontal segments on the SML plots.

To understand the significance of the interpreted zones with regard to flow performance, a modified

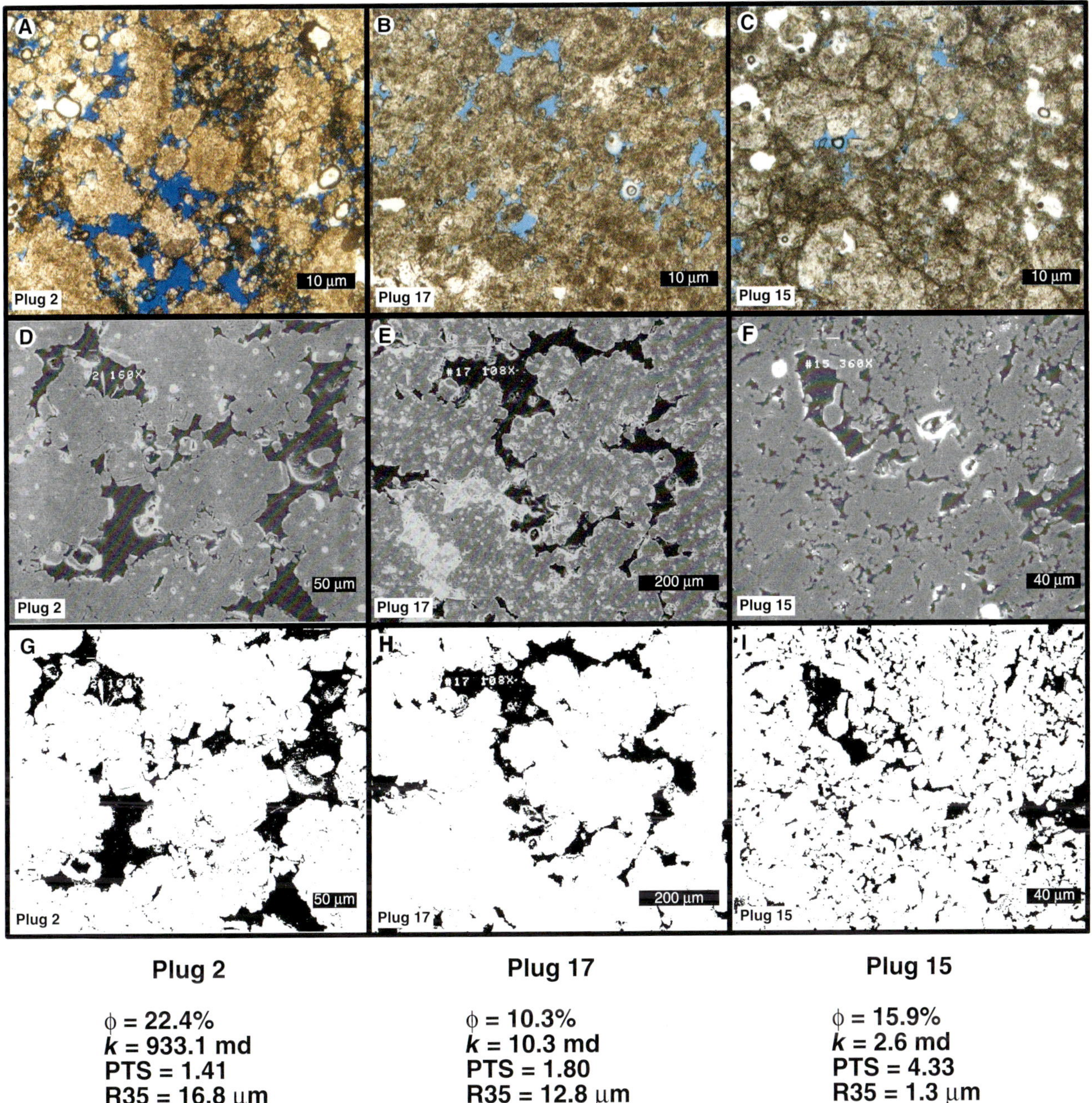

FIGURE 12. Photomicrographs of thin sections of core plugs 2, 17, and 15 obtained through an optical microscope (A–C) and with an SEM (D–F). Binary images of SEM photos, displayed as G, H, and I, illustrate the variability in pore geometry and pore-size distribution. In thin sections, blue epoxy highlights porosity; porosity on SEM and binary images is black. Reservoir quality and pore-throat sorting (PTS; higher numbers represent lower sorting) are reduced from samples 2 to 15. Although porosity values are similar in samples 17 and 15, permeability is lower for sample 15 because of the poorly sorted nature of pore throats (H and I).

Lorenz (ML) plot (Gunter et al., 1997) shows flow units arranged in order of decreasing flow capacity. The ML plot identifies which zones have a high flow potential and which zones act as barriers or baffles to fluid flow (Figure 15). In well CVU-100, 4 of the 12 flow units account for 60% of the flow capacity and only 20% of the storage capacity (Figure 15). These zones are more likely to take injected fluid, which could result in water or CO_2 cycling through high-permeability zones and bypassed pay in zones with relatively lower flow potential. Production operations that have involved selectively perforating these bypassed pay intervals within the Upper San Andres have resulted in a significant increase in oil production from those targeted wells with a very small or no increase in the volume of produced water (Scuta and Hurley, 1998). In both wells,

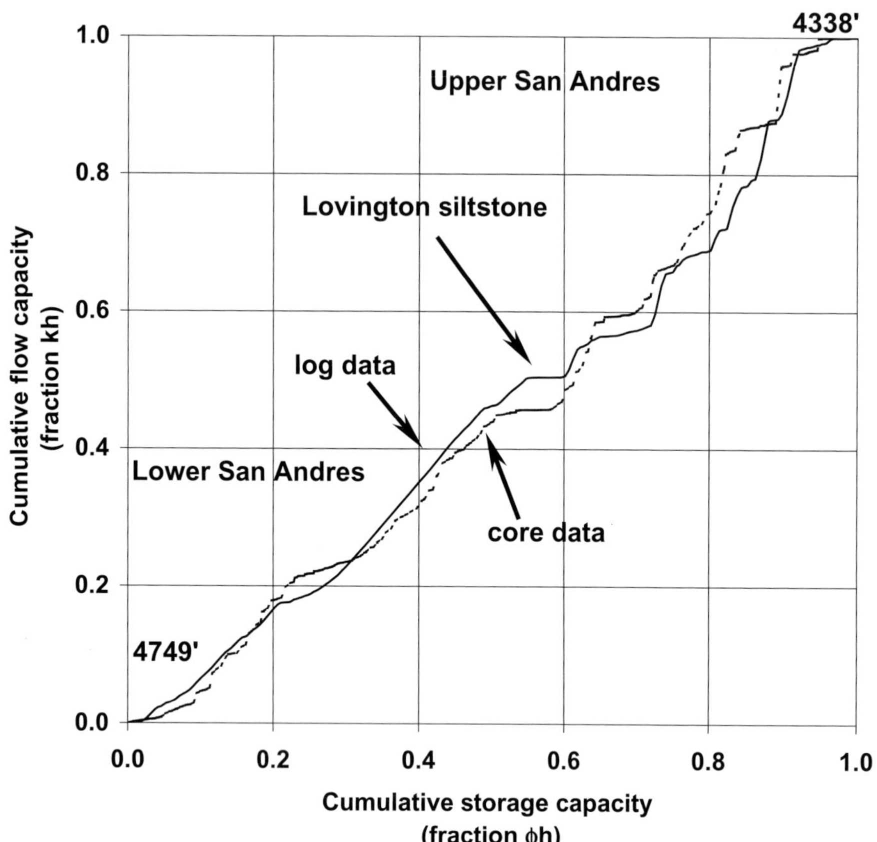

FIGURE 13. Stratigraphic modified Lorenz plot for the San Andres Formation, well CVU-345.

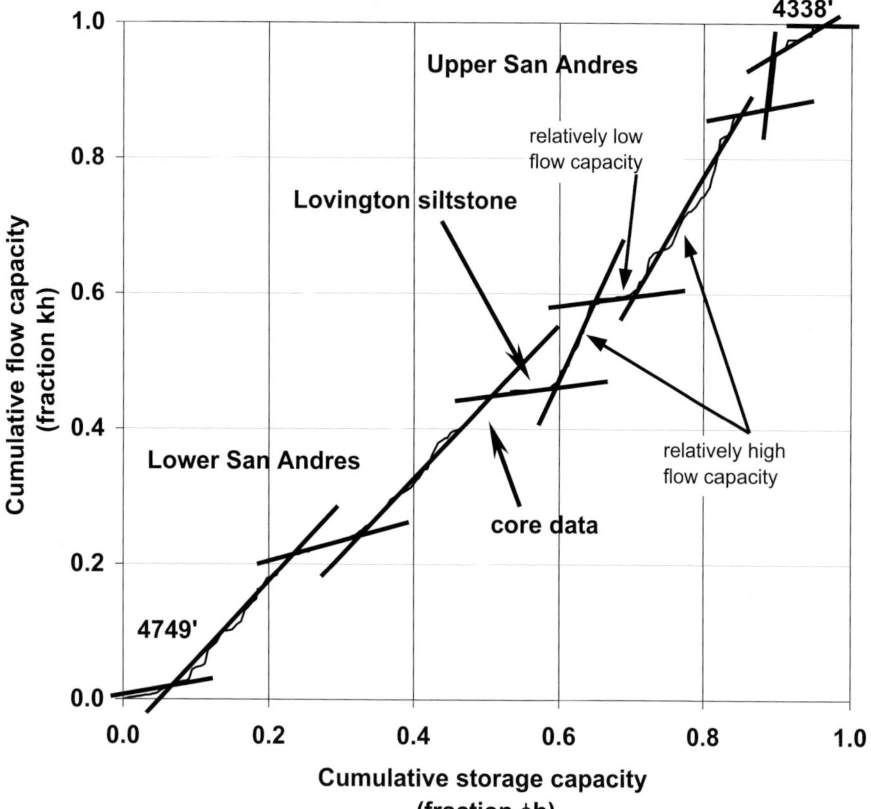

FIGURE 14. Interpreted stratigraphic modified Lorenz plot, well CVU-345. Straight-line segments are used to interpret intervals of similar reservoir quality based on changes in the slope of the curve. Each straight-line segment is inferred to be a flow unit.

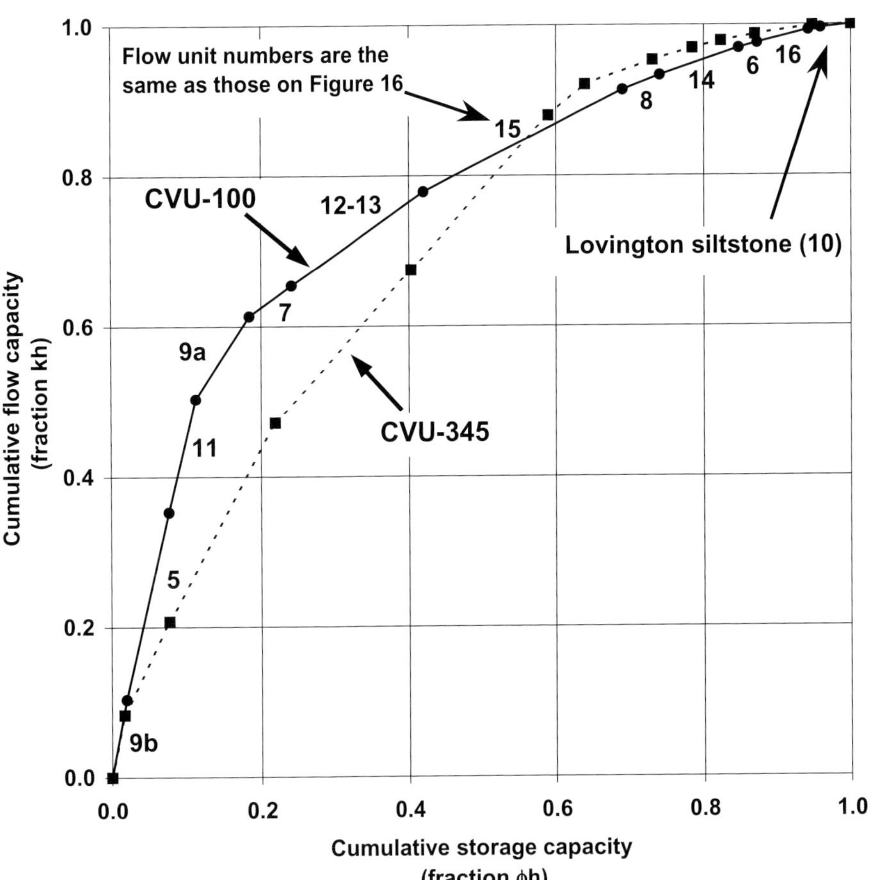

FIGURE 15. Modified Lorenz plot, wells CVU-100 and -345. Flow units defined from the stratigraphic modified Lorenz plot are plotted in order of decreasing flow capacity. Steeply dipping segments (lower left) represent flow units characterized by high flow capacity and relatively low storage capacity. Relatively flat segments (upper right) represent flow units with low flow capacity and greater storage capacity. In well CVU-100, note that 4 of the 12 flow units account for 60% of the flow capacity and only 20% of the storage capacity.

zones within the Lower San Andres tend to have greater storage capacity relative to flow capacity, whereas the opposite is observed for the Upper San Andres.

A stratigraphic flow profile for well CVU-100 (Figure 16) illustrates the flow units that were identified within the reservoir interval in that well. Flow units that were defined in cored wells were correlated across the area in noncored wells. Stratigraphic flow profiles were generated for all wells with porosity and permeability logs to correlate flow units and establish the finer-scale reservoir layering within the sequence-stratigraphic framework. This framework formed the basis for layering within the 3-D geologic model. Eighteen parasequence-scale flow units or reservoir zones were correlated across the model area within the Grayburg dolomite, Grayburg sandstone, Upper San Andres, Lovington, and Lower San Andres. Seven flow units were defined and correlated within the Upper San Andres and eight flow units were defined within the Lower San Andres. The Grayburg dolomite, Grayburg sandstone, and Lovington each comprised one reservoir zone. Two main faults interpreted from well data and compressional-wave seismic volumes were incorporated into the reservoir framework during the model-building process. These faults exhibit the greatest vertical throw and, based on production and injection response, are believed to partially compartmentalize the reservoir.

FRACTURES AND SHEAR-WAVE ANISOTROPY

The methods used to characterize fractures within the San Andres are addressed in this section. This section also presents a brief overview of shear waves and their factor in fracture characterization within the CVU.

Vertical and subvertical fractures were observed in cores from six wells and on a borehole-image log from well WS-2-26 (southwest part of study area). A combination of both open and healed (cemented) fractures was observed. Using borehole images, Scuta (1997) identified two orthogonal sets of open fractures and one set of healed fractures. The dominant orientations of open fractures are 110–140° (parallel to the present-day, maximum horizontal-stress direction) and 60–80° (parallel to the San Andres shelf margin) (Scuta, 1997). The set of healed fractures also parallels the San Andres shelf margin. Although core and borehole images confirm the presence and nature of fractures, 3-D multicomponent seismic data were necessary to spatially analyze fracture density and orientation.

When vertically propagating shear waves enter an anisotropic medium, similar to a reservoir with aligned vertical fractures, the waves are polarized, resulting in fast shear-wave (S1) particle motion parallel to the

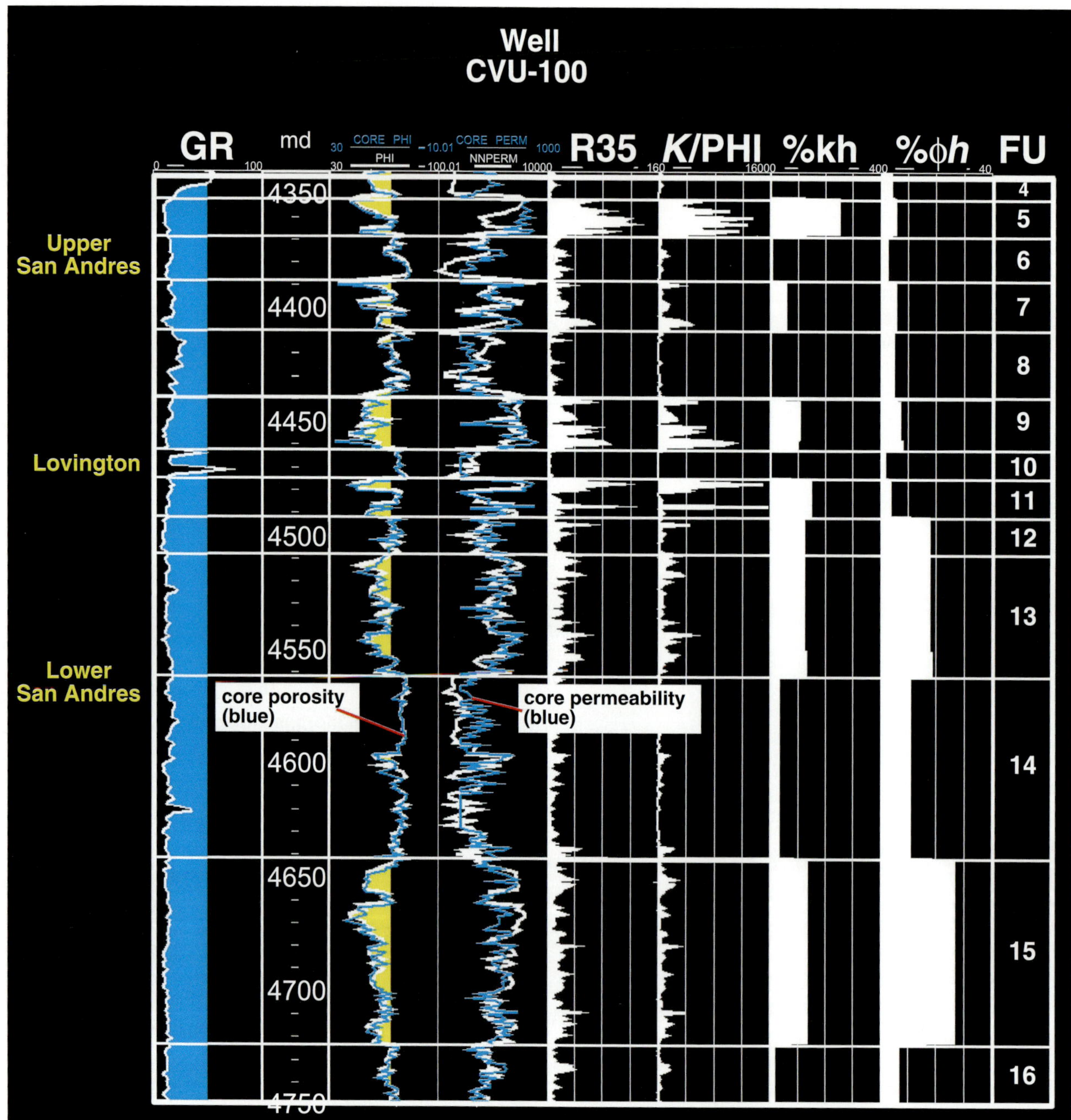

FIGURE 16. Stratigraphic flow profile, well CVU-100. Flow units (FU) are shown for the San Andres interval (labeled 5–16). Flow unit 4 is the Grayburg sandstone. This well did not penetrate the deeper flow units (17 and 18). Displayed data include gamma ray (GR), core and log porosity, core and neural-network permeability, R35, *K*/Phi, percent flow capacity (kh), percent storage capacity (ϕh), and flow unit number. Gamma-ray values less than 50 API units are shaded blue, and porosity values greater than 7% are shaded yellow.

aligned fractures and slow shear-wave (S2) particle motion perpendicular to the fractures (Winterstein, 1992). A medium (reservoir) is less rigid perpendicular to fractures, so the velocity of shear waves that are polarized in this orientation is reduced. The opposite is true regarding shear waves polarized parallel to aligned fractures. In this direction, the rigidity is greater and shear-wave velocity is greater. This difference in shear-wave velocity produces time delays between the fast and slow shear waves, known as shear-wave birefringence or splitting (Winterstein, 1992). These shear-wave properties were investigated and used to identify and characterize fractures within the San Andres reservoir.

In many petroleum reservoirs, the stress created by the overlying rock mass (vertical stress component) is generally the largest in magnitude. If differences in horizontal stress exist, maximum and minimum horizontal-stress components can be defined. Fractures within a reservoir that are oriented parallel (or near parallel) to the maximum horizontal compressive stress have a greater possibility of being open, unless filled with cement or gouge, than those fractures oriented perpendicular (or near perpendicular) to the maximum horizontal stress. Estimates of the maximum (present-day) horizontal compressive stress direction were made using borehole breakouts, VSP shear-wave polarization data, and residual-rotation analysis of surface seismic data. These analyses produced very similar results for the San Andres Formation with maximum horizontal-stress directions of 122, 123, and 118°, respectively (Mattocks, 1998; Scuta, 1997; Michaud, 1999, personal communication; Roche, 1997). Because the shear-wave surface seismic acquisition coordinate system was not the same as the natural coordinate system associated with the maximum horizontal compressive stress, the shear-wave data were rotated using Alford rotation (Alford, 1986) during seismic processing to the estimated S1 (118°) and S2 (28°) component directions. Shear-wave anisotropy is a measure of the amount of shear-wave splitting (time delay) caused by anisotropy within the reservoir, which is commonly associated with fractures and low-aspect-ratio pores (pores that are relatively flat or elongate vs. spherical). Estimates of normalized shear-wave anisotropy were computed by using isochron (time interval) differences across the reservoir interval between the processed fast (S1) and slow shear waves (S2) as

$$\frac{\Delta t_{S2} - \Delta t_{S1}}{\Delta t_{S1}}$$

where Δt_{S1} is the time interval of a given zone on the S1 seismic volume and Δt_{S2} is the time interval of the same zone on the S2 seismic volume. The amount of shear-wave splitting or anisotropy is expressed as a percent. Given variability in the local stress regime and structural complexities, it would be unrealistic to assume that all open fractures within the reservoir are exactly aligned to 118°. Likewise, the direction of the fast shear wave within the reservoir will not be aligned to 118° at all locations. Therefore, variations in open-fracture orientation from the maximum horizontal-stress direction can result in negative anisotropy values.

A shear-wave anisotropy map that covers the San Andres reservoir (Figure 17) shows the lateral variation in anisotropy across that interval. The shear-wave anisotropy reflects the anisotropy within a reservoir associated with aligned fractures and karst zones. Given how the seismic data were acquired and processed at Vacuum field, positive values of shear-wave anisotropy correspond to northwesterly aligned fractures or low-aspect-ratio pores. Positive and negative shear-wave anisotropy values computed across the reservoir interval are interpreted to coincide with two orthogonal sets of aligned vertical-fracture trends within the reservoir as observed in cores and on borehole images (Scuta, 1997). Areas of high total-fluid production also coincide with the areas of positive shear-wave anisotropy, and these areas are interpreted to exhibit a greater density of aligned vertical fractures.

INTEGRATED GEOLOGIC MODEL

A 3-D geologic model that represents the vertical and lateral heterogeneity within the reservoir was constructed for use in flow simulation to predict reservoir performance. The geologic model covers the central part of the time-lapse seismic area where the fold-of-stack is higher (range of 160–240) and there is greater confidence in the seismic data interpretation. The model area includes 50 wells. Dimensions of the model are 4070 ft (1240 m) on each side. Total thickness ranges from 681 ft (208 m) on the northwest side of the model to 868 ft (265 m) on the southeast. The model is spatially discretized into 37 rows, 37 columns ($\Delta x = \Delta y =$ 110 ft = 34 m), and 18 layers of variable thickness. The cell x and y dimensions of 110 ft (34 m) (twice the seismic bin size of 55 ft [16.75 m]) is selected to minimize the total number of model cells, maintain spatial details, and reduce the possibility of problems during simulation associated with numerical dispersion. Because the model incorporated nonvertical faults, some of the cells are divided, resulting in a total of 26,242 cells in the model (Figure 18). The formations included in the model were the Grayburg dolomite, Grayburg sandstone, and approximately 700 ft (213 m) of the upper part of the San Andres Formation (Figure 18). The Grayburg dolomite and Grayburg sandstone each comprised 1 layer, and the remaining 16 layers represented the San Andres Formation. Additional details of the 3-D geologic model are in Pranter (1999).

Parameter Estimation Using Multicomponent Seismic Attributes

Three-dimensional, multicomponent seismic data provide information to characterize lateral variations of physical rock properties, such as porosity and permeability within a reservoir. However, the vertical resolution of seismic data is much lower than the resolution provided by well information. Because individual flow units were below seismic resolution (Figure 19), map-based parameter estimation techniques using multicomponent seismic attributes were used to populate model cells with porosity and permeability.

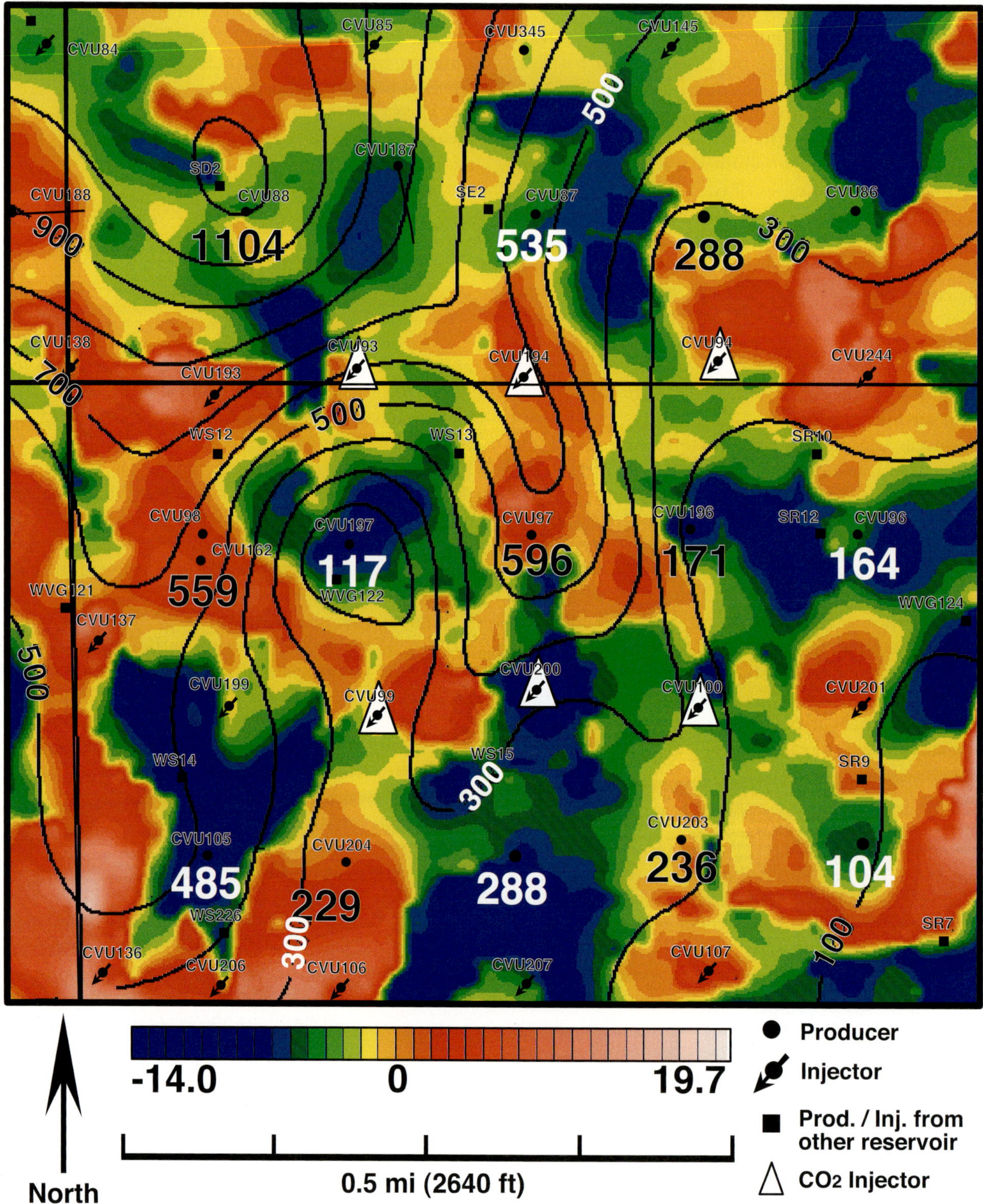

FIGURE 17. Shear-wave anisotropy in the reservoir interval. Average daily total-fluid production contours and values from the Grayburg-San Andres interval are overlain and posted in barrels of fluid per day for the period 1995–1998. Red colors indicate areas of positive shear-wave anisotropy, and blue colors indicate areas of negative shear-wave anisotropy. Yellow to orange colors reflect areas where shear-wave anisotropy is near zero.

Deterministic techniques, including multivariable regression and attribute-based rescaling, were used to estimate porosity. Well data combined with attribute dependence in SGM (Stratamodel) were used to estimate permeability. The porosity model and information on fracture density from shear-wave anisotropy were used to create a permeability model that included contributions from both matrix and fracture porosity.

Porosity Model

The lateral detail provided by the 3-D seismic data was used to distribute porosity within the geologic model. Porosity models were built separately for the Upper and Lower San Andres intervals. To estimate porosity, seismic-derived average-porosity maps are generated for the Upper and Lower San Andres. The average-porosity maps were then used as constraints to compute 3-D porosity distributions within the geologic model. Gorell (1995) used a similar method to combine reservoir-property maps with well data to populate reservoir zones that were below seismic resolution.

The process of generating average-porosity maps from seismic data involved five main steps: (1) interpret key horizons on compressional- and shear-wave seismic volumes; (2) extract seismic attributes from the intervals of interest (e.g., Upper and Lower San Andres) and identify relationships among the seismic attributes and reservoir properties from well data; (3) develop and apply a calibration function (multivariable regression equation) using the multicomponent seismic attributes; (4) compute and map residuals or differences between the actual mean porosity values at the well locations and those estimated using the calibration function; and (5) add the mapped residuals to the mean porosity map computed using the calibration function. The final product is a map of mean porosity that incorporates the lateral detail from the seismic attributes and honors well data. Schultz et al. (1994), Hinterlong et al. (1998), and Smith et al. (1998) used similar techniques to generate seismic-derived reservoir properties by integrating compressional-wave seismic data.

Numerous seismic attributes were compared to porosity data using multiple crossplots to identify meaningful relationships among the data. Average-reflection-strength (envelope) attributes provided reasonable correlations and were used for subsequent analyses. Average reflection strength is a complex trace attribute that can be thought of as amplitude independent of phase (Taner et al., 1979). Maps of Upper San Andres average porosity and reflection strength from P- and S1-wave data (Figures 20, 21) show the trends related to high porosity and low reflection strength and low porosity and high reflection strength. The well-based average-porosity maps were generated for a larger area than shown in Figures 20 and 21 and incorporate porosity data from all wells in the area.

Crossplots and statistical measures were used to quantitatively compare seismic attributes to average porosity at each well location. Average porosity for the

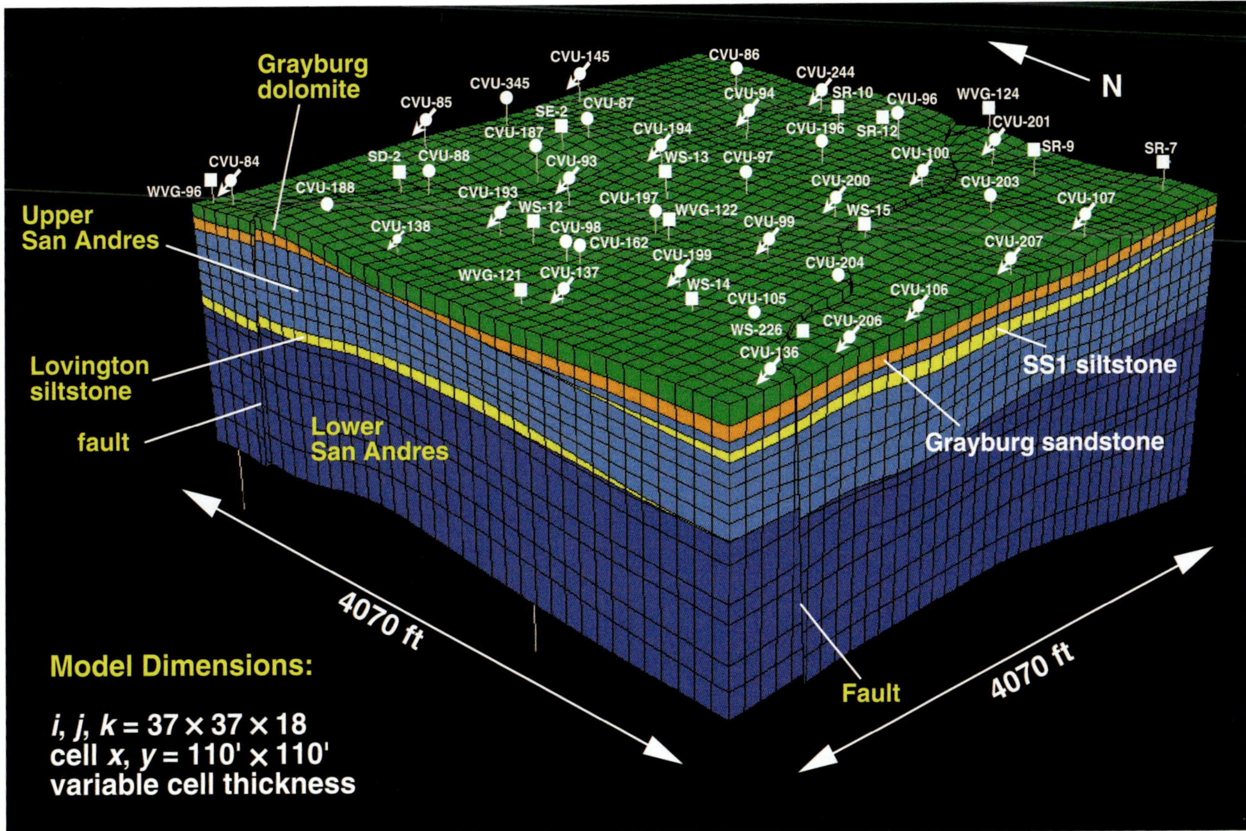

FIGURE 18. Geologic model parameters. Model framework incorporates the reservoir flow units and two main faults.

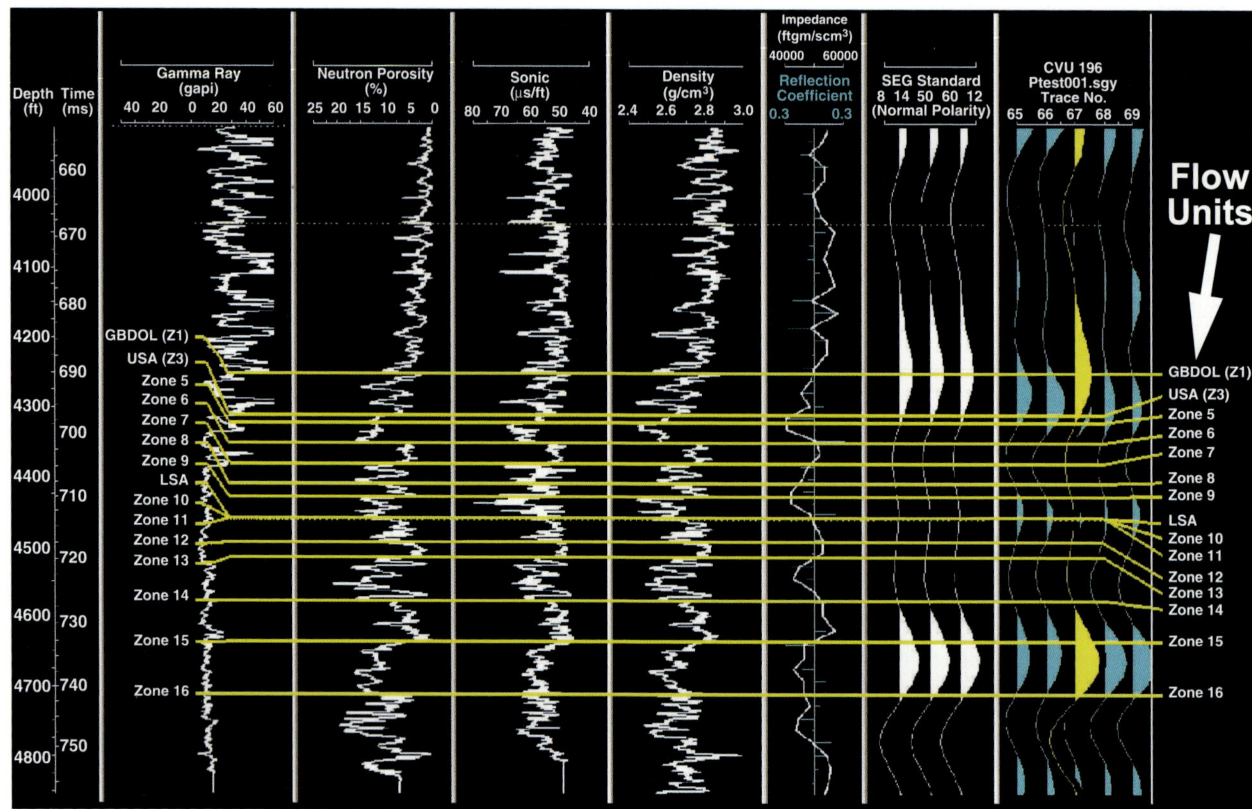

FIGURE 19. Compressional-wave synthetic seismogram, well CVU-196. Gamma ray, neutron porosity, sonic transit time, density, acoustic impedance, synthetic seismic traces, and compressional-wave surface seismic traces are displayed. Petrophysical-based flow units correspond well to vertical changes in porosity, transit time, density, and acoustic impedance. Although the Upper and Lower San Andres are resolvable on surface seismic data, individual flow units are below seismic resolution. GBDOL = Grayburg dolomite; USA = Upper San Andres; LSA = Lower San Andres.

Upper and Lower San Andres was computed from log data using average-reflection-strength values for the same intervals on the seismic data. Correlation coefficients (r) of -0.81 and -0.57 between average porosity (ϕ_{mean}) and P- (P_{ARS}) and S1-wave ($S1_{ARS}$) average reflection strength, respectively, were determined for the Upper San Andres (Figure 22). By combining these attributes using multivariable regression, the correlation coefficient increased to -0.85. The resulting calibration function

$$\phi_{mean} = 11.6372 - 0.00665504 \times (P_{ARS}) - 0.00118643 \times (S1_{ARS})$$

was used to transform the seismic data to average porosity at each bin location. For the Lower San Andres, only 10 wells penetrated the entire interval; thus, statistical estimates of correlation coefficients were considerably less reliable. However, because data from the 10 wells aligned with the trends observed for the Upper San Andres when plotted on the same graphs, the same calibration function was applied to both Upper and Lower San Andres intervals.

The calibration steps described involve a curve fit (or linear fit) to data points on the crossplots. The linear calibration curve (trend line) did not pass through all the points on the crossplot. As a result, when the seismic attributes were converted to average porosity using the model function defined by the trend line, the estimated properties did not agree exactly at most well locations (residual error). The residual error ranged from -1.75 to 1.50%, with the greatest residual error being along the margins of the model area. The residual map was added to the regression results to correct the minor differences between the estimated and actual porosity values at the well locations. The final average-porosity maps incorporate the lateral detail from the multicomponent seismic and honor well data (Figure 23). The Upper San Andres porosity maps show the southwest-to-northeast trend of higher average porosity that parallels the shelf margin. The maps also show the trend of lower average porosity to the northwest. This area of low average porosity corresponds with an area of higher total-fluid production from the San Andres that is believed to be associated with greater fracture permeability.

To estimate porosity in three dimensions, attribute-based rescaling was used. Attribute-based rescaling involves estimating porosity within each flow unit using a mapping approach (e.g., kriging, distance weighting, etc.) followed by linear rescaling of each vertical column of the geologic model (Figure 24; also see Gorell, 1995). At each location in the geologic model, the average porosity must conform to the seismic-derived average porosity. Therefore, at each x,y location, the following equation must be honored:

$$\Phi_{\text{average}}(i,j) = \frac{\sum_{k=1}^{N} \phi_k(i,j)\; h_k(i,j)}{H_{(i,j)}},$$

where Φ_{average} is the seismic-derived average porosity for the entire interval (e.g., Upper San Andres interval), ϕ_k is the porosity for layer k in the interval, h_k is the thickness of layer k in the interval, and H is the thickness of the entire interval. The estimated average porosity for the model for the Upper or Lower San Andres may not agree with the seismic-derived average-porosity map for that interval. Therefore, the model-layer porosity estimates must be modified. The following assignment was used to rescale each vertical column of the model so that the seismic-derived mean porosity map was honored:

$$\phi_{rk}(i,j) = \phi_k(i,j)\left[\frac{\Phi_{\text{average}}(i,j)H(i,j)}{\sum_{k=1}^{N} \phi_k(i,j)\; h_k(i,j)}\right],$$

where ϕ_{rk} is the rescaled porosity. Attribute-based rescaling can be viewed as the application of an area-dependent scaling factor to each of the model layers. At well locations, the seismic-derived average porosity value was equal to the average porosity from well data because of the calibration and residual correction process. Therefore, the scaling factor was equal to 1 at each well location. The final distribution of porosity within the geologic model reflects the vertical variation of porosity as defined by well data and the lateral porosity variation estimated using multicomponent seismic data.

A quantitative measure of confidence in the calibration process is useful to determine the accuracy of

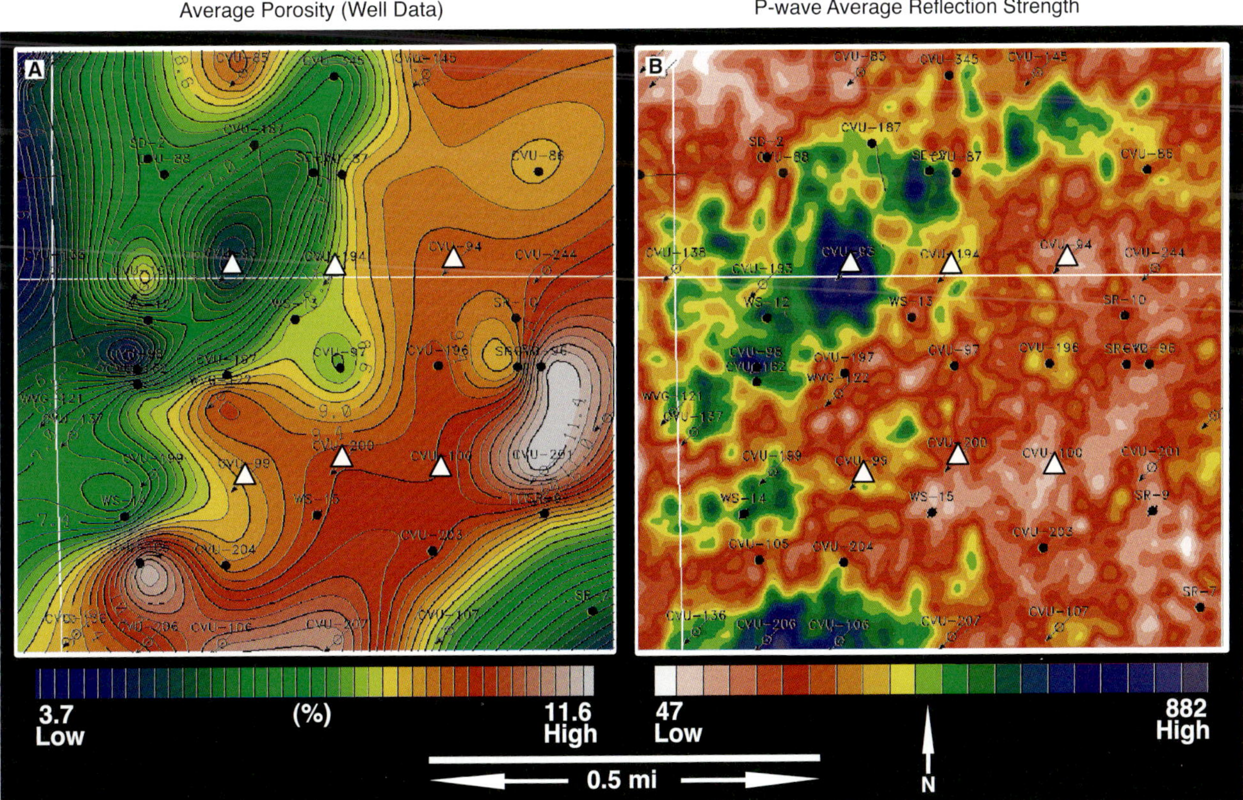

FIGURE 20. Upper San Andres maps of (A) average porosity and (B) P-wave average reflection strength. The average-porosity map is based on well data. The six CO_2 injectors are shown as triangles on each map. Areas of high average reflection strength and low porosity are blue, and areas of low average reflection strength and high porosity are red and white.

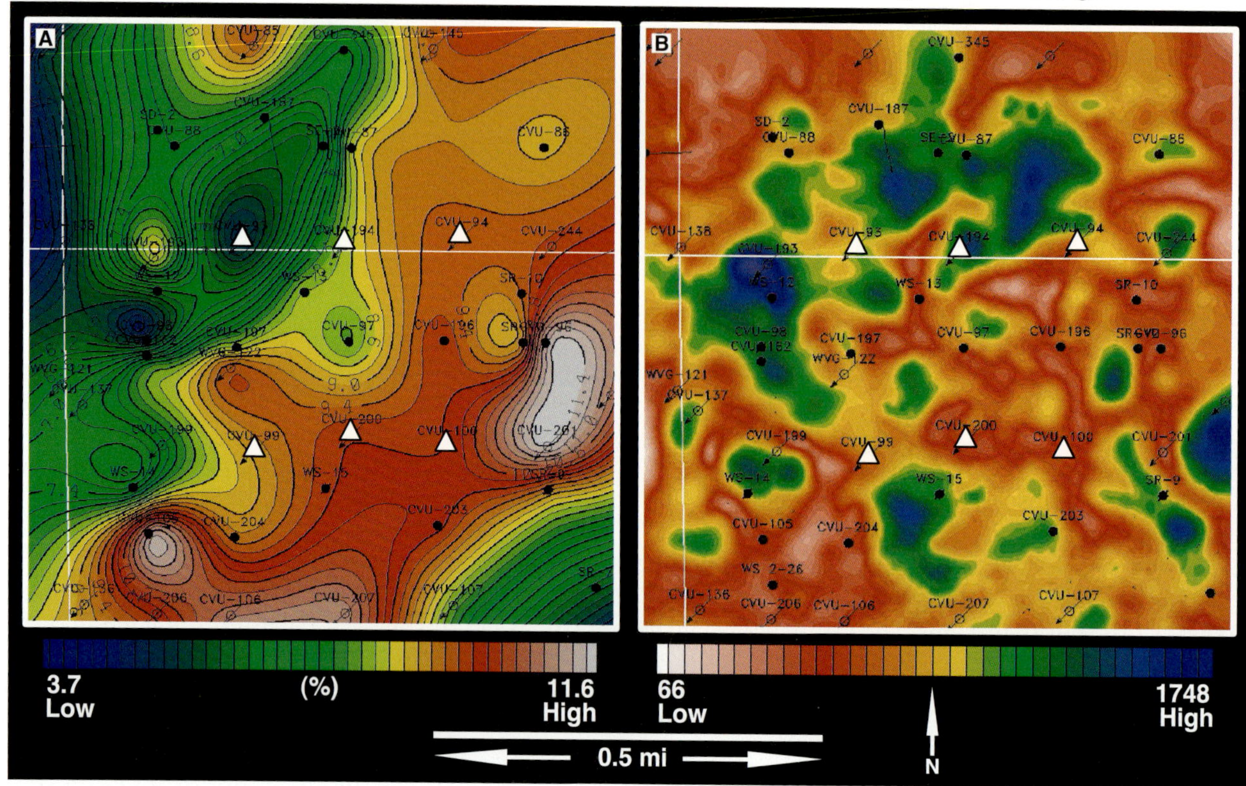

FIGURE 21. Upper San Andres maps of (A) average porosity and (B) S1-wave average reflection strength. The average-porosity map is based on well data. The six CO_2 injectors are shown as triangles on each map. Areas of high average reflection strength and low porosity are blue, and areas of low average reflection strength and high porosity are red and white.

the estimated results. We used the following method for this validation: One well was removed from the data set and a new calibration function was derived using multivariable regression. The new calibration function and residual correction were applied, and the estimated porosity value at the removed well location was compared with the actual value from the removed well. This process was repeated separately for 47 wells in the data set, and the residuals between the estimated and actual values were mapped (Pranter, 1999). A clear improvement in estimated porosity was achieved by incorporating multicomponent seismic data. Correlation coefficients (r) between estimated and actual average porosity values at well locations improved from 0.55 using well data alone to 0.89 with multicomponent seismic data (Figure 25).

Permeability Model

Permeability within the San Andres reservoir is controlled by a variety of pore types, including fractures. Within a complex reservoir like the San Andres, log-derived data typically only provide information about properties of the rock matrix. Likewise, log-derived porosity and even neural-network permeability data based on log porosity and core data do not account for the contribution of fractures to reservoir quality. Neural-network permeability values are more closely associated with matrix permeability, whereas shear-wave anisotropy from multicomponent seismic data reflects the velocity anisotropy associated with open vertical fractures and low-aspect-ratio pores associated with the rock matrix or fractures. Thus, we used the neural-network-derived permeability data and measurements of shear-wave anisotropy to estimate permeability values for the reservoir interval. Production data, borehole images, and core data were also used to address the distribution of permeability.

Permeability models were created both with and without constraints to the seismically guided porosity information. The porosity model was used to weight the distribution of neural-network permeability. This was done to account for the basic relationship that was observed between porosity and permeability and to include the lateral detail from the multicomponent seismic data that was also incorporated into the porosity model (Pranter, 1999). These steps produced a permeability model that primarily accounted for lateral

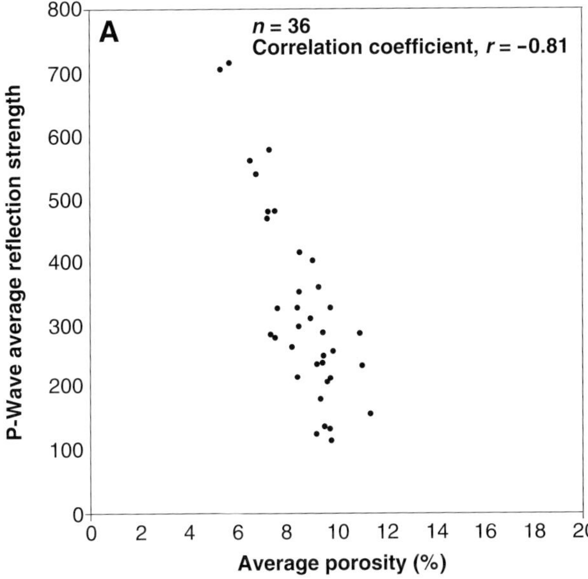

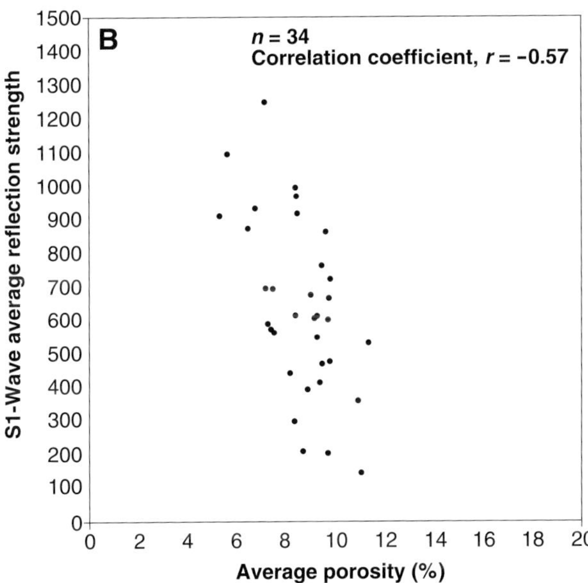

FIGURE 22. Crossplot of Upper San Andres average porosity vs. (A) P- and (B) S1-wave average reflection strength. Plotted values of average porosity and reflection strength were obtained at well locations within the model area.

heterogeneity associated with the reservoir matrix. The San Andres shear-wave anisotropy data (Figure 19) were then used to bias or modify horizontal (*kx* and *ky*) and vertical (*kz*) permeability values within each flow unit to account for the contribution of fractures to reservoir quality and reservoir performance. Permeability values in areas that exhibit positive shear-wave anisotropy were increased to account for the presence of open fractures and, therefore, higher permeability in the horizontal and vertical directions (Pranter, 1999). Figure 26 is an example of a vertical slice through the final permeability model. In general, the Upper San Andres exhibits lower horizontal permeability than the Lower San Andres. This difference in permeability within the model accounts for the observed difference in reservoir productivity between the Upper and Lower San Andres.

GEOLOGIC MODEL AND PRODUCTION DATA

Using the petrophysics-based zonation combined with multicomponent seismic data, the observed variability in permeability between and within the Upper and Lower San Andres was more accurately represented within the geologic model. Simulation results using geologic models of the study area based on well data alone did not show the preferential drainage of the Lower San Andres as observed in the field (Bard, 1999, personal communication). The discrepancy was primarily because of the similarity in permeability distributions between the two intervals that resulted from using well data alone to estimate permeability within the geologic model. Reservoir simulation models based on the modified permeability distribution more accurately accounted for the unexpectedly early CO_2-breakthrough times observed in the field than simulation models based solely on matrix properties. In addition, simulation results indicated that by using the petrophysics-based method to define reservoir zones, the need for additional upscaling was significantly reduced or eliminated, because reservoir zones for simulation are defined at the start of the modeling process rather than at the end. If necessary, additional zones or layers can be added or merged to improve the history match with production data. The sequence-stratigraphic framework must be honored to ensure that reservoir zones are properly correlated across the geologic model and simulation area.

CONCLUSIONS AND IMPLICATIONS

1) The San Andres Formation within the study area consists of stacked, shallowing-upward carbonate cycles that were deposited within subtidal, intertidal, and supratidal environments. Vertical lithofacies successions, changes in lithofacies proportions, and exposure surfaces identified in core define four high-frequency sequences within the reservoir interval. The upper two high-frequency sequences form the Upper San Andres composite sequence, and the lower two high-frequency sequences form

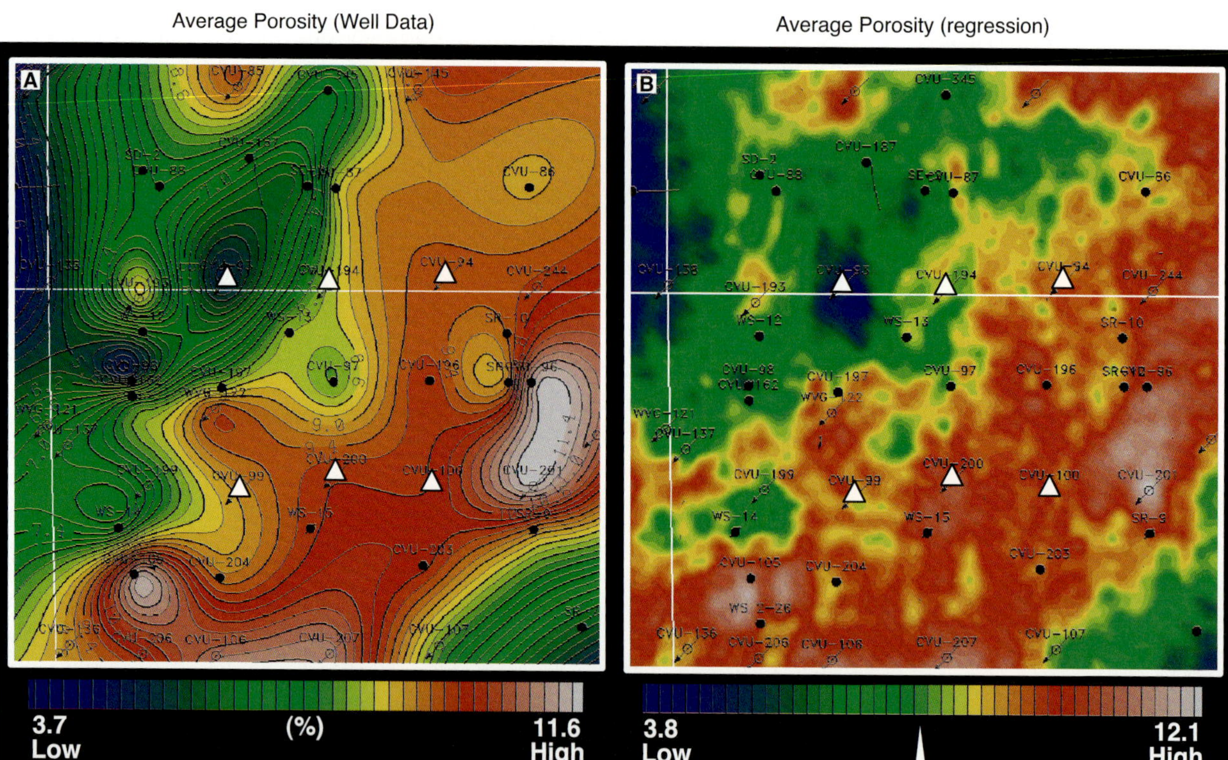

FIGURE 23. Upper San Andres maps of (A) average porosity and (B) estimated average porosity using multivariable regression. The six CO_2 injectors are shown as triangles on each map. Contour interval on map A is 0.2%. Areas of low and high average porosity are blue and red or white, respectively. The lateral variability in average porosity provided by the seismic data is incorporated in map B.

the uppermost part of the Lower San Andres composite sequence.

2) The petrophysics-based method presented here is a useful technique for characterizing reservoirs like the San Andres at Vacuum field. The method is useful to refine the gross sequence-stratigraphic interpretation when data are limited and when individual parasequences (cycles) are difficult to recognize or correlate. Likewise, it is essential that the gross sequence-stratigraphic framework be defined in advance. The petrophysics-based method would be limited in reservoirs with significant faults, complex structures, or stratigraphic complexities (e.g., shelf-slope transition) where well-to-well correlation is highly complicated. However, with additional well and core data, the petrophysics-based approach should be used in addition to proportional-, truncated-, or onlap-layer geometries to construct a more refined geologic-model framework.

3) Using a petrophysics-based method, the most significant flow units were defined within high-frequency sequences. High-frequency sequences and depositional cycles compartmentalize the reservoir vertically into alternating zones of high and low reservoir quality. Diagenetic processes, including dolomitization, anhydrite cementation and replacement, and karstification, have significantly altered the primary carbonate fabric.

4) Core porosity and permeability data, thin sections, SEM images, and PTS estimates from capillary-pressure data were used to characterize vertical variability in reservoir quality. Lithofacies and depositional cycles were difficult to directly identify in noncored wells. As a result, a petrophysics-based method was used to define flow units (reservoir zones) within high-frequency sequences. The most significant reservoir zones were identified based on vertical changes in storage (ϕh) and flow capacity (kh) and pore-throat radius (R35). Using this approach, an appropriate number of zones (layers) were defined to construct a 3-D geologic model that did not require upscaling prior to flow simulation.

5) Parameter estimation was improved by incorporating multicomponent seismic attributes with well data. Porosity was modeled using P- and S1-wave average-reflection-strength attributes. Shear-wave anisotropy measurements provided information about inferred fracture density and enhanced permeability that was not available from either core analyses or compressional-wave seismic data.

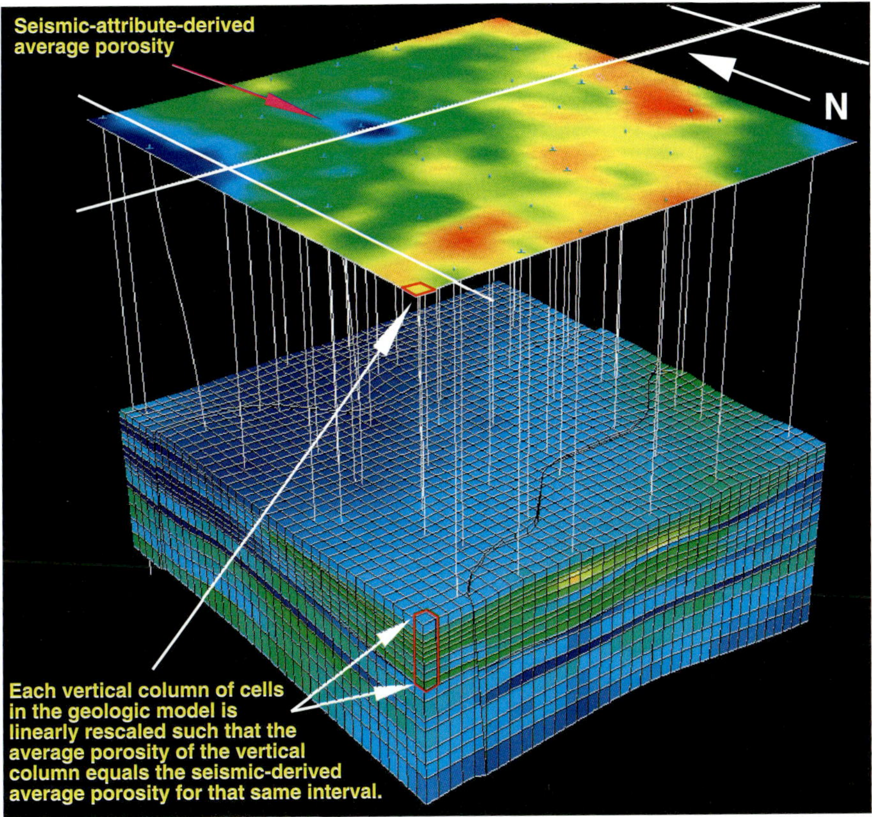

FIGURE 24. Geologic model and map of seismic-derived average porosity. To estimate porosity in each layer of the geologic model, porosity is first mapped in each layer using a mapping approach with well data. To include the lateral variability provided by the seismic data, each vertical column of cells is rescaled so that the estimated average porosity of each column honors the seismic-derived average-porosity map.

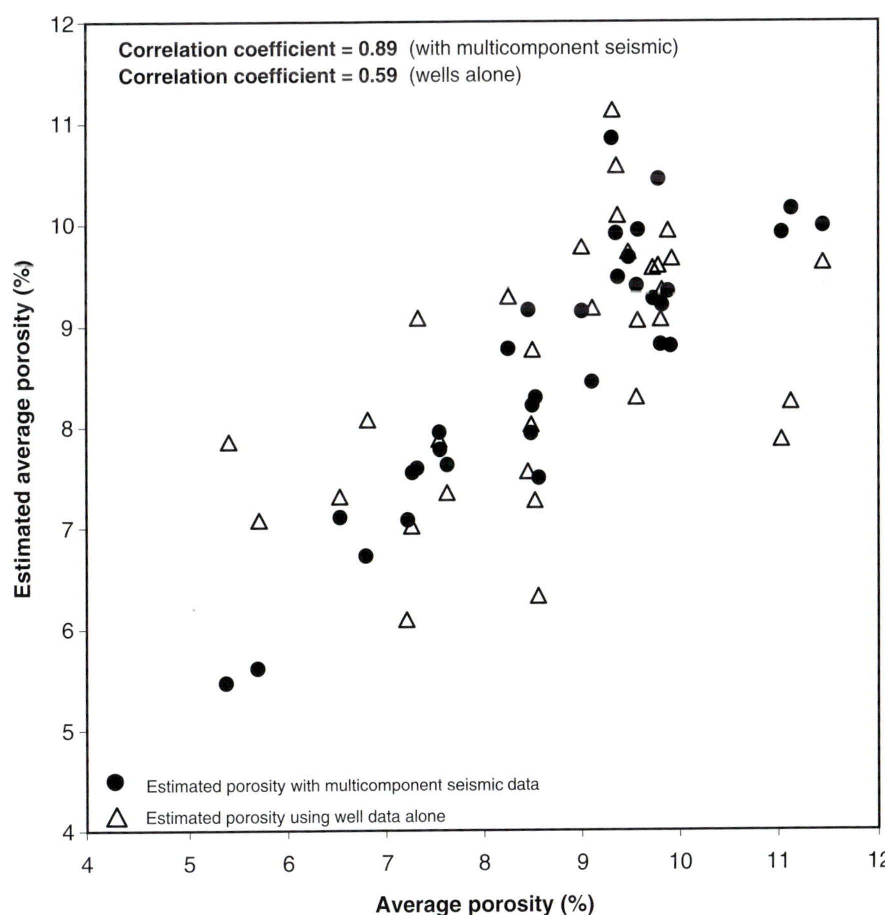

FIGURE 25. Crossplot of Upper San Andres average porosity from well data vs. estimated average porosity with and without multicomponent seismic data. Data points correspond to wells used in the validation process.

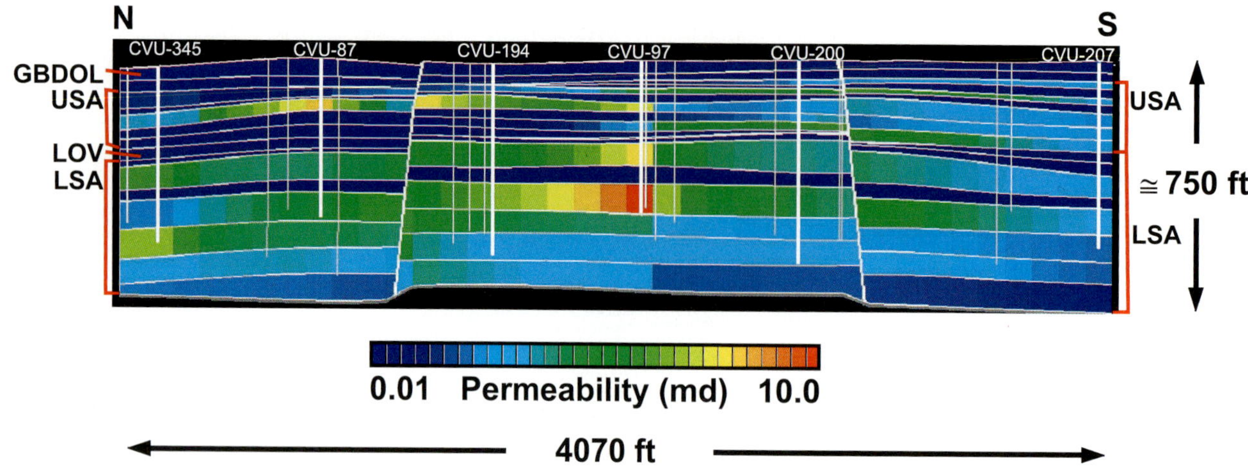

FIGURE 26. North-south permeability cross section from the 3-D geologic model. The difference in horizontal permeability between and within the Upper and Lower San Andres is shown. The difference in permeability within the model accounts for the observed difference in reservoir productivity between the Upper and Lower San Andres. GBDOL = Grayburg dolomite; USA = Upper San Andres; LOV = Lovington; LSA = Lower San Andres.

6) The petrophysics-based method can be used in other fields where porosity and estimated permeability data are available or can be determined, for example, using neural networks. The method is especially useful for fields that have limited core data and where definition of flow units based on rock fabric is difficult. Another application is to support the interpretation of reservoir zones based on rock-fabric descriptions where core data are abundant. In addition, multicomponent seismic data provide information about interwell heterogeneity that is not available when using only well data and compressional-wave seismic data.

ACKNOWLEDGMENTS

We thank the industry sponsors of the Colorado School of Mines Reservoir Characterization Project for funding and input to this study. The consortium members include AGIP, Amoco Production Company (now BP), Anadarko Petroleum Corporation, ARCO, Chevron Petroleum Technology Company, China National Petroleum Corporation, Compagnie Generale de Geophysique, Conoco Inc. (now ConocoPhillips), Dawson Geophysical Company, Exxon Production Research Company (now ExxonMobil Upstream Research Company), Gas Research Institute (now GTI), GeoQuest/Schlumberger/Geco, Golden Geophysical/Fairfield Industries, Grant Geophysical, Inc., Input/Output, Inc., INTEVEP, S.A., Japan National Oil Corporation, Landmark Graphics Corporation, Marathon Petroleum Technology Company, Maxus Energy Corporation, Nambe Geophysical, Inc., Occidental Oil and Gas Corporation, Oyo Geospace Corporation, PanCanadian Petroleum Limited (now EnCana), Phillips Petroleum Company (now ConocoPhillips), Paradigm Geophysical (formerly CogniSeis), Shell E&P Technology Company, Discovery Bay Company (now Rock Solid Images), Silicon Graphics Corporation, Solid State Geophysical, Texaco Group, Inc. (now Chevron Texaco), Union Pacific Resources Company (now Anadarko Petroleum Corporation), UNOCAL/Sprint Energy, Western Geophysical, and Veritas DGC, Inc.

The study was also supported through research grants and funding from the American Association of Petroleum Geologists, Geological Society of America, Society of Professional Well Log Analysts (now Society of Petrophysicists and Well Log Analysts), and the Department of Geology and Geological Engineering at the Colorado School of Mines. We thank Marathon Petroleum Technology Company for preparing core plugs and conducting capillary-pressure tests.

We thank Thomas A. Jones for his review of an early draft of this document. We thank AAPG reviewers Jim Weber, Susan Longacre, Gregor Eberli, and Phil Inderweisen for their helpful comments and suggestions.

APPENDIX

In the absence of core data, R35 can be estimated directly from the Winland equation:

$$\log \mathrm{R35} = 0.732 - 0.588\ \log(k) - 0.864\ \log(\phi).$$

In the Winland equation, air permeability (k) is given in millidarcies, porosity (ϕ) is given in percent, and R35 is expressed in microns. Air-brine capillary-pressure tests were conducted on 18 core plugs taken from the CVU-100 core.

Pore-throat sorting corresponds with the degree of sorting of pore throats within a rock sample (Jennings, 1987). PTS ranges from 1.0 (perfect sorting) to 8.0 (essentially no sorting) with the majority of rock samples falling between 1.2 and 5.0 (Jennings, 1987). Values of PTS were computed using the following equation adapted from a sorting coefficient equation developed by Trask (1932):

$$\text{PTS} = \sqrt{\left(\frac{\text{Third-quartile pressure}}{\text{First-quartile pressure}}\right)}$$

where the first- and third-quartile pressures are obtained directly from the capillary-pressure curve and reflect the 25% and 75% brine-saturation pressures adjusted for irreducible saturation (Jennings, 1987). For the air-brine capillary-pressure tests, brine (water) was the wetting phase.

REFERENCES CITED

Adams, S. D., 1997, Sedimentology and diagenesis of the San Andres Formation, Vacuum field, New Mexico: Master's thesis, Colorado School of Mines, Golden, Colorado, 158 p.

Ahr, W. M., 1991, Pore characteristics as surrogates for permeability in mapping reservoir flow units: Vacuum San Andres field, Lea County, New Mexico (abs.): AAPG Annual Meeting Program, v. 75, p. 532.

Alford, R. M., 1986, Shear data in the presence of azimuthal anisotropy (abs.): Expanded Abstracts from the 56th Annual International Meeting, Society of Exploration Geophysicists, p. 476–479.

Anselmetti, F. S., S. Luthi, and G. P. Eberli, 1998, Quantitative characterization of carbonate pore systems by digital image analysis: AAPG Bulletin, v. 82, p. 1815–1836.

Benson, R. D., and T. L. Davis, 2000, Time-lapse seismic monitoring and dynamic reservoir characterization, Central Vacuum Unit, Lea County, New Mexico: SPE Reservoir Evaluation & Engineering, v. 60890, p. 88–97.

Blaylock, J. J., 1999, Interpretation of a baseline 4-D multicomponent seismic survey at Vacuum field, Lea County, New Mexico: Master's thesis, Colorado School of Mines, Golden, Colorado, 121 p.

Capello de Passalacqua, M. A., 1995, Geology and rock physics of the San Andres Formation in Vacuum field, New Mexico: Master's thesis, Colorado School of Mines, Golden, Colorado, 143 p.

Craig, F. F., Jr., 1993, The reservoir engineering aspects of waterflooding: Society of Petroleum Engineers Monograph 3, p. 64.

DeVault, B., 1997, 3-D seismic prestack multicomponent amplitude analysis, Vacuum field, Lea County, New Mexico: Ph.D. dissertation, Colorado School of Mines, Golden, Colorado, 192 p.

Galarraga, M., 1999, 3-D seismic interpretation in Vacuum field area, Permian Basin, Lea County, New Mexico: Master's thesis, Colorado School of Mines, Golden, Colorado, 106 p.

Gorell, S. B., 1995, Creating 3-D reservoir models using areal geostatistical techniques combined with vertical well data: Proceedings from the Society of Petroleum Engineers Annual Technical Conference and Exhibition, SPE Paper 29670, p. 547–556.

Gunter, G. W., J. M. Finneran, D. J. Hartmann, and J. D. Miller, 1997, Early determination of reservoir flow units using an integrated petrophysical method: Proceedings of the Society of Petroleum Engineers Annual Technical Conference and Exhibition, SPE Paper 38679, p. 373–380.

Hartmann, D. J., and E. A. Beaumont, 1999, Predicting reservoir system quality and performance, *in* E. A. Beaumont and N. H. Foster, eds., AAPG Treatise of Petroleum Geology, Exploration for Oil and Gas Traps, p. 9-1–9-154.

Hills, J. M., 1984, Sedimentation, tectonism, and hydrocarbon generation in Delaware Basin, west Texas and southeastern New Mexico: AAPG Bulletin, v. 68, p. 250–267.

Hinterlong, G. D., A. R. Taylor, G. P. Watts, and K. H. Kumar, 1998, Improving flow simulation performance with a seismic-enhanced geologic model: Proceedings of the Society of Petroleum Engineers Annual Technical Conference and Exhibition, SPE Paper 39809, p. 499–504.

Jennings, J. B., 1987, Capillary pressure techniques: Application to exploration and development geology: AAPG Bulletin, v. 71, p. 1196–1209.

Kerans, C., 1995, Stratigraphic framework of the San Andres Formation, Algerita Escarpment, Guadalupe Mountains, New Mexico, *in* M. T. Moussa, ed., The San Andres in outcrop and subsurface, guidebook to the Permian Basin: SEPM 1995 Annual Field Conference, p. 7–30.

Kerans, C., and S. W. Tinker, 1997, Sequence stratigraphy and characterization of carbonate reservoirs: SEPM Short Course Notes 40, p. 30–34.

Kuester, J. L., and J. H. Mize, 1973, Optimization techniques with Fortran: New York, McGraw-Hill Book, 500 p.

Landes, K. K., 1970, New Mexico, *in* Petroleum Geology of the United States: New York, Wiley-Interscience, p. 356–368.

Leary, D. A., and J. N. Vogt, 1990, Diagenesis of the San Andres Formation (Guadalupian), Central Basin platform, Permian Basin, *in* D. G. Bebout and P. M. Harris, eds., Geologic and engineering approaches in evaluation of San Andres/Grayburg hydrocarbon reservoirs— Permian Basin: Austin, Texas, Bureau of Economic Geology, p. 21–48.

Lorenzen, R. J. L., 2000, Inversion of multicomponent time-lapse seismic data for reservoir characterization of Vacuum field, New Mexico: Ph.D. dissertation, Colorado School of Mines, Golden, Colorado, 186 p.

Lucia, F. J., 1983, Petrophysical parameters estimated from visual description of carbonate rocks: A field classification of carbonate pore space: Journal of Petroleum Technology, v. 35, p. 626–637.

Lucia, F. J., 1995, Rock fabric/petrophysical classification of carbonate pore space for reservoir characterization: AAPG Bulletin, v. 79, p. 1275–1300.

Lucia, F. J., 1999, Characterization of petrophysical flow units in carbonate reservoirs: Discussion: AAPG Bulletin, v. 83, p. 1161–1163.

Lucia, F. J., C. Kerans, and R. K. Senger, 1992, Defining flow units in dolomitized carbonate ramp reservoirs: Proceedings of the Society of Petroleum Engineers Annual Technical Conference and Exhibition, SPE Paper 24702, p. 399–406.

Martin, A. J., S. T. Soloman, and D. J. Hartmann, 1997, Characterization of petrophysical flow units in carbonate reservoirs: AAPG Bulletin, v. 81, p. 734–759.

Martin, A. J., S. T. Soloman, and D. J. Hartmann, 1999, Characterization of petrophysical flow units in carbonate reservoirs: Reply: AAPG Bulletin, v. 83, p. 1164–1173.

Mattocks, B. W., 1998, Borehole geophysical investigation of seismic anisotropy at Vacuum field, Lea County, New Mexico: Ph.D. dissertation, Colorado School of Mines, Golden, Colorado, 181 p.

Mendez-Hernandez, E., 1999, Influence of shallow heterogeneities on multicomponent 4-D seismic data at Vacuum field, New Mexico: Master's thesis, Colorado School of Mines, Golden, Colorado, 143 p.

Montgomery, S. L., 1998, Permian Clear Fork Group, North Robertson Unit: Integrated reservoir management and characterization for infill drilling, part I—Geologic analysis: AAPG Bulletin, v. 82, p. 1797–1814.

Montgomery, S. L., D. K. Davies, R. K. Vessel, J. E. Kamis, and W. H. Dixon, 1998, Permian Clear Fork Group, North Robertson Unit: Integrated reservoir management and characterization for infill drilling, part II—Petrophysical and engineering data: AAPG Bulletin, v. 82, p. 1985–2002.

Nevans, J. W., J. E. Kamis, D. K. Davies, R. K. Vessel, L. E. Doublet, and T. A. Blasingame, 1998, An integrated geologic and engineering reservoir characterization of the North Robertson (Clear Fork) Unit, Gaines County, Texas: Proceedings of the Advanced Applications of Wireline Logging for Improved Oil Recovery Workshop: U.S. Department of Energy, p. 53–70.

Pranter, M. J., 1999, Use of a petrophysical-based reservoir zonation and multicomponent seismic attributes for improved geologic modeling, Vacuum field, New Mexico: Ph.D. dissertation, Colorado School of Mines, Golden, Colorado, 366 p.

Purves, W. J., 1990, Reservoir description of the Mobil Oil Bridges State Leases (Upper San Andres reservoir), Vacuum field, Lea County, New Mexico, *in* D. G. Bebout and P. M. Harris, eds., Geologic and engineering approaches in evaluation of San Andres/Grayburg hydrocarbon reservoirs— Permian Basin: Texas Bureau of Economic Geology, p. 87–112.

Roche, S. L., 1997, Time-lapse, multicomponent, three-dimensional seismic characterization of a San Andres shallow shelf carbonate reservoir, Vacuum field, Lea County, New Mexico: Ph.D. dissertation, Colorado School of Mines, Golden, Colorado, 325 p.

Sarg, J. F., and P. J. Lehmann, 1986, Lower-middle Guadalupian facies and stratigraphy, San Andres/Grayburg Formations, Permian Basin, Guadalupe Mountains, New Mexico, *in* G. E. Moore and G. L. Wilde, eds., Lower and middle Guadalupian facies, stratigraphy, and reservoir geometries; San Andres/Grayburg Formations, Guadalupe Mountains, New Mexico and Texas: SEPM Permian Basin Section Publication 86-25, p. 1–8.

Schultz, P. S., S. Ronen, M. Hattori, and C. Corbett, 1994, Seismic-guided estimation of log properties, part 1: A data-driven interpretation methodology: Leading Edge, v. 13, no. 5, p. 305–315.

Scuta, M. S., 1997, 3-D reservoir characterization of the Central Vacuum Unit, Lea County, New Mexico: Ph.D. dissertation, Colorado School of Mines, Golden, Colorado, 274 p.

Scuta, M. S., and N. F. Hurley, 1998, Detection of bypassed pay using time-lapse log analysis, San Andres Formation, Central Vacuum Unit, New Mexico, *in* W. D. DeMis and M. K. Nelis, eds., The search continues into the 21st century: West Texas Geological Society Fall Symposium: West Texas Geological Society Publication 98-105, p. 185–189.

Smith, W. H., J. J. Reeves, and S. D. Bacon, 1998, Seismic-guided mapping of effective porosity, Grayburg/San Andres shallow shelf carbonate reservoir, Permian Basin, USA (abs.): Expanded Abstracts of the 4th Annual 3-D Seismic Symposium, Rocky Mountain Association of Geologists and Denver Geophysical Society.

Sonnenfeld, M. D., and T. A. Cross, 1993, Volumetric partitioning and facies differentiation within the Permian Upper San Andres Formation of Last Chance Canyon, Guadalupe Mountains, New Mexico, *in* R. G. Loucks and J. F. Sarg, eds., Carbonate sequence stratigraphy: Recent developments and applications: AAPG Memoir 57, p. 435–474.

Swanson, J. S. S., 1996, Four-dimensional seismic modeling of a carbon dioxide enhanced oil recovery monitoring project, Central Vacuum Unit, Lea County, New Mexico: Master's thesis, Colorado School of Mines, Golden, Colorado, 137 p.

Talley, D. J., 1997, Characterization of a San Andres carbonate reservoir using four dimensional multicomponent attribute analysis: Master's thesis, Colo rado School of Mines, Golden, Colorado, 75 p.

Taner, M. T., F. Koehler, and R. E. Sheriff, 1979, Complex seismic trace analysis: Geophysics, v. 44, p. 1041–1063.

Trask, P. D., 1932, Origin and environment of source sediments of petroleum: Houston, Texas, Gulf Publishing, 323 p.

Tucker, K. E., P. M. Harris, and R. C. Nolen-Hoeksema,

1998, Geologic investigation of cross-well seismic response in a carbonate reservoir, McElroy field, west Texas: AAPG Bulletin, v. 82, p. 1463–1503.

Voorhies IV, H. C., 1996, Application of a multicomponent vertical seismic profile to dynamic reservoir characterization at Vacuum field, Lea County, New Mexico: Master's thesis, Colorado School of Mines, Golden, Colorado, 104 p.

Wehner, S. C., and J. Prieditis, 1996, CO_2 huff-n-puff: Initial results from a waterflooded ssc reservoir: Proceedings of the Society of Petroleum Engineers Permian Basin Oil and Gas Recovery Conference, SPE Paper 35223, p. 1–10.

Winterstein, D. F., 1992, How shear-wave properties relate to rock fractures: simple cases: Leading Edge, v. 11, no. 9, p. 21–28.

4

Tinker, S. W., D. H. Caldwell, D. M. Cox, L. C. Zahm, and L. Brinton, 2004, Integrated reservoir characterization of a carbonate ramp reservoir, South Dagger Draw field, New Mexico: Seismic data are only part of the story, *in* Seismic imaging of carbonate reservoirs and systems: AAPG Memoir 81, p. 91–105.

Integrated Reservoir Characterization of a Carbonate Ramp Reservoir, South Dagger Draw Field, New Mexico: Seismic Data Are Only Part of the Story

Scott W. Tinker[1]
Bureau of Economic Geology, University of Texas at Austin, Austin, Texas, U.S.A.

Donald H. Caldwell
Marathon Oil Company, Houston, Texas, U.S.A.

Denise M. Cox
Marathon Oil Company, Denver, Colorado, U.S.A.

Laura C. Zahm
iReservoir.com, Englewood, Colorado, U.S.A.

Lisë Brinton
LithoLogic, Littleton, Colorado, U.S.A.

ABSTRACT

South Dagger Draw (SDD) field, located in southeast New Mexico, produces hydrocarbons from complex sigmoid-oblique clinoforms of the Pennsylvanian Canyon and Cisco Formations. South Dagger Draw field, a combination structural-stratigraphic trap, represents the northern extension of the Indian Basin field. Through February 2001, the Indian Basin and SDD fields together had produced nearly 23 million bbl of oil and 2 tcf of gas from Marathon Oil Company-held acreage. Vuggy porosity, formed dominantly in algal biostromes and bioherms located at the ramp-margin position of each clinoform, represents the primary reservoir. Vugs were formed by acidic hydrothermal fluids that migrated upward along joints and were baffled beneath shales, resulting in dissolution zones that are controlled by the interplay between structural joints and stratigraphic shales and carbonates.

Data used in the study include logs, cores, modern wire-line log suites, borehole image logs, and three-dimensional (3-D) acoustic impedance values from inversion of seismic data. Seismic data provide interwell information helpful for determining the present-day structure of the field but not particularly useful for interpreting the stratigraphy. High-frequency sequence-stratigraphic interpretation, guided by a depositional model derived from description of cores and outcrops, was accomplished using a necessary combination of well logs, cores, and seismic data. The sequence-stratigraphic interpretation served as input for multiple iterative seismic inversions and provided the framework for the integrated 3-D geologic model.

[1]*Present address:* Bureau of Economic Geology, University of Texas, Houston, Texas, U.S.A.

INTRODUCTION

Vertical seismic data resolution in carbonate reservoirs is significantly less than in siliciclastic reservoirs because carbonate rocks have higher velocities and greater densities. Therefore, modern three-dimensional (3-D) seismic data with dominant frequencies in the 50- to 60-Hz range must be interpreted within the context of a depositional model and integrated with all of the other available data to build a reasonable 3-D carbonate reservoir model.

South Dagger Draw (SDD) field is presented as a case study to represent a reservoir characterization workflow in a complex stratigraphic and petrophysical carbonate system. SDD field is a Pennsylvanian reservoir located in southeast New Mexico that produces from a combination structural-stratigraphic trap in the Canyon and Cisco Formations (Figure 1). The productive reservoir is represented by vuggy porosity formed mainly in algal boundstones and crinoid grainstones that form bioherms and biostromes at the steepened ramp margin. Porosity in the reservoir facies is low, ranging to as much as 6% with a few exceptions in highly vuggy intervals, and matrix permeability is also low. Permeability of the reservoir system can be moderately enhanced by fracturing and subsequent solution enhancement of the fracture network, forming small, isolated vugs, to complex, interconnected vug networks.

Through February 2001, the Indian Basin and SDD fields combined had produced nearly 23 million bbl of oil and nearly 2 tcf of gas from Marathon Oil Company acreage. The Indian Basin gas pool produces under a pressure-depletion drive on 160-acre well spacing, and the East Indian Basin and SDD field associated gas pools are on water drive with 80- and in some cases 40-acre well spacing. Complex stratigraphy and compartmentalization are confirmed by the presence of multiple oil-water and gas-oil contacts.

System	Series	Stratigraphic unit	
Permian	Guadalupian	Capitan	Tansill
		Capitan	Yates
		Capitan	Seven Rivers
		Goat Seep	Queen
		Grayburg	
		San Andres	
	Leonardian	Clear Fork	Spraberry Dean
	Wolfcampian	Wolfcamp	Horseshoe Atoll
Pennsylvanian	Virgilian	Cisco	Horseshoe Atoll
	Missourian	Canyon	Horseshoe Atoll
	Desmoinesian	Strawn	

FIGURE 1. Stratigraphic column including the Canyon and Cisco Formations in the Permian Basin of west Texas and New Mexico. Modified from Hamlin et al. (1995).

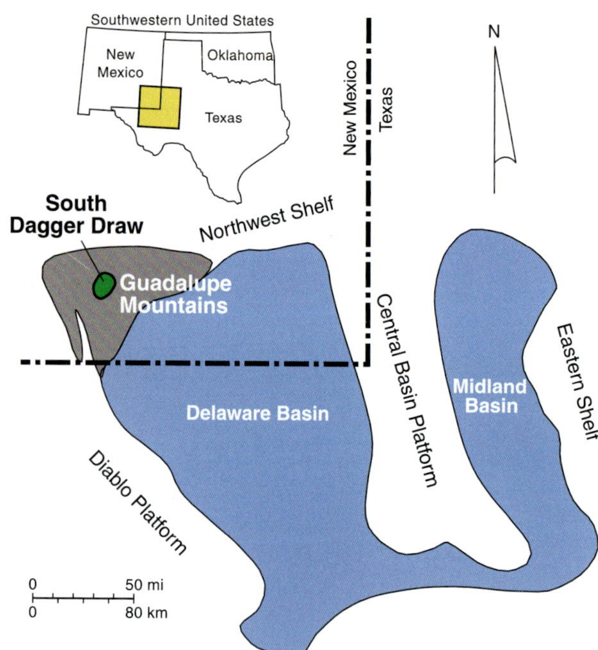

FIGURE 2. South Dagger Draw field location map.

The initial full-field sequence-stratigraphic interpretation in the SDD field was completed in 1995 using wire-line logs, cores, and one low-resolution 2-D seismic line (Tinker, 1996a, Anonymous Shelf-Margin Carbonate Example section). A second sequence-stratigraphic interpretation that integrated core, log, and 3-D seismic data was completed in 1997 (Uland et al., 1997; Tinker and Cox, 1998), and the third interpretation discussed herein incorporates additional wire-line logs from infill drilling, acoustic impedance (AI) from an iterative set of seismic data inversions, and FMI borehole image data.

GEOLOGIC SETTING

SDD field is located in the Seven Rivers Embayment of the Guadalupe Mountains in southeast New Mexico (Figure 2). This study focuses on the northern, oil-producing part of the field. The north-northwest–south-southeast–trending Huapache fault zone is located 8–16 km (5–10 mi) west to southwest of the SDD field. The Huapache fault zone, related to the Marathon-Ouachita convergent orogenic front, was active in Pennsylvanian to Early Permian (Meyer, 1966; Galley, 1968; Ross, 1986; Yang and Dorobek, 1995). Extensive glaciation of Gondwana in the Southern Hemisphere during the Pennsylvanian resulted in high-amplitude eustatic sea level changes (Peterson and Hite, 1969; Hite and Buckner, 1981; Goldhammer et al., 1994). Structural dip in the SDD field is to the northeast (Figure 3). High-amplitude eustasy, coupled with active tectonism along the Huapache fault zone during the Pennsylvanian, resulted in complex, sigmoid-oblique, prograding clinoforms (Figure 4) and created a combination structural-stratigraphic trap in the hydrocarbon-producing Canyon and Cisco Formations.

DATA

The data set used in this study includes 89 wells with various vintages of wire-line logs, 30 borehole image logs, five cored wells (~800 ft [~255 m]), a 3-D seismic volume, individual well-production data, and an outcrop analog. The nonseismic data were merged into a single SAS data set, wherein they were analyzed graphically and programmatically before being formatted for use in other steps of the workflow.

DEPOSITIONAL MODEL

Although the core data are limited, sedimentologic analysis, facies description, and stacking pattern analysis of cored wells provide valuable information to create simplified depositional models. The depositional models were used to guide the one-, two-, and three-dimensional (1-, 2-, and 3-D, respectively) stratigraphic interpretation of well logs and seismic data.

During relative sea level lowstands, terrigenous material, including clay, silt, and organic matter, were transported across the ramp and deposited in the basin, resulting in deposition of black, organic-rich shales mantling the underlying slope and toe-of-slope topography. During lowstands, carbonate sedimentation on the ramp slowed abruptly or stopped. At the major sequence boundaries, the carbonate ramp was undoubtedly subaerially exposed, but there is little recognized rock record of this exposure.

During early transgression, marine flooding of the ramp reinitiated carbonate sedimentation, resulting in deposition of argillaceous, lime mudstones in the basin and crinoidal wackestones on the ramp (Figure 5A). During late transgression and early highstand, shorelines stepped landward, most of the terrigenous material was removed from the system, and the carbonate factory was active, resulting in deposition of crinoid, fusulinid wackestones in the outer ramp, and crinoid, fusulinid packstones in the middle to inner ramp.

During mid- to late highstand, algal boundstone bioherms and biostromes and crinoid, fusulinid, packstone intermound deposits dominated the ramp-crest position, and crinoid, fusulinid wackestones to packstones were deposited in the outer ramp (Figure 5B). The algal boundstone facies is interpreted to be the shallowest water facies represented in the cores. No high-energy grainstones indicating wave base or facies that might indicate peritidal conditions were observed in the cores. Therefore, although there was most likely topography in the bioherms at the ramp margin even during late highstand, it was probably deposited in water depths of at least 10 m (>30 ft).

RESERVOIR DISTRIBUTION

An understanding of the 3-D distribution of reservoir facies resulted from integration of all data into a 3-D sequence-stratigraphic model. The results are summarized here ahead of the discussion on sequence stratigraphy and 3-D model construction. The "critical-phase" history, those processes that either create or destroy reservoir quality, of the SDD field is described below (Figure 6):

- Deposition of a series of low-angle, complex sigmoid-oblique clinoforms (distally steepened ramp) dominated by shales in lowstand and early transgressive deposits and carbonates in late transgressive and highstand deposits. The position and trend of the ramp margins is controlled largely by Pennsylvanian lineaments, and the better-quality primary reservoirs are algal boundstones formed at the ramp-margin position of each clinoform.
- Subaerial exposure was probable across high-frequency (HFS) and composite sequence (CS) boundaries, but there is little evidence recognized in core from the SDD field of prolonged exposure that would result in meteoric diagenesis and karst.
- Burial and subsequent dolomitization of the carbonate matrix, preferentially in the higher-permeability ramp-margin position.
- Acidic, hydrothermal waters sourced from below moved upward along major joints, forming irregular vugs (modified definition of Lucia, 1983), connected-vug networks, and solution-enlarged fractures. This process is similar in genesis to "oil-field karst" (Hill, 1992). The rising acidic waters were "dammed" below shales, forming a mushroom-shaped matrix-vug network preferentially in higher-permeability, ramp-margin deposits. On a different scale, similar geometries resulting from "bottom-up" processes have been described in

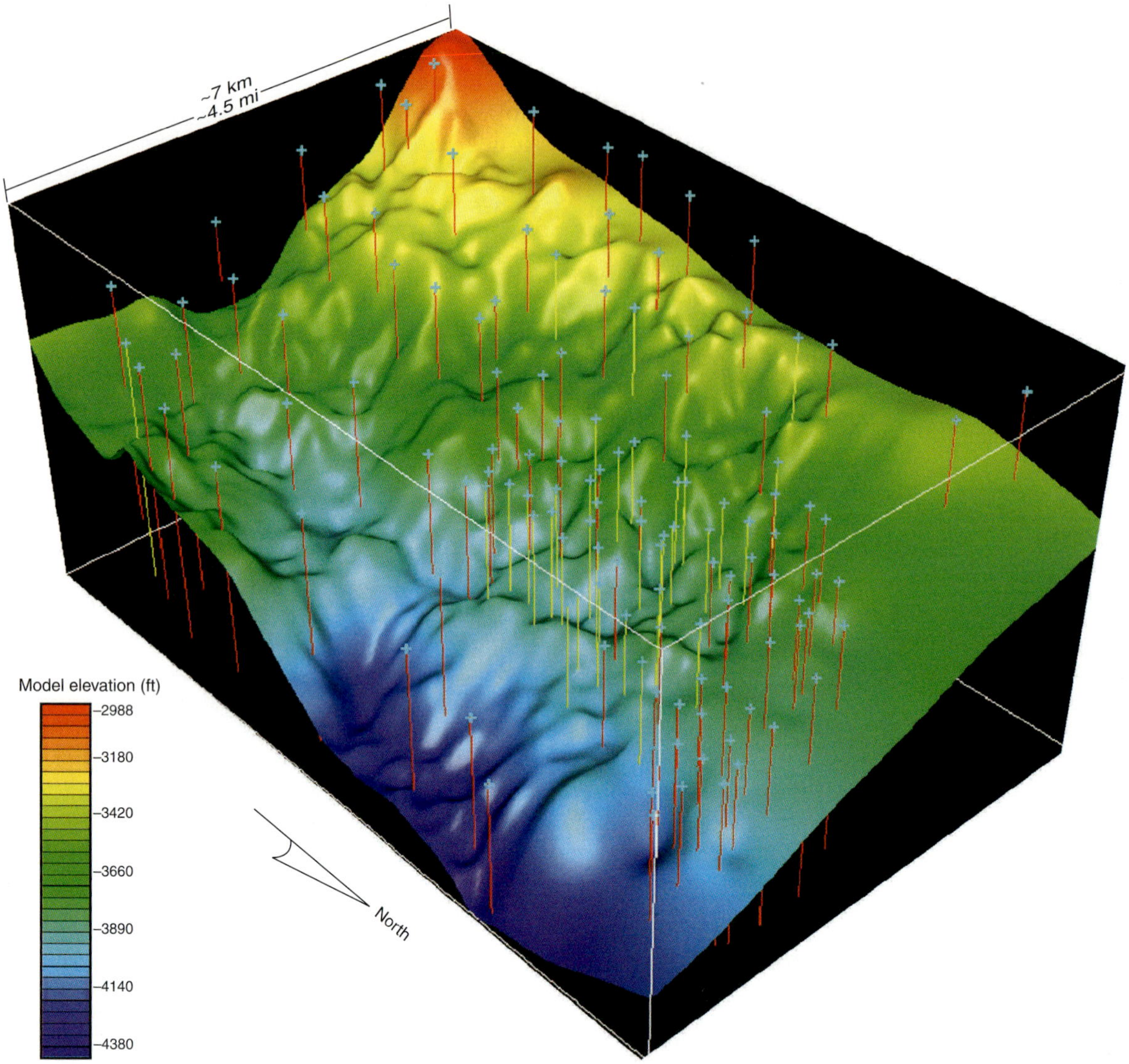

~7 km
~4.5 mi
Model elevation (ft)
–2988
–3180
–3420
–3660
–3890
–4140
–4380
North

FIGURE 3. Three-dimensional structure map based on well logs and 3-D seismic on top of the Canyon time-stratigraphic surface.

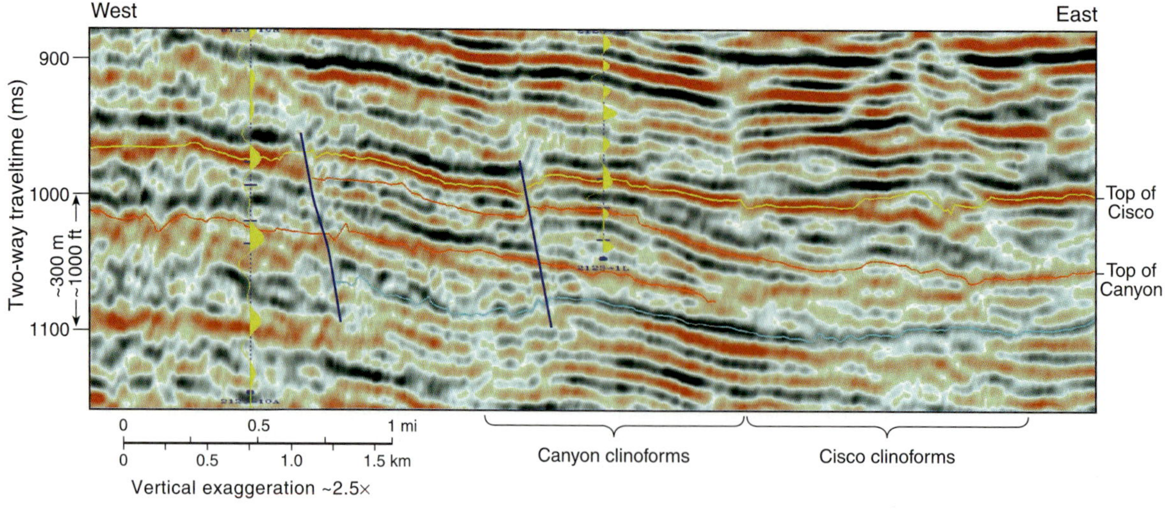

FIGURE 4. Seismic line oriented perpendicular to the Canyon margin strike. Faults parallel to Huapache are indicated.

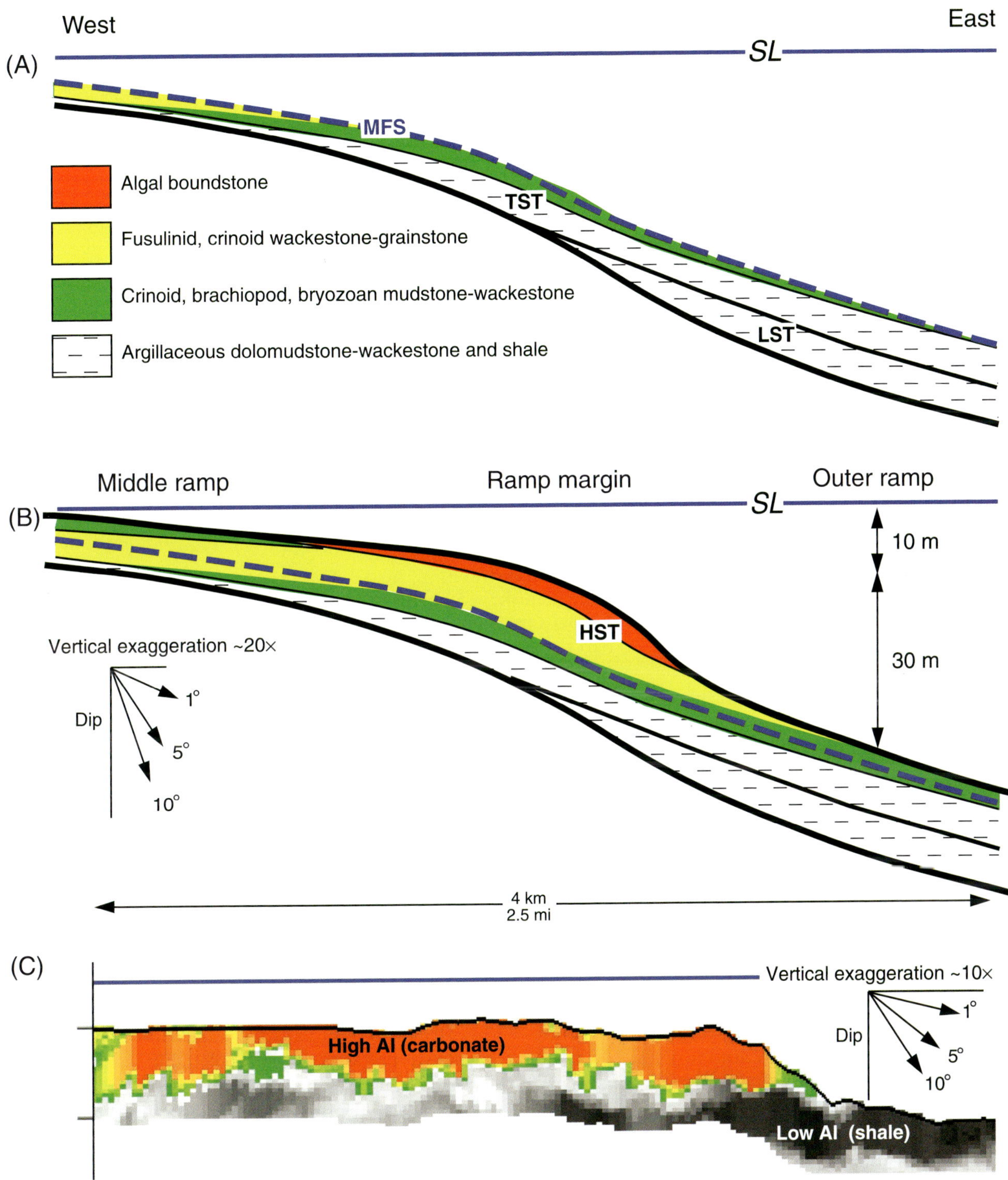

FIGURE 5. Depositional models for a high-frequency sequence. (A) Facies distributions in the lowstand systems tract (LST) and transgressive systems tract (TST). The lower sequence boundary (SB) is thick black line, the upper boundary of LST is thin black line, and the maximum flooding surface (MFS) is heavy, dashed blue line. Note the textural changes from outer to middle ramp. (B) Facies distributions in the highstand systems tract (HST). The ramp margin was most likely deposited in at least 10 m of water, depending on its position within the longer-term composite sequence. (C) Acoustic impedance (AI) cut from 3-D seismic inversion illustrating validity of depositional model. Colors represent general lithologies depicted in (A) and (B). Scale is similar, but vertical exaggeration is half of (A) and (B). Blue line represents datum on top of Cisco.

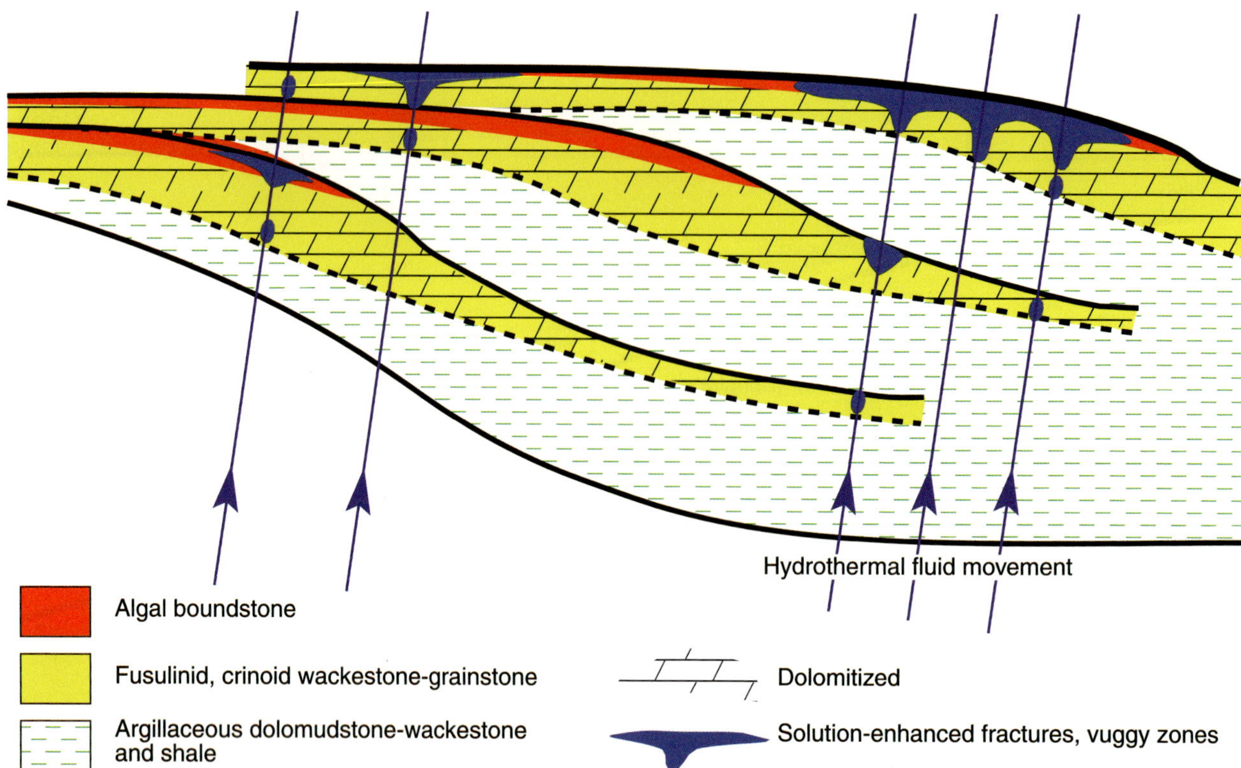

FIGURE 6. Conceptual sketch of reservoir formation and distribution. Acidic, hydrothermal fluids move up along fractures and "bank" beneath shales, forming mushroom-shaped dolomitization and dissolution (vugs and caves) regions (blue). The reservoir formation process happens preferentially where major fracture zones intersect higher primary matrix permeability ramp-crest facies (red) that are overlain by tight marine shales and mudstones (green).

outcrop (Wilson, 1990) and in the subsurface (Hurley and Budros, 1990) and have been observed by the first author in Pennsylvanian-age rocks found in the Big Hatchet Mountains of New Mexico.

- Only the ramp-margin facies that intersect a major fracture system have well-developed, vuggy porosity associated with them.
- Vugs and fractures were partially filled by baroque dolomite cement, which is interpreted as late burial cement.

The dominant *storage* is in vuggy algal boundstones and crinoid grainstones, formed where the ramp margin of a clinoform intersects a major fracture trend. Ramp margins that do not cross fracture zones are not as likely to have a well-developed vug network, and therefore will have lower storage. The partial to complete cementation of many of the vugs by late baroque dolomite cement reduces their lateral connections. Solution-enhanced fractures, which contribute minimally to storage, are more important in terms of their contribution to vertical and horizontal *permeability* of the system. Solution-enhanced fractures act to connect vugs in the ramp-crest facies, which helps to explain the high contrast between core-analysis (low) and well-test (high) permeability described by Hurley et al. (1999) for rocks in the SDD field.

SEQUENCE-STRATIGRAPHIC AND STRUCTURAL INTERPRETATION

Climate change and eustasy associated with Pennsylvanian ice-house conditions, in combination with tectonism related to the Ouachita orogeny, produced complex strtigraphy in the SDD field. All of the available data were incorporated in the stratigraphic interpretation, including cores, wire-line logs, borehole image logs, predicted facies, AI from seismic inversion, and the depositional model.

The initial full-field sequence-stratigraphic framework interpretation was completed in the SDD field in 1995 using the depositional model, stacking patterns (facies proportions, thickness variations, facies shifts, and cycle symmetry) interpreted in 89 wells, and one 2-D seismic line (Tinker, 1996a, see Anonymous Shelf-Margin Carbonate Field section). The process of 1-, 2-, and 3-D stratigraphic interpretation using well and core data is discussed in detail in Kerans and Tinker (1997) and illustrated for the SDD field in Figure 7. This initial work resulted in a stratigraphic hierarchy of three CSs, each composed of two to five HFSs, and represents a reasonable interpretation in the absence of 3-D seismic data (Figure 8).

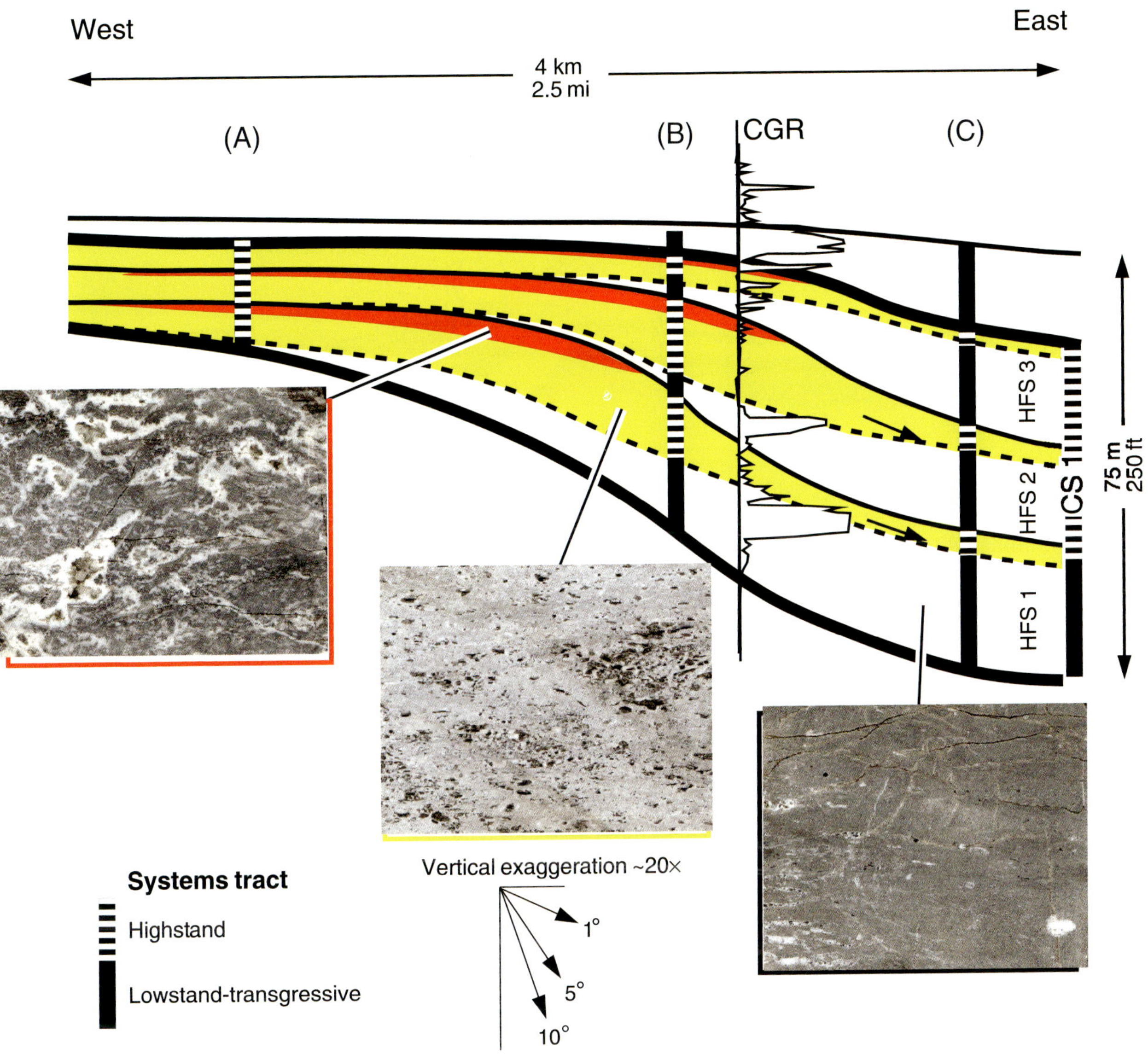

FIGURE 7. Schematic dip section illustrating stratigraphic hierarchy for composite sequence 1 (CS 1) composed of three high-frequency sequences (HFSs). At position (A), HFSs are asymmetrical, dominated by highstand systems tract (HST) facies, with little transgressive systems tract (TST) preservation. HFSs thin upward. At position (B), there are a greater proportion of TST facies in the symmetrical HFS 1 and an increasing proportion of HST facies upward in the asymmetrical HFS 2 and HFS 3. A computed gamma-ray (CGR) log from South Dagger Draw illustrates this changing proportion of facies, with a positive excursion to the right indicating high radioactivity in the shales. At position (C), all HFSs are asymmetrical, but unlike in (A), they are dominated by TST preservation. Facies proportions, symmetry, and thickness are just some of the tools that allow a stratigraphic hierarchy to be worked out from 1-D well data only. Core photographs show argillaceous lime mudstone (black border), crinoid, fusulinid, packstone (yellow border), and algal boundstones with vugs partially filled with baroque dolomite (red border). Scale across each photograph is approximately 8 cm (~3 in).

A second full-field sequence-stratigraphic framework interpretation was made that incorporated 3-D seismic data (Tinker and Cox, 1998), and a third interpretation incorporated a series of seismic inversions, a significant several additional older wire-line logs, and data from borehole image logs (bed dips, vugs, and fractures) collected programmatically from raw resistivity measurements, not image analysis. These data improved the interpretation at the SDD field, especially in terms of understanding the influential factor that structure played in stratigraphic evolution and in quantifying the significant interwell petrophysical variation.

Interpretation of seismic "surfaces" at the reservoir scale can be as misleading as lithostratigraphic correlation of well-log signatures (Tinker, 1996a), because

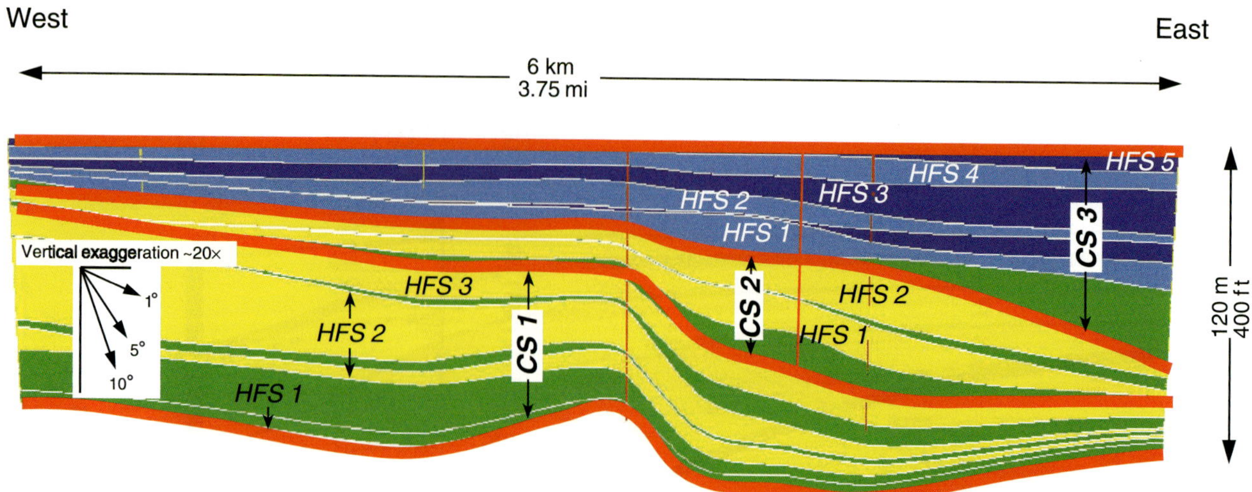

FIGURE 8. Stratigraphic framework and hierarchy worked out from 1-D well-log and core data only, before acquisition of 3-D seismic data. Picture is a dip section taken from the actual 3-D geologic model. Yellow and blue represent carbonate-dominated intervals, and green represents shale-dominated intervals.

seismic amplitude patterns can be asynchronous (Stafleau and Sonnenfeld, 1994; Tinker, 1996b). In the SDD field, reservoir-scale sequence-stratigraphic interpretation was accomplished using AI (the product of density and velocity) from seismic inversion, in conjunction with other geologic data sources, including logs, borehole images, cores, and a depositional model. Using this approach, one can correlate stratigraphic successions of genetically related rock throughout a volume, instead of correlating seismic surfaces that may actually cut across depositional surfaces (Tinker, 1996a).

Seismic Inversion

The seismic inversions in the SDD field were made with *TDROV* software developed by CGG. This software uses a 3-D volume of seismic data measurements, a 3-D stratigraphic layering framework, and simulated annealing to solve the inversion problem. Significantly, *TDROV* is not conditioned to individual well data. Four essential preliminary steps were required to perform the iterative seismic data interpretations at the SDD field: (1) the ability to reference (flatten) the seismic data volume on a stratigraphic horizon; (2) an understanding of the relationship among lithology, radioactivity, and AI; (3) depositional models; and (4) inclusion of all well logs in the seismic data volume for interpretation.

Black shales and carbonate mudstones represent lowstand and transgressive deposits and are therefore laterally continuous lithologic and near-chronostratigraphic units that drape and onlap underlying topography. Because most of the accommodation in the SDD field area was filled by the end of Cisco time, the transgressive shale across the top of the Cisco represents a relatively continuous, flat physical stratigraphic marker. For this reason, it was chosen for the datum in the SDD field area. The post-Cisco shale undoubtedly dips basinward as it fills slope topography off the Cisco margins east of the SDD field and therefore would not represent a good choice of datum to the east.

In the SDD field, stratigraphic interpretation of the 3-D seismic data required two additional pieces of information—a depositional model (discussed above) and high-resolution well logs whose responses could be related to lithology. High-resolution spectral (total radioactivity) and computed gamma-ray (uranium removed from total radioactivity) and AI logs were "hung" into the 3-D time volume on the post-Cisco shale horizon. In terms of the relationship among lithology, AI, and gamma ray, the black, organic-rich shales have the lowest AI values (low density and velocity) and the highest radioactivity values (high potassium and thorium in clays and high uranium associated with organic material). The low-porosity dolomites have the highest AI values (high velocity and density) and lowest radioactivity values (Figure 9A). Within the carbonate-dominated intervals, there is certainly no definable relationship between AI and porosity. As a crude approximation, AI values decrease as a function of increased porosity, limestone, and/or shale (Figure 9B), because substitution of any of these properties for dolomite will reduce the density and the velocity of the rock, thereby lowering the AI.

Our approach to the inversion process was highly iterative, whereby a single surface was initially interpreted in the SDD field amplitude volume and then used as input into the first inversion. A second surface was then interpreted in the AI volume, the stratigraphic layering scheme was refined, and another inversion was made. This iterative process continued for eight seismic inversions (Figure 10). Because the log and core data were not used as input into the seismic

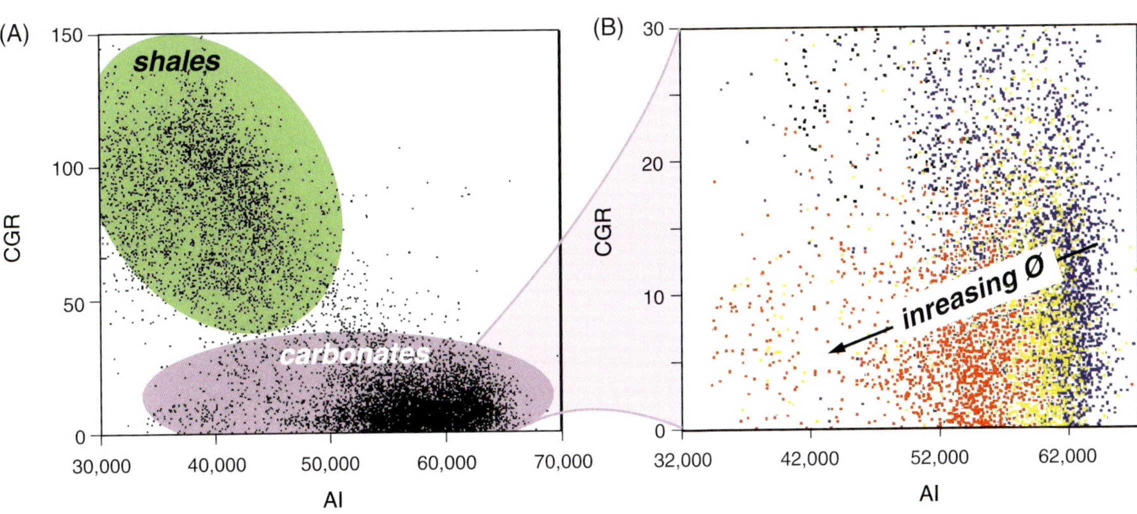

FIGURE 9. Log data illustrating the relationship between radioactivity (computed gamma-ray [CGR] log measuring potassium and thorium) and acoustic impedance (AI, product of density and velocity). (A) Shales and carbonates. (B) Carbonates only. Blue points represent less than 2% log porosity, yellow points represent 2–4% log porosity, and red points represent greater than 4% log porosity.

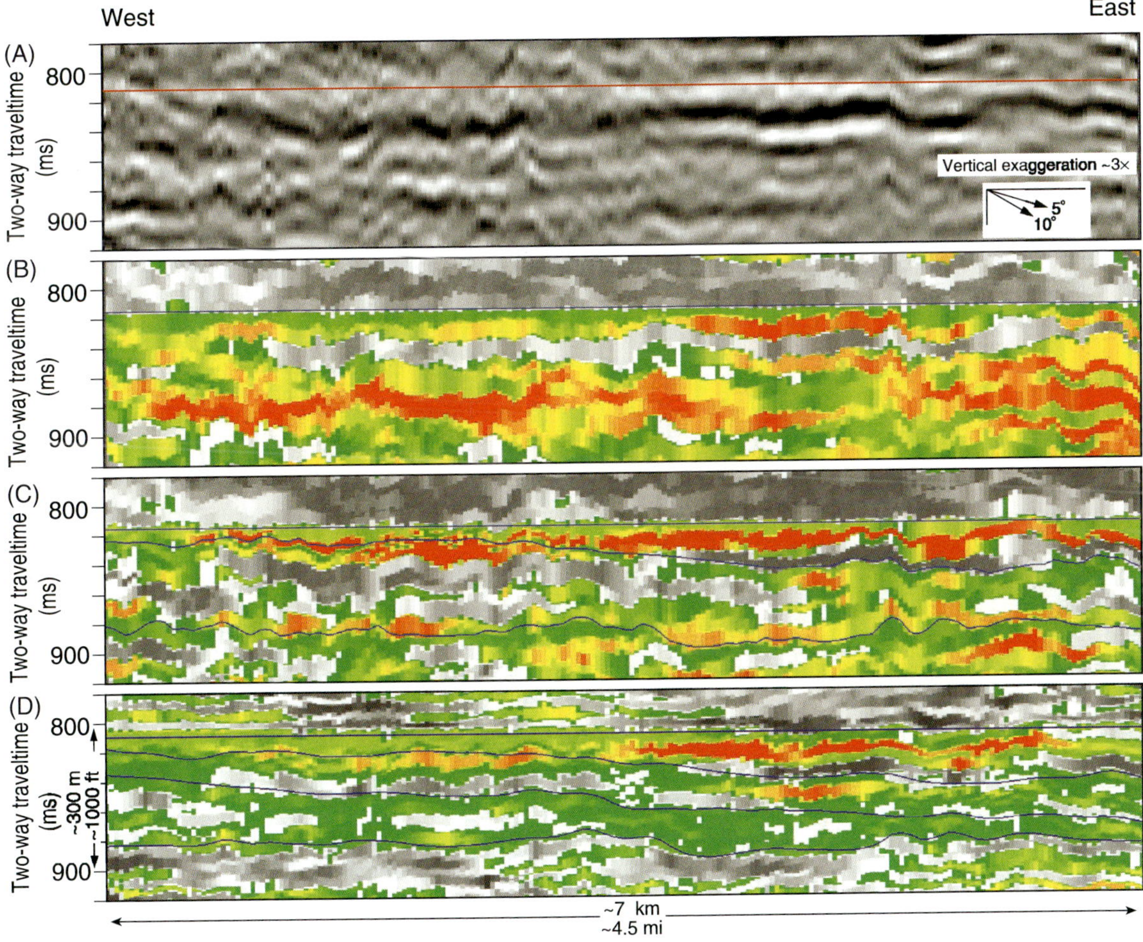

FIGURE 10. Seismic data from 3-D seismic survey. All data are from the same slice and are datumed on the shale overlying the Cisco. (A) Amplitude. (B) Acoustic impedance (AI) from the first inversion constrained by only one surface at the top of the Cisco (blue line). Gray and white represent low AI (shales), green represents midrange AI (limestones and argillaceous limestones), yellow represents moderately high AI (porous dolomites), and red represents high AI (tight dolomites). (C) Third inversion, constrained by three surfaces. (D) Eighth inversion, constrained by four surfaces. Note the change in AI from (B) to (D). This change illustrates the importance of the iterative inversion process.

inversion, data from each of the wells in the 3-D seismic data volume essentially represent a blind test. When the AI from seismic inversion matched favorably with the AI from wire-line logs in most wells (Figure 11), it was concluded that the inverted seismic data volume away from well control was reasonably accurate. An exact match between well logs and seismic data should not be expected, because well logs and seismic data measure significantly different volumes of rock. Had the inversion been derived from the well data initially, the inversion away from the initial well data would be difficult to test.

The integration of log and seismic data was necessary in the SDD field because interpretation using seismic amplitude geometry patterns alone can lead to an erroneous stratigraphic framework. For example, in many cases in the SDD field 3-D seismic data, interpreted depositional surfaces appear to cut across amplitude and AI (Figure 12; see also Figure 10). By including log data in the datumed AI volume and by using a reasonable depositional model as a guide, we were able to interpret the high-frequency sequence-stratigraphic framework with considerably greater accuracy. For example, shale that thins from the basin shelfward onto the underlying ramp may thin to below seismic data resolution (see Figure 11). However, it can still be interpreted in the seismic data volume using well logs to guide the interpretation, coupled with a depositional model to explain why the interpretation is reasonable (Figures 5–7).

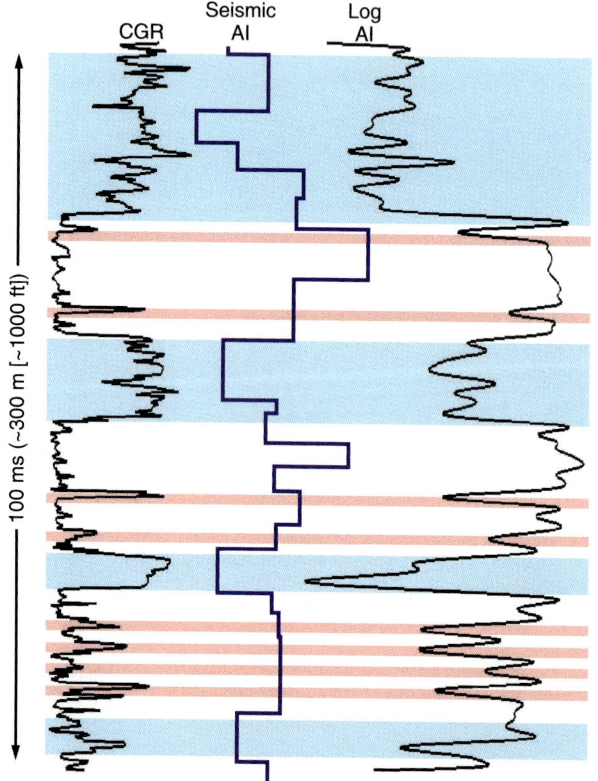

Figure 11. Illustration of benchmark testing of seismic inversion. Seismic acoustic impedance (AI) is compared with the computed gamma-ray (CGR) log and the log-derived AI. The four major shales (shown by leftward excursions of the blocky, blue-colored seismic AI curve) correspond with the high CGR values in the shales (excursions to the right) and lower values in the log AI. If the shales are too thin (<10 m [<30 ft]), then the seismic AI represents a misleading average of shale and carbonate values (red pattern).

The seismic data indicate that the position of the margins in the SDD field was highly influenced by the Pennsylvanian-age faulting associated with the Huapache fault system. A time slice from the AI volume shows the structural trends in the field area. Huapache parallel (northwest to southeast) faults and joints and conjugate northeast to southwest faults and joints control the position and orientation of the structural highs (Figure 13). When the AI volume is datumed on the shale overlying the Cisco Formation, time slices through the volume more closely represent depositional trends. A time slice from the stratigraphic (datumed) AI volume (Figure 14) shows the position of the Canyon ramp margin, the heterogeneity associated with the algal biostromes and bioherms, and the influence of Pennsylvanian-age faulting on the position of the Canyon and Cisco margins (compare with Figure 13). Unfortunately, static pictures of the 3-D AI volume do not do justice to the interpretive power of the integrated 3-D seismic data and log volume. The ability to alter color, opacity, lighting, and choice of datum, to filter out data ranges or isolate stratigraphic intervals, and to move through the AI volume while interpreting in real time, with well logs positioned properly, provides an extremely powerful visualization and interpretation tool.

THE THREE-DIMENSIONAL GEOLOGIC MODEL

Early 3-D modeling solutions in the SDD field used deterministic, inverse-distance-exponent algorithms to distribute data. Data distribution in the early 3-D model in the SDD field followed thin layers that were interpreted to represent depositional stratal geometries within a reasonable sequence-stratigraphic framework. However, because the well spacing in the SDD field is commonly greater than the scale of petrophysical variation, the resulting model did not honor the true interwell variation (Figure 15). In other carbonate fields that have broader facies tracts and greater well density, such as Yates field, this approach is perfectly adequate (Tinker and Mruk, 1995). In an attempt

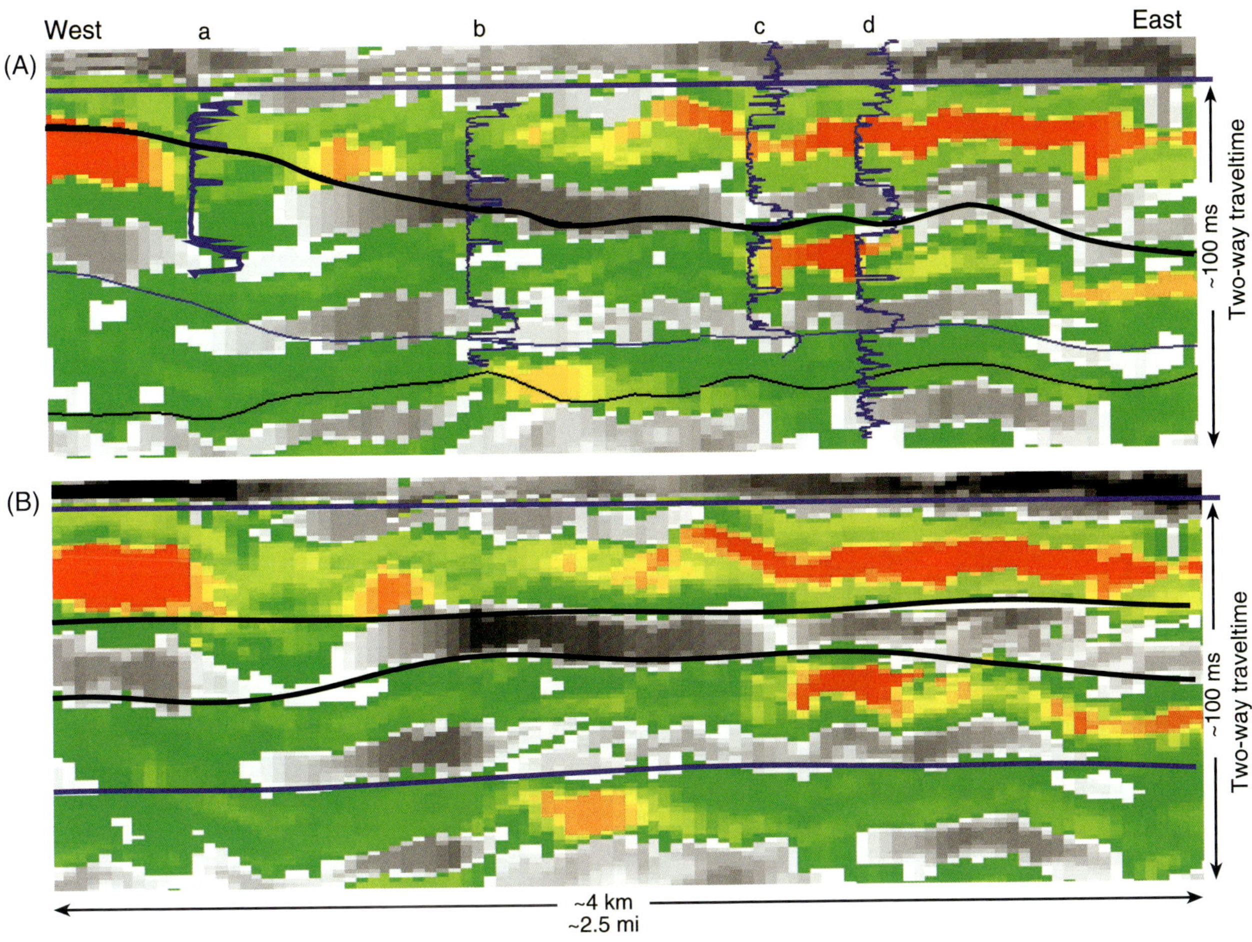

FIGURE 12. (A) Acoustic impedance from seismic inversion, shown with computed gamma-ray logs from the 3-D volume to illustrate the importance of multiple data types for stratigraphic interpretation. The shale whose base is indicated by the thick black line is thick (~15 m [~50 ft]) in well (d), thins west toward the ramp margin in wells (c) and (b), and onlaps the ramp margin in (a), where it is only 3 m (10 ft) thick. (B) An alternate (and most likely incorrect) flat-layer interpretation made using only the seismic data. The combination of a depositional model (Figure 5), an understanding of the relationship between acoustic impedance and lithology (Figure 9), and a synthetic seismic model illustrating false lateral continuity of seismic patterns in prograding systems (Tinker, 1996b) allows for the improved interpretation (A).

to introduce heterogeneity not captured by well spacing and to more closely honor population statistics for petrophysical parameters, geostatistical approaches have been employed in other fields (e.g., Journel, 1991). As with deterministic methods, the broad spectrum of geostatistical approaches will not compensate for a poor stratigraphic framework interpretation.

True interwell measurements in a reservoir come from 3-D seismic data, crosswell tomography, well tests, or tracer data. Several approaches have been described to integrate multiple-scale data, such as seismic and well data, in a 3-D model (e.g., Doyen, 1988; Araktingi and Bashore, 1992; Xu et al., 1992; A. Yang et al., 1995; C. Yang et al., 1995; Behrens et al., 1996; Doyen et al., 1997; Panda et al., 2000), and some of these approaches were used in later 3-D models at the SDD field (Figure 16).

The integrated 3-D geologic model at the SDD field has several important uses. First, volumetric calculations are improved significantly because the model is built using measurements of the complete earth volume. Second, true 3-D flow units (sensu Ebanks, 1987) can be defined, because 3-D connections based on the geologic model can be examined. In the absence of full-field models, it is possible to estimate "flow units" (Hurley et al., 1999) using Lorenz plots (Gunter et al., 1997) of cumulative storage capacity (ϕh) vs. flow capacity (kh). Although this is a useful and valid approach around a wellbore with a buildup or injection test, it does not serve to describe and predict the 3-D variability of the full, interconnected matrix, vug, and fracture volume across a field. A distinction between "well-scale flow units" and "full-field flow units" is important. The flow units identified in the model were

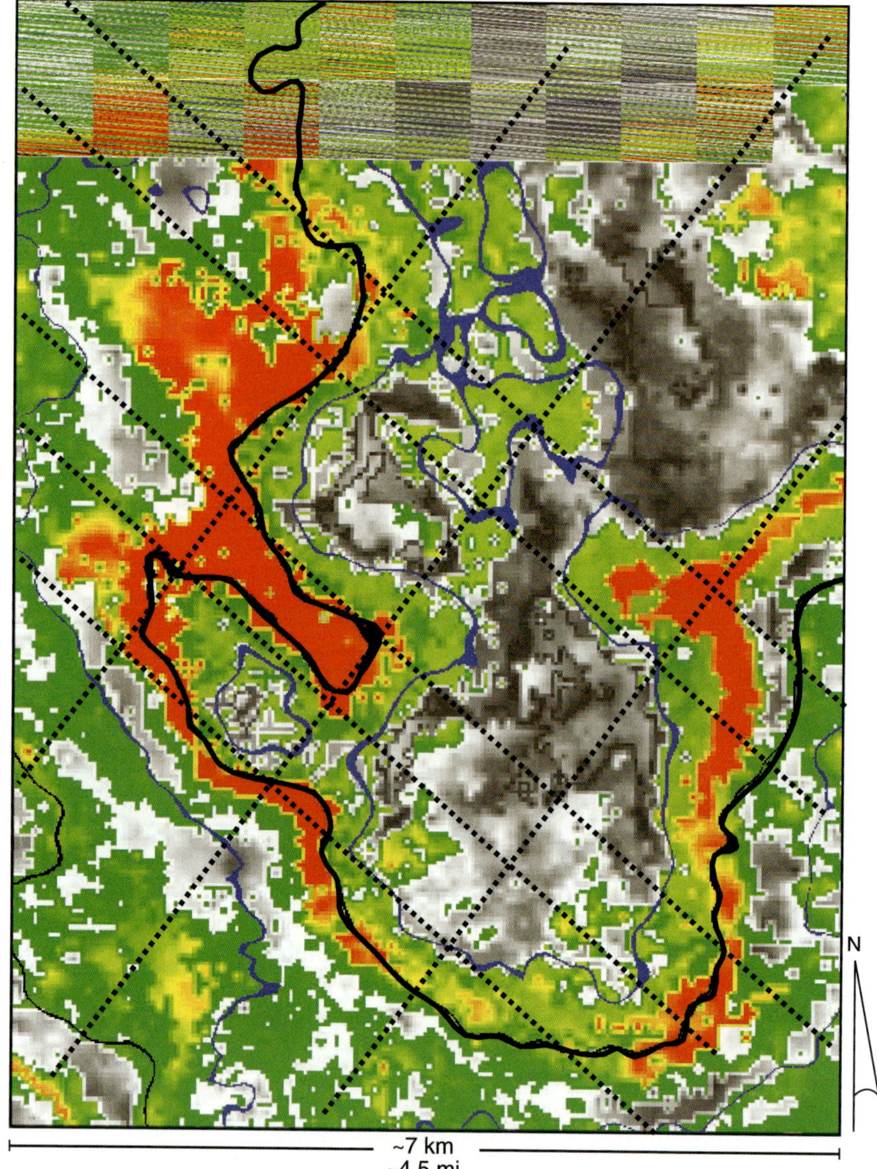

FIGURE 13. Time slice from the structural acoustic impedance volume. Interpreted structural lineaments are shown as black, dashed lines. The heavy solid line is the top of the Canyon, which represents the intersection of a depositional surface with this horizontal slice through the structural seismic volume.

used to help guide later horizontal drilling. Finally, the geologic model recognizes and incorporates all scales of measurement, including seismic measurements of the full earth volume, and provides useful, scaleable input for 3-D reservoir simulation.

CONCLUSIONS

South Dagger Draw field is presented as a case study that illustrates several generations of stratigraphic framework interpretation and 3-D modeling. SDD field and its southern neighbor, Indian Basin field, have together produced nearly 23 million bbl of oil and nearly 2 tcf of gas from vuggy porosity, formed dominantly in algal biostromes and bioherms located at the ramp-margin position of complex sigmoid-oblique clinoforms of the Pennsylvanian Canyon and Cisco Formations. Acidic, hydrothermal fluids migrated upward along joints and were held up beneath shales, causing a "mushroom effect" of dissolution in carbonate facies.

Data used in the study include logs, cores, modern wireline log suites, borehole image logs, and 3-D AI values from inversion of seismic data. The dolomite reservoirs in SDD field have high velocities and densities relative to sandstone reservoirs. Therefore, the seismic data have poor vertical resolution and are in fact misleading in regard to sequence-stratigraphic interpretation because similar facies, which are juxtaposed laterally in a prograding system, appear as horizontal reflectors on the seismic data.

Well logs and core data, used in combination with outcrop analogs, were critical to developing a reasonable depositional model, and the model was then used to guide the sequence-stratigraphic interpretation. Several iterations of seismic inversion, each guided by ever more detailed sequence-stratigraphic interpretations, resulted in an AI data volume that was incorporated into the 3-D model and used for understanding interwell variation in reservoir quality.

ACKNOWLEDGMENTS

This chapter is reprinted, with slight modification, with the permission of the Gulf Coast Section SEPM Foundation from T. F. Hentz (editor), GCS-SEPM Foundation 19th Annual Bob F. Perkins Research Conference on Advanced Reservoir Characterization for the

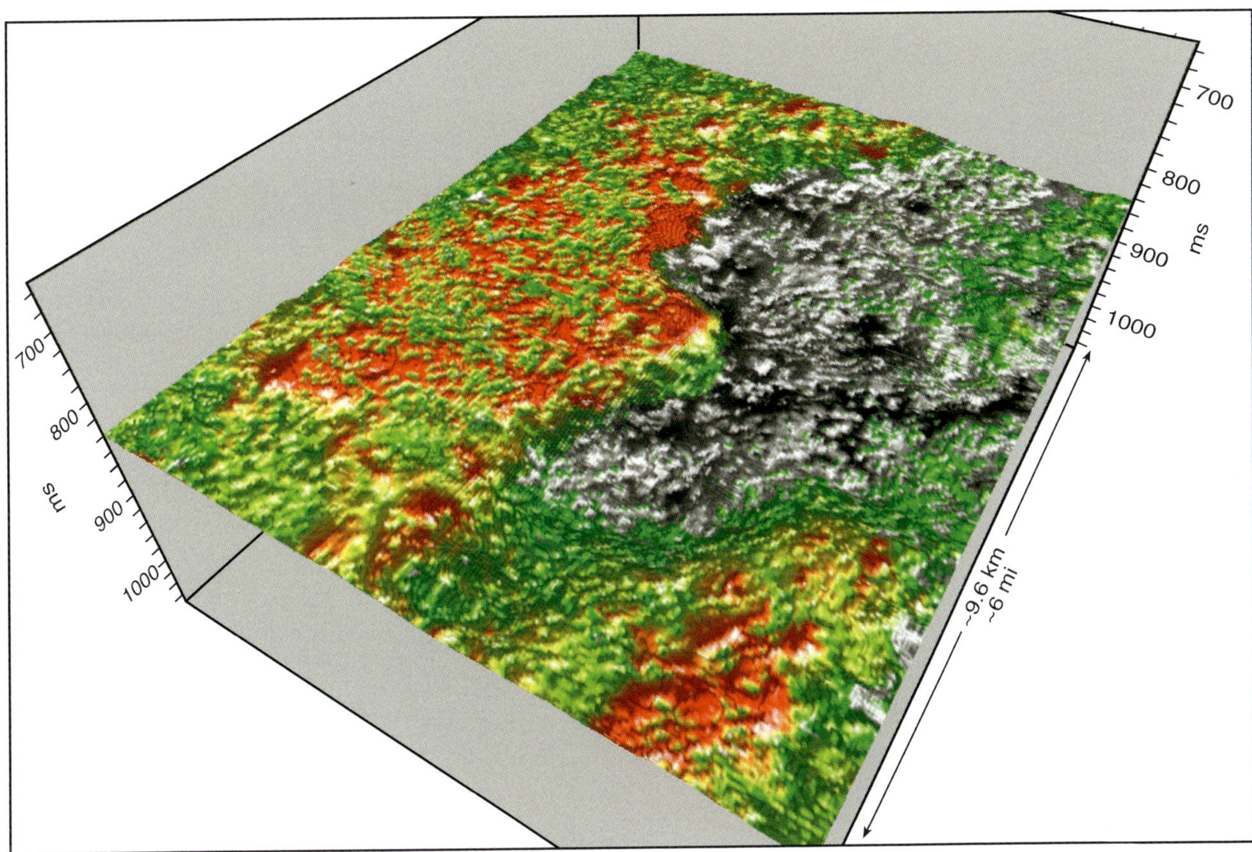

Figure 14. Three-dimensional view of the acoustic impedance (AI) volume flattened on the top of the Cisco horizon, showing the ramp margin in red (high AI) and the time-equivalent shale basin in gray (low AI). Area is the same as represented in Figure 13, and the influence of the structural lineaments illustrated in Figure 13 on the orientation of the ramp margins is apparent.

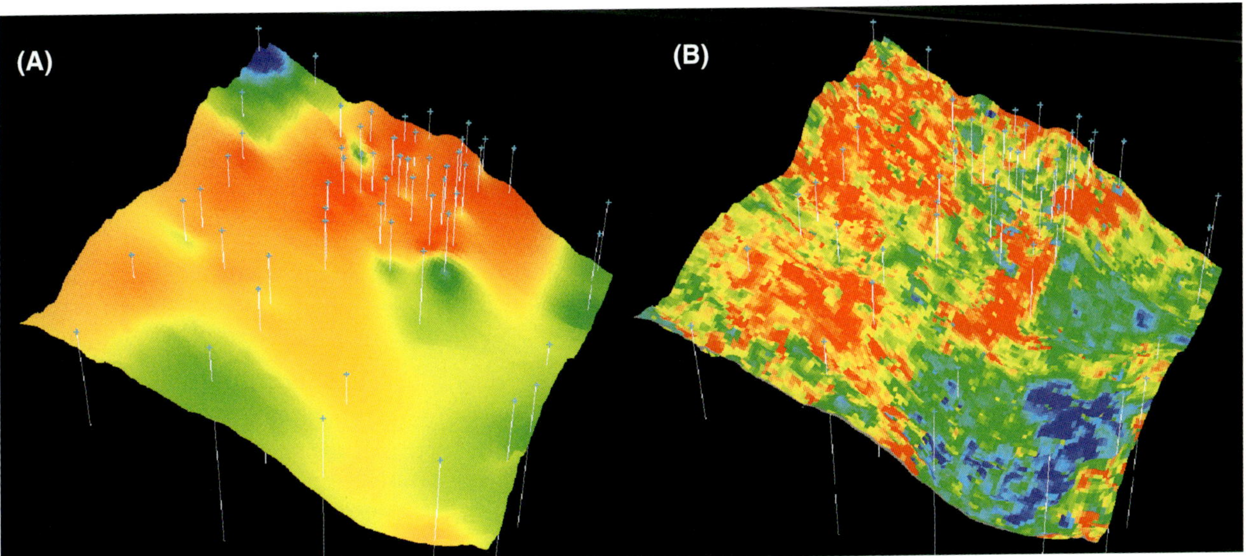

Figure 15. Porosity distribution. (A) Well logs only using inverse-distance-squared deterministic distribution algorithms. (B) Incorporating acoustic impedance from 3-D seismic data inversion. Area illustrated is approximately 7 × 7 km (~4.5 × 4.5 mi).

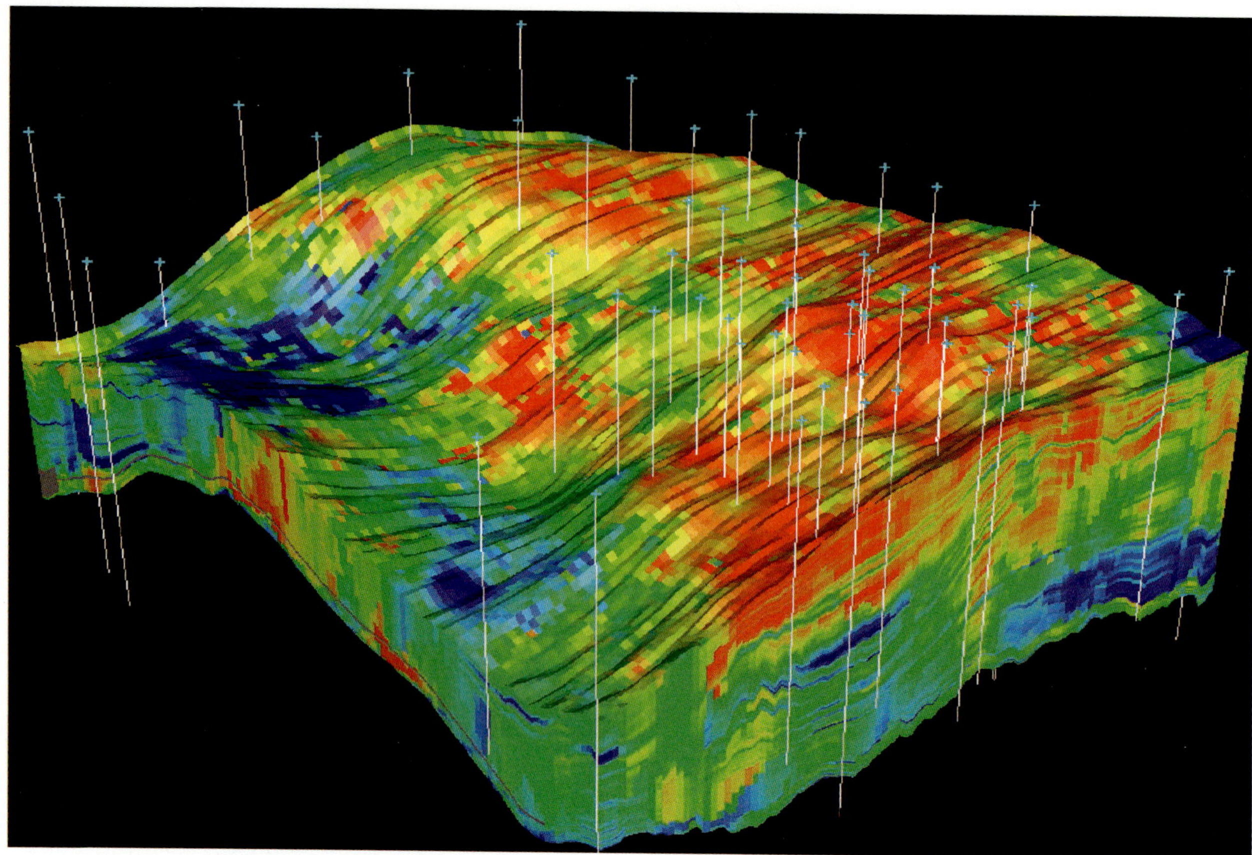

FIGURE 16. Integrated 3-D model in South Dagger Draw field. Porosity distribution is illustrated, with red and yellow representing higher values and blue and green representing lower values. Location of wells shown by vertical white lines. Area illustrated is approximately 7 × 10 km (~4.5 × 6 mi).

21st Century (1999, p. 213–232). Thanks to Art Saller, Gregor Eberli, and Mark Sonnenfeld for helping to refocus the manuscript with insightful suggestions.

REFERENCES CITED

Araktingi, U. G., and W. M. Bashore, 1992, Effects of properties in seismic data on reservoir characterization and consequent fluid flow prediction when integrated with well logs: SPE Paper 24752, p. 913–926.

Behrens, R. A., M. K. Macleod, and T. T. Tran, 1996, Incorporating seismic attribute maps in 3-D reservoir models: SPE Paper 36499, p. 31–36.

Doyen, P. M., 1988, Porosity from seismic data: A geostatistical approach: Geophysics, v. 53, no. 10, p. 1263–1275.

Doyen, P. M., D. E. Psaila, and L. D. de Boer, 1997, Reconciling data at seismic and well log scales in 3-D earth modeling: SPE Paper 38698, p. 465–474.

Ebanks, W. J. Jr., 1987, Flow unit concept-integrated approach to reservoir description for engineering projects (abs.): AAPG Bulletin, v. 71, no. 5, p. 551–552.

Galley, J. E., 1968, Some principles of tectonics in the Permian Basin, *in* W. Stewart, ed., Basins of the southwest: West Texas Geological Society Publication 1, p. 5–20.

Goldhammer, R. K., E. J. Oswald, and P. A. Dunn, 1994, High-frequency, glacio-eustatic cyclicity in the Middle Pennsylvanian of the Paradox Basin: An evaluation of Milankovitch forcing, *in* P. L. de Zboer and D. G. Smith, eds., Orbital forcing and cyclic sequences: International Association of Sedimentologists Special Publication 19, p. 243–283.

Gunter, G. W., J. M. Finneran, D. J. Hartmann, and J. D. Miller, 1997, Early determination of reservoir flow units using an integrated petrophysical method: SPE Paper 38769, p. 373–380.

Hamlin, H. S., S. J. Clift, S. P. Dutton, T. F. Hentz, and S. E. Laubach, 1995, Canyon sandstones—A geologically complex natural gas play in slope and basin facies, Val Verde Basin, southwest Texas: University of Texas at Austin, Bureau of Economic Geology Report of Investigations 232, 74 p.

Hill, C. A., 1992, Paleokarst, karst-related diagenesis, and reservoir development: Examples from Ordovician–Devonian age strata of west Texas and the Mid-continent: SEPM Publication 92-33, p. 192–194.

Hite, R. J., and D. H. Buckner, 1981, Stratigraphic correlations, facies concepts, and cyclicity in Pennsylvanian rocks of the Paradox Basin, *in* D. L. Wiegand, ed., Geology of the Paradox Basin: Rocky Mountain Association of Geologists 1981 Field Conference: Rocky Mountain Association of Geologists, p. 147–159.

Hurley, N. F., and R. Budros, 1990, Albion-Scipio and Stoney Point fields—U.S.A., *in* E. A. Beaumont and N. H. Foster, eds., Stratigraphic traps I: AAPG Treatise of Petroleum Geology Atlas of Oil and Gas Fields, p. 1–37.

Hurley, N. F., D. Pantoja, and R. A. Zimmerman, 1999, Flow unit determination in a vuggy dolomite reservoir, Dagger Draw field, New Mexico: Society of Professional Well Log Analysts Conference, Oslo, Norway, 14 p.

Journel, A. G., 1991, Geostatistics for reservoir characterization: SPE Paper 20750, p. 353–358.

Kerans, C., and S. W. Tinker, 1997, Sequence stratigraphy and characterization of carbonate reservoirs: SEPM Short Course Notes No. 40, 165 p.

Lucia, F. J., 1983, Petrophysical parameters estimated from visual description of carbonate rocks: A field classification of carbonate pore space: Journal of Petroleum Technology, v. 35, p. 626–637.

Meyer, R. F., 1966, Geology of Pennsylvanian and Wolfcampian rocks in southeast New Mexico: New Mexico Bureau of Mines and Mineral Resources Memoir 17, 123 p.

Panda, M. N., C. C. Mosher, and A. K. Chopra, 2000, Application of wavelet transforms to reservoir-data analysis and scaling: SPE Journal, v. 5, no. 1, p. 92–101.

Peterson, J. A., and R. J. Hite, 1969, Pennsylvanian evaporite-carbonate cycles and their relation to petroleum occurrence, southern Rocky Mountains: AAPG Bulletin, v. 53, p. 884–908.

Ross, C. A., 1986, Paleozoic evolution of southern margin of Permian Basin: Geological Society of America Bulletin, v. 84, p. 2851–2872.

Stafleau, J., and M. D. Sonnenfeld, 1994, Seismic models of a shelf-margin depositional sequence: Upper San Andres Formation, Last Chance Canyon, New Mexico: Journal of Sedimentary Research, v. B64, p. 481–499.

Tinker, S. W., 1996a, Building the 3-D jigsaw puzzle: Applications of sequence stratigraphy to 3-D reservoir characterization, Permian Basin, USA: AAPG Bulletin, v. 80, no. 4, p. 460–485.

Tinker, S. W., 1996b, Reservoir-scale sequence stratigraphy: McKittrick Canyon and 3-D subsurface examples, west Texas and New Mexico: Ph.D. dissertation, University of Colorado, 245 p.

Tinker, S. W., and D. M. Cox, 1998, Sequence stratigraphy and 3-D geologic models of a complex clinoform ramp-crest reservoir (abs.): AAPG Annual Convention Book of Abstracts, 2 p.

Tinker, S. W., and D. H. Mruk, 1995, Reservoir characterization of a Permian giant: Yates field, west Texas, *in* E. Stoudt and P. M. Harris, eds., Hydrocarbon reservoir characterization, geologic framework and flow-unit modeling: SEPM Short Course 34, p. 51–128.

Uland, M. J., S. W. Tinker, and D. H. Caldwell, 1997, 3-D reservoir characterization for improved reservoir management: SPE Paper 37699, 14 p.

Wilson, E. N., 1990, Dolomitization front geometry, fluid-flow patterns, and the origin of massive dolomite: The Triassic Latemar buildup, northern Italy: American Journal of Science, v. 290, no. 7, p. 741–796.

Xu, W., T. T. Tran, R. M. Srivastava, and A. G. Journel, 1992, Integrating seismic data in reservoir modeling: The collocated cokriging alternative: SPE Paper 24742, p. 833–842.

Yang, K. M., and S. L. Dorobek, 1995, The Permian Basin of west Texas and New Mexico; tectonic history of a "composite" foreland basin and its effects on stratigraphic development, *in* S. L. Dorobek and G. M. Ross, eds., Stratigraphic evolution of foreland basins: SEPM Special Publication 52, p. 149–174.

Yang, A. P., Y. Gao, and S. Henry, 1995, Reservoir characterization by integrating well data and seismic attributes: SPE Paper 30563, p. 337–361.

Yang, C. T., A. K. Chopra, J. Chu, X. Huang, and M. H. Kelkar, 1995, Integrated geostatistical reservoir description using petrophysical, geologic, and seismic data for Yacheng 13-1 field: SPE Paper 30566, p. 357–372.

5

Saller, A. H., S. Walden, S. Robertson, R. Nims, J. Schwab, H. Hagiwara, and S. Mizohata, 2004, Three-dimensional seismic imaging and reservoir modeling of an upper Paleozoic "reefal" buildup, Reinecke field, west Texas, United States, *in* Seismic imaging of carbonate reservoirs and systems: AAPG Memoir 81, p. 107–122.

Three-Dimensional Seismic Imaging and Reservoir Modeling of an Upper Paleozoic "Reefal" Buildup, Reinecke Field, West Texas, United States

Arthur H. Saller
Unocal Corporation, Sugar Land, Texas, U.S.A.

Skip Walden
Unocal Corporation, Sugar Land, Texas, U.S.A.

Steve Robertson
Pure Resources, Midland, Texas, U.S.A.

Robert Nims
Pure Resources, Midland, Texas, U.S.A.

Joe Schwab
Pure Resources, Midland, Texas, U.S.A.

Hiroshi Hagiwara[1]
Japan National Oil Company, Chiba, Japan

Shigeharu Mizohata
Japan National Oil Company, Chiba, Japan

ABSTRACT

Reinecke field is an upper Pennsylvanian to lowest Permian carbonate buildup in the southern part of the Horseshoe Atoll, west Texas, United States. The field and surrounding areas have been imaged with three 3-D seismic surveys and penetrated by many wells. Although Reinecke is commonly referred to as a reefal reservoir, deposition occurred in stratified sequences, 50–100 ft (15–30 m) thick, dominated by wackestones, packstones, and grainstones. Boundstones (mainly rich in phylloid algae) constitute only 16% of the buildup. Seismic reflectors within the buildup parallel sequence boundaries and are truncated at the margins of the buildup. Three-dimensional seismic surveys show that the top of the Reinecke buildup is highly irregular with more than 470 ft (143 m) of relief. Deep-marine shales overlie the reservoir and act as a seal for this stratigraphic trap. Reinecke's irregular, mounded morphology is the result of localized carbonate growth and erosional truncation. Much of the erosional truncation probably occurred in a deep-marine environment.

Reinecke's south dome acts a single continuous reservoir dominated by limestone (70%) with 25% dolomite. Limestone porosity is generally 5–18% (average of 11.2%)

[1]*Present address:* Japan Oil Development Company, Abu Dhabi, United Arab Emirates.

and permeability is 1–1000 md (average of 166 md). Dolomite porosity is lower (average of 8.3%), but permeability is higher (average of 894 md). Discontinuous low-permeability layers parallel to stratification serve as low-permeability baffles; however, patchy replacive dolomites cut through stratification and act as high-permeability vertical conduits. Good reservoir continuity, low-permeability baffles, and artificially enhanced bottom-water drive helped to recover more than 50% of the original oil in place. Excellent vertical reservoir continuity has allowed implementation of a crestal CO_2 flood at Reinecke field. CO_2 is being injected into the top of the structure, displacing residual and bypassed mobile oil downward for recovery in lower parts of the reservoir.

INTRODUCTION

Reinecke field is a stratigraphic trap for oil in a "reefal" buildup. Reinecke field is part of the Horseshoe Atoll, a northward-opening arc of upper Pennsylvanian carbonate buildups in the Midland Basin of west Texas, United States (Figures 1, 2). Horseshoe Atoll fields are a series of stratigraphic traps that have produced more than 2 billion bbl of oil and are characterized by high permeability and high recovery efficiencies (near 50% of original oil in place [OOIP] after waterflooding) (Galloway et al., 1983). Fields on the eastern side of the Horseshoe Atoll have stratified porosity with permeability barriers to vertical flow. Reinecke and other fields on the south and west sides of the Horseshoe Atoll have vertical permeability pathways that allow the top of the reservoir to be in pressure communication with the underlying aquifer (i.e., they produce like reefal reservoirs) (Crawford et al., 1984; Saller et al., 1999).

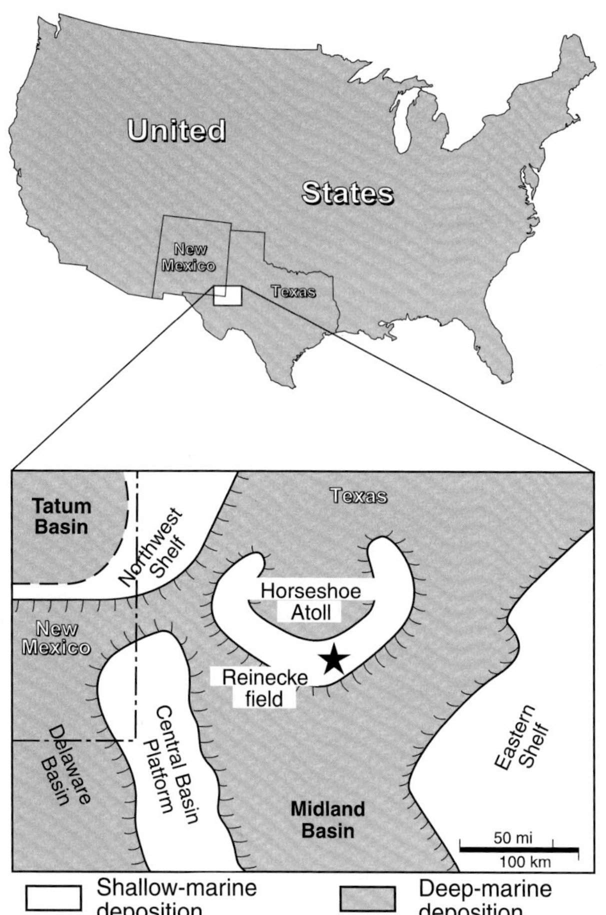

FIGURE 1. Location of Reinecke field and paleogeography during the late Pennsylvanian.

Approximately 83 million bbl of oil have been produced from Reinecke field (approximately 3 mi^2 [8 km^2]; Figure 3) since its discovery in 1950. The south dome of Reinecke field (approximately 1 mi^2 [2.6 km^2]), the focus of this chapter, has produced 46 million bbl of oil. The carbonate reservoir at the south dome of Reinecke field has approximately 470 ft (143 m) of depositional and erosional relief above its original oil-water contact at −4600 ft (−1400 m) subsea. The area has no significant tectonic deformation. The shallow-marine carbonate thins laterally and is overlain by a deep-marine shale that acts as a vertical and lateral seal. Deep-marine shales of Pennsylvanian and Early Permian age are generally assumed to be the source for the Reinecke oil.

This joint Unocal-Japan National Oil Company reservoir characterization project helped to better understand the south dome at Reinecke field to implement a crestal CO_2 flood. Major parts of the project included (1) deepening and coring of many wells that initially only penetrated the uppermost part of the reservoir, (2) acquisition and interpretation of multiple three-dimensional (3-D) seismic surveys, (3) construction of a geocellular reservoir model, (4) fluid-flow simulation and history match of the reservoir from time of discovery (1950) to the present, and (5) use that model to predict CO_2 flood response. Before the start of this project, only a few wells penetrated the entire reservoir section in the south dome of Reinecke field. In the early phases of this project, most wells were deepened to near the original oil-water contact. The database for this study includes 12 cored wells and 60 wells with wire-line logs.

Reservoir characterization included several important steps. Key stratigraphic surfaces were identified in core and then extrapolated across the rest of the field

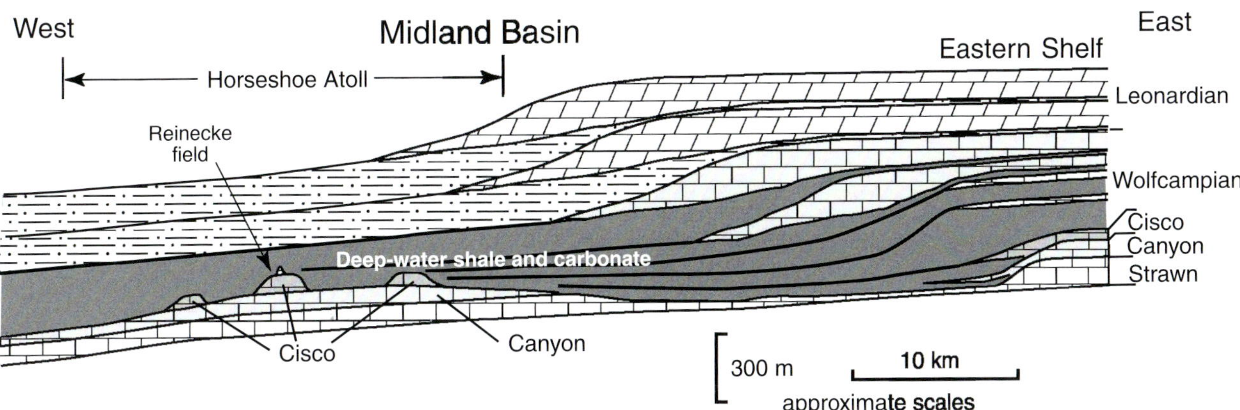

FIGURE 2. Schematic cross section across the eastern part of the Midland Basin (based in part on Galloway et al., 1983).

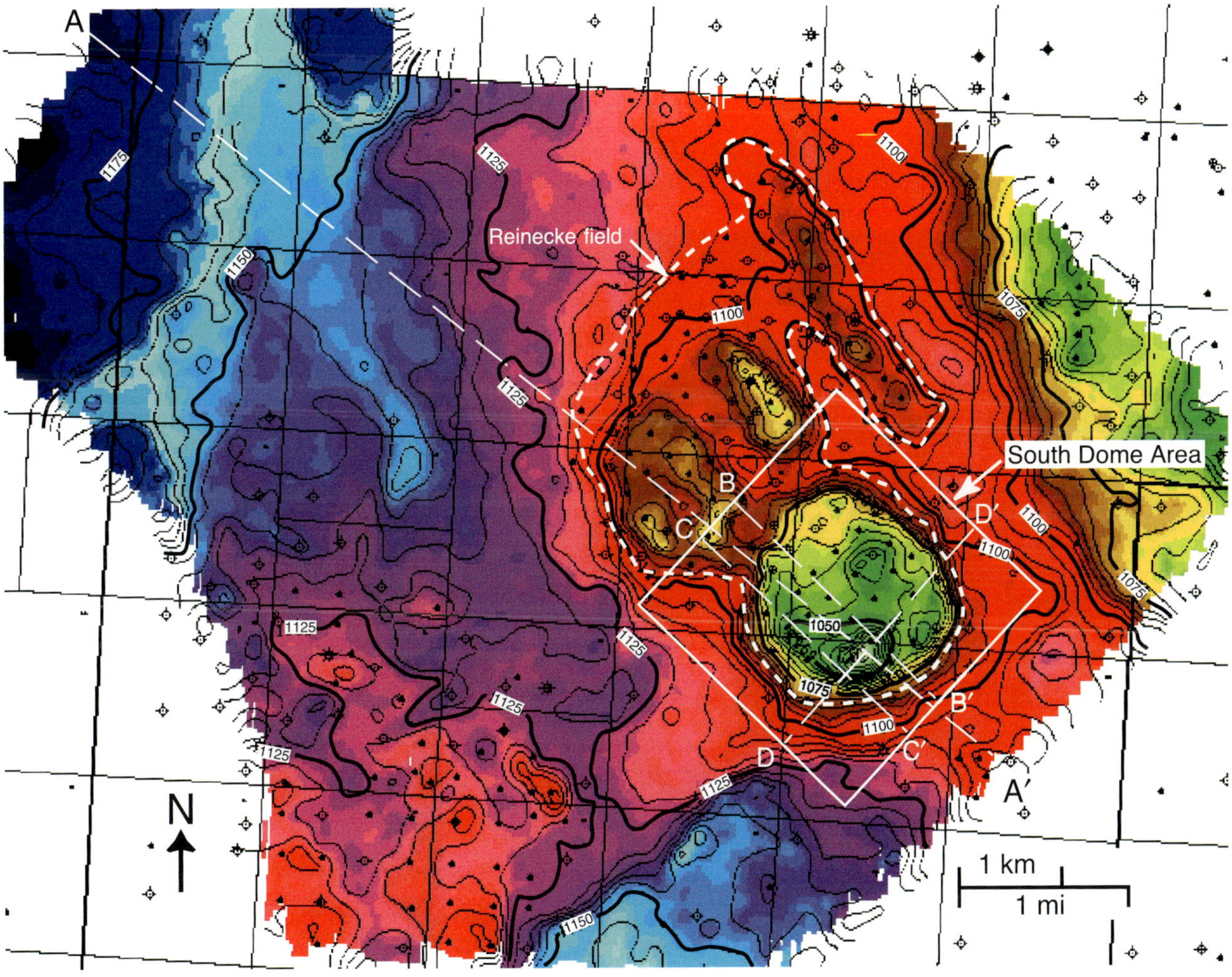

FIGURE 3. Structure map in seismic time of the upper Pennsylvanian–lowest Permian carbonate at Reinecke field and surrounding areas based on 3-D seismic data. The area has seen little tectonic deformation since the Pennsylvanian; hence, this structure represents the shape of the carbonate buildups after deposition and erosion (late Pennsylvanian to earliest Permian time). Contour interval is 5-ms two-way traveltime (TWT). Seismic section AA′ is in Figure 4, BB′ is in Figure 6, and CC′ and DD′ are in Figure 7. The south dome area was imaged in later 3-D seismic surveys (Figure 5) and modeled (Figures 12, 14).

using wire-line logs. Reservoir porosity and permeability were initially measured on core material. Porosity measured in core was used to calibrate wire-line logs. Wire-line porosity logs (mainly neutron and density logs) were then used to estimate porosity in uncored wells. Permeability in uncored wells was calculated using core-derived transforms for porosity vs. permeability. Stratigraphic tops from wells were merged with seismic surfaces in EarthVision to create stratigraphic surfaces used in the geocellular model. Porosity and permeability were extrapolated between wells by ordinary kriging and then gridded in EarthVision. A model with 35 layers was built. A RESCUE export format was used as an interface between the EarthVision geocellular model and the Eclipse flow simulator. In this chapter, we will focus on (1) describing the three-dimensional reservoir characteristics of a "reefal" buildup, i.e., the south dome of Reinecke field, (2) showing how a 3-D seismic survey better imaged this "reefal" buildup, and (3) summarizing the 3-D production history of this buildup.

THREE-DIMENSIONAL SEISMIC DATA

Three 3-D seismic surveys have been acquired at Reinecke. The first survey was shot in 1993 to locate possible undrilled pinnacles within Reinecke field and undrilled structures away from the main field. That initial, more regional survey (Figure 3) covered approximately 28 mi^2 (72 km^2) and helped in the discovery of one small nearby field. A second survey was shot in 1997 to image the south dome (approximately 5 mi^2 [13 km^2]) for reservoir characterization and act as a base line for a third survey shot in 1999 to attempt to image the crestal CO_2 flood.

The top of the Reinecke carbonate is shown regionally by the 1993 3-D seismic survey (Figure 3). Middle and upper Pennsylvanian carbonates grew on a relatively flat surface (Figure 4), and hence regional patterns of carbonate growth and erosion are shown by the structural relief on the top of the carbonate (Figure 3). The 1997 seismic survey imaged the complex top of the south dome better than the previous survey. Figure 5A shows how the south dome would be mapped by well control alone. Figure 5B shows how the top of the Reinecke reservoir would be mapped by seismic data alone. The most accurate structural map of the top of the carbonate buildup is the result of combined well and

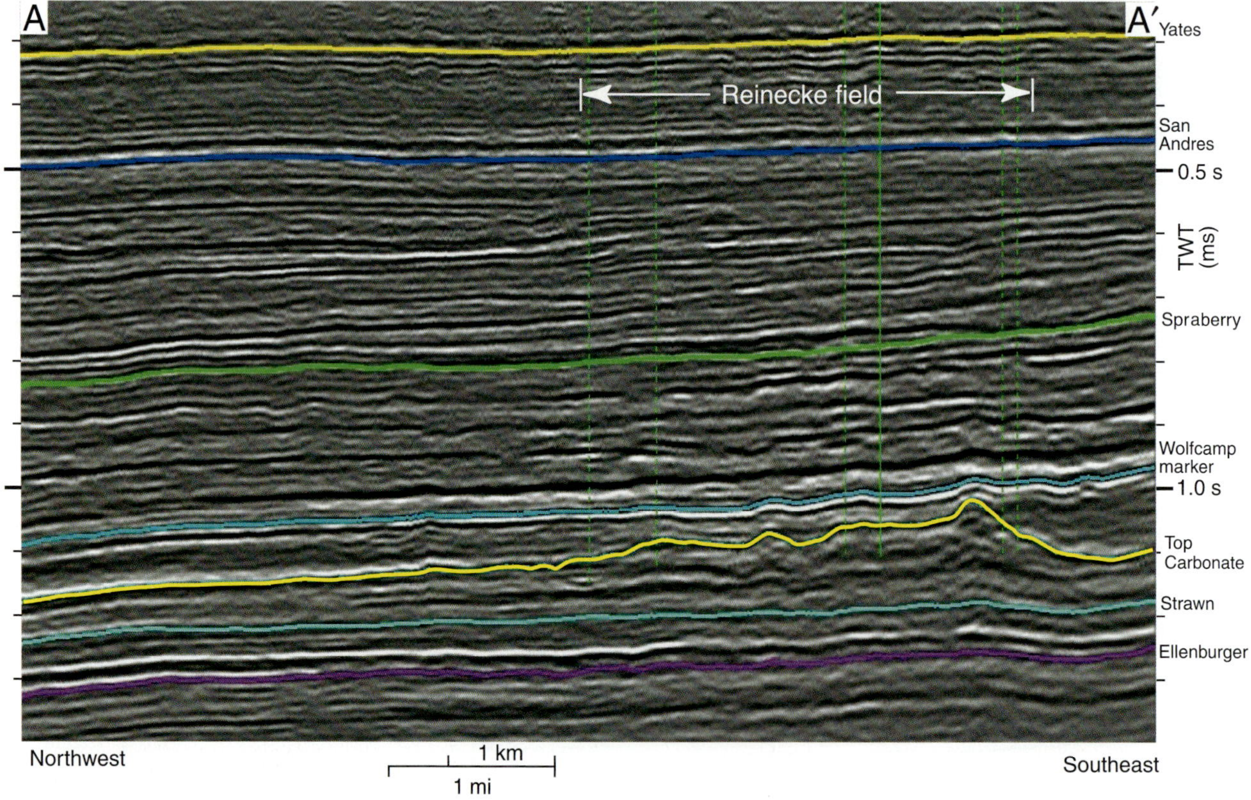

FIGURE 4. Arbitrary 3-D seismic line showing the Paleozoic section in and around Reinecke field. Note the lack of structural deformation. Selected formation tops are shown. Middle Pennsylvanian (Strawn Formation/Desmoinesian Stage), upper Pennsylvanian (Canyon Formation/Missourian Stage and Cisco Formation/Virgilian Stage), and lowest Wolfcampian (Lower Permian) shallow-marine carbonates started accumulating over a relatively flat surface. Basinal shales and carbonates of Wolfcampian age overlie the irregular upper surface of the reservoir carbonate. The Sprayberry includes basinal sand, silt, and carbonate of Leonardian age. The San Andres shelf margin prograded from east to west (right to left) across the area. The top San Andres is early Guadalupian. The Yates is late Guadalupian in age and is a shelf-interior equivalent to the middle Capitan reef. See Figure 3 for location of seismic line. TWT = two-way traveltime.

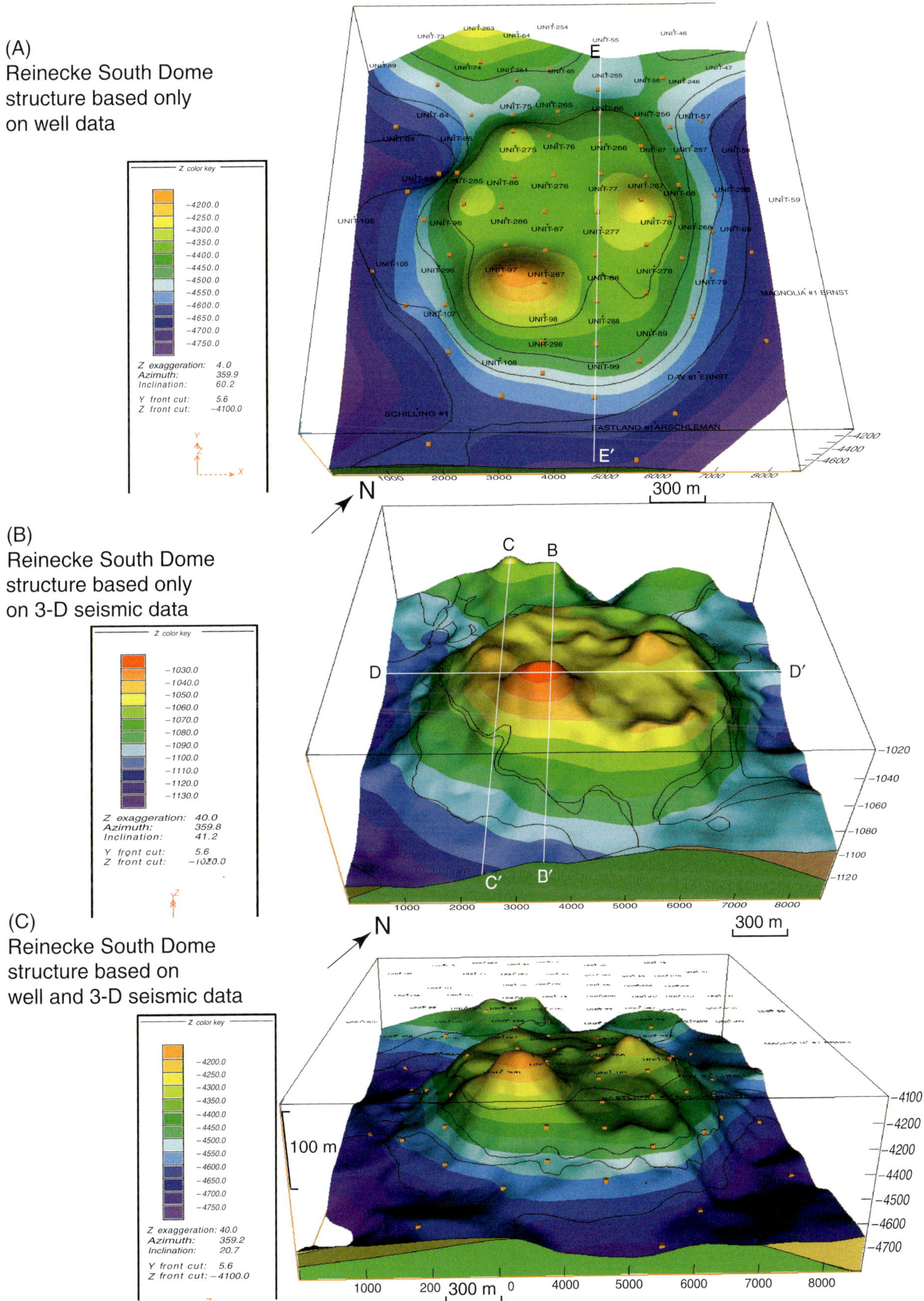

FIGURE 5. Structure of the top of carbonate, south dome of Reinecke field. All views are from the southeast. Vertical exaggeration is 4×. (A) Map (in ft subsea) based solely on well data. (B) Seismic two-way traveltime map (in ms). (C) Map (in ft subsea) based on well and 3-D seismic data.

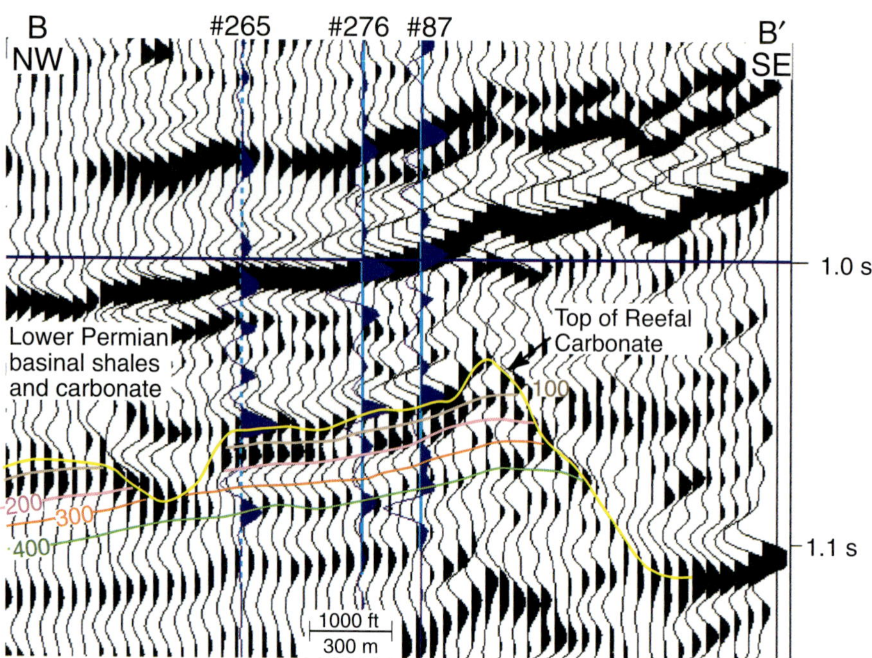

FIGURE 6. Wiggle display of an arbitrary line from the 3-D seismic survey showing sequences defined by cores. Synthetic seismograms for wells have been inserted. Note the truncation of reflectors (sequences) at the margins of the buildup. See Figures 3 and 5B for location of seismic line.

seismic data (Figure 5C). Seismic reflectors within the Reinecke carbonate are generally parallel and truncated at the margin of the Reinecke structure (Figures 6, 7). Sequence boundaries within the reservoir interval were identified in core, but could not be identified by the seismic data alone (Figure 6). Reflectors overlying the reservoir are basinal strata that downlap against the carbonate buildup with an apparent progradation toward the west (Figures 2, 7).

Two main factors have made lateral variations in reservoir characteristics (porosity, permeability, and lithology) difficult to image with the Reinecke seismic data. (1) The top of carbonate is a very irregular surface having a very large impedance contrast with the overlying shale (Figure 6). This resulted in a very high amplitude seismic response at the top of the Reinecke reservoir and an uneven penetration of seismic waves into the reservoir carbonate. Most of the reservoir is less than a seismic wavelet below the top of carbonate, resulting in most of the reservoir being in the seismic zone dominated by the shale-limestone contact. (2) Variations in limestone and dolomite give seismic impedance responses that are similar to variations in porosity within a pure limestone or pure dolomite. Because limestone-dolomite variations and porosity variations are scattered throughout the reservoir, it is very difficult to separate the two.

DEPOSITIONAL FACIES AND STRATIGRAPHY

The main reservoir at Reinecke field is an upper Pennsylvanian to lowest Permian carbonate, which is overlain and sealed by siliciclastic and carbonate mud of Early Permian age. The reservoir at Reinecke field is mainly limestone (70%) with substantial dolomite (25%) and minor amounts of mixed limestone-dolomite (20–80% dolomite; 5% of reservoir). Shale is present in trace amounts (<1%). Most of the Reinecke reservoir occurs in three main sequences (between sequence boundaries 100 and 400) that are stratified and typically 15–30 m thick (Figures 8, 9). Those three sequences are Virgilian (late Pennsylvanian) in age (G. Wilde, personal communication, 1997), and hence part of the Cisco Formation. Cores show fractures, root traces, caliches, and brecciation immediately below sequence boundaries 100, 200, 300, and 400. Small amounts of oil also occur in Wolfcampian (Lower Permian) pinnacles (above the 100 marker) and in the Canyon Formation (Missourian; below the 400 marker, but above the oil-water contact).

Deposition of these sequences occurred during "ice-house" times, and hence were subjected to high-amplitude sea level fluctuations (Veevers and Powell, 1987). Sequences typically have a thin basal part, a thick middle part, and upper part with variable thickness (Figures 8, 9). Basal parts generally contain thin shale and/or bioclastic packstone-grainstone. The middle of sequences is dominated by fossiliferous wackestone (commonly phylloid algal wackestone) and phylloid algal boundstone. Grainstone is common in the upper part of sequences (Figures 8, 9). Crinoidal grainstone occurs in the upper part of the two uppermost Pennsylvanian sequences. Oolitic grainstone is also common in the upper part of the two lower sequences (Figure 9). Nine main carbonate depositional facies are present, and boundstone comprises only 16% of this "reefal" buildup (Table 1).

Carbonates above horizon 100 are highly variable in thickness and have a sharp upper contact with the overlying shale. Wells near the edge of the

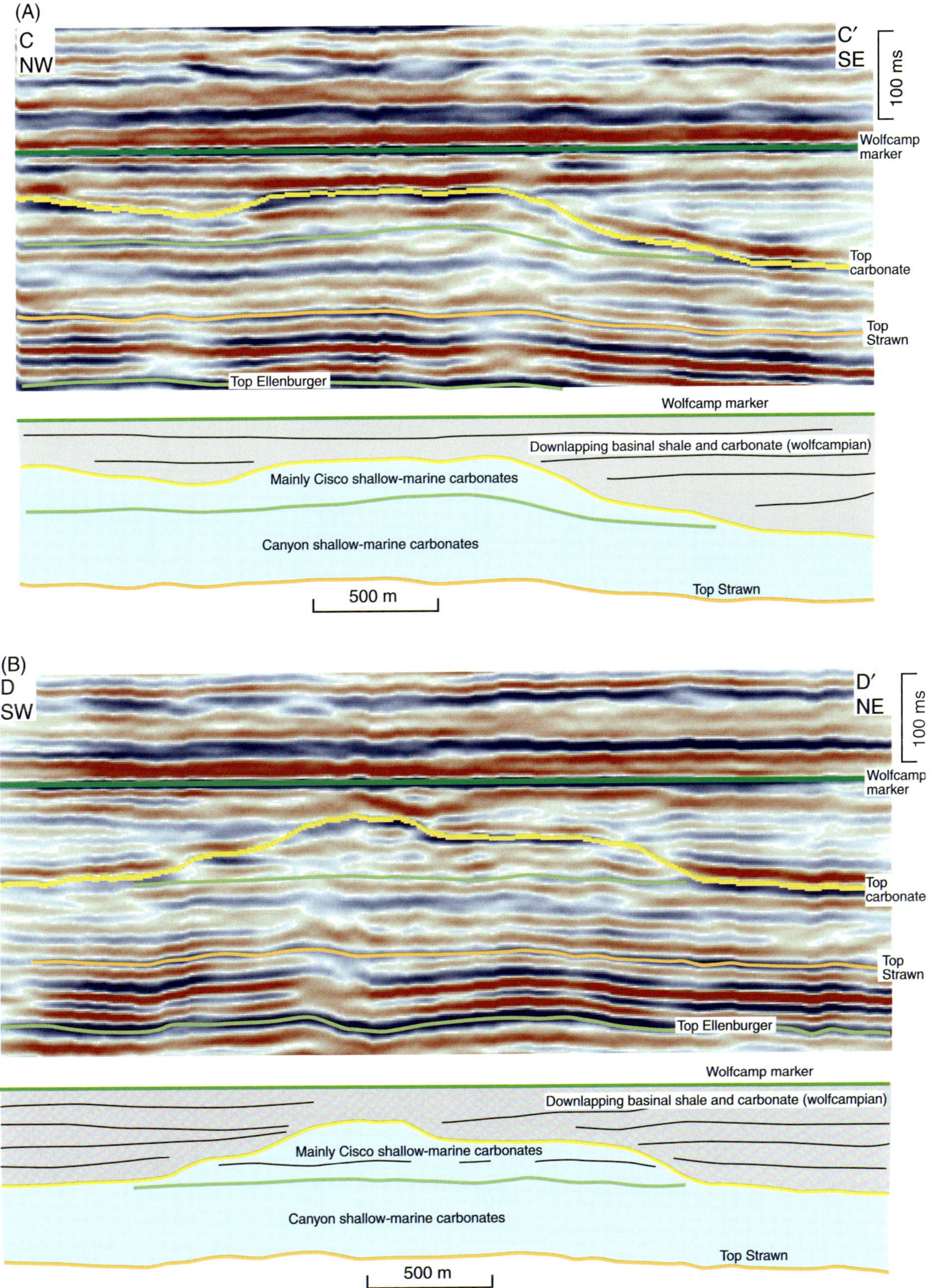

Figure 7. (A) Seismic line 237 (CC′) and (B) trace 243 (DD′) flattened on a Wolfcamp reflector in the basinal sediments overlying the Reinecke reef carbonates. Reflectors apparently onlapping the Reinecke buildup are actually basinal strata down- or baselapping against the Reinecke carbonate buildup from the east (right) with the contemporaneous shelf margin 50–100 km farther to the east (right). See Figures 3 and 5B for location of seismic line.

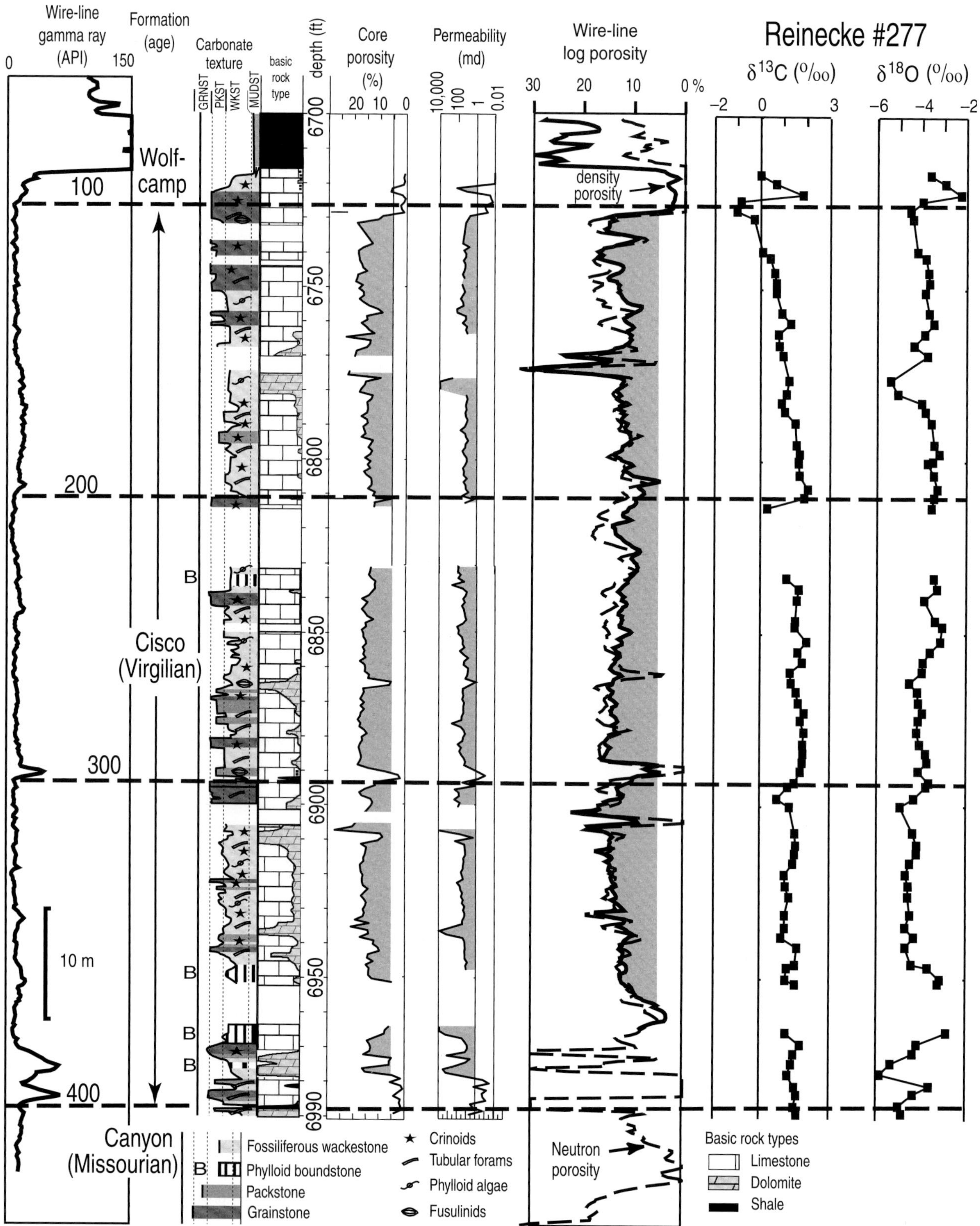

FIGURE 8. Core and wire-line logs for a "typical" south dome well, #277. Horizons 100, 200, 300, and 400 have paleosoil-related features below them and are interpreted as sequence boundaries. Thin shales are immediately above sequence boundaries 300 and 400. Stable isotope data are from whole rock samples. GRNST = grainstone; PKST = packstone; WKST = wackestone; MUDST = mudstone.

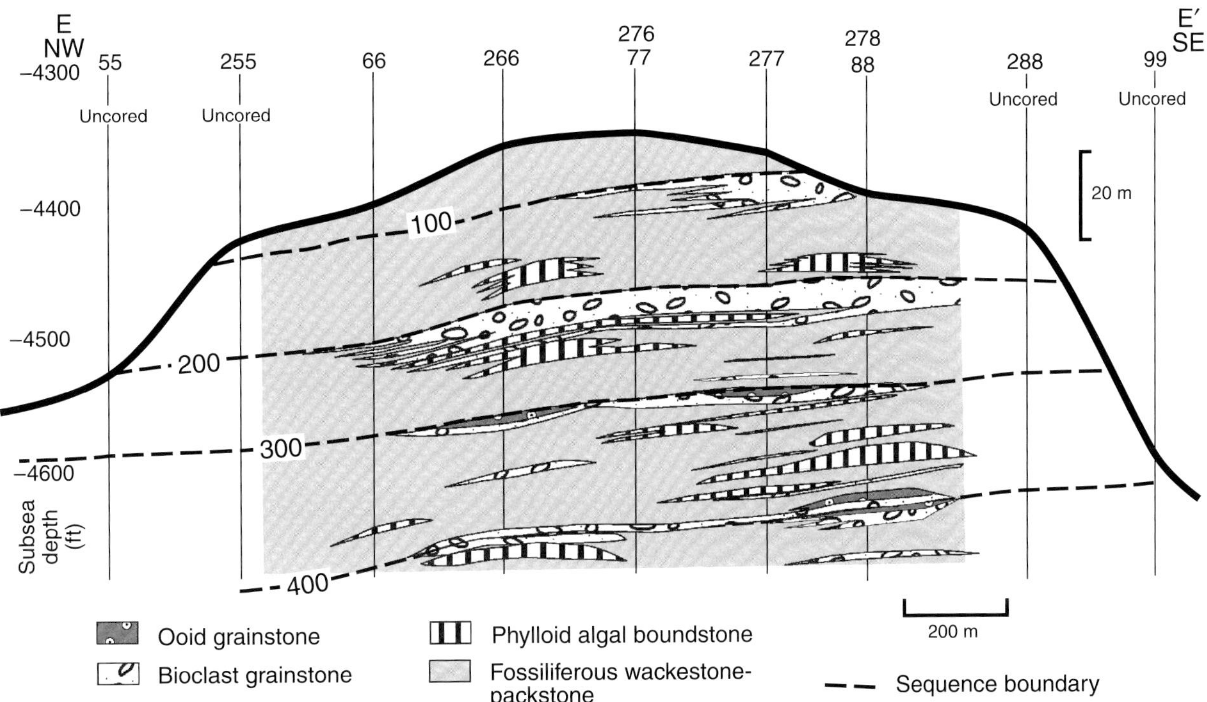

FIGURE 9. Structural cross section showing facies and stratigraphy across the south dome of Reinecke field. Facies are from core data. Horizons 100, 200, 300, and 400 are sequence boundaries with soil-related features. See Figure 5A for location of cross section.

south dome have none of this Wolfcamp interval present (Figure 9). Thin accumulations of fossiliferous packstone and grainstone occur immediately above horizon 100, and fossiliferous wackestone dominates the rest of the interval. Crinoids and phylloid algae are common. Many fractures and vugs filled with coarse, fibrous to prismatic cement are also present above horizon 100. In the #277 well, the uppermost Wolfcamp carbonate is a dark lime mudstone that is overlain by a thin spiculitic chert and then dark shale that acts as the seal for the Reinecke reservoir (Figure 8).

INTERPRETATION OF DEPOSITIONAL HISTORY

Although the south dome of Reinecke field has a mounded or "reefal" morphology, it is composed of

Table 1. Reinecke Depositional Facies— Limestone.

Facies	*Average porosity (%)*	*Average horizontal permeability (md)*	*Average vertical permeability (md)*	*Percentage of limestone facies*
1. Ooid grainstone	9.3	20.4		4
2. Bioclastic (crinoidal) grainstone	12.3	157.7	6.9	22
3. Packstone	11.8	199.6	6.0	9
4. Phylloid boundstone	12.0	690.3	82.6	13
5. Bryozoan boundstone	11.2	57.7		3
6. Phylloid wackestone-packstone	11.2	196.4	11.1	24
7. Bryozoan wackestone	12.9	20.0		2
8. Fossiliferous wackestone-packstone	10.5	29.5	12.1	22
9. Mudstone	1.4	0.79	0.01	1
Total limestone— average	11.2	165.8	11.0	100

Eight hundred and sixty-five feet of limestone were cored and analyzed.

FIGURE 10. Depositional model for a typical sequence.

Depositional model for Reinecke

1. Initial transgression and flooding of platform: Grainstones deposited

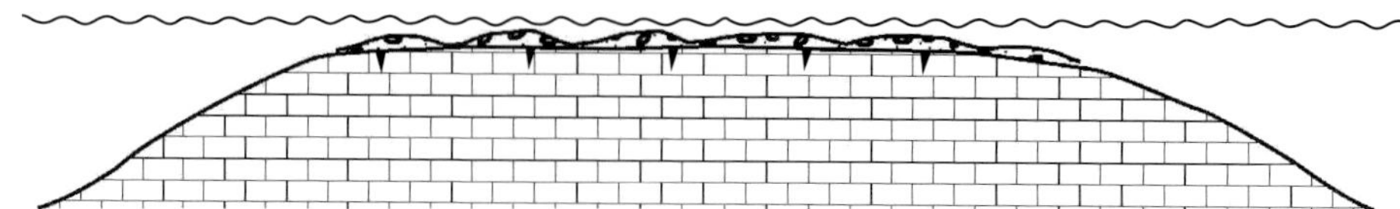

2. High sea level: subtidal environment (5–20 m deep) dominates; Fossiliferous wackestone, packstones, and phylloid algal boundstones deposited

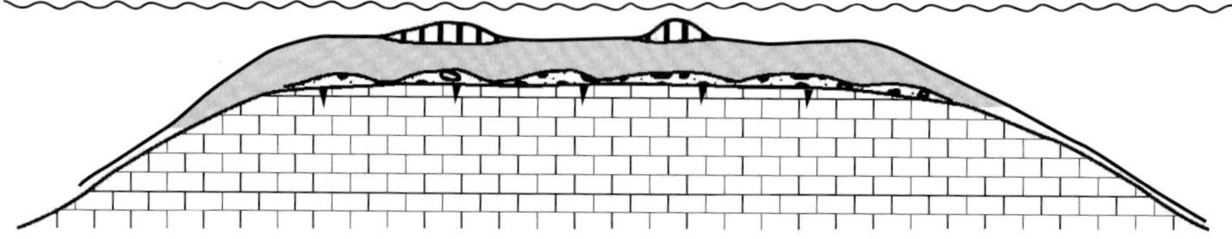

3. Shallowing of the platform: Bioclastic grainstones are deposited

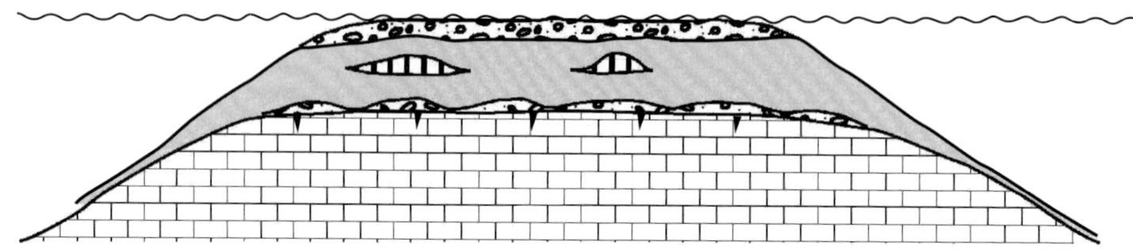

4. Sea level drops; Platform is exposed; Meteoric diagenesis rearranges pore structure and stabilizes mineralogy

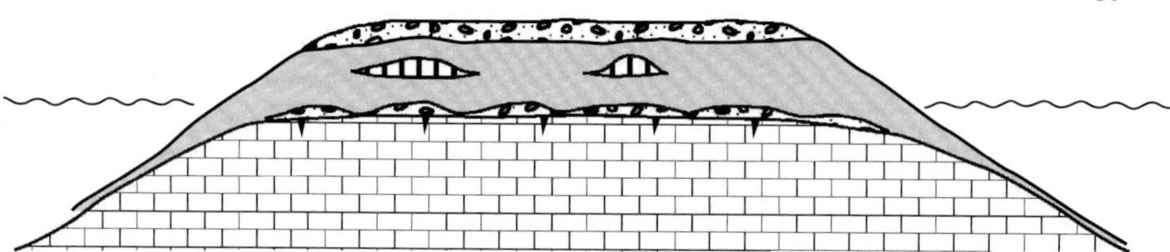

fairly stratiform sequences that apparently formed in response to variations in relative and probably eustatic sea level. Sequence boundaries are identified by soil- and exposure-related features (fractures, root traces, caliches, and brecciation) observed in core. Abrupt decreases in bulk-rock stable carbon and oxygen-isotope ratios below those surfaces support subaerial exposure (Figure 8). A depositional model for typical sequences is shown in Figure 10. The uppermost carbonate (lower Wolfcampian; above "100" marker) in the Reinecke buildup contains a no petrographic, cathodoluminescent, or stable isotopic evidence of freshwater diagenesis or subaerial exposure (J.A.D. Dickson, personal communication, 1999), suggesting that the last stage of erosion was not associated with subaerial exposure.

Relief on the Reinecke carbonate structure is probably the result of differential carbonate growth, subaerial karstification, and deep-marine erosion following deposition and drowning of the Reinecke buildup. As the 3-D seismic data show (Figures 3, 6, 7), the upper surface of the Reinecke carbonate buildup is very irregular. Several sequences (Wolfcampian and parts of two Pennsylvanian) are truncated by that surface (Figures 3, 6, 7), suggesting that the top of carbonate is, at least in part, erosional. Strata overlying Reinecke and other Horseshoe Atoll fields are deep-water shales and carbonate sediment gravity flows (Vest, 1970; Galloway et al., 1983). The apparent seismic onlap of basinal strata (Figures 6, 7) is really downlap or baselap of the distal toes of Wolfcamp sequences whose shelf margin is 50–100 km to the east (Figures 1, 2). The Reinecke area remained in a deep-marine environment from drowning in the earliest Wolfcampian (earliest Permian) through much of the Leonardian (middle Permian) (Figure 2). During that time, lowstand clastic and carbonate slope strata prograded from the east and north and largely filled the Midland Basin. Final filling of the basin occurred during deposition of the San Andres Formation (upper Leonardian and lower Guadalupian; Figure 2). Shallow-marine carbonate and evaporite deposition dominated the Reinecke area during the rest of the Guadalupian and Ochoan (middle and Late Permian; above San Andres Formation).

DISTRIBUTION OF POROSITY, DOLOMITE, AND PERMEABILITY

Most porosity at Reinecke occurs in sequences subjected to subaerial exposure and meteoric diagenesis. Meteoric diagenesis associated with subaerial

Table 2. Reinecke Depositional Facies— Dolomite.

Facies	*Average porosity (%)*	*Average horizontal permeability (md)*	*Average vertical permeability (md)*	*Percentage of dolomite facies*
1. Ooid grainstone				0
2. Bioclastic (crinoidal) grainstone	8.0	360.7	322	4.5
3. Packstone	11.8	1780.8	30.9	4.9
4. Phylloid boundstone	12.0	140.8		0.8
5. Bryozoan boundstone				0
6. Phylloid wackestone-packstone	7.9	848.4	85.0	42.3
7. Bryozoan wackestone	4.9	3.8		3.7
8. Fossiliferous wackestone-packstone	7.7	846.7	1.9	41.8
9. Mudstone	2.0	1.5	<0.01	2.0
Total dolomite— average	8.28	894	334	100
Total— all (limestone, dolomite, mixtures of both)	10.5	323	131	

Three hundred and twenty feet of dolomite were cored and analyzed. The original depositional facies of some dolomites could not be determined. One thousand and two hundred and forty feet of limestone, dolomite, and mixed limestone-dolomite were cored.

exposure apparently helped preserve intergranular porosity and microporosity in micrite by creating a lithified framework that resisted compaction during subsequent burial. Porosity is widespread in carbonates below sequence boundary 100 (Figure 8). Most carbonates above sequence boundary 100 are not porous. Calcite cements are volumetrically minor throughout the Reinecke buildup. Many pores (molds, vugs, fractures) in the Reinecke reservoir were created by leaching of biotic constituents and dolomitization (Crawford et al., 1984).

Excluding lime mudstones, depositional facies that are still limestone have similar average porosities (9.3–12.9%; Table 1). Permeability is highly variable depending on pore types within facies. Phylloid algal boundstone generally has the highest permeability. Lime wackestone and packstone with intercrystalline microporosity have lower average permeabilities (Table 1). Bioclastic grainstone with significant intergranular porosity has moderate permeability.

Dolomite is a later diagenetic phase that replaced parts of many different depositional facies during moderately deep burial (J.A.D. Dickson, personal communication, 1989). Dolomite has generally lower porosity (average of 8.3%) and higher permeability (average horizontal and vertical permeabilities of 894 and 334 md, respectively) than limestones (average porosity of 11.2%; average horizontal and vertical permeabilities of 165 and 11 md, respectively) (Table 2; Figure 10). Dolomites with substantial vuggy and coarse intercrystalline porosity have very high permeability (>500 md). Shale is present in small amounts (<1%); however, it is locally important as a barrier to vertical fluid flow.

Although porosity and permeability are variable, reservoir-grade porosity (>4%) and permeability (>1 md) are fairly continuous laterally and vertically. No fieldwide barriers to vertical or horizontal flow have been recognized in the main part of the reservoir; however, dense carbonate mudstone and shale have formed local permeability barriers or baffles. Lack of fieldwide permeability barriers is, in part, because of a lack of laterally extensive tidal flat wackestone-packstones or evaporites that effectively stratify fluid flow in many shelfal carbonates. High-permeability dolomites cut across sequences and greatly enhance vertical fluid flow. Therefore, reservoir fluids flow vertically and laterally through the Reinecke reservoir. As Figure 11 and Tables 1 and 2 show, permeability is related to reservoir rock type (limestone vs. dolomite). The 3-D distribution of porosity (Figure 12A) and dolomite

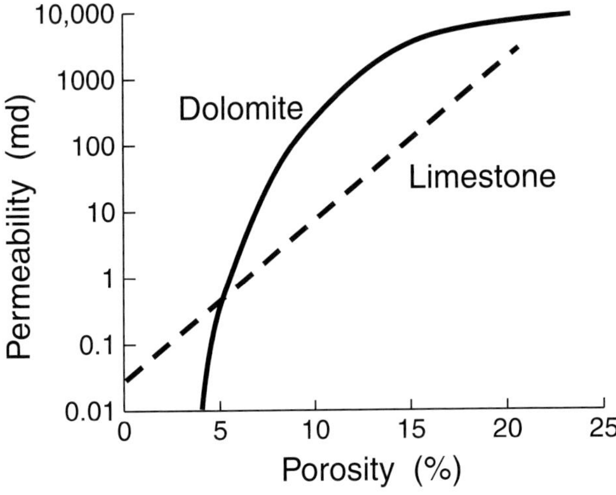

FIGURE 11. Regressions of porosity vs. permeability for limestones and dolomites in Reinecke field.

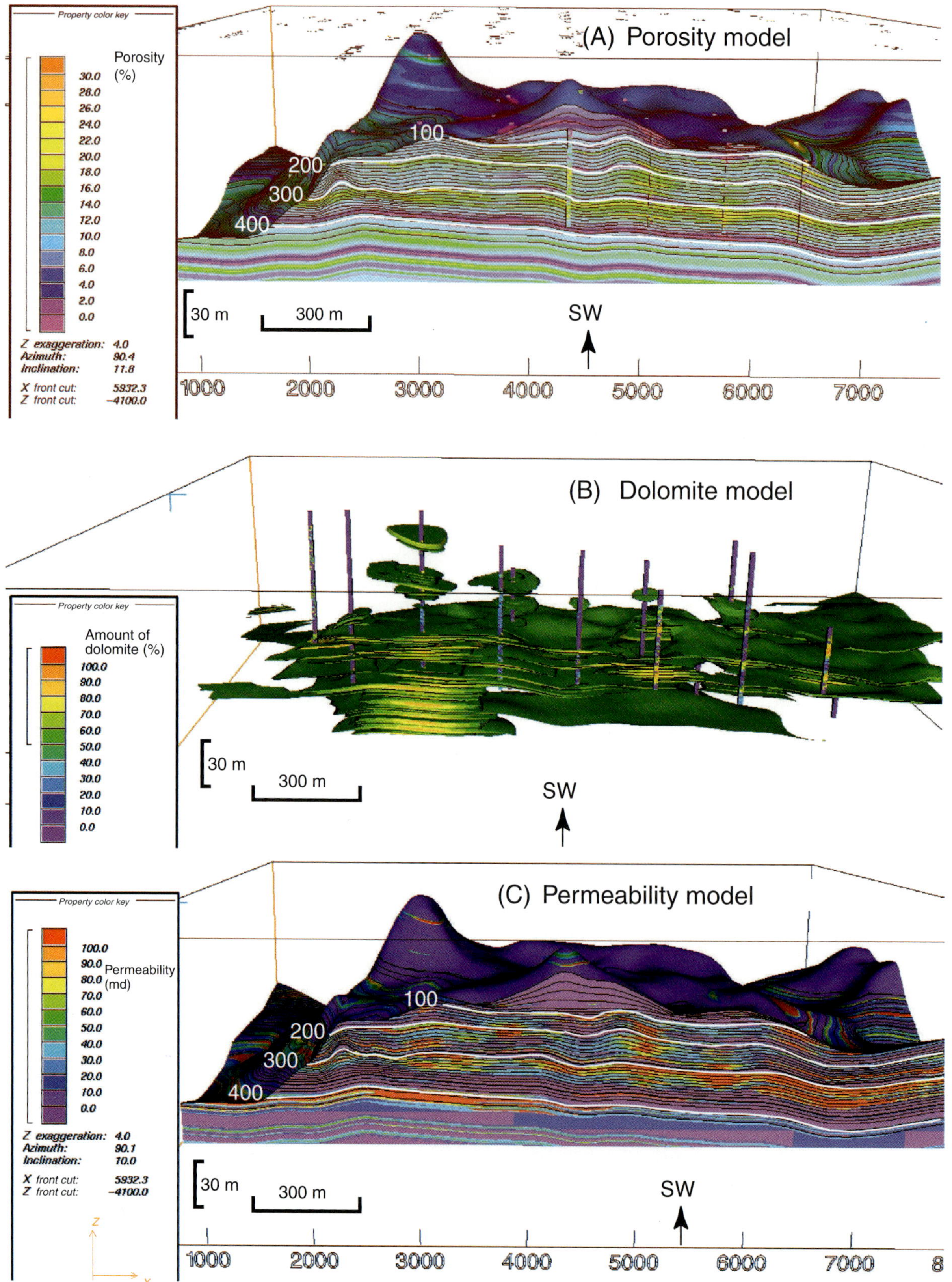

FIGURE 12. Reservoir characteristics of Reinecke's south dome looking from the northeast to the southwest (at the top). Vertical exaggerations are 4×. (A) Model for porosity distribution across the south dome of Reinecke field. (B) Model for dolomite distribution across the south dome of Reinecke field. Values are in percent dolomite. (C) Model for permeability distribution across the south dome of Reinecke field. Values are in millidarcy.

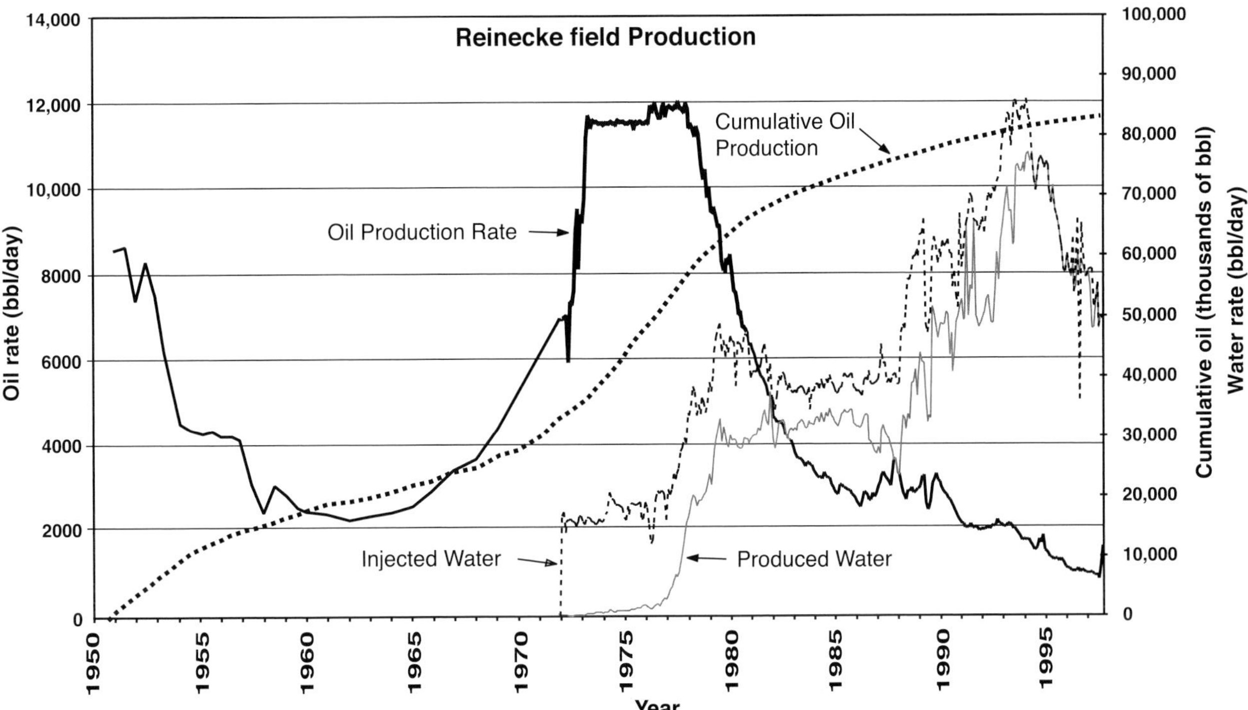

FIGURE 13. Production history of Reinecke field.

(Figure 12B) were extrapolated from well data following 3-D seismic horizons. Permeability at each location (Figure 12C) was calculated using porosity and porosity vs. permeability transforms for each lithology.

PRODUCTION HISTORY

Reinecke field was discovered in 1950 at depths of approximately 6700 ft (−4300 ft, 1310 m subsea). The oil was 42° API with a gas-oil ratio of 1266 ft^3/bbl. Coning of bottom water was recognized as a problem early in the development of the field. Consequently, most wells were completed open hole after penetrating only a few feet of porosity in the upper part of the reservoir. Original 40-ac spaced wells in Reinecke field flowed for almost 20 yr under a natural water and solution gas drive. Primary production rates at Reinecke were limited by the Texas Railroad Commission. Reinecke field produced at its top allowable rate until 1968 (Figure 13). Between 1950 and 1970, reservoir pressure fell from 3162 to 1984 psia, just below the reservoir's bubble point of 2000 psia. Water injection into the underlying aquifer began in the late 1960s as part of a pressure-maintenance program. The field was unitized in 1972, and a fieldwide pressure-maintenance program was started. Production rates and reservoir pressure climbed as water was injected below the oil-water contact (Figure 13). In the early 1970s, the Texas Railroad Commission raised allowable production rates, and production rose to approximately 11,000 bbl/day. In 1977, pumps were put on producing wells and water production increased significantly. Since 1980, water production has been increasing as oil production decreased. Drilling of 20-ac infill wells in the middle to late 1980s provided a brief increase in oil production (Figure 13). Primary, secondary, and infill production have produced more than 83 million bbl or 46% of the OOIP. In the south dome, approximately 52% of the OOIP has been produced.

This reservoir shows a classic "reefal" reservoir response to a weak bottom-water drive that was artificially enhanced. Figure 14A shows simulated hydrocarbon saturation as of 1970 with a relatively horizontal oil-water contact that had been moving up in response to production from the top of the reservoir. Between 1970 and 1977 (Figure 14B), the artificially enhanced bottom-water drive pushed much of the mobile oil to the well perforations at the top of the reservoir. By 1996 (Figure 14C), mobile water was to the top of reservoir resulting in oil being less than 2% of the fluid produced (i.e., >98% water cut).

As the model in Figure 14C shows, much residual oil (oil that will not naturally flow and is trapped in pores) remains in the Reinecke reservoir. A gravity-stable crestal CO_2 flood is being implemented at Reinecke field to recover substantial amounts of that residual oil as well as additional unswept mobile oil. CO_2 is being injected into the top of the reservoir in five wells

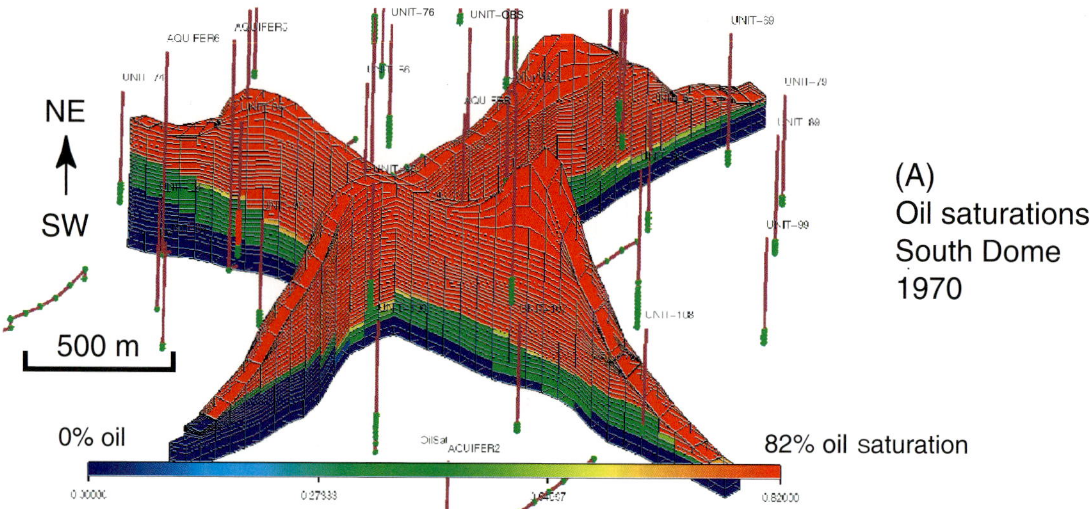
NE
SW
500 m
0% oil
82% oil saturation
(A)
Oil saturations
South Dome
1970

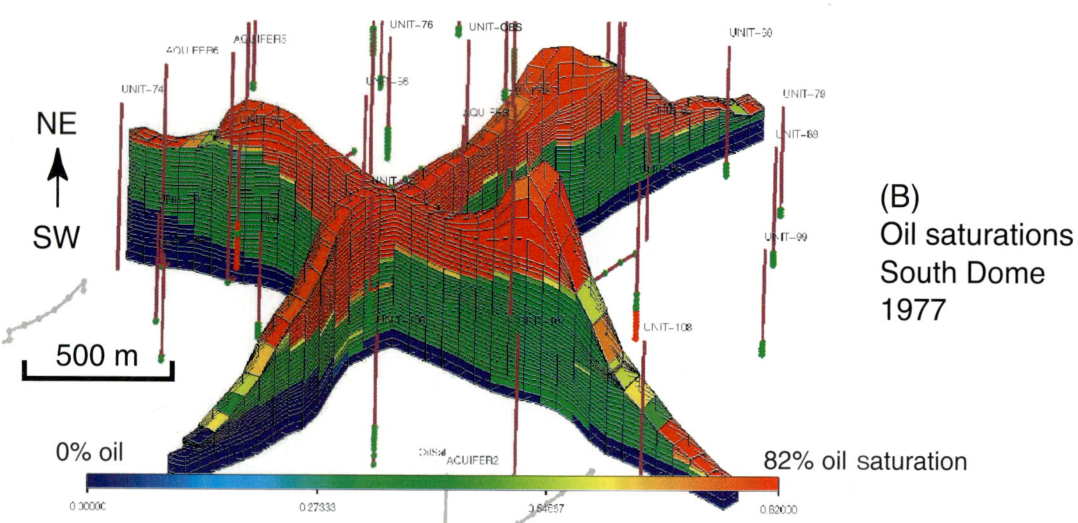
NE
SW
500 m
0% oil
82% oil saturation
(B)
Oil saturations
South Dome
1977

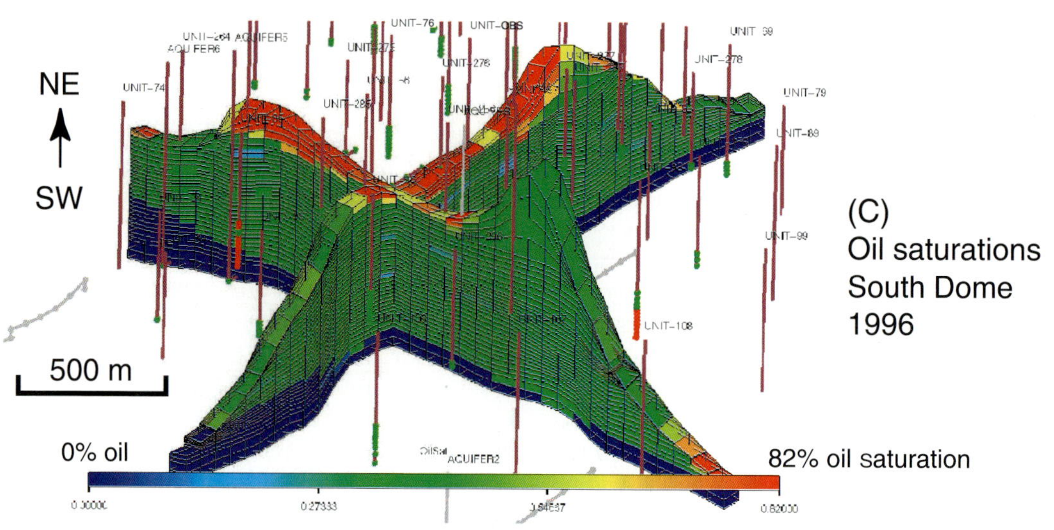
NE
SW
500 m
0% oil
82% oil saturation
(C)
Oil saturations
South Dome
1996

FIGURE 14. Reservoir simulation of the south dome showing oil saturations as of (A) 1970, (B) 1977, and 1996. Views are from the southwest to the northeast. Oil saturations are shown as a fraction of porosity. Original water saturation in the reservoir was approximately 20% (0.2) of pore space. Vertical relief is approximately 470 ft (143 m), and the south dome shown is approximately 1 mi (1.6 km) across. Between 1950 and 1970, water had moved up from the original oil-water contact (top of blue) and displaced mobile oil in the lower three to five layers (green). Water moved up through the reservoir in a fairly uniform manner displacing mobile oil between 1970 and 1977. As the water front approached the wellbores at the top of the reservoir (B–C), water began to "cone," resulting in a very heterogeneous distribution of mobile oil. Significant amounts of residual oil 25–40% of OOIP remain in the reservoir, and that is the primary target for the crestal CO_2 flood.

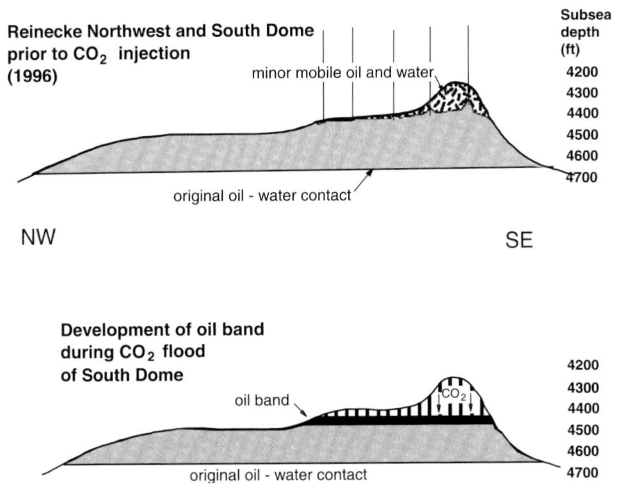

FIGURE 15. Schematic diagrams of the crestal CO_2 flood.

in the south dome (Figure 15). The overlying basinal shales should prevent CO_2 from leaking up and out of the reservoir. CO_2 should preferentially fill the highest parts of the reservoir and displace water and oil downward. A relatively horizontal oil bank is predicted to form and move downward through the reservoir as CO_2 fills the uppermost pores in the reservoir. Water injection must continue below the oil-water contact to maintain aquifer pressure. The remaining producing wells were deepened to near the original oil-water contact. Those wells will recover oil as an oil bank moves down through the reservoir below the CO_2 cap (Figure 15). This plan is feasible because of the excellent vertical and lateral continuity of porosity and permeability in the Reinecke reservoir. High-permeability streaks and lack of stratigraphic confinement of permeability within the reservoir would make a pattern flood very inefficient.

CONCLUSIONS

The reservoir at Reinecke field is a carbonate buildup that was deposited during the late Pennsylvanian and earliest Permian. Three-dimensional seismic surveys show its morphology in detail that would be impossible with well control alone. The final geometry of Reinecke's reservoir was the product of localized carbonate growth, karstification, and deep-marine erosion after the buildup was drowned. Deep marine erosion was apparently responsible for much of the very irregular top of the Reinecke carbonate. Although reefal in its morphology, the Reinecke reservoir is composed of stratiform sequences with only minor boundstone. Porosity and permeability are continuous through the Reinecke reservoir, and hence the south dome acts as a single container. Reinecke production is typical of "reefal"-type reservoirs with water from an underlying aquifer pushing oil up into perforations at the top of the reservoir. This mechanism was so efficient that 50% of the OOIP was produced by primary recovery and injection of water below the oil-water contact. A crestal CO_2 flood is currently underway and is feasible because of Reinecke's excellent reservoir continuity. The main value of the 3-D seismic surveys was to image the gross reservoir geometry in detail, which was essential for accurate volumetrics, successful reservoir simulation, and design of the crestal CO_2 flood.

ACKNOWLEDGMENTS

This chapter describes part of the results of a joint Unocal-Japan National Oil Company reservoir characterization project. Many people from both companies helped in this work, including Merle Steckel, Brian Ball, Stan Frost, John Gogas, Phil Johnston, and Tim Anderson. Constructive reviews by Steve Bachtel, Charlie Kerans, and Jose Luis Masaferro greatly improved this manuscript. We thank Japan National Oil Company and Unocal for the permission to publish this chapter.

REFERENCES CITED

Crawford, G. A., G. E. Moore, and W. Simpson, 1984, Depositional and diagenetic controls on reservoir development in a Pennsylvanian algal complex: Reinecke field, Horseshoe Atoll, west Texas: West Texas Geological Society Transactions, Southwest Section AAPG, West Texas Geological Society Publication 84-78, p. 81–90.

Galloway, W. E., T. E. Ewing, C. M. Garrett, N. Tyler, and

D. G. Bebout, 1983, Atlas of major Texas oil reservoirs: Texas Bureau of Economic Geology, 139 p.

Saller, A. H., A. W. Walden, S. Robertson, M. Steckel, J. Schwab, H. Hagiwara, and S. Mizohata, 1999, Reservoir characterization of a reefal carbonate for a crestal CO_2 flood, Reinecke field, west Texas, *in* T. F. Hentz, ed., Advanced reservoir characterization for the 21st century: Gulf Coast Section, SEPM Foundation, Nineteenth Annual Research Conference, p. 259–268.

Veevers, J. J., and C. McA. Powell, 1987, Late Paleozoic glacial episodes in Gondwanaland reflected in transgressive-regressive depositional systems in Euramerica: Geological Society of America Bulletin, v. 98, p. 475–487.

Vest, E. L., 1970, Oil fields of Pennsylvanian–Permian, Horseshoe Atoll, west Texas, *in* M. T. Halbouty, ed., Geology of giant petroleum fields: AAPG Memoir 14, p. 185–203.

Melville, P., O. Al Jeelani, S. Al Menhali, and J. Grötsch, 2004, Three-dimensional seismic analysis in the characterization of a giant carbonate field, onshore Abu Dhabi, United Arab Emirates, *in* Seismic imaging of carbonate reservoirs and systems: AAPG Memoir 81, p. 123–148.

Three-Dimensional Seismic Analysis in the Characterization of a Giant Carbonate Field, Onshore Abu Dhabi, United Arab Emirates

Peter Melville[1]
Abu Dhabi Company for Onshore Oil Operations (ADCO), Abu Dhabi, United Arab Emirates

Omar Al Jeelani
Abu Dhabi Company for Onshore Oil Operations (ADCO), Abu Dhabi, United Arab Emirates

Saeed Al Menhali[2]
Abu Dhabi Company for Onshore Oil Operations (ADCO), Abu Dhabi, United Arab Emirates

Jürgen Grötsch[3]
Abu Dhabi Company for Onshore Oil Operations (ADCO), Abu Dhabi, United Arab Emirates

ABSTRACT

A recent three-dimensional (3-D) seismic data set over an onshore oil field in Abu Dhabi, United Arab Emirates, is compared to a geologic model based on more than 200 wells. This chapter analyzes the seismic character of Upper Thamama Group (Early Cretaceous, Barremian–Aptian) carbonate reservoirs and shows how, even after more than 25 years of drilling and production, 3-D seismic data have improved the understanding of the field. A discussion of acquisition and processing techniques shows the iterative improvements in the seismic data quality, particularly in multiple suppression. The seismic provides an areal view of the structure and associated faulting. Seismic attributes were used to look at the prediction of rock properties and reservoir character.

The extensive well data in the geologic model have allowed widespread comparison with the seismic predictions, and examples are highlighted, which demonstrate the similarities and differences between the two data sets.

The chapter demonstrates that, even with a high density of well data, 3-D seismic can be a complementary data set. For this field, the combination of the two data sets

[1]*Present address:* BP, Abu Dhabi, United Arab Emirates.
[2]*Present address:* Abu Dhabi National Oil Company (ADNOC), Abu Dhabi, United Arab Emirates.
[3]*Present address:* Shell Abu Dhabi BV, Abu Dhabi, United Arab Emirates.

has allowed a better understanding of the complex wrench faulting and the behavior of the Thamama reservoir character away from the wells. Important new paleogeographic information has also been obtained from the combination of well data with the depositional geometries and patterns seen on the seismic data at Bab (Aptian age) and Asab (Late Jurassic) levels.

INTRODUCTION

The field discussed here is located in the onshore area of Abu Dhabi, United Arab Emirates. The Lower Cretaceous reservoir is an example of a layered carbonate system, characterized by cyclic alternations of high-energy skeletal-peloidal packstones-grainstones and low-energy skeletal wackestones-lime mudstones. Sweet, high-API crude oil was discovered in 1965 and further delineation wells subsequently found other hydrocarbon-bearing layers, including gas. This giant field came on stream in 1973 and is currently producing on plateau from more than 300 wells. The three-dimensional (3-D) seismic survey was shot between 1997 and 1999. The operating oil company's guidelines on publication prevent the inclusion of some specific values and types of data. (to honour confidentiality of some of the technical information, this work cites published information as a source for further detail on the area).

The field is an elongated, faulted anticline with the longer axis trending roughly northeast-southwest. It is about 30 km long and 10 km wide, with a relief of more than 180 m. Reservoir dip is 3–5° on the western flank, 2–3° on the eastern flank, and 1–2° on the northern and southern ends. The field is one of a series of anticlines formed in the Late Cretaceous by the compressional effects of the collision of Oman with Arabia. Subsequent modification of the structure during the Tertiary has occurred because of the burial compaction and also the Zagros orogeny, which has regionally tilted the area down to the north-northeast.

GEOLOGIC SETTING OF THE FIELD

During the Lower Cretaceous period, Abu Dhabi was part of a broad carbonate ramp on the southern Tethyan margin with a local intrashelf basin. This regionally extensive carbonate system contains many of the giant oil fields of the Middle East (Figure 1). In this area, the Cretaceous is divided into three groups, of which the lowermost supersequence is the Thamama Group. The Thamama is subdivided into the Habshan, Lekhwair, Kharaib, and Shuaiba-Bab Formations. The upper part of the Thamama Group contains the alphabet reservoir zones. The main hydrocarbon accumulation discussed is found in Zones A, B, and C, which are in the Shuaiba and Kharaib Formations of Barremian to Aptian age. The Thamama Zone B is the major oil-bearing and -producing reservoir in the field. A regional stratigraphic column for the field is presented in Figure 2. The detailed stratigraphy of the area has been recently modified by Granier (2000), but this present chapter uses the terminology of older articles, such as those of Murris (1980), Harris et al. (1984), and Alsharhan and Nairn (1986). An outcrop analog study has recently compared the subsurface rocks with those found in the nearby Hajar Mountains. This confirms the extensive nature of the Upper Thamama and has allowed a detailed study of the geologic sequences and their evolution (Buchem et al., 2002).

A detailed geologic model has been built for the field with 11.5 million cells at a 1-ft (0.3048-m) vertical increment and 175-m cell size. The layer correlation is based primarily on gamma-ray and porosity logs of more than 200 wells, with resistivity and finer-scale logs used where available. Well-log and core data are used for facies, porosity, and permeability data and provides the model with detailed depositional facies, sequence stratigraphy, and diagenetic understanding.

Depositional Facies

The overall depositional environment of the Thamama Group is a large carbonate ramp system on the Arabian shelf. The reservoir Zones A, B, and C, which are in the Shuaiba and Kharaib Formations of the Upper Thamama Group, comprise a cyclic sequence of shallow-marine carbonates deposited for hundreds of kilometers along the stable ramp. During the late Thamama, an intrashelf basin developed in this area; the Shuaiba shelf margin is shown on Figure 1. The sequences are predominantly carbonate with minimal siliciclastic material. Facies variations are caused by a broad regional subsidence and sea level cyclicity with related changes in depositional energy. In this field, predominantly subtidal carbonate facies associations are encountered in the porous reservoir Zones A, B, and C, which range from 18 to 58 m thick. The zones are separated by dense argillaceous limestones, A, B, and C dense layers, which range in thickness from 6

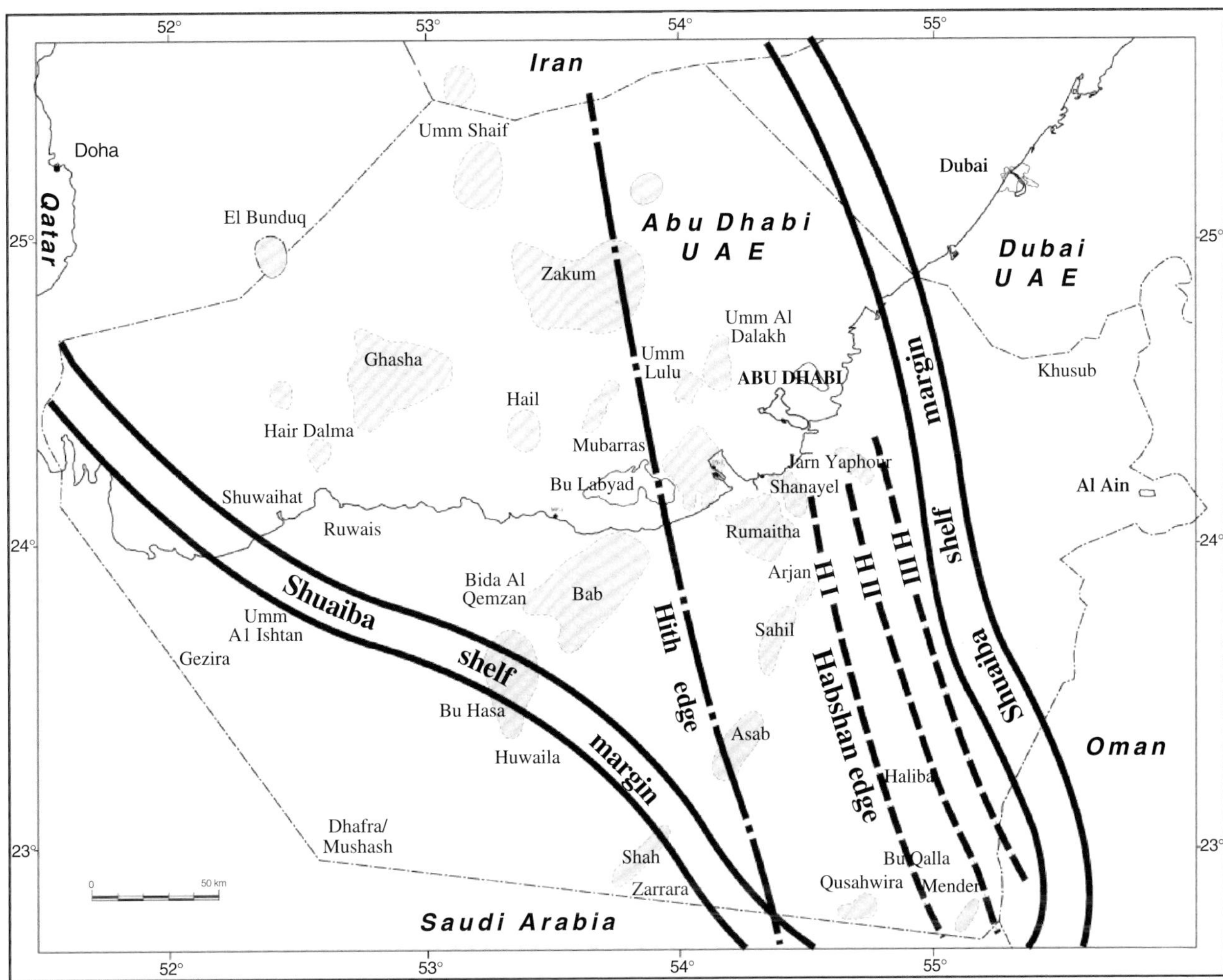

FIGURE 1. Location map of Abu Dhabi, United Arab Emirates, showing the major fields and the paleogeographic edges of the Hith (Jurassic), Habshan, and Shuaiba Formations (Lower Cretaceous).

to 14 m and give a blocky profile on the wire-line logs. Each dense zone lies immediately below its equivalent-named reservoir zone. The reservoir zones have been subdivided into layers based on sequence correlation, facies, and occurrence of stylolite-bearing dense intervals. Figure 3 is a field cross section showing the porosity of the reservoir in color with zones and subzones labeled. Six depositional lithofacies are recognized in Zone B based on their biofacies (Johnson and Budd, 1975). Zone B is subdivided into upper and lower units based on their depositional facies and reservoir properties, predominantly the change of permeability. Both units are coarsening-upward sequences. In this field, the B upper unit is typically grain supported whereas the B lower unit is lime mud supported (Alsharhan, 1993). The B upper unit consists of laterally heterogeneous bioclastic wackestones to grainstones with a peletoidal texture, as well as algal lumps, rudistid, and coarse bioclastic-bearing layers. The B lower unit consists of less-heterogeneous porous lime mudstones to wackestones with only moderate amounts of skeletal debris, which becomes more abundant toward the top of the unit (Grötsch et al., 1998). Similar depositional facies can be recognized in the Upper Thamama layers throughout the field. This suggests a rather "layer-cake" type of deposition in the reservoir Zones A, B, and C, comparable with other Thamama fields in Abu Dhabi.

Sequence-Stratigraphic Architecture

The Upper Thamama reservoir zones are built up from third-order depositional sequences (Vail et al., 1991). Published sequence-stratigraphic work is scarce and is incorporated into the recent volume by

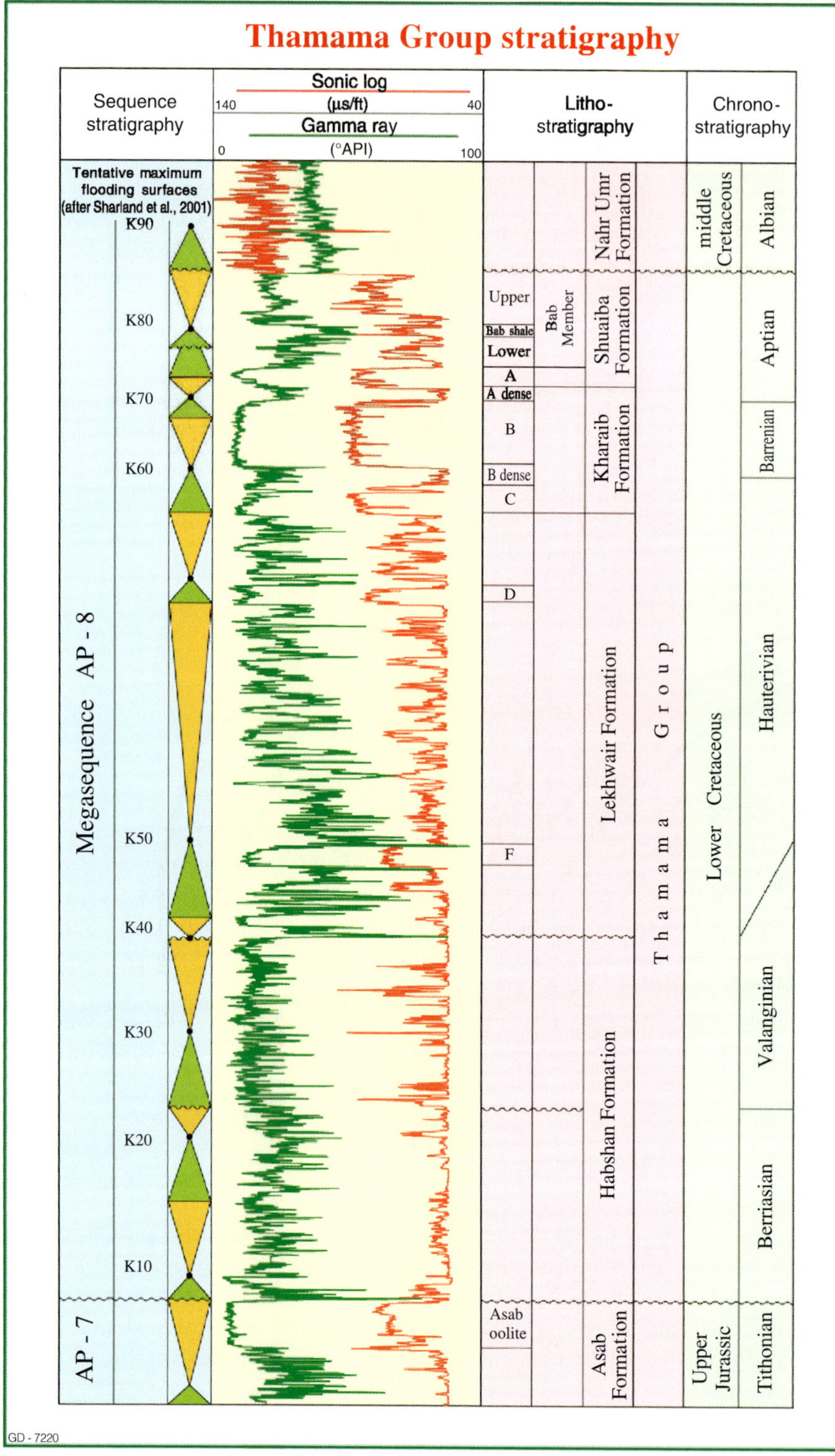

FIGURE 2. Lower Cretaceous Thamama Group stratigraphic column showing the litho-, chrono-, and sequence stratigraphy of the field.

Sharland et al. (2001); sequence boundaries using their definitions are shown on Figure 2. The third-order sequences of the Thamama A, B, and C zones are characterized by shallowing-upward trends, with a change from lower-energy deposits to higher-energy platform top deposits. This shallowing trend is mainly reflected as an overall upward increase in permeability within the cycles in accordance with the coarsening-upward facies trends. The third-order sequence can be subdivided into smaller series of shallowing-upward fourth-order cycles that can be subaerially exposed at their tops. These fourth-order cycles are made up of meter to submeter cycles interspersed with storm-deposited bioclastic streaks. The main dense layers between the reservoir zones represent the maximum flooding surfaces of the third-order cycles, the Zone A dense layer being the regional equivalent of the Hawar shale. Examples of the sequence technique applied to the Kharaib and Shuaiba of offshore fields in Abu Dhabi are shown in Azer and Toland (1993) and Boichard et al. (1994), where the TH II Zone offshore compares with the Thamama B Zone onshore.

Diagenesis

Although similar depositional facies can be recognized in the Upper Thamama all over the field, they can have significantly different properties of porosity and permeability because of differences in diagenetic overprinting, either early (cementation, local dolomitization) or middle to late (compaction, stylolitization). These processes are well described in Burchette and Britton (1985) for a nearby middle Cretaceous carbonate reservoir system.

Early diagenesis is related to paleotopography and consequent subaerial exposures, meteoric water and water table effects. These effects can develop at different timescales, i.e., controlled by third-, fourth-, or even fifth-order sequence effects. As an example, the main third-order shallowing trend of Zone B is reponsible

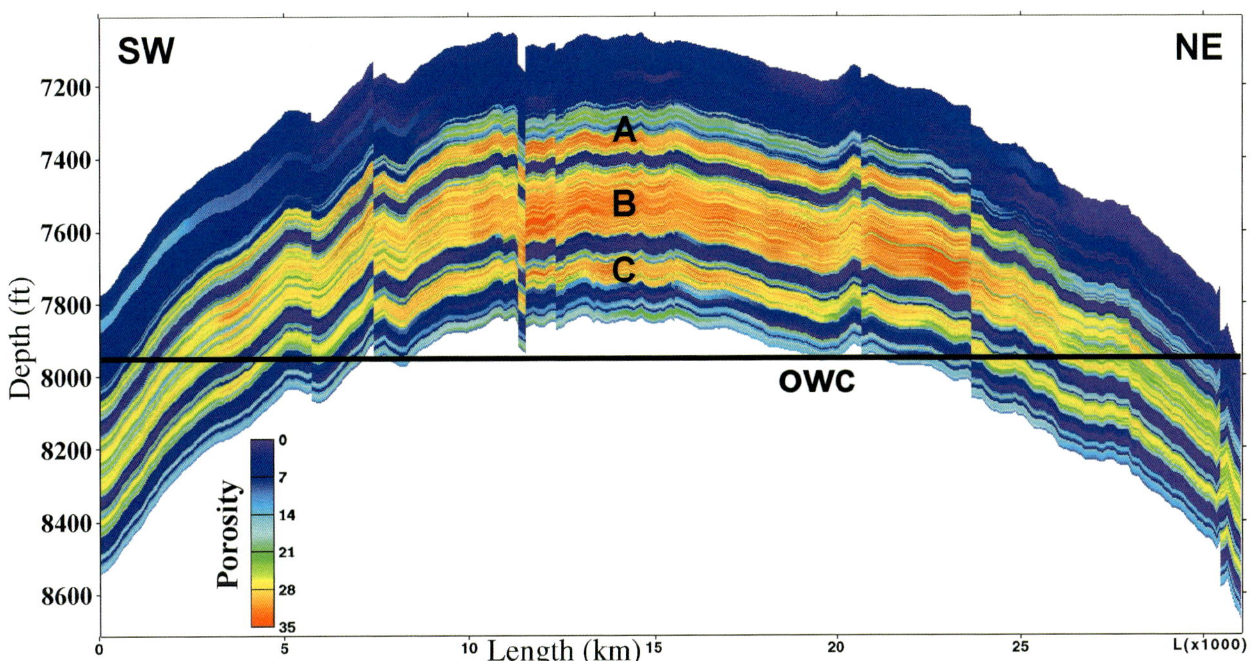

FIGURE 3. A cross section of the geologic model showing porosity of the reservoir zones. Note the decreasing porosity downflank with an increase of dense bodies in the middle of Zone B. OWC = oil-water contact.

for an intense leaching, followed by a pervasive cementation phase, leading to the development of a so-called dense layer, DB1. This thin fieldwide dense layer has been noted by Koepnick (1987) to be characterized by the absence of oil stain.

Middle to late diagenesis is much more related to burial compaction and is evidenced through clear trends of thickness and porosity reduction from crest to flank. These downflank diagenetic changes mostly occur where the fourth- and fifth-order maximum floodings have deposited slightly clay-enriched fine-grained sequences that are more favorable for pressure-solution processes within an homogeneous carbonate facies fabric. The resultant features of such compactional effects are the stylolites, forming most of the intra-reservoir dense layers, whose number and thickness increase downflank. The downflank distribution of the stylolitic dense layers has changed the original overall "layer-cake" properties of the reservoir toward a more depth-related property. This change in reservoir property is seen to affect thickness and porosity within the field that both decrease with increasing depth (Figure 4). The Late Cretaceous paleodepth gives the best fit of porosity to depth of burial (Grötsch et al., 2000).

Evidence suggest that the stylolitization, faulting, hydrocarbon migration, and trap formation are contemporaneous and Santonian to Campanian in age (Fiqa Formation). The distribution and shape of the thin stylolitic dense layers is somewhat irregular, with some of them extending continuously to the flank, whereas others are of limited areal extent. Comparison with the fault pattern derived from 3-D seismic shows that the faulting has had some effect, with some dense bodies ending at faults and others being localized along fault trends (Grötsch et al., 2000). This suggests that the main phase of burial diagenesis is contemporaneous with or postdates the faulting of Santonian to Campanian in age (Fiqa Formation).

Discussion of the stylolite cementation process in Koepnick (1987) documents the diagenesis of this Zone B and shows that most stylolites are oil stained and formed in the presence of oil. The marked increase of stylolite cements from crest to flank suggests that the diagenesis was progressively inhibited by the exclusion of water for solute transport as the oil-water contact (OWC) moved deeper. Hence, the decreasing porosity downflank would have been caused by the increased impact of diagenesis in the non-hydrocarbon-bearing areas during hydrocarbon infill of the structure. Paleo-reconstruction shows that trap formation started during the Turonian, with the majority of the present-day closure formed in the Santonian to Campanian. The timing for the hydrocarbon fill is also Santonian to Campanian, as suggested by source-rock generation studies. Hence, the diagenesis, which occurred during hydrocarbon fill, must be of Santonian to Campanian age.

In the adjacent field to the north, Oswald et al. (1995) show how a similar timing for the stylolitization has been established from paleodepth vs. porosity plots, paleo-OWC, and thermal history. Hawas and Takezaki (1994) show oil typing and generation timing for other United Arab Emirates fields.

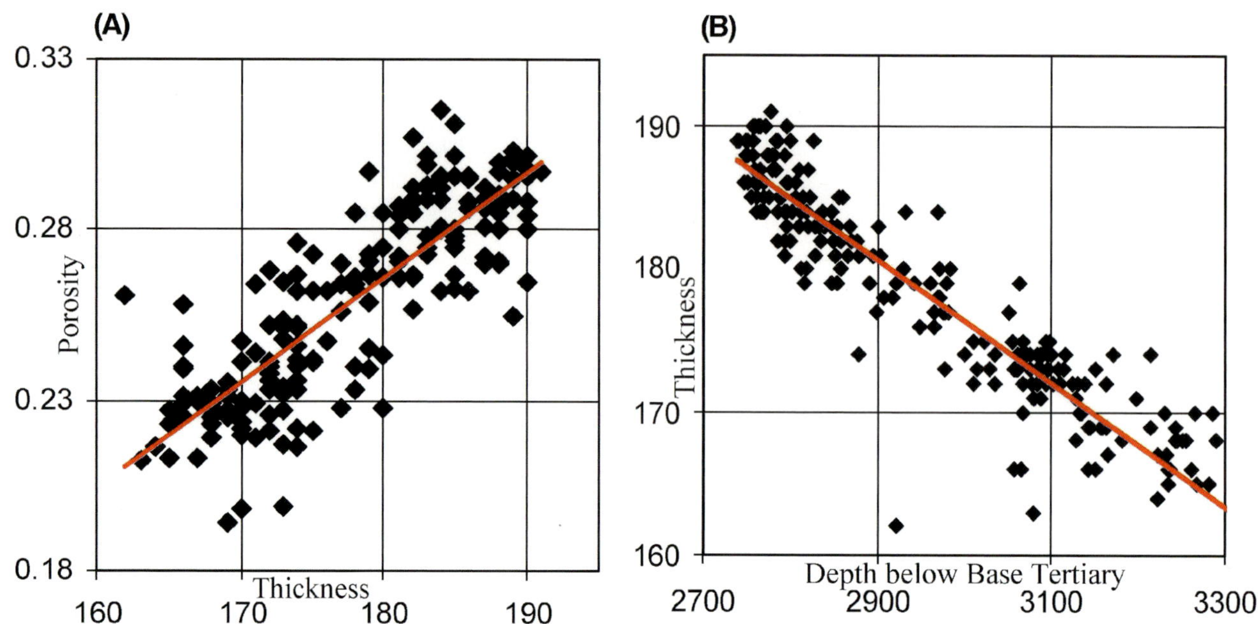

FIGURE 4. Two crossplots of well data. (A) This shows the good relationship between average Zone B porosity and thickness. (B) This shows that there is also a good relationship between Zone B thickness and depth. The depth shown here is below Base Tertiary because this gives a better fit than the present-day depth. Similarly, a better fit of porosity to depth of burial is found to be of Late Cretaceous age rather than present day (Grötsch, 2000).

TWO-DIMENSIONAL SEISMIC AND WELL-BASED STRUCTURE

The two-dimensional (2-D) seismic data over the field was shot mostly in the late 1950s and early 1960s. Mapping of dynamite seismic lines in 1962 showed a large four-way dip closed structure with an amplitude anomaly at the Thamama level (Figure 5). Only a few additional 2-D lines have been shot since the discovery well in 1965 and these provided a sparse grid with spacing ranging from 3 × 4 to 4 × 6 km (field size is 10 × 30 km).

The majority of the field appraisal and development maps was based on the well data, and in 1984, after the primary development drilling of more than 170 wells, the field was considered to be essentially unfaulted. A few seismic lines crossed the field in the early 1990s as part of a regional 2-D grid and these showed that faults could be imaged by the seismic data, although they could not be correlated between the sparse lines. Faults have been known to exist since 1966 when an early well encountered a 16-m missing section in the Bab Formation. Fault cutouts have subsequently been encountered throughout the Cretaceous section with throws of as much as 41 m. No fault cutouts are seen in the Simsima or in the overlying Tertiary sequences. With the horizontalization of wells to manage water influx (Hassan et al., 1990), several faults were encountered, offsetting the target layer at the base of Zone B. A study was made in 1996 of the vertical wells to identify fault cutouts, and fault orientations were identified in those horizontal wells with Formation MicroImager (FMI) data. These data were used to create a view of the expected fault pattern in the field and Figure 6 shows the best pre-3-D field map tying the 2-D seismic data, wells, and the locations of the known faults.

THREE-DIMENSIONAL SEISMIC

The 3-D survey acquisition commenced in 1997 with an initial 2-D test line, shot to confirm the parameter choice to be used for the full 3-D. The 3-D survey covered some 700 km^2, and its acquisition and processing were completed in 1999, 26 years after the first oil production.

The acquisition was carried out using 18 swaths, each 2 km wide. The receivers were laid out with an increment of 50 m in a northwest-southeast direction as 16 lines, 250 m apart and 6 km long. They were then moved along the in-line direction until being moved sideways by eight lines for the next swath. The shot lines were in a north-south direction at an angle to the receivers and also 250 m spacing. The vibrator points were every 56 m along the shot line in the central eight lines of the receiver swath. This provided a final 3-D trace spacing of 25 m with a nominal 96 fold and a cross-line direction along the long axis of

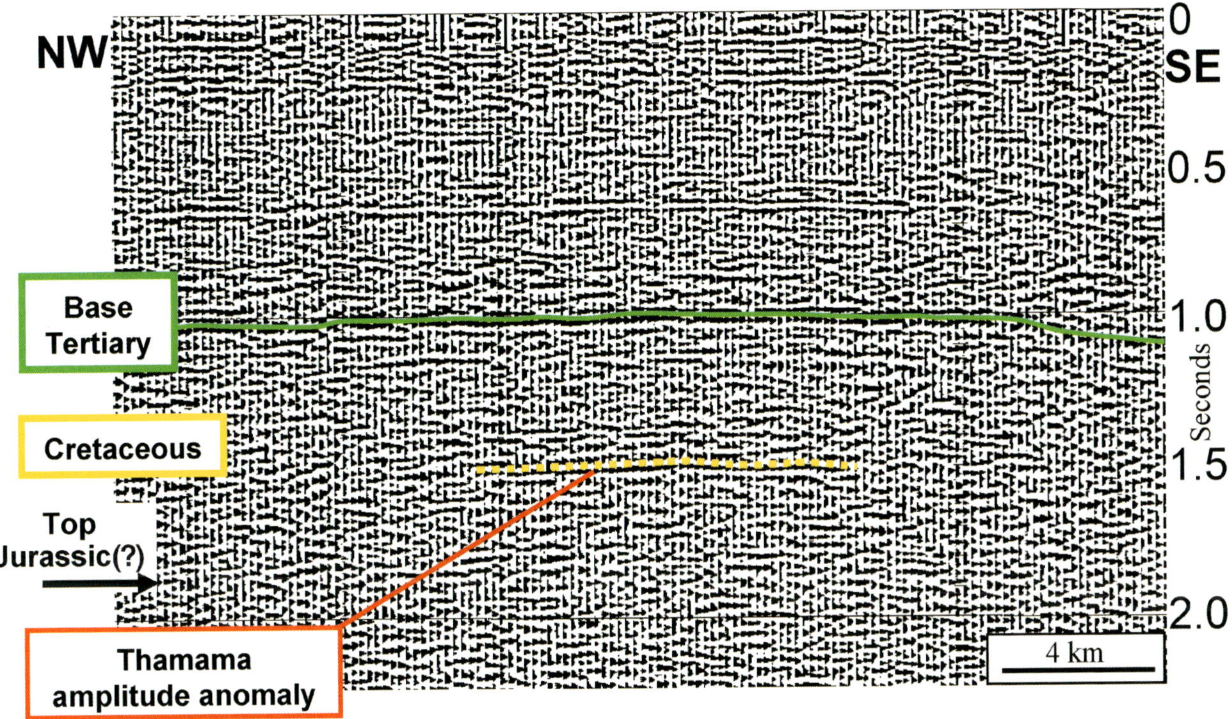

FIGURE 5. Early appraisal seismic section from 1966, northwest-southeast across the field, showing a high-amplitude anomaly at Cretaceous Thamama Zone F level.

the field. Acquisition was complicated by a harsh, hot, desert environment and significant field infrastructure with more than 300 wellheads.

Processing of the data had to deal with some key issues of the area. The large sand-dune topography and near-surface weathering both generated statics and reverberated noise. The subsurface geologic sequence of Abu Dhabi is highly layered, of high velocity, and also contains strong velocity contrasts, with several near inversions in the velocity profile. This results in seismic data that have events with a train of multiples generated at several levels, with little moveout discrimination.

Four versions of an axial line show the evolution in data quality of the 2-D test line, field-processed 3-D, priority 3-D cube, and the final 3-D data set (Figure 7). On these sections, a fault zone is progressively sharpened and noise, multiples, and statics are reduced. The 2-D test line shot prior to the 3-D shows a lot of banding resulting from interbed multiples and undulations resulting from near-surface statics (Figure 7A). For the 3-D, some processing was done in the field

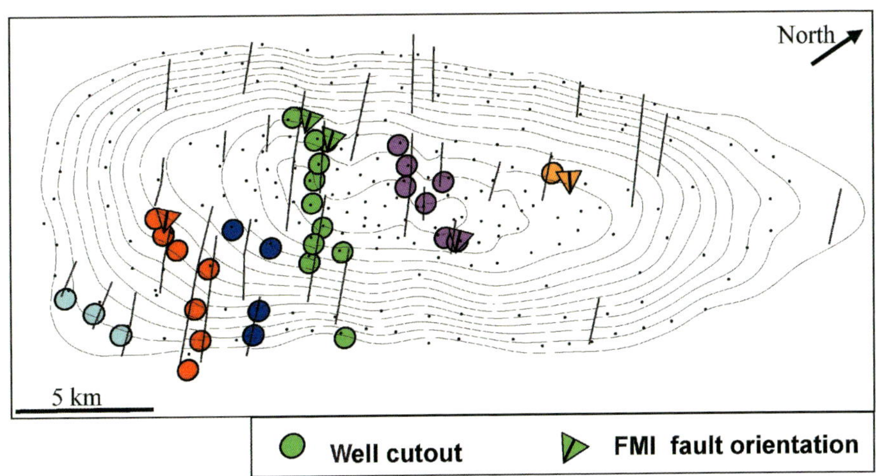

FIGURE 6. Pre-3-D Top Thamama Zone B depth map, 1998, contoured at a 25-ft (7.62-m) increment, based on more than 200 wells (black dots) and limited 2-D seismic data. The fault pattern was interpreted from geologic cutouts observed in vertical wells (circles), horizontal well Formation MicroImaging (FMI) data (triangles) and 2-D seismic data. The color coding represents these points grouped into the different wrench trends revealed by the later 3-D.

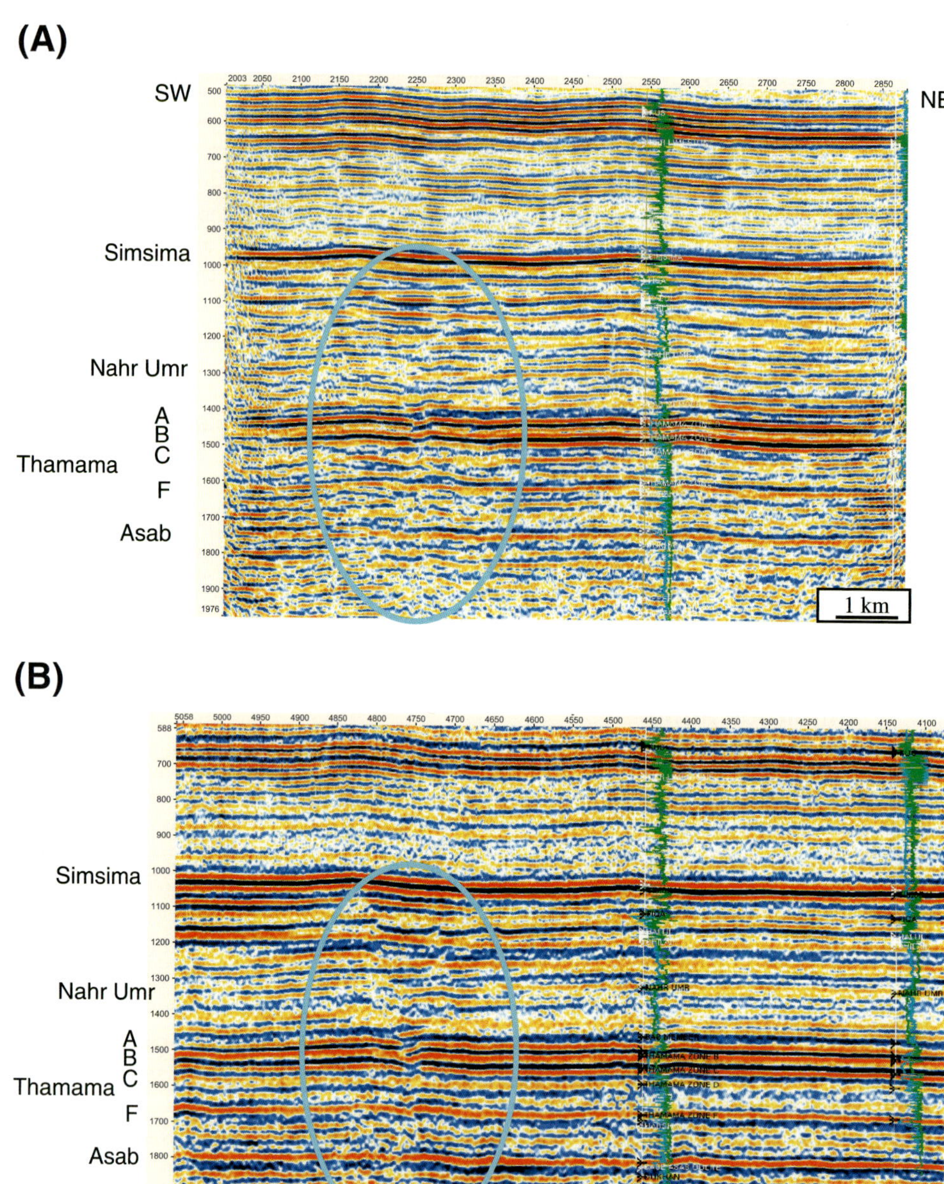

Figure 7. (A) A seismic test line, shot as 2-D for 3-D acquisition parameter evaluation. The line runs along the crest of the field and shows the lower Tertiary and Cretaceous section with a well tie. The 2-D seismic data suffer badly from long period statics, appearing as undulations, and multiples, which run through the poorly imaged fault zone (blue oval). (B) A 3-D cross-line from the field-processed volume in similar position as 2-D test line. This 3-D cross-line shows the improved fault imaging over the 2-D. However, the regular processing sequence, applied by the acquisition crew, leaves undulations resulting from statics and interfering multiples. (C) The same 3-D cross-line from a priority-processed volume. The application of a near-surface velocity model has improved the statics. However, even with careful velocity picking and a sharp FK demultiple, the multiple interference remains. (D) The same 3-D cross-line from the final volume. The use of longer offset data in the stack with AAMO has allowed better multiple reduction revealing the wrench fault zone and the more geologic-looking layering.

through to stack sections. Simple elevation statics were applied and a velocity field was derived from a limited number of velocity analyses, then a noise-reducing filter in the temporal frequency wave number (FK) domain and signal shaping with a gap deconvolution. This processing revealed good signal at the target (Figure 7B), with better imaging than the 2-D, but confirmed that the problem of statics and multiples was not to be solved by 3-D acquisition alone. The multiples were seen more obviously as horizontal banding on the flanks of the structure where they cut across the dipping data. A priority 3-D cube was processed through a conventional sequence using refraction statics, a prestack 95% FK demultiple, and deconvolution to attack the multiple. To improve the statics, a deeper uphole program (of around 150 m) was acquired to tie the base weathering refractor beneath the sand, observed in early processing. Although this approach assisted the demultiple algorithms to reduce some of the ringing, it did not remove the multiple interference at the target zone (Figure 7C). For the full cube processing, a technique to use the longer offsets was developed using a higher order moveout routine, apparent anisotropic moveout (AAMO, Western Geophysical). This technique has significantly increased the proportion of the longer offsets available to the stack at target level and hence provided better multiple discrimination and improved the signal-to-noise ratio (Figure 7D). However,

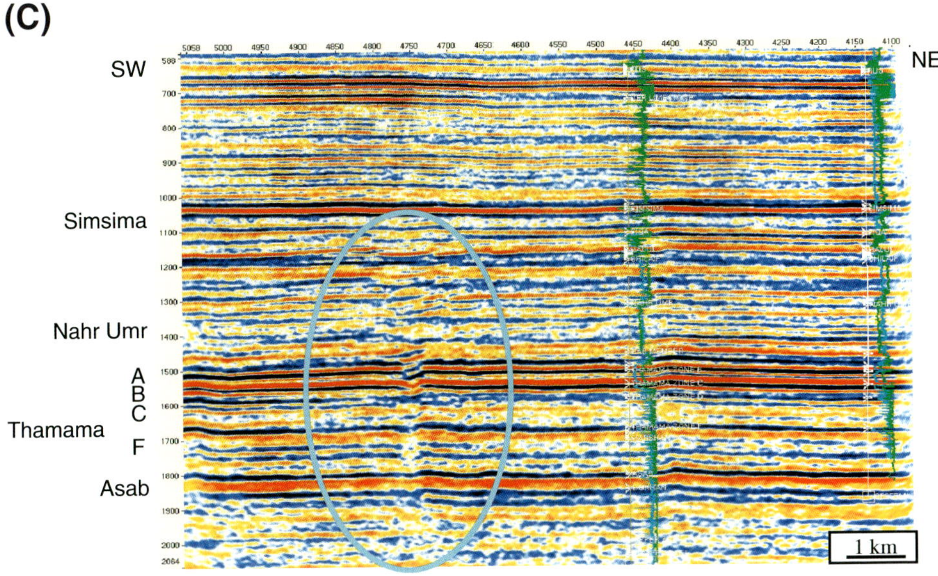

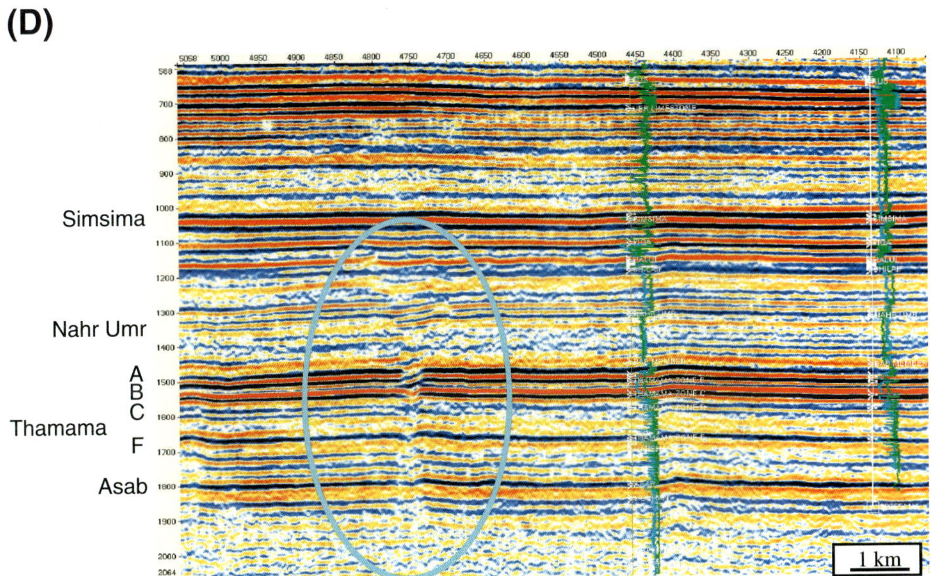

FIGURE 7. (cont.).

the inclusion of the long offsets has slightly reduced the high frequencies at target.

The improvement in the final data was most readily seen on a cross-line as a reduction in the multiple bands on the flanks of the structure compared to the priority cube data (Figure 8). The processing sequence was very sensitive to the velocity used, and horizon-consistent velocities were picked on analyses incrementing less than 1%. Other processing techniques were applied to the full data set, including an FK filter poststack to successfully deal with the subtle survey "footprint" seen in several horizon-amplitude maps of intermediate processing results. The final data show no sign of "footprint" at these levels, and the Top Asab Formation discussed later is considered to be "footprint-free." Residual noise attenuation was also used to reduce the random noise not removed by the stack process.

Through the use of these techniques, the final data quality was considered to be very good. However, for future surveys, it was suggested that source effort should be more disseminated to reduce surface noise and a higher frequency sweep be used to boost frequency at target to help resolve the interference between some of the relatively thin, porous/dense carbonate reservoir layers.

Velocity Control Wells

Vertical seismic profile (VSP) surveys are available in several wells, providing a reasonable coverage of geophysical well ties across the field. For a well tie, a synthetic seismic trace from logs is compared to the seismic data traces around a well and the VSP corridor stack trace. Figure 9 is the synthetic tie for a deep well and shows a good match among panels 3 and 4 from the left, the well synthetic seismic with VSP trace, and the 3-D final seismic trace at the well location. The average velocity curve on the far right shows two near inversions in the velocity profile, in the Fiqa and Nahr Umr sequences. Earlier comparisons of field data and priority cube data with the wells had significant remnant multiple in the data, typically showing up as events within the Nahr Umr sequence, which should be a quiet zone. The synthetics confirm the horizon ties at the wells and show that the final polarity of the seismic for the Top Zone B event (increase in porosity) is a decrease in acoustic impedance (AI), a negative number, and a trough. Well synthetics also showed that the packages of events in the Thamama were tuned at the seismic frequencies available at target (8–60 Hz).

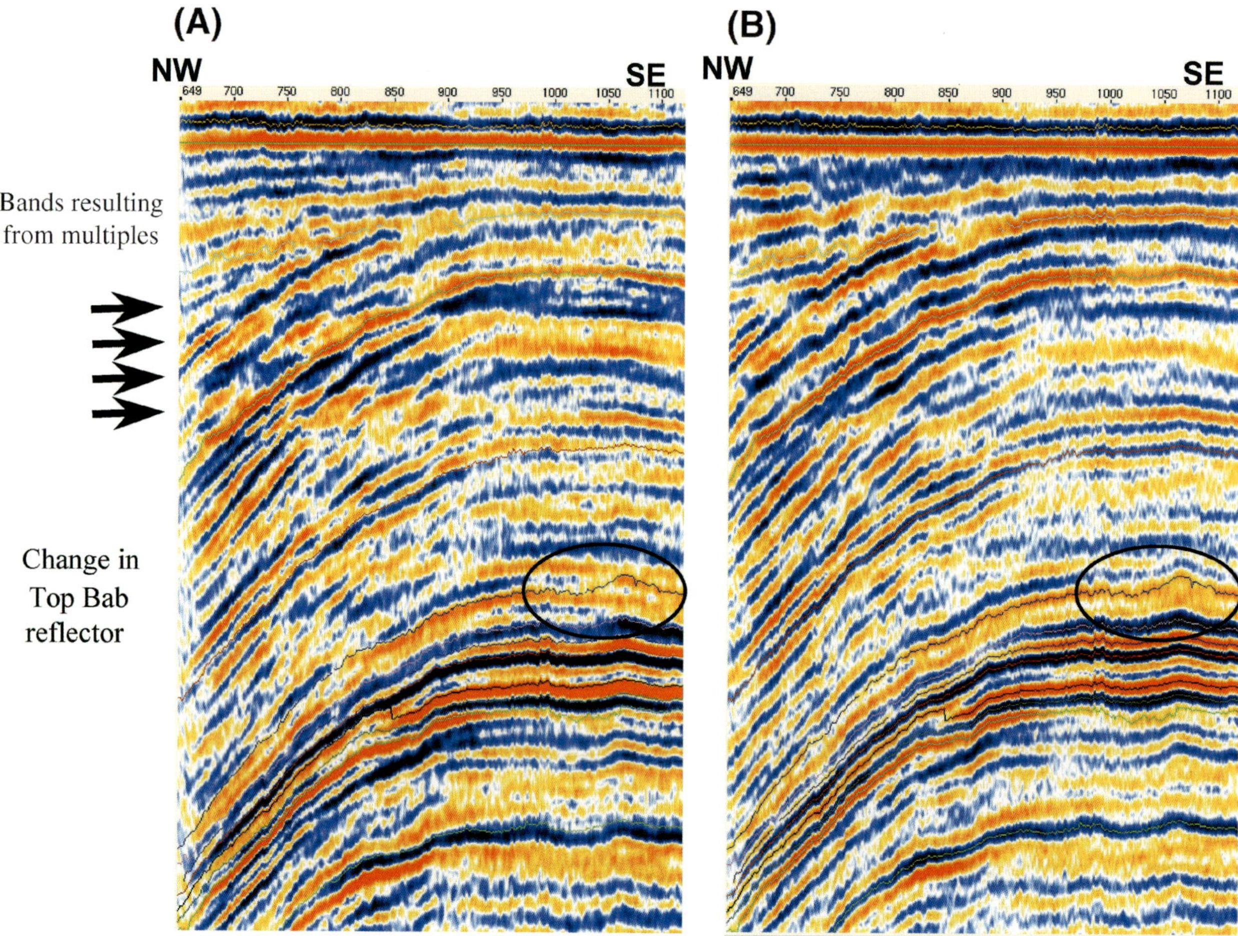

FIGURE 8. A 3-D seismic section from the crest to the flank with the (A) conventional (NMO) and (B) longer offsets processing (AAMO). The section is flattened at Base Tertiary to make the shallow-generated multiples appear near flat. (A) Bands resulting from multiple interference can be seen to cut across the 5° dip of the Upper Cretaceous reflectors of the field's flank. On the crest, these multiples are parallel to the primary data and they interfere and confuse the picture. (B) The multiples are much reduced, with the effect that on the crest the primary reflectors are now better imaged and can be mapped. Horizons from the AAMO data are displayed on the NMO data.

THREE-DIMENSIONAL SEISMIC INTERPRETATION

Conventional horizon tracking was used to map the key layers in the zone of interest. Many different attribute maps were made to highlight fault lineaments and some of these were also displayed with a false illumination to highlight features in a particular orientation. Attributes can be calculated from the values associated with the picked surface itself or from intervals including or parallel to the surface. The key attributes used for a surface are amplitude, dip, azimuth, and curvature, and integrated amplitude, reflection heterogeneity, and correlation coefficient for interval attributes. In the case of the latter, this can also be considered similar to the horizon slice through a variance or coherency cube, which have also been generated for this data set.

Structure and Faulting

The 3-D data have shown significant wrench faulting cutting through the anticline. An axial line (Figure 10) shows the Cretaceous section with near-vertical faults cutting up to the top of the Fiqa Formation. The Top Thamama Zone B seismic two-way traveltime (TWT) map (Figure 11) clearly demonstrates that the large anticline is cut by dextral wrench faults containing typical small pull-apart and pop-up blocks and horsetail fault patterns. The gross structural shape is similar to the map based on wells (Figure 6), but the amount and type of faulting had not been predicted from the wells. A striking picture of the structure can be made using "illumination" that uses angled light to highlight-shade structure. This can reveal features not seen on a dip map. Figure 12 shows an illuminated 3-D perspective view colored by time, showing the relationship of the faulting to the relief. The

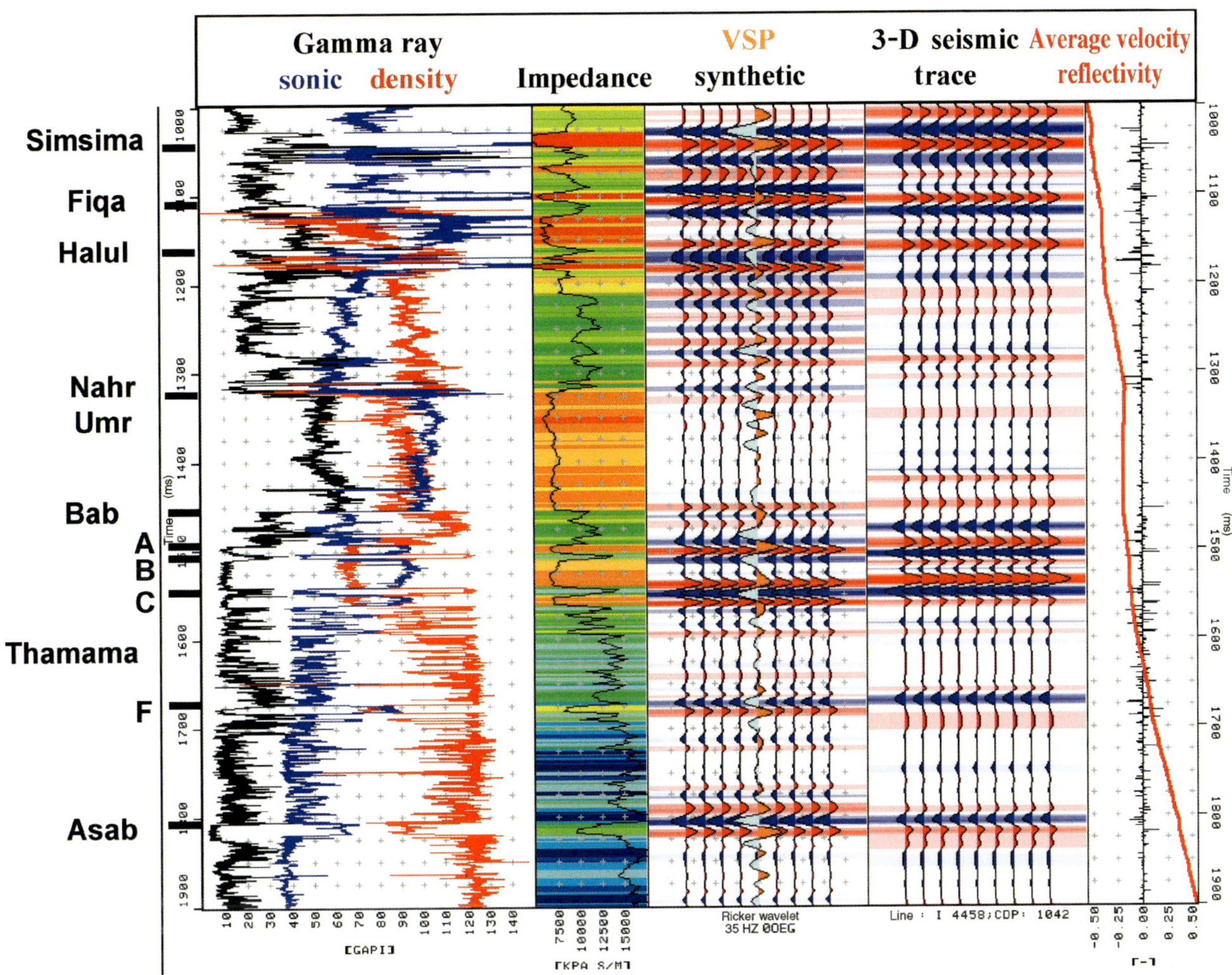

FIGURE 9. A representative tie of seismic to well for the field. This shows conventional well logs in the left panel and well-derived AI in the second panel. The third panel shows a synthetic seismic trace, displayed as intensity and wiggle formats, derived using the well impedance and a wavelet. The single overlain trace is the recorded vertical seismic profile (VSP). The third panel is a good comparison to the fourth, the trace from the 3-D seismic at the well location. The fifth panel shows the well reflectivity and average velocity with a velocity inversion in the Nahr Umr Formation.

image reveals some of the gentler ripples in the anticlinal fold, with this illumination from the south showing north-northwest–south-southeast ripples. Other illumination angles also show ripples parallel to the fold axis in a northeast-southwest direction. Attribute maps as well as conventional surfaces were used to map some of the minor faulting. The map of principal curvature (Figure 13) looks at one of the components of the dip of the Top Thamama Zone B surface. It emphasizes the subtle changes in the dip, revealing a striking pattern (in red) of the fault edges and (in green) the edges of the smaller folds-flexures. Because this is a surface-generated attribute, the regional structure (anticline) is removed.

The faulting in the field is seen as high-angle wrench faults that form flower structures in the Upper Cretaceous section (Figure 10). In the main wrench zones, many faults at the Upper Cretaceous Halul level coalesce downward into a near-vertical system at Top Thamama Zone F. The throws on the faults vary both vertically and laterally, because of both dip slip and strike slip, with the result that a single fault trace can have normal and reverse behavior. Faulting can sometimes be more visible in the Upper Cretaceous and traced downward to anomalies at reservoir level. Several papers on wrench tectonics in Abu Dhabi have discussed the effects of the Oman orogeny of Late Cretaceous to Paleocene. In Marzouk and Sattar's (1994) work, wrench faults in a northwest-southeast direction were proposed as being sinistral; however, this new seismic data show that the deep-seated west-northwest–east-southeast wrench trends in this field are dextral in motion. In Silva et al.'s (1996) work, the east-west compression of the area is discussed, with

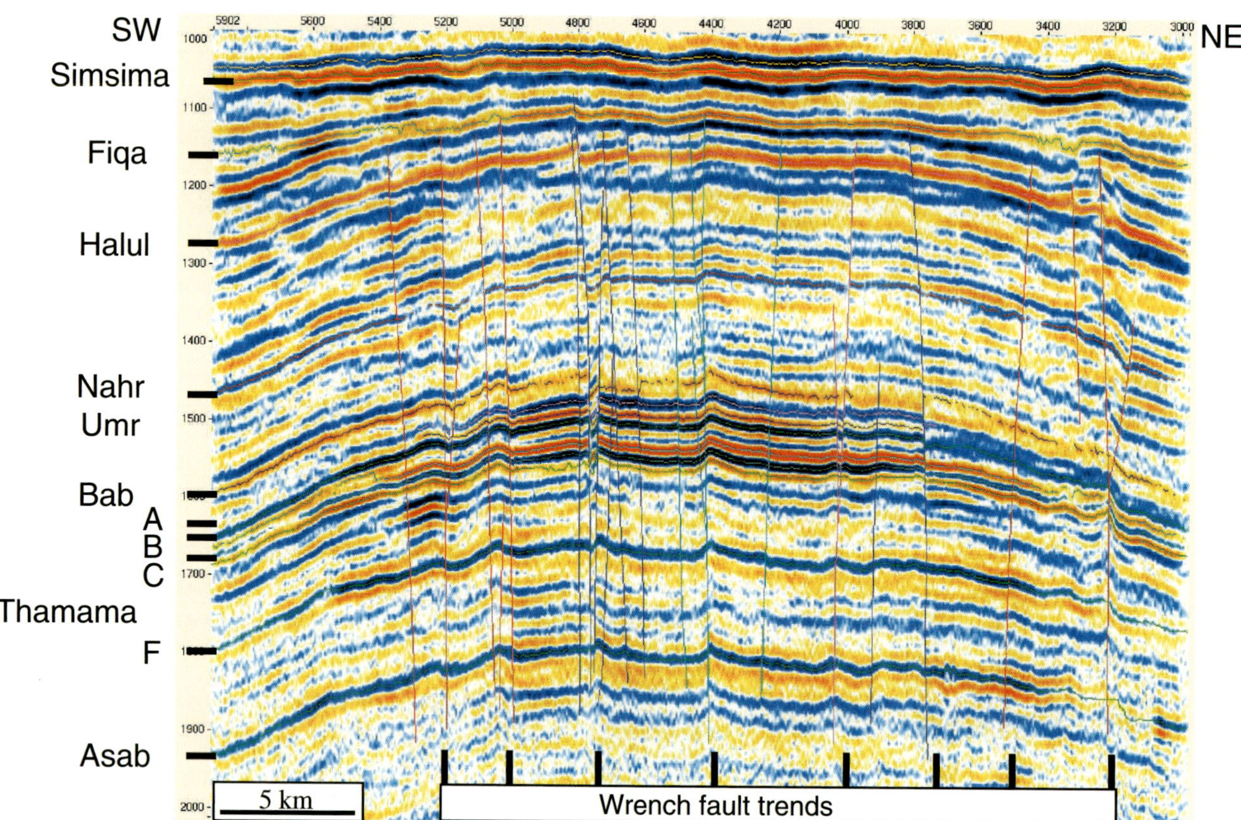

FIGURE 10. A seismic line along the crest of the field (anticline axis) showing the Cretaceous section with interpreted horizons. The large anticline is cut by near-vertical Upper Cretaceous-aged wrench faults.

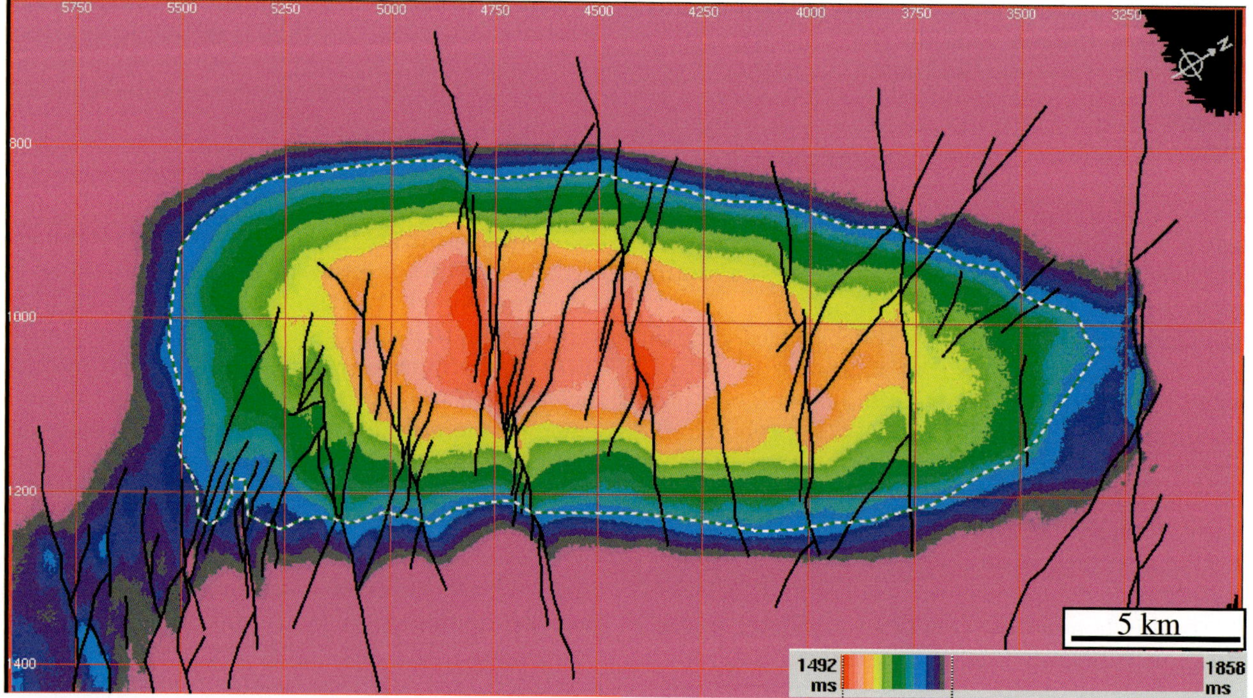

FIGURE 11. The workstation TWT map of Top Thamama Zone B, with dashed outline of Zone B OWC, shows the broad anticline cut by wrench faults.

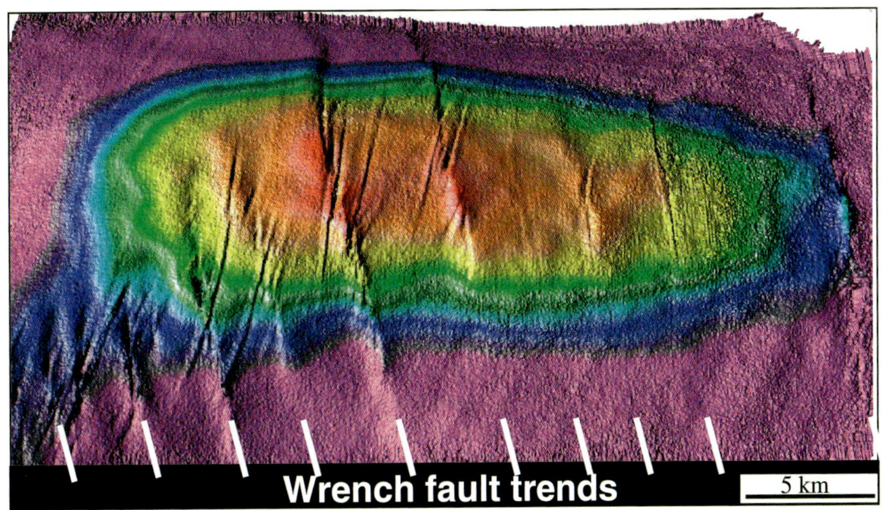

FIGURE 12. A 3-D visualization of Top Thamama Zone B time surface, colored by time and illuminated from the south, shows the fault pattern in perspective. The wrench faults are larger and more complicated in the south of the field, and fault links can be seen between the wrench zones. The in-line parallel stripes in the north of the field are where the B event loses amplitude and quality.

dextral wrench faulting proposed for this field, which is now confirmed by the seismic data.

To demonstrate the magnitude of the faulting, a piece of seismic across a wrench zone can be compared to a section of Thamama outcrop from Oman (Figure 14). The outcrop layering is a good match to the seismic, with similar layer thickness, confirming the extensive regional nature of this formation, as noted by Buchem et al. (2002) among others. The faults on the seismic show how the different zones would connect via juxtaposition. The ADCO office tower is shown for everyday scale and Zone B is 58 m thick at this crestal location. The 3-D seismic has revealed the complex fault structure of the field with zonal juxtaposition among Thamama Zones A, B, and C, contrary to previous published models (e.g., Alsharhan, 1993). This fault connectivity has now been incorporated into the production reservoir model and gives an improved history match of the zonal reservoir pressures.

The combination of the 3-D seismic data and the well data to make a depth map has revealed a very good relationship of velocity to depth at Top Thamama Zone

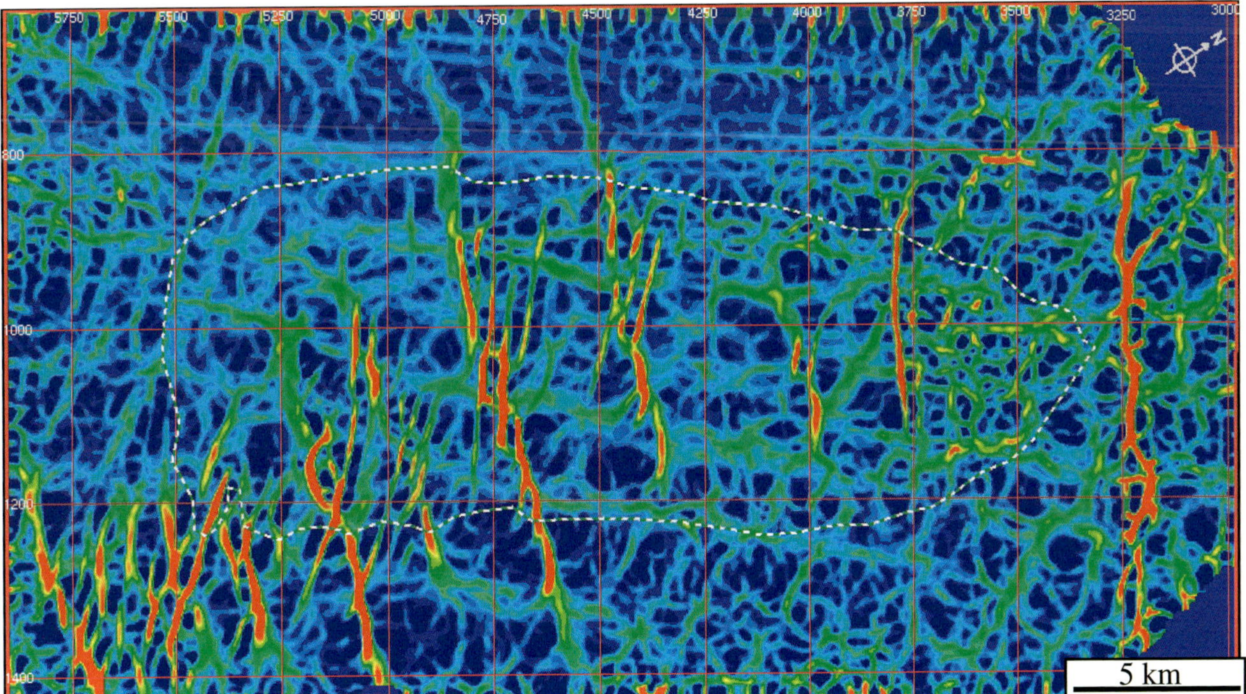

FIGURE 13. The principal curvature of Top Thamama Zone B time surface. The color range shows the magnitude of change in surface curvature such that larger faults appear as red and the smaller faults and flexures appear as green.

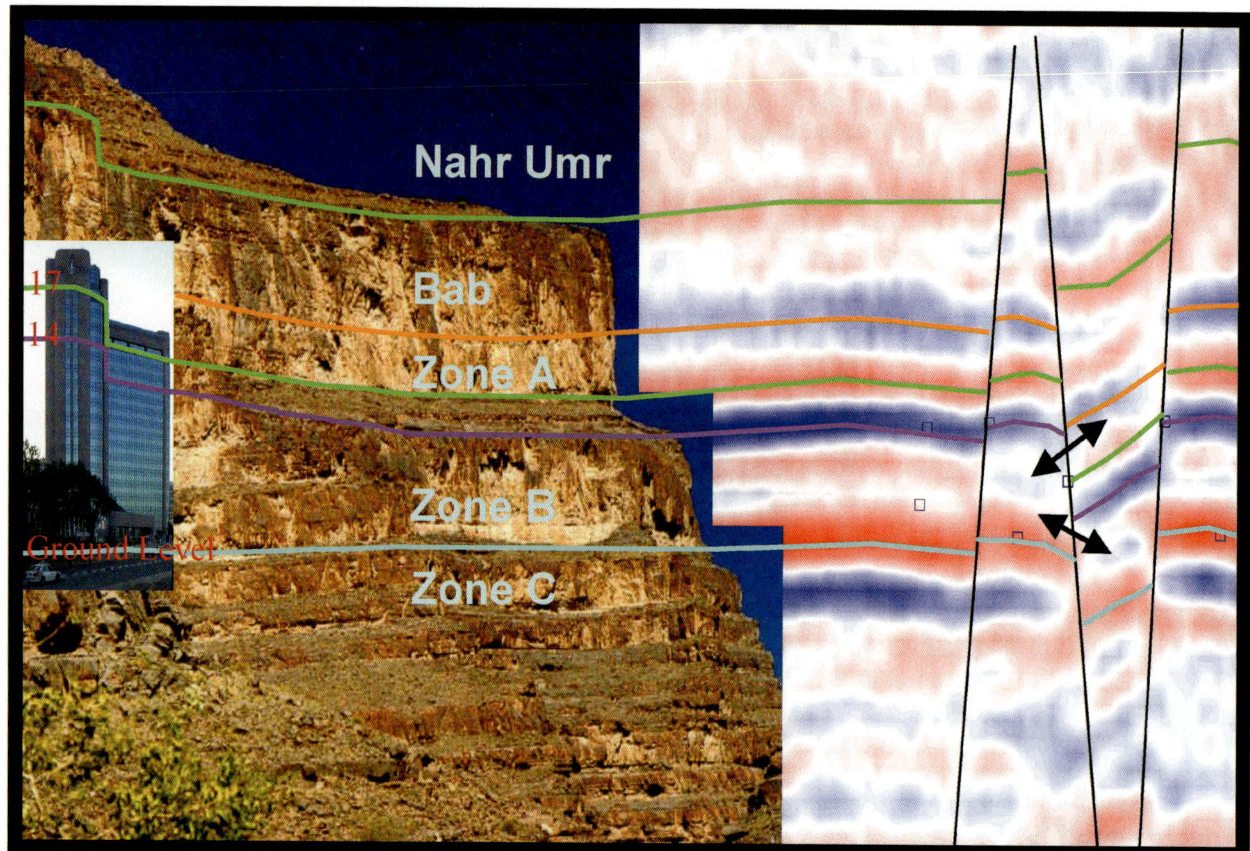

FIGURE 14. Comparison of Upper Thamama Group outcrop in Wadi Guhl, Oman, with a 3-D seismic section. The seismic and outcrop layering are of similar thickness, and the ADCO office tower provides an everyday scale for the 58-m-thick Zone B (14 floors). It is to be noted that the seismic only resolves the large-scale layering and that the fine-scale geology from a core or logs of such a section are not resolved by the seismic. However, the seismic can see faults in the main layers down to as small as 5 m throw. This seismic section shows the reservoir zones juxtaposed by a 25-m wrench fault (connections marked with arrows).

B, with a single linear pseudovelocity function tying the wells to within 15 m. However, at the detailed scale, this residual can be seen to be a subtle variation of velocity in the wrench trends, which are some 1% faster than the adjacent, more stable, blocks. This variation in velocity in the fault trends was also noticed on the processing velocities when horizon-consistent interval and average velocity maps were made. For small, wrench-related structures elsewhere, without wells, care in depth conversion would be required to avoid this "pull-up" effect.

Reservoir Character

The 3-D seismic has revealed much about the structure of the field and is also being used for reservoir characterization. On the axial line (Figure 10), the seismic events in the Thamama Group can be seen to weaken and the layers thin downflank, suggesting a marked decrease in the reservoir properties toward the water leg. A southwest-northeast line along the flank, outside the field, flattened at Top Thamama Zone B (Figure 15), highlights the highly layered nature of the Thamama A, B, and C zones. There is only a change to progrades in the uppermost part of the sequence, in the Upper Bab.

The observed seismic TWT thickness of Thamama Zone B (Figure 16) shows a marked thinning of the Zone B porous interval from crest to flank, but with little thinning outside the field area. Comparing this to the thickness of the Zone B porous interval based on well data, which ranges from 58 m in the crestal area to 51 m downflank (Figure 17), it can be seen that there is a good match within the field area, but the thinning seen by the wells is incorrectly extrapolated downdip.

The well data show a good relationship of impedance against porosity, suggesting that seismic amplitude should be related to porosity. It is therefore a surprise to see that the shape of the Top Thamama Zone B amplitude, with low values in the north (Figure 18), does not match the Thamama Zone B average porosity from well data, which ranges from 34% in the crestal

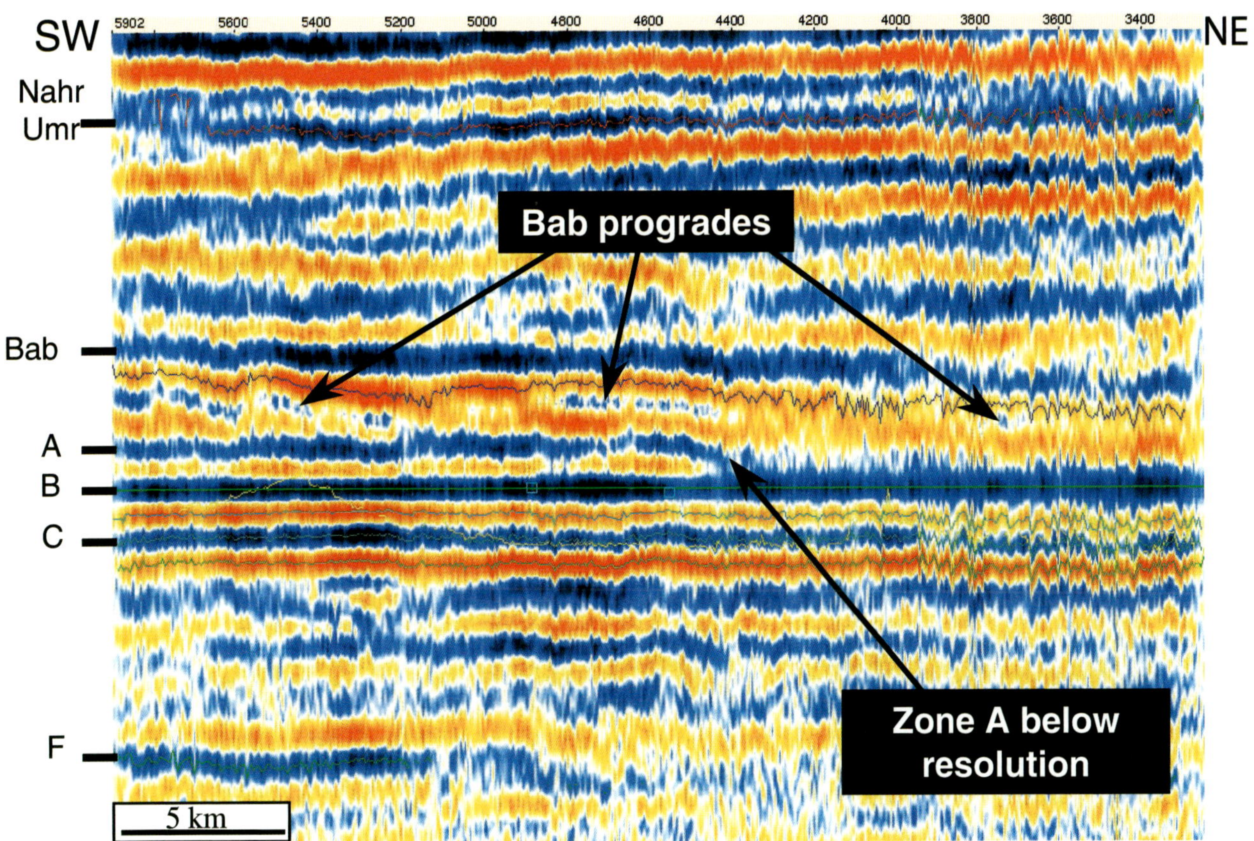

FIGURE 15. A seismic line parallel to field axis, outside of closure. The section is flattened at Top Zone B and shows the layered A, B, and C zones. The overlying Bab Formation shows northerly prograding clinoforms, steeper in the south. Zone A thins to the north and seems to disappear where it goes below seismic resolution.

area to less than 20% downflank (Figure 19). From the well data, the dense layer immediately above Top Thamama Zone B, the A dense layer, is observed to be present everywhere in the field but thins to the north. It is this thinning of A dense layer to below seismic resolution that has led to the apparent seismic pinch-out of Thamama Zone A, as seen on the flattened cross-line Figure 15. Similarly, this thinning is also the cause of the poorer amplitudes of the Top Thamama Zone B seismic event in the north, which, if there were no wells, could have the potential for a misleading prediction of low porosity from amplitude. An integrated amplitude map of near Top Thamama Zone B (a trace attribute of the upper part of the B porous section rather than its boundary) (Figure 20), improves the match of the map to the average porosity.

The Base Thamama Zone B amplitude shows a clearer anomaly (Figure 21) than Top Thamama Zone B. However, this anomaly is a little smaller than the Zone B field outline. A comparison of the amplitude map with the wedge zone, where the layers pass through the transition zone, shows that it is only as the B lower is filled with oil that the base B amplitude increases. The well data show that late cementation increases rapidly into the water leg and thus the seismic signal is responding to the lower reflection contrast of the lower-porosity Thamama Zone B with the underlying B dense layer.

For the thinner intervals, the seismic data could also be used to predict properties, such as thickness and porosity, but as these intervals are tuned, there is a risk in this. For example, the seismic time isopach for B dense layer might be interpreted to suggest that it is thicker over the crest of the field. However, inspection of the well data shows that this would be an incorrect interpretation, as the B dense layer gently thickens to the northeast except for fault cutout points.

Seismic Impedance

To help remove some of the tuning effects of the wavelet, the seismic data were converted to AI. The conversion used a sparse spike technique, focused on the Thamama section. An axial line of AI (Figure 22) shows the low-impedance reservoir layers over the crest of the field increasing in impedance and decreasing in thickness toward the flanks. A more detailed

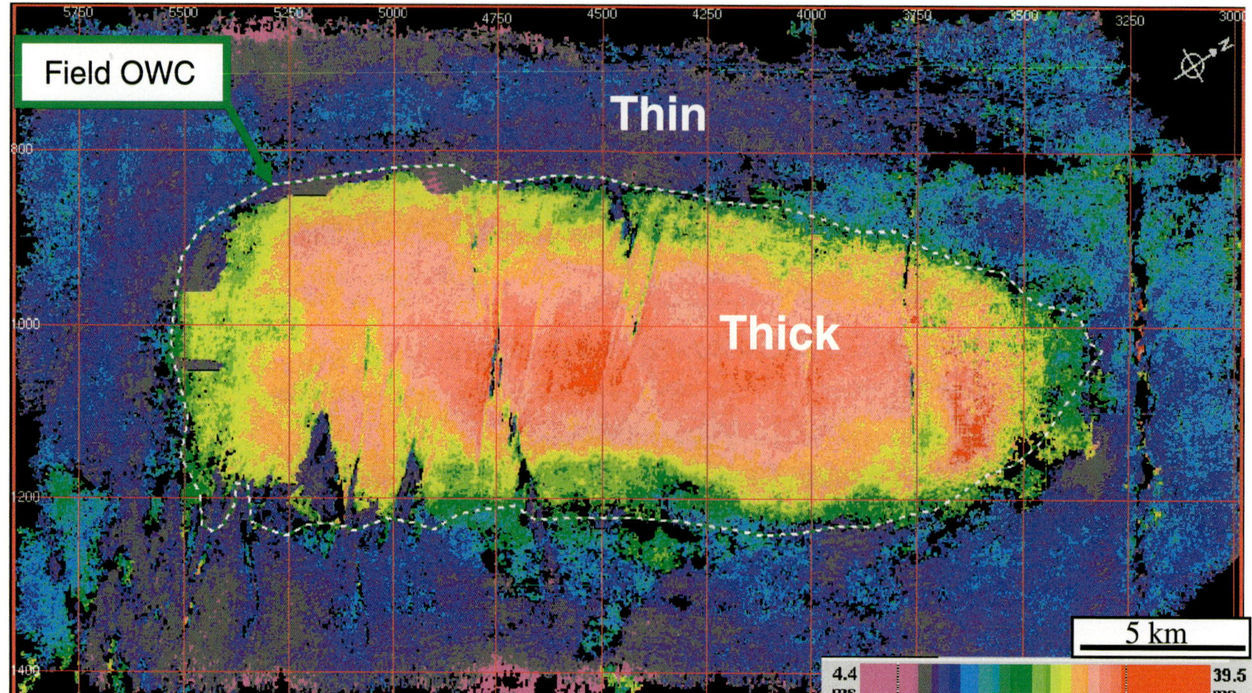

FIGURE 16. The seismic time isopach of Thamama Zone B reservoir shows that the reservoir thins toward the oil-water contact (OWC) with little further change in the water leg. The faulting clearly affects the thickness, with sharp changes across the faults, caused by the downthrown block undergoing greater diagenesis. The sharp changes at some places on the edge of the field are where seismic tuning effects make the horizon picks jump.

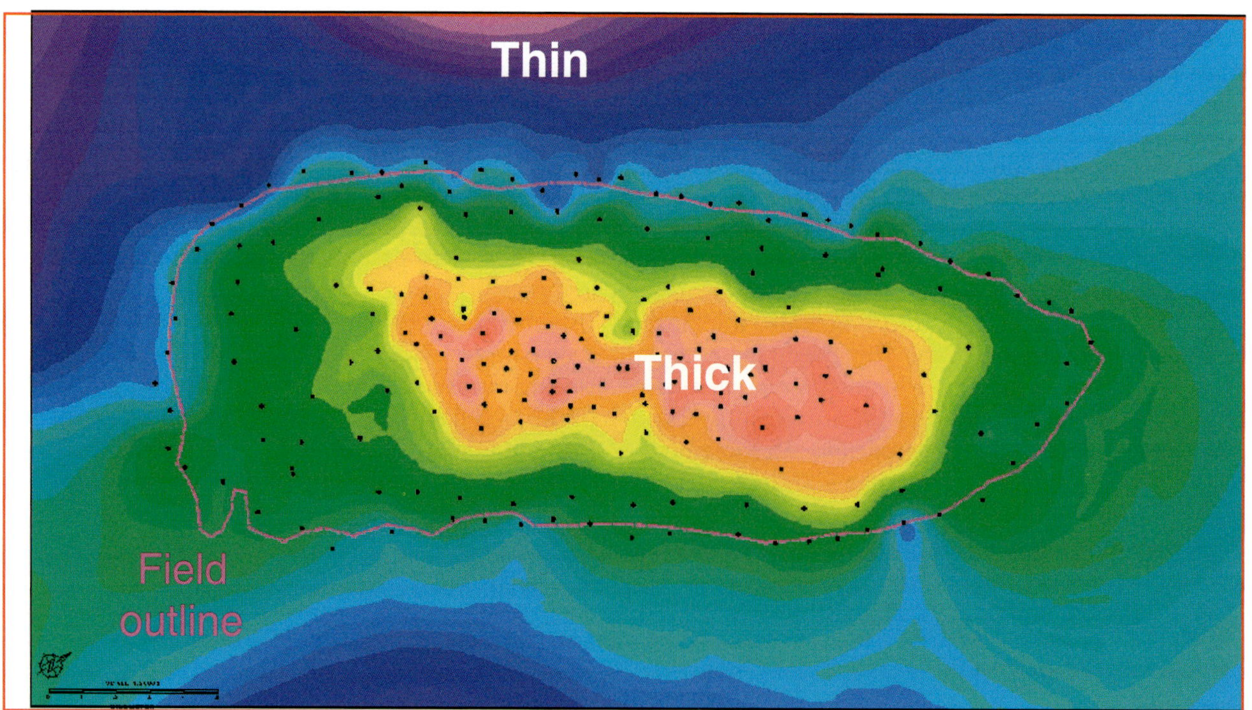

FIGURE 17. The Thamama Zone B reservoir thickness from wells shows the thinning caused by increasing diagenesis from crest to flank. Although similar to the seismic time isopach, the well-only model continues the thinning outside of the field, away from well control. Fault cutouts appear as spots rather than the thin lineations seen on the seismic isopach.

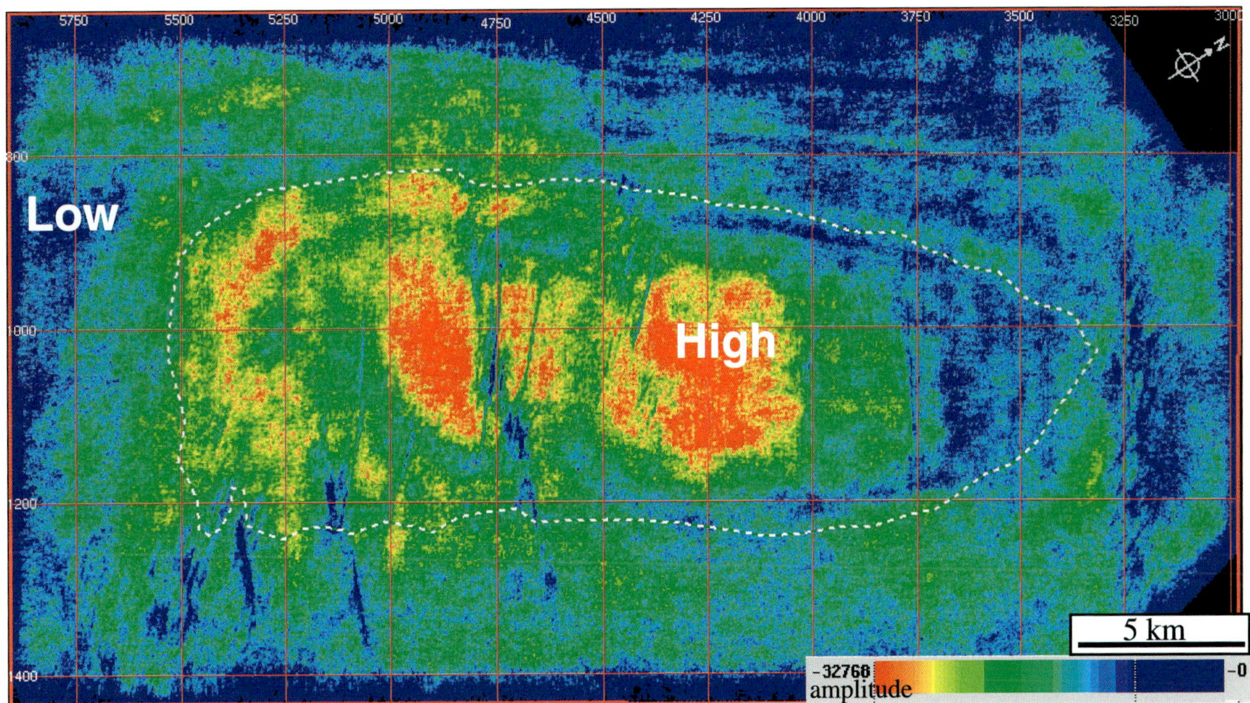

FIGURE 18. The Top Thamama Zone B amplitude shows a patchy anomaly within the field area. This anomaly does not match the true field shape because of a weak seismic event in the north and rings caused by interference from remnants of strong multiples in the data.

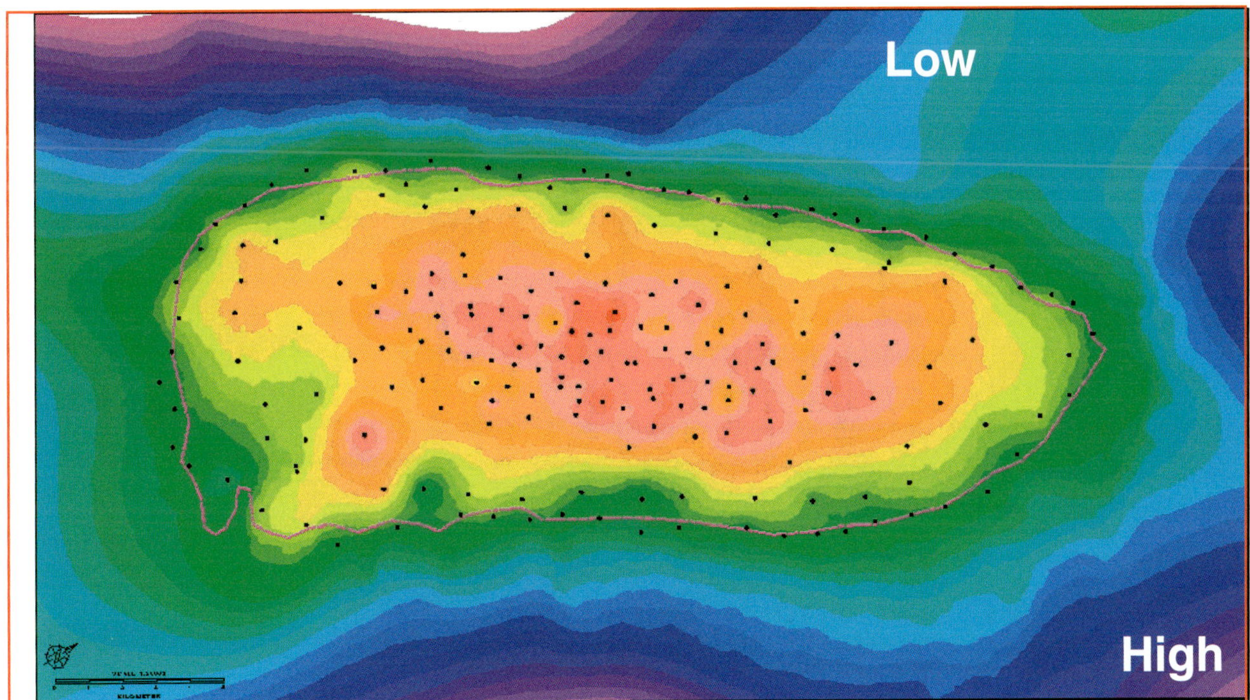

FIGURE 19. Thamama Zone B reservoir porosity from wells showing a reduction from crest to flank. Although there is a good relation between porosity and thickness, this map is not quite the same shape as the thickness map from well data, but a better match to the seismic time isopach map, which responds to both velocity and thickness.

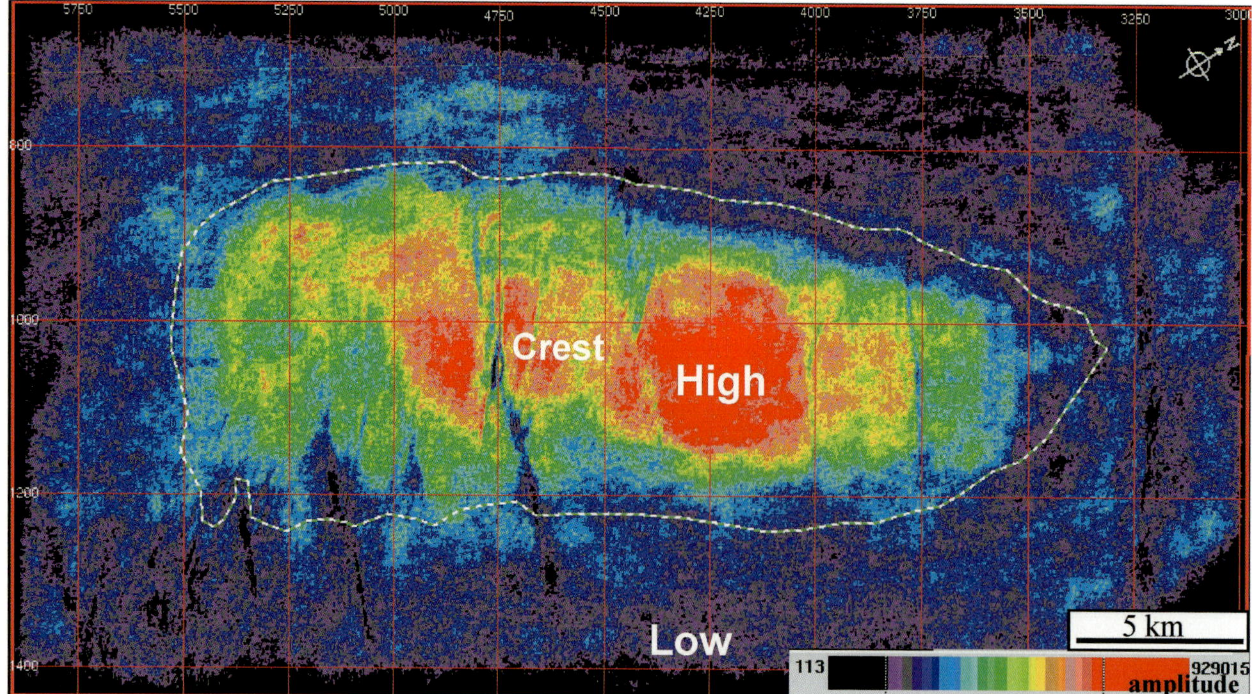

Figure 20. Integrated amplitude of upper part of Thamama Zone B showing anomaly inside field area with high amplitudes, slightly offset from structural crest. This thick-amplitude map (Top Zone B ±8 ms) shows more clearly than the simple horizon amplitude, the amplitude changes at top B. Note gradational amplitude toward contact.

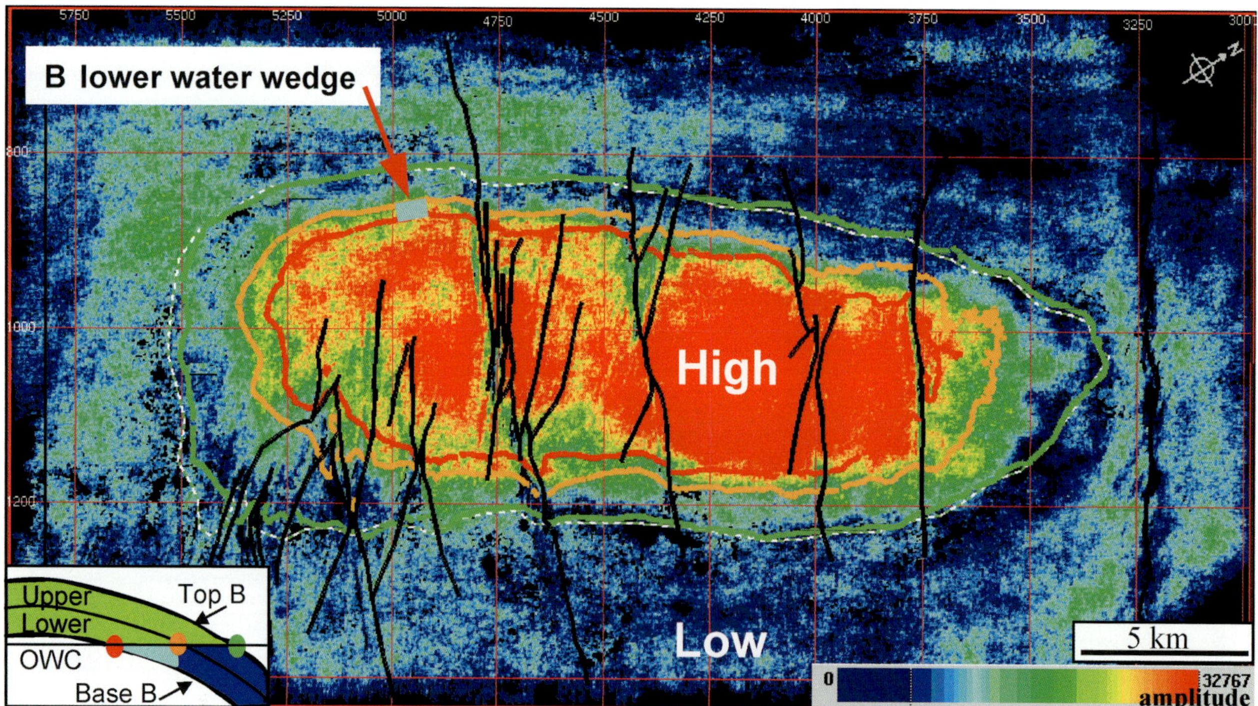

Figure 21. Seismic amplitude of Base Thamama Zone B (Top Zone B dense layer). This amplitude is responding to porosity changes of the lower part of Thamama Zone B. The three structural contours, green, orange, and red lines, show where the Top Zone B, Top Lower Zone B, and Base Zone B intersect the oil-water contact (OWC). The amplitude anomaly increases between the Top Lower Zone B and Base B OWC where porosity has been preserved by hydrocarbon fill.

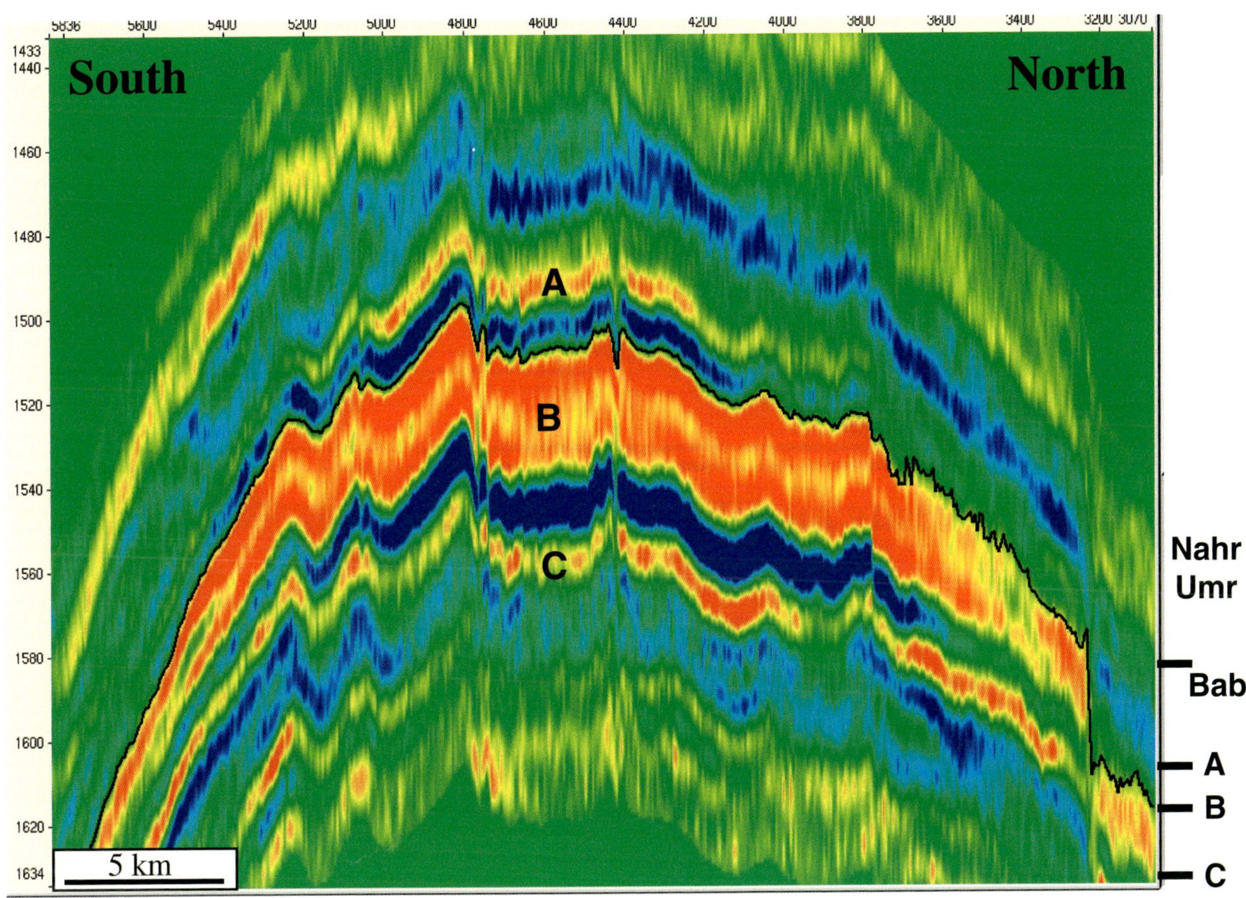

FIGURE 22. A seismic line along the anticline axis converted to AI showing the Upper Thamama reservoir Zones A, B, and C. The red end of the spectrum shows low AI, equivalent to soft porous limestone; the dark blue is high AI, i.e., hard/dense limestone. The reservoir layers can be seen to become higher impedance, lower porosity, toward the flanks of the field.

depth section on the flank (Figure 23) compares the seismic impedance with the porosity model. This shows that the center of Thamama Zone B increases in impedance, separating B upper and lower. In the geology model, there is also a subzone in the wells that has increased stylolites and becomes "denser" toward the edge of the field. The B upper and lower also increase in impedance toward the edge of the field, in agreement with the well data and the model, which show that the B upper and lower are of lower porosity at a greater depth on the flanks of the field. Figure 24 shows sections of impedance along a traverse that zigzags between wells along the field, back and forth from crest to flank. Figure 24A is an interpolation of impedance from these wells between 0 and 125 Hz. This section was high-cut filtered down to 60 Hz (Figure 24B), which gives a good match to the seismic impedance along this traverse (Figure 24C). It is interesting to note how much resolution and definition of the layers has been lost when filtering the well interpolation and what might be seen if much higher seismic frequencies could be recorded at target.

Paleostructure

The deeper gas-filled Thamama Zone F has an amplitude anomaly (Figure 25), which suggests the shape of the closure is slightly different in the north compared to the outline of the upper levels. Inspection of the Zone B to F time isopach shows a thick anomaly in the field area, with a maximum thickness in the northern half of the area (Figure 26). An axial line along the crest, flattened at top C (Figure 27), clearly shows the greater preserved thickness of Upper Thamama within the field area and can be compared to a line outside the field that has no thickening (Figure 15). The asymmetry seen in this thickening-preservation results from the changes in the structural closure during the Late Cretaceous and Tertiary, affecting diagenesis during and after hydrocarbon fill. A similar history is discussed for an adjacent field to the north in Oswald et al. (1995). The axial section from the geology model colored by porosity (Figure 3) also suggests this subtle secondary diagenetic overprint, with dense layers a little farther up the south

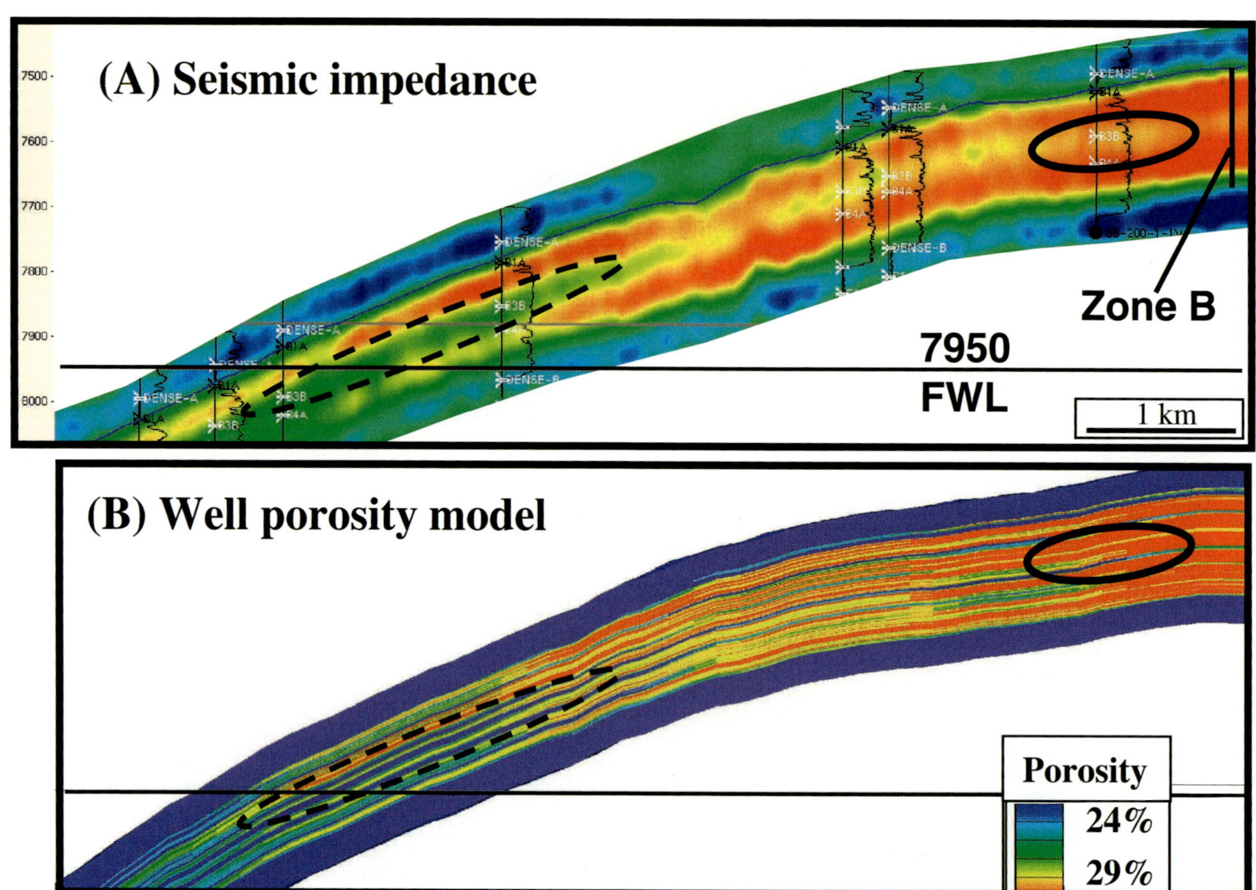

FIGURE 23. This compares a depth section of the seismic impedance and a well-derived porosity section for Zone B on the flank of the structure. (A) The low-impedance reservoir section is progressively divided toward the flank by an increased impedance in the middle of Zone B. There is also an increase of impedance in B upper and lower in the transition zone. These changes match porosity changes in (B) where there is a downflank decrease of porosity near to the middle of the reservoir resulting from an increase in stylolites. FWL = free-water level.

flank than the north, with better porosities developed a little to the north of the crest.

Depositional Geometry and Paleogeography

The Upper Bab Member of the Shuaiba Formation can clearly be seen to be a progradational sequence on the seismic (Figure 15). Mapping shows that these prograding clinoforms change their direction from northeast to north with prograde angle decreasing from south to north across the field area. Well data show the youngest sequences to be present only in the north. These northern low-angle progrades are predominantly tight lime mudstone, whereas the older, steeper progrades contain reservoir-quality carbonates. This is in line with the Bab sequence-stratigraphic model of Buchem et al. (2002), where the Upper Bab (their sequence IV) is a regressive phase and hence would give rise to basinward-stepping progrades in front of the Shuaiba shelf margin (Figure 1).

A deeper reservoir level in the field lies at the top of the Jurassic, where some wells penetrate Asab Formation oolites and grainstones of Late Jurassic age (late Kimmeridgian to Tithonian). These are overlain by the Habshan Formation, which was deposited by the major Cretaceous transgression of the area. The correlation of these Asab oolites with the Arab and Hith Formations, which are found regionally to the south and west, is problematic, although a well in the south of the field area contains Hith evaporites at the level equivalent to the Asab oolites. From regional well data, it is apparent that this field lies at the seaward edge of the main evaporitic Hith deposition (Matos, 1994), which is shown to pinch out in an east-northeast direction with a north-northwest–south-southeast edge in this area (Figure 1). The overlying Habshan in the wells can be correlated with the sequences discussed by Landmesser and Saydam (1996) and Aziz and Abd El-Sattar (1997), which show a good seismic delineation of the younger Habshan shelf edges and basin, which developed to the east of this field, also with a north-northwest–south-southeast trend (Figure 1).

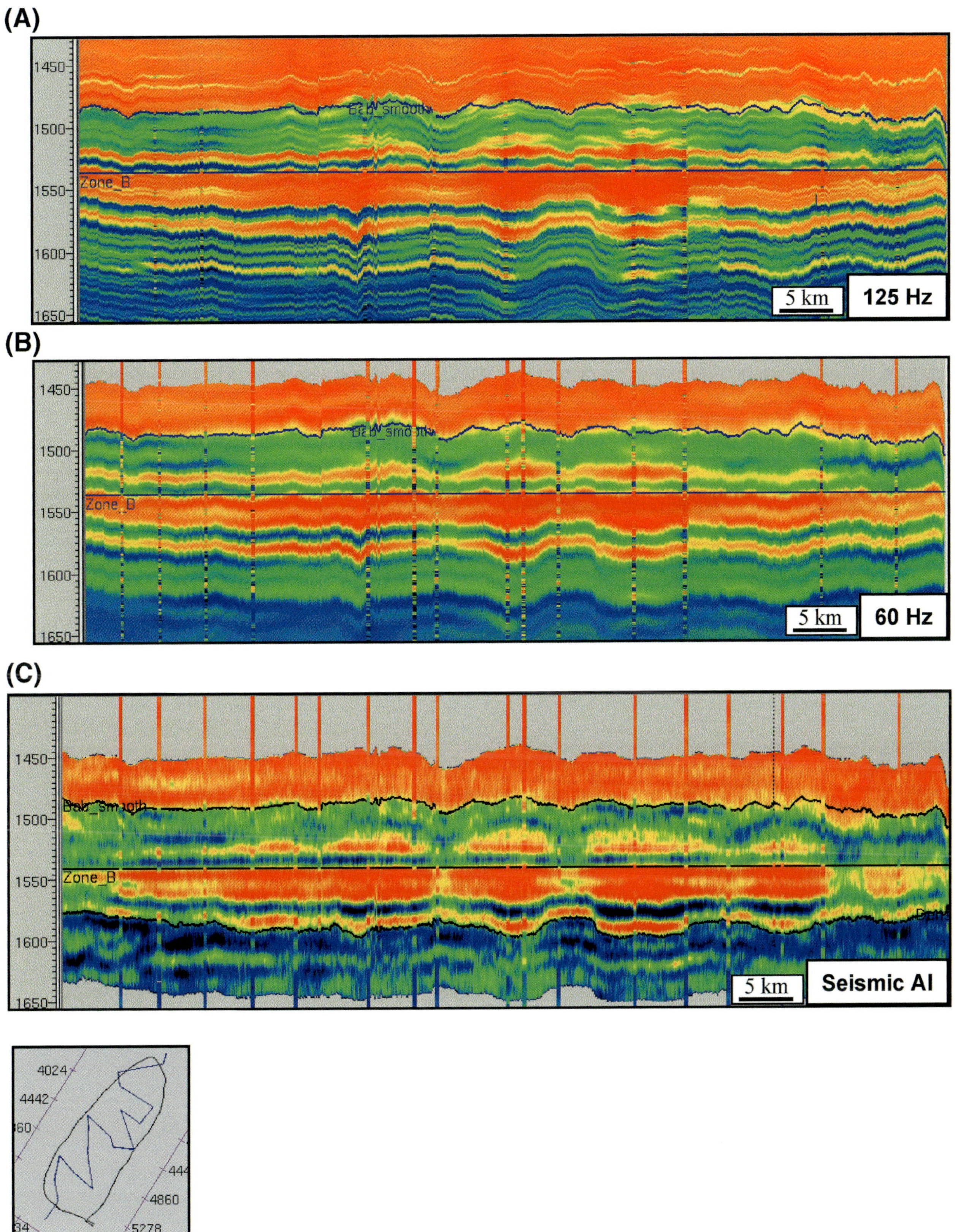

FIGURE 24. (A–C) A time-section of acoustic impedance (AI) for the Upper Thamama Zones A, B, and C on a zigzag traverse along the field. (A, B) Interpolated well AI, with the well AI logs shown as thin columns. (A) The frequency range is between 0 and 125 Hz. (B) The frequencies are filtered to below 60 Hz. (C) Seismically derived AI. This shows a reasonable match to B, although in places lower-frequency content causes the reservoir zones to break up and disappear.

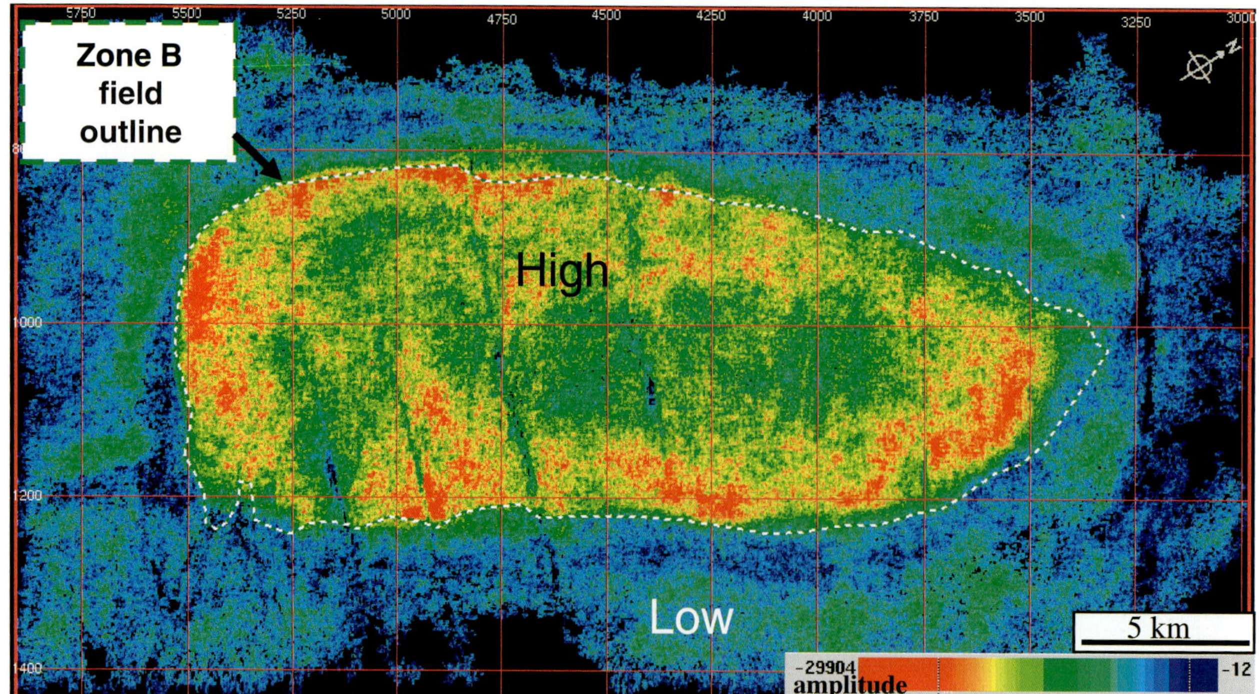

FIGURE 25. The Thamama Zone F amplitude map clearly shows the anomaly of the Zone F reservoir. At this separate reservoir level, the anomaly has a different shape from the shallower Zone B field outline.

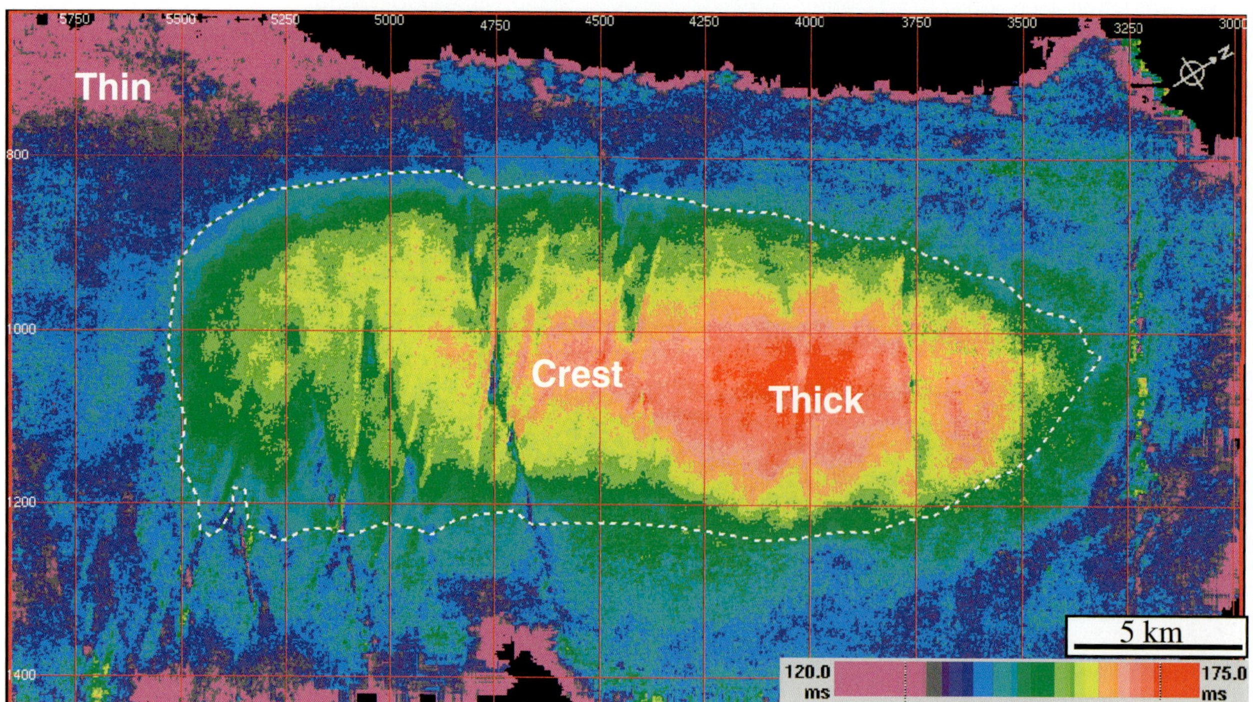

FIGURE 26. Thamama Zone B to F time isopach shows an offset of thickness from the present-day crestal location resulting from asymmetric diagenetic preservation. Thickness difference between crest and contact is about 20 ms (12%).

The seismic horizon of the Top Asab level shows an east-northeast progradation in the far northeast of the field, suggesting that in this area the Asab Formation is shallow marine. The illuminated dip map (Figure 28) reveals that beneath the main field area there are long linear ridgelike anomalies running north-northwest–south-southeast, which are different from the main wrench trends that are in a

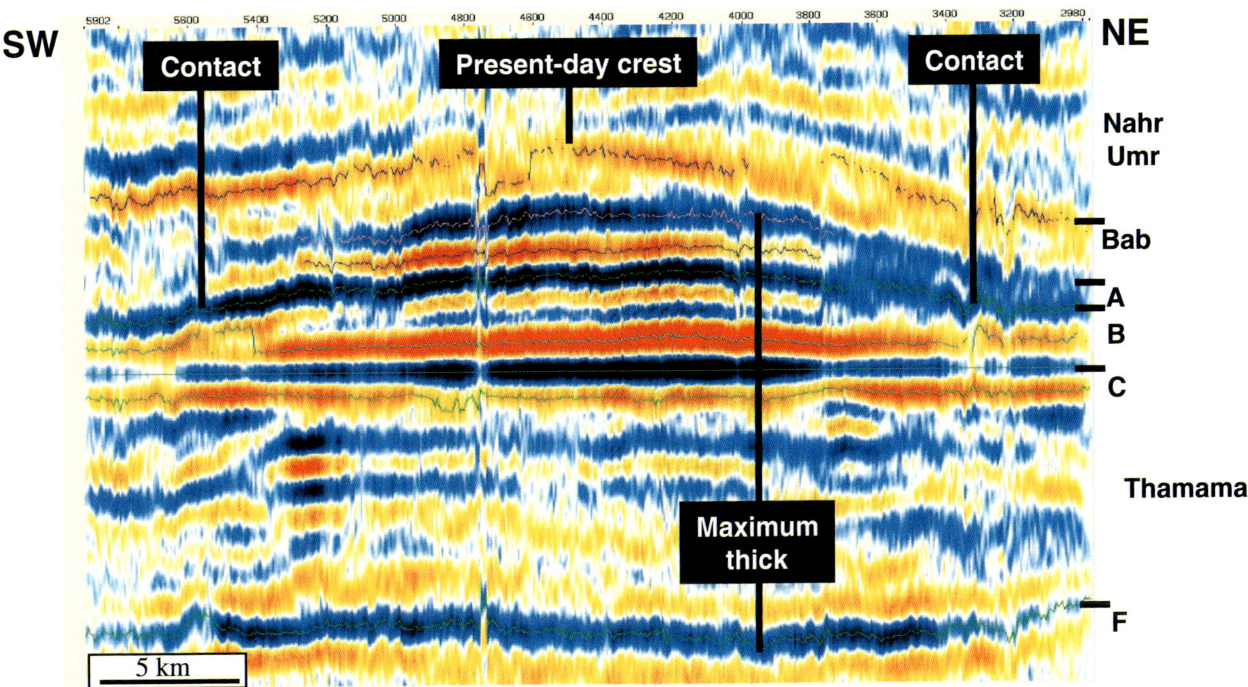

FIGURE 27. Seismic line along the crest of the field, flattened at top C. This shows thickening of the Thamama section resulting from diagenetic preservation, with maximum thickness offset from present-day crest.

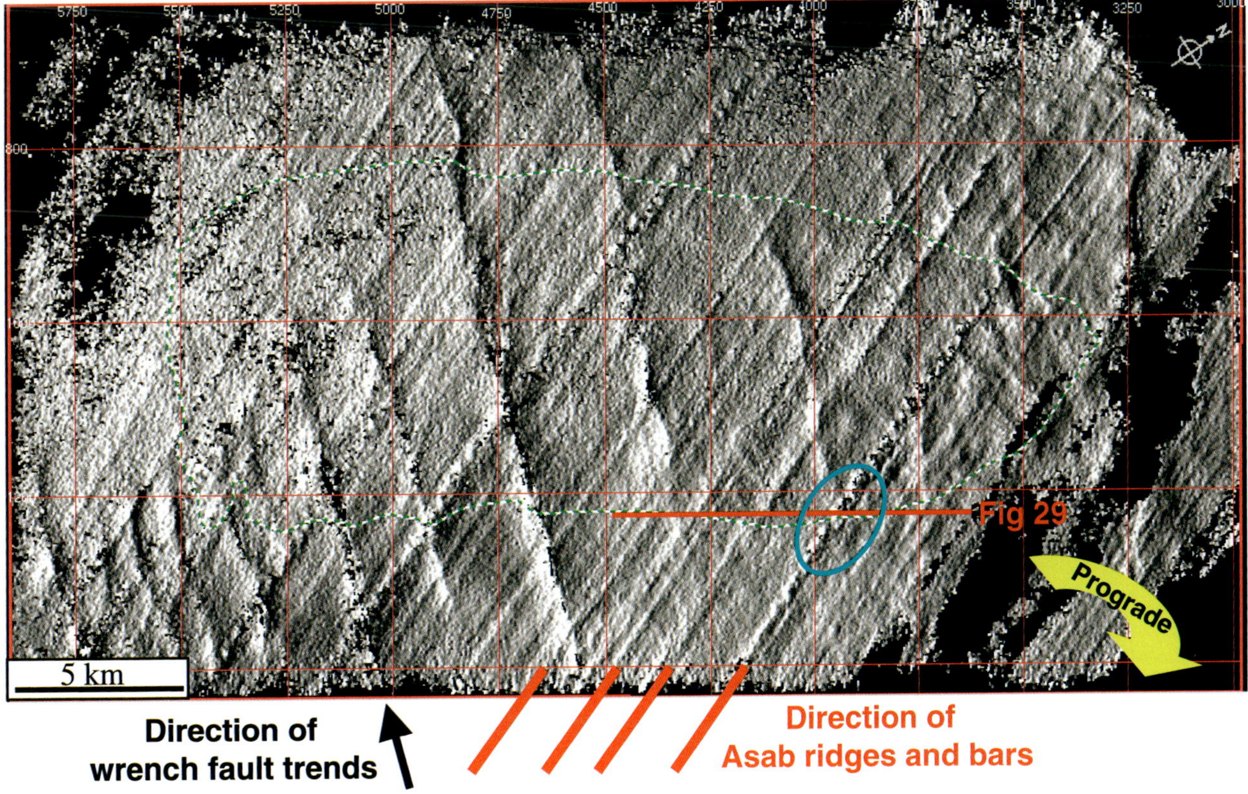

FIGURE 28. Dip map of Top Asab Formation, Top Jurassic, illuminated from the south. This shows both wrench fault trends and depositional lineaments. To the south of a prograde in the north are linear bars and ridges with a north-northwest–south-southeast grain. Larger ridges can be seen to have intermittent gaps.

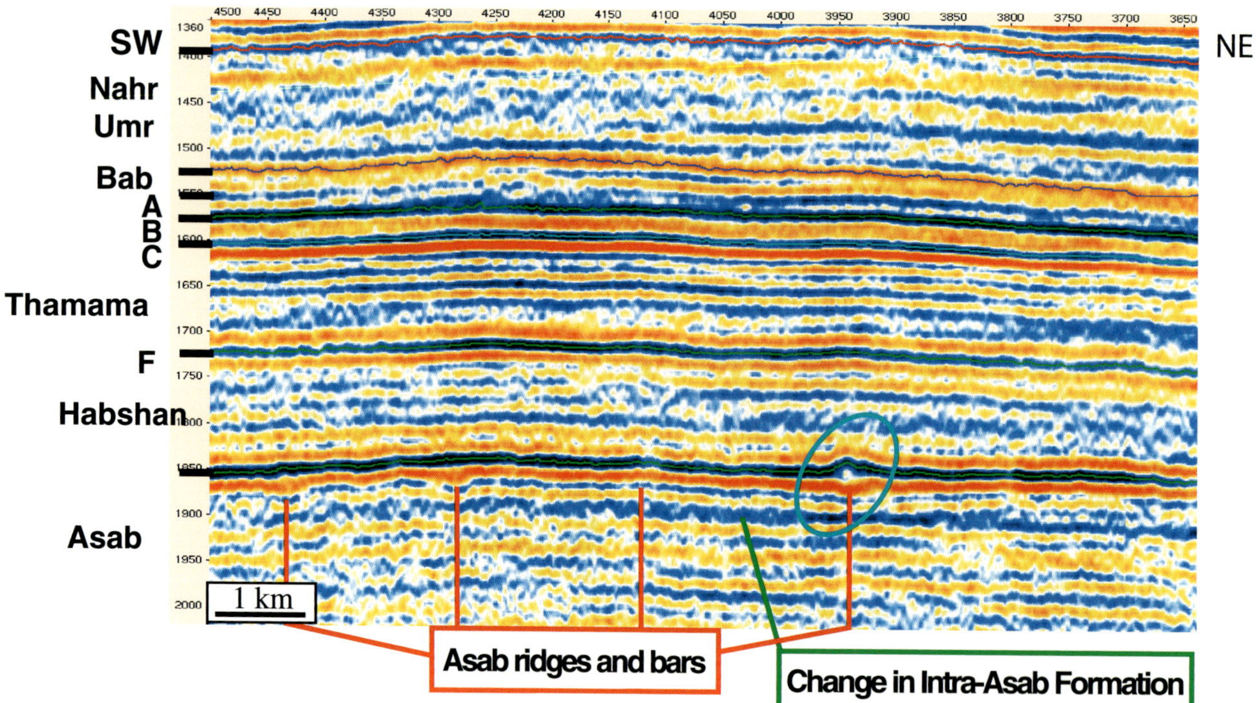

FIGURE 29. A detailed section of seismic line along flank of field showing lens-shaped anomalies at Top Asab corresponding to lineaments seen on the dip map. These lens features do not occur in the shallower section and are not "footprint." Beneath the Top Asab, an older edge can also be seen, where the character of a reflector changes from strong to weak and hummocky.

west-northwest–east-southeast direction. Initially, the lineations were thought to be seismic "footprint," but no other surfaces show a similar pattern. Careful analysis shows that the features are not completely parallel or linear, nor are they oriented correctly to the acquisition pattern to be "footprint." After careful study, similar ridgelike features have subsequently been found elsewhere on other 3-D data along the trend of the Hith edge. A northeast-southwest seismic section (Figure 29) on the northeast flank of the field across these ridges shows that they are caused by bodies at the top of the Asab Formation that are lens-shaped in cross section. They are only occasionally associated with deeper faults and a few can be seen to be draped by shallower surfaces. Some of the larger ridges can be seen to be discontinuous along their length and are considered to be shallow to marginal marine barrier-beach bars with intersecting channels. The ridges lie shoreward of the prograde, become smaller, and die away farther to the southwest, where Hith evaporites are found in a well. The seismic interpretation of these ridges as shoreline features matches with earlier interpretation of the well data by Matos (1994), who suggested that because of the absence of marine cements, the oolites are emergent shoals in a wave-dominated barrier island complex. Such shoreline features are rarely seen by seismic and have been preserved here by the rapid Jurassic to Cretaceous transgression. A character change can also be seen in the underlying Upper Jurassic Asab section (Figure 29), from a strong reflector in the north to a weak, hummocky reflector in the south. This is inferred to be an earlier shoreline edge set back and beneath the Top Asab edge.

CONCLUSIONS

It is clear that neither seismic nor well data provide a full picture of the subsurface strata. The seismic data can provide an areal picture with lateral detail, but lack the fine vertical detail of the wells. The wells have hard data, such as core, for detailed sedimentology and 6-in. sampled log data for petrophysics, but lack the lateral sampling away from the borehole. Together, the wells and seismic data form complementary data sets for a better understanding of the subsurface geology.

In this field, the combination of the two data sets has allowed major improvements to be made to the reservoir model with better modeling of the behavior of the reservoir layers between the wells. In particular, there is now a better understanding of the downflank stylolitic dense layers. The reservoir zones are now recognized as connecting through the complex wrench faults. Significant paleogeographic information has also been obtained from the depositional geometries seen on the seismic data at Bab and Asab levels, which may have an impact on continuing exploration.

ACKNOWLEDGMENTS

The authors thank ADCO and ADNOC for permission to publish this chapter and their colleagues, past and present, in ADCO for their ideas and unpublished work. Thanks are also due to the editors of this volume for their constructive suggestions and to Alison Melville for advice on English usage and linguistic style.

REFERENCES CITED

Alsharhan, A. S., 1993, Asab field— United Arab Emirates: Rub Al Khali Basin, Abu Dhabi, *in* N. H. Foster and E. A. Beaumont, comps., Structural traps VIII: AAPG Treatise of Petroleum Geology, Atlas of Oil and Gas Fields, p. 69–97.

Alsharhan, A. S., and A. E. M. Nairn, 1986, A review of the Cretaceous formations in the Arabian Peninsula and Gulf; Part I. Lower Cretaceous (Thamama group) stratigraphy and paleogeography: Journal of Petroleum Geology, v. 9, p. 365–392.

Azer, S. R., and C. Toland, 1993, Sea level changes in the Aptian and Barremian (Upper Thamama) of offshore Abu Dhabi: Middle East Oil Technical Conference and Exhibition, Bahrain, SPE Paper 25610, p. 141–154.

Aziz, S. K., and M. Abd El-Sattar, 1997, Sequence stratigraphic modelling of the Lower Thamama group, east onshore Abu Dhabi, United Arab Emirates: GeoArabia, v. 2, p. 179–202.

Boichard, R. A. P., A. S. Al-Suweidi, and H. Karakhian, 1994, Sequence boundary types and related porosity evolutions: Example of the Upper Thamama Group in field "A" offshore Abu Dhabi, United Arab Emirates, *in* M. I. Al-Husseini, ed., Geo'94, Middle East Petroleum Geosciences Conference 1: Bahrain, Gulf Petrolink, p. 191–201.

Buchem, F. S. P. Van, B. Pittet, H. Hillgärtner, J. Grötsch, A. I. Al Mansouri, I. M. Billing, H. H. J. Droste, and W. H. Oterdoom, 2002, High-resolution sequence stratigraphic architecture of Barremian/Aptian carbonate systems in northern Oman and the United Arab Emirates (Kharaib and Shu'aiba formations): GeoArabia, v. 7, p. 461–500.

Burchette, T. P., and S. R. Britton, 1985, Carbonate facies analysis in the exploration for hydrocarbons: A case study from the Cretaceous of the Middle East, *in* P. J. Brenchley and B. P. J. Williams, eds., Sedimentology: Recent developments and applied aspects: Geological Society of London, p. 311–338.

Granier, B. R. C., 2000, Lower Cretaceous stratigraphy of Abu Dhabi and the United Arab Emirates— A reappraisal: Ninth Abu Dhabi International Petroleum Exhibition and Conference, ADIPEC 2000, Abu Dhabi, Conference Proceedings, ADIPEC 0918, p. 526–535.

Grötsch, J., O. Al-Jeelani, and Y. Al-Mehairi, 1998, Integrated reservoir characterisation of a giant Lower Cretaceous oil field, Abu Dhabi, U.A.E.: Eighth Abu Dhabi International Petroleum Exhibition and Conference, ADIPEC 1998, Abu Dhabi, Conference Proceedings, SPE 49454, p. 77–86.

Grötsch, J., P. Melville, O. Al-Jeelani, K. Leyrer, and M. S. Efnik, 2000, Integrated 3-D reservoir characterisation in a giant Lower Cretaceous carbonate reservoir, Abu Dhabi, 4th Middle East Geoscience Conference, Bahrain, March 25–27: GeoArabia, v. 5, p. 97–99.

Harris, P. M., S. H. Frost, G. A. Seiglie, and N. Schneiderman, 1984, Regional unconformities and depositional cycles, Cretaceous of the Arabian peninsula, *in* J. S. Schlee, ed., Interregional unconformities and hydrocarbon accumulation: AAPG Memoir 36, p. 67–80.

Hassan, A., F. Youssef, and M. Ayoub, 1990, Controlling water flood advance in Asab Thamama B reservoir: Fourth Abu Dhabi International Petroleum Exhibition and Conference, ADIPEC 1990, Abu Dhabi, Conference Proceedings, SPE 701, p. 292–306.

Hawas, M. F., and H. Takezaki, 1994, A model for migration and accumulation of hydrocarbons in the Thamama and Arab reservoirs in Abu Dhabi, *in* M. I. Al-Husseini, ed., Geo '94, Middle East Petroleum Geosciences Conference 2: Bahrain, Gulf Petrolink, p. 483–494

Johnson, J. A. D., and S. R. Budd, 1975, The geology of the Zone B and Zone C Lower Cretaceous limestone reservoirs of Asab field, Abu Dhabi: Ninth Arab Petroleum Congress (Dubai), Paper 109 (B-3), p. 1–24.

Koepnick, R. B., 1987, Distribution and permeability of stylolite-bearing horizons within a Lower Cretaceous carbonate reservoir in the Middle East: SPE Formation Evaluation, v. 2, p. 137–142.

Landmesser, P., and A. S. Saydam, 1996, Seismostratigraphic interpretation of Lower Thamama/Habshan in SE Abu Dhabi: Seventh Abu Dhabi International Petroleum Exhibition and Conference, ADIPEC 1996, Abu Dhabi, Conference Proceedings, SPE 36204, p. 244–254.

Marzouk, I., and M. A. Sattar, 1994, Wrench tectonics in Abu Dhabi, United Arab Emirates, *in* M. I. Al-Husseini, ed., Geo'94, Middle East Petroleum Geosciences Conference 2: Bahrain, Gulf Petrolink, p. 655–668.

Matos, J. E. de, 1994, Upper Jurassic–Lower Cretaceous stratigraphy: The Arab, Hith and Rayda formations in Abu Dhabi, *in* M. D. Simmons, ed., Micropalaeontology and hydrocarbon exploration in the Middle East: London, Chapman & Hall, p. 81–101.

Murris, R. J., 1980, Middle East: Stratigraphic evolution and oil habitat: AAPG Bulletin, v. 64, p. 597–618.

Oswald, E. J., H. W. Mueller, D. F. Goff, H. Al-Habshi, and S. Al-Matroushi, 1995, Controls on porosity evolution in Thamama Group carbonate reservoirs in Abu

Dhabi, U.A.E.: Middle East Oil Show, Bahrain, Conference Proceedings, SPE 029797, p. 251–265.

Sharland, P. R., R. Archer, D. M. Casey, R. B. Davies, S. H. Hall, A. P. Heward, A. D. Horbury, and M. D. Simmons, 2001, Arabian plate sequence stratigraphy: GeoArabia Special Publication 2: Manama, Gulf Petro-Link, 371 p.

Silva, F. P., J. F. Rodrigues, C. Maciel, and S. Alomari, 1996, NE Abu Dhabi— New evidences of wrench tectonics associated with the development of the Oman mountains foredeep: Seventh Abu Dhabi International Petroleum Exhibition and Conference, ADIPEC 1996, Abu Dhabi, Conference Proceedings, SPE 36276, p. 255–262.

Vail, P. R., F. Audemard, S. A. Bowman, P. N. Eisner, and C. Perez-Cruz, 1991, The stratigraphic signatures of tectonics, eustasy and sedimentology, *in* G. Einsele, W. Ricken, and A. Seilacher, eds., Cycles and events in stratigraphy: Berlin, Springer-Verlag, p. 617–659.

7

Heubeck, C., K. Story, P. Peng, C. Sullivan, and S. Duff, 2004, An integrated reservoir study of the Liuhua 11-1 field using a high-resolution three-dimensional seismic data set, *in* Seismic imaging of carbonate reservoirs and systems: AAPG Memoir 81, p. 149–168.

An Integrated Reservoir Study of the Liuhua 11-1 Field Using a High-Resolution Three-Dimensional Seismic Data Set

Christoph Heubeck
Department of Geosciences, Freie Universität Berlin, Berlin, Germany

Kenneth Story
DDD Energy, Houston, Texas, U.S.A.

Pat Peng
BP, Houston, Texas, U.S.A.

Claire Sullivan
BP, Houston, Texas, U.S.A.

Stuart Duff
Independent contractor, Wellington, New Zealand

ABSTRACT

Liuhua 11-1 field, located in the Pearl River Mouth Basin offshore south China, consists of diagenetically altered Miocene limestone comprising a shallow-water carbonate bank. This bank forms the topmost and youngest interval of a larger, extensively karsted, buried carbonate platform. A three-dimensional (3-D) seismic survey of Liuhua field yielded a very high-resolution data set (>200 Hz), allowing a spatial resolution less than 5 m. This data set was subsequently used to produce a reservoir model that closely linked petrophysical, log, and seismic data.

The carbonate stratigraphy suggests several subaerial exposure events that significantly modify primary stratification of the carbonate bank through diagenesis. These include freshwater leaching, burial compaction, cementation, and late diagenetic flushing of the bank. The combined diagenetic changes had three principal effects: (1) exacerbation of primary facies-dependent differences in porosity through a series of dissolution-reprecipitation steps; (2) widespread incipient carbonate collapse at or below the scale of seismic resolution; and (3) formation of numerous regionally occurring karst sinkholes of as much as 400 m diameter shortly before final drowning of the platform. Incipient collapse of the friable carbonate framework is expressed seismically by a reduction in amplitude.

Carbonate dissolution appears to be ongoing because sagging continues to affect all strata overlying the reservoir to the seafloor. Subsurface dissolution may be a result of either flushing of the carbonate platform by cold, undersaturated marine waters or may be a result of active biodegradation of the hydrocarbons along the oil-water contact and the concomitant release of acids.

INTRODUCTION

The Pearl River Mouth Basin forms the passive Atlantic-type margin of south China related to the Tertiary opening of the South China Sea (Fulthorpe and Schlanger, 1989; Dorobek, 1997). In this basin, a geographically restricted carbonate platform formed on top of the subsiding Dongsha horst block during rifting of the South China Sea in the middle Miocene (Figure 1). The topmost, geographically most restrictive, and youngest part of this platform now forms an elevated horst block containing the Liuhua 11-1 field (Christian and Tyrrell, 1991; Erlich et al., 1990; Moldovanyi et al., 1995; Turner and Hu, 1990, 1991; Turner, 1990; Tyrrell and Christian, 1992). The reservoir is comprised of middle Miocene bioclastic rhodolith packstones and foram-algal grainstones of the Zhujiang Formation. They overlie fluviodeltaic sandstones and shales of the Miocene Zhuhai Formation, which, in turn, unconformably overlie granitic basement of the Dongsha block (Figure 2). The carbonate bank (Zhujiang Formation) is sharply overlain by more than 1000 m of Miocene–Holocene marine shales of the Hanjiang Formation across a flooding surface. This flooding surface was described as an example of a "drowning unconformity" by Erlich et al. (1990, 1991), Tyrrell and Christian (1992), and Erlich et al. (1993).

The reservoir, containing approximately 1.3 BBOIP, produces biodegraded, 16–22° API oil from shallow depths with a very strong bottom-water drive. The

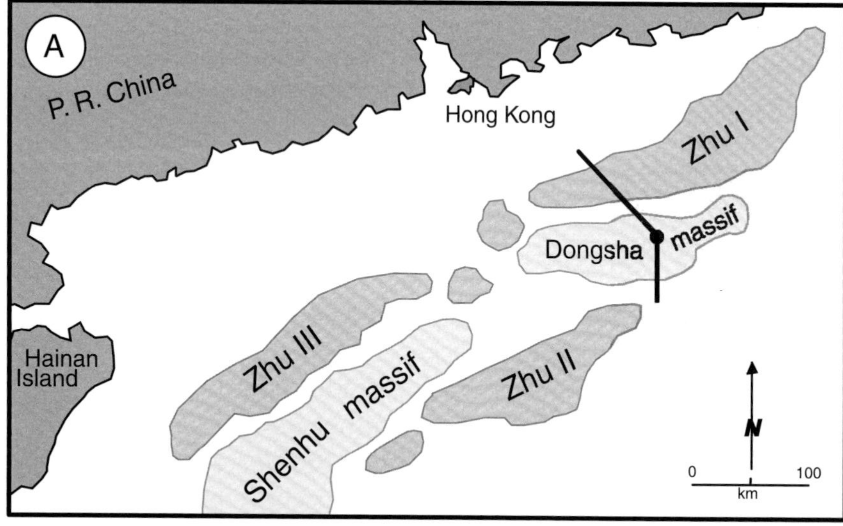

FIGURE 1. (A) Basin map of Pearl River Mouth Basin. Liuhua field is located approximately 210 km southeast of Hong Kong in a water depth of 300 m in platform carbonates that formed in the structurally highest part of the Dongsha horst block during Miocene rifting of the South China Sea between the Zhu I and Zhu II turbidite-filled basins. (B) Schematic regional geologic cross section. Modified from Tyrrell and Christian (1992).

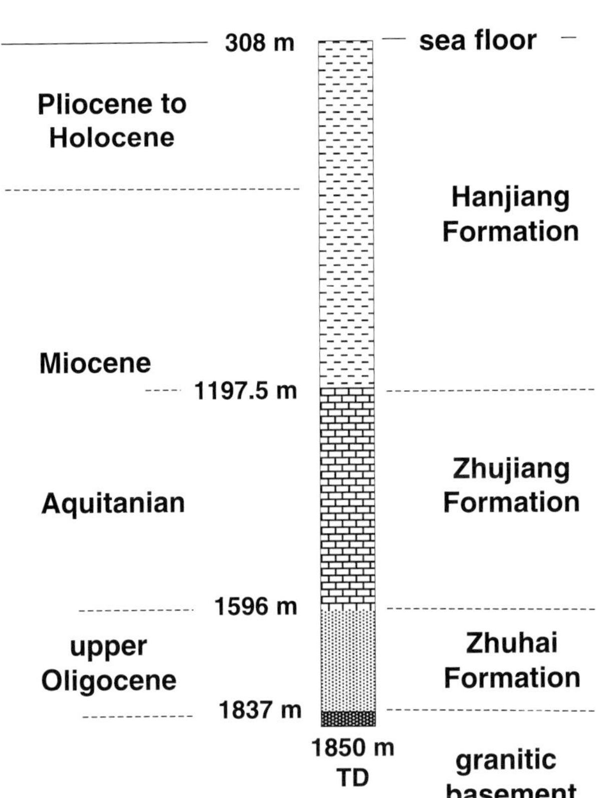

FIGURE 2. Stratigraphic column encountered in well LH 11-1-1A, the discovery well for the Liuhua field (modified after Tyrrell and Christian, 1992). The sedimentary sequence overlying the crystalline basement of the Dongsha horst block is a typical rift-drift sequence of increasing water depth. The fluviodeltaic deposits of the Zhuhai Formation are overlain by Miocene platform carbonates of the Zhujiang Formation whose upper part contains the Liuhua reservoir. The Zhujiang Formation is, in turn, unconformably overlain by deep-marine shales of the Hanjiang Formation. TD = total depth.

combination of heavy oil, shallow reservoir, and strong water drive requires the accurate placement of horizontal production wells.

This chapter describes the principal insights gained from the high-resolution seismic survey over Liuhua field into distribution and origins of reservoir heterogeneity through the integration of geophysical, geological, and petrophysical data, thereby allowing insights in subaerial and subsurface karst processes on field scale. It also briefly describes the construction of a reservoir model from the seismic data through integration of production data from wells with petrophysical parameters from core and logs, visualization, and finally reservoir modeling, including predictive flow simulation.

RESERVOIR GEOLOGY

Field History, Depositional Environment, and Diagenesis

The Liuhua 11-1 field is located 210 km southeast of Hong Kong in a water depth of approximately 300 m (Figure 1). It was discovered by Amoco in January 1987. Significant technical challenges to field development included the water depth, its location in the "typhoon alley," and the presence of heavy oil in a large but shallow and relatively thin reservoir interval with a very strong bottom-water drive. Several geologic aspects of Liuhua and the Dongsha horst have been described in detail in previous publications, including its regional setting and exploration history (Tyrrell and Christian, 1992; Wu et al., 1997), its stratigraphy and karst features from seismic data (Moldovanyi et al., 1995), and the drowning of the carbonate platform (Erlich et al., 1990). Field development was authorized in March 1993 and production from the first of a total of 25 horizontal extended-reach wells started in March 1996 (Gu and Ye, 1992). To optimize production through the quantification of reservoir compartmentalization and heterogeneity, a three-dimensional (3-D) survey was acquired over the field in July 1997. Geologic and seismic acquisition parameters enabled the acquisition of an extremely high resolution seismic data set.

The field is a combination stratigraphic-structural trap in which the primary carbonate bank buildup formed on a structural high. Continued faulting and gentle anticlinal folding of the bank provided a secondary trapping mechanism after deposition of the Hanjiang Shale seal.

Although the bioclastic grainstones and packstones had excellent primary intergranular porosity, most of the reservoir porosity is diagenetic in origin. Petrographic and isotopic data from Liuhua cores indicate that the reservoir potential of Zhujiang carbonates has been modified (from oldest to youngest) by the following processes (Boutell and Moldovanyi, 1992; Moldovanyi et al., 1995; Wagner et al., 1995):

1) Repeated subaerial exposure of the upper part of the carbonate bank and leaching of metastable grains by meteoric waters (James and Choquette, 1990)
2) Corresponding precipitation of calcite cement at various levels near the paleo-ground water table
3) Selective brecciation of low-strength intervals resulting from compaction
4) Postburial leaching, particularly of coarse-grained, high-porosity intervals of sufficient mechanical

Table 1. Sequence of diagenetic events affecting the reservoir of Liuhua field.

Event number (earliest to latest)	*Diagenetic event*	*Impact on porosity development*
1	Subaerial exposure; leaching of metastable grains by meteoric water	⇑
2	Precipitation of calcite cement at various levels near the paleoground water table	⇓⇓
3	Brecciation of low-strength intervals resulting from compaction	⇓⇓
4	Postburial leaching, particularly of coarse-grained, high-porosity intervals of sufficient mechanical strength	⇑⇑
5	Solution compaction, affecting primarily fine-grained intervals	⇓⇓
6	Burial cementation by equant calcite spar	⇓
7	Mechanical compaction affecting primarily intervals of lower strengths	⇓
8	Leaching by compaction water and fluids	⇑⇑⇑

strength, such as red algal boundstones, by cold, undersaturated marine waters

5) Solution compaction affecting primarily fine-grained intervals
6) Burial cementation by equant calcite spar
7) Mechanical compaction affecting primarily intervals of lower strengths
8) Leaching by compaction water and fluids

The effects of these processes varied through time (Table 1). Some processes, such as meteoric leaching, were most important during marine and shallow burial diagenesis, whereas others, such as solution compaction and compaction-water leaching, became more important during later stages of burial and may be active even today. The reservoir is free of dolomite and of terrigenous input, such as clay minerals, based on log evaluation and core data. This monomineralic $CaCO_3$ composition facilitated seismic interpretation and petrophysical modeling.

As a result of the extensive diagenetic overprint, low-porosity, highly cemented ("tight") streaks of occur commonly within the Liuhua reservoir. They are laterally extensive, but not contiguous. Core inspection, petrography, and log analysis of these zones show that their thickness ranges from a few centimeters to a few tens of meters. Turner and Hu (1991) and Moldovanyi et al. (1995) describe their petrographic and isotopic characteristics in detail. These authors (as well as P. Wagner, 1995, personal communication) conclude that the low-porosity zones are principally the result of several events of subaerial exposure of the carbonate bank and precipitation of dissolved calcite by meteoric fluids near the water table and were accentuated by subsequent diagenetic events.

The principal low-porosity zones are sufficiently continuous, thick, or well-cemented to affect fluid flow through the reservoir and are, in part, above the seismic resolution of approximately 4 m. Therefore, they can be characterized in the 3-D reservoir description and the numerical reservoir model. The low-porosity zones have a double effect on well productivity: They protect the wellbore from early water breakthrough and reduce pressure support from the strong bottom-water drive. Consequently, development wells drilled into a high-porosity lower zone with only a moderately tight subjacent zone had very high productivity but watered out rapidly.

Reservoir Stratigraphy

Seismic stratigraphy of the Liuhua bank shows evidence of being predominantly primary in some places and predominantly diagenetic in origin in others. Core inspection and thin-section petrography show that the reservoir stratigraphy on a seismic scale is a result of postdepositional (i.e., diagenetic) porosity enhancements and reductions; however, the cores also show that these diagenetic changes follow primary grain-size differences and therefore outline primary depositional facies. Seismic stratification appears to outline primary depositional features, such as mounds, bank fronts, and platform flats (Figure 4). We therefore postulate that most porosity stratification in Liuhua field represents actual time surfaces that bound depositional units (Moldovanyi et al., 1995), representing a diagenetic enhancement of an original facies-dependent primary stratigraphy.

The detailed reservoir stratigraphy of Liuhua field has been outlined, based on the largely cored discovery and appraisal wells, by Tyrrell and Christian (1992) and Moldovanyi et al. (1995) and will not be repeated here. Wagner et al. (1995) described its petrographic and isotopic composition. The stratigraphic description below essentially confirms these descriptions but adds additional detail gained through the drilling and geologic interpretation of the development wells (Stark, 1991; Heubeck, 1998).

Liuhua field comprises (from top to bottom) the following units (Figures 3, 4):

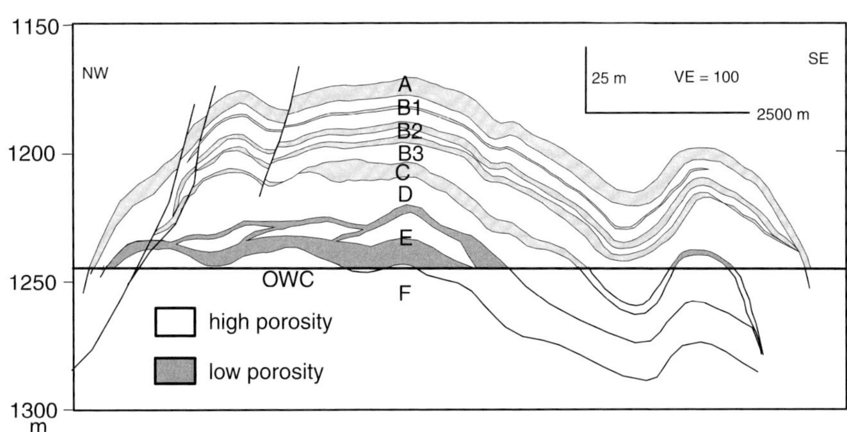

FIGURE 3. Simplified geologic cross section across Liuhua field. The stratigraphic architecture of Liuhua 11-1 field alternates between thin zones of high and low porosity with varying degrees of fieldwide continuity. These variations constitute largely highly varying degrees of cementation, but may also, in part, accentuate primary fabric and grain-size variation of the different facies in the carbonate bank depositional environment. Note the high vertical exaggeration (VE). OWC = oil-water contact.

- The A Zone, uppermost unit of the Zhujiang Formation on the Liuhua structure, is dominantly a firm rhodolith-foraminifera packstone approximately 4.3–5.5 m thick, infiltrated by marine clays. Porosity is mostly interskeletal and averages about 13%. The A Zone is overlain by a thin, discontinuous glauconitic sand at the base of the overlying Hanjiang Formation that buried the drowned bank.
- The B1 Zone is a firm, friable rhodolith-foraminiferal packstone, 7.9–8.5 m thick (Figure 3). Bioclastic and skeletal material include algae, calcareous benthonic foraminifera, and indeterminate skeletal material that possibly includes bryozoans, coral fragments, and radiolaria. Intraskeletal and framework porosity ranges between 25 and 33%. This unit is the uppermost high-porosity zone within the reservoir. Extensive development drilling in this zone showed that a 1- to 1.5-m-thick "lower"-porosity (17–24%) streak (termed the *B12 subzone*) occurs at a fixed position within the B1 Zone throughout the development area. This slightly more cemented streak serves as an excellent marker during directional drilling operations.
- The B2 Zone is approximately 13.7 m thick and lithologically identical to the B1 Zone but is generally more cemented by opaque calcite cement (Figures 3, 5). In comparison to the B1 Zone, voids, cavities, and spaces in the B2 Zone are greatly reduced in size and frequency. This zone is highly stratified internally with seven alternating high- (16–24%) and low- (6–12%) porosity streaks, each about 0.3–2.5 m thick. Thickness and porosity values of these streaks are consistent throughout the 2- × 4-km area accessed by the development wells. Therefore, the detailed stratification within the B2 Zone, expressed in porosity logs of individual wellbores, can be readily correlated across the field.
- The B3 Zone, approximately 9 m thick, is lithologically similar to the B1 Zone but shows even higher visual and log porosity, ranging from 23 to 34%. Moldic, interskeletal, and intraskeletal porosity are common. A single development well drilled into this zone reached high productivity, probably because of better pressure support from the water leg, but watered out quickly.
- The C Zone is predominantly a milky white foraminiferal-skeletal packstone, tightly cemented by white opaque calcite, with only minor visual porosity in cuttings and core (Figure 5). Fracture porosity, however, is common. The C Zone is the principal "tight" zone within the reservoir. On 3-D seismic data, the C Zone grades from a strong and continuous seismic reflector on the west side of the field with approximately 14% log-calculated porosity in a seismically discontinuous, 6-m-thick unit of approximately 27% porosity to the east. No development well has penetrated the C Zone or any of the underlying zones.
- The D Zone was penetrated only twice by appraisal wells. It appears internally complex on seismic (Figures 3, 4) but is overall a high-porosity zone with a mean log porosity 28%. Low-porosity streaks within the D Zone are visible on seismic sections. These may outline possible depositional mounds.
- The E Zone forms a low-porosity zone of foraminiferal packstone, approximately 17 m thick, at the base of the oil-filled reservoir. The low porosity may be a result of calcite cementation along the former oil-water contact (OWC) by degradation of oil by oxygen-rich water. Its average porosity is 16% but varies widely. Subsequent deformation caused the E Zone to bend into a very open anticline. At the bank margins, the E Zone dips below the OWC. It is, therefore, not effective in protecting overlying oil and horizontal wellbores from water coning.
- The F Zone is a highly porous zone of leached, friable skeletal packstone and grainstone below the OWC. Interpretation of core texture and thin

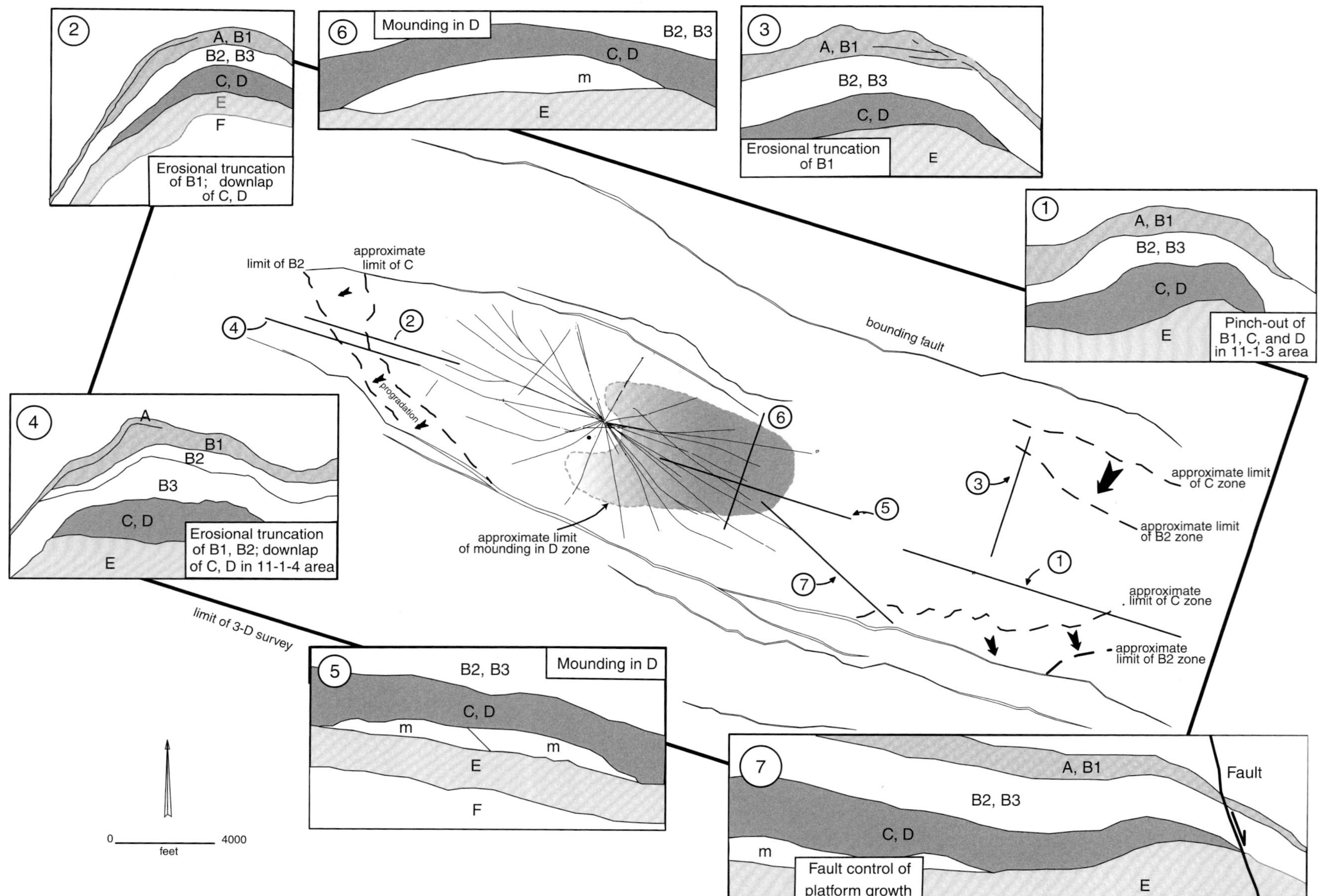

FIGURE 4. Liuhua bank stratigraphic architecture from 3-D seismic data, based on selected seismic sections through parts of the field. Central map shows area of 3-D survey, location of faults bounding the platform, development wells, and profiles 1 through 7 (numbered insets) taken from seismic time sections. These insets show suspected mounding in the D Zone and limited seaward (southward) progradation of the bank margin in its upper part (B and C Zones).

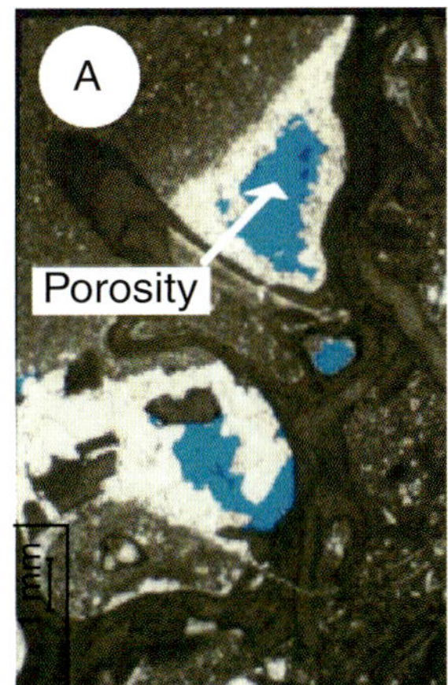

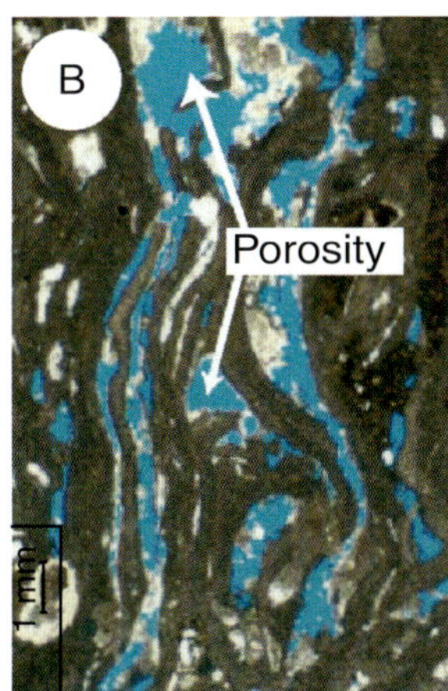

FIGURE 5. Thin-section photomicrographs and core slabs of the Zhujiang Formation in Liuhua field. (A) Thin section showing vuggy and moldic porosity in the well-cemented, low-porosity C Zone. (B) Algal boundstone from the high-porosity B2 subzone showing interskeletal porosity. (C) False-color core-slab photograph showing friable rhodolith packstone, also from the high-porosity B2 Zone. Core width is approximately 12 cm.

sections suggests that this zone was exposed to flushing by cold seawater undersaturated with respect to $CaCO_3$, whereas the strong water drive implies that this flushing may extend to the present.

SEISMIC GEOPHYSICS

Rationale for Three-Dimensional Seismic Data

Development wells had delivered an abundance of information about the composition, structure, and fine-scale stratigraphy of the A, B1, B2, and B3 Zones. Well production data, however, had made it clear that individual well performance was also strongly affected by the continuity, permeability, and stratigraphy of the zones underlying the producing sections of the wellbores (Figures 3, 4). Little was known about these zones aside from the partially cored appraisal wells and the two-dimensional (2-D) seismic lines, thus preventing proper estimates about expected well productivity and risk of premature water coning. Seismic source and acquisition parameter modeling and testing on a reprocessed 2-D line suggested that a 3-D survey could clarify internal reservoir heterogeneity to a high degree and identify high-porosity well targets.

Seismic Acquisition and Inversion

The seismic acquisition was conducted during calm seas with short 1500-m streamers and shallow 3.5-m-tow depths. The 180-Hz field data, enhanced during processing, produced peak frequencies of 240 Hz. Temporal and spatial resolution of this data set is approximately 4 m; detectability is about 1.5 m. Approximately 4 million traces were processed at a bin spacing of 5×5 m over a 100-km^2 area (Figure 6).

Seismic stratigraphy and wellbore data were used to constrain an acoustic impedance inversion of the time data. Because the relationship of acoustic impedance to core porosity is nearly linear for Liuhua, because of the monomineralic mineralogy of the reservoir limestone, fieldwide acoustic impedance was readily calculated, converted to porosity through a quadratic regression equation related to core-plug data, and matched to core-calibrated log data from the vertical wells penetrating the reservoir. Ultimately, petrophysical data and modeling coupled with the seismic inversion were used together to create a spatial distribution of porosity, permeability, and saturation.

Seismic Interpretation

Interpretation of the regional 2-D seismic data has identified the sequence-stratigraphic framework of the Dongsha Platform and recognized the regional distribution of (poorly resolved) karst features (Moldovanyi et al., 1995). Over the bank buildup of Liuhua field, another 2-D seismic survey, spaced 2 and 1 km apart, has identified the general reservoir architecture of the field and recognized its problematic stratigraphic heterogeneity (Moldovanyi et al., 1995). The newly acquired 3-D data offered the opportunity to greatly expand the level of detail in the structural and stratigraphic interpretation. The interpretation focused on (1) quantifying the relative contributions of stratigraphic trapping vs. folding, faulting, and fracturing in trap definition; (2) the influence of damage halos near the bounding faults in constraining productive wellbore lengths; (3) which "major" faults penetrated the reservoir to the OWC and served as conduits for water coning and which "minor" faults did not; and (4) the impact of karst collapse structures on reservoir hydraulics and field development.

Five stratigraphic tops (A, B2, C, E, and F) and the principal bounding faults of the Liuhua horst block

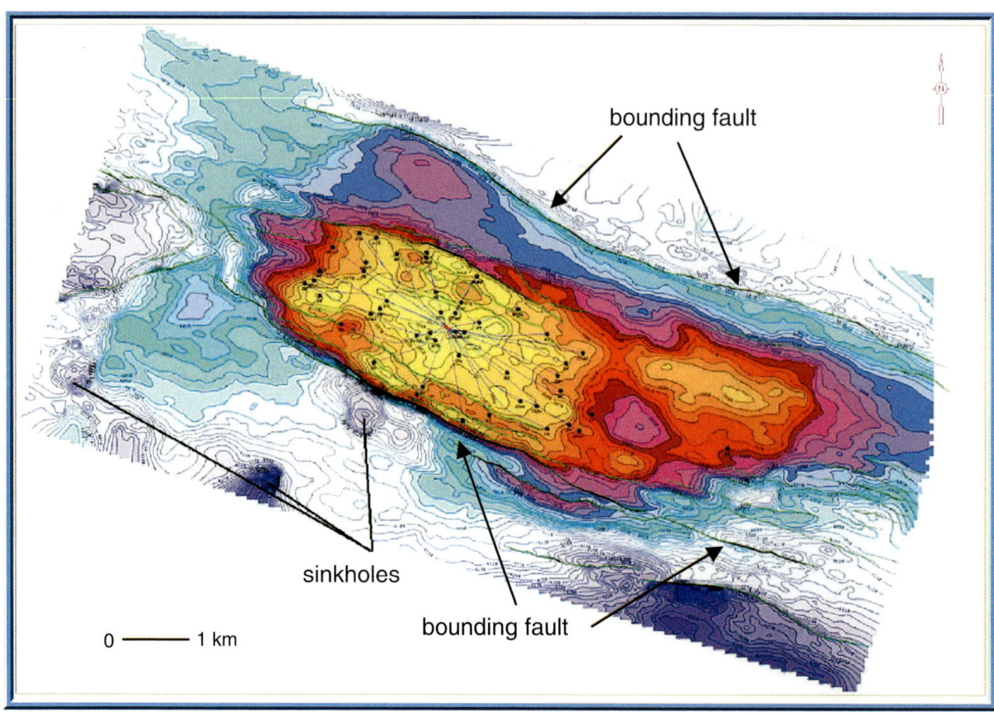

FIGURE 6. Depth-structure map for the top of the Liuhua carbonate platform, also top of the reservoir (top of zone A). Development wells are shown as thin black lines, extending from the platform at the central and structurally highest part of the carbonate bank. Note the west-northwest–east-southeast–striking faults bounding the horst structure and large several circular depressions interpreted as carbonate-collapse features (karst-related sinkholes) just to the southwest of the field.

were mapped on the seismic data set. Structural interpretation grouped structural elements into one of five classes: bounding faults, major faults, minor faults, chaotic zones, and gas sag margins. The depth structure map for the top unit of the carbonate complex (the Top A surface) is shown in Figure 6.

The Liuhua horst block between the major bounding faults is gently folded into a broad, northwest-southeast–trending, doubly plunging anticline (Figure 7). Total relief on this anticline at the top of the reservoir between the top of the bounding faults and the structural crest is approximately 11 m over a distance

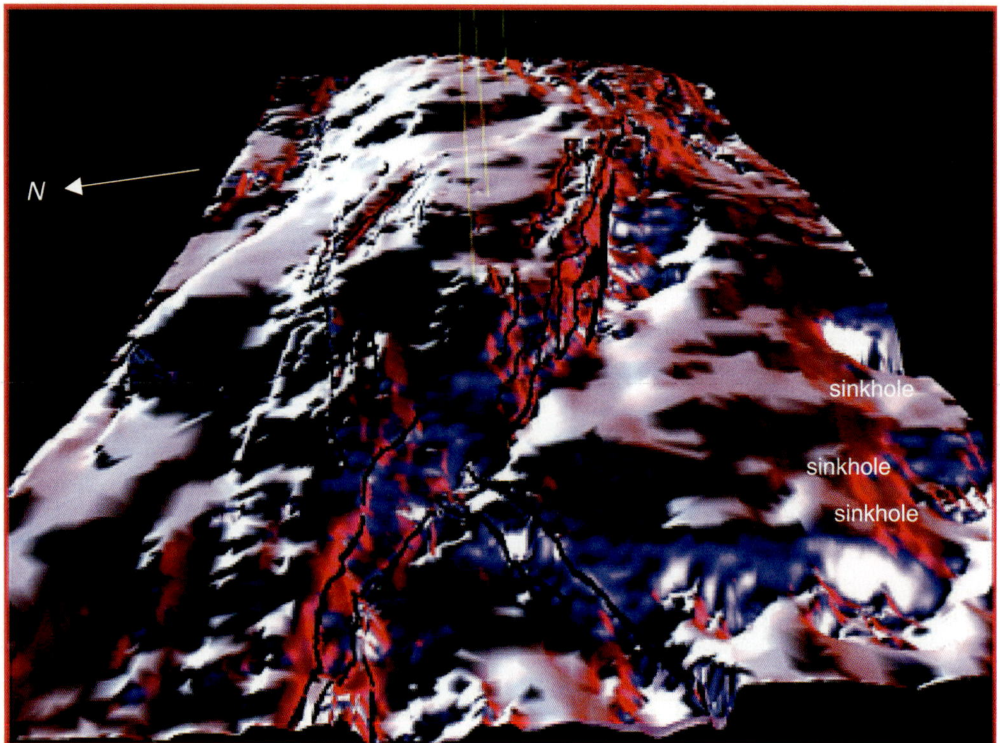

FIGURE 7. Perspective view of the same surface as in Figure 6, the top surface of the reservoir. View is from the northwest along the long axis of the gently folded and moderately faulted Liuhua horst block. Vertical yellow lines show wellbore traces of the wells LH 11-1-4 (foreground, appraisal), LH 11-1-1A (middle ground, discovery well), and LH 11-1-3 (background, appraisal). Several karst sinkholes are partly or completely imaged in the right foreground. Vertical exaggeration = 20.

of approximately 1.2 km. Maximum relief between the structural crest and either down-plunge end of the anticline is approximately 75 m over a distance of 5 km. The maximum elevation difference between the OWC and the crest of the structure is also 75 m.

Bathymetric relief between bank edge and top, associated with original deposition, was probably on the order of approximately 10 m, judged by the bank-margin depositional geometry, the lack of cut-and-fill features throughout, and the uniformity of primary stratification across the bank interior (Figures 3, 4). Facies changes related to present-day structural elevation are not apparent. Therefore, much of the Liuhua folding was probably postdepositional and was probably closely related to the major faulting along the bounding faults near the bank margin. Only some faulting near the bank edge appears to affect zone thicknesses and was, therefore, pre- or syndepositional, at least in part.

Two groups of "bounding faults" on the north and south sides of the Liuhua reservoir delimit the production area, defining an intervening horst block. Maximum dip-slip displacement on the southern and northern bounding faults is 49 and 24 m, respectively. Displacement diminishes along strike because of transfer to subsidiary faults and folds and because of minor changes in strike. In seismic cross sections, the bounding faults can be traced from crystalline basement to the seafloor, which is blanketed by Holocene sediments (Figure 8). The bounding faults may, therefore, still be active. "Gas sags" on seismic time sections occur along strike on both bounding faults. Damage halos associated with bounding faults are approximately 300 m wide on the southern fault but distinctly narrower on the northern bounding faults, possibly because of the reduced offset. Within each damage halo, subsidiary normal faults with offsets generally less than 5 m parallel the bounding faults.

Several "major faults" penetrate the reservoir to the aquifer. The faults typically show a minimum dip-slip displacement of 5 m in their midsection and extend for greater than 300 m in map view. These faults are water conduits and must be strictly avoided by the wellbores. The common "minor faults" represent in part en echelon–stepping segments of suspected strike-slip fault zones. Several wellbore trajectories penetrated minor faults (Figure 10) but minor evidence of faulting was found in only one well log. Individual well productivity does not appear to be adversely affected by penetration of minor faults.

"Chaotic zones" of seismic time sections are zones of diminished and discontinuous amplitude reflectivity and low, distributed coherence in the individual low-porosity zones B2, C, and E. No chaotic zone was cored in Liuhua but several of these zones were penetrated by production wells. Cuttings and well logs through these zones showed no deviation from their appearance in other wellbores. The "chaotic zones" are likely zones of stratigraphic brecciation resulting from carbonate dissolution and probably serve as major channels for water coning toward the production wellbores.

"Gas sags" appear as continuous, straight, or curved zones of low coherency. They appear downfaulted on seismic time sections (Figure 9). However, direct evidence from logs and drill-bit cuttings shows that these seismic discontinuities are artifacts caused by an abrupt lateral variation of seismic traveltime and are not, or only to a minor degree, caused by actual structural displacement. Because seismic reflectors appear offset on seismic time-domain sections, the resulting amplitude discontinuity is computed as a zone of low coherence and resembles a fault on a coherence map (Figure 10).

Conventional structural fracturing appears to be unimportant on a fieldwide scale and therefore for reservoir flow simulation studies in Liuhua. Microfracturing, however, may be more significant. Two production wells were logged by microfracture logging tools but showed only a few hairline cracks along with one open and one closed (cemented) fracture each.

Karst Features

Moldovanyi et al. (1995) correctly interpreted the "hummocky seismic facies" at the top of the Zhujiang Limestone on the regional 2-D seismic lines across the Dongsha Platform as a regional extensive karst field of individual and nested sinkholes (dolines). Of the region surveyed by 2-D seismic, seven large and many small sinkholes fall within the area of the 3-D survey; in addition, the high resolution of the seismic data allows the interpretation of other karst-related features that will be discussed below.

The sinkholes are imaged best with a coherency map at the top of the reservoir (Figure 10). They are typically 100 to approximately 500 m across and show a relief of as much as 15 m. Similar karst sinkholes associated with former sea level lowstands are common features on modern reefs and carbonate banks, e.g., in the Bahamas and Belize (Gascoyne et al., 1979; Mylroie and Carew, 1995) and in the Maldives (Purdy and Bertram, 1993); they are commonly known as "blue holes." Their surface expression is circular because of the dissolution over a cave system at depth and cave-ceiling collapse. Karst dissolution features may form particularly early in regions of tropical cyclone activity (Bourrouilh, 1998). In cross section (Figures 8, 11, 12), at least two additional observations can be made: (1) columns of low amplitude exist within and above the collapse zones and also near the bounding faults; they extend to the sea floor; and (2) concentric structural sag above

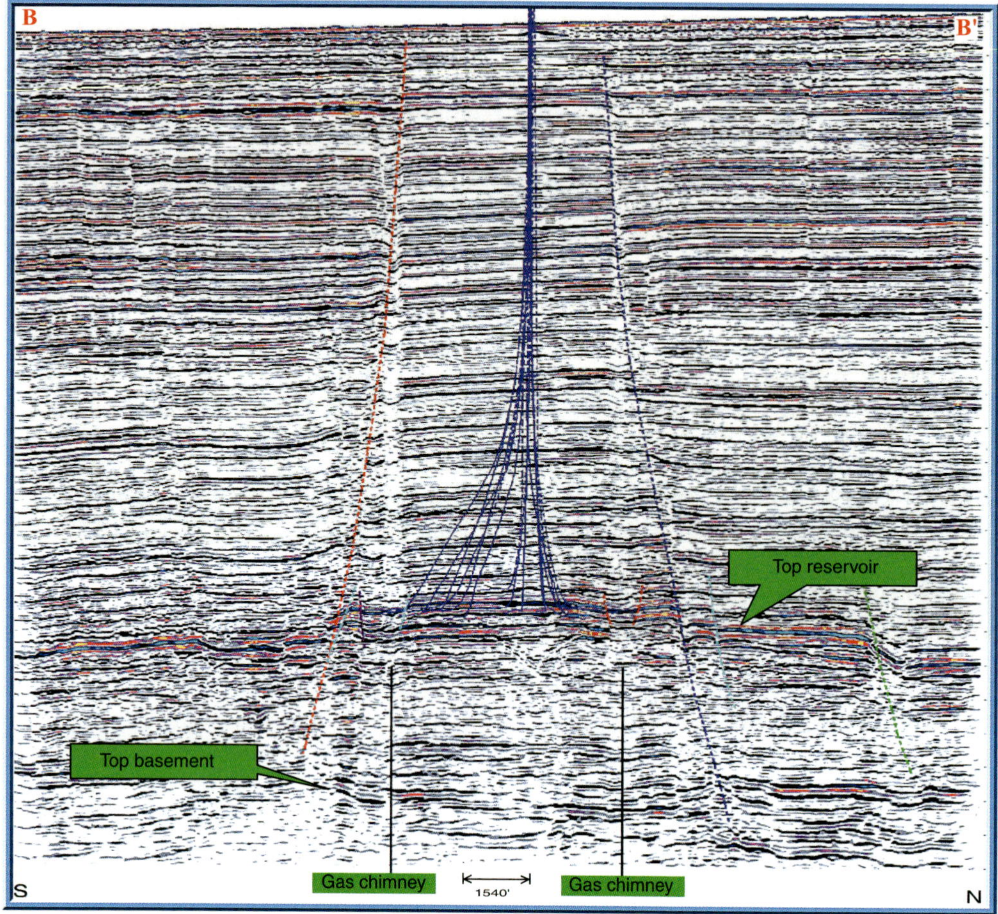

Figure 8. Seismic time section BB′ (for location, see Figure 10) across Liuhua field. Bounding faults can be traced from well below the reservoir nearly to the sea floor, possibly also offsetting Holocene sediments, and suggest that the bounding faults may still be active. Two gas chimneys emanating from the reservoir are imaged as diffuse vertical zones of reduced amplitude and time sag.

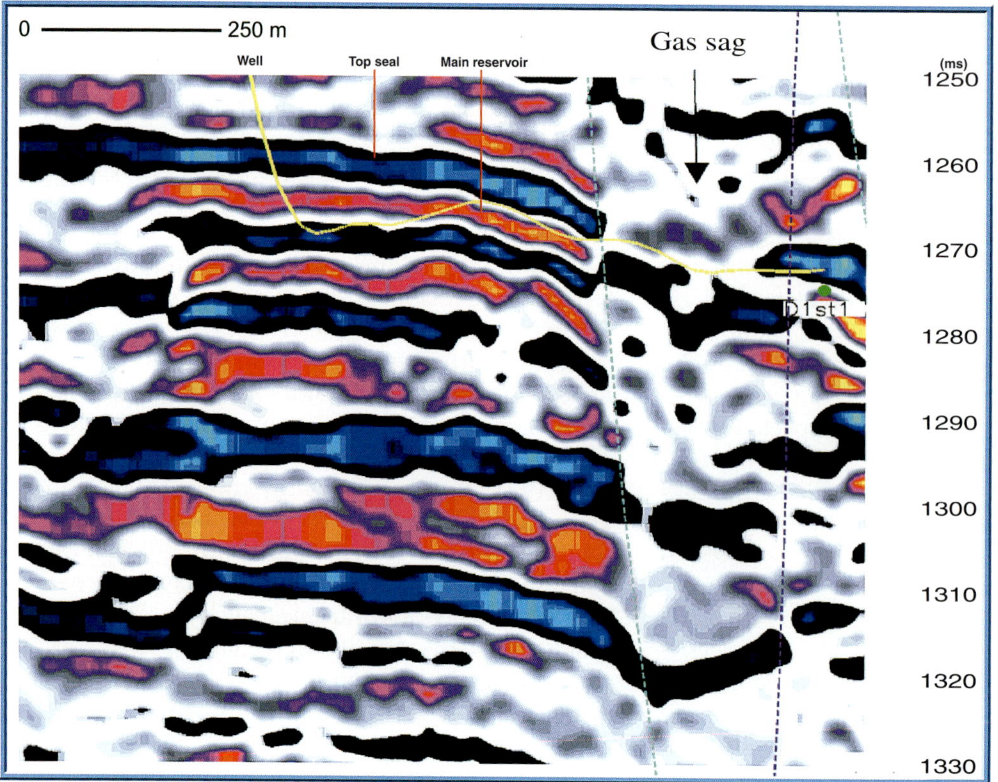

Figure 9. Detail of "gas sag" at the top of the reservoir in seismic time section. Gas sag is imaged through reduced amplitude and a laterally sharply defined zone of seismic sag, thus resembling a structural graben. The wellbore (thin yellow line) crossing the time sag, however, drilled continuously within the B Zone of reservoir, without showing evidence of any structural offset. Horizontal width of image is approximately 2 km. Vertical scale: 1 ms ≈ 2 m.

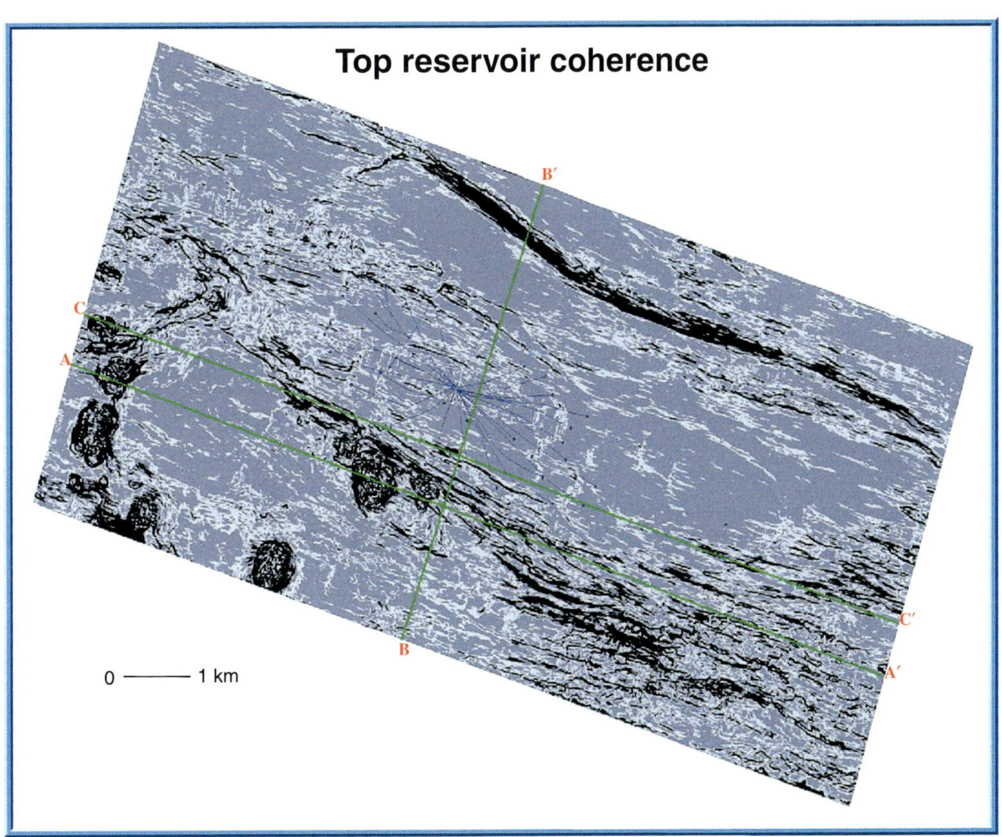

FIGURE 10. Seismic map of the top of the reservoir showing coherence attribute over the survey area of approximately 7 × 12 km. Traces of horizontal wellbores are shown in blue. Notice cluster of circular, partly intersecting sinkholes to the southwest of the reservoir, in part associated with fault zones bounding the reservoir. Cross section AA′ is shown in Figure 12, cross section BB′ is shown in Figure 8, and cross section CC′ is shown in Figure 11.

the collapse zones affects the entire overburden. The east-west seismic section in Figure 12 demonstrates these characteristics over the large sinkhole along the southern fault margin. Apparently, a gas-generating process is occurring near the base of the reservoir level; the upward migration of gas is facilitated by the sinkhole. We believe that the apparently active subsurface limestone dissolution, the associated discrete and diffuse carbonate collapse, and gas-chimney formation are related and can be attributed to either or both of two processes: (1) ongoing biodegradation of the oil and related release of acids or (2) continued flushing of the Dongsha Platform by cold marine waters undersaturated with respect to $CaCO_3$.

The first alternative, termed *oil-field karst*, has been proposed for various worldwide examples (Hill, 1992, 1995). In the case of the Liuhua reservoir, bacteria at the OWC may decompose hydrocarbon into carbon dioxide, methane, and hydrogen sulfide. CO_2 released into the seawater forms carbonic acid and dissolves the carbonate substrate, causing sagging and collapse in the overlying strata. Methane, hydrogen sulfide, excess carbon dioxide, and a large amount of water rise up along the bounding and major faults and through the collapse pipes and leave behind an amplitude dim zone in the microfractured marine shale associated with the gas chimneys. The telltale signs of inactive oil-field karst, gypsum, however, have not been reported from Liuhua cores. It is possible that it may have been dissolved.

Alternatively, cold marine waters may also contribute to carbonate dissolution. The Zhuhai sandstone aquifer, underlying the Zhujiang Limestone, crops out on the seafloor at the fault-bounded edge of the Dongsha Platform, resulting in a strong bottom-water drive. The chalky, highly porous, and leached texture of the reservoir F Zone below the OWC testifies that carbonate dissolution was active in the past. In what may have been the first step in a two-stage process, compaction-water leaching may have also initially contributed to the alteration of the F Zone by moving water vertically downward, then horizontally outward.

From a field-production standpoint, the ongoing solution collapse is important in that it will create vertical zones for water encroachment both outside of and within the productive area of the reservoir. Zones of reduced reflection strength (“dim zones”) not only overlie clearly collapsed, typically circular off-structure sinkholes, but are also associated with subtle fracture swarms paralleling the bounding faults within the reservoir. These smaller features are generally only faintly visible on the coherency map (Figure 10). There are also nonlinear partial-dissolution features that are not completely collapsed into cylindrical patterns. Only the irregular edges of some of these features are discernible

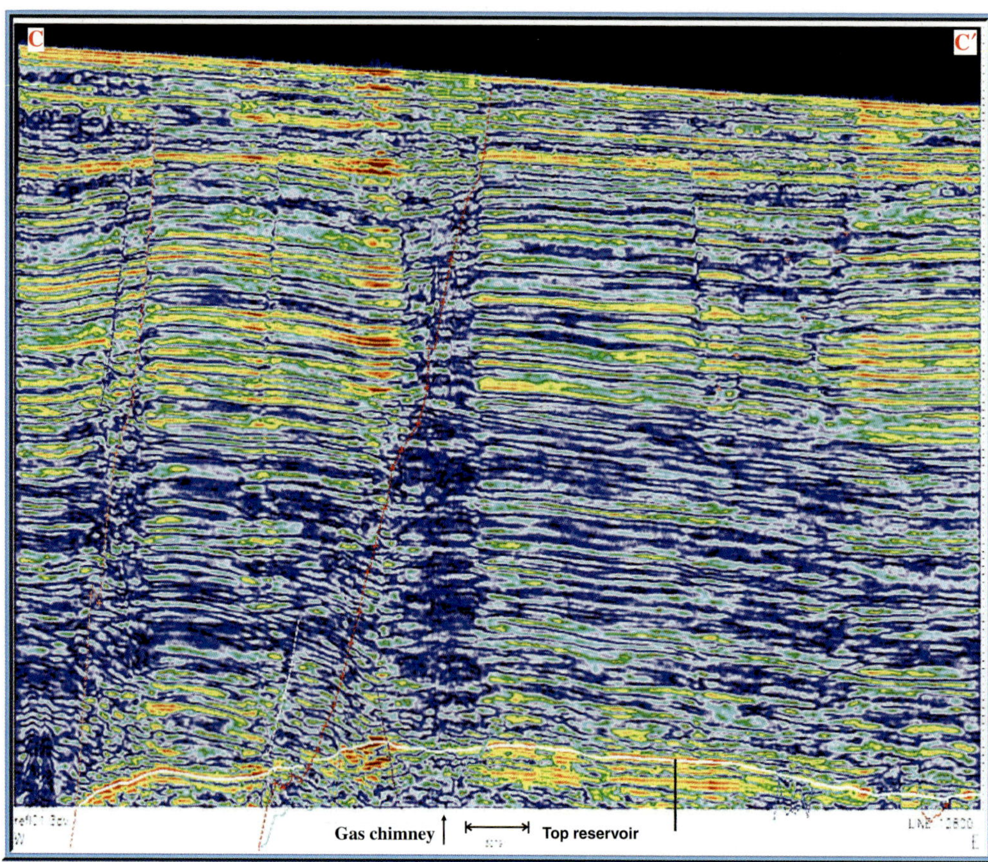

FIGURE 11. Seismic time section CC′ along strike of the Liuhua horst block, crossing the Liuhua field slightly off-center. Width of image is approximately 12 km. Top of the reservoir is shown as the gently anticlinally folded and faulted bright reflector near the bottom of the image. This section obliquely crosses a zone of gas sag presumably related to minor and diffuse faulting in the central, structurally flat part of the reservoir, and is associated with a major gas chimney represented by a column of low amplitude (see Figure 10 for location). High amplitudes next to this gas chimney but well above the reservoir may be escaped gas, charging turbiditic sandstones of Pliocene–Pleistocene age.

(Figure 10). Several isolated, partial-dissolution "dimples" that are suspected to exist within the core producing area are associated with these subtle attribute irregularities.

Core description can only partially confirm the widespread evidence of various degrees of exposure from seismic data. The principal geologic evidence for repeated exposure events derives from thin section petrography (meniscus cements, early moldic porosity; Wagner et al., 1995) and oxygen-isotope stratigraphy on the Liuhua cores (Boutell and Moldovanyi, 1992) that indicate a vadose zone diagenetic environment and a periodic influence of meteoric water. However, there is a lack of macroscopic diagnostic fabrics, such as collapse breccias (McMechan et al., 1998; Loucks, 1999), pink or laminated cave-fill facies (Cocozza and Gandin, 1990), radial fibrous or blocky calcite, tension cracks and fissures, or chaotic stratigraphy (Kerans, 1988), that are typical characteristics of karsted reservoirs and which would fill a "mesoscale" data gap between the microscopic petrographic data and the macroscopic seismic evidence for exposure and karsting. The lack of exposure-diagnostic mesoscopic fabrics is largely because of the fact that production wellpaths were carefully designed to avoid apparently structured zones (because of the danger of early water coning) and zones of low porosity (because of the evident relationship with reduced production volumes). However, where amplitude "dim zones" were penetrated by an (uncored) development well, no unusual fabrics or lithologies were reported from cuttings descriptions either. Therefore, the features responsible for seismic amplitude reduction are probably at or below the seismic detection limit and are not evident in the cuttings; they may be brecciated areas of incipient carbonate collapse ("crackle breccia" of Loucks, 1999). Production data suggest that the "dim zones" may be responsible for higher-than-expected water production from individual wells. The suspected microfracturing associated with both the major and incipient solution collapse was modeled successfully during the reservoir hydraulic simulation with local adjustments to the vertical water transmissibility.

Within the limits of the 3-D survey, the large sinkholes are preferentially associated with the subvertical bounding faults. The fact that karst processes were controlled by these faults suggests that these faults were major vertical fluid conduits already early in the diagenetic process and may have been a factor in bounding the extent of carbonate platform growth,

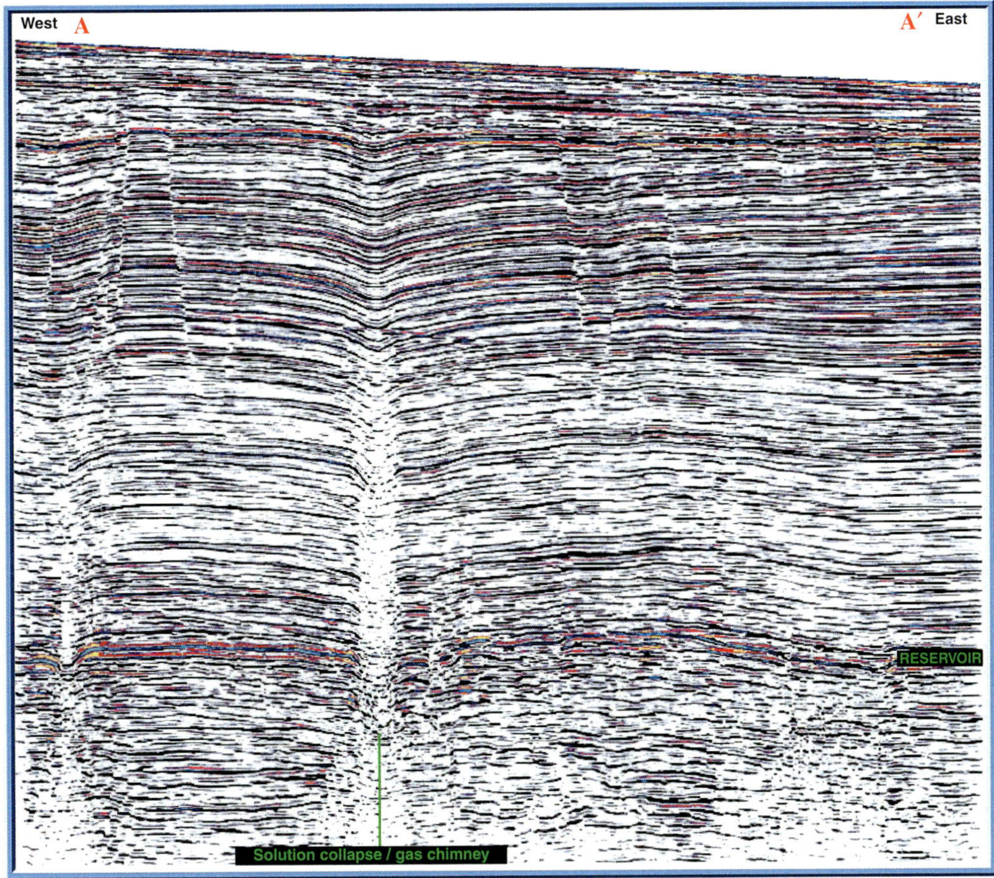

FIGURE 12. East-west–oriented seismic time cross section AA′, crossing the largest karst sinkhole of the 3-D survey, located just outside the major faults bounding of the development area. The concentric structural sag above the collapse zones affects the entire overburden.

similar to their distribution in the west Texas Yates field (Tinker and Hruk, 1995). While none of the large karst sinkholes are located within the development area, there does appear to be a strong relationship between the poor production of the western field area and the large sinkhole cluster along the southern bounding fault nearby (Figures 6, 10).

Porosity degradation in the western field area, suggested by a lack of contiguous acoustic resonance in the reef reservoir complex, can also be demonstrated in a spectral decomposition map (Figure 13). This frequency-analysis attribute indicates that the western area has poor acoustic values, that the core producing area has high values, and that the eastern, undeveloped area shows arcuate structures of alternating high and low values. Oils from the latter two reservoir segments also have a different composition and viscosity. This attribute can probably be interpreted as a measure of petrophysical matrix continuity, or coherence, in the frequency domain. The spectral decomposition values can be used to delineate zones of increasing matrix discontinuity that may either result from variable primary depositional patterns (e.g., carbonate facies of different water depths) and/or various degrees of carbonate dissolution and microfracturing.

Reservoir Characterization and Visualization

The objective of the reservoir characterization was to produce a detailed and accurate depth-converted porosity and permeability model of the reservoir useful for identification of contiguous high-porosity trends and for wellbore planning, history matching of producing wells, and predictive reservoir simulation. Nine additional surfaces from well control were added to the five surfaces picked from seismic control and interpreted using a sequence-stratigraphic framework to improve simulation resolution. The inversion values (porosity) range from 10 to 34% and demonstrate the spatial variability of the porosity zones (Figure 14).

The thin porosity layers in the reservoir and the localized effects of "time sag" in the solution collapse zones complicated the structural depth conversion of the model. Because the main producing zone averages only approximately 8 m in thickness (approximately 4 ms), the depth conversion required high precision to ensure that the 25 horizontal wellbores plotted in the correct reservoir zones as interpreted from logs. Log-calculated porosity data from vertical and horizontal wells showed an excellent agreement with the

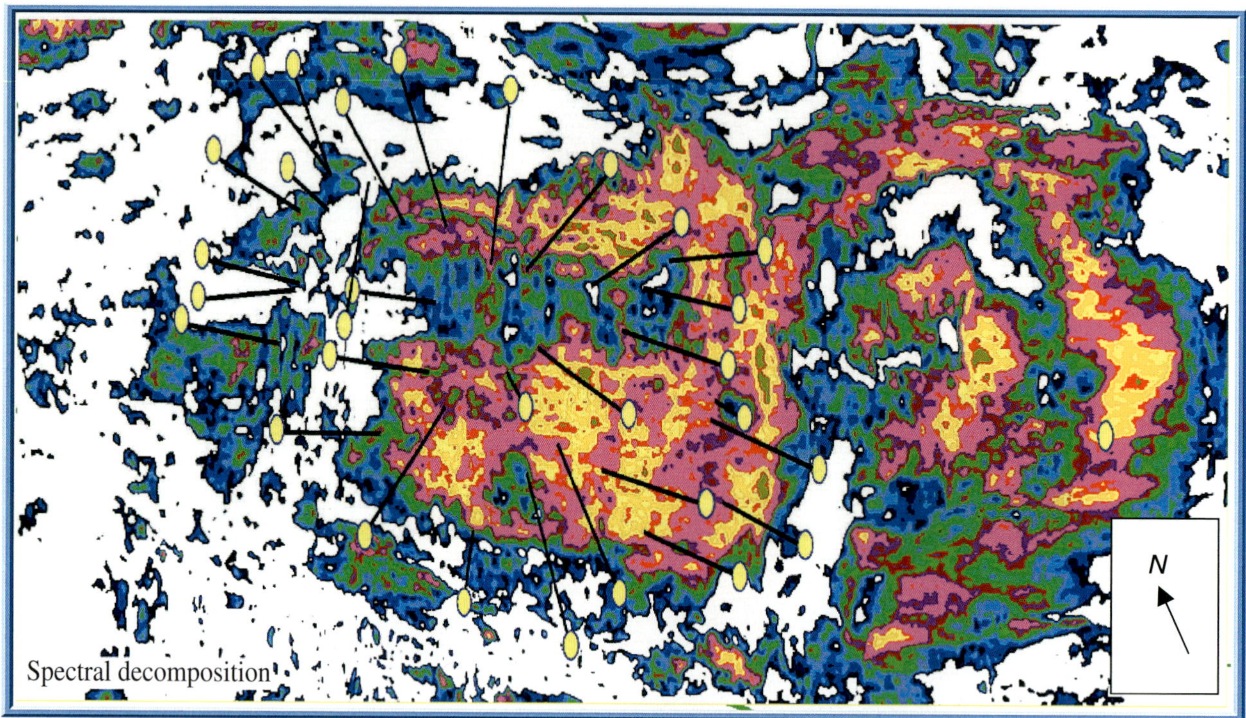

Figure 13. Spectral decomposition attribute map of the B Zone of the Liuhua reservoir. A lack of contiguous acoustic resonance (white and blue colors) in the western field area is interpreted as reduced porosity resulting from the destruction of continuity in trace-to-trace similarity. Spectral decomposition attribute is indicative of primary matrix competence, which in the case of Liuhua field is probably a primary depositional, facies-dependent parameter, modified during karsting and carbonate dissolution.

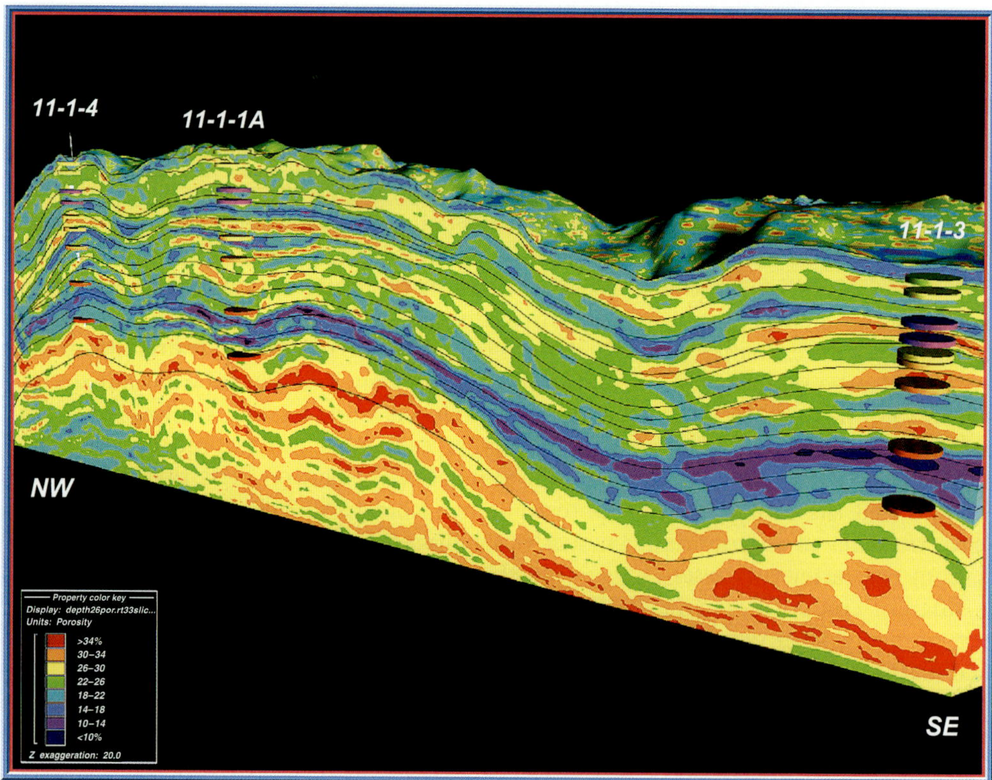

Figure 14. Porosity volume of Liuhua field cut along a northwest-southeast cross section. This orientation along the long axis of the reservoir shows tops of internal reservoir zones (black lines) picked from seismic, tops from vertical wellbores LH 11-1-4, LH 11-1-1A, and LH 11-1-3 (disks), and porosity calculated from inversion of acoustic impedance using core-calibrated porosity values (coloring). Zones of high porosity are shown in red, whereas zones of low porosity are shown in blue. Vertical exaggeration = 20. Note the linear but discontinuous zones of alternating high and low porosity. Extracted low-porosity volumes colored blue and violet are shown in Figure 16.

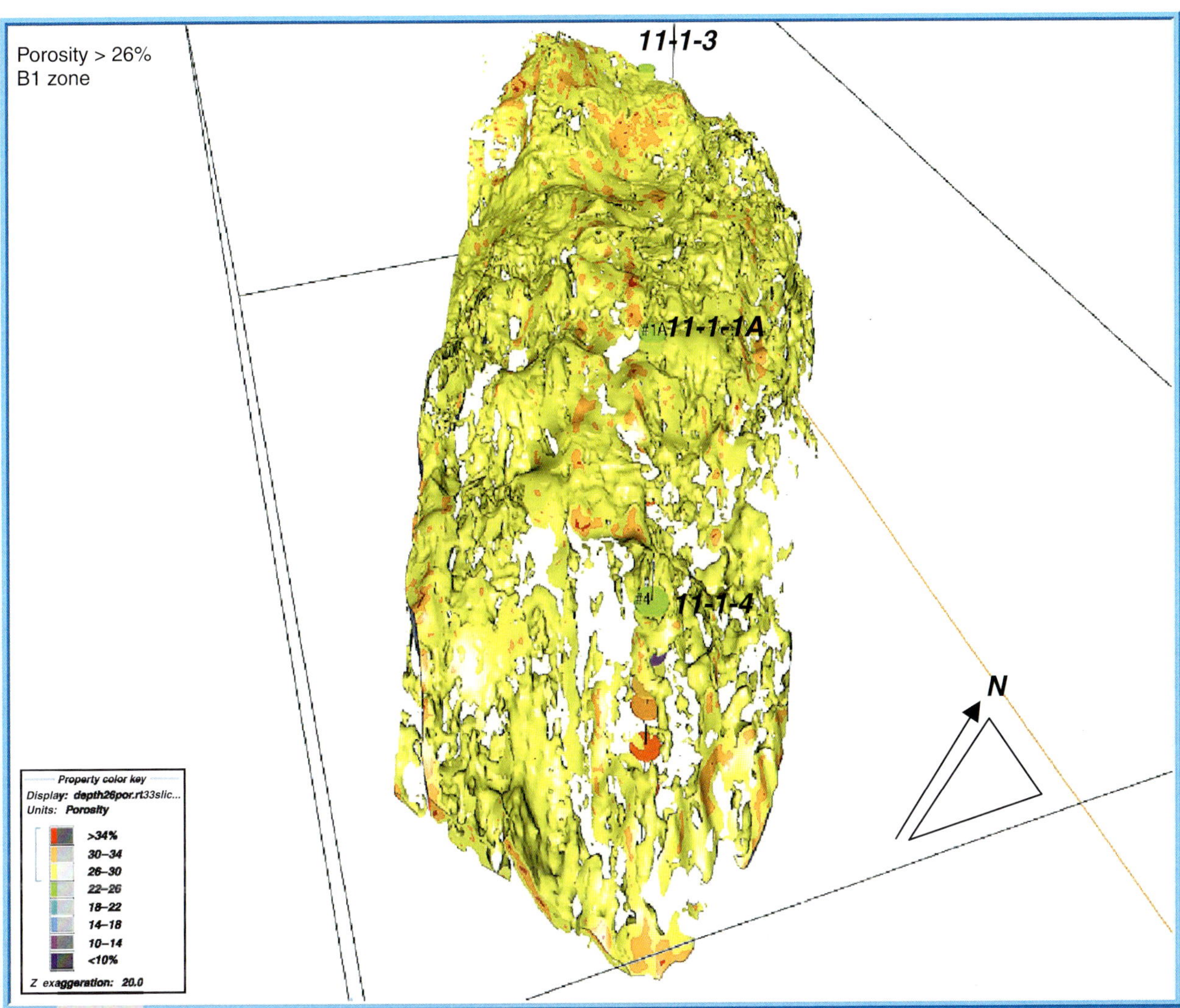

FIGURE 15. Extracted reservoir volume showing only porosity values greater than 26% (high-quality reservoir) from the topmost and most densely drilled producing zone, B1. View is from northwest to southeast, similar to the viewpoint of Figure 7. Location and tops of the three vertical wells LH 11-1-4 (foreground), LH 11-1-1A (middle ground), and LH 11-1-3 (background) are shown as green disks. Images of this kind are useful to assess fieldwide porosity or tightness of individual reservoir zones. The penetration of continuous zones of high porosity by production wellbores is a critical prerequisite for high flow rates and economic well lifespan. Note the more contiguous and slightly higher porosity in the southeastern part of the field (background), contrasting with the "patchy" distribution of high-quality reservoir in the foreground. Reduction in porosity there is probably associated with physical collapse of porous carbonate framework through faulting and dissolution, stylolitization, increased cementation along selected faults, and reduced effectiveness of diagenetic dissolution processes.

depth-converted porosity data calculated from seismic data; this served as a critical quality check of the conversion accuracy.

The reservoir porosity characterization allows the display of contoured porosity volumes, either associated with specific zones or with defined porosity value intervals to visualize regional distribution or the degree of tightness of individual reservoir zones (Figures 15, 16). The extracted volume of high-quality reservoir (>26% porosity) within the main target horizon indicates that high porosity is reasonably continuous in the central development area and the southeastern, undeveloped extension area. Stratification in the western area, however, which is accessed by several development wells, is considerably more discontinuous than originally anticipated because of porosity heterogeneity. The extent of the "low-porosity" zones (<14% porosity, Figure 16) is far from continuous, illustrating the major production problem of Liuhua field: There are no contiguous aquitards to protect the topset wellbores from early water breakthrough. The results demonstrate that Liuhua reservoir is not a simply stratified stack of alternating tight and porous strata, as envisioned in the original development plan. Rather, the principal low-porosity zones

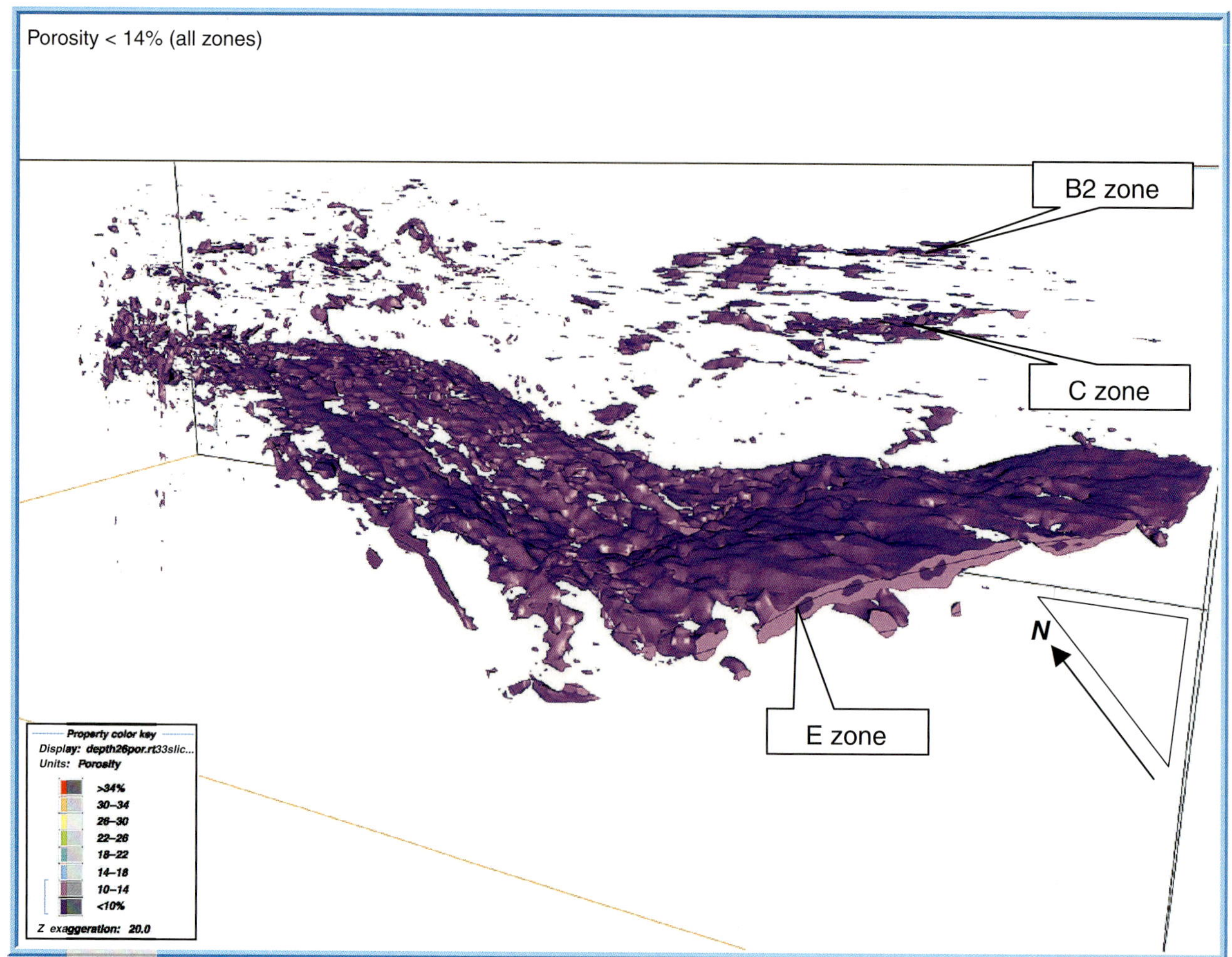

FIGURE 16. Partial volume extracted from porosity volume of Figure 14, showing only porosity values less than 14% ("tight" zones) from all zones of Liuhua. The volume displays the high variability in thickness and continuity of low-porosity zones within the reservoir. The original assumption of fieldwide continuity for tight zones is met only for E Zone. However, this zone dips below the OWC at the field margins and in the gentle syncline in the middle foreground. It therefore offers only limited wellbore protection against premature water breakthrough. Structurally higher low-porosity zones (C, B2) are highly discontinuous.

(C, E) are penetrated by numerous gaps of high porosity. Overall, stratigraphic discontinuity of the low porosity zones (principally C, but also B2 and, in places, E) resulting from any of the processes discussed above will greatly weaken their role of protecting the topset production wellbores (mostly in B1) from early water breakthrough.

RESERVOIR SIMULATION

A final reservoir simulation model honored the geoscience reservoir model. The zonal-porosity averages for all 14 units in the reservoir characterization model were transferred directly into the simulator. The geologic correlations and petrophysical rock-type modeling were incorporated with permeability values calculated from porosity and integrated with water saturation values in a depth-converted porosity model. This integration allowed the definition of petrophysical rock types and hydraulic flow units. For simulation purposes, the original structure and porosity model, which had been gridded at 75′ by 75′ horizontally by 5′ vertically, was upscaled to reach an acceptable maximum number of grid cells for the reservoir simulator while still adequately representing the internal reservoir heterogeneity. This required a 3- to 10-fold reduction in vertical resolution. The final simulation model used 15 layers to represent internal reservoir geometry and fluid distribution, with the lowest two layers representing the aquifer. Overall, the model comprised 89,100 grid blocks (Figure 17).

Permeability and water saturation were calculated by regressing core porosity vs. core permeability (Figure 18) and log-calculated effective porosity vs. log-calculated water saturation. Steady-state measurements on 16 core plugs provided data for relative permeability. The simulation model used four different relative permeability curves representing four petrophysical rock types. The model calculated the rock type of each grid block based on its permeability-to-porosity ratio. Thus, once the rock type was determined, the relative permeability value for the grid block could be calculated for any given water saturation.

History Matching

Even after incorporating the information from 3-D seismic data, the large scatter in the permeability-vs.-porosity relationship (Figure 18) and the unknown impact of faults and discontinuities on flow behavior imposed significant uncertainties in the reservoir description, thereby limiting the predictive ability of the reservoir model. A useful method to improve this shortcoming is history matching, in which simulated reservoir productivity is compared with historic field-wide and well-specific production performance. Reservoir parameters can then be adjusted to obtain a reasonable match. We adjusted, where necessary, vertical-to-horizontal permeability ratio, permeability value, and effective length of the wellbore. Because of the variability of the historical data caused by pump-operating conditions, temporary shut-ins, and gauge accuracy, we focused on matching the monthly average performance instead of the transient variations of data.

History matching suggests that production performance is controlled by water coning through the formation and by water channeling through faults, fractures, dissolution microfractures, and gaps in the

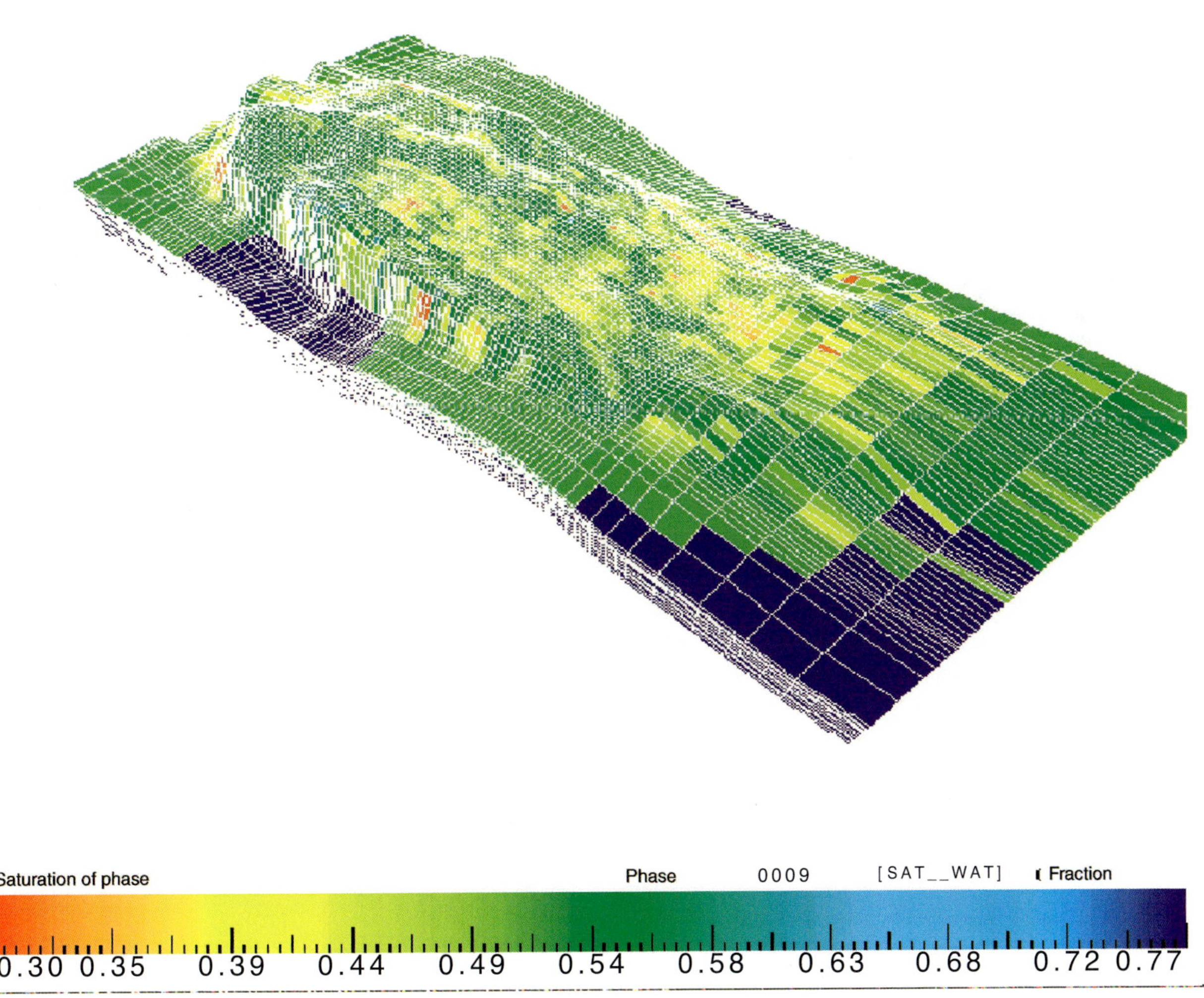

FIGURE 17. Reservoir simulation model of Liuhua field showing initial fieldwide water saturation in principal reservoir zone B1.

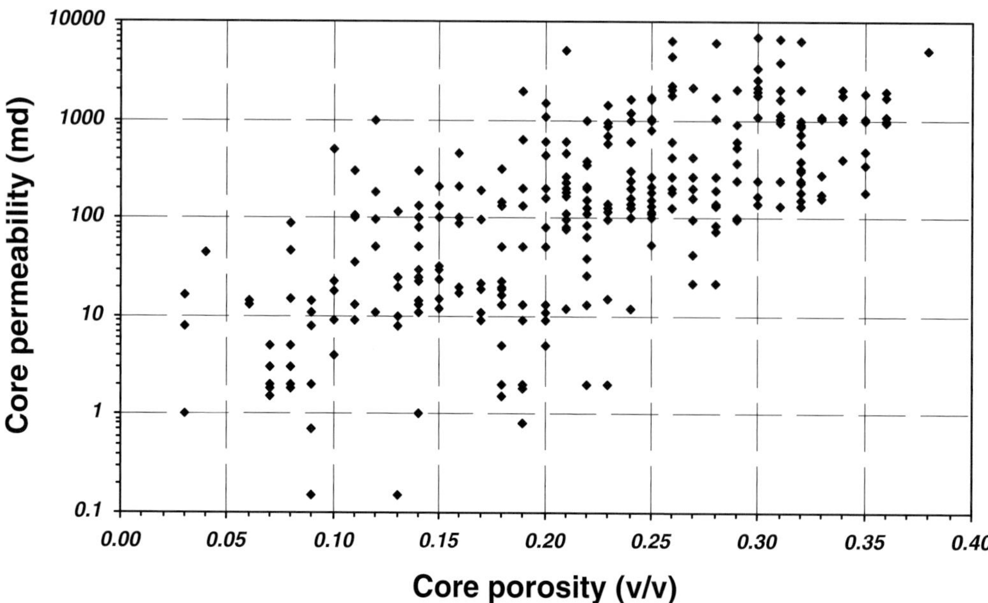

FIGURE 18. Permeability-vs.-porosity crossplot from selected plugs of the three vertical, cored wells through Liuhua field. The large scatter indicates that the rock types include a range of pore types (intergranular, intragranular, moldic, vuggy, fracture) and have experienced multiple porosity-modifying events.

low-porosity zones. The 3-D porosity model and long-term production tests support such a model. History matching also allows a model in which low-porosity zones block water coning, leaving only bounding and internal faults for water encroachment. Such a model predicts significant zones of high oil saturation and bypassed pay but is inconsistent with the bulk of the findings from the reservoir characterization. Results also indicate that uncertainties in water saturation and porosity (±5%) and horizontal permeability in the lower zones should only have a limited impact on future production performance.

Good history matches of individual well performance were achieved for a range of simulation descriptions. Modification of the geologic variables (porosity, permeability, thickness) were kept to an absolute minimum and made rapid solutions for individual well production histories possible. The reservoir characterization process, which provided high-quality detail on number and thickness of layers, fieldwide binned porosity, permeability, and water-saturation data constrained the history-matching process considerably so that the combined characterization and simulation process took ultimately significantly less time than would have been required without a reservoir characterization.

CONCLUSIONS

The very high resolution of the 3-D seismic data revealed a degree of insight into small-scale stratification and internal heterogeneity that is normally unattainable for most reservoirs. Detailed images of carbonate-collapse zones, high-resolution porosity stratification, and 2- to 4-m offsets on minor faults became readily visible. Major faults are surrounded by a halo of minor faults that accommodate both normal and strike-slip displacement along en echelon sets and that may have major impact on well productivity and reservoir fluid flow. Diffuse carbonate solution, microbreccia formation, and gas chimneys were found to be important factors affecting reservoir hydraulics. Even the 4-m resolution attained by this survey could not unequivocally identify the nature of amplitude "dim zones" associated with high water production; it is likely that these represent zones of incipient carbonate collapse. As a result of the high-resolution reservoir characterization, much of the prior geological and geophysical understanding of the reservoir was revised significantly.

In the Liuhua reservoir, the shallow, low-relief trap combined with heavy oil under strong water drive made the high-precision characterization an economic necessity. Similar high-resolution surveys for other reservoirs will likely bring out surprises that will allow improvements in reservoir management through extended field life, sustained production increase, or reduced costs. Our results suggest that significant aspects of reservoir characteristics may be missed in surveys of lower resolution.

ACKNOWLEDGMENTS

This work was performed while the first four authors were with Amoco (now BP), Houston, Texas. The authors would like to thank the management of BP Amoco, Kerr McGee, and China National Offshore Oil for permission to publish this work. Shane Pelecnaty, Gregor Eberli, and Mike Grammer improved the manuscript through helpful reviews. The senior author thanks Mrs. Britta Ernst and Mrs. Martina Grundmann for help with the figures.

REFERENCES CITED

Bourrouilh, L. J. F. G., 1998, The role of high-energy events (hurricanes and/or tsunamis) in the sedimentation, diagenesis and karst initiation of tropical shallow water carbonate platforms and atolls: Sedimentary Geology, v. 118, p. 3–36.

Boutell, R. D., and E. P. Moldovanyi, 1992, Sedimentology and diagenesis of the Miocene Liuhua carbonate platform, Pearl River Mouth Basin, South China Sea (abs.): AAPG Annual Meeting Program, v. 1, p. 12.

Christian, H. E., Jr., and W. W. Tyrrell Jr., 1991, Exploration history of the Liuhua 11-1-1A discovery, Pearl River Mouth Basin, China: Proceedings, Offshore Technology Conference, v. 23, p. 93–100.

Cocozza, T., and A. Gandin, 1990, Carbonate deposition during early rifting: The Cambrian of Sardinia and the Triassic–Jurassic of Tuscany, Italy, *in* M. E. Tucker, J. L. Wilson, P. D. Crevello, J. R. Sarg, and F. Read, eds., Carbonate facies: International Association of Sedimentologists Special Publication 9, p. 9–38.

Dorobek, S. L., 1997, Miocene carbonate platforms of the South China Sea region: Tectonic controls on platform inception and termination (abs.): AAPG Annual Meeting Program, v. 6, p. A29.

Erlich, R. N., S. F. Barrett, and B. J. Guo, 1990, Seismic and geologic characteristics of drowning events on carbonate platforms: AAPG Bulletin, v. 74, p. 523–1537.

Erlich, R. N., S. F. Barrett, and B. J. Guo, 1991, Drowning events on carbonate platforms: A key to hydrocarbon entrapment?: Proceedings, Offshore Technology Conference, v. 23, p. 101–112.

Erlich, R. N., A. P. Longo Jr., and S. Hyare, 1993, Response of carbonate platform margins to drowning: Evidence of environmental collapse, *in* R. G. Loucks and J. F. Sarg, eds., Carbonate sequence stratigraphy: Recent developments and application: AAPG Memoir 57, p. 241–266.

Fulthorpe, C. S., and S. O. Schlanger, 1989, Paleo-oceanographic and tectonic settings of early Miocene reefs and associated carbonates of offshore Southeast Asia: AAPG Bulletin, v. 73, p. 729–756.

Gascoyne, M., G. J. Benjamin, H. P. Schwarcz, and D. C. Ford, 1979, Sea-level lowering during the Illinoian glaciation: Evidence from a Bahama "blue hole": Science, v. 205, p. 806–808.

Gu, K. J., and Y. Q. Ye, 1992, Tests verify advantages of horizontal well in offshore China oil field: Oil & Gas Journal, v. 90, (42), Oct. 19, 1992, p. 76–80.

Heubeck, C., P. Peng, C. Story, and Z. J. Yuan, 1998, Detailed carbonate stratigraphy from log interpretation of horizontal wells assists in reservoir management (Liuhua field, offshore China) (abs.): AAPG Annual Meeting Program and Extended Abstracts, v. 1, p. A295.

Hill, C. A., 1992, Sulfuric acid oil-field karst, *in* M. P. Candelaria and C. L. Reed, eds., Paleokarst, karst-related diagenesis, and reservoir development: Examples from Ordovician–Devonian age strata of west Texas and the Mid-Continent: SEPM, Permian Basin Chapter Publication 92-33, p. 192–194.

Hill, C. A., 1995, H_2S-related porosity and sulfuric acid oil-field karst, *in* D. A. Budd, A. H. Saller, and P. M. Harris, eds., Unconformities and porosity in carbonate strata: AAPG Memoir 63, p. 301–306.

James, N. P., and P. W. Choquette, 1990, The meteoric diagenetic environment, *in* I. A. McIlreath and D. W. Morrow, eds., Diagenesis: Geoscience Canada Reprint Series 4, p. 35–74.

Kerans, C., 1988, Karst-controlled reservoir heterogeneity in Ellenburger Group carbonates of west Texas: AAPG Bulletin, v. 72, p. 1160–1183.

Loucks, R. G., 1999, Paleocave carbonate reservoirs: Origins, burial-depth modifications, spatial complexity, and reservoir implications: Journal of Sedimentary Research, v. 83, p. 1795–1834.

McMechan, G. A., R. G. Loucks, X. X. Zeng, and P. Mescher, 1998, Ground penetrating radar imaging of a collapsed paleocave system in the Ellenburger dolomite, central Texas: Journal of Applied Geophysics, v. 39, p. 1–10.

Moldovanyi, E. P., F. M. Wall, and J. Y. Zhang, 1995, Regional exposure events and platform evolution of the Zhujiang Formation carbonates, Pearl River Mouth Basin: Evidence from primary and diagenetic seismic facies, *in* D. A. Budd, A. H. Saller, and P. M. Harris, eds., Unconformities and porosity in carbonate strata: AAPG Memoir 63, p. 125–140.

Mylroie, J. E., and J. L. Carew, 1995, Geology and karst geomorphology of San Salvador Island, Bahamas: Carbonates and Evaporites, v. 10, p. 293–206.

Purdy, E. G., and G. T. Bertram, 1993, Carbonate concepts from the Maldives, Indian Ocean: AAPG Studies in Geology 34, p. 1–56.

Stark, P. H., 1991, Horizontal drilling: Overview of geologic aspects and opportunities: AAPG Bulletin, v. 75, p. 1399.

Tinker, S. W., and D. H. Hruk, 1995, Reservoir characterization of a Permian giant: Yates field, west Texas, *in* E. L. Stoudt and P. M. Harris, eds., Hydrocarbon reservoir characterization: Geologic framework and flow unit modeling: SEPM Short Course Notes, p. 51–128.

Turner, N. L., 1990, The lower Miocene Liuhua carbonate reservoir, Pearl River Mouth Basin, offshore People's Republic of China (abs.): AAPG Bulletin, v. 74, p. 1006–1007.

Turner, N. L., and P. Z. Hu, 1990, The lower Miocene Liuhua carbonate reservoir, Pearl River Mouth Basin, offshore People's Republic of China: AAPG Bulletin, v. 74, p. 781.

Turner, N. L., and P. Z. Hu, 1991, The lower Miocene Liuhua carbonate reservoir, Pearl River Mouth Basin, offshore People's Republic of China: Proceedings, Offshore Technology Conference, v. 23, p. 113–123.

Tyrrell, W. W., Jr., and H. E. Christian, 1992, Exploration history of the Liuhua 11-1-1A Miocene carbonate discovery, Pearl River Mouth Basin, China: AAPG Bulletin, v. 76, p. 1209–1223.

Wagner, P. D., D. R. Tasker, and G. P. Wahlman, 1995, Reservoir degradation and compartmentalization below subaerial unconformities: Limestone examples from west Texas, China, and Oman, *in* D. A. Budd, A. H. Saller, and P. M. Harris, eds., Unconformities and porosity in carbonate strata: AAPG Memoir 63, p. 177–194.

Wu, X. C., P. H. Li, and P. Z. Hu, 1997, Reservoirs of chalkified Miocene reef-banks on Dongsha Platform in South China Sea, *in* Z. C. Sun, T. B. Wang, D. L. Ye, and G. J. Song, eds., Proceedings of the 30th International Geological Congress 18A, p. 239–253.

8

Neuhaus, D., J. Borgomano, J.-C. Jauffred, C. Mercadier, S. Olotu, and J. Grötsch, 2004, Quantitative seismic reservoir characterization of an Oligocene–Miocene carbonate buildup: Malampaya field, Philippines, *in* Seismic imaging of carbonate reservoirs and systems: AAPG Memoir 81, p. 169–183.

Quantitative Seismic Reservoir Characterization of an Oligocene–Miocene Carbonate Buildup: Malampaya Field, Philippines

Dietmar Neuhaus[1]

Shell Philippines Exploration BV, Alabang, Muntinlupa, Philippines

Jean Borgomano

Shell E&P Technology and Applied Research, Volmerlaan 8, Rijswijk, The Netherlands

Jean-Claude Jauffred

Shell E&P Technology and Applied Research, Volmerlaan 8, Rijswijk, The Netherlands

Christophe Mercadier

Shell E&P Technology and Applied Research, Volmerlaan 8, Rijswijk, The Netherlands

Sam Olotu[2]

Shell E&P Technology and Applied Research, Rijswijk, The Netherlands

Jürgen Grötsch

Shell Abu Dhabi BV, Abu Dhabi, United Arabian Emirates

ABSTRACT

The complex reservoir architecture of the Malampaya carbonate buildup offshore Palawan, Philippines, was initially controlled by a rugged clastic basement morphology, which was overgrown by atoll structures during the Oligocene and early Miocene. Additional factors with major impact on reservoir quality are frequent and high-amplitude relative sea level fluctuations, ocean currents, and prevailing wind directions. Primary depositional reservoir-quality distribution has been overprinted by diagenetic events, primarily as a result of repeated platform-top exposure and submarine cementation. Inherent noise within the previous seismic data introduced by the complex overburden and buildup morphology has resulted in inconsistent seismic attribute distribution. Therefore, earlier reservoir modeling efforts used seismic horizon and volume interpretation only, coupled with the sequence- and cyclostratigraphic architecture and the concept of reservoir rock types for field development planning.

Prior to gas-development drilling, another attempt was made to extract direct reservoir-quality information from the reprocessed three-dimensional (3-D) seismic data to validate the earlier deterministic reservoir models. Improved 3-D prestack depth

[1]*Present address:* Nederlandse Aardolie Maatschappij, Assen, The Netherlands.
[2]*Present address:* Shell Petroleum Development Company, Lagos, Nigeria.

migration based on a new velocity model has been the foundation of the quantitative seismic analysis for reservoir characterization, static modeling, reserves estimation, and optimized gas development and oil appraisal well targeting. High-porosity areas in the upper part of the reservoir were identified using top-reservoir reflection amplitudes. This provided the tool to minimize penetration of low-porosity, fractured zones prone to mud losses in the gas development wells. A series of acoustic-impedance inversions were applied to create reservoir porosity cubes from seismic and to target wells in good reservoir areas. Porosity cubes are also essential for a correct time-depth conversion of the static model, using a linear porosity-velocity relationship in clean carbonates, which was derived from well data. Several static model realizations were created using the porosity cubes from seismic as a backdrop combined with 3-D seismic facies analysis and a depositional model based on well data and analogs. The results of the five gas-development wells have confirmed the modeled reservoir-quality distribution within the lagoonal part of the northern Malampaya accumulation. Early production performance following first gas in October 2001 is indicative of excellent lateral pressure communication in this area of the buildup, in accordance with earlier dynamic models.

Porosity-height realizations created from the different seismic porosity cubes proved valuable to visualize uncertainty in reservoir-quality distribution within the Malampaya oil rim and formed the basis for targeting a horizontal appraisal well. The MA-10 horizontal oil-rim appraisal well drilled at the end of 2001 confirmed the forecasted facies distribution and reservoir properties as derived from the model.

Based on the new quantitative seismic reservoir characterization, additional areas of potentially good reservoir quality have been identified in the southern Malampaya culmination and on the western flank of the northern culmination. Both areas were previously considered to contain low-porosity reservoir caused by pervasive early marine cementation.

INTRODUCTION

The Malampaya field is located in deep-water (850–1200 m) offshore Palawan, Philippines. Discovered in 1989, the field contains a gas column as much as 650 m thick and a 56-m-thick oil rim (API 29.4) within two Oligocene–Miocene carbonate buildups at a depth of approximately 3000 m subsea (Figure 1). Since October 2001, the gas is being produced through a subsea manifold and five deviated wells, whereas the oil rim will be further appraised by a horizontal well.

The Nido Limestone was described earlier by Wolfart et al. (1986) and Wiedicke (1987). It contains several small hydrocarbon accumulations in the offshore Palawan area (Longman, 1981). However, the Malampaya buildup shows significant differences with respect to age, morphology, depositional-facies distribution, reservoir architecture, and hydrocarbon volumes (Grötsch and Mercadier, 1999).

In 1991, the Malampaya field was covered with a 25- × 25-m bin size three-dimensional (3-D) seismic survey. The data quality is affected by streamer feathering caused by surface currents coupled with positioning uncertainties, resulting in uneven offset distributions and data holes. The complex overburden containing high-velocity imbricate conglomerate channels and the rugged seabed topography introduce large ray-bending effects, which in turn give rise to nonhyperbolic residual moveout. High-frequency attenuation in the overburden and the effects of residual moveout limit the frequency spectrum at Nido level, resulting in a dominant frequency within the reservoir of about 20 Hz and a vertical resolution of about 80 m.

In 1994, the world's first industry application of 3-D prestack depth migration (PSDM) was applied to the Malampaya 3-D seismic survey. This significantly improved structural positioning; however, the amplitude content was found to be unreliable for quantitative reservoir predictions. Previous reserve estimates and initial development well targeting were based on an integrated petroleum engineering study, which used seismic horizon and volume interpretation, coupled with sequence- and cyclostratigraphic architecture and the concept of reservoir rock types. The 3-D velocity model as input for seismic processing was identified as one key remaining issue for improving the quality of

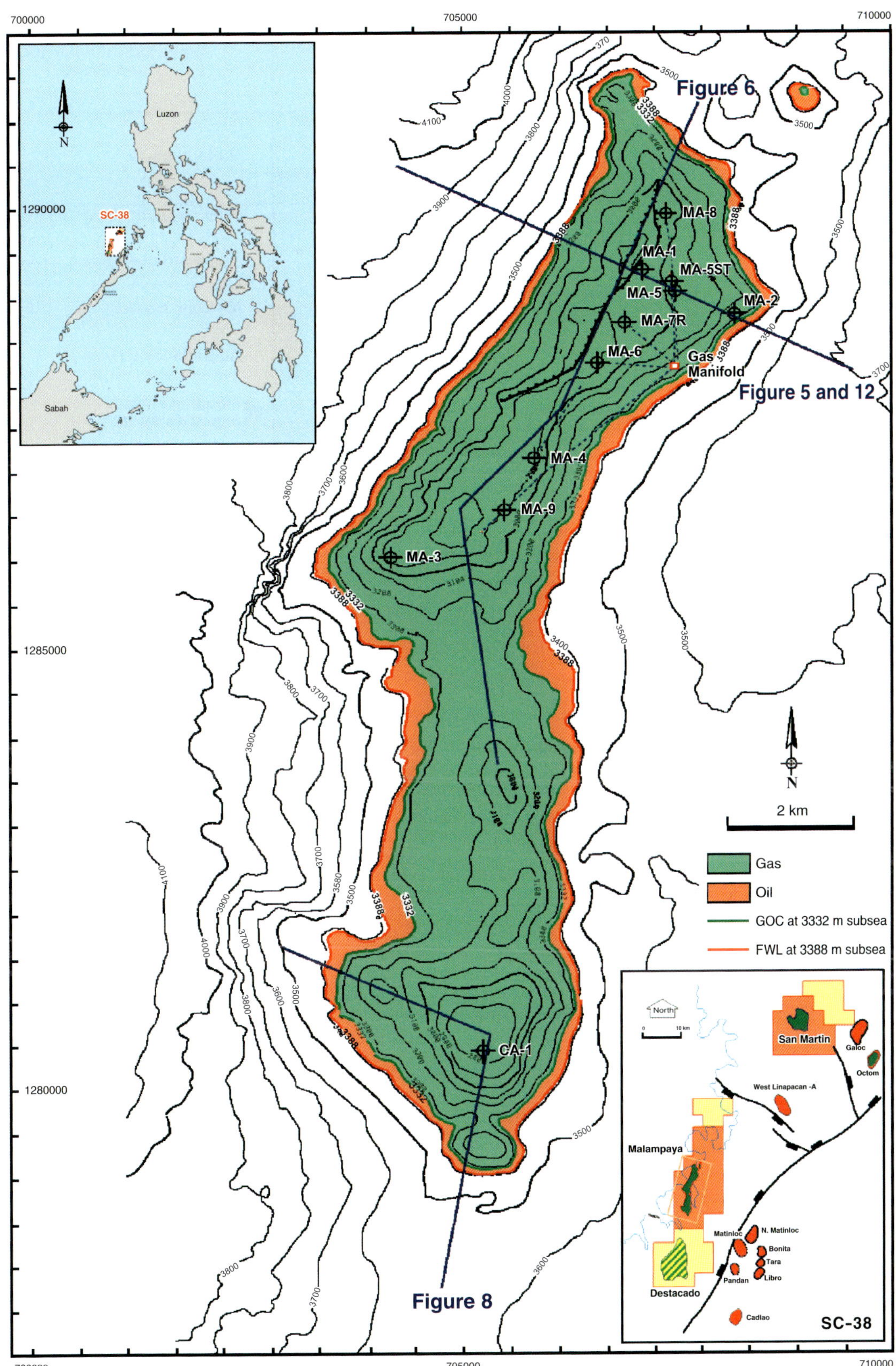

FIGURE 1. Top Nido reservoir depth map of the Malampaya buildup located within block SC-38, offshore northwest Palawan in the Philippines. Note the 1000-m water-depth contour in the inset map on the lower right corner. The five exploration and appraisal wells (CA-1, MA-1 to MA-4) and the five gas-development wells (MA-5 to MA-9) are indicated, as is the gas-oil contact (GOC) at 3332 m subsea and the free-water level (FWL) at 3388 m subsea. The seismic sections are shown as black lines.

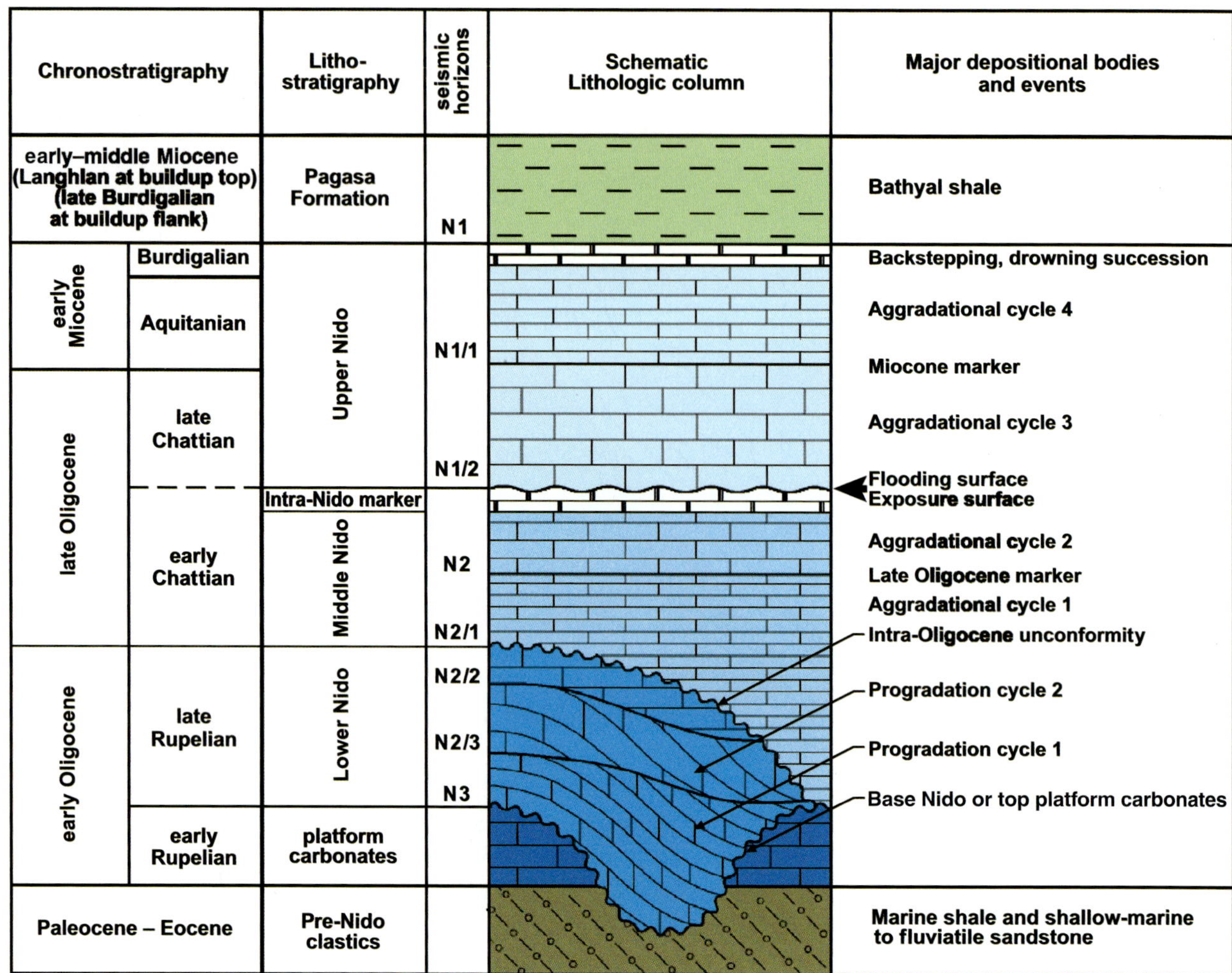

FIGURE 2. Chronostratigraphy and lithostratigraphy, main horizons mapped from 3-D seismic, schematic lithology, and major depositional cycles and events during the buildup growth. (From Grötsch and Mercadier, 1999).

the available seismic data (Grötsch and Mercadier, 1999).

In 2000, the seismic data were reprocessed using the latest PSDM technology and improved velocity modeling to validate and optimize the development well locations and reserves estimates. This resulted in a significant reduction of nonhyperbolic moveout, enhancement of reflection continuity, and an improvement of true amplitude preservation. Hence, the reprocessing greatly improved the applicability of 3-D seismic data as direct input for reservoir characterization and modeling purposes. The results from this work are documented in this chapter.

BUILDUP GROWTH HISTORY AND DEPOSITIONAL MODEL

Extensive gathering of stratigraphic data, both by Sr isotopes and biostratigraphy, allow reconstruction of the tectonic and growth history of the Malampaya and Camago buildups, which in turn enables reconstruction of relative sea level history by using the two atoll structures as dipsticks in the ocean (Grötsch and Mercadier, 1999). After the initial opening of the South China Sea in the Paleocene–Eocene, the Nido Limestone was deposited during the late Eocene to early Miocene period (Figure 2; Grötsch and Mercadier, 1999). The regional distribution of the Nido Limestone is predominantly controlled by the underlying northeast-southwest–trending extensional basement faults related to the rifting. The resulting basement morphology was a prime controlling factor in the development of complex reservoir geometries (Figures 3–5). The initial platform carbonates of late Eocene–early Oligocene age onlap onto a rugged morphology, which forms in places a Pre-Nido paleohigh in the core of the Malampaya structure as penetrated by well MA-4 (Figures 4, 6). The drifting phase in the South China Sea continued with the deposition of the first progradational phase of platform-derived slope carbonates during the early

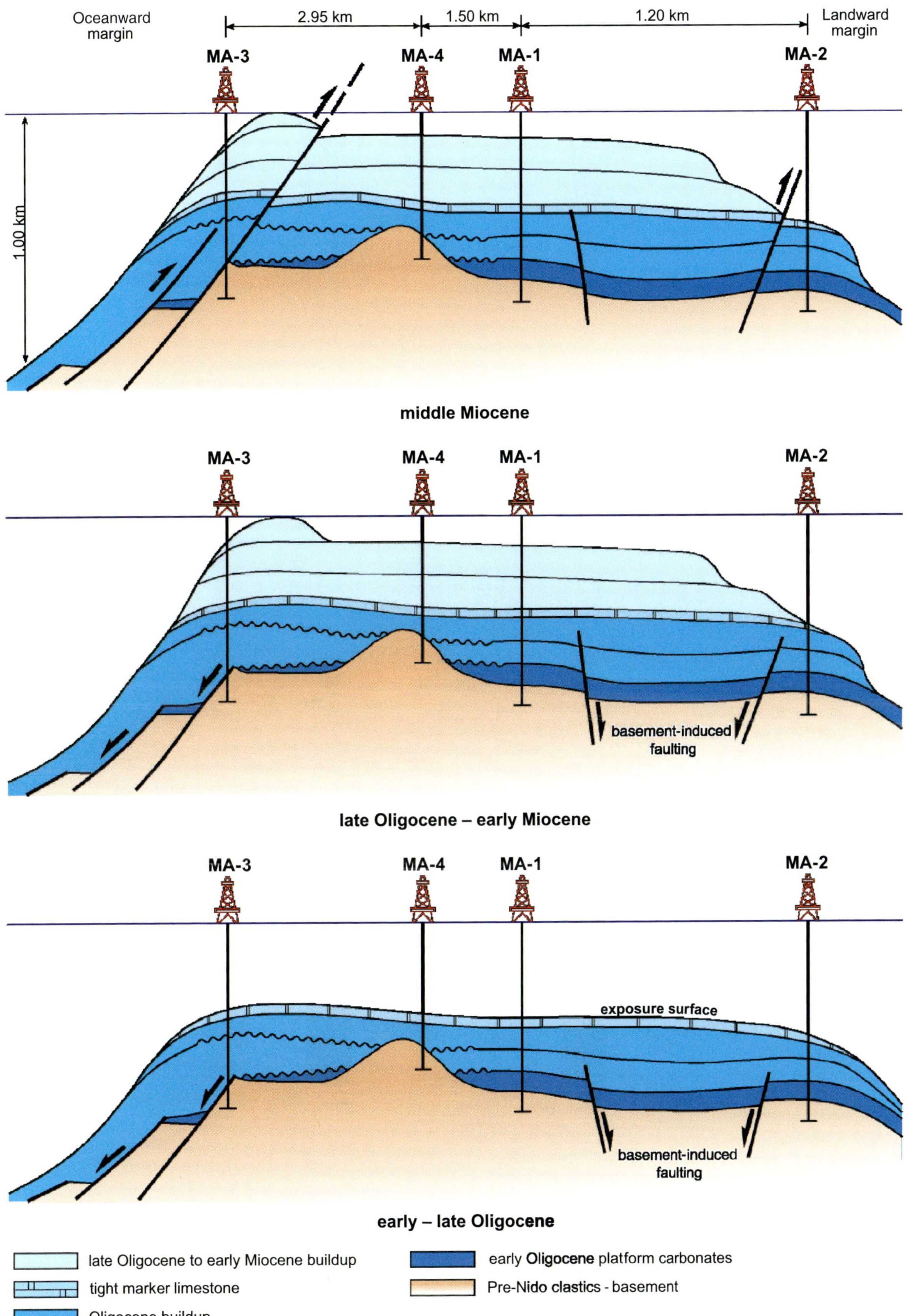

FIGURE 3. Schematic drawing showing a composite section that illustrates the tectonic and sedimentary history of the Malampaya buildup (adapted from Grötsch and Mercadier, 1999).

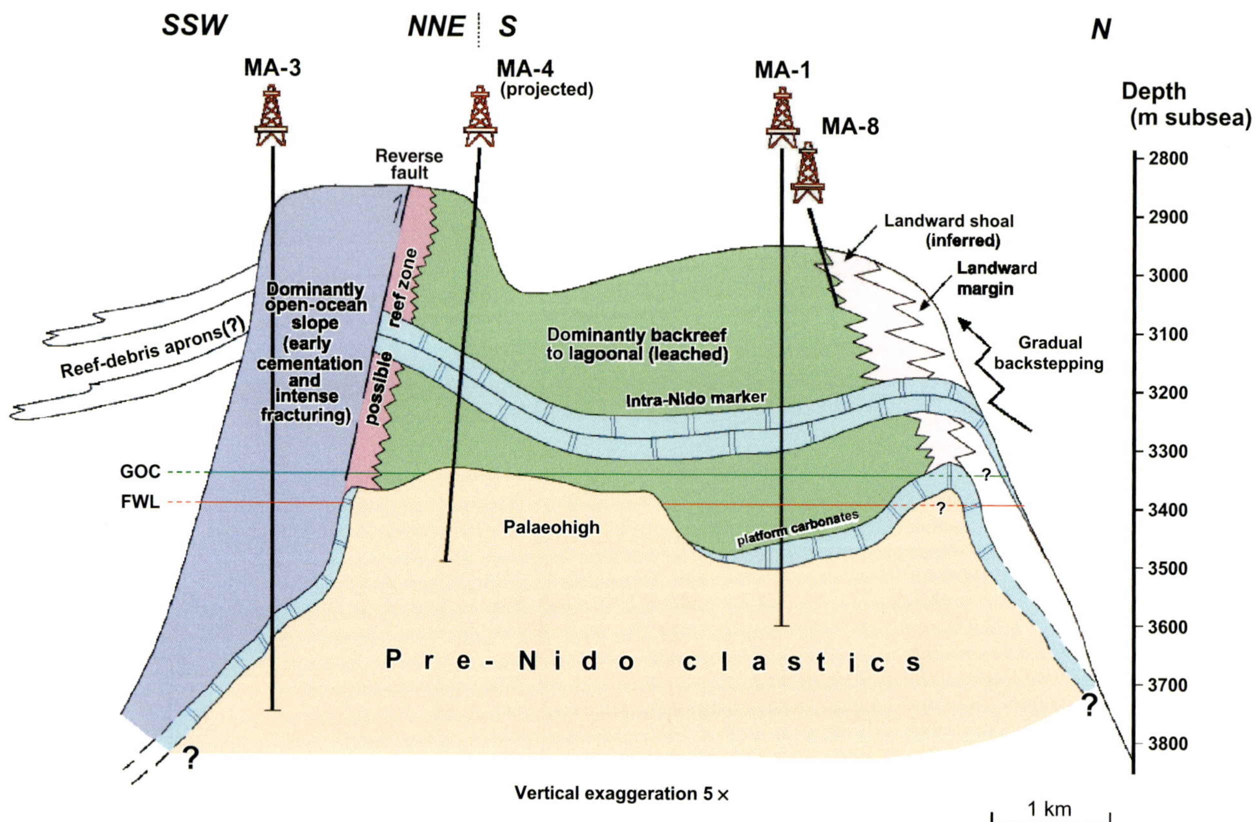

FIGURE 4. Schematic geologic cross section and facies distribution in the Malampaya buildup. GOC = gas-oil contact; FWL = free-water level.

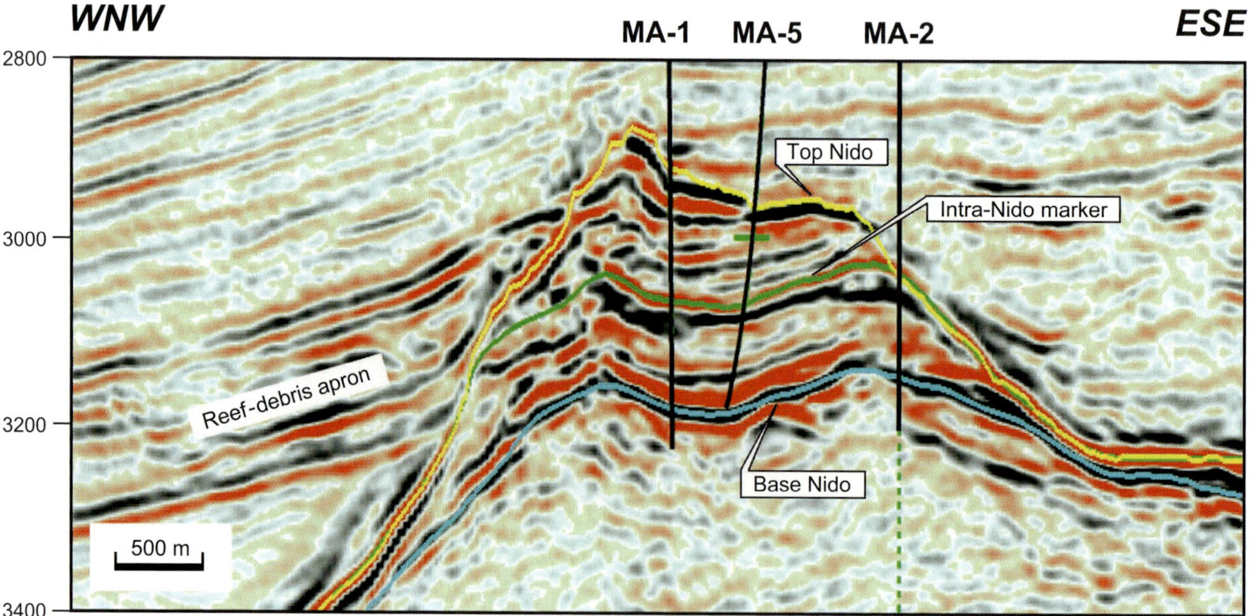

FIGURE 5. West-northwest–east-southeast 3-D PSDM seismic section across the Malampaya field. Wells MA-1 and MA-2 are exploration and appraisal wells, well MA-5 is a gas-development well deepened to appraise the oil rim. The high-amplitude events onlapping on the oceanward side are interpreted as potential reef-debris deposits, the high-amplitude unit east of MA-2 is interpreted as a turbidite channel within the deep-marine Pagasa shales cut perpendicular to the depositional axis.

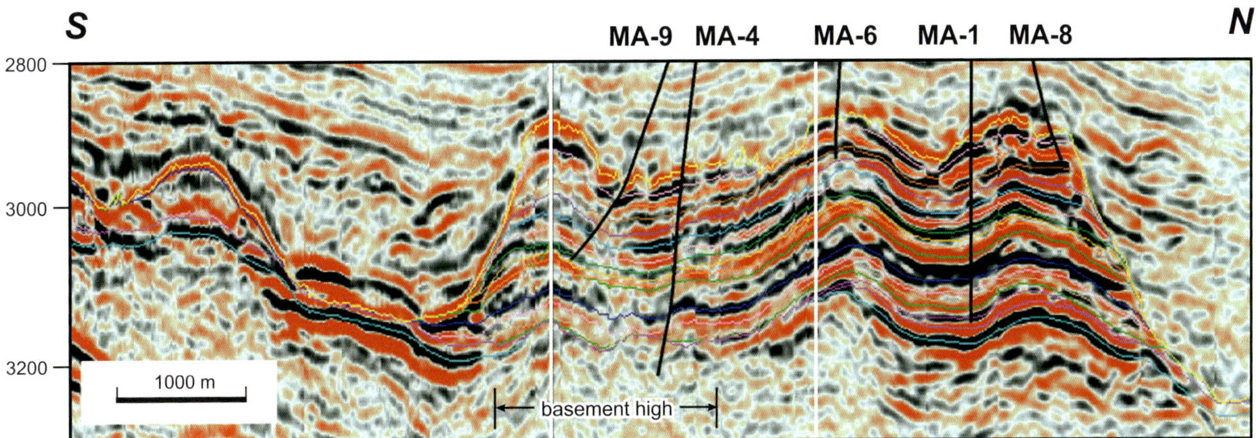

FIGURE 6. South-north 3-D PSDM seismic section showing the onlap between MA-1 and MA-6 of the Eocene–Oligocene platform carbonates at the bottom of the Nido sequence onto a pre-Nido paleohigh penetrated by MA-4. Note the steep backstepping north of MA-8. The detailed horizon interpretation was used for the static modeling.

Oligocene as seen on seismic in the southwestern part of Malampaya and Camago. A middle Oligocene unconformity marks the transition from progradational to postdrift aggradational growth. For information on the tectonic evolution of the South China Sea, see also Holloway (1982).

During the late Oligocene and early Miocene, buildup aggradation and subsequent backstepping of the eastern margin occurred as a result of successive rapid high-amplitude falls and rises of relative sea level (Grötsch and Mercadier, 1999). The depositional-facies distribution during this main growth phase was controlled by the basement morphology, relative sea level fluctuations, ocean currents, and prevailing wind direction. The oceanward-facing margins of many carbonate platforms are exposed to strong current and wave action and, therefore, commonly have the highest potential for reef growth (Bosscher and Schlager, 1993). On the Malampaya buildup, the major reef framework and debris shedding are observed along its western margin facing the open ocean of the South China Sea. The actual reef zone has not been penetrated by any Malampaya wells and is likely to have a subseismic scale of only a few tens of meters wide (Grötsch and Mercadier, 1999). However, the amount of reef debris observed in wells MA-1 and MA-7 suggests that these wells are located close to a highly productive reef zone. MA-7, for example, penetrated some 246 m of Upper Nido Limestone, of which 14.8 m were cored. A 4.5-m section within this core was interpreted to belong to a reef-debris system predominantly comprising broken and rolled coral and calcisponge fragments. Interpretation of borehole images and wire-line logs suggests that 30–40% of the drilled intervals in MA-1 and MA-7 are formed by such reef debris.

High-relief pinnacle reefs developed predominantly on the steep flanks of the southern Malampaya culmination (Figure 7). High-amplitude parallel events originating at the oceanward side of the Malampaya buildup and dimming toward the deeper basin are interpreted as probable reef-debris deposits (Figures 5, 7). The reservoir potential of these sediments is untested to date.

The Malampaya buildup drowned in the early Burdigalian (late Miocene) and was covered by the deep-marine shale sequence of the Pagasa Formation. Periodic uplift of the Palawan Islands toward the east resulted in the influx of coarser material through turbidite channels on the landward-eastern side of Malampaya (Figure 5).

SEISMIC VOLUME ANALYSIS

The boundary between the reef-zone/open-ocean slope on one side and the back reef/lagoon on the other side has also been imaged with Shell proprietary software for seismic facies classification and seismic volume segmentation (Figure 8). The method uses a supervised neural network approach and volume attributes to classify the seismic volume into seismic facies of interest. The neural network is trained on polygons of the seismic data, which are selected to represent the seismic character of the facies to be classified. The (volumes) attributes are either calculated within the program or input from previously generated volumes. The result is a volume in which every sample is assigned to one or the others of the selected facies. The Malampaya seismic volume was first segmented into two facies: reef-zone/open-ocean slope and back reef/lagoon. The attributes used in the first pass (Figure 8A) were a combination of amplitude, measures of continuity-coherence, and dip and azimuth. This resulted in a clear fieldwide distinction of the reef-zone/open-ocean slope and the back reef/lagoon. The shapes of these two seismic facies were

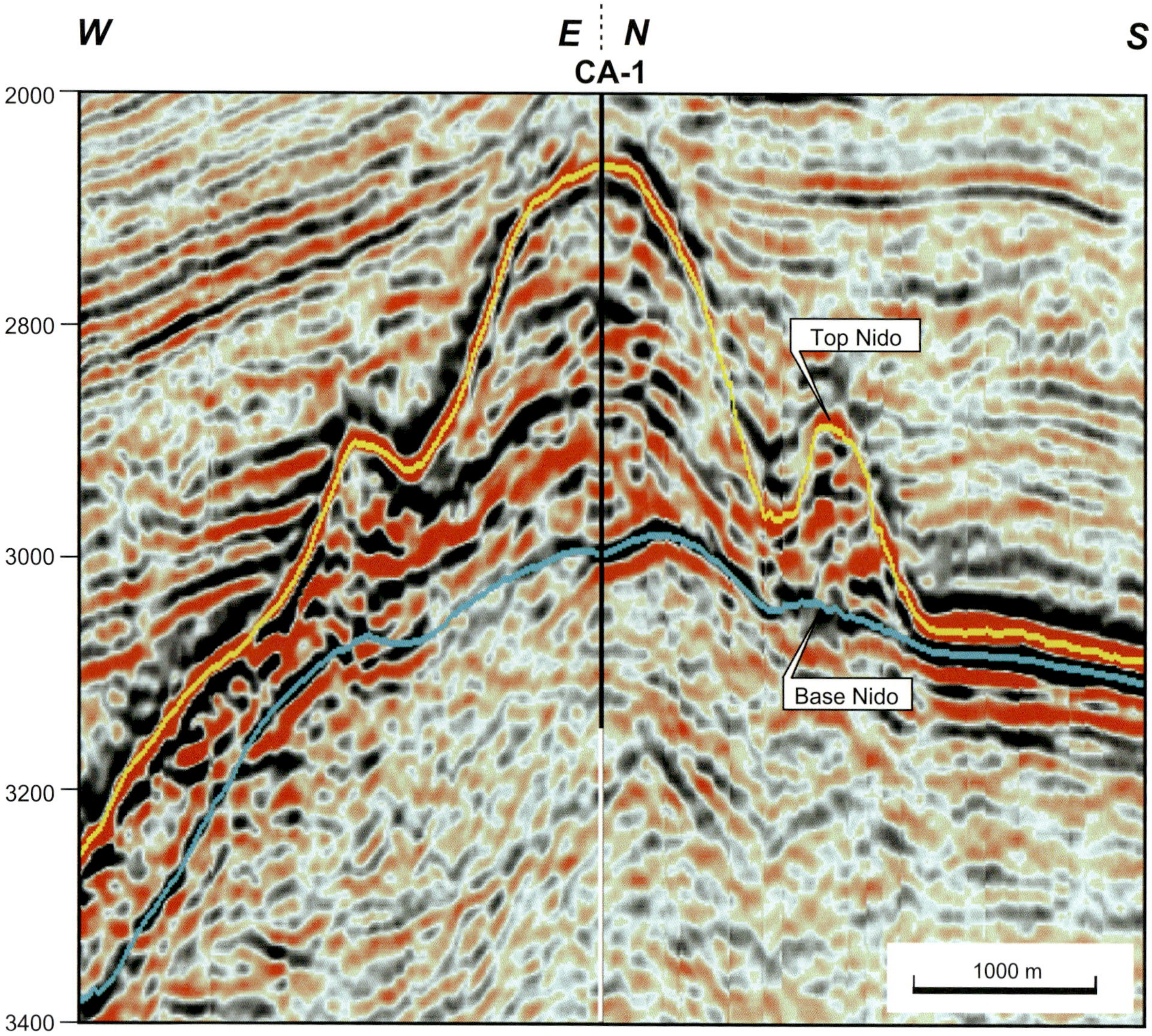

FIGURE 7. Pinnacle reefs on the slopes of the southern Malampaya culmination.

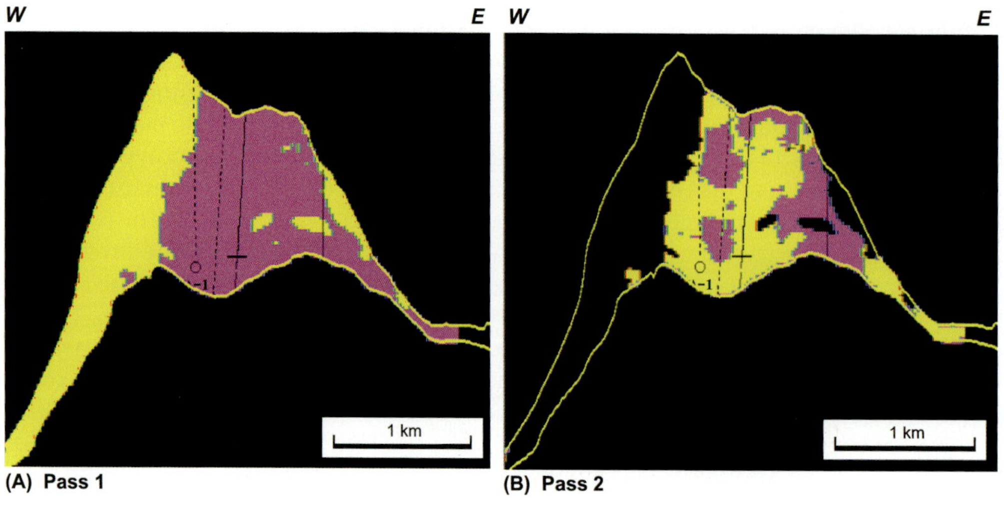

FIGURE 8. Neural network–based multi-attributes volume segmentation and seismic facies analysis along the seismic line shown in Figure 5. (A) Pass 1 volume segmentation results. The yellow colors highlight the reef-zone/ocean slope, whereas the purple color represents the back reef/lagoon. (B) Pass 2. The back reef/lagoon was further segmented to locate potential patch reefs, landward margin, and landward shoal (purple).

subsequently used in the static modeling to constrain, in particular, the extent of marine cementation.

Once a volume has been classified in main facies units, it is possible to further classify a unit into subfacies during a second-pass filtering. This was applied to Malampaya for the back-reef/lagoon facies. Again, a supervised neural network approach (with new training sets) was used with a similar set of attributes as applied previously with the exception of dip and azimuth. This allowed further segmentation of the lagoonal facies to locate potential patch reefs, the landward margin, and landward shoal.

From a sequence-stratigraphic point of view (Schlager, 1999), the evolution of the Malampaya carbonate buildup can be split into three major system tracts: first, a transgressive system tract that reflects the installation of the carbonate platform and the backstepping of its margins (late Eocene–Oligocene); second, an "empty-bucket" system tract (early Miocene) that corresponds to the drowning of the reef interior and the aggradation of the oceanward reef; and third, a drowning system tract (late Miocene) that marks the demise of the Nido Limestone Platform. This stepwise disappearance of rift-related shallow-marine carbonates during the Tertiary is also described as an example in the Gulf of Aden (Borgomano and Peters, 2004). The scarcity of identified "flat tops" in the Malampaya seismic could suggest that highstand and lowstand system tracts are not very well developed within the Malampaya buildup.

IDENTIFICATION OF HIGH-POROSITY AREAS FOR GAS-DEVELOPMENT DRILLING

The Malampaya carbonate buildup is overlain by the dominantly deep-marine shale sequence of the Pagasa Formation (Figure 5). The Pagasa exhibits a clear compaction-related acoustic-impedance depth trend, which is locally disturbed by thin turbiditic siltstone-sandstone layers. In contrast, carbonate acoustic impedance is strongly related to porosity (Figure 9). A crossplot of the Pagasa and Nido acoustic-impedance depth trends provides a tool to predict the reservoir porosity within the upper part of the buildup from top-reservoir amplitude (Figure 10).

The variable character of the Top Nido reflection can be observed in Figure 11. West of the MA-1 well, the Top Nido (yellow horizon) is represented by a very strong negative (red) loop, indicating the presence of low-porosity reservoir at the top. West and east of MA-5, the Top Nido pick is at the zero-crossing from the negative (red) to the positive (black) loop, confirmed

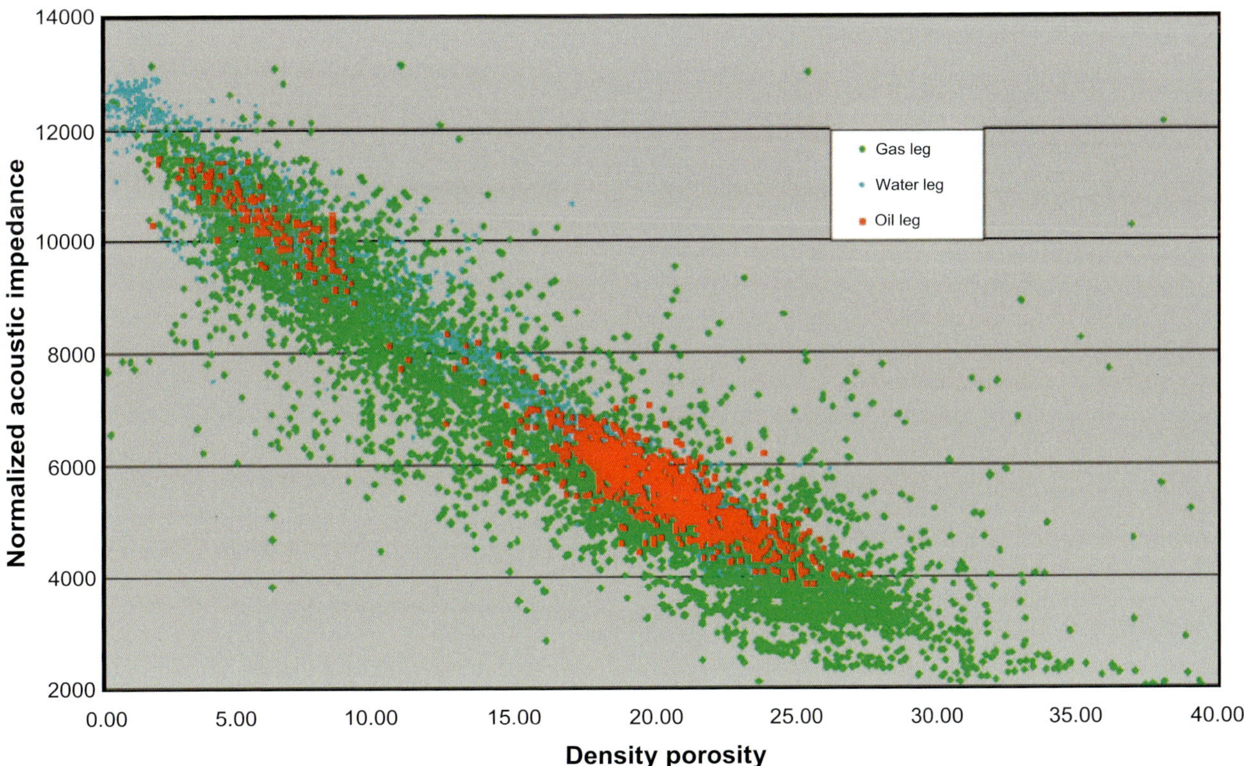

FIGURE 9. Acoustic impedance vs. density (compensated formation density log) porosity and porefill. The porosity distribution in the oil and water leg is bimodal, reflecting the low-porosity early-Oligocene platform carbonates and ocean slope material and the higher-porosity late-Oligocene aggradational units (see Figure 2).

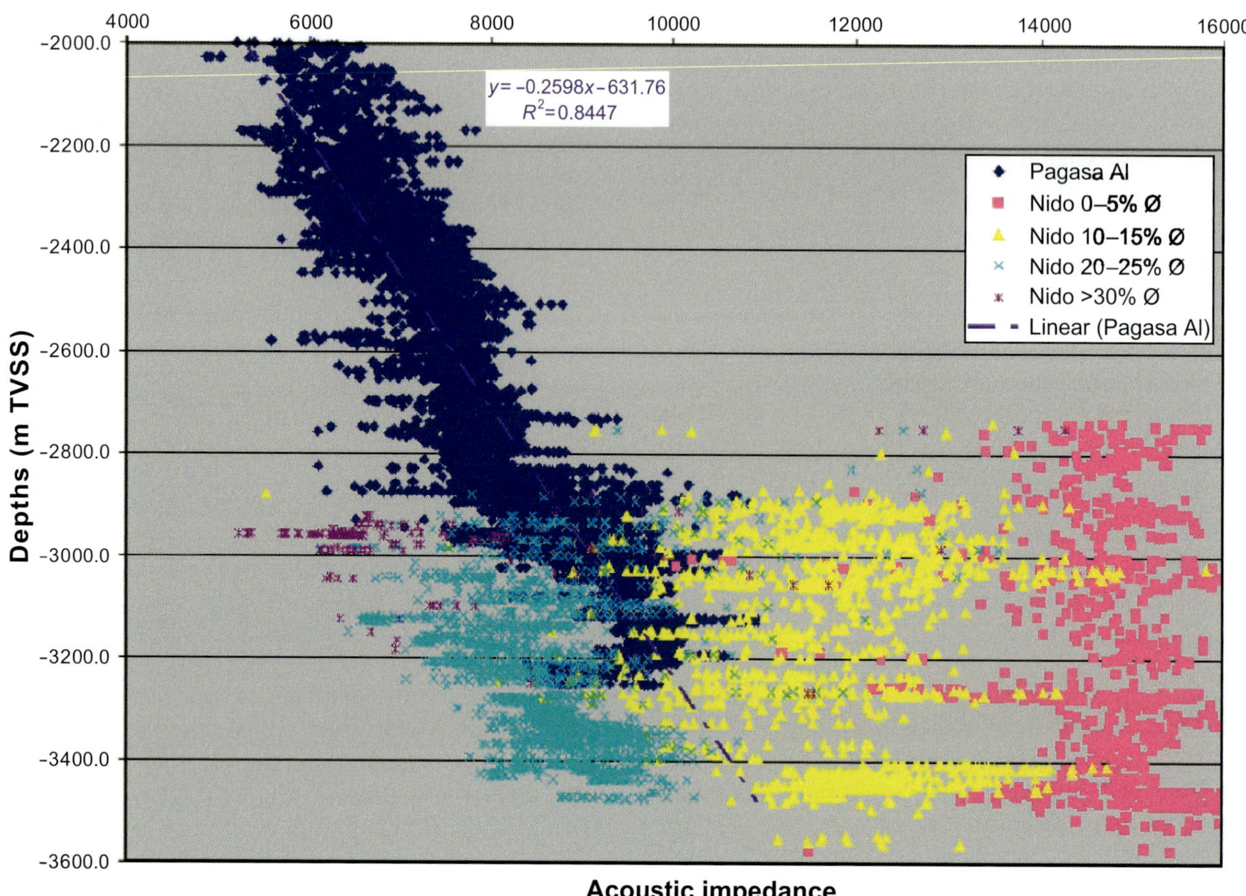

FIGURE 10. Reservoir and overburden acoustic-impedance trends vs. depth and reservoir porosity. A positive acoustic-impedance contrast is observed at the interface between the Pagasa shale and low-porosity (approximately 0–15%) Nido Limestone, resulting in a strong negative (red) loop at Top Nido using Society of Exploration Geophysicists normal polarity. In case the Pagasa is overlying high-porosity (≳25%) Nido Limestone, a weak positive (black) loop can be observed.

by a vertical seismic profile (VSP) in MA-5. The Top Nido pick halfway between MA-5 and MA-2 then falls onto a black (positive loop). At this location, a high-porosity (>25%) reservoir interval is interpreted to subcrop the Pagasa. MA-2 intersected the tight Intra-Nido marker (between green and blue horizons) at top reservoir. Without consideration for the relationship between Top Nido reservoir porosity and reflection character, the seismic interpretation between MA-5 and MA-2 could be considered an interpretation artifact.

Based on VSP and synthetic seismogram analysis, the maximum negative amplitude within a ±10-ms window around the mapped Top Nido reflection was found to best characterize the Pagasa-Nido interface. Considering the large vertical relief of the Malampaya buildup, the resulting raw amplitudes were depth corrected to remove the Pagasa acoustic-impedance depth trend overprint. The final amplitudes were used in two ways.

First, the analysis indicates that the highest Top Nido porosities are located along the eastern side of the field (Figure 12). This distribution is in agreement with the diagenetic model of the field described by Grötsch and Mercadier (1999) and analogous to modern reef settings (Purser, 1980). Thereby, early phreatic leaching enhanced reservoir-matrix porosities predominantly in the back reef, lagoon, and landward shoal (in the east), whereas early marine cementation destroyed most of the porosity in the fore-reef slope and oceanward reef (in the west) setting. The new seismic analysis, however, suggests the potential presence of high-porosity reservoir on parts of the eastern flank in the southern culmination. This area had previously been assumed to be of poor reservoir quality based on the results of the CA-1 well.

Secondly, operationally, the data were used to predict the presence of potential mud-loss zones at top reservoir. In Malampaya, open fractures and associated mud losses are restricted to low-porosity intervals. This observation is based on log data (especially from Formation MicroImager and dipole shear imaging logs), core data, detailed mud-loss monitoring, and geomechanical fracture modeling. Therefore, identification of low-porosity zones at Top Nido provides a tool to avoid mud losses on entering the reservoir— a condition that could result in drilling problems. For example, well MA-9 was repositioned to avoid such a predicted loss zone.

THREE-DIMENSIONAL TIME-TO-DEPTH CONVERSION USING ACOUSTIC-IMPEDANCE DATA

Seismic velocity in pure carbonates (like Malampaya) is predominantly a function of matrix porosity, whereas porefill has only a limited effect (Figure 9). Therefore, the time-to-depth conversion of a 3-D seismic–derived static reservoir model requires the knowledge of reservoir porosity in three dimensions. The newly generated Malampaya 3-D PSDM data were converted to acoustic impedance, and then porosity, using Jason Geoscience Sparse Spike Inversion, Jason Stochastic Inversion (see Shanor et al., 2001), and PROMISE (Shell proprietary probabilistic inversion software; see Leguijt, 2001). Eighteen horizons were interpreted fieldwide from reflectivity (Figures 6, 11) and impedance data and converted to depth using velocities derived from inversion-based average interval porosities. After calibrating the depth grids at the well locations, the resulting frame was infilled with the depth-stretched porosity cube in GEOCAP (Shell proprietary object-based geologic modeling tool) and DEPSIM (Shell proprietary layer-based geologic modeling tool) (Figure 13).

SEISMIC INVERSION, MULTIPLE RESERVOIR REALIZATIONS, AND VOLUMETRICS

The procedure described in the previous paragraph was repeated for various Top Nido depth grids and porosity-cube realizations, resulting in six static model scenarios with the following input:

where
Method 1: Base-case Top Nido depth grid, Jason Sparse Spike Inversion porosity cube;
Method 2: Low-case Top Nido depth grid (considering horizon time pick and overburden velocity uncertainty), Jason Sparse Spike Inversion porosity cube;
Method 3: High-case Top Nido depth grid (considering horizon time pick and overburden velocity uncertainty), Jason Sparse Spike Inversion porosity cube;
Method 4: Base-case Top Nido depth grid, Jason Statmod Mean porosity cube created from 35 porosity-cube realizations;

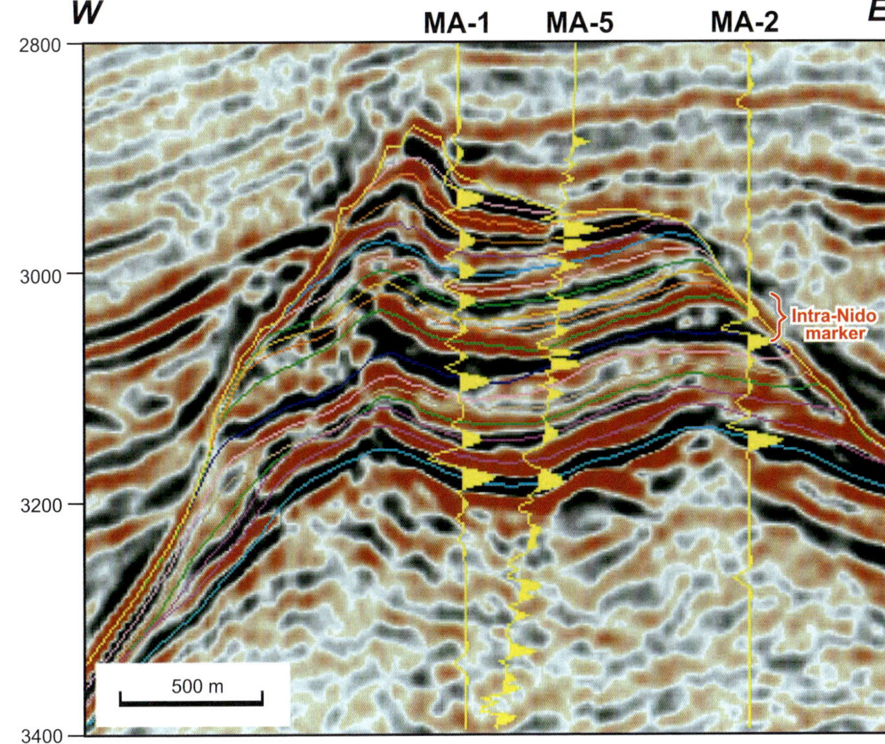

FIGURE 11. Detailed horizon interpretation used for time-depth stretch of porosity cubes and static modeling. West of the MA-1 well, the Top Nido (yellow) is represented by a very strong negative (red) loop, indicating the presence of low-porosity reservoir at the top. West and east of MA-5, the Top Nido pick is at the zero-crossing from the negative (red) to the positive (black) loop, confirmed by a VSP in MA-5. The Top Nido picks halfway between MA-5 and MA-2 then falls onto a black (positive loop). At this location, a high-porosity (>25%) reservoir interval is interpreted to subcrop the Pagasa. Note that this is not an interpretation artifact. MA-2 intersected the tight Intra-Nido marker (between green and blue horizons) at top reservoir.

Method 5: Base-case Top Nido depth grid, Jason Statmod low-case porosity cube created by subtracting the standard deviation uncertainty porosity cube from the mean porosity cube;

Method 6: Base-case Top Nido depth grid, Jason Statmod high-case porosity cube created by adding the standard deviation uncertainty porosity cube to the mean porosity cube.

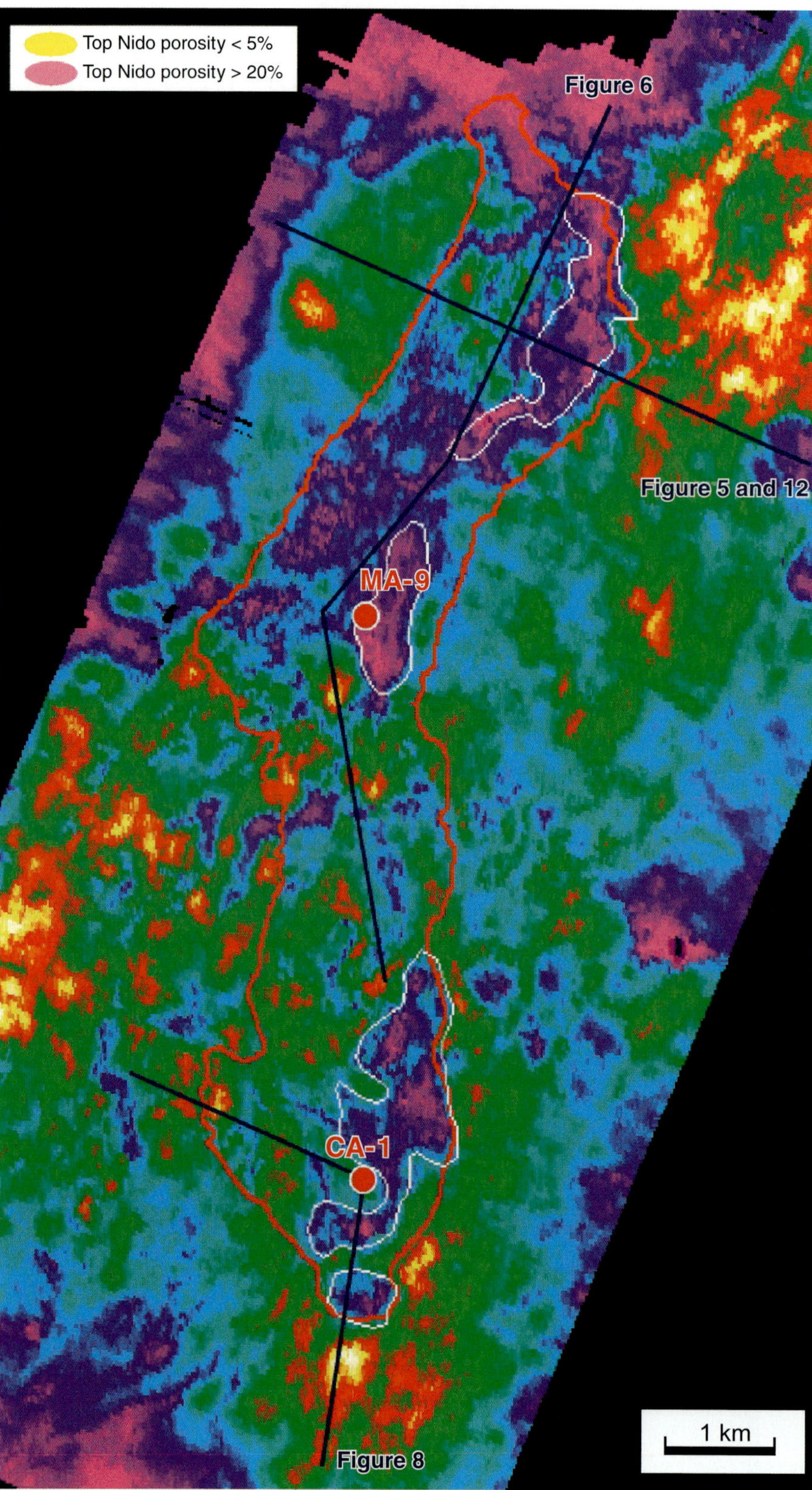

FIGURE 12. Top-reservoir porosity from amplitude analysis (maximum negative amplitude in a ±10-ms window around Top Nido), depth corrected. High Top Nido porosity areas are concentrated on the eastern side of the field (white polygons), where the reservoir quality was enhanced through diagenesis in a meteoric environment.

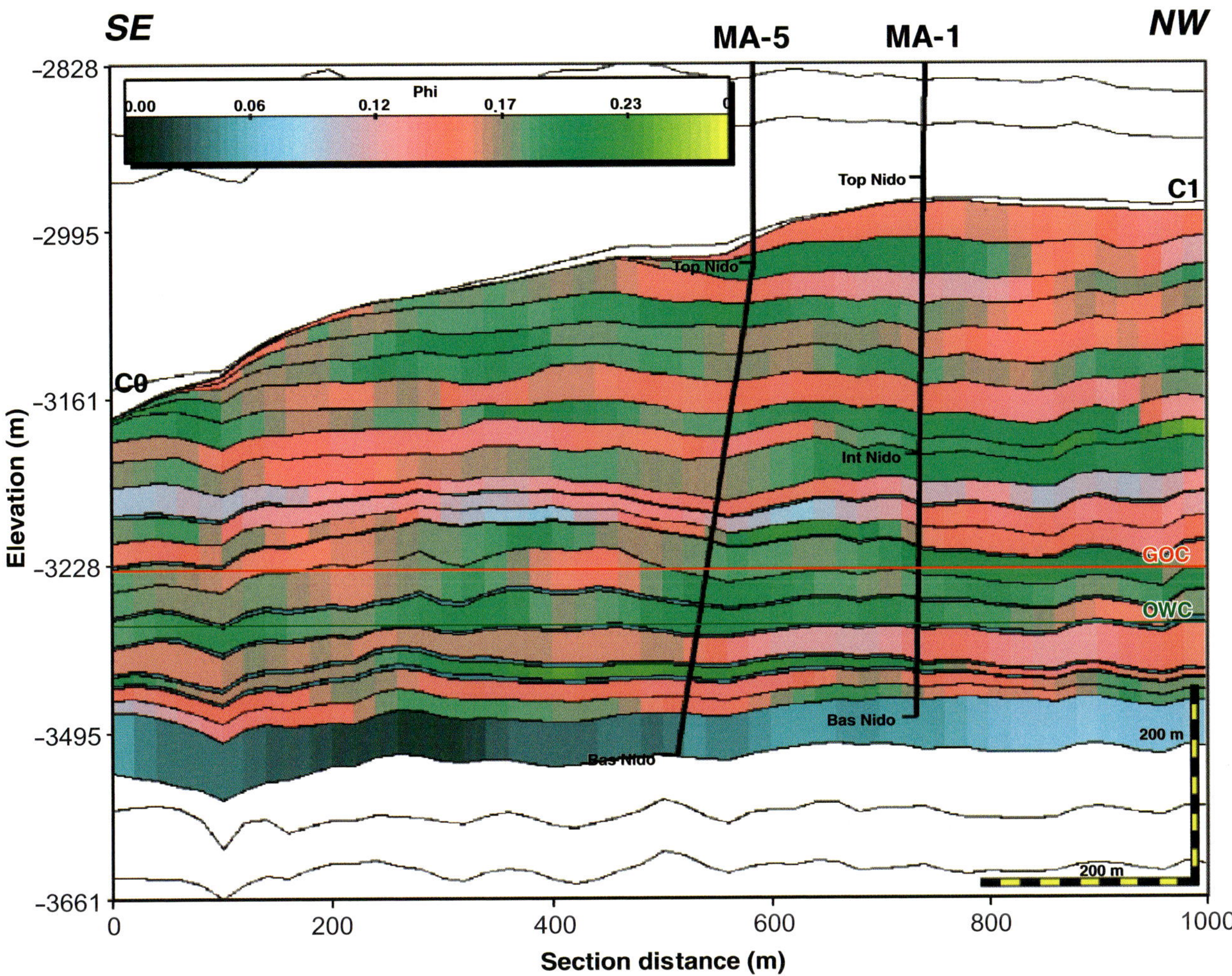

FIGURE 13. Vertical section through static reservoir model in depth. The porosity infill of the reservoir layers is derived from PROMISE acoustic-impedance inversion. Some thin, very low porosity layers at reservoir unit boundaries in the lower part have been manually added. These layers are too thin to be resolved by the 3-D seismic, but they have been recognized as correlatable layers in the wells. The low-porosity layers could form seals, in case they are not fractured, or high permeability layers, in case they are highly fractured. GOC = gas-oil contact; OWC = oil-water contact.

The resulting range of in-place volumes from these predominantly seismic-driven models was found to be in agreement with previous estimates using the entirely different approach described in Grötsch and Mercadier (1999). The static models were then further enhanced in GEOCAP/DEPSIM by adding deterministic information below seismic resolution (like low- and high-permeability streaks observed in wells, Figure 13) and reservoir rock-type classification for porosity-permeability and saturation modeling before export to dynamic reservoir simulation software (MoReS).

HORIZONTAL DRILLING FOR OIL-RIM APPRAISAL

The porosity volume cubes from seismic and different static reservoir model realizations can be used to visualize hydrocarbon distribution given the gas-oil contact at 3332 m subsea and the free-water level at 3388 m subsea. In Figure 14, the distribution of oil-in-place within the 56-m-thick oil column was visualized by constructing porosity-height maps (= oil-rim thickness × oil-rim average porosity). From this, it becomes apparent that in all scenarios the largest oil volumes are located in the northern part of the field, whereas in the center, the oil is restricted to a narrow rim around a basement high (Figure 4). The combined displays are useful tools for identifying reservoir sweet spots for different realizations and for targeting development wells.

CONCLUSIONS

Field development planning, including an integrated reservoir characterization and modeling effort

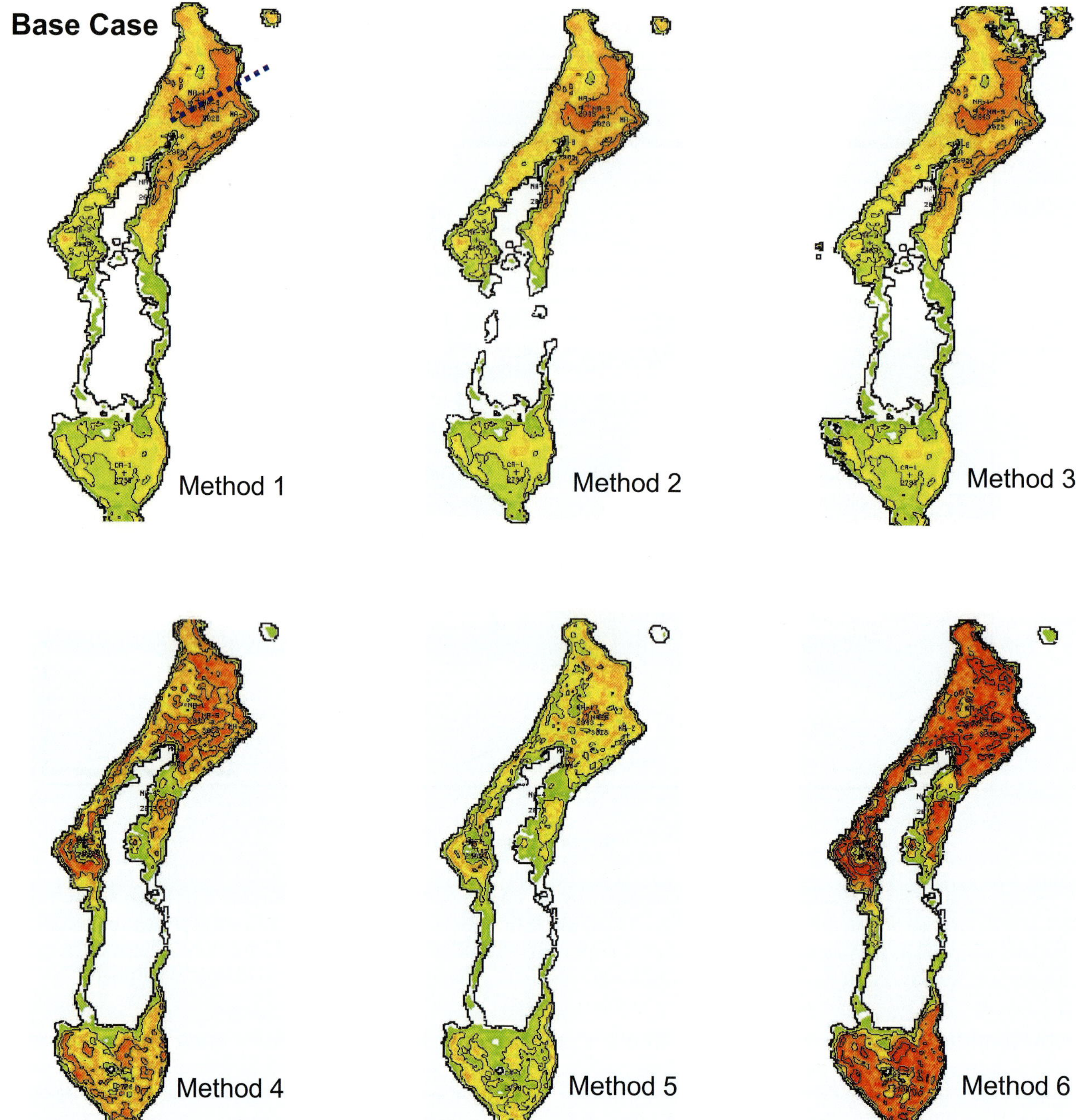

Figure 14. Porosity-height scenarios for the Malampaya oil rim. The base case is shown at top left. Green = low-porosity-height values; red = high-porosity-height values.

during 1994–1996, highlighted the importance of the 3-D velocity model for seismic processing and time-to-depth conversion in the Malampaya 3-D survey (Grötsch and Mercadier, 1999). The renewed 3-D PSDM presented here used such an updated and improved 3-D velocity model with respect to the 1991 Malampaya 3-D seismic data set. This reprocessed seismic data set formed the input for quantitative seismic-analysis tools used for a second-phase reservoir characterization, static modeling, reserves estimation, and optimized gas development and oil-appraisal well targeting.

Top Nido reflection amplitudes were used to predict matrix porosity within the upper part of the carbonate reservoir. In line with previous reservoir models, high-porosity areas are predominantly present in the central lagoonal area, where meteoric leaching has enhanced the reservoir quality, and on the eastern, landward side of the reef, which was protected from early marine cementation processes. These areas were preferentially targeted in gas-development drilling. In contrast, low-porosity zones known to be prone to fracturing were avoided to minimize the risk of mud losses, which are potential related drilling problems.

Various acoustic-impedance inversion techniques were used to create reservoir-porosity realizations. Considering the linear porosity-velocity relationship in clean carbonates, porosity cubes are essential for a correct time-depth conversion of carbonate static reservoir models. Several static model realizations were created using the porosity cubes as a backdrop, combined with 3-D seismic facies analysis and a depositional model based on well data and analogs. The models were used to calculate in-place volumes and to target wells in good reservoir areas. Gas-development well results and early production performance support the modeled reservoir distribution within the lagoonal part of the northern Malampaya accumulation.

Porosity-height scenarios created from the different porosity cubes were used to visualize reservoir-quality distribution within the Malampaya oil rim and formed the basis for targeting a horizontal appraisal well. The MA-10 horizontal oil-rim appraisal well drilled at the end of 2001 confirmed the forecasted facies distribution and reservoir properties.

ACKNOWLEDGMENTS

This article is largely based on work performed during the Malampaya gas-development and oil-appraisal drilling campaign in 2000 and 2001 and the contribution of the subsurface team in Shell Philippines Exploration BV (SPEX), namely, G. Davies, J. Esquito, G. Loftus, and O. Tosun is greatly appreciated. Seismic processing, special seismic studies, and reservoir modeling were performed at Shell Exploration and Production Technology and Research (SEPTAR) in Rijswijk, The Netherlands. Contributions by A. van den Berg, T. Carlson, J. Leguijt, L. Mieles-de Pina, E. Sims, and T. Tjan are highly appreciated. Comments by the reviewers Bruce Hart and Gregor Eberli helped considerably to improve the manuscript. We are grateful to SPEX, Texaco Philippines Inc., and PNOC-EC for authorizing publication of these data.

REFERENCES CITED

Borgomano, J. R. F., and J. Peters, 2004, Outcrop and seismic expressions of coral reefs, carbonate platforms and adjacent deposits in the Tertiary of the Salalah Basin, south Oman, *in* G. P. Eberli, J. L. Masaferro, and J. F. Sarg, eds., Seismic imaging of carbonate reservoirs and systems: AAPG Memoir 81, p. 251–266.

Bosscher, H., and W. Schlager, 1993, Accumulation rates of carbonate platforms: Journal of Geology, v. 101, p. 345–355.

Grötsch, J., and C. Mercadier, 1999, Integrated 3-D reservoir modeling based on 3-D seismic: The Tertiary Malampaya and Camago buildups, offshore Palawan, Philippines: AAPG Bulletin, v. 83, p. 1703–1728.

Holloway, N. H., 1982, North Palawan Block, Philippines—Its relation to the Asian mainland and role in evolution of the South China Sea: AAPG Bulletin, v. 66, p. 1355–1383.

Leguijt, J., 2001, A promising approach to subsurface information integration: 63rd EAGE Conference and Technical Exhibition, Amsterdam, The Netherlands.

Longman, M. W., 1981, Fracture porosity in reef talus of a Miocene pinnacle-reef reservoir, Nido-B field, the Philippines, *in* P. C. Roehl and P. W. Choquette, eds., Carbonate petroleum reservoirs: New York, Springer-Verlag, p. 549–560.

Purser, B. H., 1980, Sedimentation et diagenese des carbonates neritiques recents. Publication de l' IFP, tome 1: Paris, Editions Technip, 366 p.

Schlager, W., 1999, Sequence stratigraphy of carbonate rocks: Leading Edge, v. 18, p. 901–907.

Shanor, G., M. Rawanchaikul, M. Sams, R. Muggli, G. Tiley, and J. Ghulam, 2001, A geostatistical inversion to flow simulation workflow example: Makarem field, Oman: 63rd EAGE Conference and Technical Exhibition, Amsterdam, The Netherlands.

Wiedicke, M., 1987, Stratigraphie, Mikrofazies und Diagenese tertiärer Karbonate aus dem Südchinesischen Meer (Dangerous Grounds— Palawan, Philippines): Facies, v. 16, p. 195–302.

Wolfart, R., P. Cepek, F. Grahmann, E. Kemper, and H. Proth, 1986, Stratigraphy of Palawan Island, Philippines: Newsletters on Stratigraphy, v. 16, p. 19–48.

9

Droste, H., and M. Van Steenwinkel, 2004, Stratal geometries and patterns of platform Carbonates: The Cretaceous of Oman, *in* Seismic imaging of carbonate reservoirs and systems: AAPG Memoir 81, p. 185–206.

Stratal Geometries and Patterns of Platform Carbonates: The Cretaceous of Oman

Henk Droste[1]
Petroleum Development Oman, Muscat, Sultanate of Oman

Mia Van Steenwinkel
Petroleum Development Oman, Muscat, Sultanate of Oman

ABSTRACT

Extensive carbonate platforms covered the eastern part of the Arabian Plate during Mesozoic times. The interior parts of these platforms are commonly visualized as undifferentiated, extensive shallow-water areas, where carbonates accumulate by aggradation. This view is based on the fact that individual shallowing-upward carbonate packages are laterally extensive.

The improvement of seismic quality and resolution, however, reveal internal geometries within the carbonates. The Cretaceous carbonate platform of Oman, for example, shows a complex internal architecture, rather than a "layer-cake" configuration. Recognition of these stratal geometries has important implications for prospective hydrocarbon discovery and development in these sequences.

The aim of this study is a better understanding of the internal architecture of carbonate platforms, which can guide exploratory-play evaluation and at the same time field-scale reservoir modeling and production performance of carbonates in general. Two examples— and two scales— have been chosen to highlight the internal complexity of these carbonate systems:

(1) Habshan Formation: large-scale Arabian Plate margin configuration (approximately 300 m thick, progradation of more than 250 km);
(2) Natih Formation: smaller-scale intrashelf carbonates configuration (approximately 50–100 m thick, progradation of more than 50–60 km).

These examples are based on good-quality, high-density seismic data, closely spaced wells and excellent outcrop exposures.

Arabian Plate Margin Configuration (Habshan Formation): The Cretaceous carbonate platform was initiated in central Oman during the major transgression over the base

[1]*Present address:* JVR Centre for Carbonate Studies, Sultan Qaboos University, Al Khod, Sultanate of Oman.

Cretaceous unconformity. After a rapid progradation of some 250 km to the north and northeast, the platform edge aggraded, leading to the development of a 700-m-thick platform succession.

A well-developed clinoform complex occurs in the prograding lower part of the platform sequence, represented by the "Habshan system." This clinoform complex consists of a series of forward-built stratigraphic packages of 10–20 km width, each showing a change from low- to high-angle clinoforms. These packages are thought to represent third-order sea level cycles superimposed on a second-order regressive trend. The variation in clinoform dip angle is interpreted to reflect changes in accommodation space available during relative sea level lowstands and highstands. It is also associated with variations in sediment fabric. The high-angle clinoforms developed during platform aggradation in times of sea level highstand. They are composed of thick sequences of porous and permeable shallow-water-derived grainstones and packstones. The low-angle clinoforms represent forced regressive wedges formed during lowering sea level. They are composed of muddy, deeper-water calcareous shales, with platform-derived porous units.

Intrashelf Carbonates Configuration (e.g., Natih Formation): In the younger platform interior carbonates, seismic data reveal the presence of similar but smaller-scale clinoform complexes and the occurrence of intraplatform basins. Mapping of clinoform belts and directions of progradation in the platform interior Natih E shows that this extensive carbonate member consists of several separate platforms, which merged by lateral accretion. The platforms started to grow in areas with relatively low subsidence rates, such as basement highs and salt domes, following a regional rise in sea level. Merging of the platforms was not always complete and in the intervening areas relict intraplatform basins developed, which were later filled with shales. Similar to the large Habshan clinoform system, the clinoforms of the younger platform interior show cyclic variations in slope angle, associated with variations in sediment composition and related to relative change in sea level.

In addition to the internal features, the top of the Natih A Platform is characterized by uplift, karstification, and erosion. An extensive system of deeply incised meandering fluvial channel systems is observed on three-dimensional seismic.

Generally, the integration of seismic and well data is crucial to the recognition of inclined stratal geometries and the diachronic character of lithologic units. This recognition can have important implications for hydrocarbon prospectivity and reservoir development. At exploration scale, it allows the definition of stratigraphic trapping potential. In addition, risks and opportunities for reservoir and seal can be better evaluated. At development scale, it will guide well-log correlations away from the "layer-cake" model. This way, the understanding and prediction of reservoir heterogeneities, continuity, sweep efficiency, early high water cut, and water flood can be improved and field development plans optimized.

INTRODUCTION

During Cretaceous times, the eastern part of the Arabian Plate was covered by extensive carbonate platforms, which currently contain major hydrocarbon accumulations (Figure 1; Murris, 1980; Harris and Frost, 1984; Jordan et al., 1985). The interior parts of these platforms are commonly visualized as extensive shallow-water areas, where carbonates accumulate by aggradation.

Previous seismic studies in Oman (Haan et al., 1990) and the United Arab Emirates (Landmesser and Saydam, 1996; Aziz and El-Sattar, 1997) have shown that the lower Cretaceous platform edge consists of

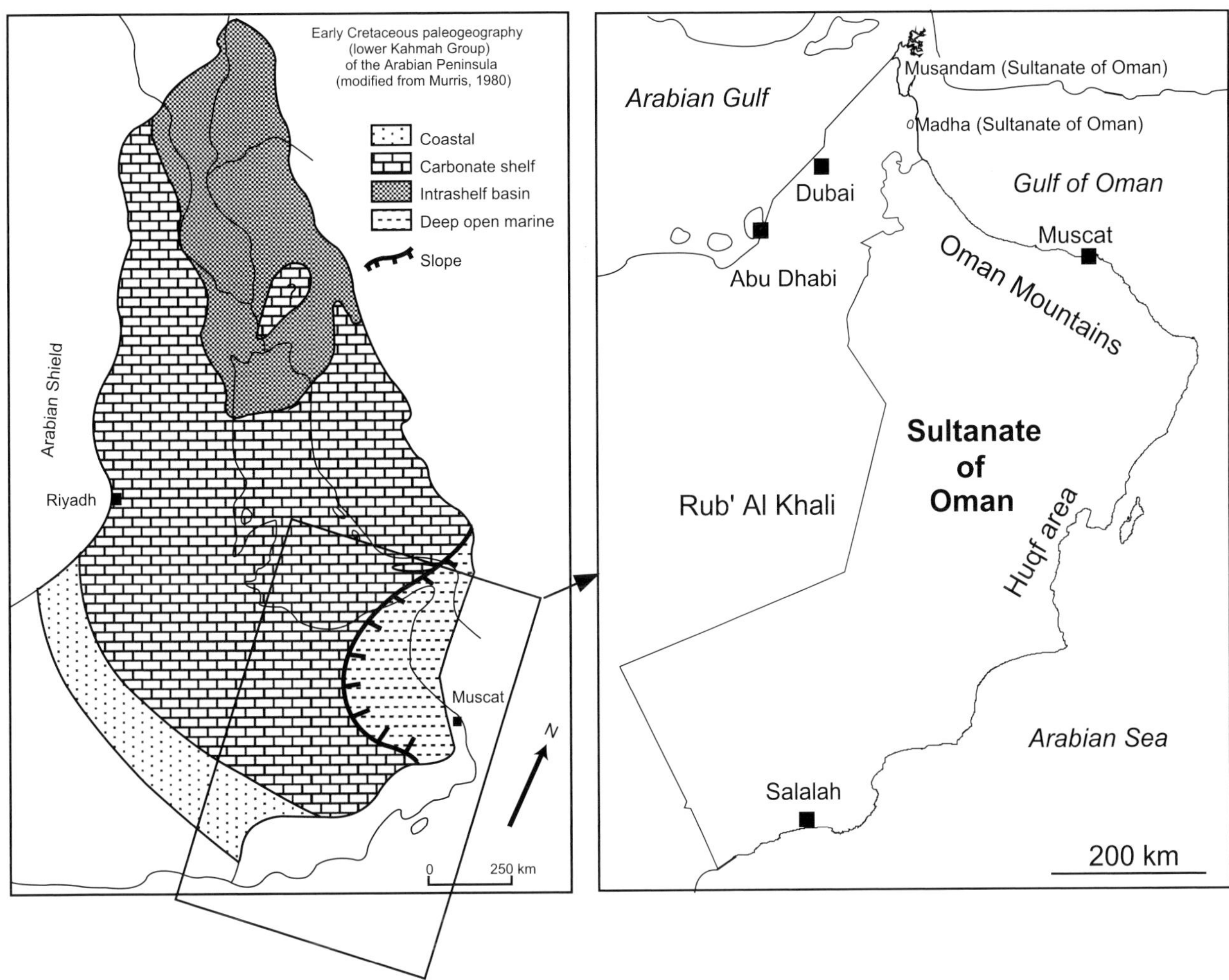

FIGURE 1. Simplified paleogeographic map of the Arabian Peninsula during the Early Cretaceous and the present-day geography of Oman.

well-defined progradational clinoform belts. The platform interior also seems to consist of complex internal geometries. This has been shown by high-resolution sequence stratigraphy studies based on well and outcrop data of the Shuaiba (Van Steenwinkel, 1992, 1996; Witt and Gökdag, 1994; van Buchem et al., 2002b) and Natih (van Buchem et al., 1996, 2002a) formations in Oman as well as of the lateral equivalent of the Natih Formation in the United Arab Emirates, the Mishrif Formation (Burchette, 1993; Pascoe et al., 1995).

This study aims at guiding exploratory-play evaluation and at the same time field-scale reservoir modeling and production performance of carbonates in general by better understanding the internal architecture of carbonate platforms. The Cretaceous of north Oman is used as an example, which has good-quality, high-density seismic data, closely spaced wells, and continuous outcrop.

Two- and three-dimensional seismic data, well logs, core material, and outcrop data have been taken into account for this study. Excellent outcrops of Mesozoic carbonates occur in the Oman Mountains (Figure 1). The latter formed by collision of the Arabian and Eurasian plates, associated with the obduction of oceanic crust onto the Arabian Plate during the Late Cretaceous (Glennie et al., 1974). Wadis (canyons) cutting into the uplifted and folded Mesozoic strata reveal the architecture and evolution of the northern edge of the Mesozoic carbonate platform (Pratt and Smewing, 1990, 1993a, b; van Buchem et al., 1996, 2002a; Masse et al., 1997; Immenhauser et al., 2002; Hillgärtner et al., 2002).

South of the Oman Mountains, Mesozoic outcrops are scarce, because approximately 2 km of Tertiary sediments cover the Cretaceous in central and south Oman. However, this rocky desert area has been the focus of hydrocarbon exploration for nearly 50 years, and the lack of outcrops is compensated by an extensive seismic and well database. Two examples have been chosen to highlight the internal complexity of carbonate systems:

1) Habshan Formation: large-scale, Arabian Plate margin evolution (approximately 300 m thick, progradation of more than 250 km);

2) The Natih Formation: smaller-scale, intrashelf carbonates configuration (approximately 50–100 m thick, progradation of more than 50–60 km).

GEOLOGIC SETTING

During the Mesozoic, the Sultanate of Oman was located at the northeastern edge of the Arabian Plate (Figure 1) and was part of an extensive carbonate platform that covered most of the Middle East (Murris, 1980). The platform edge was located just north of the present-day Oman Mountains, in the north of Oman (Pratt and Smewing, 1990). Pulses of clastic sediment onto this platform were derived from the exposed Arabian Shield and fringing exposures of Paleozoic sediments in the southwest. The carbonate province consists of a series of stacked carbonate platforms separated by major stratigraphic breaks (Figure 2). These breaks represent regional tectonic unconformities, related to the break-up of Gondwana and subsequently, the collision of the Eurasian and African-Arab plates during the Late Cretaceous and Tertiary (Loosveld et al., 1996).

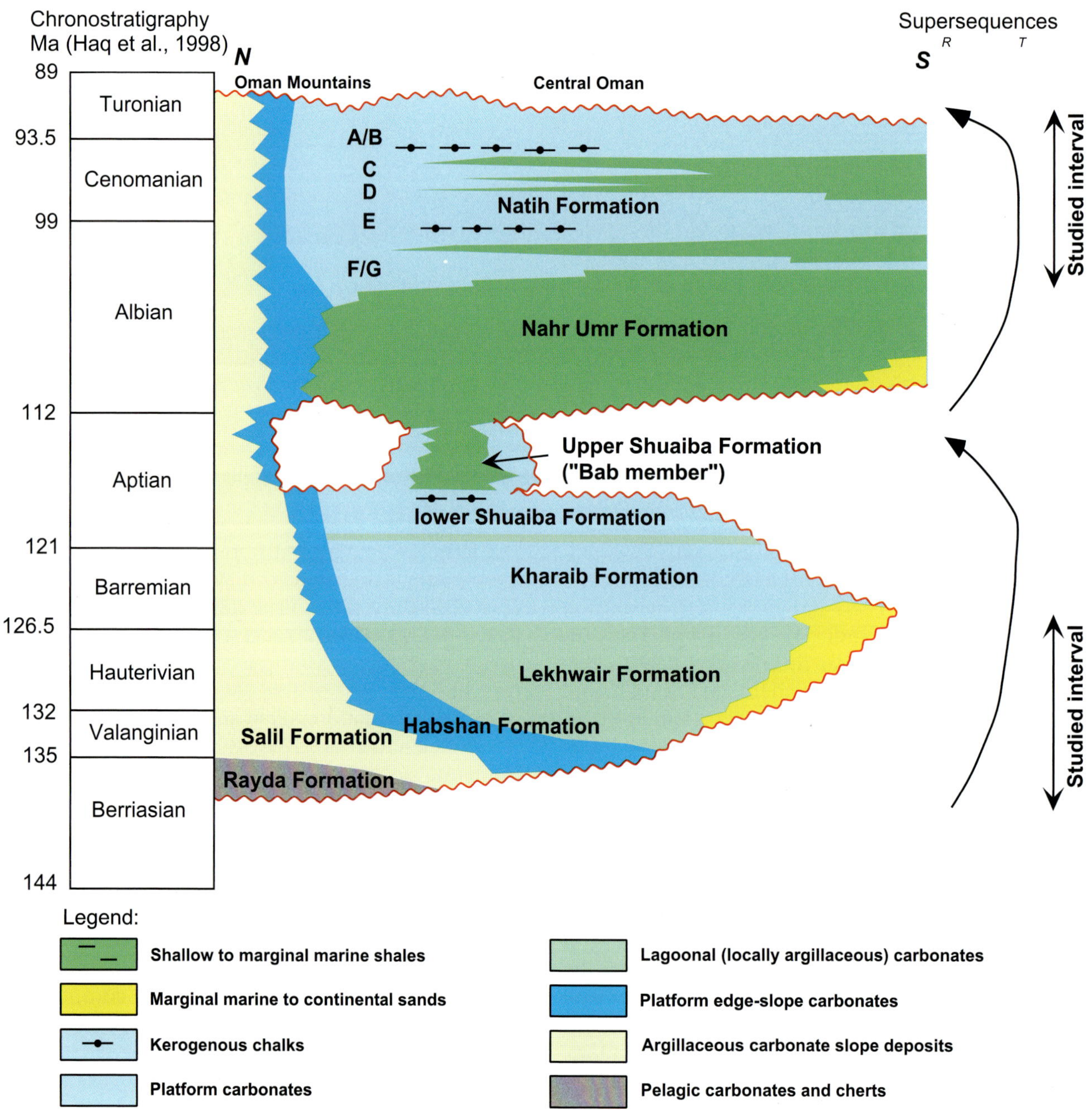

FIGURE 2. Cretaceous stratigraphy of Oman.

The Cretaceous carbonate-platform complex in Oman is as much as 1200 m thick and 1000 km wide. It unconformably overlies Jurassic and older strata, which were tilted, uplifted, and eroded in the late Jurassic. Along the platform margin in the north, the uplift was followed by a major downwarping in the Berriasian (Rabu et al., 1990; Pratt and Smewing, 1990, 1993a, b). The uplift and the subsequent foundering of the Jurassic carbonate-platform edge resulted in a drowned shelf. Consequently, the platform edge retreated some 300 km to the south onto the Arabian Plate.

During Berriasian to Turonian times, carbonate deposition was regularly interrupted by subaerial exposure and the influx of fine-grained clastics from the south. These events are attributed to tectonic movements of the Arabian Shield (Sharief et al., 1989). The largest one was a relative sea level fall in the late Aptian (Figure 2), which can be recognized as a regional unconformity over the entire Middle East (Harris et al., 1984). In Oman, this unconformity is associated with major karstification and erosion. It was followed by a widespread influx of fine-grained clastics of the Nahr Umr Formation (Witt and Gökdag, 1994). The Cretaceous platform was terminated by a regional phase of uplift and subaerial exposure in the Turonian. This is thought to be related to the formation of a peripheral foreland bulge during the initial phases of the collision between Eurasia and the Arabian margin (O'Connor and Patton, 1986; Warburton et al., 1990).

STRATIGRAPHY

The focus of this chapter is on the Habshan system in the Kahmah Group and the Natih E Member in the Wasia Group (Figure 2). Both groups are shortly summarized below (see also Glennie et al., 1974; Hughes Clarke, 1988).

Kahmah Group

The Kahmah Group ranges from late Berriasian to Aptian and consists of six formations (Figure 2). They are in stratigraphic order:

- *Rayda Formation:* pelagic lime mudstones and cherts;
- *Salil Formation:* argillaceous lime mudstone and wackestone (slope deposits);
- *Habshan Formation:* oolitic and bioclastic grainstones to packstones deposited in a platform margin to slope setting;
- *Lekhwair Formation:* cycles consisting of restricted-marine, argillaceous limestones and more open-marine, skeletal wackstones to peloidal-skeletal lime packstones and grainstones with rudist biostromes;
- *Kharaib and lower Shuaiba Formations:* deepening and shallowing upward cycles of restricted-marine, argillaceous lime wackestones to packstones, microbial boundstones, open-marine packstones and grainstones with rudist biostromes (van Buchem et al., 2002b);
- *Upper Shuaiba Formation:* argillaceous lime mudstones interbedded with deeper-water shales and redeposited grainstones and packstones, locally coarse-grained shallow-water packstones to grainstones. This formation is only locally developed in the northwest of Oman.

Biostratigraphic data indicate that most formation boundaries in the Kahmah Group are diachronous (Scott, 1990; Simmons, 1994). The Rayda, Salil, Habshan, and Lekhwair Formations are Berriasian to Hauterivian–Barremian in age. The Kharaib and the Shuaiba Formations are more isochronous units: late Barremian to early Aptian for the Kharaib, early Aptian for the lower Shuaiba, and late Aptian for the upper Shuaiba (Witt and Gökdag, 1994). The upper Shuaiba Formation is restricted to the Bab intraplatform basin (located in northwestern Oman and the United Arab Emirates), where it was deposited following a relative drop in sea level of several tens of meters, spanning the early to late Aptian boundary (Van Steenwinkel, 1992; van Buchem et al., 2002b).

Wasia Group

The Wasia Group is Albian to early Turonian in age and consists of the Nahr Umr Formation at the base and the Natih Formation at the top.

- The Nahr Umr Formation is a laterally extensive unit of shallow-marine, calcareous shales, grading into lime mudstones toward the north. This formation becomes sandy to the south, especially where it onlaps pre-Cretaceous clastics.
- The Natih Formation consists of mainly mud-supported and some grain-supported limestones with local rudist development, alternating with calcareous shales. It contains two organic-rich chalk levels with source-rock potential deposited in intraplatform basins (Hughes Clarke, 1988; van Buchem et al., 1996, 2002a).

The Nahr Umr Formation is late Aptian to late Albian (Witt and Gökdag, 1994; Immenhauser et al., 1999). It conformably overlies upper Shuaiba infill of the Bab Basin, whereas outside the Bab Basin, a mid- to late Aptian stratigraphic hiatus occurs between the Nahr Umr and the lower Shuaiba (Figure 2; Witt and

Gökdag, 1994). Farther south, the hiatus encompasses more time, as the Nahr Umr progressively onlaps truncated older stratigraphic units. Based on the above and on regional sequence-stratigraphic correlations in Oman and the United Arab Emirates (Van Steenwinkel, 1992; van Buchem et al., 2002b), we suggest that the "base Nahr Umr unconformity," which occurs at the top Shuaiba level outside the intraplatform Bab Basin, corresponds to the sequence boundary at the base of the upper Shuaiba in the Bab Basin. In this case, both the upper Shuaiba and the Nahr Umr belong to the same genetic unit that is progressively onlapping the unconformity. This implies that this unconformity occurs at the early to late Aptian boundary.

The Natih Formation conformably overlies the Nahr Umr. It is late Albian to early Turonian in age (Simmons and Hart, 1987; Smith et al., 1990; Scott, 1990; Kennedy and Simmons, 1991; Philip et al., 1995; van Buchem et al., 1996).

STRATIGRAPHIC ARCHITECTURE

Two scales of platform geometries are seen (Figure 3). The large-scale geometry refers to the general configuration of the Cretaceous platform, the relative position of its platform edge (the Habshan Formation), and its overall change from progradation to aggradation through geologic time. The smaller-scale geometries refer to packages of several tens to a hundred meters in thickness in the Kharaib, Shuaiba, and Natih Formations of the platform interior.

Arabian Plate Margin Configuration (Habshan Formation)

Figure 3 shows a schematic geologic cross section, based on seismic, well, and outcrop data, through the Cretaceous platform of Oman. The map in Figure 4 shows the distribution of the main elements of the platform.

The large-scale platform started to grow in central Oman after a major transgression over a structurally collapsed Jurassic carbonate platform. From here, the edge of the platform prograded 250 km to the north in approximately 15 Ma (Berriasian to late Hauterivian). This corresponds to an average progradation rate of 17 km/m.y. This prograding carbonate belt is referred to as Habshan Formation. During this phase, bioclastic and oolitic sands dominated the platform edge. The water depths in front of the platform were in the order of a few hundred meters. The Salil and Rayda Formations are interpreted as slope and deep-sea deposits that are isochronous with the prograding carbonate platform (see also Conally and Scott, 1985).

This progradational phase was followed by a mainly aggradational trend, which lasted approximately 35 Ma

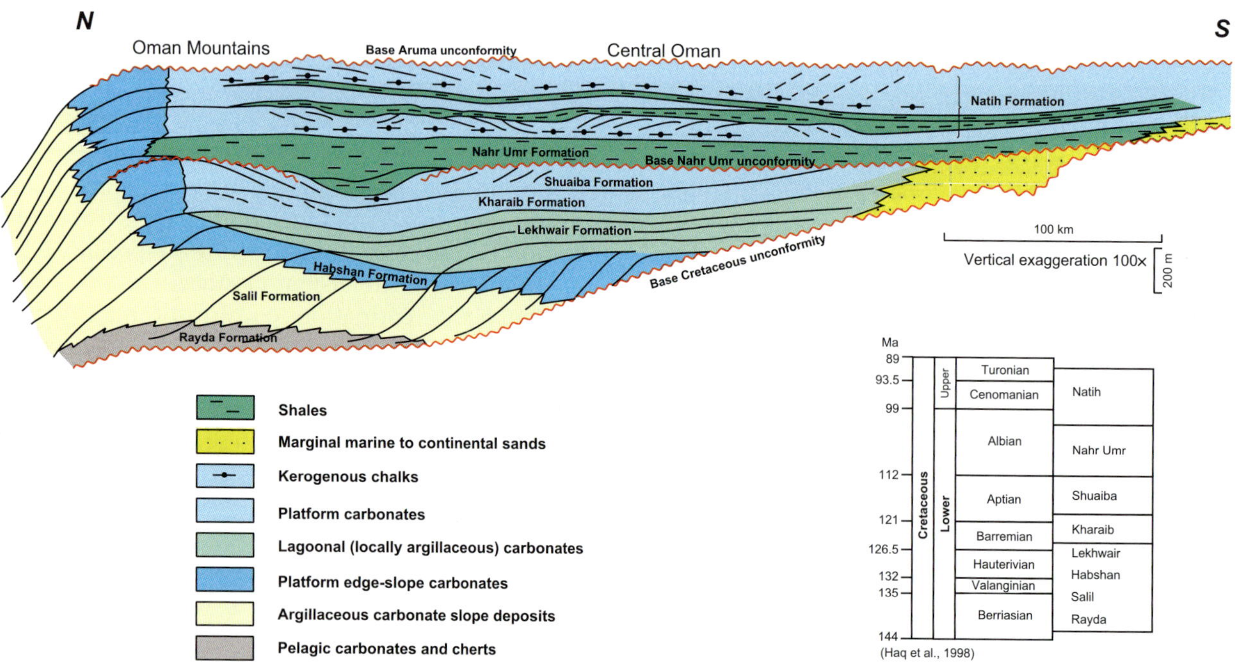

FIGURE 3. North-south schematic geologic cross section through the Cretaceous carbonate platform of Oman (see Figure 4 for the location of the cross section).

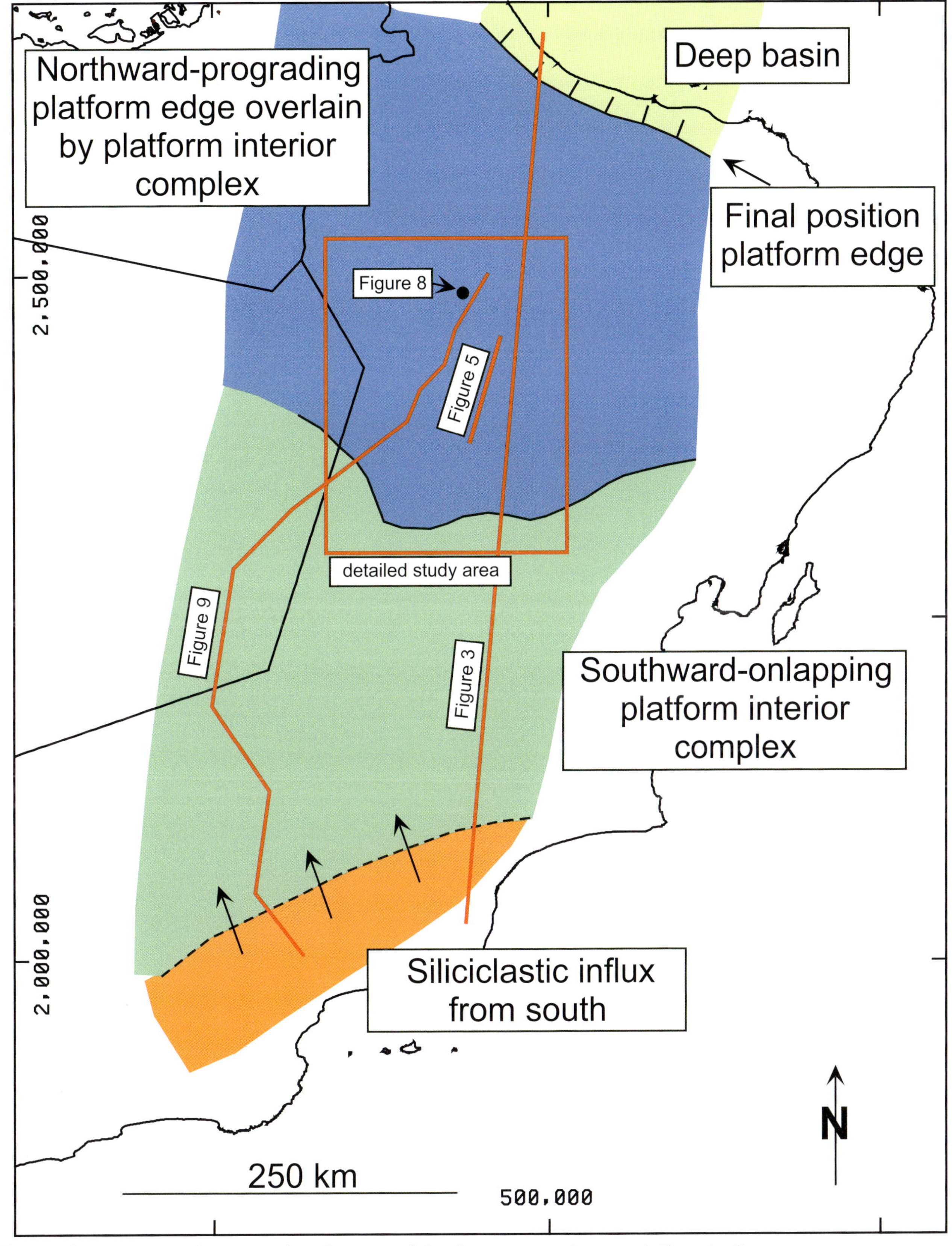

FIGURE 4. The main paleogeographic elements of the Cretaceous carbonate platform in Oman.

(Barremian to early Turonian). Only minor shifts in the position of the platform margin occurred. The 700-m-thick platform interior that developed behind the platform margin forms the classic "layer-cake" platform, which contains the smaller-scale geometries described below.

During this period, the platform interior carbonates also started to onlap the base Cretaceous unconformity toward the south, commonly above a thin, transgressive lag of siliciclastics. Locally thick clastics were deposited in salt-withdrawal basins of southern Oman. Subcropping Paleozoic clastics in this area are thought to be an important source of the coarser-grained clastics.

Figure 5 shows a seismic line covering a 75-km-long section through the central part of the Cretaceous platform (see Figure 4 for location). At about 1500 ms (above "B Kahmah" marker), well-defined clinoforms of the "Habshan system" (Habshan-Salil-Rayda Formations) document northward progradation (see also Haan et al., 1990; Landmesser and Saydam, 1996; Aziz and El-Sattar, 1997). This prograding belt represents the initial stage of the large-scale Cretaceous platform development (see also Figure 3). Well data show that facies shallow upward within a clinoform belt from basinal (Rayda Formation) to slope (Salil Formation) into shallow-marine oolitic and bioclastic shoals (Habshan Formation). The Habshan clinoforms show wide variations in inclination. Dips range from less than 1° to 20°. The seismic line (Figure 5) runs parallel to the direction of progradation. Seismic mapping of the clinoforms confirms that variations in dip are real and not related to different progradation directions. Cores show that the difference in inclination is associated with variations in sediment fabric, a phenomenon that has also been recognized in other case studies (Kenter, 1990). Intervals with high-angle clinoforms consist of thick sequences with platform-derived grainstones and packstones deposited on the shelf margin. The low-angle clinoforms are characterized by open-marine, mud-dominated lime turbidites and marls, with intercalated platform-derived wedges.

Subsurface cores show that the high-angle seismic reflections are caused by the impedance contrast across dolomitized hardgrounds within the grainy slope

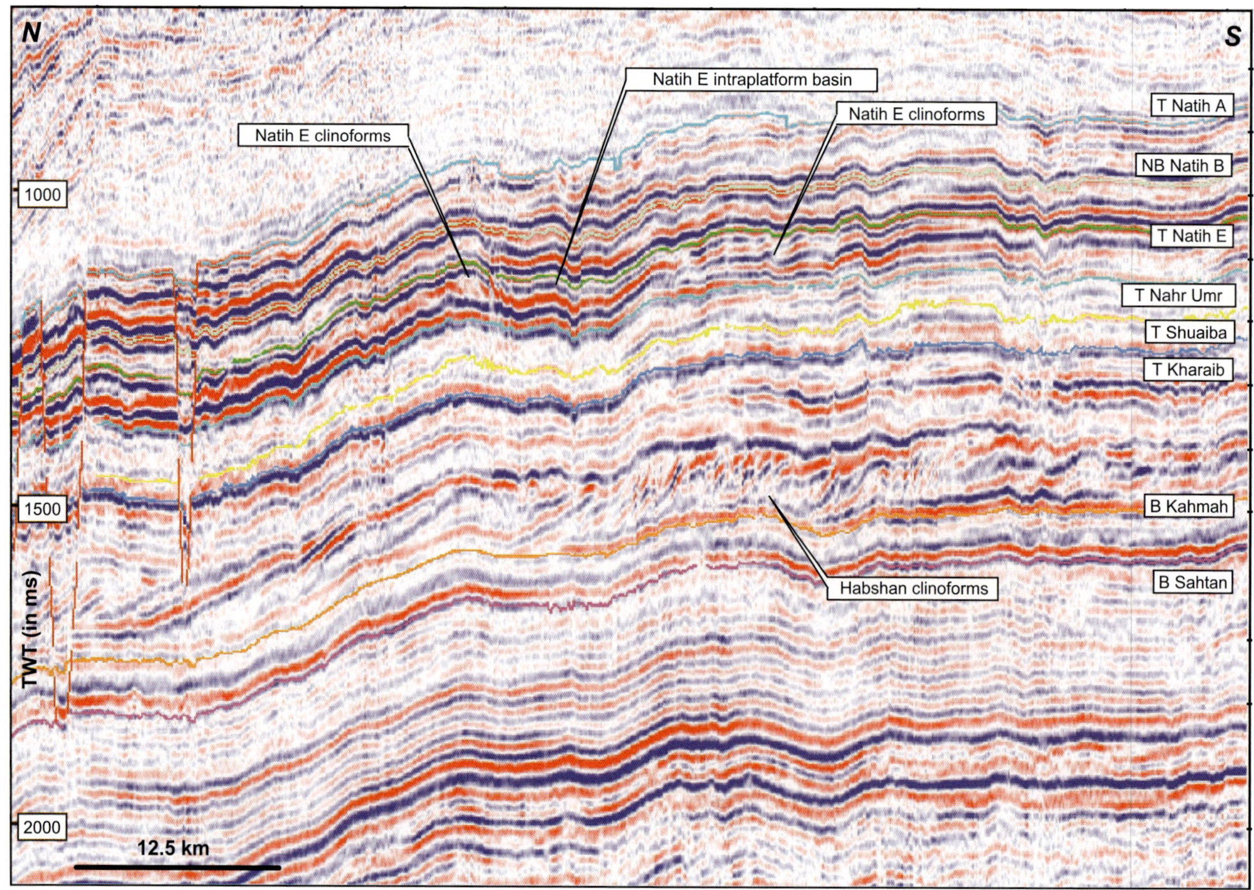

FIGURE 5. Regional seismic line through the north Oman Cretaceous carbonate platform (see Figure 4 for the location of the line). TWT = Two-way traveltime.

deposits. The low-angle clinoforms in the finer-grained sediments are reflected by the impedance contrast between submarine cemented lime-packstone turbidites and the surrounding marly limestones.

The clinoform interval in Figure 5 is overlain by a 700-m-thick package of strong, more or less parallel, horizontal reflections (from 1400 ms up to the "T Natih a" marker). This interval contains the platform interior carbonate complex (Lekhwair to Natih Formations) as described below.

Figure 6 shows the "Habshan clinoform system" in more detail. The system consists of the Habshan carbonate platform and upper slope facies, the Salil slope to basinal facies, and the Rayda deep-sea facies. It is internally organized as packages that are 10–20 km wide. Each package shows a change from low- to high-angle clinoforms. High-angle clinoforms are interpreted to have been formed during aggradation of the Habshan carbonate platform during relative sea level rise, whereas low-angle clinoforms result from forced regressions during relative sea level fall. At least six of these cycles are seen on the available seismic data, which only covers part of the platform. Interestingly, seismic lines through a time-equivalent carbonate-platform complex in the Neuquén Basin of Argentina show very similar variations in stacking patterns and clinoform geometries (Mitchum and Uliana, 1985). The geometries in the Habshan system strongly resemble the high-resolution architectural patterns in Miocene reefs of Mallorca (Spain), described by Pomar (1991, 1993), and the Permian Capitan Reef of west Texas (Sonnenfeld and Cross, 1993; Osleger and Tinker, 1999). These reefs show a hierarchical stacking of accretionary units comprising aggrading, prograding, and offlapping units. The seismic resolution at Habshan level does not allow a similarly detailed high-resolution analysis the larger-scale trends can be detected.

Being part of an overall progradation lasting 15 m.y, the duration of each Habshan cycle is estimated to be circa 1–3 m.y. It is here suggested that these clinoform packages were formed by third-order eustatic sea level cycles superimposed on a second-order regressive trend.

The interpretation of the high- to low-angle packages of the Cretaceous platform edge is illustrated in Figure 7, which represents a schematic conceptual model of one third-order cycle. During early sea level highstands, carbonate growth was more rapid on the platform edge, which built up to a steep slope. In lowstand periods when the rate of sea level fall exceeded the rate of subsidence, large parts of the platform were exposed. Less carbonate production and the influx of fine clastics resulted in the shedding of fine-grained argillaceous sediments into the basin. Low-angle, offlapping clinoforms (forced-regression type) result.

During the subsequent increasing rise in sea level, the previous lowstand clinoforms started to be flooded again. This triggered a gradually larger area of carbonate (oolite) production, whereas the clastic influx decreased. As more accommodation space became available, the stacking pattern changed from progradation to aggradation. The newly created platform edge, being farther seaward compared to the previous highstand margin, formed a barrier to the lagoonal area behind

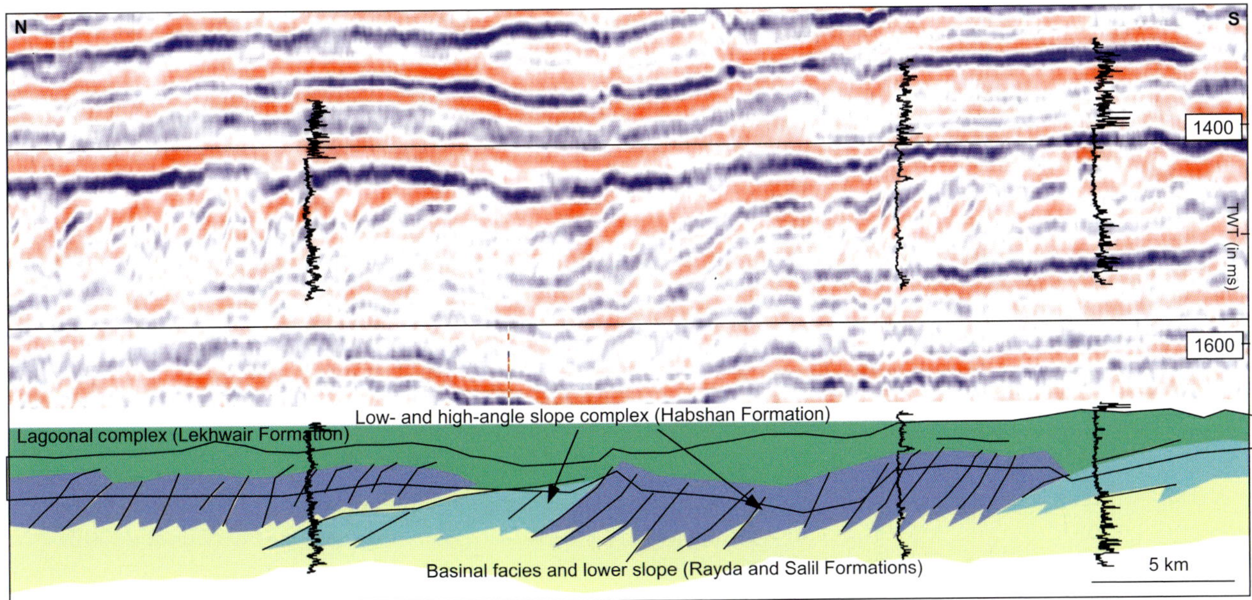

Figure 6. Enlarged view of the central part of Figure 5 showing details of seismic geometries in the basal part of the platform. The prograding platform complex consists of laterally stacked low- and high-angle clinoform wedges. TWT = Two-way traveltime.

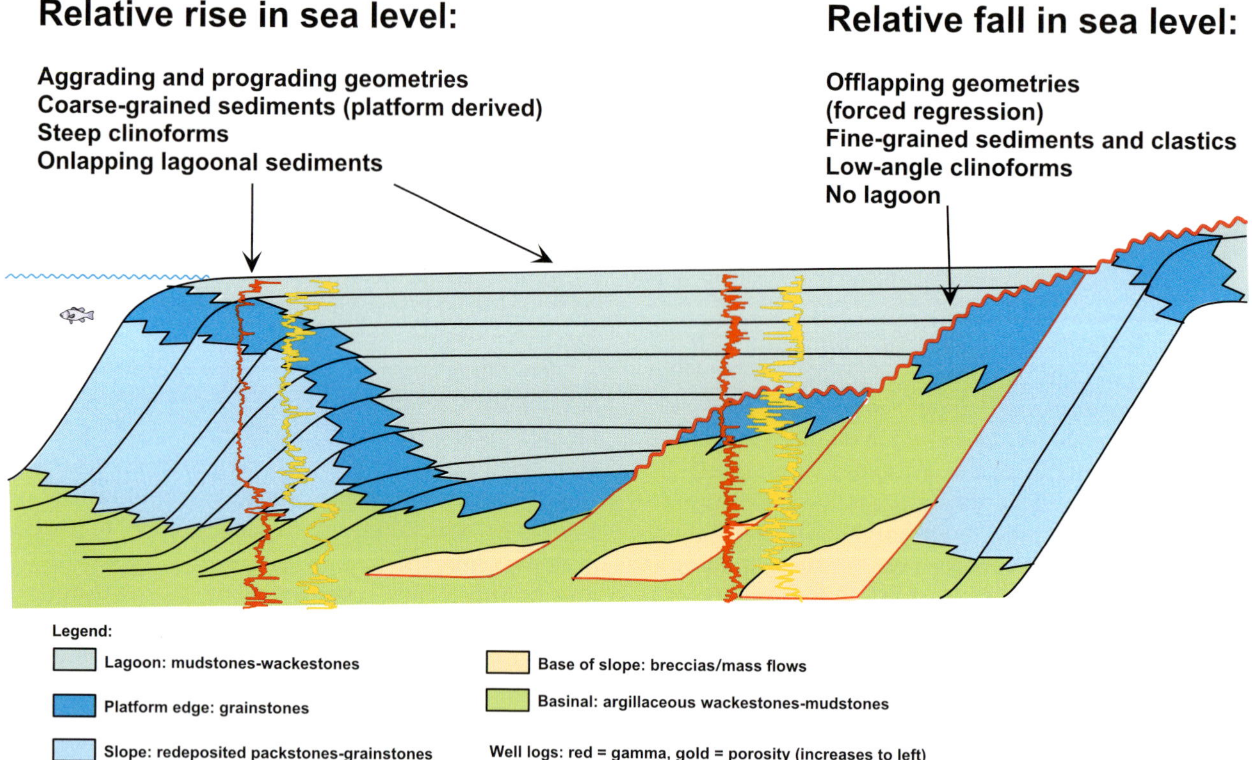

FIGURE 7. Schematic model of one third-order platform-edge cycle superimposed on an overall regressive, second-order trend. The variations in relative sea level impact stacking patterns and composition of the slope sediments and their inclination.

it. The lagoonal sediments onlap the older, low-angle wedges but transgression of the previous high-angle platform edge has not been observed.

The overall progradation of the Habshan system in the Berriasian to Hauterivian indicates that the third-order sea level rises were buffered by the overall, second-order regressive trend. Carbonate production was always faster than the increase in accommodation space. A transgressive systems tract with a landward shift of the platform edge is generally missing. The changeover from lowstand to highstand at the inflection point of an eustatic sea level rise is represented by a relatively high aggradation/progradation ratio, leading to a steep platform edge. Additional factors that control clinoform geometries are climatic changes, which impact clastic influx and carbonate production, and differential compaction (Hunt and Fitchen, 1999).

The resulting depositional sequence is represented by an alternation of low-angle, offlapping, lowstand clinoforms (forced-regression type) and high-angle highstand clinoforms. The lowstand clinoforms are primarily composed of fine-grained pelagic carbonates and fine clastics (low net-to-gross ratio). The highstand clinoforms contain shallow-water, coarse-grained platform or platform-derived carbonates, without clastic influx (high net-to-gross ratio).

Platform-Interior Carbonate Geometries (Natih Formation)

The main phase of carbonate deposition in the platform interior occurred during the Barremian to Turonian aggradational phase (Figure 3). Carbonate deposition was interrupted several times by subaerial exposure and the influx of fine-grained clastics. The internal architecture of platform-interior carbonates is illustrated based on the Natih E Member.

Facies Succession

A log through the platform-interior Natih Formation shows repetitive sedimentary cycles up to several tens of meters thick (Figure 8). The cycles start with a thin mixed carbonate-shale that is overlain by a thick carbonate. These cycles have historically been used for lithostratigraphic subdivisions (Natih A–G Members). Similar cycles have also been recognized in other formations within the platform interior, e.g., the Shuaiba and Kharaib Formations (Pittet et al., 2002).

Each cycle consists of a relatively thin deepening and a much thicker shallowing-upward interval. The shales at the base of the cycles are marginal marine and contain a restricted marine fauna. They show a rapid deepening-upward trend into pelagic carbonates

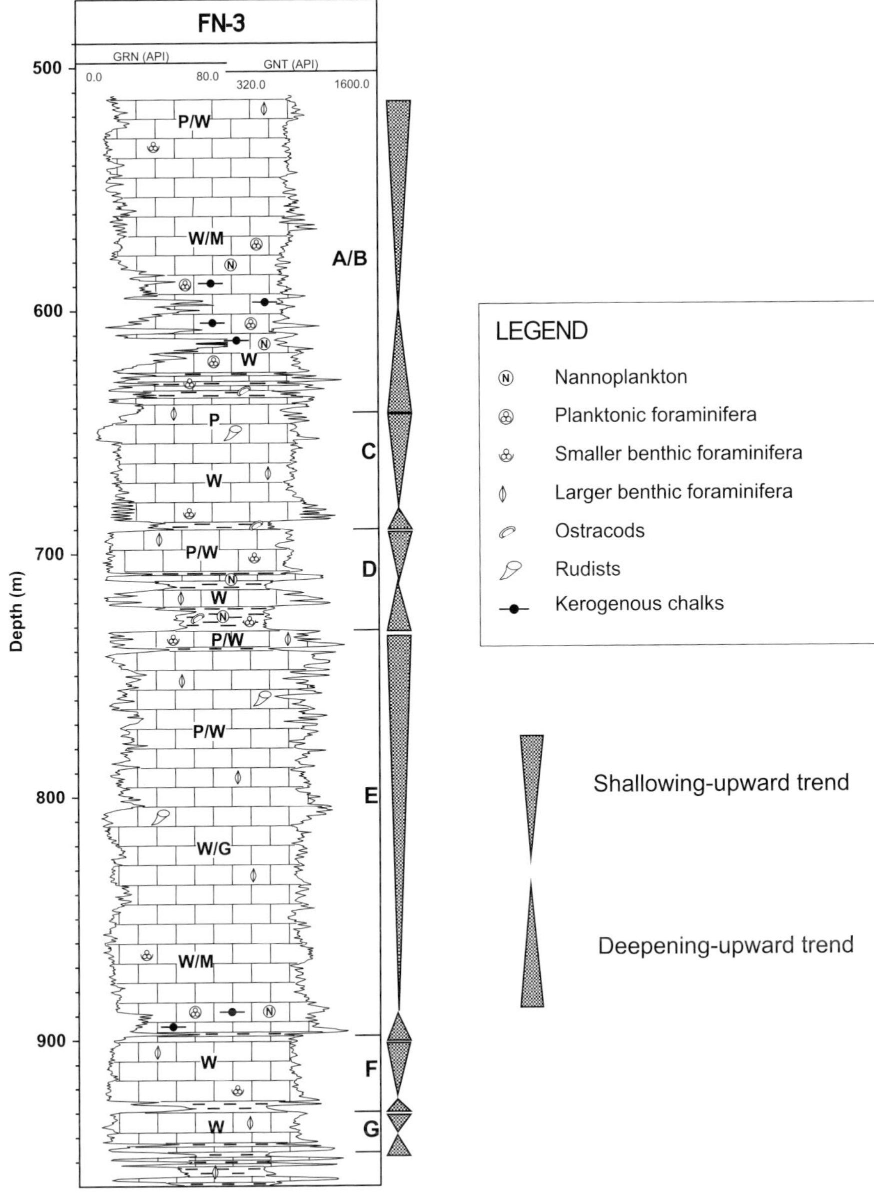

FIGURE 8. Example of intraplatform carbonate cycles: Type log for the Natih Formation from well Fahud North-3 (FN-3) (see Figure 4 for the location of the well). P = packstones; W = wackestones; M = mudstones; G = grainstones.

and sometimes organic-rich chalks (e.g., Natih E and B). Carbon contents of up to 15% have been measured in these intervals (Terken, 1999), indicating the presence of anoxic bottom waters. The main part of the cycle consists of a shallowing-upward trend through bioclastic lime wackestones to packstones, rudist shoals, lagoonal wackestones-packstones, and intertidal grainstones. The coarser-grained sediments at the top of the cycles have commonly been subaerially exposed and leached. The maximum water depths on the platform are estimated to be in the order of several tens of meters and unlikely to exceed 100 m.

Geometries

A regional cross section of the Natih Formation, based on well data (Figure 9; see Figure 4 for location), highlights that the intervals of clastic influx can be correlated over several hundreds of kilometers, which suggests a layer-cake geometry. That impression is further enhanced by the wireline log patterns of carbonates, which are rather monotonous and undifferentiated and, apart from the top and the bottom of the units, show no markers that can be used for lateral correlation.

As a result, the carbonates are commonly visualized as aggrading units. However, as indicated by the seismic expression (see below), the carbonates internally consist of inclined packages, similar to, but at a smaller scale than the ones described above for the platform edge Habshan Formation.

At first sight, the seismic data of the platform interior also seem to suggest a layer-cake stratigraphy (Figure 5, between 900 and 1400 ms). This is because the reflectivity is primarily caused by the strong impedance contrasts between the carbonates and the laterally extensive argillaceous beds. However, on an enlarged view of the central part of this line (shown in Figure 10), subtle but clear inclined geometries can be recognized in the Natih E, which is one of the thicker Natih Members (see Figures 8, 9). These clinoforms show different directions and delineate prograding carbonate-platform systems, which are separated by (shale-filled) intraplatform basins.

Figure 11 shows the mapped Natih E clinoform belts in central Oman (location of study area is shown on Figure 4). The hatching indicates the direction of progradation. Consistent high dips from dipmeter data have been plotted as well. The different progradation directions on the map indicate that the Natih E was internally composed of separate prograding carbonate platforms that amalgamated through time (e.g., Al Huwaisah and Yibal area). As in the case of the Natih E, the merging of these platforms was not always complete, leaving intraplatform "basins" of several tens of

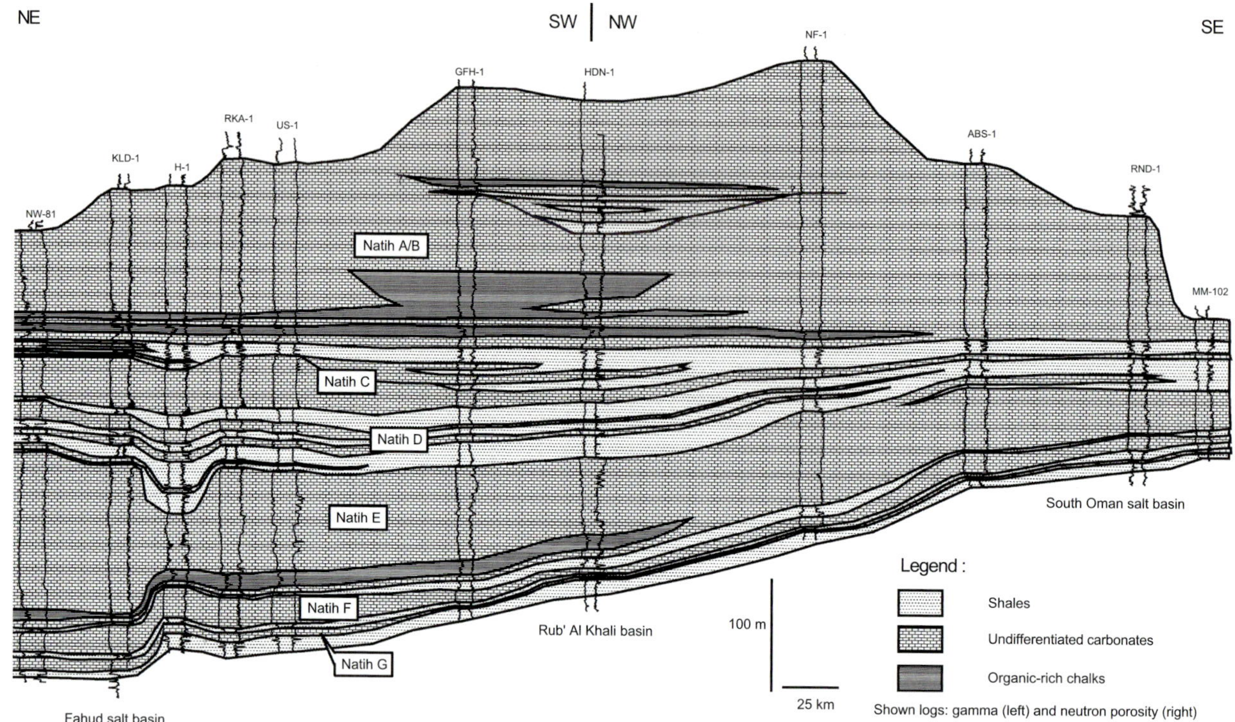

FIGURE 9. Regional cross section through the Natih Formation along the eastern edge of the Rub' Al Khali Basin based on well-log correlations (see Figure 4 for the location of the line).

kilometers wide, filled with clays. Platform initiation and the location of intraplatform basins are structurally controlled, e.g., by fault zones, basement highs, salt doming, and salt withdrawal. The coalescence of smaller platforms has also been described for large, steep isolated platforms, such as the northwestern Great Bahama Bank (Eberli and Ginsburg, 1987, 1989).

Figure 12 shows low-angle offlapping to high-angle aggrading units in the platform interior, similar to, but at a smaller scale than the ones of the Habshan platform edge.

The seismic lines in Figures 10 and 12 furthermore show that the horizontal reflectors should be correlated with caution. High-frequency relative sea level changes created complicated offlap and onlap relationships with the clinoform complex, as well as internal hiatuses (see also Pomar, 1991, 1993).

Figure 13 shows a well-correlation panel along the seismic section of Figure 12, superimposed with the clinoform dips plotted to scale. As indicated by the integrated well and seismic data, the dip angle of the clinoforms is here also related to the grain fabric.

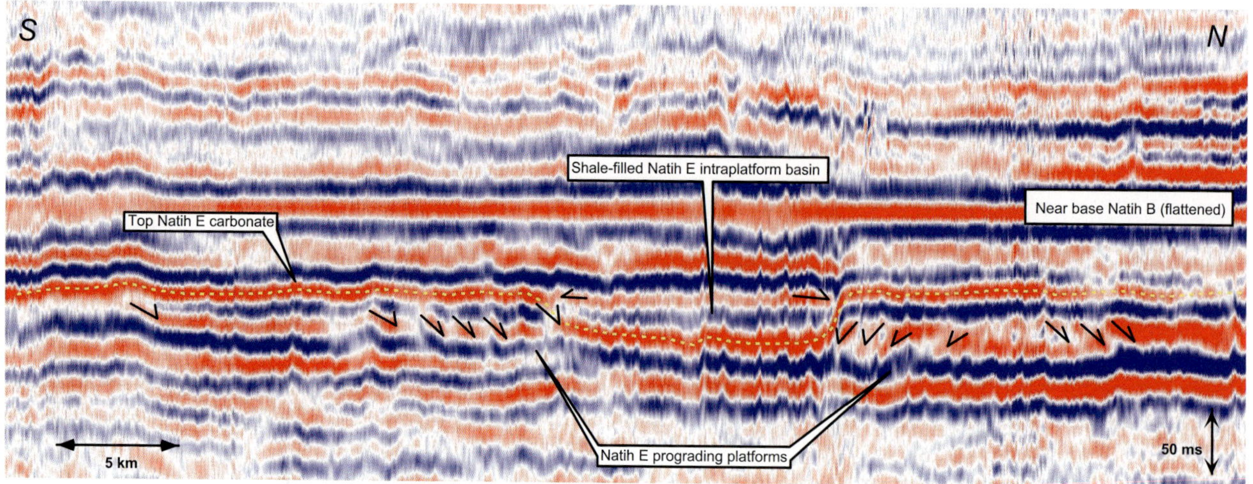

FIGURE 10. Detail of seismic line in Figure 5, flattened along the near base Natih B reflector, showing stratal geometries within the Natih E platform carbonates.

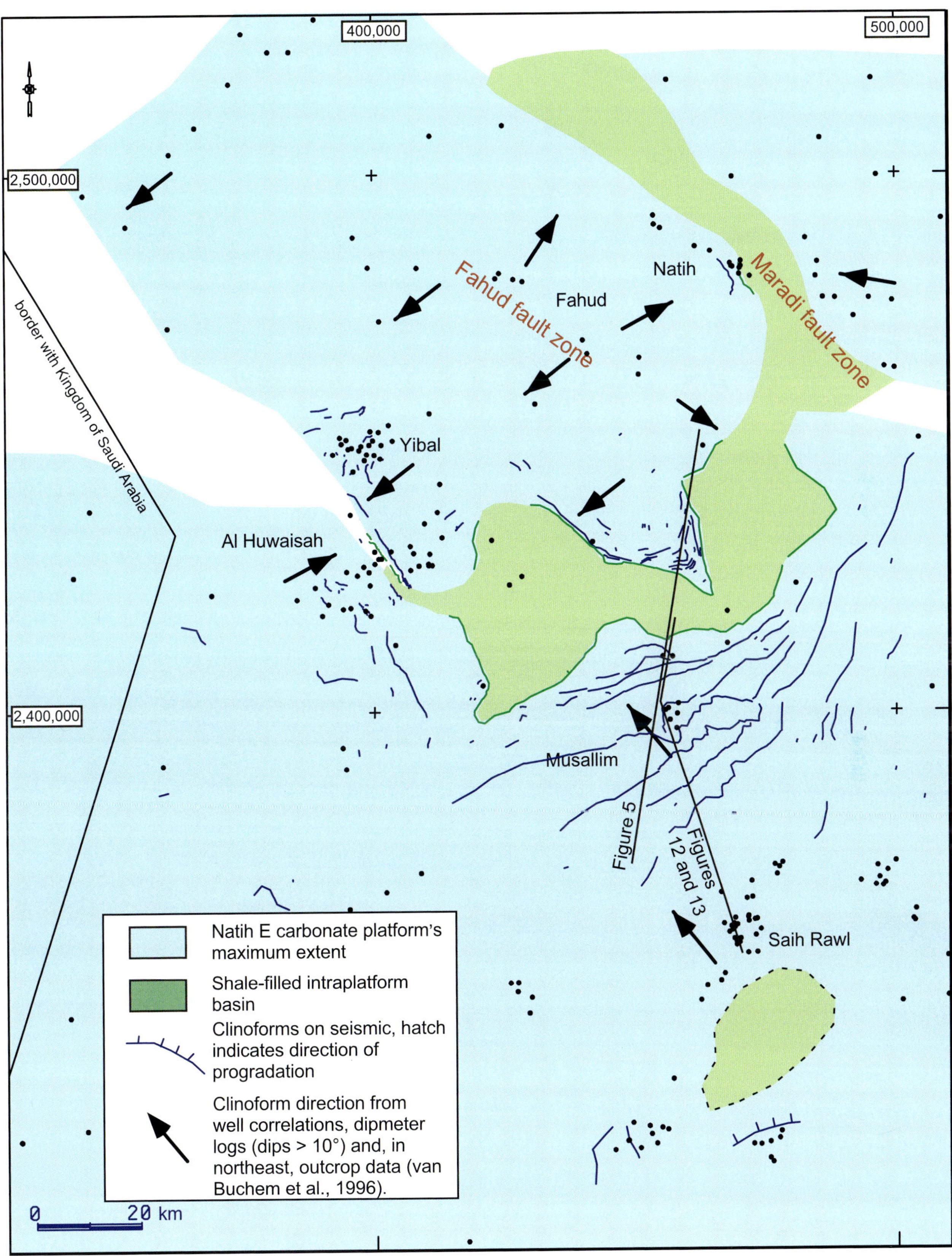

FIGURE 11. Paleogeographic map of the Natih E Platform carbonate and shale-filled remnant intraplatform basins in north Oman (see Figure 4 for the location of the study area).

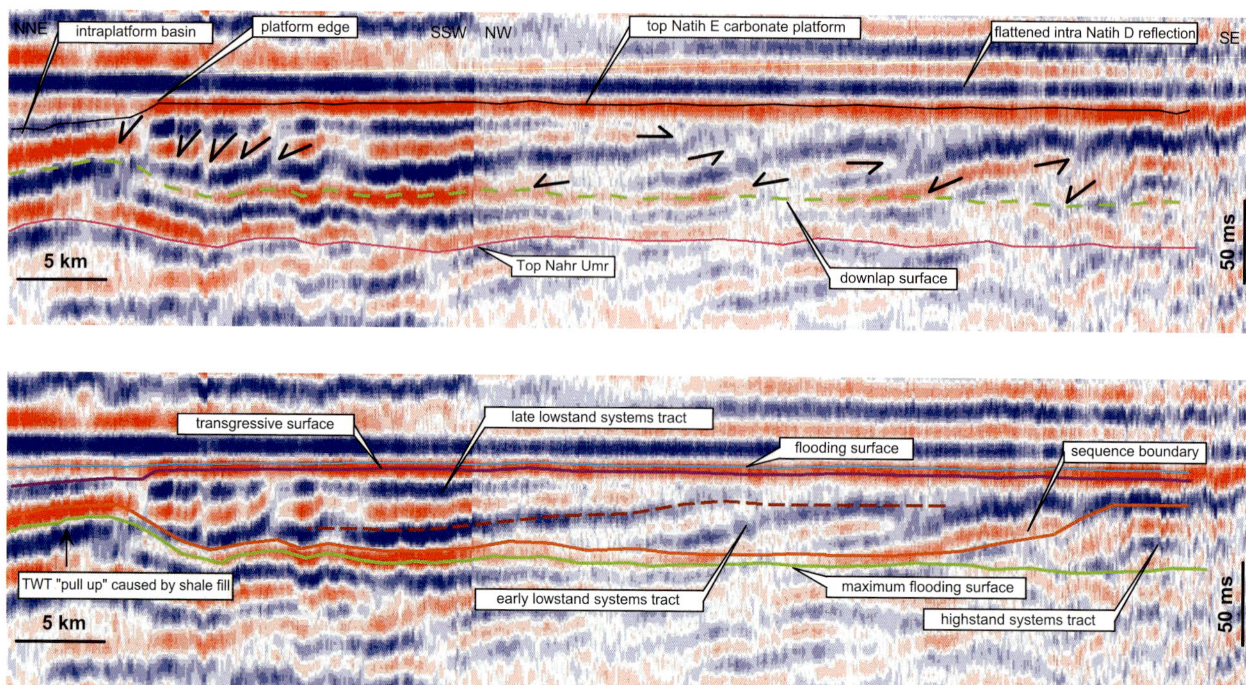

FIGURE 12. Seismic line cutting perpendicular through the clinoform belt at the northern edge of Musallim Natih E Platform (see Figure 11 for the location of the line). TWT = Two-way traveltime.

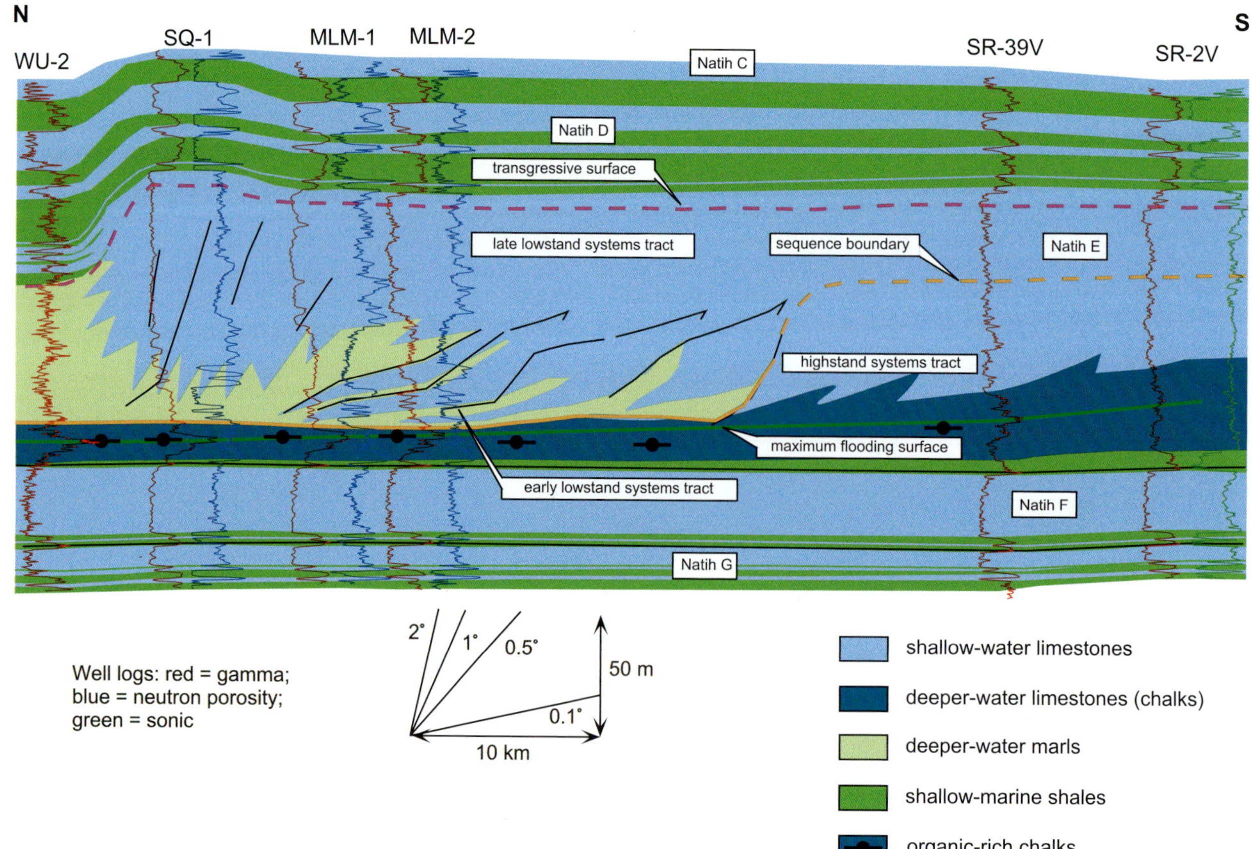

FIGURE 13. Well-correlation panel along the seismic line shown in Figure 12 with a projection of the seismically defined clinoforms.

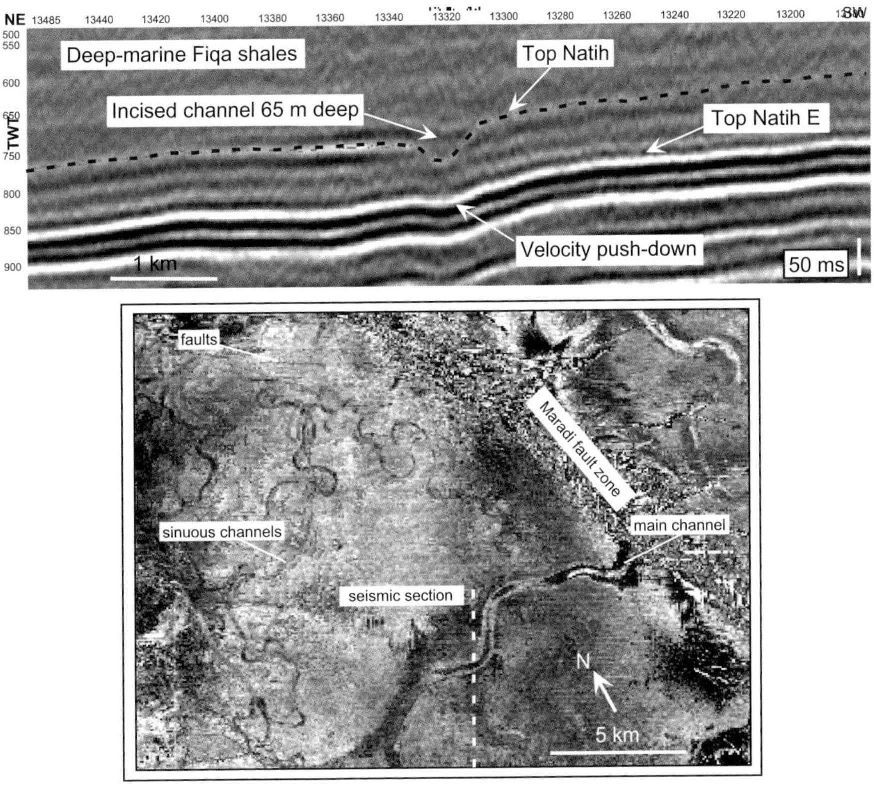

FIGURE 14. Channel complex cut into Top Natih Formation. Top: Seismic line through channel. Bottom: Amplitude map along time slice 44 ms below Top Natih horizon in north Oman showing a complex of narrow sinuous channels and wider northeast-southwest–trending main channel. TWT = Two-way traveltime.

Low-angle clinoforms occur in the marly units and high-angle clinoforms are composed of platform-derived grainstones.

This figure also shows that the overall dips observed on this seismic line are very small: The "high-angle" dips are only 2°, whereas the low-angle dips are less than 0.5°. The relationship with the seismic data indicates that, despite the gentlest dips, wells that are 10 km apart (in dip direction) are hardly correlatable. The correlation is further complicated by the fact that these low dips cannot be identified as clinoforms in cores or image logs. This difficulty of correlations has important implications for reservoir modeling and development (see below).

The intraplatform carbonates can prograde out over large distances in a short geologic time. The Natih E Platform progradation is estimated to have occurred at a rate of at least 27 km/m.y. This estimate is based on a measured 50 km of progradation during the Cenomanian, which lasted 5.5 m.y. (Figure 2). During this time, three depositional sequences (Natih C, D, and E) were deposited. Assuming that the 50-km-wide Natih E progradational belt was deposited over a third of the Cenomanian time (1.8 Ma) and ignoring the time involved to deposit the intervening shales, 27 km/m.y. is a conservative estimate. This intraplatform progradation is faster than the progradation of the Habshan Platform edge (17 km/m.y.) and also faster than the one estimated for the Miocene platform edge in Mallorca (>10 km/m.y.; Pomar, 1993). This is probably because intraplatform carbonates do not have much vertical space to build upward, and therefore spread laterally more quickly.

Within the intraplatform succession, the inclined stratal geometries are seismically best expressed in the Natih E, as it is a thick cycle of nearly 100 m. Thus, the internal configuration is within seismic resolution. Clinoforms have also been observed in other Natih cycles, as well as in the Shuaiba and Kharaib Formations. This observation strongly suggests the presence of a similar stratigraphic architecture as that of the Natih E and, consequently, the potential for stratigraphic traps in the platform complex.

The Top Natih Platform Incision

The Cretaceous carbonate platform was terminated by a regional phase of uplift and subaerial exposure in the Turonian. The uplift is thought to be related to the formation of a wide peripheral foreland bulge during the initial phases of the collision of the Arabian and Eurasian plates (O'Connor and Patton, 1986; Warburton et al., 1990). As a result, the Top Natih has been faulted, truncated, karstified (Montenat et al., 1999), and incised by extensive channel systems.

Figure 14 illustrates the seismic expression of these channels in central Oman. Several generations of highly sinuous channels are visible. Some incisions cut more than 150 m into the underlying carbonates and reach a width of several hundred meters. The period of emersion only lasted a few million years (Scott, 1990) and was followed by a rapid deepening in the Coniacian–Santonian, as the foreland basin started to develop (e.g., Burchette, 1993). During this drowning, the paleotopography of the Top Natih was covered by shales of the basal Fiqa Formation (Aruma Group). Well data show that the lower part of this shale package is lignitic and contains thin streaks of siltstone and fine sandstone, as well as traces of glauconite and marine fauna. This suggests a marginal marine (estuarine) to lagoonal setting. The top of the lagoonal shales is marked by an iron-oolitic glauconitic claystone, which is overlain by hemipelagic and pelagic mudstones.

This Top Natih A configuration of incised reservoir units, laterally sealed with shale-filled channels opens up a new play opportunity in the Oman Cretaceous.

GEOLOGIC MODEL FOR THE INTRASHELF CRETACEOUS CARBONATES

A geologic model for the intrashelf Cretaceous carbonates is based on the Natih E study and presented in Figure 15. Carbonate growth in the intrashelf basin initially developed small buildups in areas where subsidence was relatively low, e.g., near basement highs and salt domes. These buildups developed into intrashelf platforms, which merged in some cases as they prograded out. Cyclic variations in the composition and the angle of the slope sediments occurred in these platforms in response to sea level variations. Regional variations in sediment fabric and the amount of progradation are also inferred to have occurred as a result of different orientations of the platform margins with respect of the predominant wind direction as well as variations in subsidence.

In the intrashelf basins between these platforms, sediments predominantly consist of pelagic carbonates, which are commonly rich in organic content. Where the coalescence of the platforms was complete, a continuous carbonate unit results, although with a strong lateral variation in rock properties. However, in cases where platforms were not completely merged, "remnant" or unfilled intraplatform basins remained. This could be because of a decrease or stop in carbonate production during a relative sea level fall and/or to increased clastic influx. Such basins were later filled with shale (as in the Natih E), with mixed carbonate-shale (e.g., upper Shuaiba, Witt and Gökdag, 1994; Pittet et al., 2002), or with evaporites during the subsequent lowstand (e.g., Jurassic Hanifa-Jubaila Formation of Qatar, Droste, 1990). The size of such basins can range from 20–30 km (Natih E) to a few hundred

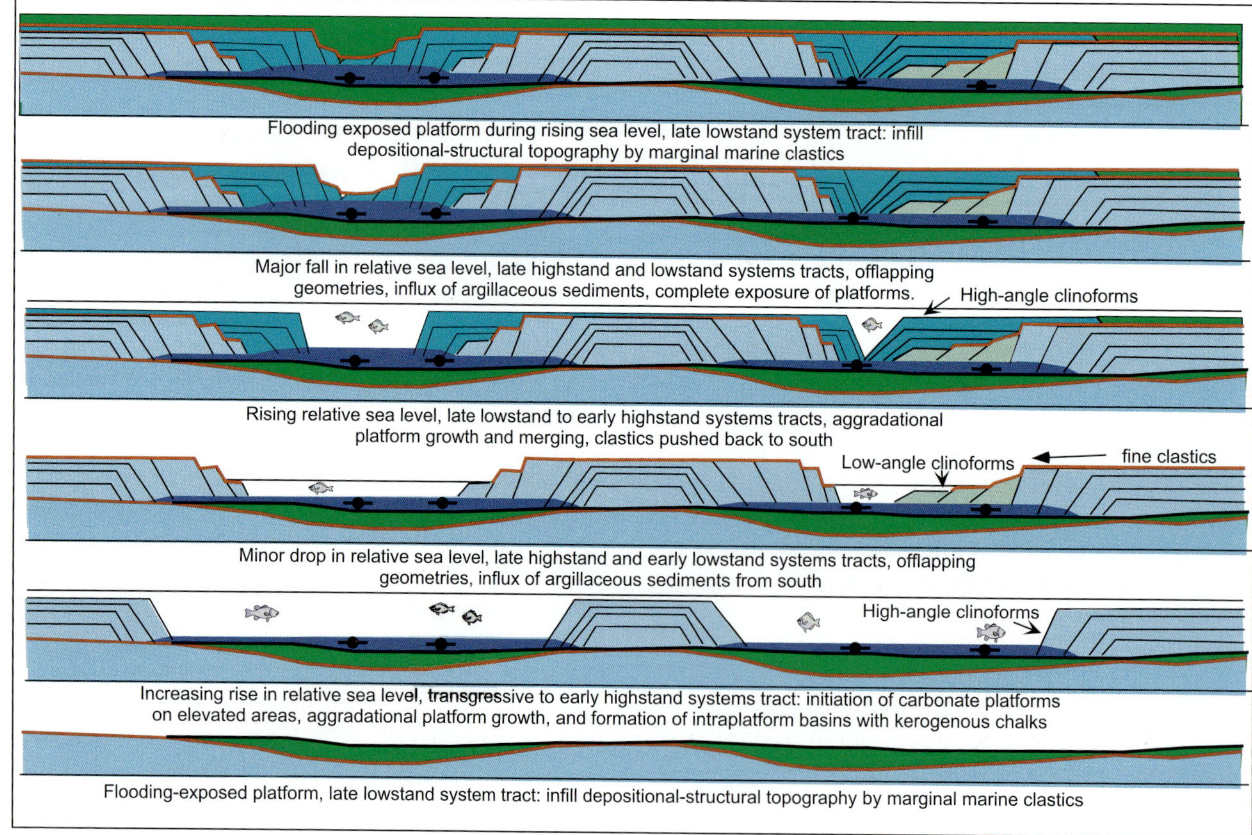

FIGURE 15. Geologic model for the evolution of intrashelf Cretaceous carbonates based on a study of the Natih E Member.

kilometers (observed in other stratigraphic intervals, see Murris, 1980). The fine-grained sediments that fill the seaways may form good seals.

Inclined surfaces and progradation are a common feature within the platform interior carbonates. Similar to the Habshan platform-edge progradation, "steep" and "gentle" dips occur in the intrashelf carbonates. The steeper slopes (5–35°) are related to highstand carbonate-platform progradation, whereas the gentle (<1°) slopes are related to lowstand prograding wedges (forced-regression type).

CONCLUSIONS AND IMPLICATIONS

High-resolution seismic data of the Cretaceous carbonate platform in Oman show a complex internal architecture with abundant inclined stratal geometries. Recognition of these geometries and the diachronic character of lithologic units has important implications for hydrocarbon prospectivity and reservoir development. At exploration scale, it allows a mappable risk and opportunity evaluation in hydrocarbon play analysis and a refined definition of stratigraphic trapping potential. At development scale, the understanding and prediction of reservoir heterogeneities, sweep efficiency, early high water cut, water flood, etc. will help to optimize field development plans.

Implications for Hydrocarbon Exploration

Mappable Play Evaluation

The geometries seen on seismic and the relationship of sedimentary facies with clinoform dip angles allow reservoir and seal risks, as well as net-to-gross ratios to be mapped using seismic expression. Reservoir risk is low and net-to-gross ratios are high in the grain-dominated, high-angle units. Reservoir risk is higher in the mud-dominated, low-angle units, which have lower net-to-gross ratios. Seal risk, however, is higher in high-angle units and lower in low-angle units.

Stratigraphic Trap Recognition

Recognition of stratigraphic trapping potential is considerably increased. Several types of stratigraphic traps are envisaged (Figure 16):

- *Combined structural/stratigraphic traps in the platform interior.* The irregular outline of the intrashelf carbonate platform edges (e.g. Natih Formation), with the onlapping intra-platform shales as top and lateral seal. This requires tectonic tilt in the opposite direction of the foreset progradation.
- *Lowstand grainstone wedges at the foot of a foreset belt in the platform interior* are another candidate for stratigraphic trapping. They are sealed by regional shales at the top and by fine-grained carbonates at the base.
- *Habshan grainstone belts sealed by Lekhwair lagoonal carbonates.* This trapping mechanism relies on the lateral extent of good-quality seal in the Lekhwair Formation and on good-quality seat seals provided by the Salil slope facies.
- *Base-of-slope carbonates offshore of the Habshan foresets* are sealed at top and base by deep-water (Salil) mudstones.
- *Truncation traps below the Nahr Umr shales in the Kahmah or older formations.* Lagoonal shales or tight carbonate mudstones of the Kahmah Formation may provide seat seals.
- *Paleotopographic traps* at the Top Natih level are incised by meandering channels several hundred meters wide, with depths of as much as 150 m, and filled with shales. The shale-filled channels can act as a lateral seal in combined structural-stratigraphic traps.

Implications for Hydrocarbon Development

Implications for hydrocarbon development are all related to the lateral correlation of the inclined geometries. Figure 17 shows an example of the Natih E, which suggests that dips of 35° in a platform of 100-m-thick downlap over an horizontal distance of only 143 m. In carbonate platforms with such slope angles, intracarbonate correlations between wells only a few hundred meters apart are not possible. Slope angles of 1° downlap over a distance of 5.7 km.

As the Natih E example suggests, the predominance of dipping geometries implies that if based on well data alone, there is limited hope for high-resolution correlations within such carbonate systems. However, a geologic model with layer-cake correlations will leave many aspects of production behavior poorly understood. Integration with seismic data is therefore of crucial importance for understanding reservoir behavior.

- *"Unexpected" or "random" reservoir quality variations:* Because each clinoform represents a smaller-scale depositional cycle, it may internally be formed of facies with different reservoir characteristics. The

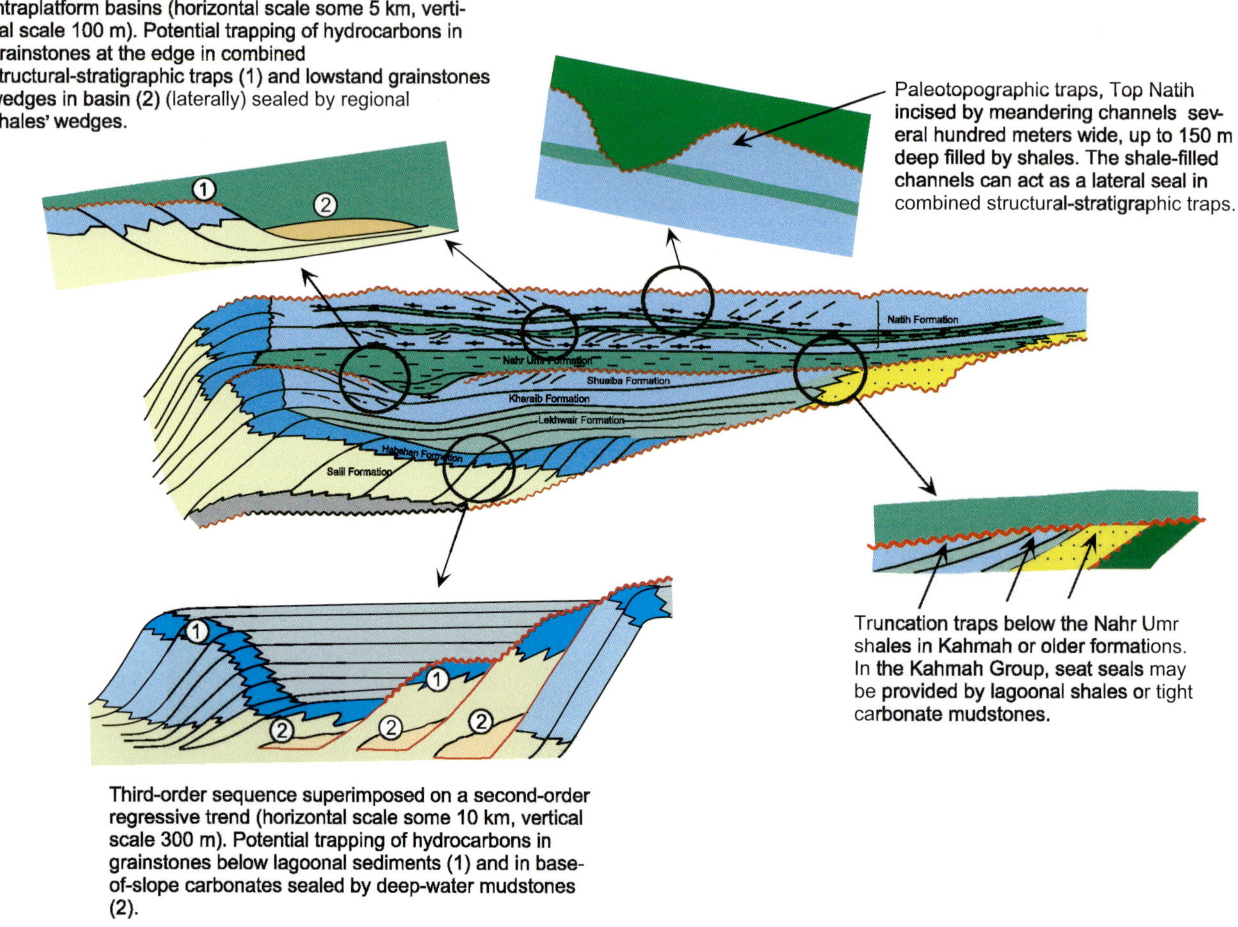

FIGURE 16. Concepts for stratigraphic traps in the Cretaceous carbonate-platform complex of Oman.

vertical variability may be picked up from log and core data, but the lateral variability can be highly underestimated if layer-cake correlations are assumed.

- Hardgrounds and cemented zones along clinoforms may further enhance reservoir heterogeneities. Although sometimes very thin (only centimeters in size), these zones should be taken into account in dynamic reservoir modeling.
- At any scale, the stacking pattern of depositional cycles (e.g., progradation vs. aggradation) determines the direction in which facies variations are predominant. This facies distribution has implications for Kv/Kh ratios and therefore also for the sweep efficiency in a reservoir. In progradational systems, the facies belts shift horizontally, which leads to better lateral facies correlatability and higher vertical variability. Consequently, the horizontal permeability is better than the vertical (Kv < Kh). In aggradational systems, facies belts are stacked on top of each other, which leads to lower vertical variability and higher lateral variability. Consequently, the vertical permeability is better than the horizontal (Kv > Kh).
- The different dip angles also relate to Kv/Kh variations.
- The *sweep efficiency*, being directly related to the Kv/Kh ratio and variability, can be more optimally planned for if stacking patterns are understood and variations in dip angle taken into account in the reservoir model.
- Water fingers and differential high water-cut development may be the result of high-permeability facies in foresets that dip into the water leg and act as conduits for water. This was interpreted for the Shuaiba Formation of the Al Huwaisah field (Figure 18, Van Steenwinkel, 1996). The wells in this zone produce with high gross rates, early high water cut but rapidly declining oil rates, high well-test permeabilities, and very good water-drive support. This zone is also characterized by a poor sweep efficiency and remaining oil pockets in low permeability zones. Infill-drilling wells in these zones are designed to target the remaining oil in the low-permeability pockets.

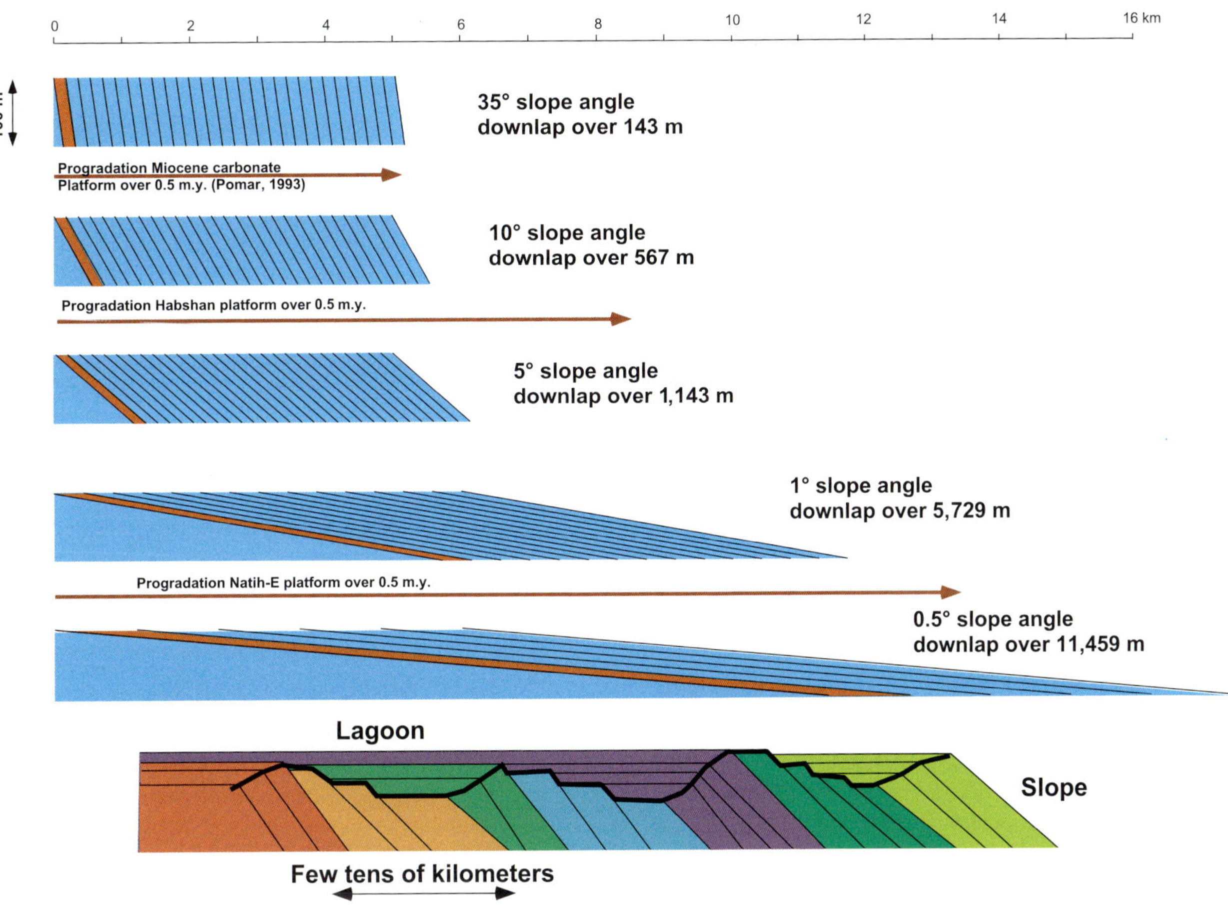

Figure 17. Time-line geometries in carbonate platforms and their impact on high-resolution lateral correlations.

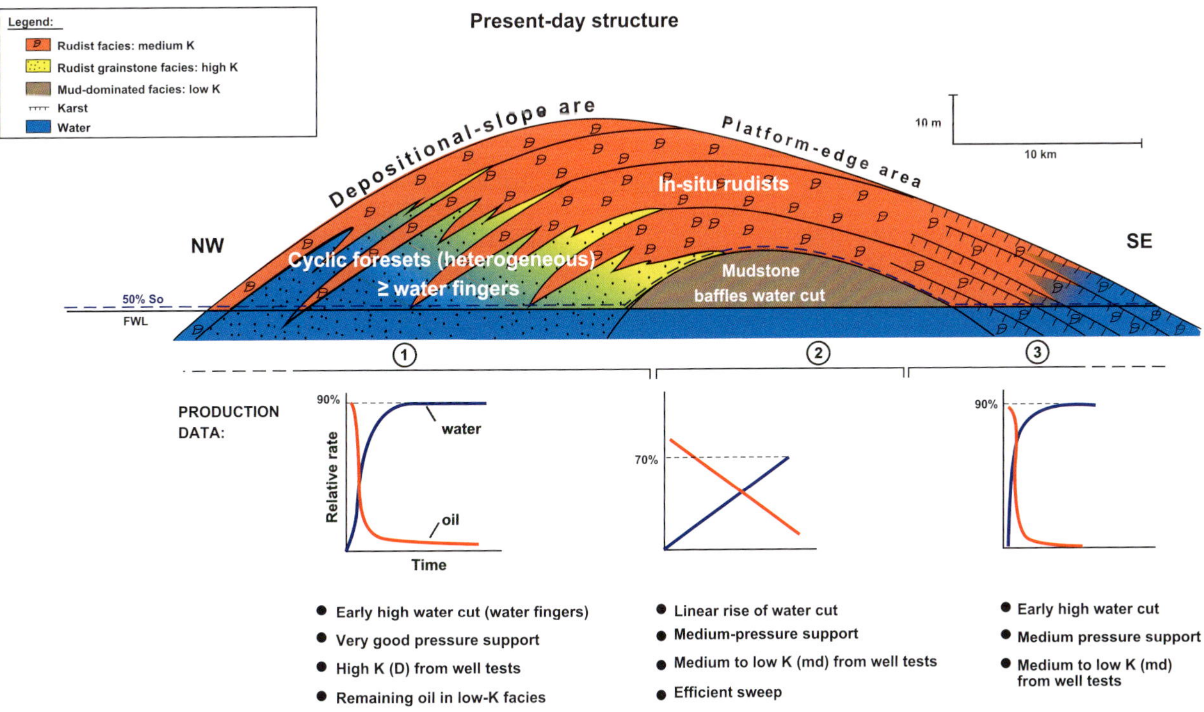

Figure 18. Example of the impact of clinoform geometries on the production performance from the Shuaiba Formation in the Al Huwaisah field.

ACKNOWLEDGMENTS

This study is based on an integration of published outcrop data and unpublished seismic and well data of the Cretaceous of Oman, kindly made available by Petroleum Development Oman. We thank the Ministry of Oil and Gas and Petroleum Development Oman for their permission to publish this chapter. We would like to acknowledge the work of present and former Petroleum Development Oman geologists reported in various unpublished company reports, which provided many of the data for this chapter. We are grateful to Mark Partington for making available Figure 14. This manuscript greatly benefited from the reviews by Steve Bachtel and John Mitchell and from helpful suggestions and informal reviews by Gregor Eberli and Frans van Buchem. We thank them for their time and efforts.

REFERENCES CITED

Aziz, K. S., and M. M. A. El-Sattar, 1997, Sequence stratigraphic modeling of the lower Thamama Group, east onshore Abu Dhabi, United Arab Emirates: GeoArabia, v. 2, no. 2, p. 179–202.

Burchette, T. P., 1993, Mishrif Formation (Cenomanian–Turonian), southern Arabian Gulf: Carbonate platform growth along a cratonic basin margin, *in* J. A. T. Simo, R. W. Scott, and J. P. Masse, eds., Cretaceous carbonate platforms: AAPG Memoir 56, p. 185–199.

Conally, T. C., and R. W. Scott, 1985, Carbonate sediment-fill of an oceanic shelf, Lower Cretaceous, Arabian Peninsula, *in* P. D. Crevello and P. M. Harris, eds., Deep water carbonates: SEPM Core Workshop 6, p. 266–302.

Droste, H. J., 1990, Depositional cycles and source rock development in an epeiric intra-platform basin: The Hanifa Formation of the Arabian Peninsula: Sedimentary Geology, v. 69, p. 281–296.

Eberli, G. P., and R. N. Ginsburg, 1987, Segmentation and coalescence of Cenozoic carbonate platforms in northwestern Great Bahama Bank: Geology, v. 15, p. 75–79.

Eberli, G. P., and R. N. Ginsburg, 1989, Cenozoic progradation of northwestern Great Bahama Bank, a record of lateral platform growth and sea-level fluctuations, *in* P. D. Crevello, J. L. Wilson, J. F. Sarg, and J. F. Read, eds., Controls on carbonate platform and basin development: SEPM Special Publication 44, p. 339–351.

Glennie, K. W., M. G. H. Boeuf, M. W. Hughes Clarke, M. Moody-Stuart, W. F. H. Pilaar, and B. M. Reinhardt, 1974, The geology of the Oman Mountains: Verhandelingen van het Koninklijk Nederlands Geologisch Mijnbouwkundig Genootschap, v. 31, 423 p.

Haan, E. A., S. G. Corbin, M. W. Hughes Clarke, and J. E. Mabillard, 1990, The lower Kahmah Group of Oman: The carbonate fill of a marginal shelf basin, *in* A. H. F. Robertson, M. P. Searle, and A. C. Ries, eds., The geology and tectonics of the Oman region: Geological Society Special Publication 49, p. 109–125.

Harris, P. M., and S. H. Frost, 1984, Middle Cretaceous carbonate reservoirs, Fahud field and northwestern Oman: AAPG Bulletin, v. 68, p. 649–658.

Harris, P. M., S. H. Frost, G. A. Seiglie, and N. Schneidermann, 1984, Regional unconformities and depositional cycles, Cretaceous of the Arabian Peninsula, *in* J. S. Schlee, ed., Interregional unconformities and hydrocarbon accumulation: AAPG Memoir 36, p. 67–80.

Hillgärtner, H., F. S. P. van Buchem, F. Gaumet, P. Razin, B. Pittet, J. Grötsch, and H. Droste, 2003, The Barremian–Aptian evolution of the eastern Arabian carbonate platform margin (northern Oman): Journal of Sedimentary Research, v. 73, no. 5, p. 756–773.

Hughes Clarke, M. W. H., 1988, Stratigraphy and rock unit nomenclature in the oil-producing area of interior Oman: Journal of Petroleum Geology, v. 11, no. 1, p. 5–60.

Hunt, D., and W. M. Fitchen, 1999, Compaction and the dynamics of carbonate-platform development: Insights from the Permian Delaware and Midland Basins, southeast New Mexico and west Texas, U.S.A., *in* P. M. Harris, A. H. Saller, and J. A. Simo, eds., Advances in carbonate sequence stratigraphy: Applications to reservoirs, outcrops and models: SEPM Special Publication 63, p. 75–106.

Immenhauser, A., W. Schlager, S. J. Burns, R. W. Scott, T. Geel, J. Lehmann, S. van der Gaast, and L. J. A. Bolder-Schrijver, 1999, Late Aptian to late Albian sea-level fluctuations constrained by geochemical and biological evidence (Nahr Umr Formation, Oman): Journal of Sedimentary Research, v. 69, no. 2, p. 434–446.

Immenhauser, A., B. van der Kooij, A. van Vliet, W. Schlager, and R. W. Scott, 2002, An ocean-facing Aptian–Albian carbonate margin, Oman: Sedimentology, v. 48, p. 1187–1207.

Jordan, C. F., T. C. Connally, and H. A. Vest, 1985, Middle Cretaceous carbonates of the Mishrif Formation, Fateh field, offshore Dubai, *in* P. O. Roehl and P. W. Choquette, eds., Carbonate petroleum reservoirs: New York, Springer-Verlag, p. 425–442.

Kennedy, W. J., and M. D. Simmons, 1991, Mid-Cretaceous ammonites and associated microfossils from the central Oman Mountains: Newsletter in Stratigraphy, v. 25, p. 127–154.

Kenter, J. A. M., 1990, Carbonate platform flanks: Slope angle and sediment fabric: Sedimentology, v. 37, p. 777–794.

Landmesser, P., and A. S. Saydam, 1996, Seismostratigraphic interpretation of lower Thamama/Habshan in SE Abu Dhabi: SPE Paper 36204, p. 244–254.

Loosveld, R. J. H., A. Bell, and J. J. M. Terken, 1996, The tectonic evolution of interior Oman: GeoArabia, v. 1, no. 1, p. 28–51.

Masse, J. P., J. Borgomano, and S. Al-Maskiry, 1997,

Stratigraphy and tectonosedimentary evolution of a late Aptian–Albian carbonate margin: The northeastern Jebel Akhdar (Sultanate of Oman): Sedimentary Geology, v. 113, p. 269–280.

Mitchum, R. M., and M. A. Uliana, 1985, Seismic stratigraphy of carbonate depositional sequences, Upper Jurassic–Lower Cretaceous, Neuquén Basin, Argentina, *in* O. R. Berg and D. G. Woolverton, eds., Seismic sequence stratigraphy II: AAPG Memoir 39, p. 255–274.

Montenat, C., H.-J. Soudet, P. Barrier, and A. Chereau, 1999, Karstification and tectonic evolution of the Jabal Madar (Adam Foothills, Arabian Platform) during the Upper Cretaceous: Bulletin Centre Recherche Elf Exploration Production, v. 22, p. 161–183.

Murris, R. J., 1980, Middle East: Stratigraphic evolution and oil habitat: AAPG Bulletin, v. 64, p. 597–618.

O'Connor, S. J., and T. L. Patton, 1986, Middle Cretaceous carbonate reservoirs, Fahud field and northwestern Oman: Discussion: AAPG Bulletin, v. 70, p. 1799–1801.

Osleger, D. A., and S. W. Tinker, 1999, Three dimensional architecture of Upper Permian high frequency sequences, Yates Capitan shelf margin, Permian Basin, U.S.A, *in* P. M. Harris, A. H. Saller, and J. A. Simo, eds., Advances in carbonate sequence stratigraphy: Applications to reservoirs, outcrops and models: SEPM Special Publication 63, p. 169–185.

Pascoe, R. P., N. P. Evans, and T. L. Harland, 1995, The generation of unconformities within the Mishrif and Laffan Formations of Dubai and adjacent areas: Applications to exploration and production, *in* M. I. Al-Husseini, ed., Geo'94, Middle East Petroleum Geosciences Conference, Gulf Petrolink, Bahrain, v. 2, p. 749–760.

Philip, J., J. Borgomano, and S. Al Maskiry, 1995, Cenomanian–early Turonian carbonate platform of Northern Oman: Stratigraphy and palaeo-environments: Palaeogeography, Palaeoclimatology and Palaeoecology, v. 119, p. 77–92.

Pittet, B., F. S. P. van Buchem, H. Hillgärtner, P. Razin, J. Grötsch, and H. Droste, 2002, Ecological succession, paleoenvironmental change and depositional sequences of the Barremian–Aptian shallow water carbonates in northern Oman: Sedimentology, v. 49, no. 3, p. 555–581.

Pomar, L., 1991, Reef geometries, erosion surfaces and high-frequency sea-level changes, upper Miocene Reef Complex, Mallorca, Spain: Sedimentology, v. 38, p. 243–269.

Pomar, L., 1993, High-resolution sequence stratigraphy in prograding Miocene carbonates: Application to seismic interpretation, *in* R. G. Loucks and J. F. Sarg, eds., Carbonate sequence stratigraphy: AAPG Memoir 57, p. 389–407.

Pratt, B. R., and J. D. Smewing, 1990, Jurassic and Early Cretaceous platform margin configuration and evolution, central Oman Mountains, *in* A. H. F. Robertson, M. P. Searle, and A. C. Ries, eds., The geology and tectonics of the Oman region: Geological Society Special Publication 49, p. 69–88.

Pratt, B. R., and J. D. Smewing, 1993a, Early Cretaceous platform margin, Oman, eastern Arabian Peninsula, *in* T. Simo, R. W. Scott, and J. P. Masse, eds., Cretaceous carbonate platforms: AAPG Memoir 56, p. 201–212.

Pratt, B. R., and J. D. Smewing, 1993b. Early Cretaceous platform-margin configuration and evolution in the central Oman Mountains, Arabian Peninsula: AAPG Bulletin, v. 77, p. 225–244.

Rabu, D. J. Le Metour, F. Bechennec, M. Beurrier, M. Villey, and C. Bourdillon-Jeudy de Grissac, 1990, Sedimentary aspects of the Eo-Alpine cycle on the northeast edge of the Arabian Platform (Oman Mountains), *in* A. H. F. Robertson, M. P. Searle, and A. C. Ries, eds., The geology and tectonics of the Oman region: Geological Society Special Publication 49, p. 49–68.

Scott, R. W., 1990, Chronostratigraphy of the Cretaceous carbonate shelf, southeastern Arabia, *in* A. H. F. Robertson, M. P. Searle, and A. C. Ries, eds., The geology and tectonics of the Oman region: Geological Society Special Publication 49, p. 89–108.

Sharief, F. A., K. Magara and H. M. Abdulla, 1989, Depositional system and reservoir potential of the middle Cretaceous Wasia Formation in central-eastern Arabia: Marine and Petroleum Geology, v. 6, p. 303–315.

Simmons, M. D., 1994, Micropalaeontological biozonation of the Kahmah Group (Early Cretaceous), Central Oman Mountains, *in* M. D. Simmons, ed., Micropalaeontology and hydrocarbon exploration in the Middle East: British Micropalaeontological Society Publication Series: London, Chapman & Hall, p. 177–220.

Simmons, M. D., and M. B. Hart, 1987, The biostratigraphy and microfacies of the Early to middle Cretaceous carbonates of Wadi Mi'aidin, central Oman Mountains, *in* M. B. Hart, ed., Micropalaeontology of carbonate environments: Chichester, UK, Ellis Horwood, p. 176–207.

Smith, A. B., M. D. Simmons, and A. Racey, 1990, Cenomanian echinoids, larger foraminifera and calcareous algae from the Natih Formation, central Oman Mountains: Cretaceous Research, v. 11, p. 29–69.

Sonnenfeld, M. D., and T. A. Cross, 1993, Volumetric partitioning and facies differentiation within Permian upper San Andres Formation of Last Chance Canyon, Guadalupe Mountains, New Mexico, *in* R. G. Loucks and J. F. Sarg, eds., Carbonate sequence stratigraphy recent developments and applications: AAPG Memoir 57, p. 435–474.

Terken, J. M. J., 1999, The Natih petroleum system of Oman: GeoArabia, v. 4, p. 157–180.

van Buchem, F. S. P., P. Razin, P. W. Homewood, J. M. Philip, G. P. Eberli, J.-P. Platel, J. Roger, R. Eschard, G. M. J. Desaubliaux, T. Boisseau, J.-P. Leduc, R. Labourdette, and S. Cantaloube, 1996, High resolution sequence stratigraphy of the Natih Formation (Cenomanian/Turonian) in northern Oman: Distribution of source rocks and reservoir facies: GeoArabia, v. 1, p. 65–91.

van Buchem, F. S. P., P. Razin, P. W. Homewood, H.

Oterdoom, and J. Philip, 2002a, Stratigraphic organization of carbonate ramps and organic-rich intrashelf basins: Natih Formation (middle Cretaceous) of northern Oman: AAPG Bulletin, v. 86, p. 21–54.

van Buchem, F. S. P., B. Pittet, H. Hillgärtner, J. Grötsch, A. Al Mansouri, O. Al-jeelani, I. Billing, M. Van Steenwinkel, H. Droste, and H. Oterdoom, 2002b, High resolution sequence stratigraphic architecture of Barremian/Aptian carbonate systems in northern Oman and the United Arab Emirates (Kharaib and Shu'aiba Formations): GeoArabia, Bahrain, v. 7, no. 3, p. 461–500.

Van Steenwinkel, M., 1992, Sequence stratigraphy of the Shuaiba Formation, Oman: Tentative correlations: Shell International Internal Report, 11 p.

Van Steenwinkel, M., 1996, Al Huwaisah Shuaiba reservoir geological model: Petroleum Development Oman Internal Report, 18 p.

Warburton, J., T. J. Burnhill, R. H. Graham, and K. P. Isaac, 1990, The evolution of the Oman Mountains foreland basin, *in* A. H. F. Robertson, M. P. Searle, and A. C. Ries, eds., The geology and tectonics of the Oman region: Geological Society Special Publication 49, p. 419–427.

Witt, W., and H. Gökdag, 1994, Orbitolinid biostratigraphy of the Shuaiba Formation (Aptian), Oman—Implications for reservoir development, *in* M. D. Simmons, ed., Micropalaeontology and hydrocarbon evaluation in the Middle East: London, Chapman & Hall, p. 221–243.

Eberli, G. P., F. S. Anselmetti, C. Betzler, J.-H. Van Konijnenburg, and D. Bernoulli, 2004, Carbonate platform to basin transitions on seismic data and in outcrops: Great Bahama Bank and the Maiella Platform margin, Italy, *in* Seismic imaging of carbonate reservoirs and systems: AAPG Memoir 81, p. 207–250.

Carbonate Platform to Basin Transitions on Seismic Data and in Outcrops: Great Bahama Bank and the Maiella Platform Margin, Italy

Gregor P. Eberli
University of Miami, Miami, Florida, U.S.A.

Flavio S. Anselmetti
Swiss Federal Institute of Technology (ETH), Zurich, Switzerland

Christian Betzler
Geologisch-Palaeontologisches Institut, Hamburg, Germany

Jan-Henk Van Konijnenburg
Sarawak Shell Bhd, Sarawak, Malaysia

Daniel Bernoulli
University of Basel, Basel, Switzerland

ABSTRACT

The comparison of seismic and core data from the western Great Bahama Bank with the exhumed Maiella Platform margin and its adjacent slope in the Apennines of Italy relates the seismic facies to depositional facies and processes. Both platforms evolved similarly from an escarpment-bounded, aggrading platform in the Cretaceous to a prograding platform in the Tertiary. This comparison helps to improve seismic interpretation of isolated carbonate platform systems.

Platform interior deposits are typically horizontally layered cycles of shallow-water carbonates, but the seismic sections from Great Bahama Bank are dominated by a chaotic to transparent seismic facies. Synthetic seismic sections of the Maiella Platform margin demonstrate that the chaotic to transparent seismic facies is a product of low-impedance contrasts in the platform carbonates. Both platforms were bounded in the Cretaceous by an escarpment that separated the platform from onlapping basinal and slope sediments. This juxtaposition of facies is recorded in the seismic facies by the lateral change from chaotic platform to inclined continuous reflections of the slope. The outcrops of the Maiella Platform margin help assess the processes that formed these escarpments. Small concave scallops and associated megabreccias in the basinal section document episodic erosion during the platform growth, indicating that the escarpment was growing simultaneously with the platform.

Both platforms prograde after burial of the escarpment by basinal sediments. On the western margin of the Great Bahama Bank, progradation started in the middle Miocene and advanced the platform margin approximately 25 km westward to its present position. Progradation is documented on the seismic data by clinoform geometry and the expansion of the interpreted platform seismic facies. The prograding system of western Great Bahama Bank consists of sigmoidal clinoforms with foresets that are approximately 600 m high. The foresets are characterized by reflections with variable amplitude and continuity. Discontinuous high-amplitude packages are interrupted by low-amplitude, nearly transparent units of periplatform ooze. Channels of variable size dissect the entire slope but deep incisions with a persistent cut-and-fill geometry occur preferentially at sequence boundaries. These incised submarine canyons are oriented downslope perpendicular to the strike of the platform margin. Most of the gravity-flow deposits bypassed the upper and middle slope and are deposited on the lower slope and on the toe-of-slope. These redeposited carbonates are seismically characterized by discontinuous to chaotic high-amplitude reflections that suggest a heterogeneous environment of depositional lobes.

Core data indicate that a tripartite facies succession of slope, reef margin, and platform interior deposits forms the topsets of the prograding clinoforms on Great Bahama Bank. This facies succession is also found in the Maiella Platform margin that prograded across the underlying slope during Eocene time. Synthetic seismic sections show that the reefal units appear as transparent zones on the seismic data, corroborating the calibration made by a core-to-seismic correlation in the Bahamas.

Along the Maiella Platform margin, incised slope canyons are exposed, revealing the lithologies of the channel fills. The Maiella canyons are filled with coarse, fining-upward mass gravity-flow deposits that fine upward. The outcrops in the Gran Sasso area display the heterogeneity of the toe-of-slope environment that is characterized by small, amalgamated lobes with feeder channels in largely pelagic background sediment.

INTRODUCTION

A thorough seismic analysis includes the interpretation of depositional environments and facies and the documentation of stratigraphic and structural relationships from seismic data. A common first step in seismic interpretation is to relate seismic facies, which includes seismic reflection configuration, shape, amplitude, and continuity, to lithology and depositional environments (Mitchum et al., 1977; Ramsayer, 1979; Brown and Fisher, 1980). With the advance of three-dimensional (3-D) seismic data, these seismic facies parameters have been expanded to include other attributes and the third dimension (Weimer and Davis, 1996; Brown, 1999). Despite the technical advances in acquisition, processing, and display of seismic data, the prediction of subsurface lithologies still relies on the correlation of the seismic facies to the interpreted depositional facies. Not all seismic facies do, however, uniquely represent depositional facies. A chaotic seismic facies might be caused by very heterogeneous deposits or by seismic imaging problems. Likewise, a depositional facies might be altered by postdepositional processes, potentially resulting in a nonunique acoustic character of a depositional environment. This is especially true in carbonates where impedance is the combined product of sedimentation and diagenesis (Anselmetti and Eberli, 1993, 1997). In carbonates, diagenetic processes can easily change original mineralogy, porosity, and density. These changes are difficult to detect on seismic sections unless they affect a large volume of rocks (Wagner, 1997). In short, the nonunique acoustic character of carbonates makes their interpretation a challenge.

Generally, two methods are used to overcome this challenge. The first approach is to examine carbonate depositional environments and to produce plausible seismic criteria for recognizing the facies on seismic sections (Fontaine et al., 1987; Macurda, 1997). This approach qualitatively relates the depositional sedimentary bodies to the seismic facies. For example, the homogeneous micritic limestone (chalk) of the basinal setting is likely to result in a low-amplitude to transparent seismic facies (Fontaine et al., 1987). The second technique to evaluate the relationship between rock bodies and their seismic response is seismic modeling. Modeling includes one-dimensional borehole modeling, forward modeling of subsurface data, and two-dimensional (2-D) modeling of outcrop data. Neidell and Poggiagliolmi (1977) and Fagin (1991) discuss techniques, basic principles, and several case studies. In the context of seismic facies and stratigraphy, 2-D outcrop modeling can help calibrate a seismic facies and bridge the gap between the detail of geologic sections and low-resolution seismic sections (e.g., Rudolph et al., 1989; Biddle et al., 1992; Stafleu and Schlager, 1993; Stafleu and Sonnenfeld, 1994; Bracco Gartner and Schlager, 1999; Schwab and Eberli, 2000).

In this chapter, we use a combination of both methods to evaluate the seismic facies of platform-basin transects across isolated carbonate platforms. We compare the seismic facies across the margin of western Great Bahama Bank with the seismic-scale outcrops of the Maiella Platform margin. What is here called the *Maiella Platform margin* is a 2000-m-thick section of the Apulian Platform that is exposed in the Montagna della Maiella (Abruzzi, central Italy), where a platform-basin transition is exposed along a north-south transect of approximately 30 km and a width of 10–15 km (Figure 4; Bally, 1954; Accarie, 1988; Eberli et al., 1993; Sanders, 1994; Vecsei et al., 1998). The Maiella Platform margin, ranging in age from Jurassic to Miocene, has a similar evolution to the evolution of the Great Bahama Bank and is an excellent ancient analog for the modern platform. Both platforms evolved from isolated vertically aggrading platforms with steep, marginal escarpments into progradational platforms (Eberli and Ginsburg, 1987; Eberli et al., 1993). In particular, two platform-basin transects, one over the western edge of the Great Bahama Bank and one across the Maiella Platform margin, reveal striking similarities. The excellent outcrops in the Maiella Mountains are of seismic scale and suitable for a comparison of seismic and depositional facies of the platforms and their slopes (Eberli et al., 1993; Bernoulli et al., 1996; Anselmetti et al., 1997; Van Konijnenburg et al., 1999).

The comparison is organized along the three seismic facies belts that are recognized from the platform interior to the basin on Great Bahama Bank: the aggrading platform bounded by an escarpment, the prograding margin, and the slope and basin. The objectives of the chapter are to (a) document the seismic characteristics (seismic facies and geometries) of these three environments, (b) relate the seismic facies of each environment to the depositional facies using the synthetic seismic sections, core information, and/or outcrop analogs from the Maiella Platform margin, and (c) discuss the depositional and diagenetic processes responsible for the sedimentary architecture and seismic characteristics in each environment.

GEOLOGIC SETTING OF THE TWO PLATFORMS

The isolated carbonate platforms of the Bahamas are commonly used as a modern analog for the interpretation of ancient carbonate deposits. Precise description of the different facies belts of the Bahamas-Florida region by, among others, Illing (1954) and Newell and Rigby (1957) laid the foundation for carbonate-facies analysis and helped interpret ancient shallow-water carbonates. Similarly, the Lower Cretaceous to upper Miocene part of the Apulian Platform, which is exposed in the Maiella Mountains (Italy), is an ancient analog of the Great Bahama Bank. Other parts of Apulian Platform are exposed today in Apulia, the Gargano peninsula, and the Ionian islands and are known from the subsurface of southern Italy and the Adriatic Sea. This platform has a similar evolution to the evolution of the Great Bahama Bank. The similarities start with the initiation of the two platforms. Both were established on rifted margins about at the same time during the Early to Middle Jurassic— the Maiella Platform margin in the Mesozoic Tethys ocean and Great Bahama Bank in the central Atlantic (Figure 1).

Great Bahama Bank is the largest isolated platform of the Bahamian archipelago. It is an impressive edifice of approximately 600 km in length (northwest-southeast) and 350 km in width (east-west) situated between 22° and 28° north of the equator. The platform shape is irregular with a deep basin (Exuma Sound) and troughs (Northeast and Northwest Providence Channels, Tongue of the Ocean) cutting into the bank (Figure 2). An older trough, the Straits of Andros that divided the northern Great Bahama Bank into two banks (Bimini Bank and Andros Bank), was filled and incorporated into the present-day Great Bahama Bank (Figure 3; Eberli and Ginsburg, 1987, 1989). Similar infilled intraplatform depressions and troughs also occur in the southern part of the modern Great Bahama Bank (Masaferro and Eberli, 1999).

The Bahamian platforms experienced phases of tectonic segmentation and periods of platform expansion and coalescence (Eberli and Ginsburg, 1987, 1989; Austin et al., 1988; Masaferro and Eberli, 1999). Platform

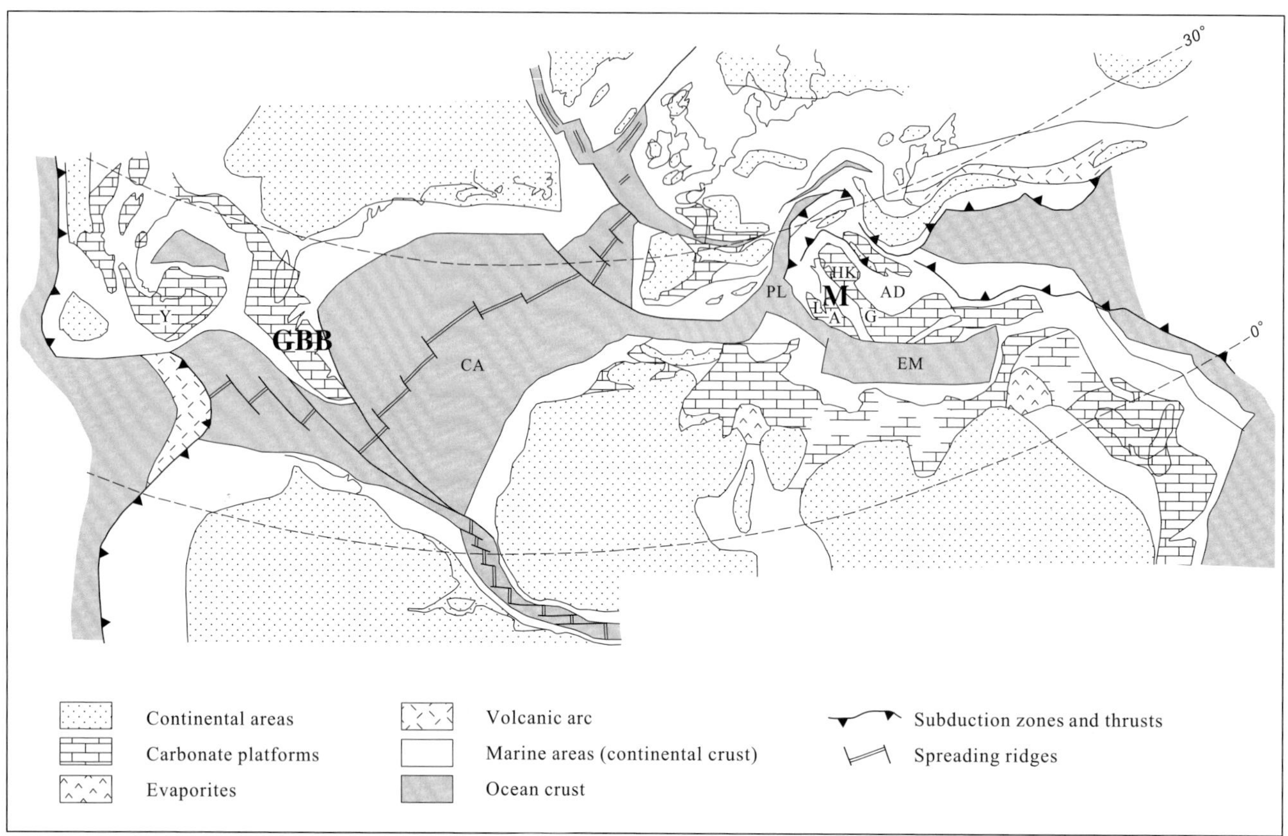

FIGURE 1. Late Cretaceous (94–92 Ma) paleogeography and the location of the Great Bahama Bank (GBB) and the peri-Adriatic Platforms (M) within the frame of the Cretaceous Tethys ocean. A = Apulian Platform; CA = central Atlantic; EM = eastern Mediterranean; G = Gavrovo; GBB = Great Bahama Bank; HK = high karst; L = Lazio-Abruzzi Platform; M = Maiella; PL = Piemont-Liguria ocean; Y = Yucatan; AD = Adria.

expansion in the Early Cretaceous buried parts of the rift topography to form what became known as the "megabank" (Sheridan et al., 1981, Ladd and Sheridan 1987). During the middle Cretaceous, a phase of tectonic segmentation and drowning of this "megabank" occurred. During the Late Cretaceous and Paleogene, the Great Bahama Bank consisted of several vertically aggrading platforms that later coalesced to form the modern Great Bahama Bank (Eberli and Ginsburg, 1987, 1989; Masaferro and Eberli, 1999). In the area of what is today the northwestern part of the Great Bahama Bank, two vertically aggrading platforms (Bimini Bank and Andros Bank) were identified on the seismic lines. The seismic facies of these aggrading platforms is mostly transparent to chaotic that abruptly terminate toward inclined reflections (Figure 3). This abrupt seismic facies change from the platform to adjacent slope deposits indicates that both platforms were bounded on the western margin by an escarpment (Figure 3; Eberli and Ginsburg, 1987, 1989; Bernoulli et al., 1996). The relative quiescence in the Neogene led to platform progradation and the coalescence of the smaller banks to form the modern Great Bahama Bank (Figure 3; Eberli and Ginsburg, 1987). At the western side of the Great Bahama Bank. Onlapping sediments buried the escarpment transforming the margin into a low-angle and, since the middle Miocene, into a prograding platform (Eberli and Ginsburg, 1989). Progradation advanced the margin approximately 25 km into the Straits of Florida (Figure 3).

The Maiella Platform margin is situated on Herzynian continental basement. Crustal extension preceding the formation of the Tethys ocean broke up the large Late Triassic to Early Jurassic megaplatform (Hauptdolomit-Calcare Massiccio Formation) into a series of isolated platforms and intervening basins, partly mimicking the rift morphology (Bernoulli and Jenkyns, 1974; Winterer and Bosellini, 1981; Bice and Stewart, 1990). These platforms display large-scale similarities in size, morphology, facies, and internal architecture with the platforms of the Bahamas archipelago (Bernoulli, 1972; d'Argenio et al., 1975; Bosellini, 1989). Furthermore, deeper basins, comparable to the present-day troughs in the Bahamian archipelago, separated the peri-Adriatic platforms from each other. One of the largest of these isolated Mesozoic to middle Tertiary carbonate platforms was the Apulian Platform that was more than 400 km long (750 km?). The Maiella Platform margin was

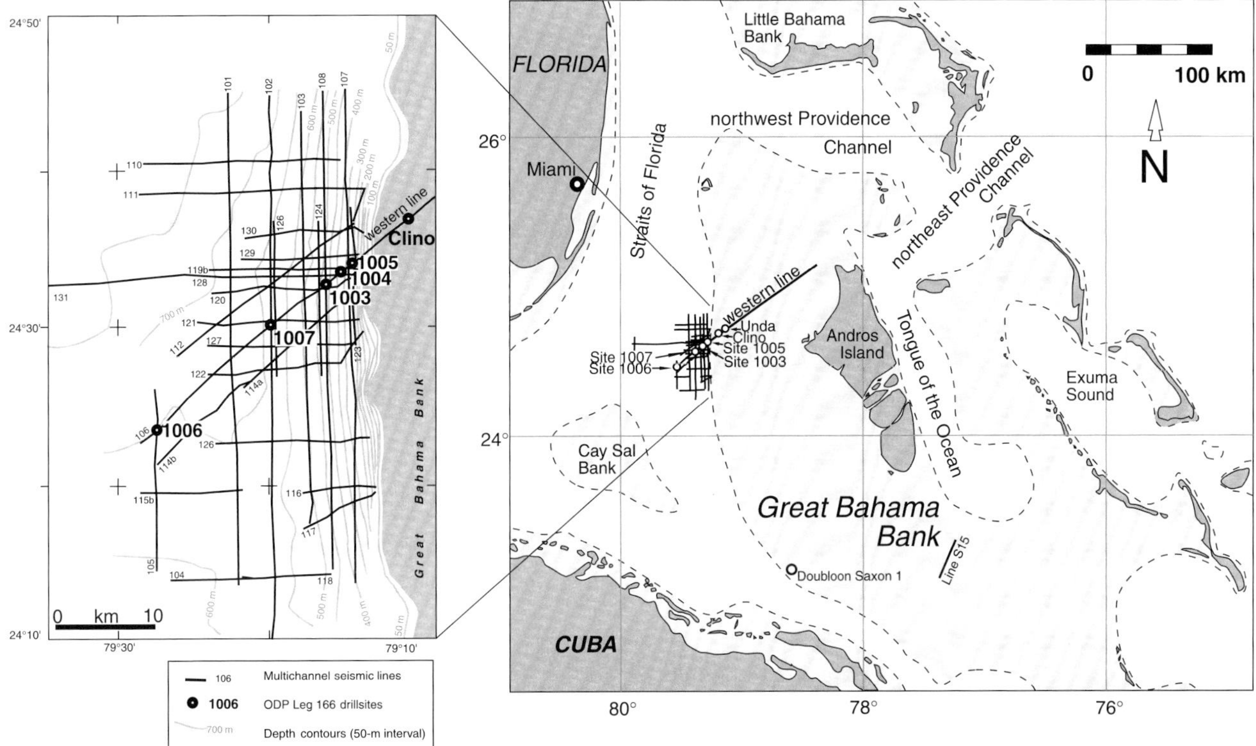

FIGURE 2. Map of the Great Bahama Bank with location of wells and seismic data set used in this study. Inset of western margin of the Great Bahama Bank with the location of the drill sites of Ocean Drilling Program (ODP) Leg 166 and the high-resolution multichannel seismic lines acquired with the R/V *Lone Star*.

situated at a paleolatitude of approximately 10–30° N in a subtropical, warm, and only seasonally humid climate (Accarie and Deconinck, 1989). In the Gran Sasso area, approximately 50 km north of the Maiella Mountains, time-equivalent base-of-slope deposits are exposed in a set of thrust sheets (Figure 4; Van Konijnenburg, 1997; Van Konijnenburg et al., 1999).

The Maiella Platform margin was aggradational during most of the Cretaceous and progradational during the latest Cretaceous and the Tertiary. During most of the Cretaceous, an escarpment separated a shallow-water platform margin rimmed by rudist biostromes in the south from a pelagic facies with intercalations of gravity-flow deposits to the north (Figure 5A). In the middle Cretaceous, the platform was exposed and slightly tilted. Humid conditions during the exposure in the Aptian to Cenomanian are indicated by extensive bauxite horizons (d'Argenio, 1970). Toward the end of the Campanian, onlapping basinal sediments buried the escarpment and the platform started to prograde (Orfento Formation, Supersequence 2). This platform expansion is coeval with the development of a basin-floor fan at the base of slope in the Gran Sasso area (Figure 6). The progradation was halted by renewed emergence of the platform during the latest Maastrichtian and early Danian. Coeval basin sediments consist mostly of pelagic limestones (uppermost Monte Corvo Formation).

A northward shift in latitude from the Late Cretaceous to the Tertiary, together with climatic changes during the Tertiary, led to a gradual disappearance of corals and to cooler and slightly deeper water benthic communities dominated by bryozoans, coralline algae, and large foraminifera (Carannante et al., 1988; Mutti et al., 1999). Basinal breccias that contain Cretaceous and lower Tertiary lithoclasts record platform margin erosion and retrogradation in the late Paleocene and Eocene (Sanders, 1994; Van Konijnenburg, 1997; Vecsei et al., 1998). Shallow-water carbonate deposition continued up to the late Miocene when platform growth was terminated by the Messinian salinity crisis (Eberli et al., 1993). Subsequently, the platform interior and its northern margin were incorporated into the fold belt of the southern Apennines. In the Maiella Mountains, the former platform margin is magnificently exposed in the frontal anticline of the last thrust sheet.

SEISMIC AND CORE DATA

Sufficient seismic data from Great Bahama Bank are available to image various depositional environments and architectural elements that make up the modern platform (e.g., Hine et al., 1981; Mullins et al., 1984; Eberli and Ginsburg, 1987, 1989; Austin et al., 1988;

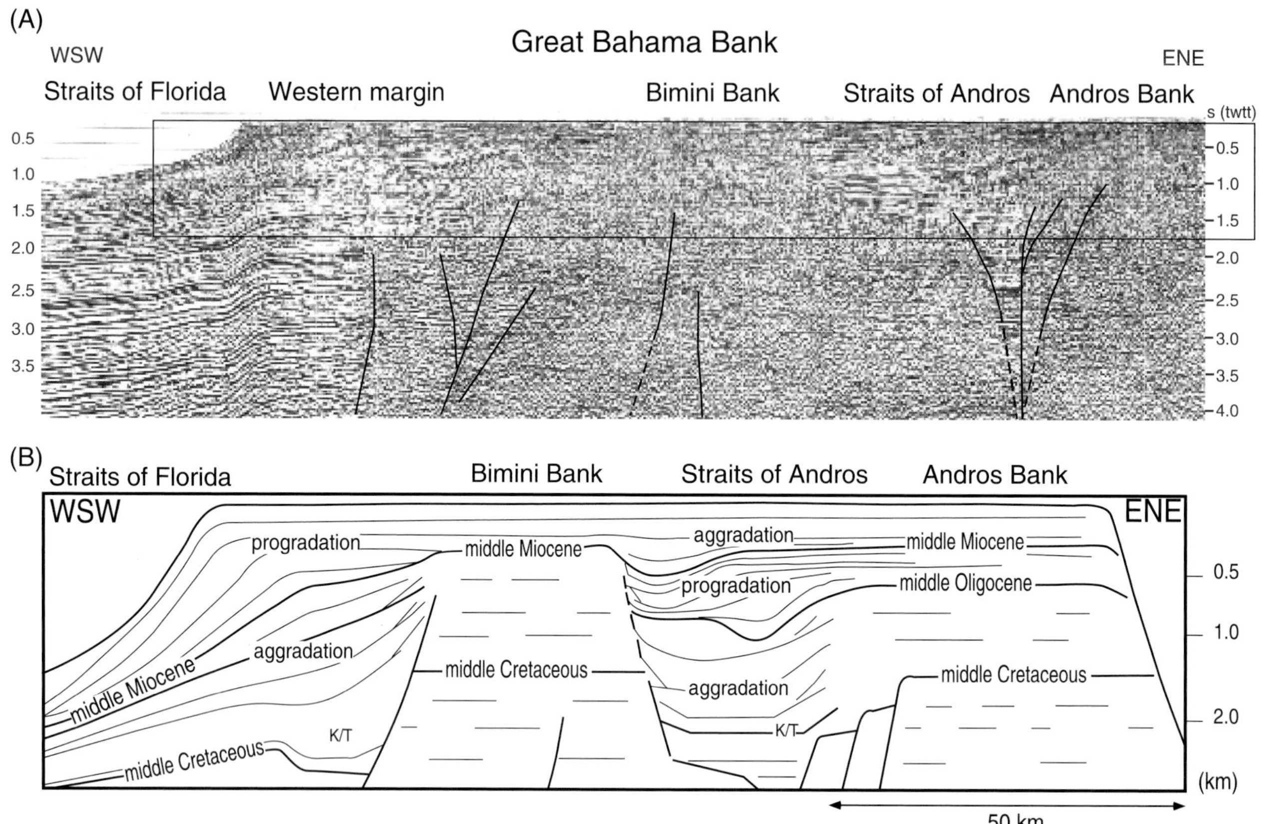

FIGURE 3. (A) Squeezed seismic line (western line in Figure 2) across Great Bahama Bank shows deep-seated faults of the two nuclear banks and a buried intraplatform seaway that coalesced to form the modern Great Bahama Bank. The base of platform edifice is not imaged seismically. (B) Schematic cross section of the upper part across Great Bahama Bank (box of panel A) approximately along the western seismic line with interpreted ages of the Upper Cretaceous to Holocene strata (modified from Eberli and Ginsburg, 1987).

Masaferro and Eberli, 1999; Anselmetti et al., 2000). In particular, several vintages of multichannel seismic profiles across Great Bahama Bank and its margin provide seismic images of the internal architecture of the largest platform of the archipelago (Eberli and Ginsburg, 1987, 1989; Schlager et al., 1988; Masaferro and Eberli, 1999; Anselmetti et al., 2000). The seismic data sets used in this chapter are of two vintages. Two seismic data sets across the top of the Great Bahama Bank were shot in the 1980s and consist of approximately 1500 km of mostly unmigrated, multichannel seismic profiles. Ten to 12 air guns with variable volumes (720–2280 in.^{3} [11,798–37,362 cm^3]) were used to acoustically penetrate the platform; shotpoint interval was 25 m and occasionally 50 m. The signal was sampled at intervals of 2 ms and deconvolved before stacking. Resolution in these seismic data is approximately 30 m in the upper and 40 m in the lower part of the seismic sections.

The slope and western margin of the Great Bahama Bank are imaged with approximately 1000 km of multichannel seismic lines (Figure 2). The sections were shot using a 45/105-in^3 GI air gun operated at 2200 psi and 2 m water depth onboard R/V *Lone Star*, which is operated by Rice University. A 600-m-long, 24-channel streamer received the signals in a water depth of 3–4 m. The acoustic signal of the GI air gun is characterized by a frequency spectra between 20 and 500 Hz, with a central frequency of approximately 100–200 Hz. For data processing, a time-variable high-pass filter was applied, ranging from 40–50 Hz in the shallow parts to 20–30 Hz in the deeper zones, while the low-pass filter was set throughout to 450–500 Hz. After common depth-point stacking (12-fold), migrating, and filtering, an automatic gain control (AGC) was applied for the display of all sections with a window length of 100 ms. Because of the sharp amplitude peak of the GI air gun, which is hardly disturbed by a bubble pulse, the seismic resolution can reach 5 m, particularly in low-velocity deposits that show a highly coherent seismic facies.

Several recent, deep drill holes provide information about the lithology of the imaged environments (Austin et al., 1988; Harwood and Towers, 1988; Eberli et al., 1997; Eberli, 2000). The exploration well Great Isaac well 1 at the northwestern tip of the bank (Schlager et al., 1988)

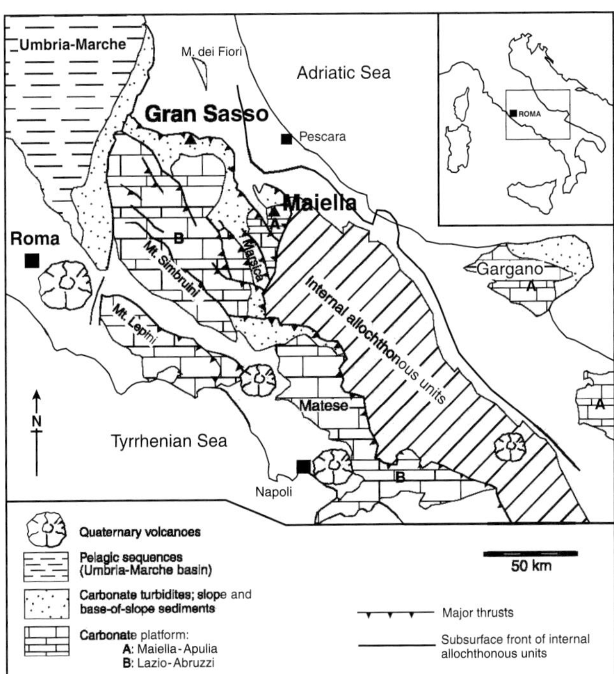

FIGURE 4. Location map of the Maiella Platform margin and the base-of-slope setting in the Gran Sasso area and the present-day position of the Apulia (A) and Lazio-Abruzzi-Campania (B) Platforms. Facies distribution refers to the Upper Cretaceous strata.

and the Doubloon Saxon well in the southwest (Walles, 1993) give lithostratigraphic and biostratigraphic information for the Cretaceous and Paleogene parts of the banks. In this chapter, we concentrate on the cross-bank profile (western line, Figure 2), along which seven sites were drilled in connection with the Bahamas Transect. The cores of the Bahamas Transect drill sites provide the lithologic, stratigraphic, and petrophysical information for the Neogene part of the western margin of the Great Bahama Bank. These cores are correlated to the seismic data by using the time-depth conversion from interval velocities obtained from check-shot surveys in each hole (Anselmetti et al., 2000; Eberli et al., 2001). In addition, synthetic seismic profiles were constructed at each drill site (Bernet, 2001; Eberli et al., 2001).

The synthetic seismic model of the Maiella Platform margin was calculated from geometries and petrophysical data retrieved from outcrops (Anselmetti et al., 1997). To define a layered impedance model, velocities and densities of 186 minicores representing all major outcropping lithologies were determined in the laboratory. The impedance model was converted into synthetic seismic data by applying a computer-simulated modeling procedure that uses the normal incidence ray-tracing method at variable frequencies, amplitude gains, and noise levels. The impedance model, 14 km in width and 1.5 km in depth, was covered by a virtual survey of 400 shotpoints that are identical with 400 common midpoints (CMPs) and 400 receiver points. The resulting CMP distance equals 35 m, which is a reasonable value for true seismic lines. The time sample interval for all synthetic sections is 2 ms. The purpose of seismic modeling is to evaluate the seismic response of a geologic setting by computer-simulating a seismic survey (Anselmetti et al., 1997).

COMPARISON OF SEISMIC AND DEPOSITIONAL FACIES

General Overview

The similar geologic evolution from an escarpment-bounded aggrading platform to a prograding platform of the western Great Bahama Bank and the Maiella Platform margin results in similar seismic facies of the two platforms (Figures 5, 7). In both transects, an aggrading platform interior produces three seismic facies. The Cretaceous parts of both platforms are characterized by a mostly transparent to chaotic seismic facies with local high-amplitude horizons. A high-amplitude seismic reflection horizon caps this facies. The Tertiary parts of both platforms are imaged with more continuous seismic reflections. In contrast, the slope and basinal sections consist of inclined reflections of variable amplitude and continuity. During the aggrading phase, this facies terminates abruptly against the platform seismic facies along an escarpment. After burial of the escarpment by onlapping slope and basinal sediments, both platforms began to prograde. Horizontal reflections of platform top facies are expanding over inclined reflections of the slope. In Great Bahama Bank, the two facies are typically separated by narrow zone of transparent seismic facies that is correlated to marginal reefal units. Similarly, in the Maiella Platform margin, reefal units advance over slope deposits during progradation.

The slope sections of the Great Bahama Bank are seismically imaged by reflections of variable amplitudes and continuity. Thick low-amplitude to transparent-upper-slope reflections are intercalated by downcutting high-amplitude reflections outlining channel incisions. Toward the lower-slope and toe-of-slope reflection, continuity decreases as small-scale channeling and mounded reflection geometries increase. These seismic geometries and the redeposited carbonates recovered in cores document small channel and levee complexes and overlapping lobes on the lower slope. Outcrop geometries of redeposited carbonates at the toe of slope and the basin floor in the Gran Sasso area are similar to the geometries observed on the seismic sections. In the following, these three elements—escarpment-bounded aggrading platform, prograding margin, and slope—will be described in detail and correlated and compared to the outcrops.

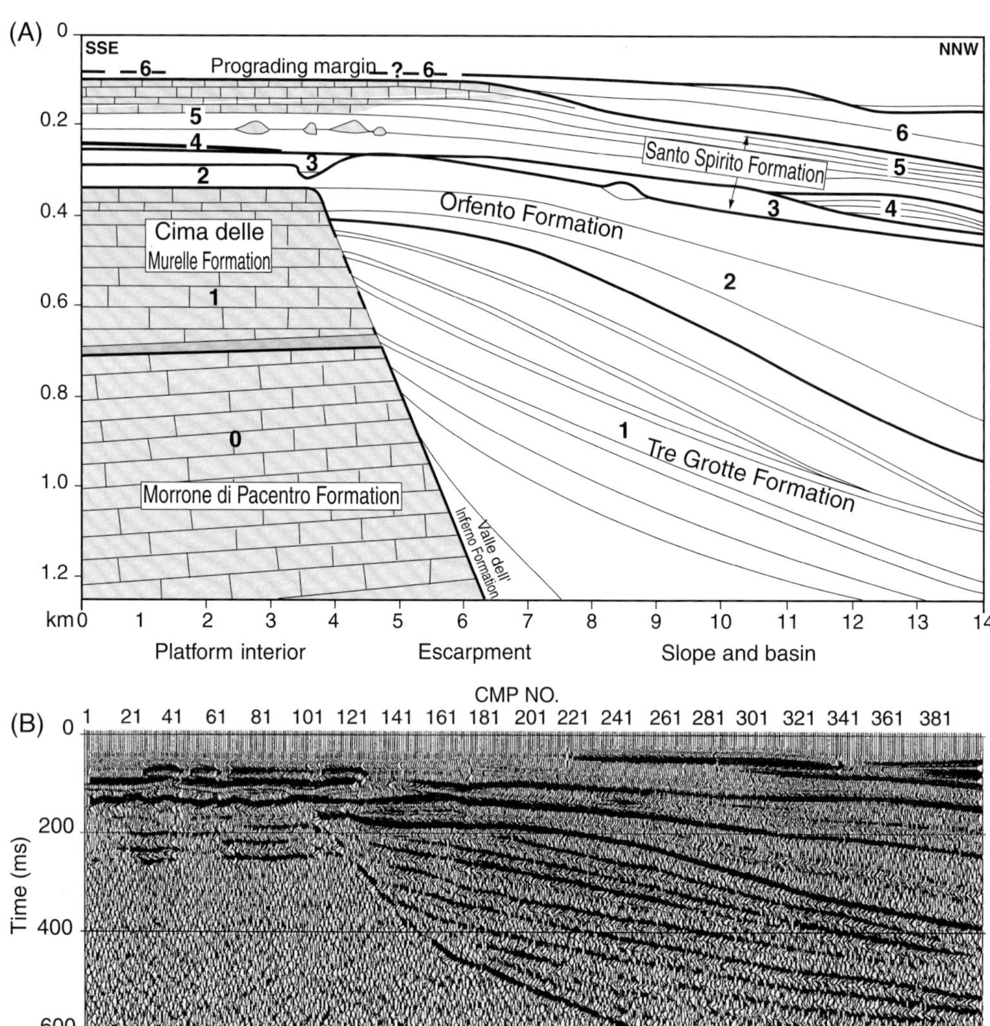

FIGURE 5. (A) Schematic cross section of the Maiella Platform margin. Seven second-order supersequences (0–1) can be distinguished (reprinted by permission of Vecsei, 1991). Supersequences 0 and 1 represent the Cretaceous platform. Upper Cretaceous Supersequence 2 buries the relief along the platform. Paleogene Supersequences 3–5 are characterized by erosion and repeated flooding of the platform, redeposited megabreccias (Supersequences 3 and 4), progradation of reefal units (Supersequence 5), and resedimented and pelagic carbonates on the slope. Miocene Supersequence 6 was deposited on a wide and shallow shelf. Bold lines are supersequence boundaries; shaded areas mark shallow-water carbonates. (B) Synthetic seismic section across the Maiella Platform margin displaying the seismic facies of the Cretaceous platform interior, the escarpment that separates the platform from the adjacent basin, and the slope and basin sediments. The platform interior is seismically nearly transparent, with some discontinuous reflections in the upper part and capped by a high-amplitude reflection. The escarpment is imaged as a continuous reflection marking the onlap surface for the continuous reflections of the basin and slope facies. Horizontal scale is the same as in schematic cross section above (modified from Anselmetti et al., 1997). CMP = common midpoints.

Aggradational Platform and Escarpment

Aggrading platforms are generally constructed by the consecutive deposition of shallow-water carbonates that are usually interrupted by subaerial exposure horizons. These successions are related to the rhythmic creation of accommodation space as a result of high-frequency sea level fluctuations (e.g., Goldhammer et al., 1990; Strasser et al., 1999). The slope angles of vertically aggrading platforms vary widely but there is a trend of increased slope angle with increased slope height, and escarpment margins along carbonate platforms are relatively common throughout the stratigraphic record (Schlager and Ginsburg, 1981). Escarpments are produced by a combination of aggradational growth and erosion, arising from the ability of carbonate platforms to construct very steep slopes and from the erosional processes that affect these slopes (Hurst and Surlyk, 1984; Freeman-Lynde and Ryan, 1985; Paull and Neumann, 1987; Mullins and Hine, 1989; Eberli et al., 1993). They have approximately 40° or more declivity and a cumulative height of more than 500 m. They have a stable foundation, thereby aggrading over time in a single location and with long duration. The escarpment may be partially buried during growth, resulting in a smaller escarpment height during any one depositional time (Surlyk and Ineson, 1992; Eberli et al., 1993). Large aggradational/erosional escarpments are fundamentally different from the modern, steep constructional margins, variously named erosional escarpment or the wall (Grammer, 1991), fore-reef escarpment (James and Ginsburg, 1979), submarine cliff (James and

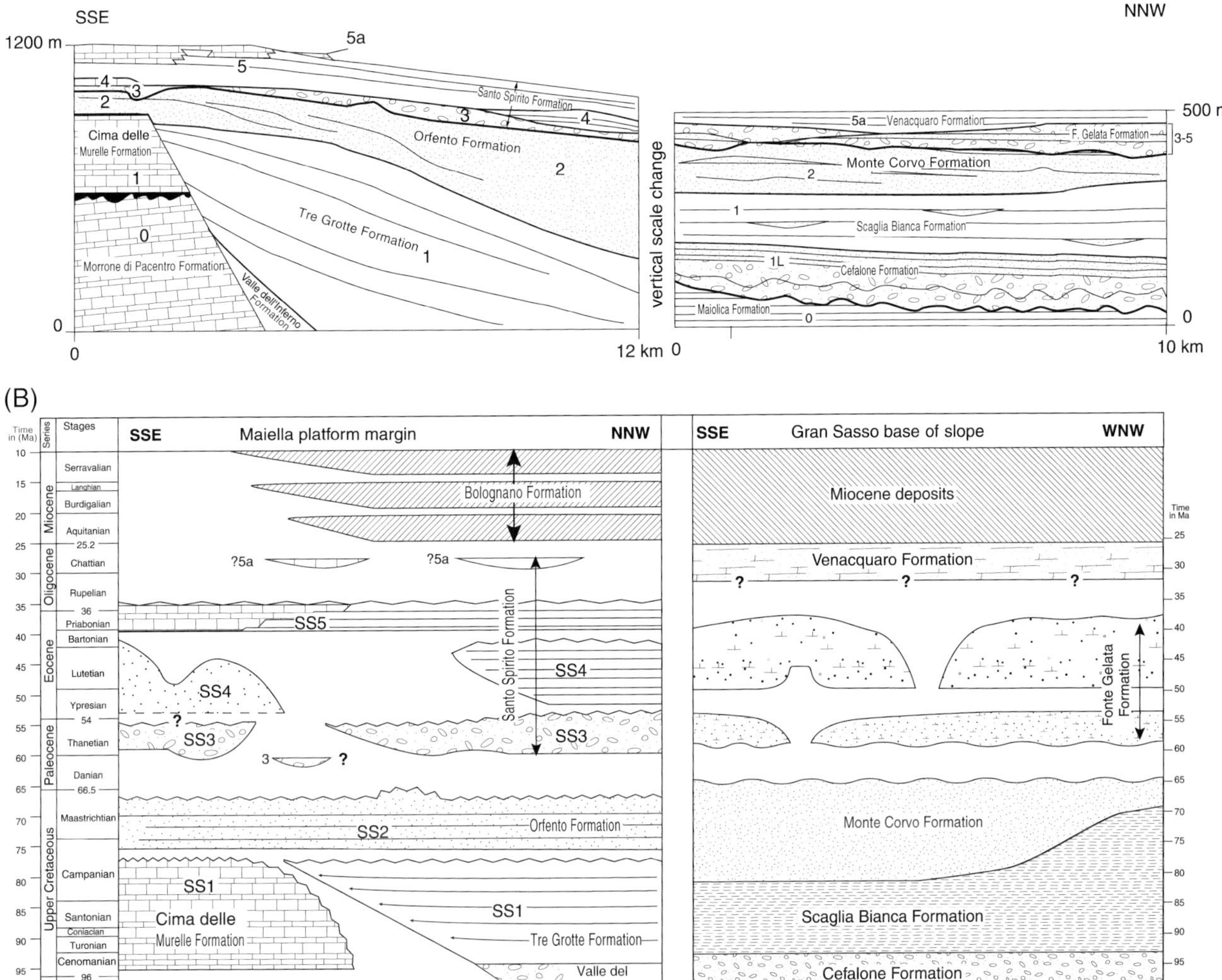

FIGURE 6. (A) Stratigraphic architecture and (B) chronostratigraphy and time-space diagram of the Maiella Platform margin and the Gran Sasso base of slope. (Reprinted by permission of Vecsei (1991), Eberli et al. (1993), Sanders (1994), Mutti et al. (1996), and Van Konijnenburg (1997)). The two areas are not physically connected and correlation is based on biostratigraphy and facies evolution. Numbers refer to the supersequences defined in the Maiella Platform margin and the correlative units in base-of-slope setting in the Gran Sasso area (modified from Van Konijnenburg et al., 1999).

Ginsburg, 1979). These smaller features are part of the depositional processes associated with the progradational nature of the carbonate margins.

The seismic facies of escarpment-bounded platforms is characterized by the coexistence of the seismic facies of the platform interior and the near-vertical juxtaposition of onlapping reflections of the adjacent deep-water sediments (Corso et al., 1988). Such a facies transition is buried in the subsurface of western Great Bahama Bank but both seismic facies have never been drilled. In the Maiella Mountains, the platform with the escarpment and the adjacent basin sediments are exposed and a synthetic seismic section across this transition is compared to the seismic image of western Great Bahama Bank (Vecsei, 1991; Eberli et al., 1993; Anselmetti et al., 1997; Stössel, 1999).

Seismic Facies of the Platform Interior of the Great Bahama Bank

Three seismic facies are observed in the aggrading platform interior of the Great Bahama Bank. The most

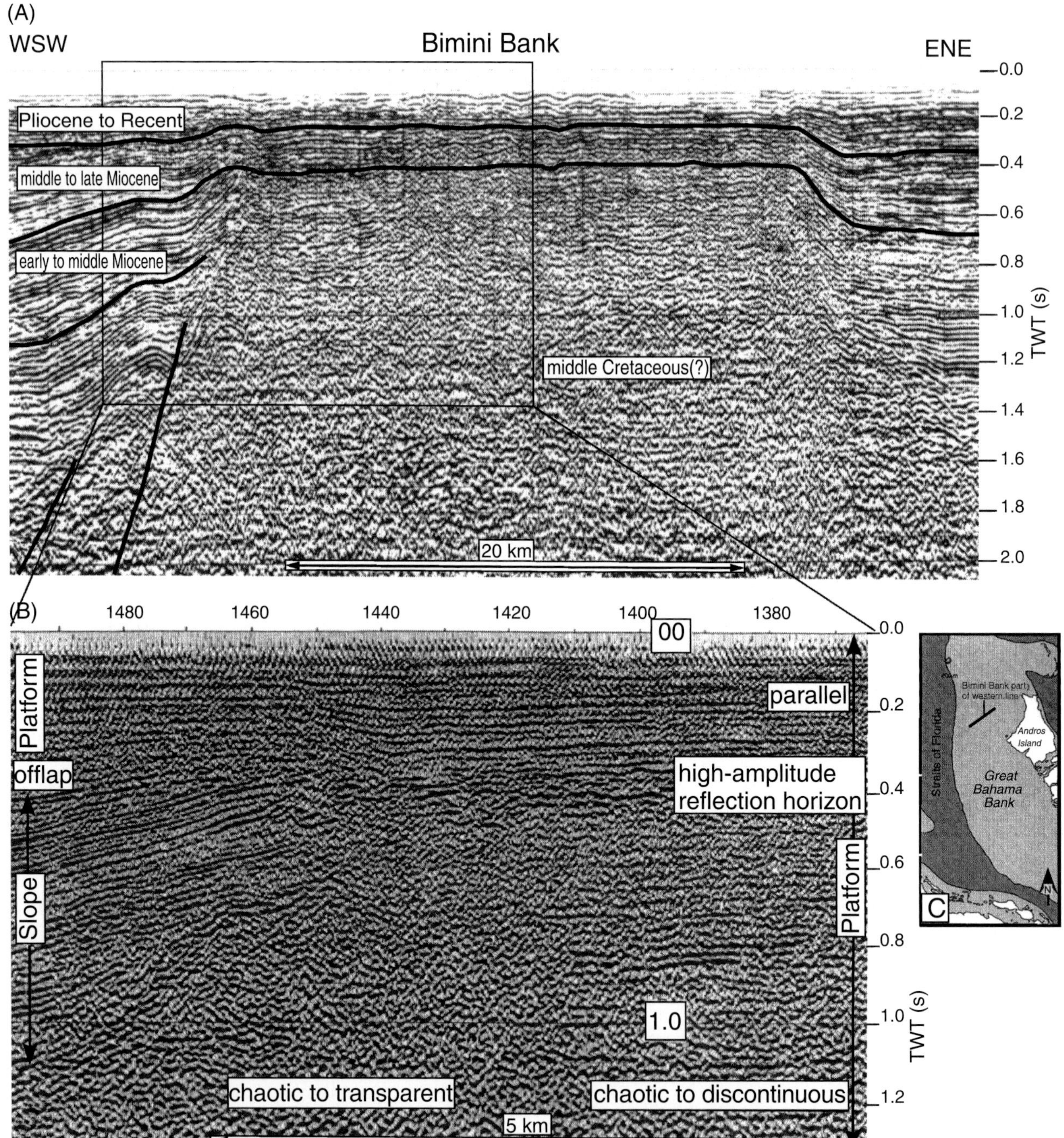

FIGURE 7. Seismic sections across Bimini Bank displaying the seismic facies of platform interior. (A) Squeezed section across the entire Bimini Bank shows the steep-sided morphology of the bank and the seismic facies difference to adjacent slope facies. The platform interior is characterized by two seismic facies. The lower part of the bank is imaged as transparent to chaotic seismic facies but a few high-amplitude reflection horizons can be distinguished. The Tertiary top part of the bank is imaged by horizontally layered continuous reflections. (B) Close-up of the western margin of Bimini Bank. The parallel continuous reflections at the top are separated from the chaotic to transparent facies by a high-amplitude reflection horizon that is early middle Miocene in age. Short and discontinuous high-amplitude reflections in the chaotic facies are generally horizontal. Platform progradation in the middle Miocene expands the platform interior over the inclined continuous high amplitude reflections of the slope sediments. TWT = two-way traveltime.

dominant, both in thickness and lateral extent, is the transparent to chaotic seismic facies (Figure 7). The transparent to chaotic seismic facies is approximately 2 s two-way traveltime (TWT) or 5 km thick. It overlies a group of regional high-amplitude reflections interpreted as the Triassic to Middle Jurassic(?) volcaniclastics, evaporites, and carbonates (Figure 8; Masaferro and Eberli, 1999). In the northwestern part of the bank, the chaotic facies grades laterally into more continuous low-amplitude reflections (Eberli and Ginsburg, 1989). Throughout the rest of the Great Bahama Bank, more continuous high-amplitude seismic reflections occur locally within the chaotic seismic facies that vary in continuity and strength. The most prominent of these reflection events is observed at about 1-s-TWT depth in the northern Great Bahama Bank, where it is seen as discontinuous, partially hummocky interval of approximately 0.1 s TWT. Toward the western margin along the western seismic line, the horizon is reduced to one reflection (Figures 3, 7).

The correlation of a time-depth-converted seismic line to the lithologies of the nearby Doubloon Saxon well shows that the dominant transparent to chaotic seismic facies of the platform interior is composed of dolomites and limestones (Figure 8; Walles, 1993; Masaferro and Eberli, 1999). The more continuous seismic horizons in the Cretaceous part (Barremian–middle Aptian) correlate in the core to an interlayering of evaporites with dolomites and limestones. Obviously, the density and velocity differences between carbonates and evaporites cause enough impedance contrast to generate high-amplitude reflections.

The transparent to chaotic seismic facies is bounded at the top by an interval of high-amplitude reflections that form a second seismic facies in the aggrading platform. These reflections change laterally from continuous to discontinuous and more hummocky toward the margin (Figure 7B). In the Great Isaac area of northwestern Great Bahama Bank, a high-amplitude reflection marks the top of a Cretaceous platform succession, while the underlying platform carbonates 200 m thick are imaged as a discontinuous to chaotic seismic facies. High-amplitude reflections below this seismic facies can be correlated with carbonate-anhydrite alternations observed deeper in the well bore (Schlager et al., 1988). At this location, the Cretaceous shallow-water carbonates

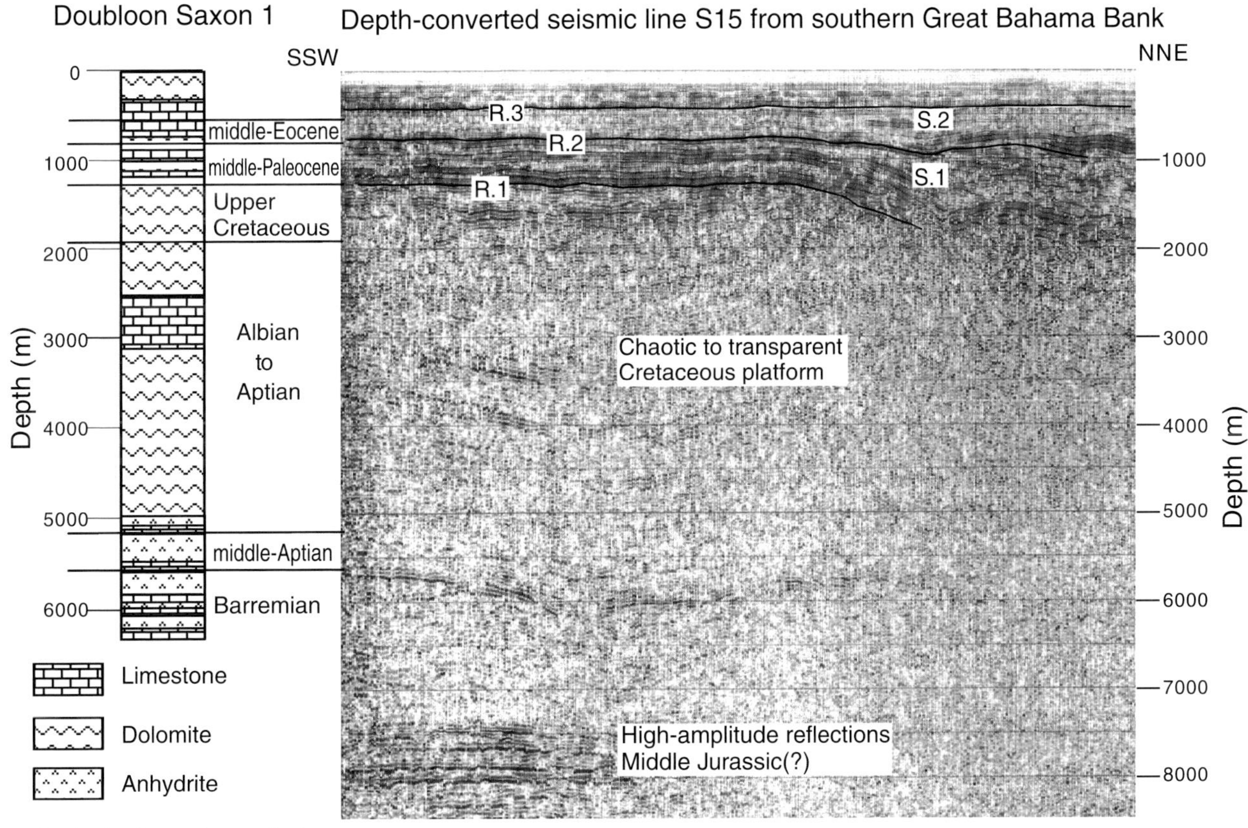

FIGURE 8. Seismic facies of platform interior displayed in depth-converted seismic section from southern Great Bahama Bank correlated to the lithology and ages of the Doubloon Saxon well described by Walles (1993). High-amplitude continuous seismic reflections (Middle Jurassic(?)) form the acoustic basement. Chaotic to transparent seismic facies are Cretaceous carbonates. The continuous reflections within this facies correlate to anhydrite horizons or limestone-dolomite transitions. R.1–R.3 = high-amplitude continuous correlation horizons; S.1 = high-amplitude continuous seismic reflections (Upper Cretaceous–middle Paleocene); S.2 = transparent low-amplitude seismic facies (upper Paleocene–Holocene). For location, see Figure 2 (modified from Masaferro and Eberli, 1999).

and evaporites are overlain by deep-water chalks that are Late Cretaceous to Miocene in age (Schlager et al., 1988).

A high-amplitude horizon typically separates the chaotic seismic facies from the third seismic facies, which is characterized by continuous low-amplitude seismic reflections, which form approximately the top 0.4 s TWT of the seismic section of the Great Bahama Bank (Figures 7, 8). Within this top unit, intraplatform depressions occur locally that are usually filled by prograding clinoforms (Eberli and Ginsburg, 1987, 1989; Masaferro and Eberli, 1999). In the Bimini Embayment, these clinoforms are sigmoidal, whereas in the Straits of Andros and in the southern Great Bahama Bank, the fill is mostly complex sigmoid-oblique (Figure 8; Eberli and Ginsburg, 1988, 1989; Masaferro and Eberli, 1999).

Buried Escarpment of Western Great Bahama Bank

Buried beneath the top two seismic facies of the Great Bahama Bank occurs an abrupt seismic facies transition from the transparent to chaotic seismic facies to inclined reflections with relatively high amplitude and continuity (Figure 9). In the lower part, the transition is wider and both slope and platform seismic facies are nearly chaotic. Occasional horizontal reflections in the chaotic platform facies indicate that the horizontally layered platform strata reach this facies boundary (Figure 9). In the upper part, the inclined reflections onlap the chaotic platform facies.

This near-vertical facies transition is similar in seismic expression to buried escarpments in the Bahamas-Florida region (Bryant et al., 1979; Corso et al., 1988; Denny et al., 1994; Masaferro, 1997). Thus, the boundary between the chaotic platform facies and inclined slope reflections is interpreted to be the seismic image of a buried escarpment. The height of the escarpment is approximately 0.5 s TWT, which converts to approximately 800 m, assuming a 3.2-km/s interval velocity of the platform carbonates. The escarpment is situated above a deep-rooted fault that displaces the older strata down to the west (Figure 3). The escarpment existed along western Great Bahama Bank throughout the Late Cretaceous, Paleogene, and the earliest part of the Neogene (Figures 3, 9). After the burial of the escarpment

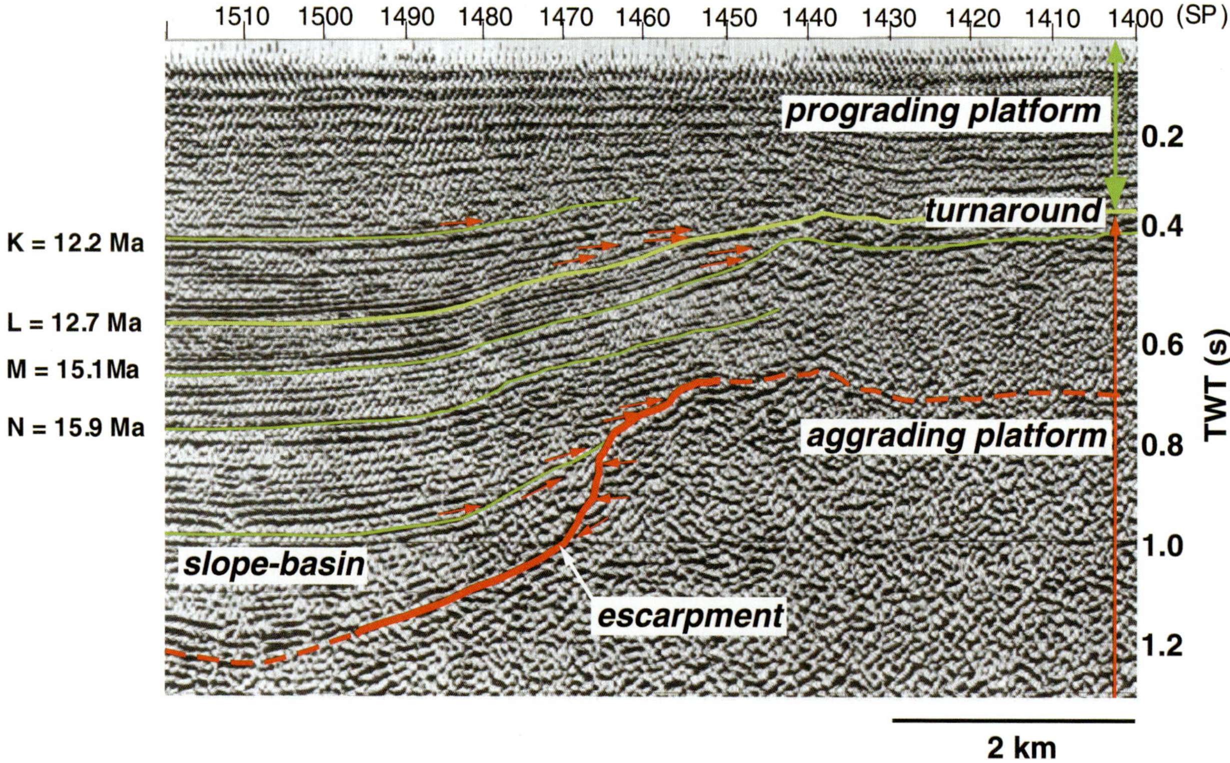

FIGURE 9. Part of seismic section within Great Bahama Bank displaying the buried platform escarpment and a constructional margin that leads to the turnaround from aggradation to progradation of the Great Bahama Bank. The escarpment was approximately 800 m high and separates the chaotic seismic facies of the bank from the continuous seismic reflections of the slope. In contrast, in the constructional margin, the seismic facies transition from platform to basin is more gradual. Letters to the left indicate seismic sequence boundaries. SP = shotpoints; TWT = two-way traveltime.

in the earliest Miocene, a less pronounced but still steep constructional margin persisted along the vertically aggrading platform before the platform prograded in the middle Miocene (Figure 9).

Correlation of Seismic and Lithologic Facies of the Great Bahama Bank Platform

Lithologies in the six deep cores on Great Bahama Bank indicate a general succession of limestones and dolomites in the Lower Cretaceous and few evaporites in the Upper Cretaceous part. Density and velocity differences between the evaporites and the carbonates obviously produce the reflectivity for high-amplitude reflections in the otherwise chaotic to transparent seismic facies of the platform interior. The low reflectivity of large parts of the shallow-water platform seems to be related small impedance contrasts within the platform succession.

The seismic character of the third seismic facies, the low-amplitude parallel reflections at the top of the Great Bahama Bank, is, however, not caused by carbonate-anhydrite alternations. No evaporites were recovered in the shallow drill holes through the Neogene section of the Great Bahama Bank (Beach and Ginsburg, 1980; Beach, 1982; Manfrino and Ginsburg, 2001). Boreholes Unda and Clino, which are located on the western seismic line, correlate these low-amplitude parallel reflections to vertically stacked parasequences of shallow-water carbonates, each capped by a horizon of subaerial exposure (Manfrino and Ginsburg, 2001; Eberli et al., 2001). In this highly porous platform succession (Lucayan Formation), most of these parasequences have a coarse skeletal grainstone or molluskan rudstone base overlain by shallow-marine lithologies that vary widely in their composition (Beach and Ginsburg, 1980; Beach, 1993; Kievman, 1998). Furthermore, the compositional variations, in conjunction with subsequent diagenesis, produce large vertical variations of porosity and velocity that cause the impedance contrasts for continuous low-amplitude reflections (Eberli et al., 2001).

The truncation of the platform interior facies at the escarpment compares well to the unconformable nature of the modern Bahamas Escarpment where erosional processes have removed the frontal parts of the marginal facies and beds are exposed with interior and backreef facies (Freeman-Lynde et al., 1981; Freeman-Lynde and Ryan, 1985; Paull et al., 1991).

Platform Interior Facies of the Maiella Platform Margin

The aggrading platform interior of the Maiella Platform margin is subdivided into Lower and Upper Cretaceous deposits (Figure 5A; Crescenti et al., 1969; Accarie, 1988; Vecsei, 1991; Eberli et al., 1993; Sanders, 1994; Stössel, 1999). They are separated by an irregular surface of subaerial exposure with evidence of intense karstification. Karst holes were filled with breccia, large speleothems, and, locally, bauxite. The duration of this middle Cretaceous hiatus is late Albian to middle Cenomanian in age (Crescenti et al., 1969; Accarie, 1988). A discordance of about 5° between the Lower and Upper Cretaceous deposits indicates slight tilting in addition to relative uplift.

The facies types of the Lower Cretaceous platform limestones (Morrone di Pacentro Formation) indicate low-energy platform interior and higher-energy platform margin environments (Crescenti et al., 1969). Bedding is indistinct, but the arrangement of beds in packages as much as 20 m thick suggests cyclic deposition. Each cycle typically starts with bioturbated or massive subtidal wackestones followed by laminated or cross-bedded peloidal grainstones and peritidal fenestral mudstones. The top of a cycle is formed either by a horizon of reworked lithoclasts and black pebbles or a thin clayey layer (Stössel, 1999).

During the Late Cretaceous up to the early Campanian (Cima delle Murelle Formation), the platform margin was nonrimmed and vertically aggrading (Figures 10, 11). At the outer part of the platform, rudist biostromes and bioclastic grainstone dominate the facies assemblage (Crescenti et al., 1969; Accarie, 1988; Sanders, 1994; Stössel, 1999). The inner and more protected part of the platform is made up of finer-grained sediment and carbonate mud (Sanders, 1994; Stössel, 1999). The western flank of the Cima delle Murelle, located at the northern edge of the Maiella Platform margin, displays an approximately 1-km-long cross section through the platform interior (Figure 10). At this location, the Cima delle Murelle Formation comprises approximately 250 m of coarse-grained bioclastic grainstone-rudstone that are intercalated with several rudist biostromes. The succession is cyclic and consists of bioclastic grainstone that grade upward into coarse, poorly sorted rudstone to floatstone, capped by rounded and cross-bedded bioclastic grainstone and rudstone (Sanders, 1994; Stössel, 1999).

A major sea level fall in the late Campanian exposed the platform. Calcrete formed and extensive vadose and phreatic meteoric diagenesis caps the cyclic platform deposits (Mutti, 1995). During the subsequent transgression, the platform margin backstepped, and slope deposits of the Orfento Formation overlie the platform succession (Figures 5, 11). These sediments, called *Calcare Critallini* by Italian authors, are composed of well-sorted bioclastic sand of mainly rudist debris with high porosity and little cementation (Anselmetti et al., 1997). The Orfento Formation forms a wedge that thickens into the basin and shows an overall progradational geometry (Figures 5, 11; Eberli et al., 1993; Mutti et al., 1996).

(A)

(B)

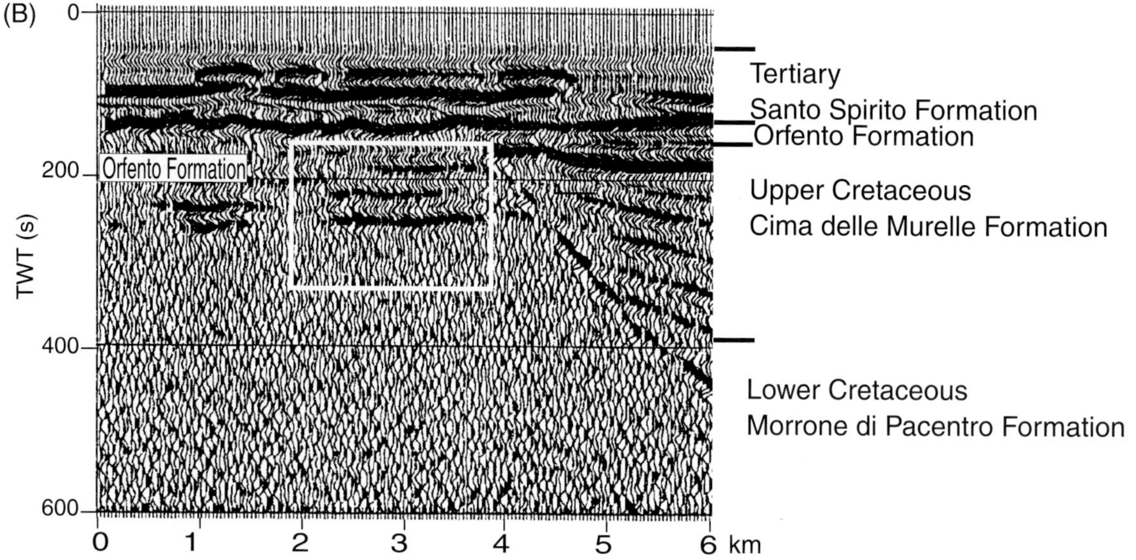

FIGURE 10. (A) Photograph of Cima delle Murelle displaying the cyclic sedimentation of shallow-water carbonates of the Cenomanian to lower Campanian Cima delle Murelle Formation. The cliff is approximately 270 m high (modified from Stössel, 1999). (B) Part of the synthetic seismic section of the platform interior and the adjacent slope. White rectangle outlines approximate position of outcrop. The horizontally bedded Upper Cretaceous strata are imaged as three reflections in a seismic transparent zone. TWT = two-way traveltime.

Maiella Escarpment

The Maiella Platform margin was bordered at its northern edge by an escarpment that, at the end of the Early Cretaceous, was about 1000 m high (Figures 11–13). The Maiella escarpment strikes east-west with an irregular undulating surface and a north dip of approximately 35° (Figure 12; see Crescenti et al., 1969; Accarie, 1988; Eberli et al., 1993). Horizontally bedded platform carbonates are truncated at the escarpment. The Lower Cretaceous beds in the escarpment wall lack a pronounced rim facies but contain back-reef facies with peritidal carbonates and oolithic grainstones (Vecsei, 1991). In the Upper Cretaceous part of the escarpment, rudist biostromes and related facies are exposed (Figure 11).

The escarpment is onlapped and buried by three stratigraphic units. From the bottom going up, these units are the Valle dell'Inferno, Tre Grotte, and Orfento Formations. The Valle dell'Inferno Formation is a wedge of talus breccia (middle to late Cenomanian) that onlaps the eastern part of the Maiella escarpment (Figures 5, 13). It is a 240-m-thick unit of steeply inclined beds (30–50°) that form overlapping lobes of coarse grainstone and

Figure 11. (A) View of the western side of Valle dell'Orfento displaying major features of the Maiella Platform margin. (B) In the foreground is the Cretaceous platform, which is onlapped along the escarpment by slope sediments. The overlying uppermost Cretaceous sediments level the relief between platform and slope. The top of the mountain is formed by upper Eocene slope and lower Oligocene reefal units, which prograde over slope and shelf deposits. Paleocene and lower–middle Eocene sediments are present only as small relics not distinguished in detail. Vertical relief from escarpment at lower right corner to the mountaintop is approximately 1200 m (from Anselmetti et al., 1997).

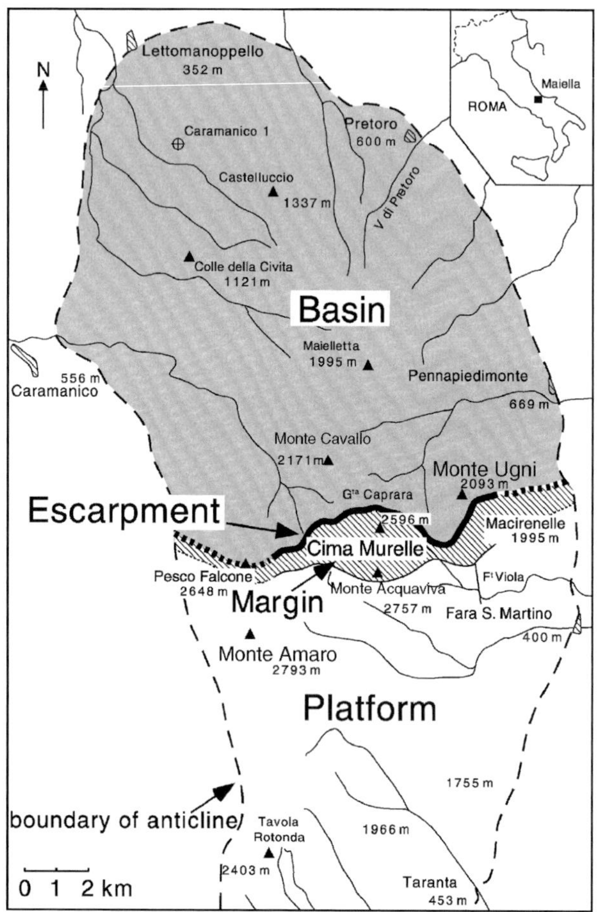

FIGURE 12. Map of the Maiella Platform margin escarpment illustrating the undulating nature of the escarpment. The east-west–striking Cretaceous escarpment separates the Maiella Platform margin from the basinal section to the north (modified from Accarie et al., 1986).

breccia (Vecsei, 1991). The grainstone is mostly fragments of rudists and admixtures of orbitilinids and ooids, whereas the breccia components consist of platform rocks of Early Cretaceous age (Vecsei, 1991). Onlapping this wedge and most of the rest of the escarpment are megabreccias, calcareous turbidites, and periplatform sediments of the Tre Grotte Formation (Turonian–late Campanian; Accarie, 1988; Vecsei, 1991). The beds slope from a few degrees to approximately 10°. The bulk of the basinal deposits are calcareous turbidites and pelagic deposits, but, within the 800-m-thick package, six megabreccia units occur that range in thickness from 10 to 50 m. Components of the matrix-poor, commonly amalgamated breccias include fragments of rudist biostromes and debris and minor amounts of micritic limestone with fauna from the back-reef environment (Vecsei, 1991). The top 150 m of the escarpment is buried by the upper Campanian–Maastrichtian Orfento Formation that has a detrital character (Vecsei, 1991; Mutti et al., 1996). Lime mudstones and wackestones form the fine-grained background sediment in which bioclastic, calcareous turbidite beds and breccias are deposited. The components of the megabreccias contain lithoclasts and bioclastic detritus from the Orfento Formation itself and subordinately from outer shelf facies of the older platform (Cima delle Murelle Formation; Mutti et al., 1996).

Synthetic Seismic Facies of the Maiella Platform Interior and Escarpment

Synthetic seismic sections across the Maiella Platform margin explain the seismic facies of the carbonate-platform margin system (Figures 5, 14; Anselmetti et al., 1997). The overall horizontally bedded and cyclic facies of the Cretaceous Maiella Platform margin are expected to generate a synthetic seismic section with a series of horizontal reflections. The resulting synthetic seismic sections, however, display a mostly transparent to chaotic seismic facies in the platform, which is onlapped along the escarpment by a succession of high-amplitude slope reflections (Figure 14B). The reason for the transparent to chaotic seismic facies of the platform interior is the low reflection coefficient in the platform carbonates that is masked by the noise (10%) in the synthetic seismic section. Recognition of a reflection through the noise requires a certain amplitude level. Jones and Nur (1982) assume that a reflection coefficient higher than ±0.05 causes a reflection with amplitude that can be seen on most seismic sections. Modern seismic techniques with high signal-to-noise ratios might even image layer boundaries with a coefficient of only ±0.02. The Lower Cretaceous platform section, however, shows a succession of layers with velocity values that range between 6250 and 6400 m/s and densities between 2.68 and 2.7 g/cm^3, resulting in reflection coefficients of ±0.016 (Anselmetti et al., 1997). These low-impedance contrasts and reflection coefficients cause reflections with very low energy. As a consequence, the layers cannot be detected when noise is present on the synthetic sections (Figure 14D). The result is a seismically transparent to chaotic Cretaceous platform (Figure 14B, C), despite the considerable vertical and lateral extent of the layers.

The Upper Cretaceous platform, which has somewhat higher impedance contrasts because rudist biostromes with high porosity and low velocity intercalate with grainstone with lower porosity and lower velocity. Rudist biostromes have an average porosity of 14.2% and an average velocity of 5566 m/s, whereas the other platform carbonates have an average porosity of 4.6% and a velocity of 6044 m/s. These slight impedance differences produce some low-coherence reflections that are partly visible through the randomly generated noise (Figures 10B, 14). The middle Cretaceous unconformity,

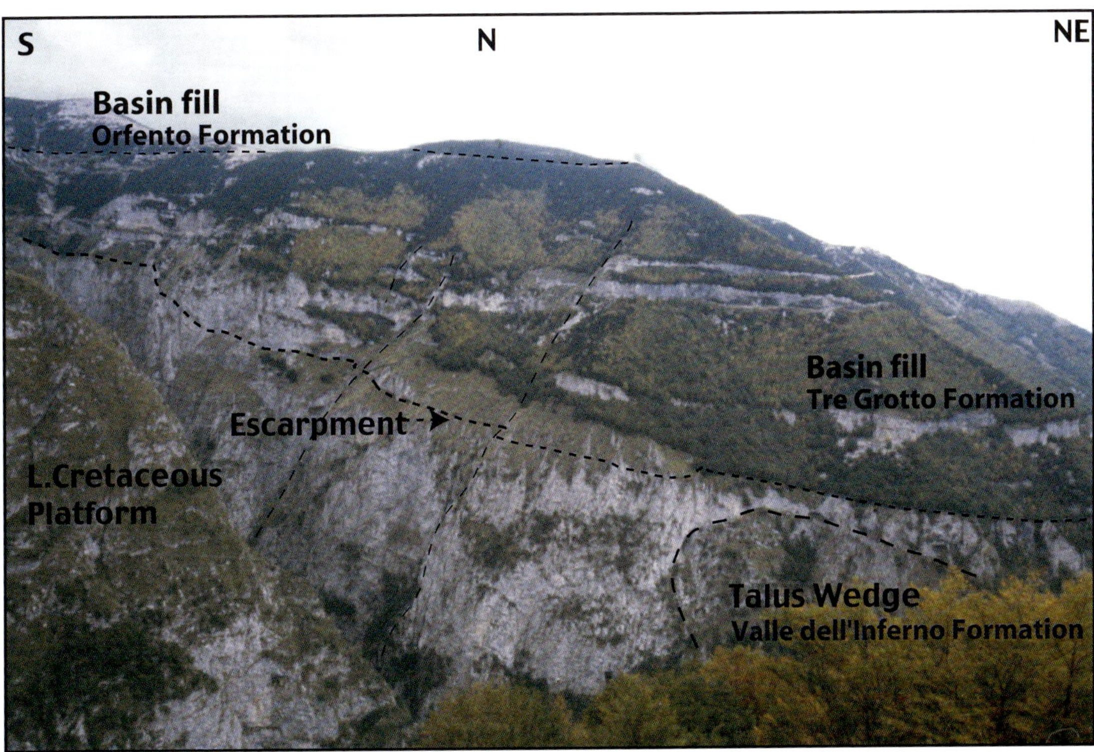

FIGURE 13. Lower part of the Maiella escarpment at Monte d'Ugni. The escarpment is onlapped by a talus wedge (Valle dell'Inferno Formation) and two formations consisting of megabreccias, turbidites, and hemipelagic sediments (Tre Grotte and Orfento Formations). The amalgamated megabreccias form cliffs in the forested slope. Horizontal view is approximately 1.8 km; vertical view is approximately 1 km.

marked by pockets of low-velocity rocks, is also imaged by a noncontinuous reflection. At dominant frequencies above 40 Hz, the reflection from this unconformity has significantly higher amplitude than reflections from the rudist biostromes (Figure 14). The top of the platform, which is overlain by the low-velocity Orfento Formation, forms a major contrast in acoustic impedance. This contact yields a high-amplitude reflection on the synthetic sections. The modeled karstic nature of this contact is visible on the seismic section by a wavy, but still continuous, reflection (Figures 5, 14).

Despite its irregular and steeply dipping geometry, the platform escarpment is imaged clearly on all synthetic sections (Figures 5B, 14, 15). Regardless of the chosen frequency, polarity, or gain, the escarpment produces its own reflection and is recognized by the termination of onlapping slope reflections. It separates high-velocity platform rocks from lower-velocity slope sediments and is characterized by a high-amplitude reflection (Figures 14, 15). The ray tracing of all reflections from the escarpment shows that most of the escarpment is imaged by rays that have angles of incidence between 10 and 45° to the vertical (Anselmetti et al., 1997). The irregular surface produces bundling and divergence effects, in particular at concave and convex niches in the platform wall that, on real seismic sections, is likely to be the origin of numerous diffractions. These diffractions are not simulated properly with the modeling program. The oblique nature of the ray paths also results in a relatively large offset between the position of the escarpment on the unmigrated seismic section and its true position. On real 2-D seismic sections, a correct migration of such a setting would be rather difficult, because the relief of the steep escarpment is quite complicated (Anselmetti et al., 1997).

Comparison of Synthetic Seismic Facies of the Aggrading Parts of the Maiella Platform Margin to Those of the Great Bahama Bank

The seismic facies of the Maiella Platform margin correlates well with the seismic facies observed on seismic lines across Great Bahama Bank (Figures 3, 7, 8). In particular, the nearly transparent to chaotic seismic facies of Lower Cretaceous part is similar. In addition, the middle Cretaceous correlation horizon has a similar high-amplitude discontinuous character. In both cases, a high-amplitude reflection marks the top of the Cretaceous platform part, although the overlying strata also consist of carbonates. Furthermore, the Tertiary parts in both platforms generally have more continuous seismic reflections (Figure 15). The continuous shallow-water deposition was interrupted in both areas

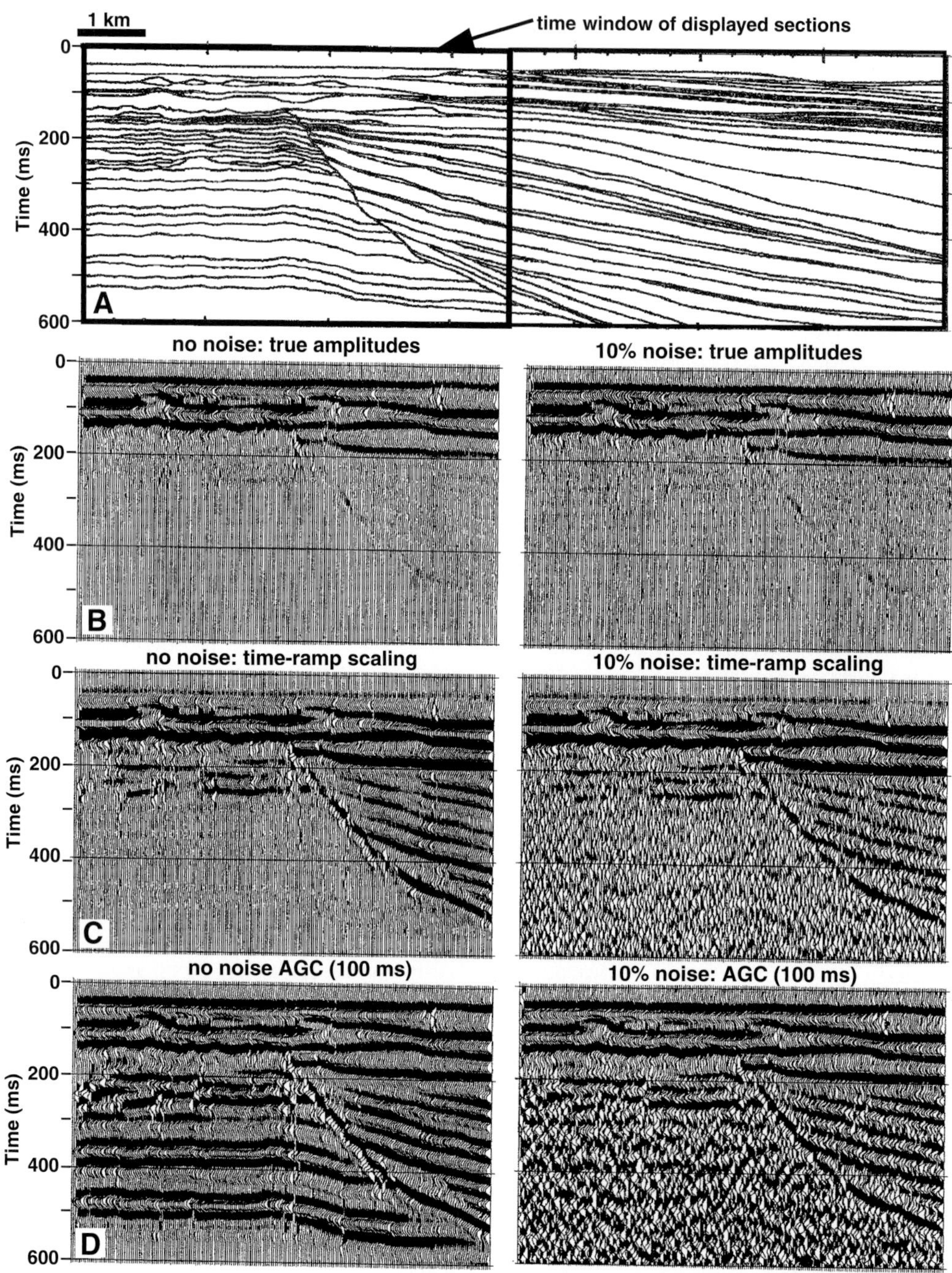

FIGURE 14. Synthetic seismic section across Maiella Platform margin with a comparison of different noise and amplitude-gain options. (A) Impedance model converted to time. The frame marks the displayed synthetic sections of B–D. The sections on the left are seismograms with no noise, whereas those on the right are disturbed by randomly added noise (10% of average amplitude). (B) True amplitude sections without amplitude gain. Both synthetic sections are similar, and only shallow reflections can be recognized. (C) Sections with time-ramp scaling. The power function with an exponent of 1.2 results in a realistic image; the high-amplitude reflections on the slope are well displayed, whereas the Lower Cretaceous platform is relatively transparent, with or without added noise. The Upper Cretaceous platform shows some internal structure (rudist biostromes). (D) Sections with applied automatic gain control (AGC) of 100 ms. The AGC keeps the total amplitude within a time-window constant. Without noise, the extremely low-impedance contrasts of the Lower Cretaceous platform ($R = \pm 0.016$) are amplified distorting the seismic image (from Anselmetti et al., 1997).

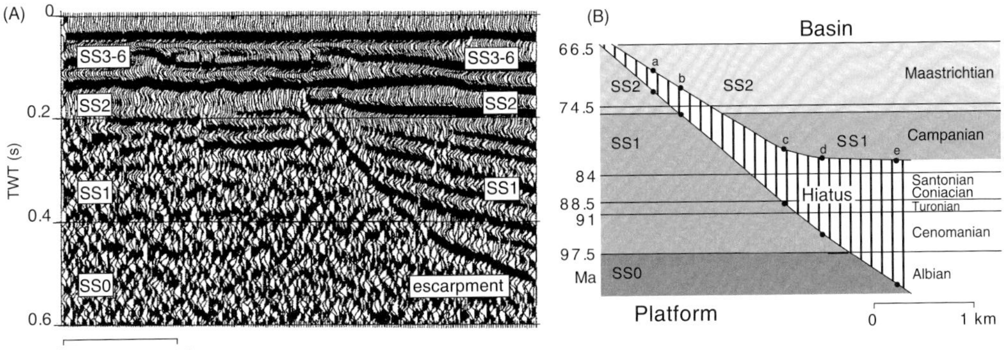

FIGURE 15. (A) The synthetic seismic section displays the escarpment as a continuous reflection, separating transparent to chaotic platform interior seismic facies from the inclined continuous slope facies. (B) Stratigraphic relationship between platform growth and onlapping sequences along the Maiella escarpment. The occurrence of coeval Cretaceous platform and basin sediments documents the growth of the escarpment throughout the Cretaceous. The decrease of the hiatus from the bottom to the top of the escarpment is a result of the progressive burial of the escarpment by basinal sediments (modified from Vecsei, 1991). TWT = two-way traveltime.

in the middle Cretaceous. The Maiella Platform margin became exposed, and bauxite formed in a humid climate before the platform reestablished (Eberli et al., 1993; Anselmetti et al., 1997). In the Bahamas, during the middle Cretaceous event, several platforms drowned and margins stepped back as a result of environmental and tectonic stress (Schlager et al., 1988; Eberli, 1991). Although the sedimentary expression of this event differs in both areas, it obviously resulted in the vertical juxtaposition of lithologies with different impedance contrasts, creating a regional seismic reflection.

A nearly transparent facies of the Cretaceous platform part is also characteristic for the Florida Platform and the drowned platforms along the Campeche escarpment (Corso et al., 1988; Macurda, 1988, 1997). Based on the modeling results of the Maiella Platform margin, we speculate that porosity and velocity variations of these platform carbonates are too small to produce an impedance contrast that is sufficient to overcome the signal-to-noise ratio. In the Middle East, shales that produce high-impedance contrasts separate the low-amplitude Cretaceous carbonate shelf packages. In the Bu Hasa field, for example, the Shuaiba Formation in the aggrading southern area is characterized by weak, discontinuous reflections that become chaotic toward the shelf margin. The prograding northern area in contrast shows shingled clinoforms with alternating weak and strong reflections (Fitchen, 1997).

In the Tertiary section of both platforms, the seismic reflection continuity is higher. In the Maiella Platform margin, this increased reflectivity is a result of the vertical juxtaposition of slope, reefal, and platform facies with highly variable porosity and velocity values. Cores from the western margin of the Great Bahama Bank also show a high degree of facies variability caused by backstepped margins and subsequent progradation (Eberli et al., 2001). These shifts in facies and diagenetic belts ultimately result in a more heterogeneous vertical facies succession with different velocity and porosity that produce higher reflectivity and more continuous seismic facies. Drowned Tertiary platforms in southeast Asia also show generally continuous reflections in the platform interior (Epting, 1989; Grötsch and Mercadier, 1999). Nevertheless, areas of chaotic facies are often observed in the aggrading parts of such platforms (Bachtel et al., 2004).

Comparison of the Bahamas-Florida and Maiella Escarpments

A chaotic seismic facies is a common characteristic of steep carbonate margins (Schlager, 1992). The chaotic facies can be related to the occurrence of reefs at the platform edge or, in the case of escarpments, to the energy dispersion of the seismic energy by the bundling of the seismic rays on the irregular escarpment wall and on the steep onlapping talus wedge (Dillon et al., 1988). The effect of dispersion is strongest on unburied escarpments whereas escarpments that are buried completely by basinal sediments tend to produce a discernible reflection. Examples for both situations are found along the Florida Escarpment (Corso et al., 1988). In many cases, the talus wedge produces chaotic seismic facies that makes it difficult to draw the precise boundary between an escarpment and the platform (Corso et al., 1988; Dillon et al., 1988). The Great Bahama Bank escarpment is such an example.

With regard to geometry and facies, the Maiella escarpment is very similar to modern exhumed escarpments. The undulating nature, steep slope angle, and irregular shape are characteristic features of escarpments. Dredge samples along modern escarpments record platform interior facies (bank-interior, back-reef, or restricted peritidal-lagoonal facies) outcropping at the escarpment (Paull and Dillon, 1980; Freeman-Lynde et al., 1981; Freeman-Lynde and Ryan, 1985; Paull et al., 1990, 1991). The facies in the Maiella escarpment also document the erosion of the marginal facies, reaching occasionally the lagoonal facies. The absence of high-energy margin or reef-belt facies in the escarpment is commonly interpreted to indicate various amounts of margin erosion, in the case of the Blake Plateau, as much as 15 km of headward erosion (Paull and Dillon, 1980; Corso et al., 1988). The problem with this interpretation is the lack of documented debris in the adjacent deep-water areas. Along the Maiella escarpment, the erosional products are found in the basal talus wedge and the megabreccias. Although spectacular as beds, volumetrically, these breccias are a relatively small part of the basin fill. The small volumes of platform material imply that the margin facies are very narrow along steep escarpment-bounded platforms.

The age correlation between platform and basin indicates that the platform grew coeval with the sedimentation of the onlapping basinal sediments (Figure 15; Vecsei, 1991). The occurrence of breccias containing platform clasts in all three onlapping units also suggests that erosion continuously shaped the escarpment during its growth. The component assemblages document that back cutting of the platform margin occasionally reached the back-reef environment. The main center of deposition of the erosional products is a narrow belt along the platform. Boreholes, 20 km basinward of the escarpment, contain no coarse-grained redeposited beds (Crescenti et al., 1969).

Erosion of the platform appears to be caused primarily by defacement along scallops (Eberli et al., 1993). Accarie et al. (1986) proposed synsedimentary faulting as a trigger mechanism but the escarpment is clearly not a fault. No fault breccias in the escarpment or as components in the megabreccias are present. The lithologic and stratigraphic evidence indicates that the escarpment was maintained and grew with approximately the same declivity throughout the Late Cretaceous by a combination of platform aggradation and erosional processes.

Prograding Platform Margin

Isolated platforms can vary between an aggrading to a prograding mode depending on interplay between relative sea level rise and rate of sediment production (Schlager, 1992; Bosellini, 1984; Read, 1985; Eberli, 1991). The Maiella Platform margin underwent this transition in the Eocene (Eberli et al., 1993). At Great Bahama Bank, the turnaround from aggradation to progradation of the occurred at approximately 12.7 Ma (Figures 9, 16). Since then, the margin has advanced

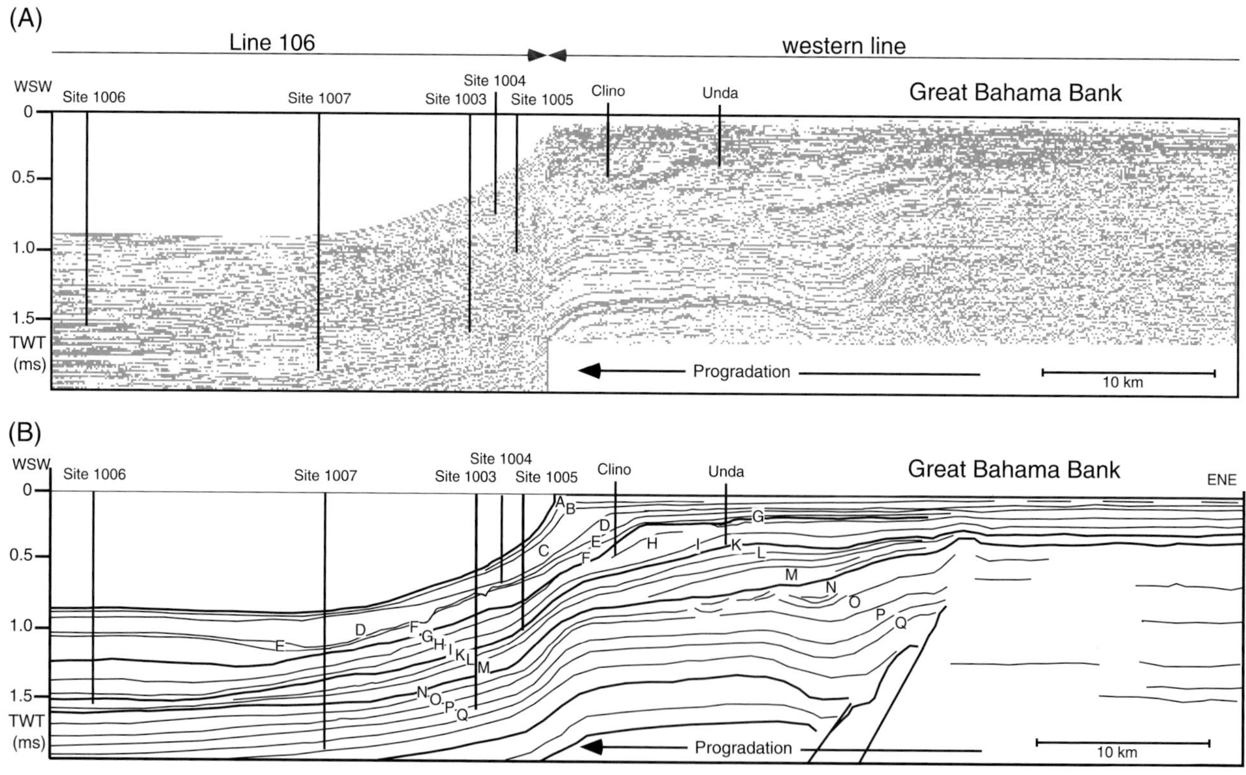

FIGURE 16. (A) Composite seismic line (western seismic line and seismic line 106 to the west) of prograding margin of the Great Bahama Bank with boreholes drilled along this transect during the Bahamas Drilling Project (Unda and Clino) and the ODP Leg 166 (Sites 1003–1007). Progradation started in the late Miocene and advanced the margin approximately 27 km into the Straits of Florida. (B) Line drawing of the major reflections and seismic sequences. Note how the platform margin steepens during sequences F–A (modified from Eberli et al., 1997). TWT = two-way traveltime.

more than 25 km into the Straits of Florida (Figures 3, 16). On the seismic section, this turnaround can be identified by the expansion of the chaotic seismic facies westward over continuous reflections of the upper-slope facies (Figure 9).

Prograding Margin of the Great Bahama Bank

Figure 16 displays the seismic image of the prograding margin of western Great Bahama Bank. On a large scale, the onset of the progradation is identified with the development of clinoform geometries over inclined slope reflections. At the platform margin, the turnaround from aggradation to progradation is indicated by offlapping reflections and the expansion of the chaotic platform facies across continuous slope reflections (Figure 9). Although the progradation is an impressive 25 km over the last 12.7 m.y., unconformities and onlap patterns subdivide the prograding clinoforms into sequences, indicating pulsed progradation. The prograding sequences display a generally sigmoidal shape with increasing steepening of the foresets in the younger sequences (Figure 16). Internally, the prograding sigmoids show horizontal high-amplitude topset reflections, chaotic to discontinuous or low-amplitude reflections in the uppermost part of the foresets, and high-amplitude reflections in the rest of the foresets. The uppermost foreset is the platform margin, which shows the seismically diverse reflections geometries. Near-horizontal reflections alternate with steep-dipping reflections of various lengths and amplitudes (Figure 17A). The geometry becomes concave downslope, reflection packages thin and some downlap geometries. These foreset reflections are generally high in amplitude but have intercalations of low-amplitude packages. These low-amplitude packages also have low reflection strength (Figure 17B). The modern platform margin is the latest of prograding clinoforms and consists of a near-vertical margin or wall (Ginsburg et al., 1991) with onlapping highstand deposits (Figure 17C). A high-resolution seismic line across the margin reveals the geometry of the youngest prograding pulse (Figure 17C). The onlapping wedge is a nearly transparent seismic facies that thins basinward as some reflections downlap onto the underlying high-amplitude reflection. This high-amplitude slope reflection seems to be connected to the low-amplitude reflection that images the wall. Below the high-amplitude slope reflection, the couplet of near-transparent seismic facies overlaying a high-amplitude slope reflection is repeated (Figure 17C). This lower-slope reflection can be followed underneath the wall into the bank.

Cores at Ocean Drilling Program (ODP) Sites 1005, 1008, and 1009 calibrate this seismic image. The onlapping transparent package consists of Holocene periplatform ooze (Eberli et al., 1997; Malone et al., 2001). At ODP Site 1009, the high-amplitude slope reflection corresponds to a submarine hardground with borings and encrusting organisms. At this site, seven successions of well-lithified layers separated by periplatform sediments were encountered and can be related to the alternation of high- and low-amplitude reflection patterns (Eberli et al., 1997). Age dating indicates that the well-cemented layers formed during glacial lowstand, whereas the less-cemented intervals were deposited during sea level highstands (Malone et al., 2001). The cemented lowstand units can reach velocities of 4–5 km/s, whereas the highstand sediments have much lower velocities that start at 1.6 km/s and increase downcore (Eberli et al., 1997). On the seismic image, the platform-derived-material highstand deposits form low-amplitude to nearly transparent highstand wedges that thin basinward and are separated by high-amplitude seismic reflections of well-cemented slope intervals.

Farther in the platform, core borings Unda and Clino allowed for facies calibration with the prograding seismic sequences (Figures 16, 18). The prograding clinothems in both cores consist of a tripartite succession of slope deposits, overlain by a reefal zone, which is overlain by shallow-water carbonates. The tripartite facies succession is correlated to the seismic facies in Figure 18 at the Clino well. In Clino, the top succession of shallow-water facies of the platform interior (21.6–98.45 meters below sea floor, or mbsf) is seismically imaged by three to four horizontally layered medium-amplitude seismic reflections. Lithologically, this seismic facies is composed of 10 vertically stacked parasequences of shallow-water carbonates, each of which is capped by a subaerial exposure horizon (Manfrino and Ginsburg, 2001).

The seismic facies of the reefal zone is low-amplitude, discontinuous to chaotic and transparent (Figure 18). In this reefal unit (197.44–98.45 mbsf), there is an upward progression from deep-reef/foreslope to fore-reef/reef crest (173 m) and finally to back reef. The major development of reef growth began on unconsolidated upper-slope sediments (Kenter et al., 2001; Manfrino and Ginsburg, 2001). At borehole Unda, which is located 8.5 km farther to the east, Pliocene corals are common, whereas the corals in Clino are largely Pleistocene to Holocene in age (Manfrino and Ginsburg, 2001). The younger ages at the more distal site, Clino, corroborate the prograding nature of the margin inferred from the geometries on the seismic section.

The fore-reef/reefal zone is deposited over a 479.3-m-thick succession of slope deposits (Kenter et al., 2001). The slope facies is composed of monotonous fine-sand to silt-sized skeletal and peloidal grains with some coarser intercalations. The interval between 197.44 and 367 m consists of alternating layers, decimeters to meters thick, of lime mudstone, peloidal packstone to

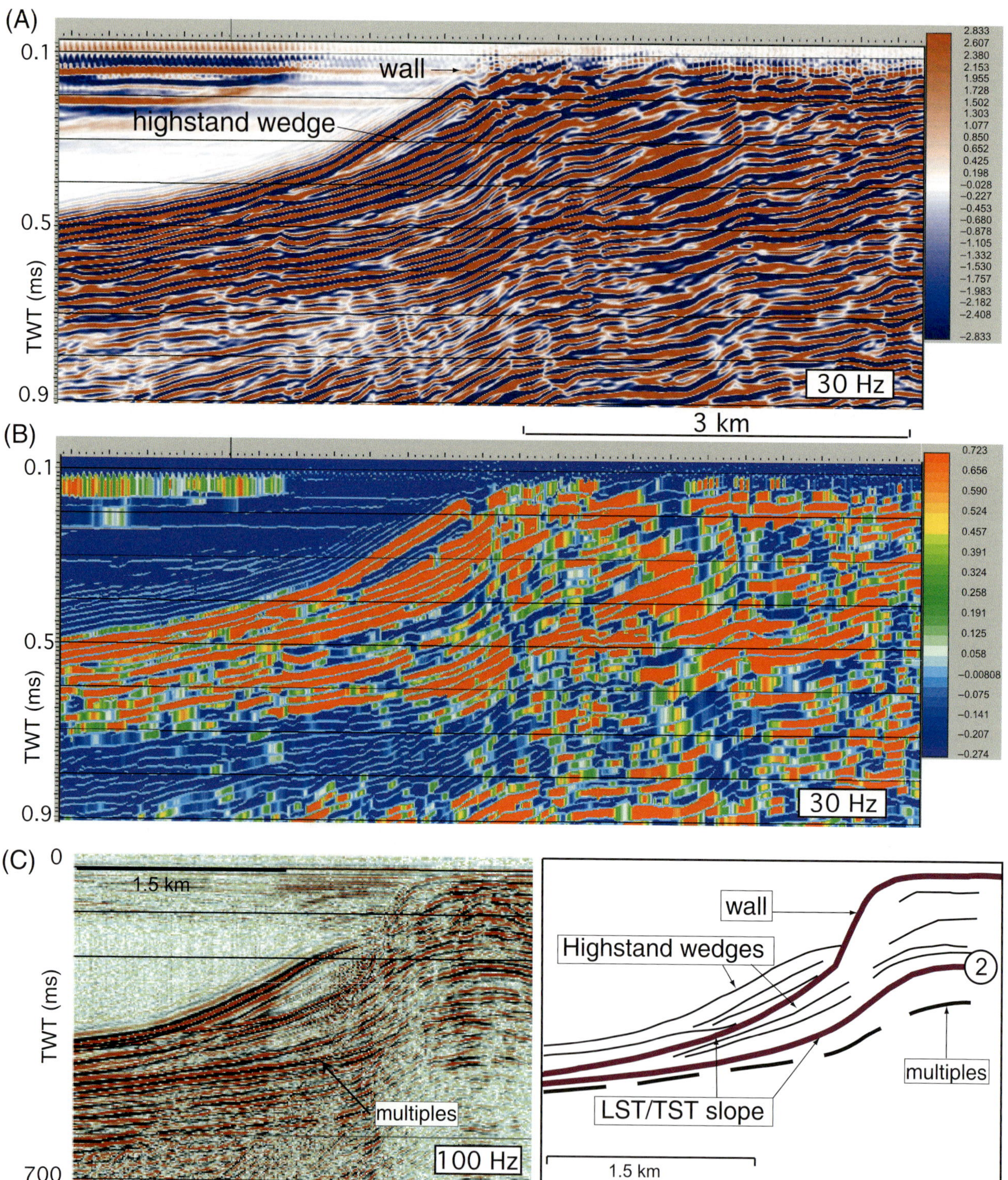

FIGURE 17. The prograding margin of the Great Bahama Bank. (A) The amplitude display (30 Hz, western line) of the margin reveals the sigmoidal shape of the prograding clinoforms. Horizontally layered high-amplitude reflections of the topset grade into discontinuous inclined reflections of the upper foreset and farther downslope into continuous high-amplitude reflections. (B) The display of the reflection strength reveals the intercalation of low-reflection-strength packages (dark blue) between high-amplitude upper-slope reflections. (C) High-resolution seismic data (~100 Hz) images the low-amplitude packages and between strong slope reflections and the vertical wall of the modern platform edge. Pull-up of inclined slope reflections (2) indicates high-velocity rocks at the modern platform edge. LST = lowstand systems tract; TST = transgressive systems tract.

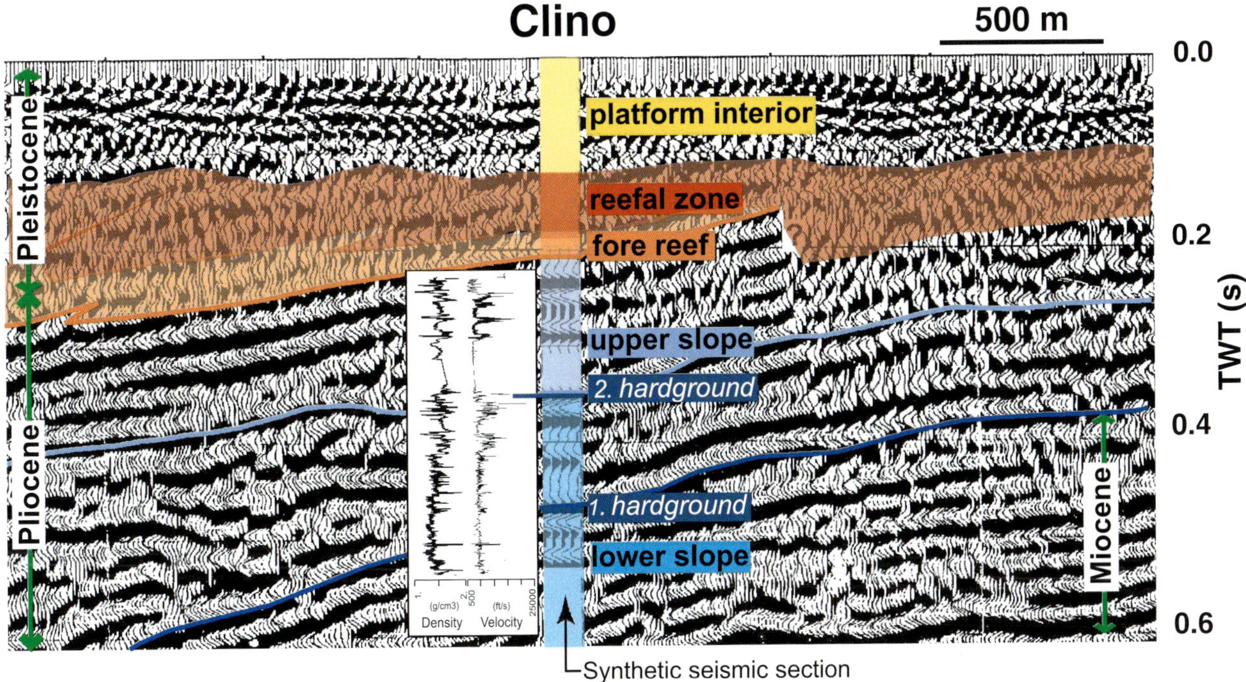

FIGURE 18. Close-up of seismic line with lithologies of borehole Clino and interpreted outline of the reefal zone away from the well. Inclined high-amplitude, continuous reflections characterize the slope, transparent to discontinuous seismic facies correlates with a reefal zone, and the shallow-water facies of the platform interior are imaged as horizontally layered continuous reflections. TWT = two-way traveltime.

grainstone, and minor skeletal packstone to grainstone. Below 367.03 m, the section is made up of meter-scale, alternating laminated wackestone to packstone, and porous to tightly cemented packstone to grainstone. Grain composition consists of mixed peloids and skeletal grains, with 5–10% planktonic foraminifera. Twelve intervals of coarser-grained skeletal packstone to grainstone are intercalated in the fine-grained slope facies. Two of these intercalations overlay discontinuity surfaces that indicate significant periods of nondeposition and/or erosion on the slope at 536.33 and 367 m, respectively (Kenter et al., 2001). The discontinuity surface at 536.33 m is a phosphatic hardground ("1. hardground" in Figure 18); the one at 367 m is a partially eroded hardground ("2. hardground" in Figure 18). The upper-slope deposits contain few gravity mass flows. In the lowermost part of the slope sections, graded beds are interpreted as turbidites, but most of the deposition on the slope is from fallout of offbank-transported sediment (Kenter et al., 2001).

Seismically, the slope sections are characterized by inclined, high-amplitude, and continuous reflections. Several reflections can be related to zones of intense cementation rather than to significant changes of facies. The highest-impedance contrasts are observed at sequence boundaries where sediment composition changed and diagenetic alteration occurred, for example, at fine-grained sediments with hardgrounds or firm grounds that are overlain by coarse-grained sediments. Thus, differences in cementation and slight variations in composition cause the high reflectivity of these relatively homogenous slope sediments (Eberli et al., 2001).

Prograding Margin of the Maiella Platform

The Maiella Platform margin experienced two phases of progradation; one in the latest Cretaceous and one in the late Eocene to Oligocene. The Cretaceous progradation occurred after burial of the escarpment in the Late Campanian–Maastrichtian (Mutti et al., 1996; Vecsei, 1998; Vecsei et al., 1998). A prograding wedge (Orfento Formation, Supersequence 2) transformed the Maiella Platform margin into a distally steepened ramp (Figure 5). Within the prograding wedge, shingled sigmoidal sequences composed mostly of rudist debris are separated by erosional unconformities. Prograding units are dominated by grainstone and packstone of the upper and lower shoreface environments that overlay mass gravity-flow deposits like megabreccias and turbidites (Mutti et al., 1996; Vecsei, 1998). Seismically, the Cretaceous prograding unit is imaged as a basinward-thickening wedge, in which the sigmoidal nature of the individual sequences is not resolved (Figure 19C).

Progradation of the Maastrichtian platform was terminated by subaerial exposure at the end of the Cretaceous and during part of the Paleocene (Accarie, 1988; Vecsei, 1991). No extensive shallow-water platform is preserved on the former platform from the Latest Maastrichtian through "middle" to late Ypresian. However, small coralgal reefs of Danian to early Thanetian and of late Thanetian age are observed as large olistoliths on the slope about 4 km seaward of the former escarpment (Moussavian and Vecsei, 1995). Great volumes of Thanetian reef clasts also fill younger channels and document that shallow-water reefs on top of the Cretaceous platform were eroded (cf. Figure 26; Vecsei et al., 1998). Middle(?)–late Ypresian to Bartonian intervals consist of alveolinid rudstone and floatstone, but subsequent erosion nearly removed them (Bally, 1954).

The deeply eroded platform was rapidly flooded in the Middle Eocene, resulting in an extensive backstepping of the platform margin. In the late Eocene and Oligocene, reefs developed and prograded approximately 4 km basinward (Supersequence 5, upper part of Santo Spirito Formation). In the exposed part of the Maiella Platform margin, this progradation is recorded in a 180-m-thick supersequence of skeletal sands with several small patch reefs of corals and hydrozoans (Figure 19A). The bioclastic sands are grainstone to packstone that contain benthic foraminifera, algae, bryozoans, and other skeletal fragments (Figure 19A; Vecsei, 1991; Sanders, 1994). In the early Oligocene, coral reefs (Pesco Falcone Formation) bloomed and eventually prograded approximately 4 km basinward over these fine carbonate sands (Eberli et al., 1993; Vecsei et al., 1998). Sea level fall in the Chattian exposed these reefs. To the north of the former escarpment, the reefs are capped by Aquitanian cross-bedded calcareous sands, consisting of bryozoan grainstone with glauconite and a variety of benthic foraminifera onlap (Figure 19B; Crescenti, 1969; Mutti et al., 1999). This change to temperate carbonates in the Mediterranean, which began approximately at 20 Ma, predates the major Neogene global-cooling step between 14 and 12 Ma (Mutti et al., 1999).

On the synthetic seismic section, the slope part of the Tertiary prograding unit is imaged by gently dipping, continuous, high-amplitude reflections (Figure 19C). The patch reefs, although small in size, are well visible as discontinuous reflections. This seismic character is a result of the strong impedance contrast between the reefs and the surrounding slope carbonates (Anselmetti et al., 1997). In contrast, the thick prograding lower Oligocene reefs exhibit internally a low-impedance contrast and produce a transparent facies above the slope sections (Figure 19C; Anselmetti et al., 1997).

Comparison of Prograding Margins

The tripartite succession of slope, reef, and shallow-water facies in the prograding western margin of the Great Bahama Bank is remarkably similar to the Tertiary progradation of the Maiella Platform margin. In particular, the rapid transition from slope to massive fore-reef/reef facies that is observed in the cores Unda and Clino is present over long distances along the Maiella Platform margin. The importance of corals in platform progradation is also documented by previous shallow borings on the leeward side of the bank (Beach and Ginsburg, 1980; Beach, 1982) and by single-channel seismic sections of the Pleistocene–Holocene platform margin of Little Bahama Bank (Hine et al., 1981) and into facies associations and seismic expressions of other prograding carbonate shelves (e.g., Pomar, 1993; Pomar et al., 1996; Harris and Saller, 1999; Osleger and Tinker, 1999). In all these examples, the tripartite succession of slope, reef, and shallow-water facies produces a characteristic seismic facies pattern. The slope facies is generally of high amplitude and continuous. Individual clinoform packages thin basinward. The reflection that marks the top of the lowstand package is a downlap surface for prograding clinoforms (Eberli and Ginsburg, 1989; Pomar, 1993). The fore-reef/reef interval is usually a seismically transparent zone because the well-cemented interval is petrophysically a massive unit with small impedance contrasts. Occasionally dipping reflections outline the bedding surfaces within these reef-dominated margins (Sarg, 1989; Tyrell and Davis, 1989). The transparent zone separates the horizontal reflections of the platform top from inclined slope reflections (Pomar, 1993; Harris and Saller, 1999). The shallow-water platform facies is imaged as high-amplitude seismic reflections in the marginal areas where a high-impedance contrast exists between the reef and the shallow-water facies.

FIGURE 19. Tertiary progradation of the Maiella Platform margin. (A) One hundred and eighty-meter-thick Eocene to Oligocene prograding unit at Pesco Falcone. The partly eroded Maastrichtian unit of rudist grainstones is overlain by middle to upper Eocene (Bartonian–Priabonian) slope deposits with local patch reefs. The slope is overlain by a prograding reefal belt that expanded the platform approximately 4 km during the Oligocene (early–middle Rupelian). (B) The distal reefs of the Oligocene unit are overlain by cross-bedded, Miocene grainstone containing bryozoan, red algae, forminifera, and phosphatic grains. (C) Seismic facies of the two prograding units on the synthetic seismic section. The Campanian–Maastrichtian prograding unit forms a basinward-thickening wedge with low-amplitude internal reflections. In the prograding Tertiary unit, patch reefs generate short high-amplitude reflections and the Oligocene reefal unit is mostly transparent. The coeval slope deposits are imaged by continuous high-amplitude reflections (modified from Anselmetti et al., 1997). TWT = two-way traveltime; HST = highstand systems tract; LST = lowstand systems tract.

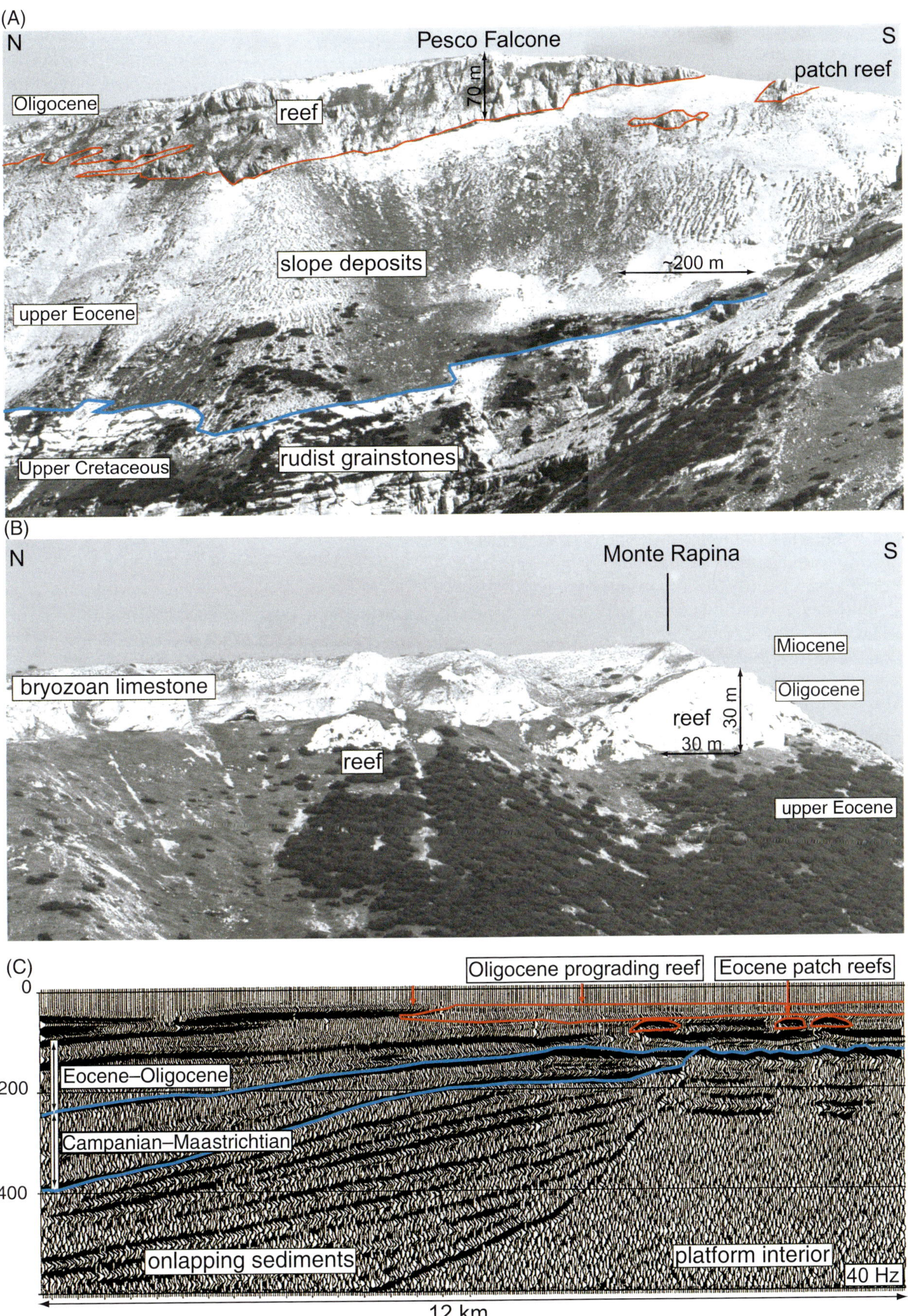
(A)
N
Pesco Falcone
S
patch reef
Oligocene
reef
70 m
slope deposits
~200 m
upper Eocene
Upper Cretaceous
rudist grainstones
(B)
N
Monte Rapina
S
Miocene
bryozoan limestone
Oligocene
30 m
reef
30 m
reef
upper Eocene
(C)
Oligocene prograding reef
Eocene patch reefs
0
200
400
Time (ms)
Eocene–Oligocene
Campanian–Maastrichtian
onlapping sediments
platform interior
40 Hz
12 km

Toward the platform, the amplitude usually decreases (Eberli and Ginsburg, 1989; Pomar, 1993). In cases where the back-reef/shelf interior facies contains evaporites and/or clastics (attached shelf), topset reflections maintain their amplitude (Harris and Saller, 1999).

Western Great Bahama Bank displays a characteristic steepening of the margin during the progradation (Betzler et al., 1999). The steepening is probably related to the increase of the amplitude of sea level change resulting from the onset of Northern Hemisphere glaciation during the Pliocene. During the transgression, the platform margin kept up with sea level and built a near-vertical margin in excess of 100 m (Hine and Mullins, 1983; Ginsburg et al., 1991; Grammer et al., 1993). Highstand sediments bypass the vertical wall and form characteristic onlapping highstand wedges on the upper slope (Glaser and Droxler, 1991; Grammer et al., 1993).

Slope

Slopes of isolated carbonate platforms are the main depositional areas for platform-derived sediment. Sediment composition, sedimentation rate, marginal erosion, and ocean currents control the morphology and slope declivity (Cook and Mullins, 1983; Kenter, 1990; Adams and Schlager, 2000). Mullins et al. (1984) made the first comprehensive description of a modern carbonate slope on the northern slope of Little Bahama Bank and related the depositional environments to the seismic facies. Their subdivision of an upper slope with canyons and a lower slope with hummocky morphology and chaotic seismic facies has proven to be a general pattern of carbonate slopes. Slopes receive their sediments by two mechanisms: fall-out from the water column of both pelagic and platform-derived material and deposition from mass gravity flows. On the upper slopes, fine-grained, offbank-transported periplatform ooze accumulates with a high sedimentation rate (Mullins et al., 1984; Eberli et al., 1997). Incised canyons act as funnels for mass gravity-flows that bypass this part of the slope. The gravity flows are deposited in shingled lobes at the middle to lower slope and as more continuous deposits at the toe of slope and the adjacent basin floor (Crevello and Schlager, 1980; Mullins et al., 1984; Betzler et al., 1999; Anselmetti et al., 2000). Along Great Bahama Bank and the Maiella Platform margin, these morphologic elements are imaged on seismic data and exposed in outcrops, respectively.

The slopes of the prograding system of western Great Bahama Bank are approximately 600 m high and show a concave-upward geometry with a maximum inclination of 8° (Adams and Schlager, 2000). The slope consists of a variety of seismic facies that are related to various facies and depositional processes along the slope profile (Figures 16, 20; Eberli and Ginsburg, 1989; Betzler et al., 1999; Adams and Schlager, 2000). The slope reflections begin at the massive transparent to chaotic seismic facies of the platform margin. In the Miocene to lower Pliocene part of the seismic lines, the slope angles decrease to less than 4° at the proximal position and to less than 2–3° in the distal part (Betzler et al., 1999). On the upper slope, high-amplitude reflections separate transparent to low-amplitude intervals and channel incisions produce downcutting high-amplitude reflections within a general low-amplitude seismic facies. In the middle to lower slope, continuity of seismic reflections is short, resulting in a discontinuous to chaotic seismic facies. The discontinuous character is partly caused by abundant small-scale channeling (Figure 20). At the toe of slope and farther basinward, continuity of the reflections increases and the more continuous reflections often have a slightly mounded geometry (Figure 20). In the following, three slope environments are described in detail and compared to the outcrop geometries of Maiella Platform margin and basin.

Upper Slope and Canyons of the Great Bahama Bank

On the upper slope of the Great Bahama Bank, the inclined reflections are variable in amplitude and continuity. High-amplitude reflections delineate sedimentary packages that downlap or thin basinward. A low-amplitude, nearly transparent unit about 100 ms thick that forms the early Pliocene slope is intercalated into these high-amplitude packages (Figures 21, 22). Along depositional strike, channels of variable size (Figure 22) dissect the upper and middle parts of the slope. The channels are preferentially aligned along stratigraphic horizons that are identified as sequence boundaries (Figures 16, 21; Anselmetti et al., 2000). For example, at the top of the seismically almost transparent lower Pliocene section, spectacular incisions with depths more than 100 m occur at the early–late Pliocene sequence boundary E/F (Figure 22). Many of the channels display a persistent cut-and-fill geometry (Figure 22). Although the early–late Pliocene boundary marks the most prominent period of channeling, many other Neogene sequence boundaries are also characterized by canyonlike incisions. These canyons are sometimes arranged in a complicated stacking pattern, resulting from lateral migration through time (Figure 22B). Other canyons can remain stable through time and simply stack vertically (Figure 22C). Even in sections in which major canyons are absent, the seismic facies of the slope deposits is characterized by frequent smaller-scale incisions, yielding incoherent and irregular seismic reflection patterns (Figure 21).

These seismic facies document the combined effect of erosional and depositional processes operating on

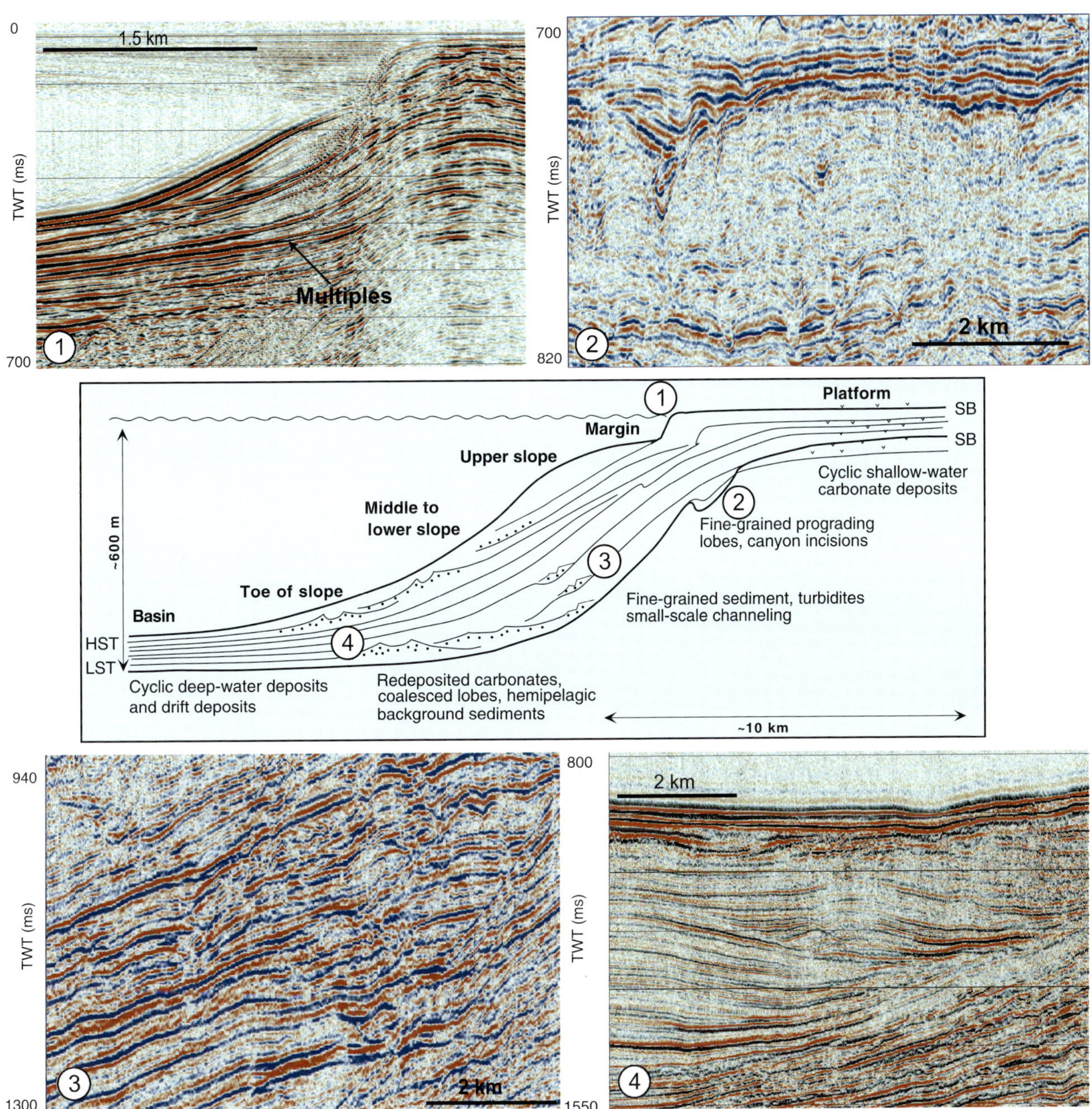

FIGURE 20. Overview of geometries and facies of the margin and slope of western Great Bahama Bank. Centerpiece is a schematic cross section from the margin and slope subenvironments to the basin floor. Four seismic images relate these environments to the seismic record. (1) Modern margin with the characteristic near-vertical wall and the cemented upper slope is onlapped by transparent facies of unconsolidated Holocene slope deposits. The strong reflection below the first transparent package is the cemented late Pleistocene surface. (2) Upper slope: Transparent facies of fine-grained slope deposits with high-amplitude channel incisions. (3) Middle to lower slope: Discontinuous high-amplitude reflections on channelized slope. (4) Basin-floor fan: Mounded external geometry with moderate amplitudes on the flanks (levees) and chaotic to transparent facies on top (feeder channel).

the slopes. At the modern margin of western Great Bahama Bank, a steep cemented slope with a declivity of approximately 40° extends seaward from the foot of the wall. The cemented slope is onlapped by a thick wedge of Holocene sediments that rapidly decreases in declivity to approximately 3.5° (Wilber et al., 1990; Betzler et al., 1999; Anselmetti et al., 2000). Upper-slope sediment is mostly periplatform ooze with few thin calcareous turbidites. At ODP Site 1005, the most proximal site drilled on the upper slope during ODP Leg 166, the sedimentary succession consists of unlithified to partially lithified wackestone and slightly coarser-grained intervals consisting of packstone and grainstone (Figure 21). This pattern of cemented uppermost slope covered by

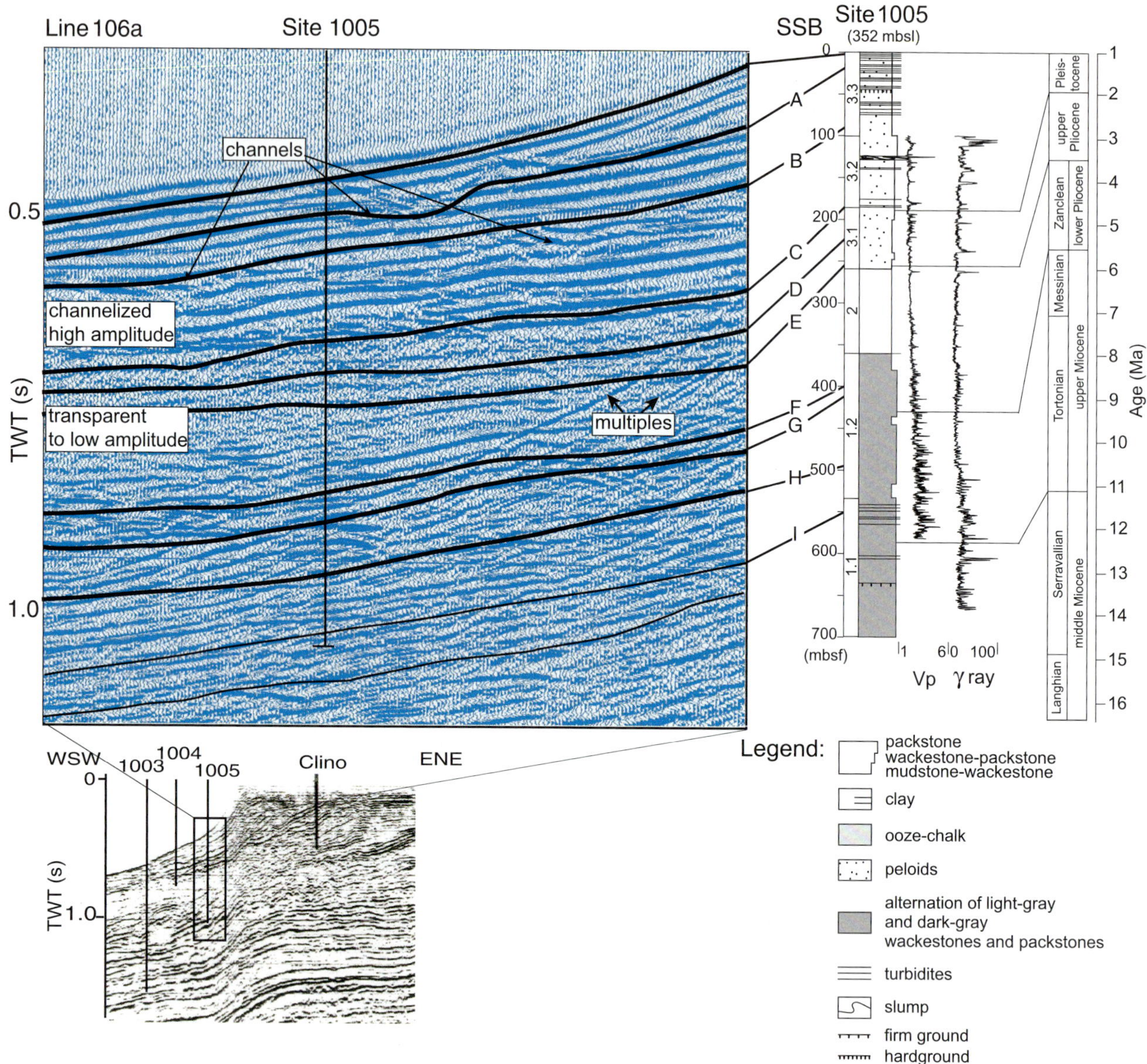

FIGURE 21. Close-up of dip line of the upper slope off western Great Bahama Bank correlated to the lithologies recovered at ODP Site 1005. The slope strata contain three seismic facies. Low-amplitude to transparent facies image homogeneous fine-grained, periplatform ooze packages. Continuous to discontinuous high-amplitude reflections reflect slope intervals with various small-scale channels. High-amplitude reflections forming downcutting unconformities are the base of submarine channels. Vp = velocity log; γ ray = gamma-ray log.

a thick wedge of periplatform ooze results from high-frequency sea level fluctuations. Compositional variations document an alternating pattern of bank flooding, concomitant shedding to the slope with periods of exposed banks, a shutdown of shallow-water carbonate production, and largely pelagic sedimentation (Eberli et al., 1997). During sea level lowstands when sedimentation rates on the slope are low, submarine cementation forms hardgrounds on the uppermost slopes (Grammer et al., 1993; Malone et al., 2001). Subsequent flooding of the bank and concomitant shedding onto the slopes allow for the deposition of thick wedges of platform-derived material (periplatform ooze and fine-grained turbidites) during sea level highstands (Droxler and Schlager, 1985; Grammer et al., 1993).

The canyons originate close to the platform edge. They extend to the base of the slope and are oriented perpendicular to the slope (Anselmetti et al., 2000). The increased rate of incisions under seismic sequence boundaries indicates that the currents that form these canyons are more vigorous during sea level lowstands. The canyon fill was not penetrated by cores on the western margin of the Great Bahama Bank, but a short core (16.5 m) into a similar incision was drilled on northern Little Bahama Bank at ODP Site 629. The facies of the fill consists of sandy carbonate ooze, lime

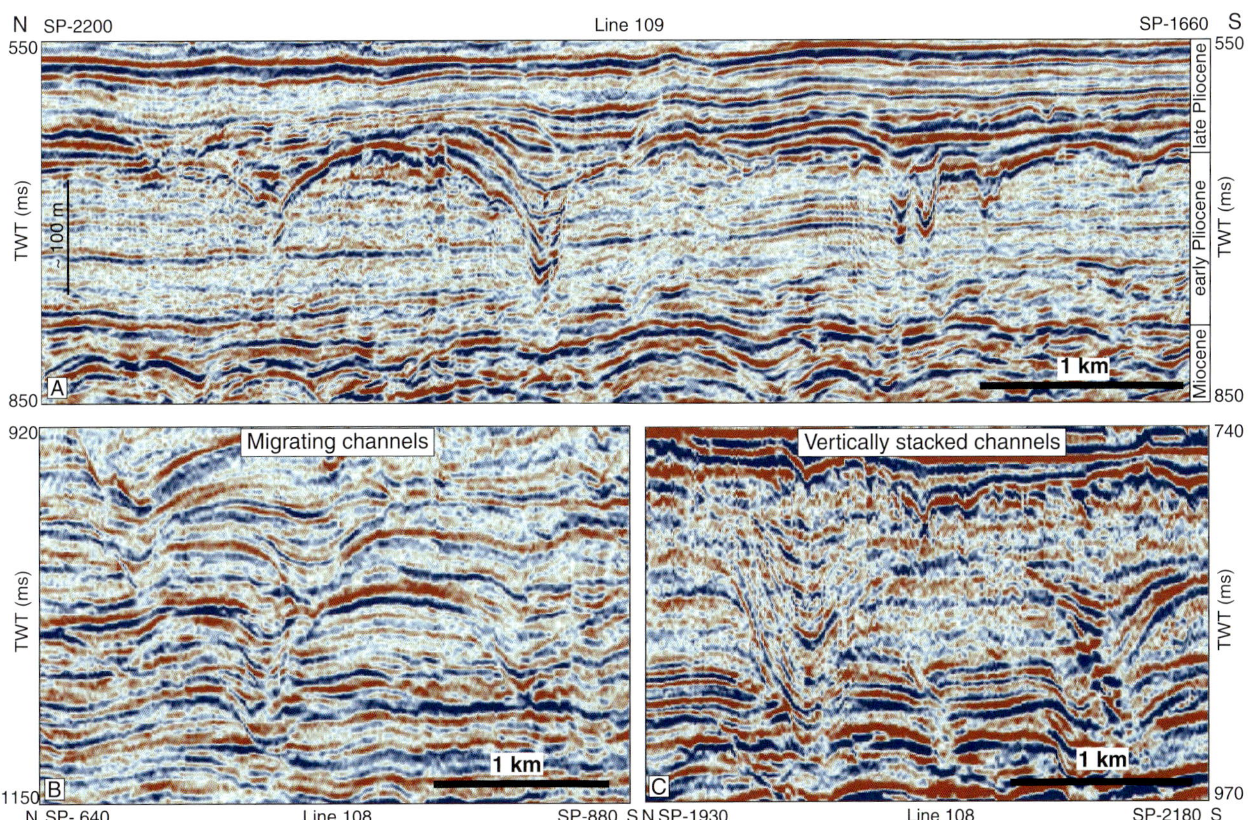

FIGURE 22. Strike seismic lines display the submarine channels that originate on the upper slope and run to the toe-of-slope of western Great Bahama Bank. (A) Submarine channels at the early-to-late Pliocene boundary cut into low-amplitude to transparent upper-slope sediments. The channels are spaced at approximately 1 km and have a maximum depths of 150 m. The channel fill shows a cut-and-fill geometry. An older level of incisions (~830 ms TWT [two-way traveltime]) occurs at the Miocene–Pliocene boundary. (B) A series of migrating channels of early Pliocene age results in a complex vertical and lateral stacking pattern. (C) Vertically stacked channels cut into low-amplitude to transparent facies of the early Pliocene slope deposits (modified from Anselmetti et al., 2000). TWT = two-way traveltime.

sand and rubble, and fragments of friable limestone, all of late Quaternary age (Austin et al., 1988). This composition indicates that the canyon fill consists of coarse carbonate debris derived from the platform and upper slope.

Middle-Lower to Toe-of-Slope Facies at Great Bahama Bank

The declivity of the lower and middle slope is approximately 2–3°. The seismic facies of the lower slope is characterized by partly short, discontinuous high-amplitude reflections and intercalations of low-amplitude packages. On the middle slope, small downcutting high-amplitude reflections that interrupt continuous reflections indicate abundant small-scale channeling (Figures 20, 23). In the lower slope and at the toe-of-slope, slightly mounded and shingled geometries are observed on both dip and strike lines (Figures 23, 24). Small-scale faults are also typical for the middle to lower slope (Figure 23). These faults are probably caused by slope instability and downslope movements of sediment. Such slope adjustment processes were also described from the Miocene of the northern flank of Little Bahama Bank (Harwood and Towers, 1988).

ODP Site 1003 recovered sediments from the middle slope and ODP Site 1007 penetrated the toe-of-slope and basin floor. These cores help to correlate the seismic with the lithologic facies (Figures 23–25). The dominant sediment of the Oligocene to lowermost Pliocene slope section is an alternation of light- and dark-gray wackestone and packstone. The light-gray wackestone and packstone contain shallow-water allochems and planktonic foraminifera. In contrast, the dark-gray wackestone and packstone contain planktonic, benthic foraminifera, and clay, but no shallow-water bioclasts. Siliciclastic content is between 2 and 10%, and at Site 1007, between 10 and 20% (Eberli et al., 1997; Betzler et al., 1999; Frank and Bernet, 2000). Cyclicity was driven by high-frequency sea level changes. Light-gray layers containing shallow-water bioclasts were

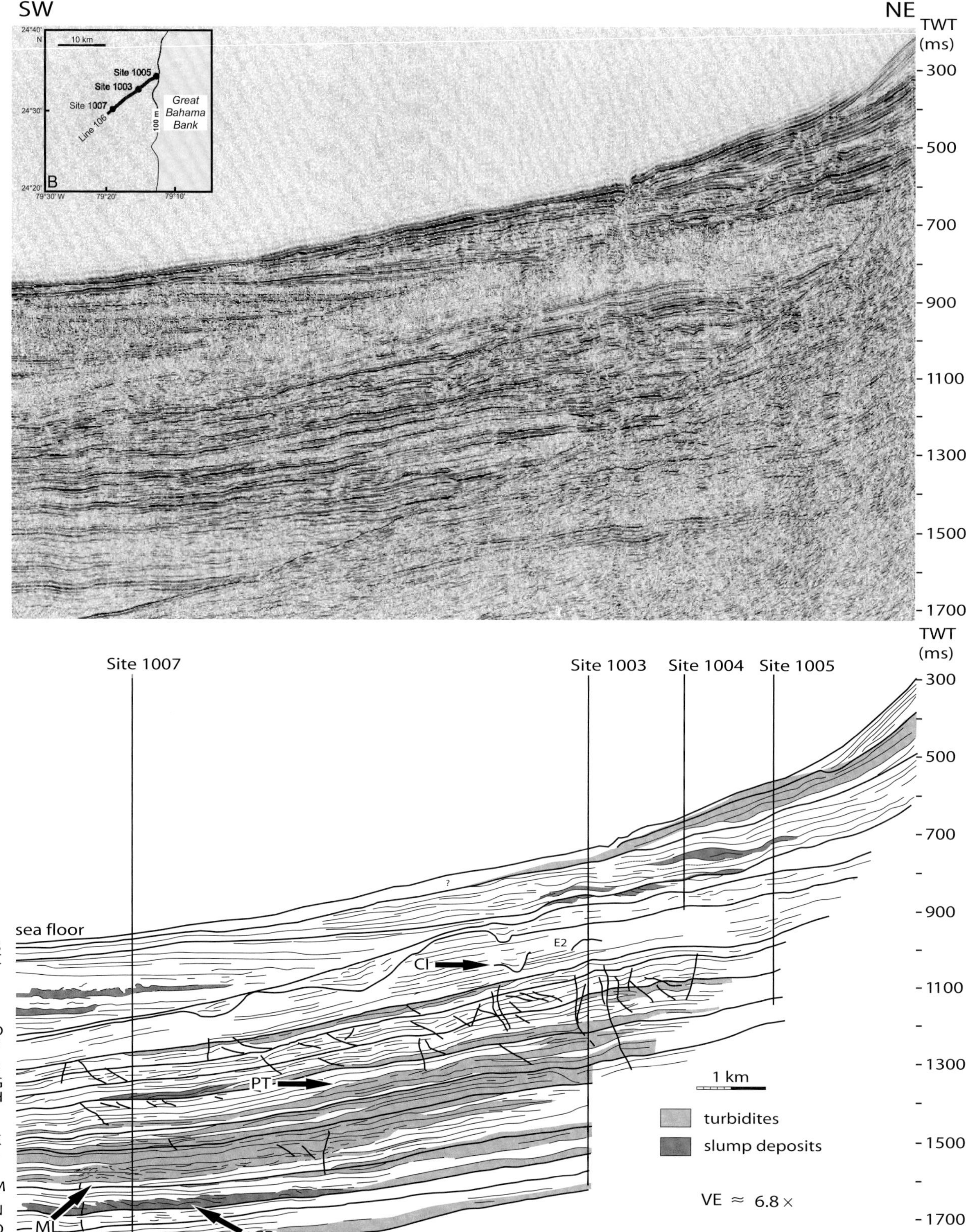

FIGURE 23. Proximal slope and toe of slope of seismic line 106 with line drawing showing interpretation of geometries of turbidite and slump packages. ODP Sites 1003 and 1007 provide information of turbidite and slump packages. Faults probably represent detachment surfaces produced by downslope movements of sediment. The arrows in the line drawing point to features discussed in the text. CI = canyon incision in Sequence F; PT = prograding turbidite package in Sequence K; ML = mounded lobes in Sequence M; CR = convoluted reflections in slump of Sequence N. VE = vertical exaggeration; TWT = two-way traveltime. Reprinted by permission of Betzler et al. (1999).

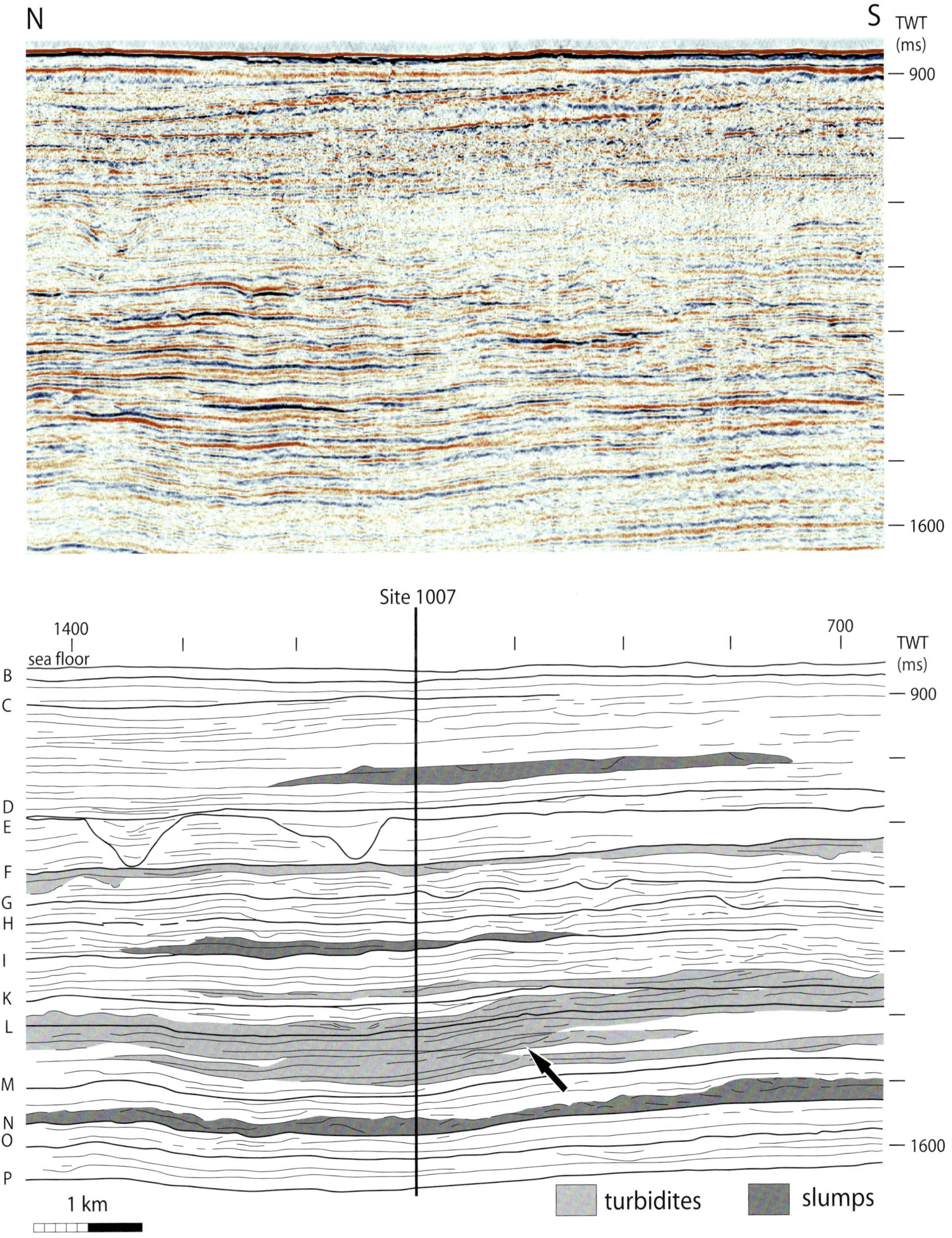

FIGURE 24. Strike line (seismic line 102) parallel to platform margin across ODP Site 1007 and interpretation. Turbidite packages pinch out laterally, documenting the overlapping lobes of redeposited carbonates. In Sequence E channels cut into the near-transparent facies of Sequence F. Arrow indicates shingled internal geometry of a turbidite package. TWT = two-way traveltime. Reprinted by permission of Betzler et al. (1999).

formed during sea level highstands, whereas the dark-gray layers are dominantly pelagic and deposited during sea level lowstands. Intercalated in these deposits are calciturbidites and some slump horizons (Figure 25). The turbidite packstone to floatstone contain shallow-water bioclasts and planktonic foraminifera. Slump deposits occur as isolated intervals. The slumps consist of contorted periplatform ooze and calciturbidites (Betzler et al., 1999).

On the middle to lower slope, sedimentation is variable (Site 1003), consisting of fine-grained, more neritic sediments, pelagic deposits, or turbidite successions. The variability in facies and the small-scale channeling is reflected in the seismic facies that is characterized by discontinuous seismic reflection pattern (Figures 23, 24). The seismic data indicate that channels and incisions are abundant but small. Low-amplitude, nearly transparent seismic packages that interrupt the discontinuous high-amplitude packages correlate to homogeneous platform deposits. For example, a 100-ms-thick early Pliocene transparent unit (sequences F and E) consists of non-lithified to partially lithified lime mudstone to wackestone with planktonic and benthic foraminifera, calcareous nannoplankton, minor diatoms, and minor radiolarians.

Most of the redeposited carbonates accumulate at the toe of slope where the reflections in the seismic lines show mounded to lobate geometries (Site 1007) (Figures 23, 24; Betzler et al., 1999; Bernet et al., 2000; Eberli, 2000). The turbidite lenses are discontinuous within the marl and limestone alternations. A good example of this arrangement is provided in Sequence M, southwest of Site 1007 (Figures 23, 24). At this site, Sequence M consists of a lower turbidite-poor and an upper turbidite-rich interval. The turbidite succession can be split up into three individual bodies. On a seismic scale, these bodies pinch out updip from Site 1007. The lower two bodies are mounded structures as much as 1 km wide and 20–30 m thick. Reflections in these mounded zones are discontinuous. In the central part of the mounds, flat to slightly concave reflections occur. Shingled reflection geometries indicate that individual turbidite bodies stack laterally. In addition to down- and updip terminations, turbidite bodies also pinch out parallel to strike, for example, 1.5 km south of Site 1007 (Figure 24). The same depositional geometries also occur in the turbidite systems of Sequences L and K. The dominant direction of lateral accretion of these bodies is toward the north; only very minor south-directed package-internal dips occur (Betzler et al., 1999). Slump deposits occur in different positions in the lower slope and the toe of slope (Figure 23). The slump in Sequence N at Site 1007 appears as a zone with curved and convoluted reflections. The same type of geometric expression of slumped intervals is in Sequence D at Site 1007 and in Sequences D and C at Sites 1003–1005. The slumps in the basal part of Sequence I at Site 1007 show a different geometry (Figure 24). This interval is characterized by sigmoidal, downlapping reflections.

West of the slope, a drift is deposited in the basin axis that is seen on the seismic lines as a large mound with continuous low-amplitude internal reflections (Figures 16, 23; Betzler et al., 1999; Anselmetti et al., 2000). The drift deposits interfinger and onlap the slope deposit. This mound formed a positive relief and generated a depression along the base of slope. The depression is filled by mound-shaped deposit that parallels the axis of the depression. The seismic facies is continuous low-amplitude reflections on the side of the mound, whereas the top is transparent (Figure 20). The mound is interpreted as a channel-levee deposit in which the seismically transparent top is the feeder channel of the fan. If this is correct, the mass gravity-flow deposits were deflected along the depression into a large submarine fan to form a channel-levee system.

Slope and Canyons of the Maiella Platform Margin

Slope strata and submarine canyons are exposed in the Maiella Platform margin at several stratigraphic levels (Accarie, 1988; Sanders, 1994; Mutti et al., 1996; Bernoulli, 2001). Incised channels are several hundreds meters wide and tens of meters deep (Accarie, 1988). The oldest channel system is pre-Campanian in age and cuts into lagoonal deposits. It is filled with stacked channelized breccias, composed of platform-derived lithoclasts in a matrix of bio- and lithoclastic calcareous sand. A younger series of channel incisions is Campanian to Maastrichtian in age and cuts into the slope sediments of the prograding part of the Orfento Formation. One of the best-exposed channel systems (Cima dell'Altare location; Sanders, 1994) consists of five amalgamated incisions that crosscut each other with minor lateral migration of the channel axis. The entire system is approximately 80 m thick and approximately 1400 m wide. Each fill thins and fines upward from bioclastic and lithoclastic megaconglomerates and breccias upward into coarse bioclastic-lithoclastic grainstone and siltstone (Sanders, 1994). The channel system is laterally and vertically imbedded in lime mudstone, calcisiltstone, and calcarenite that are part of a slope that prograded toward the east-southeast in layers inclined approximately 15°. Maastrichtian beds consisting mainly of rudist sand conformably overlie it.

Near the top of the Maastrichtian prograding system, a channel system occurs at the base of prograding lobes during a time of forced regression (Mutti et al., 1996). In this system, the 20- to 40-m-thick channels are limited to the uppermost slope and can be traced for only for a few kilometers. They have an erosive base and are filled with

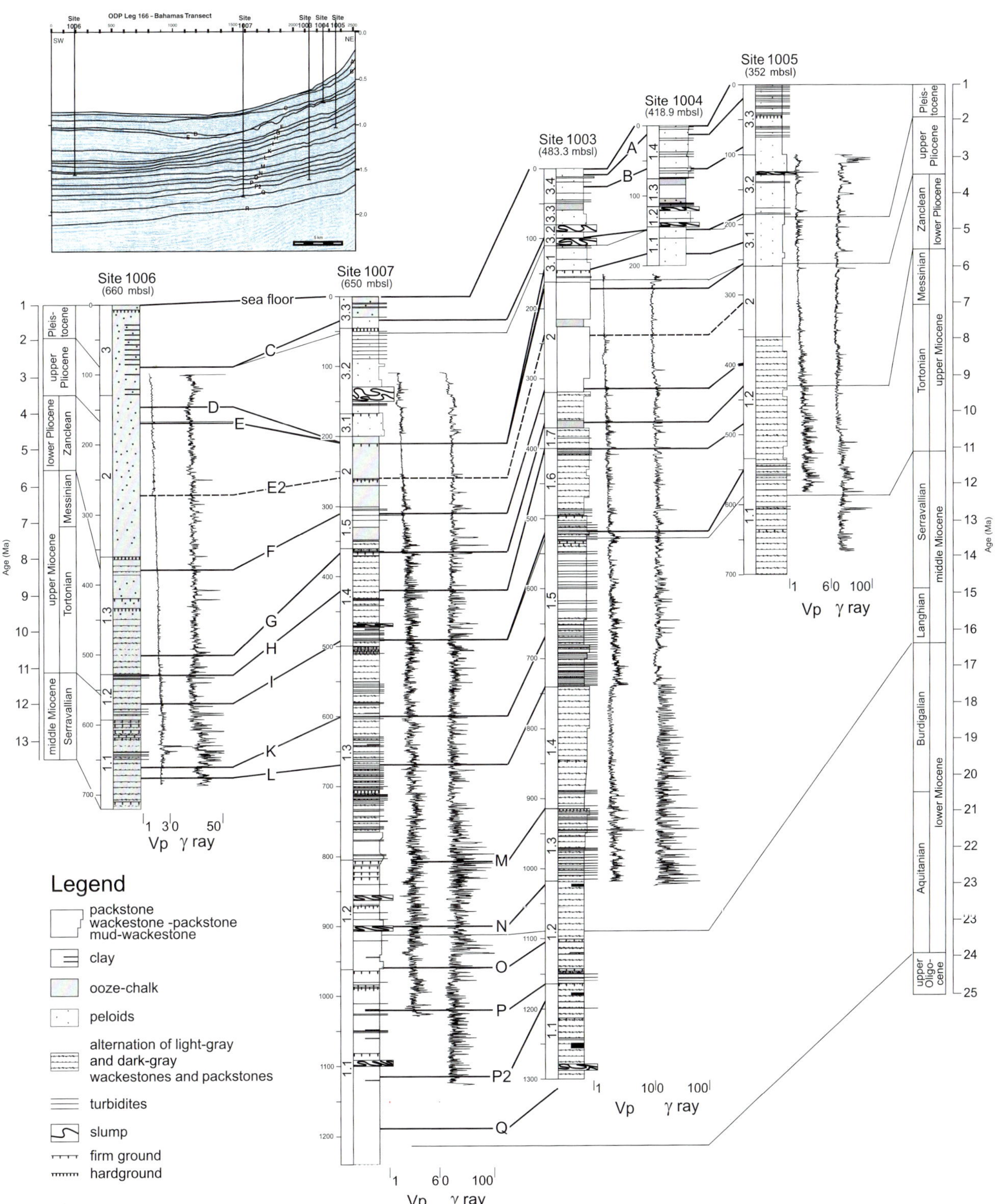

FIGURE 25. Summary of lithologies and ages of the slope and basin part along the Bahamas Transect drilled during ODP Leg 166. Numbers refer to sedimentary units, letters to sequence boundaries. Vp = velocity log (km/s); γ ray = gamma-ray log (CPS units). Age assignments are after Eberli et al. (1997) (modified from Betzler et al., 1999).

coarse, poorly- to medium-well-sorted rudstone that fines upward into intraclastic and skeletal grainstone.

Another series of channels formed in the late Paleocene and early Eocene during a time of minor aggradation of the platform (Bernoulli, 2001). These channels are multistoried in outcrops at Monte San Angelo (Figure 26). The first incision cuts into the Maastrichtian grainstone and is dated as late Paleocene (Sanders, 1994).

The fill consists of large boulders and rudstone of reworked Maastrichtian and lower Paleocene shallow-water bioclastic material (Figure 26C). A younger, also late Paleocene incision followed the same channel axis. It partly eroded the older fill and widened the channel. During a third incision, the axis shifted farther to the west. This fill consists of lithobioclastic conglomerates that grade upward into coarse bioclastic and lithoclastic rudstone to grainstone. Calcareous siltstones form the top of the channel fill. The channel dimensions increase with time. The first incision is approximately 400 m wide, whereas the younger ones widen it to approximately 1700 m. The depths of the channels are between 50 and 100 m. The dimension and internal architecture of this channel system are very similar to those previously described concerning the slopes of the Great Bahama Bank (Figure 26D).

The channels of the Maiella Platform margin document that the platform was at times subject to deep erosion, during which large parts of the platform were removed mechanically. All channel fills show a thinning- and fining-upward trend. The basal, clast-supported megaconglomerate in places contains clasts as much as 10 m in diameter that were probably deposited by rock fall or debris flows. The somewhat finer and more matrix-rich conglomerates show sedimentary structures

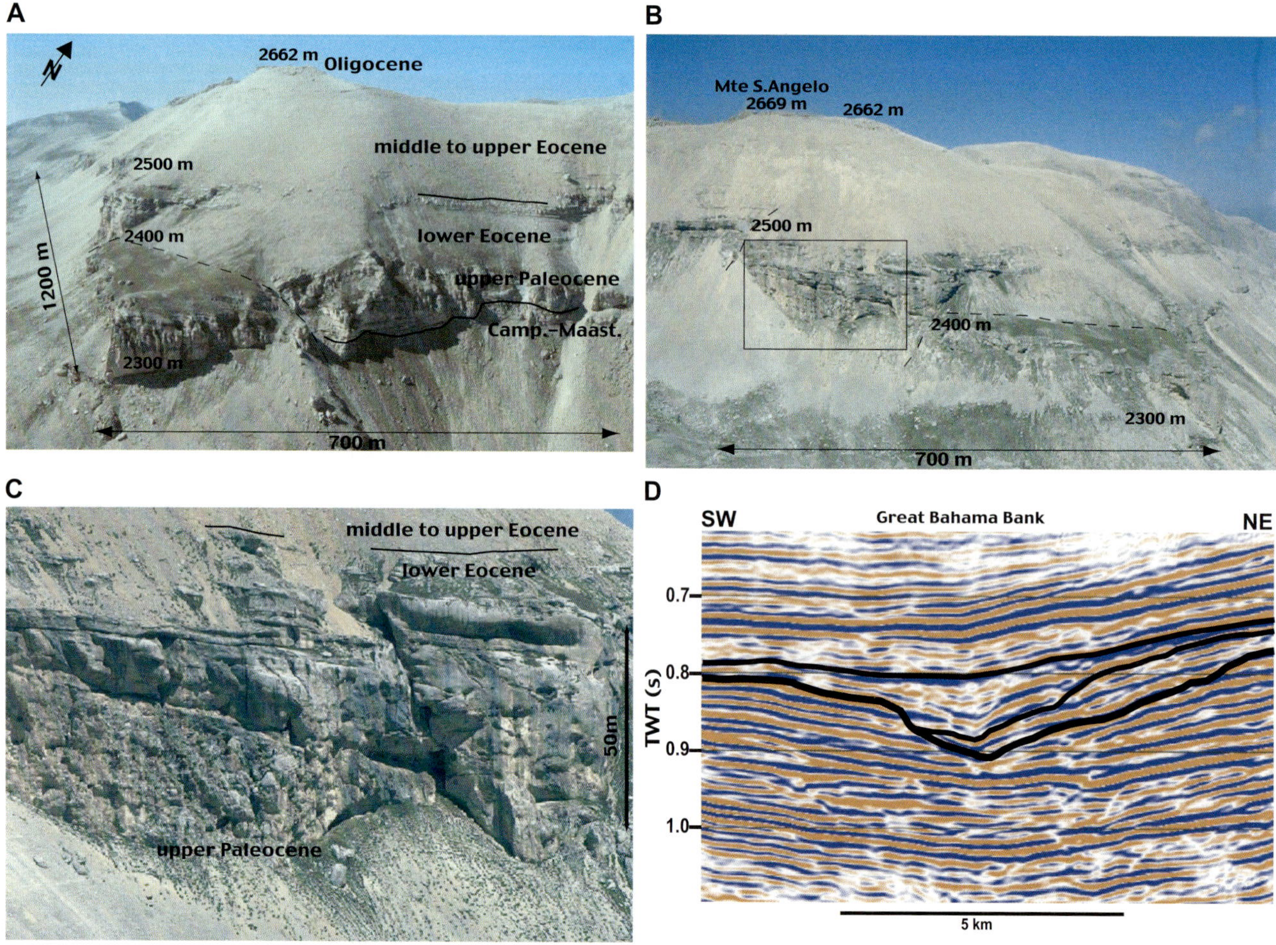

FIGURE 26. (A) Upper Paleocene–lowermost Eocene (upper Thanethian–upper Ilerdian) channel cutting into the Upper Cretaceous slope of the Maiella Platform margin. Channeling occurred in stages, producing a cut-and-fill geometry. Thickness in axis is approximately 100 m; width is approximately 1.5 km. The channel is buried by upper Eocene (Priabonian) slope deposits that form a recessive scree slope. Massive cliffs at the top of the slope are Oligocene reefs. Dashed line indicates an Alpine fault that downthrows the southern part of the channel by about 60 m. (B) View from southwest into the channel displaying the cut-and-fill geometry of the infilling strata. The erosive base of the channel is concave-downward, whereas the top is nearly flat because calcirudites and grainstones fill the channel. (C) Close-up of channel fill displaying the main facies types. A carbonate megaconglomerate with boulders as much as 5 m in diameter of Cretaceous platform clasts fills the base of the channel. The overlying mass-gravity-flow deposits thin upward. A truncation surface separates the channel from upper Eocene bioclastic limestones. (D) Middle Miocene (15.1 Ma) channel in the western slope of the Great Bahama Bank with cut-and-fill geometry similar to the channel exposed in the Maiella Platform margin. TWT = two-way traveltime.

that indicate deposition from debris flow and high-density turbidity currents. Grading in many beds indicate that turbidity currents were a dominant depositional mechanism. None of the beds of the channel deposits show evidence of subaerial exposure, although some clasts within the conglomerates had been altered by meteoric diagenesis prior to redeposition in the channel (Mutti et al., 1996). The general absence of meteoric diagenesis and their position in the slope indicates that these channels are submarine in origin. As in the Bahamas, the channels are either stacked vertically or laterally migrating. Figure 27 is a schematic line drawing of the migrating channel system at Monte San Angelo, illustrating the crosscutting relationships, facies distribution, and the migration of the incisions.

The channel incisions are concentrated at certain stratigraphic horizons where facies shifts indicate events of major base-level lowering. Sequence boundaries are postulated at the base of incised channel-fill deposits similar to those on the Great Bahama Bank (Sanders, 1994; Mutti et al., 1996; Vecsei et al., 1998). Increased channeling during relative sea level fall indicates that base-level lowering in carbonates significantly affect mechanical platform erosion. The result of this base-level fall is large volumes of mass-gravity-flow deposits in slope canyons and at the base of slope (Sarg, 1988).

Toe-of-Slope Section of the Maiella Platform Margin

North of the Maiella Mountains in the Gran Sasso area, large seismic-scale outcrops display Lower Cretaceous to Oligocene base-of-slope deposits in an almost 3-D framework. The succession was deposited on the slope and base of slope adjacent to the Lazio-Abruzzi Platform and south of the Apulian Platform and merges with the pelagic successions of the Umbria-Marche basin to the north (Figure 4). The sedimentary succession of the Gran Sasso area is subdivided into six formations. The Monte Corvo Formation, which is of early Campanian (or Maastrichtian) to earliest Danian age, is compared to the toe-of-slope deposits along the western Great Bahama Bank. This time interval corresponds to the Orfento Formation in the Maiella region, when the escarpment of the Maiella Platform margin was buried, and a distally steepened ramp was established during progradation of the platform (Mutti et al., 1996; Van Konijnenburg et al., 1999).

The Monte Corvo Formation consists of fine- to medium-grained skeletal packstone or grainstone beds that are intercalated with thin (2–5 cm thick) layers of pelagic lime mudstone and wackestone. The packstone and grainstone beds were deposited in broad, flat, sheetlike, or slightly channelized beds, hundreds to more than a thousand meters wide. Figure 28 shows the characteristics of these deposits. The packstone and grainstone beds show graded bedding and are interpreted to result from deposition by turbidity currents. They consist of fine- to medium-grained skeletal fragments, derived dominantly from rudists, inoceramids, echinoderms, and larger benthic foraminifera (*Orbitoides*, *Siderolites*). Bed surfaces are flat or slightly undulating and typically eroded. Amalgamation of beds is common and obscures individual layers (Figure 28C). In places, minor internal unconformities are associated with channelized deposits. Bedsets with a positive relief and associated onlap and offlap geometries are also observed, even along the same stratigraphic level.

Clast-supported, intraclastic breccia beds occur mainly in the upper part of the formation. The breccia beds fill broad, 100- to 200-m wide channels (Figure 28A, B), and overbank deposits extend laterally for several hundred meters. They contain intraclasts of fine- to medium-grained packstone and grainstone and pelagic lime mudstone that are from 2 cm to over 1 m in diameter. Many of these intraclastic breccias are graded and all have erosive bases, in many places with substantial downcutting (Figure 28B). In the upper part of the formation, the frequency and thickness of the breccias

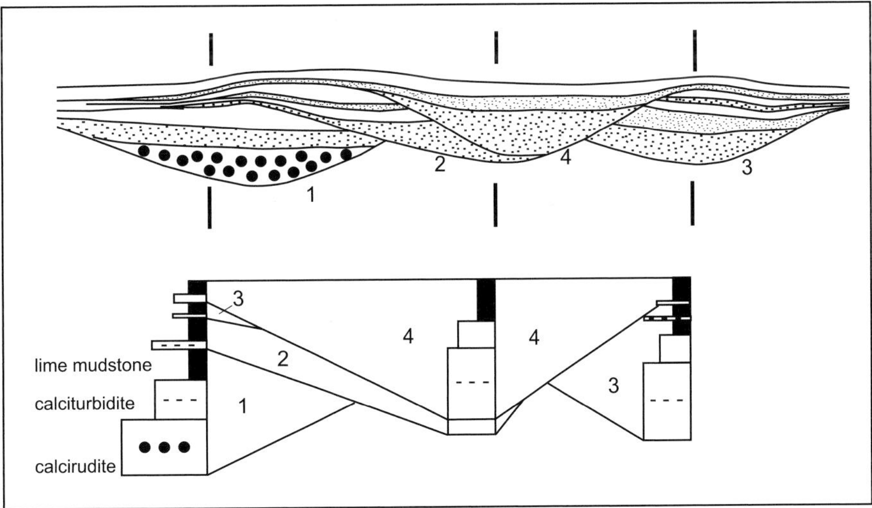

FIGURE 27. Line drawing and schematic sections of an exposed Eocene channel complex in the Maiella Platform margin. Four nested channels are distinguishable. Each of these channels cut and used a different thalweg, producing a migrating-channel complex. The channel-fill deposits fine upward (redrawn from Sanders, 1994).

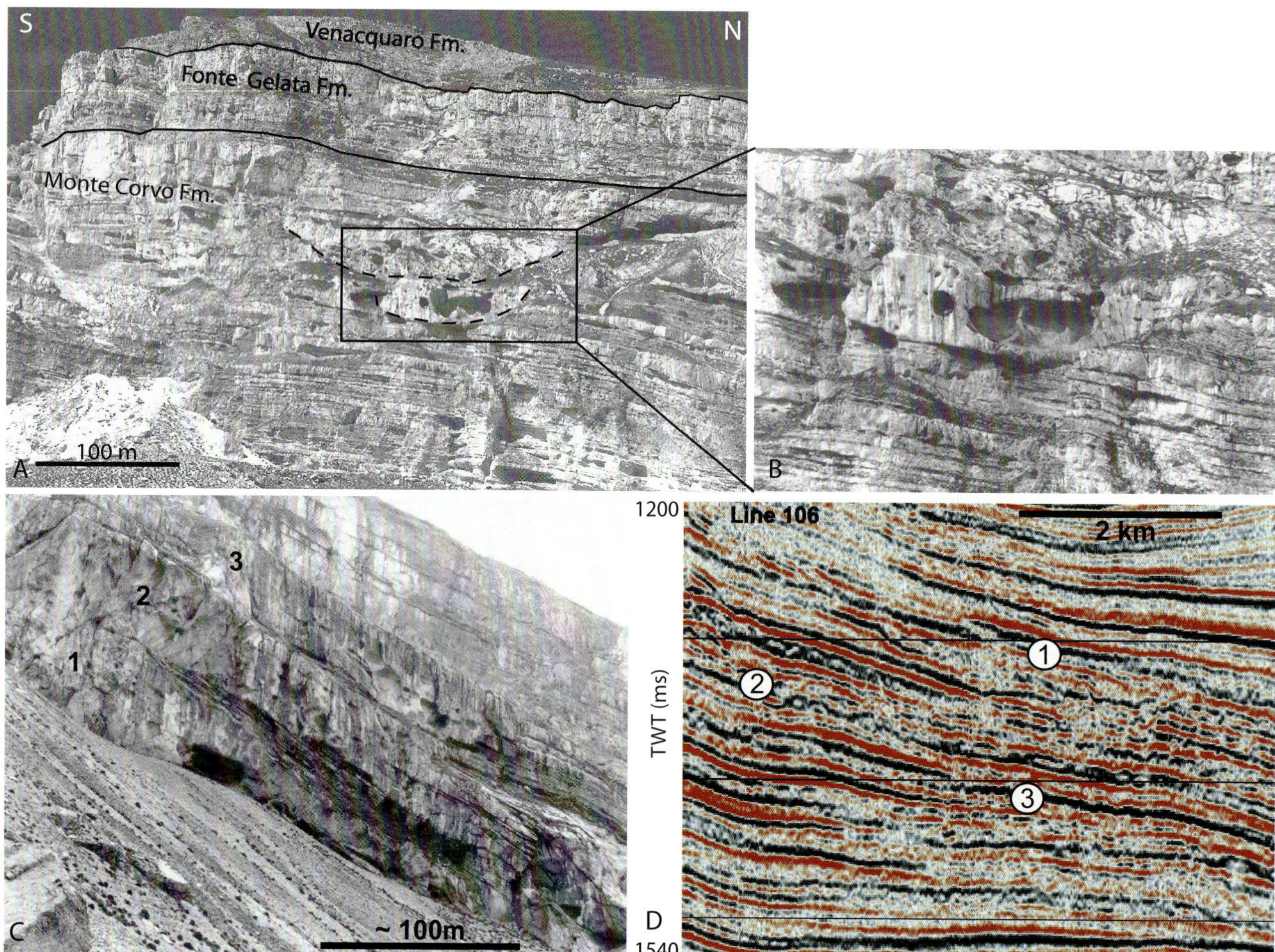

FIGURE 28. Outcrops in the Gran Sasso area display the size and geometries of the Upper Cretaceous lower-slope and basin-floor deposits north of the Maiella Platform margin. (A) Monto Corvo east side: The lower part of the Monte Corvo Formation consists of thick-bedded bioclastic turbidites with a low-angle positive relief and sheetlike intraclastic breccias. The upper part of the formation consists of stacked, channelized intraclastic breccias and calcarenite turbidites. (B) Close-up of two stacked channels filled with conglomerates in the Monto Corvo Formation. The two channels are approximately 100 (top) and 200 m (bottom) wide. (C) Three stacked small lobes of redeposited carbonates on the basin floor exposed at Monte Fosso. (D) Part of seismic line 106 in the Bahamas of the basin floor and lower slope with stacked channels (1), small scale channeling (2), and lobes (3) that are comparable in size and geometry with the outcrop equivalents in the Gran Sasso area. TWT = two-way traveltime.

increase considerably in the center of the Gran Sasso (around Sella dei Grilli and Monte Corvo). This increase in breccia deposition started during the late Maastrichtian. Laterally, to the west and east, this trend is not observed and breccia deposits remain uncommon. The increase of channelized debris-flow deposits in the central part of the Gran Sasso area is explained by progradation of a channel-levee complex. The youngest sediments of the Monte Corvo Formation are pelagic limestones of earliest Danian age (planktonic foraminifera biozone P1a/c). They overlie the redeposited sediments or, at one location (Rio Arno), pelagic limestones of latest Maastrichtian age. Debris-flow and turbidite-filled channels with erosive bases, flat-based turbiditic calcarenites with a positive relief, and onlap and offlap geometries are characteristic of channel-levee complexes typical of basin-floor submarine fans (Mullins and Cook, 1986; Coniglio and Dix, 1992).

The seismic facies and the fine-scale geometries seen on the seismic line 106 are remarkably similar to these outcrop deposits (Figure 28). On the seismic lines, broad, stacked channels 1–2 km wide and some tens of meters deep are comparable in size with the channelized breccia units in the upper part of the Monte Corvo Formation (Figure 28D-1). Smaller-scale channeling is seen as disrupting the continuity of reflections (Figure 28D-2). Lobe-shaped reflections with onlap geometries are probably the seismic expression of the lobate turbidite deposits (Figure 28D-3). The increase of channels in the upper part of the Monte Corvo Formation indicates progradation of a complex channel-levee complex. Progradation of the turbidite packages drilled at Site 1007 is documented by the shingled and downlapping reflections within the slightly mounded units. The low resolution of the seismic image does not, however, give enough credit to the complex nature of these toe-of-slope deposits with their vertical and lateral facies heterogeneity.

DISCUSSION AND CONCLUSIONS

The comparison of the seismic sections of the western margin of the Great Bahama Bank with depositional facies that crop out in Maiella Mountains and the Gran Sasso area adds to the interpretation of the architecture and facies distribution of isolated platforms. Figure 29 summarizes this comparison of the five major depositional environments. The seismic facies of the platform interior of the Great Bahama Bank lacks a calibration with cores, but this can be understood with the help of synthetic seismic sections. It shows how horizontally layered strata can produce a chaotic to transparent seismic image when the impedance contrast between beds is very low. Impedance is the product of density times velocity, density is inversely correlated to porosity. In siliciclastic rocks, density and porosity are most important for the determining the impedance. In carbonates, however, velocity is equally important as porosity for impedance. The reason is that velocity in carbonates is the combined result of porosity and pore type (Anselmetti and Eberli, 1993). Low impedance is pronounced in the Cretaceous platform strata that produced chaotic to transparent seismic facies in the platform interior. Intercalations of evaporites might, however, produce strong reflection horizons in the otherwise chaotic seismic facies. The Tertiary parts of both platforms display a higher amplitude and more continuous reflections.

The changes in rock physics and the resulting difference in seismic character between the Cretaceous and Tertiary parts of the platforms may be the result of changes in carbonate facies resulting from differences in climate between the Cretaceous and the Tertiary. The Cretaceous was a time of stable and warm climate without major buildups of ice caps. In this greenhouse world, subsidence and low-amplitude eustatic sea level changes resulted in the vertical aggradation of carbonate cycles without large-facies shifts. The succession of similar types of rocks, although cyclic in nature, produces low-impedance contrasts and reflectivity that is below the noise level.

In the Tertiary section of both platforms, the vertical succession consists of varied lithologies. The lithologic variability may be directly related to the climate with extremely hot climates in the Paleocene and Eocene and high-frequency glaciations throughout the Neogene (Miller et al., 1991; Zachos et al., 1994). These Neogene glaciations caused high-amplitude sea level fluctuations that significantly affected the carbonate environments, leading to periodic platform exposure and incipient drowning and margin backstep in both platforms (Vecsei, 1991; Sanders, 1994; Kievman, 1998; Eberli et al., 2001; Kenter et al., 2001). These high-frequency sea level changes led to large facies and diagenetic changes. Consequently, impedance contrasts are prone to be higher. As a result, the Tertiary platform interiors are imaged by more continuous reflections.

In both platforms, the margins are characterized by the juxtaposition of different lithologic and seismic facies, in particular where the margin is steep as is the case of the Pliocene–Pleistocene Great Bahama Bank. Although these vertical and lateral facies changes are easy to recognize in outcrop, the seismic data in many places do not resolve these changes accurately, because diffractions obscure the signal.

Sedimentologic and stratigraphic data from the outcrops of the Maiella escarpment indicate that the construction of such escarpments results from the interplay of platform growth and erosion. Similar findings from the modern Bahamas Escarpment suggest that erosion of marginal facies by margin failure is a common process in maintaining the escarpment (Freeman-Lynde et al., 1988; Mullins and Hine, 1989). The escarpment separates basin and slope facies from the eroded platform facies. This facies separation is seen on seismic sections as the lateral transition from the chaotic seismic facies of the platform interior to the more continuous dipping reflections of the slope (Bryant et al., 1979; Corso et al., 1988; Denny et al., 1994; Masaferro, 1997). Diffractions on the steep escarpment and on irregularities of the escarpment obscure this transition on seismic lines. The deep part of the seismic data in Great Bahama Bank images a fault that reaches the base of the escarpment, providing evidence that the initiation of this escarpment was fault controlled (Figure 3).

On the seismic data from the slopes of the western Great Bahama Bank, all the elements that have been described from the modern slopes (e.g., Little Bahama Bank; Mullins et al., 1984) can be found at several stratigraphic levels. According to Mullins et al. (1984), the upper slope is characterized by parallel, low-amplitude to transparent seismic facies that is cut by deep canyons trending perpendicular to the margin. Cores from ODP Leg 166 sites help calibrate the parallel seismic facies to alternations of periplatform ooze with better-cemented intervals (Eberli et al., 1997; Malone et al., 2001). The exposed canyons in the Maiella Platform margin show that the canyon-fill facies is very coarse at the base with a general thinning- and fining-upward trend. Depositional mechanisms include rock fall, debris flows, and

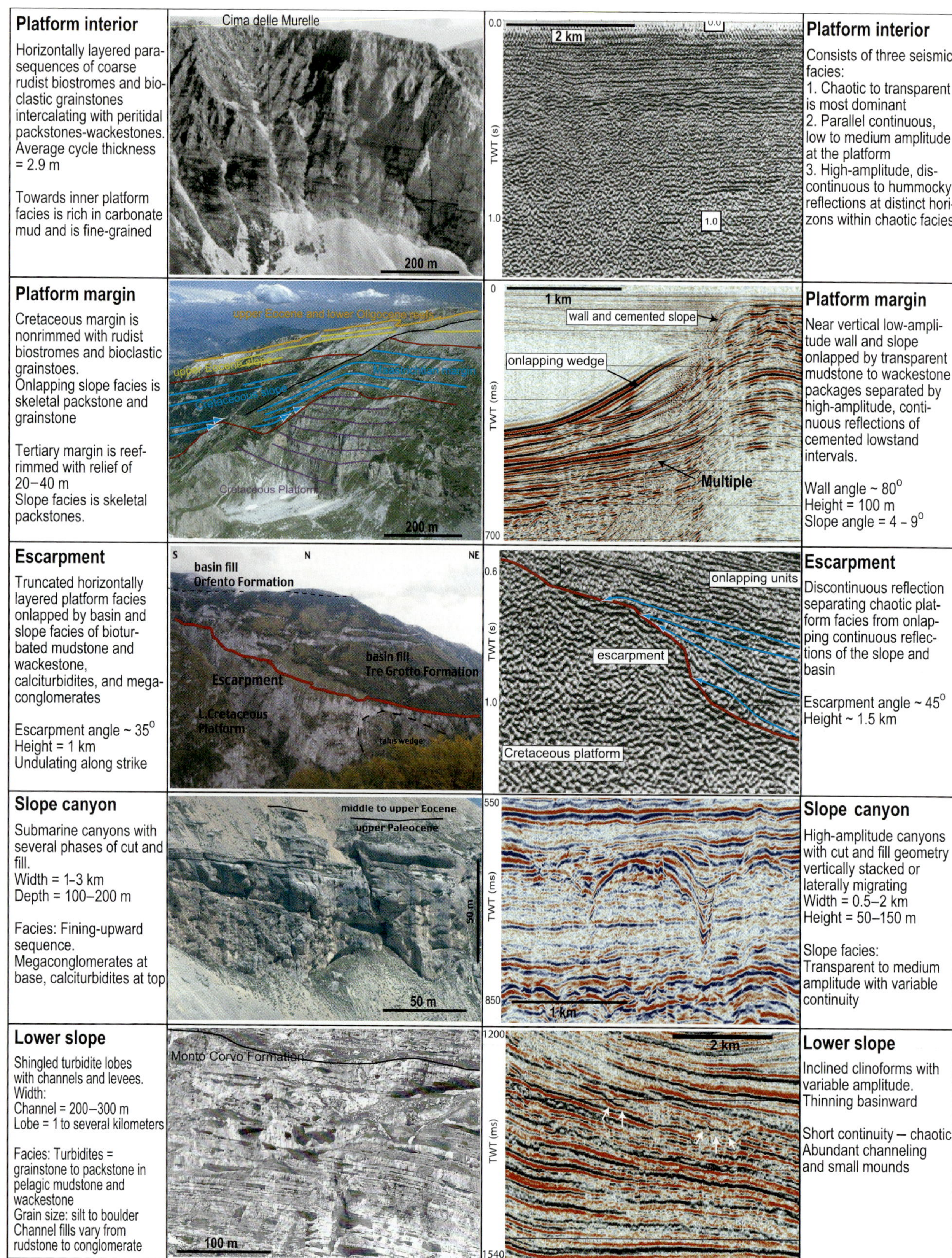

FIGURE 29. Comparison of lithologic and seismic facies from the Great Bahama Bank and the Maiella Platform margin. TWT = two-way traveltime.

high- and low-density turbidity currents. Stratigraphic information documents the long-lived nature of these canyons and the repeated incision and filling that can only be inferred from the seismic image (Figure 29). Canyons are more frequent during times of lower sea level, indicating that sea level lowstands result in significant mechanical erosion of the platform.

The lower slope and the toe of slope are seismically characterized by high-amplitude reflections that are discontinuous and sometimes chaotic. Piston cores and cores from the ODP sites indicate that this discontinuous seismic facies is formed by redeposited carbonates that preferentially accumulate at the lower slope and the toe of slope (Crevello and Schlager, 1980; Mullins et al., 1984; Anselmetti et al., 2000). Betzler et al. (1999) showed that the redeposited carbonates are arranged in laterally shingled lobes, which shift in concert with sea level changes farther basinward or more updip. This architecture of shingled lobes has previously been documented along the steep-sided platform (Cook and Mullins, 1983). A similar geometry is here recognized on the Miocene low-angle slopes when Great Bahama Bank had a more ramplike geometry. The outcrops at the Gran Sasso area demonstrate the complex facies distribution in these shingled lobes with small-scale channeling and mounded features on an outcrop scale. Although the toe-of-slope is the depositional center for the redeposited carbonates, the sedimentation rates in these lobes are surprisingly low. In the Gran Sasso area, considerably less sediment accumulated on the base of slope than on the Maiella Platform margin (Van Konijnenburg et al., 1999). Similarly, the clinoforms of western Great Bahama Bank thin toward the basin, documenting that the sedimentation rate of fine-grained sediment transported offbank and preferentially deposited upslope outpaces the redeposition of carbonates in the lower-slope and toe-of-slope areas.

ACKNOWLEDGMENTS

A lot of the data for this comparison between Great Bahama Bank and the Maiella Platform margin was assembled in two large research efforts. The Maiella Platform margin was studied over 10 yr under the guidance of Daniel Bernoulli in projects funded by the Swiss National Science Foundation (grants 20-20-27 457.89, 20-45131.95, and 20-359007.92). Many of the ideas were discussed and tested by the students and postdocs who worked in the Maiella. We thank them all for their input. The Bahamas Transect was also a long-term research effort that started with donations of the seismic lines from Texaco Inc. and Western Geophysical to the Comparative Sedimentology Laboratory at the University of Miami. Two cores along the western seismic line that were drilled in the Bahamas Drilling Project, under the guidance of Robert Ginsburg, provided the first calibration of the seismic transect. Funds for the drilling, coring, and logging were provided by a National Science Foundation (NSF) grant (OCE-8917295) and contributions from the industrial associates to the Comparative Sedimentology Laboratory and the Swiss National Science Foundation. Subsequent funds for laboratory analyses of cores and logs came from grants from the NSF (OCE-9204294) and the Department of Energy (DE-FG05-92ER14253). The seismic data acquisition and interpretation of the seismic data along the western margin of the Great Bahama Bank was funded by NSF grant OCE-9314586 (to G. P. Eberli, D. F. McNeill, and P. K. Swart). This grant included a subcontract to Rice University for data acquisition on the R/V *Lone Star*. André Droxler and John Anderson tailored the acquisition system to fit our needs. An upgrade of the seismic processing facility by NSF grant OCE-9615141 (to C. Scholz and G. P. Eberli) improved the data presentation.

Ground truthing of the seismic data was provided by the cores recovered during Ocean Drilling Program Leg 166. We thank the scientific shipboard party and the technical staff on the Joides Resolution for all their input. Funding from JOI-USSAC grant 166F000330 (to G. P. Eberli) and the Industrial Associates of the Comparative Sedimentology Laboratory paid for subsequent core and seismic analyses. Christian Betzler received grants from the Deutsche Forschungsgemeinschaft (projects 1272/5 and 1272/6) for his analysis of the slope deposits.

The manuscript benefited significantly from the careful reviews, comments, and suggestions of Associate Editor J. Frederick "Rick" Sarg and the two reviewers, Steve Bachtel and Jim Weber.

REFERENCES CITED

Accarie, H., 1988, Dynamique sédimentaire et structurale au passage platform/bassin. Les faciès Crétacés et Tertiaries: Massif de la Maiella (Abruzzi, Italie): Ecole des Mines de Paris, Mémoires des Sciences de la Terre 5, p. 1–158.

Accarie, H., and J.-F. Deconinck, 1989, Relation entre présence de palygorskite et périodes d'anoxie à l'Albien supérieur et au Turonien inférieur à moyen dans le massif de la Maïella (Abruzzes, Italie): Comptes Rendus de l'Académie des Sciences (Paris), tome 308/II, p. 1267–1272.

Accarie, H., B. Beaudoin, R. Cussey, Ph. Joseph, and S. Triboulet, 1986, Dynamique sédimentaire et structurale au passage plateforme/bassin. Les faciès carbonatés crétacés du massif de la Maiella (Abruzzes, Italie): Memorie della Società Geologica Italiana, v. 36, p. 217–231.

Adams, E. W., and W. Schlager, 2000, Basic type of

submarine slope curvature: Journal of Sedimentary Research, v. 70, p. 814–828.
Anselmetti, F. S., and G. P. Eberli, 1993, Controls on sonic velocity in carbonate rocks, *in* R. C. Liebermann, ed., Pure and Applied Geophysics, v. 141, no. 2/3, p. 287–323.
Anselmetti, F. S., and G. P. Eberli, 1997, Sonic velocity in carbonate sediments and rocks, *in* F. J. Marfurt and A. Palaz, eds., Carbonate seismology: SEG Geophysical Developments Series 6, p. 53–74.
Anselmetti, F. S., G. P. Eberli, and D. Bernoulli, 1997, Seismic modeling of a carbonate platform margin (Montagna della Maiella, Italy): Variations in seismic facies and implications for sequence stratigraphy, *in* F. J. Marfurt and A. Palaz, eds., Carbonate seismology: Society of Exploration Geophysicists Geophysical Developments Series 6, p. 373–406.
Anselmetti, F. S., G. P. Eberli, and Z.-D. Ding, 2000, From the Great Bahama Bank into the Straits of Florida: A margin architecture controlled by sea level fluctuations and ocean currents: Geological Society of America Bulletin, v. 112, p. 829–844.
Austin, J. A., Jr., W. Schlager, et al., 1988, Proceedings of the Ocean Drilling Program, Scientific Results: Ocean Drilling Program, Texas A&M University, College Station, Texas, v. 101, 501 p.
Bachtel, S. L., R. D. Kissling, D. Martono, S. Rahardjanto, P. A. Dunn, B. A. MacDonald, 2004, Seismic stratigraphic evolution of the Miocene–Pliocene Segitiga Platform, East Natuna Sea, Indonesia: The origin, growth, and demise of an isolated carbonate platform, *in* G. P. Eberli, J. L. Masaferro, and J. F. Sarg, eds., Seismic imaging of carbonate reservoirs and systems: AAPG Memoir 81, p. 309–328.
Bally, A., 1954, Geologische Untersuchungen in den SE-Abruzzen: Ph.D. thesis, Zürich University, 289 p.
Beach, D. K., 1982, Depositional and diagenetic history of Pliocene–Pleistocene carbonates of northwestern Great Bahama Bank: Evolution of a carbonate platform: Ph.D. dissertation, University of Miami, 600 p.
Beach, D. K., 1993, Submarine cementation of subsurface Pliocene carbonates from the interior of Great Bahama Bank: Journal of Sedimentary Petrology, v. 63, p. 1059–1069.
Beach, D. K., and R. N. Ginsburg, 1980, Facies succession, Plio-Pleistocene carbonates, northwestern Great Bahama Bank: AAPG Bulletin, v. 64, p. 1634–1642.
Bernet, K. H., 2001, Hierarchies of sea level changes along the Bahamas Transect: Ph.D. dissertation, University of Miami, p. 204.
Bernet, K. H., G. P. Eberli, and A. Gilli, 2000, Turbidite frequency and composition in the distal part of the Bahamas Transect, *in* P. K. Swart, G. P. Eberli, M. J. Malone, and J. K. Sarg, eds.,Proceedings of the Ocean Drilling Program, Scientific Results: Ocean Drilling Program, Texas A&M University, College Station, Texas, v. 166, p. 45–60.
Bernoulli, D., 1972, North Atlantic and Mediterranean Mesozoic facies, a comparison, *in* C. D. Hollister, J. I. Ewing, eds., Initial reports of the Deep Sea Drilling Project 11: Washington, DC, U.S. Governement Printing Office, p. 801–871.
Bernoulli, D., 2001, Mesozoic–Tertiary carbonate platforms, slopes and basins of the external Apennines and Sicily, *in* G. B. Vai and I. P. Martini, eds., Anatomy of an orogen: The Apennines and adjacent Mediterranean Basins: Great Britain, Kluwer Academic Press, p. 307–326.
Bernoulli, D., and H. C. Jenkyns, 1974, Alpine, Mediterranean and central Atlantic Mesozoic facies in relation to the early evolution of the Tethys, *in* R. H. Dott Jr. and R. H. Shaver, eds., Modern and ancient geosynclinal sedimentation: SEPM Special Publication 19, p. 129–160.
Bernoulli, D., F. S. Anselmetti, G. P. Eberli, M. Mutti, J. S. Pignatti, D. G. K. Sanders, and A. Vecsei, 1996, Montagna della Maiella: The sedimentary and sequential evolution of a Bahamian-type carbonate platform of the South-Tethyan continental margin: Memorie della Società Geologica Italiana, v. 51, p. 7–12.
Betzler, C., J. J. Reijmer, K. H. Bernet, G. P. Eberli, and F. S. Anselmetti, 1999, Sedimentary variations in space and time along the leeward flank of Great Bahama Bank (ODP Leg 166): Sedimentology, v. 46, p. 1127–1145.
Bice, D. M., and K. G. Stewart, 1990, The formation and drowning of isolated carbonate seamounts: Tectonic and ecological controls in the northern Apennines, *in* M. E. Tucker, J. L. Wilson, P. D. Crevello, J. F. Sarg, and J. F. Read, eds., Carbonate platforms: Facies, sequences and evolution: Special Publication of the International Association of Sedimentologists 9, p. 145–168.
Biddle, K. V., W. Schlager, K. W. Rudolph, and T. L. Bush, 1992, Seismic model of a progradational carbonate platform, Picco di Vallandro, the Dolomites, Northern Italy: AAPG Bulletin, v. 76, p. 14–30.
Bosellini, A., 1984, Progradation geometries of carbonate platforms: Examples from the Triassic of the Dolomites, northern Italy: Sedimentology, v. 31, p. 1–24.
Bosellini, A., 1989, Dynamics of Tethyan carbonate platforms, *in* P. D. Crevello, J. L. Wilson, J. F. Sarg, and J. F. Read, eds., Controls on carbonate platform and basin development: SEPM Special Publication 44, p. 3–13.
Bracco Gartner, G. L., and W. Schlager, 1999, Discrimination between onlap and lithologic interfingering in seismic models of outcrops: AAPG Bulletin, v. 83, p. 952–971.
Brown, A. R., 1999, Interpretation of 3-dimensional seismic data: AAPG Memoir 42, 5th ed., 341 p.
Brown, L. P., and W. L. Fisher, 1980, Seismic stratigraphic exploration and petroleum exploration: AAPG Continuing Education Course Notes Series No. 16, 116 p.
Bryant, W. R., A. A. Meyerhoff, N. K. Brown, M. A. Furrer, T. E. Pyle, and J. W. Antoine, 1979, Escarpments, reef trends, and diapiric structures, eastern Gulf of Mexico: AAPG Bulletin, v. 53, p. 2506–2542.
Carannante, G., M. Esteban, J. D. Milliman, and L. Simone, 1988, Carbonate lithofacies as paleolatitude indicators: Problems and limitations: Sedimentary Geology, v. 60, p. 333–346.

Coniglio, M., and G. R. Dix, 1992, Carbonate slopes, *in* R. G. Walker and N. P. James, eds., Facies models: Response to sea-level change: Geological Association of Canada, p. 349–374.

Cook, H. E., and H. T. Mullins, 1983, Basin margin environment, *in* P. A. Scholle, D. G. Bebout, and C. H. Moore, eds., Carbonate depositional environments: AAPG Memoir 33, p. 540–617.

Corso, W., R. T. Buffler, and J. A. Austin Jr., 1988, Erosion of the southern Florida Escarpment, *in* A. W. Bally, ed., Atlas of seismic stratigraphy, v. 2: AAPG Studies in Geology No. 27, p. 149–151.

Crescenti, U., A. Crostella, G. Donzelli, and G. Raffi, 1969, Stratigrafia della serie calcarea dal Lias al Miocene nella regione Marchigiano-Abruzzese (Parte II— Litostratigrafia, biostratigrafia, paleogeografia). Memorie della Società Geologica Italiana, v. 8, p. 343–420.

Crevello, P. D., and W. Schlager, 1980, Carbonate debris sheets and turbidites, Exuma Sound, Bahamas: Journal of Sedimentary Petrology, v. 50, p. 1121–1148.

D'Argenio, B., 1970, Central and southern Italy Cretaceous bauxite stratigraphy and paleogeography: Annales Instituti Geologici Publici Hungarici, v. 54, p. 221–233.

D'Argenio, B., P. de Castro, C. Emiliani, and L. Simone, 1975, Bahamian and Apenninic limestones of identical lithofacies and age: AAPG Bulletin, v. 59, p. 524–533.

Denny, W., J. A. Austin, and R. T. Buffler, 1994, Seismic stratigraphy and geologic history of mid-Cretaceous through Cenozoic rocks, southern Straits of Florida: AAPG Bulletin, v. 78, p. 461–487.

Dillon, W. P., A. M. Trehu, P. C. Valentine, and M. M. Ball, 1988, Eroded carbonate platform margin— The Blake Escarpment off southeastern United States, *in* A. W. Bally, ed., Atlas of seismic stratigraphy, v. 2: AAPG Studies in Geology No. 27, p. 140–148.

Droxler, A. W., and W. Schlager, 1985, Glacial versus interglacial sedimentation rates and turbidite frequency in the Bahamas: Geology, v. 13, p. 799–802.

Eberli, G. P., 1991, Growth and demise of isolated carbonate platforms: Bahamian controversies, *in* D. W. Müller, J. A. McKenzie, and H. Weissert, eds., Controversies in modern geology: Academic Press, London, p. 231–248.

Eberli, G. P., 2000, The record of Neogene sea-level changes in the prograding carbonates along the Bahamas Transect— Leg 166 synthesis, *in* P. K. Swart, G. P. Eberli, M. J. Malone, and J. K. Sarg, eds., Proceedings of the Ocean Drilling Program, Scientific Results: Ocean Drilling Program, Texas A&M University, College Station, Texas, v. 166, p. 167–177.

Eberli, G. P., and R. N. Ginsburg, 1987, Segmentation and coalescence of platforms, Tertiary, NW Great Bahama Bank: Geology, v. 15, p. 75–79.

Eberli, G. P., and R. N. Ginsburg, 1988, Aggrading and prograding infill of buried Cenozoic seaways, Northwestern Great Bahama Bank, *in* A. W. Bally, ed., Atlas of seismic stratigraphy, v. 2: AAPG Studies in Geology Series No. 27, p. 97–103.

Eberli, G. P., and R. N. Ginsburg, 1989, Cenozoic progradation of northwestern Great Bahama Bank, a record of lateral platform growth and sea level fluctuations, *in* P. D. Crevello, J. L. Wilson, J. F. Sarg, and J. F. Read, eds., Controls on carbonate platform and basin development: SEPM Special Publication 44, p. 339–351.

Eberli, G. P., D. Bernoulli, D. Sanders, and A. Vecsei, 1993, From aggradation to progradation: The Maiella Platform margin (Abruzzi, Italy), *in* J. T. Simo, R. W. Scott, and J.-P. Masse, eds., Cretaceous carbonate platforms: AAPG Memoir 56, p. 213–232.

Eberli, G. P., P. K. Swart, M. Malone, and Scientific Party, 1997, *in* P. K. Swart, G. P. Eberli, M. J. Malone, and J. K. Sarg, eds., Proceedings of the Ocean Drilling Program, Scientific Results: Ocean Drilling Program, Texas A&M University, College Station, Texas, v. 166.

Eberli, G. P., F. S. Anselmetti, J. A. M. Kenter, D. F. McNeill, R. N. Ginsburg, P. K. Swart, and L. A. Melim, 2001, Calibration of seismic sequence stratigraphy with cores and logs, *in* R. N. Ginsburg, ed., Subsurface geology of a prograding carbonate platform margin, Great Bahama Bank: Results of the Bahamas Drilling Project: SEPM Special Publication 70, p. 241–266.

Epting, M., 1989, Miocene carbonate build-ups of central Luconia, offshore Sarawak, *in* A. W. Bally, ed., Atlas of seismic stratigraphy, v. 2: AAPG Studies in Geology No. 27, p. 168–173.

Fagin, S. W., 1991, Seismic modeling of geologic structures: Society of Exploration Geophysicists Geophysical Development Series 2, 269 p.

Fitchen, W. M., 1997, Carbonate sequence stratigraphy and its application to hydrocarbon exploration and reservoir development, *in* F. J. Marfurt and A. Palaz, eds., Carbonate seismology: Society of Exploration Geophysicists Geophysical Developments Series 6, p. 121–178.

Fontaine, J. M., R. Cussey, J. Lacaze, R. Lanaud, and L. Yapaudjian, 1987, Seismic interpretation of carbonate depositional environments: AAPG Bulletin, v. 71, p. 281–297.

Frank, T. D., and K. Bernet, 2000, Isotopic signature of burial diagenesis and primary lithological contrasts in periplatform carbonates (Miocene, Great Bahama Bank): Sedimentology, v. 47, p. 1119–1134.

Freeman-Lynde, R. P., and W. B. F. Ryan, 1985, Erosional modification of Bahamas Escarpment: Geological Society of America Bulletin, v. 96, p. 481–494.

Freeman-Lynde, R. P., M. B. Cita, F. Jadoul, E. L. Miller, and W. B. F. Ryan, 1981, Marine geology of the Bahama escarpment: Marine Geology, v. 44, p. 119–156.

Freeman-Lynde, R. P., W. R. McClain, and K. C. Lohmann, 1988, Deep-marine origin of equant spar cements in Oligocene–Miocene perireef boundstones, Leg 101, Sire 635, Northeast Providence Channel, Bahamas, *in* J. A. Austin Jr., W. Schlager, et al., Proceedings of the Ocean Drilling Program, Scientific Results: Ocean Drilling Program, Texas A&M University, College Station, Texas, v. 101, p. 255–261.

Ginsburg, R. N., P. M. Harris, G. P. Eberli, and P. K. Swart, 1991, The growth potential of a bypass margin, Great

Bahama Bank: Journal of Sedimentary Geology, v. 61, no. 6, p. 976–987.

Glaser, K. S., and A. W. Droxler, 1991, High production and highstand shedding from deeply submerged carbonate banks, northern Nicaragua Rise: Journal of Sedimentary Petrology, v. 61, p. 128–142.

Goldhammer, R. K., P. A. Dunn, and L. A. Hardie, 1990, Depositional cycles, composite sea level changes, cycle stacking patterns, and the hierarchy of stratigraphic forcing: Examples from platform carbonates of the Alpine Triassic: Geological Society of America Bulletin, v. 102, p. 535–562.

Grammer, G. M., 1991, Formation and evolution of Quaternary carbonate foreslopes, Tongue of the Ocean, Bahamas: Ph.D. dissertation, Coral Gables, Florida, Univ. of Miami, p. 1–314.

Grammer, G. M., R. N. Ginsburg, and P. M. Harris, 1993, Timing of deposition, diagenesis, and failure of steep carbonate slopes in response to a high-amplitude/ high-frequency fluctuation in sea level, Tongue of the Ocean, Bahamas, *in* R. G. Loucks and J. F. Sarg, eds., Carbonate sequence stratigraphy: AAPG Memoir 57, p. 107–131.

Grötsch, J., and C. Mercadier, 1999, Integrated 3-D reservoir modeling based on 3-D seismic: The Tertiary Malampaya and Camago buildups, offshore Palawan, Philippines: AAPG Bulletin, v. 83, p. 1703–1728.

Harris, P. M., and A. H. Saller, 1999, Subsurface expression of the Capitan depositional system and implications for hydrocarbon reservoirs, northeastern Delaware Basin, *in* A. H. Saller, P. M. Harris, B. L. Kirkland, and S. J. Mazzullo, eds., Geologic framework of the Capitan Reef: SEPM Special Publication 65, p. 37–49.

Harwood, G. M., and P. A. Towers, 1988, Seismic sedimentologic interpretation of a carbonate slope, north margin of Little Bahama Bank, *in* J. A. Austin Jr., W. Schlager, et al., Proceedings of the Ocean Drilling Program, Scientific Results: Ocean Drilling Program, Texas A&M University, College Station, Texas, v. 101, p. 263–277.

Hine, A. C., and H. T. Mullins, 1983, Modern carbonate shelf-slope breaks, *in* D. J. Stanley and G. Moore, eds., The shelf break: Critical interface on continental margins: SEPM Special Publication 33, p. 169–188.

Hine, A. C., R. J. Wilber, J. M. Bane, A. C. Neumann, and K. R. Lorenson, 1981, Offbank transport of carbonate sands along open, leeward bank margins: Northern Bahamas: Marine Geology, v. 42, p. 327–348.

Hurst, J. M., and F. Surlyk, 1984, Tectonic control of Silurian shelf margin morphology and facies, North Greenland: AAPG Bulletin, v. 68, p. 1–17.

Illing, L. V., 1954, Bahamian calcareous sands: AAPG Bulletin, v. 38, p. 1–95.

James, N. P., and R. N. Ginsburg, 1979, The seaward margin of Belize barrier and atoll reefs: International Association of Sedimentologists Special Publication 3: Oxford, Blackwell, 191 p.

Jones, T., and A. Nur, 1982, Seismic velocities and anisotropy in mylonites and the reflectivity of deep crustal fault zones: Geology, v. 10, 260–263.

Kenter, J. A. M., 1990, Carbonate platform flanks: Slope angle and sediment fabric: Sedimentology, v. 37, p. 777–794.

Kenter, J. A. M., R. N. Ginsburg, and S. R. Troelstra, 2001, Sea-level-driven sedimentation patterns on the slope and margin, *in* R. N. Ginsburg, ed., Subsurface geology of a prograding carbonate platform margin, Great Bahama Bank: Results of the Bahamas Drilling Project: SEPM Special Publication 70, p. 61–100.

Kievman, C. M., 1998, Match between late Pleistocene Great Bahama Bank and deep-sea oxygen isotope records of sea level: Geology, v. 26, p. 635–638.

Ladd J. W., and R. E. Sheridan, 1987, Seismic stratigraphy of the Bahamas: AAPG Bulletin, v. 7, p. 719–736.

Macurda, D. B., Jr., 1988, Seismic stratigraphy of carbonate platform sediments, southwest Florida, *in* A. W. Bally, ed., Atlas of seismic stratigraphy, v. 2: AAPG Studies in Geology 27, p. 159–161.

Macurda, D. B., Jr., 1997, Carbonate seismic facies analysis, *in* F. J. Marfurt and A. Palaz, eds., Carbonate seismology: Society of Exploration Geophysicists Geophysical Developments Series 6, p. 95–119.

Malone, M. J., N. C. Slowey, and G. M. Henderson, 2001, Early diagenesis of shallow-water periplatform carbonate sediments, leeward margin, Great Bahama Bank (Ocean Drilling Program Leg 166): Geological Society of America Bulletin, v. 113, p. 881–894.

Manfrino C., and R. N. Ginsburg, 2001, Pliocene to Pleistocene depositional history of the upper platform margin, *in* R. N. Ginsburg, ed., Subsurface geology of a prograding carbonate platform margin, Great Bahama Bank: Results of the Bahamas Drilling Project: SEPM Special Publication 70, p. 17–39.

Masaferro, J. L., 1997, Interplay of tectonism and carbonate sedimentation in the Bahamas foreland basin: Ph.D. dissertation, University of Miami, 147 p.

Masaferro, J. L., and G. P. Eberli, 1999, Jurassic–Cenozoic structural evolution of southern Great Bahama Bank, southern Bahamas, *in* P. Mann, ed., Caribbean Basins, sedimentary basins of the world 4: Amsterdam, Elsevier, p. 167–193.

Miller, K. G., J. D. Wright, and R. G. Fairbanks, 1991, Unlocking the ice house: Oligocene–Miocene oxygen isotopes, eustasy, and margin erosion: Journal of Geophysical Research, v. 96, p. 6829–6848.

Mitchum, R. M, P. R. Vail, and J. B. Sangree, 1977, Stratigraphic interpretation of seismic reflection patterns in depositional sequences, *in* C. E. Payton, ed., Seismic stratigraphy— Applications to hydrocarbon exploration: AAPG Memoir 26, p. 117–133.

Moussavian, E., and A. Vecsei, 1995, Paleocene reef sediments from the Maiella carbonate platform, Italy: Facies, v. 32, p. 213–222.

Mullins, H. T., and H. E. Cook, 1986, Carbonate apron models: Alternatives to the submarine fan model for paleoenvironmental analysis and hydrocarbon exploration: Sedimentary Geology, v. 48, p. 37–79.

Mullins, H. T., and A. C. Hine, 1989, Scalloped bank margins: Beginning of the end of carbonate platforms: Geology, v. 17, p. 30–33.

Mullins, H. T., K. C. Heath, H. M. Van Buren, and C. R. Newton, 1984, Anatomy of a modern open-ocean carbonate slope: Northern Little Bahama Bank: Sedimentology, v. 31, p. 141–168.

Mutti, M., 1995, Porosity development and diagenesis in the Orfento Supersequence and its bounding unconformities (Upper Cretaceous, Maiella Mountains, Italy), *in* D. A. Budd, A. Saller, and P. M. Harris, eds., Unconformities and porosity in carbonate strata: AAPG Memoir 63, p. 141–158.

Mutti, M., D. Bernoulli, G. P. Eberli, and A. Vecsei, 1996, Facies associations and timing of progradation with respect to sea level cycles in an Upper Cretaceous platform margin (Orfento Supersequence, Maiella, Italy): Journal of Sedimentary Research, v. 66, p. 781–799.

Mutti, M., D. Bernoulli, S. Spezzaferri, and P. Stille, 1999, Lower and middle Miocene carbonate facies in the Central Mediterranean: The impact of paleoceanography on sequence stratigraphy, *in* P. M. Harris, A. H. Saller, and J. A. Simo, eds., Advances in carbonate sequence stratigraphy: Application to reservoirs, outcrops, and models: SEPM Special Publication 63, p. 371–384.

Neidell, N. S., and E. Poggiagliolmi, 1977, Stratigraphic modeling and interpretation— Geophysical principles and techniques, *in* C. E. Payton, ed., Seismic stratigraphy— Applications to hydrocarbon exploration: AAPG Memoir 26, p. 389–416.

Newell, N. D., and J. K. Rigby, 1957, Geological studies on the Great Bahama Banks: SEPM Special Publication 5, p. 15–79.

Osleger, D. A., and S. W. Tinker, 1999, Three-dimensional architecture of Upper Permian high-frequency sequences, Yates-Capitan shelf margin, Permian Basin, U.S.A., *in* P. M. Harris, A. H. Saller, and J. A. Simo, eds., Advances in carbonate sequence stratigraphy: Application to reservoirs, outcrops, and models: SEPM Special Publication 63, p. 169–185.

Paull, C. K., and W. P. Dillon, 1980, Erosional origin of the Blake Escarpment: An alternative hypothesis: Geology, v. 8, p. 538–542.

Paull, C. K., and A. C. Neumann, 1987, Continental margin brine seeps: Their geological consequences: Geology, v. 15, p. 545–548.

Paull, C. K, F. N. Spiess, J. R. Curray, and D. C. Twichell, 1990, Origin of Florida Canyon and the role of spring sapping on the formation of submarine box canyons: Geological Society of America Bulletin, v. 102, p. 502–515.

Paull, C. K., R. Freeman-Lynde, T. J. Bralower, J. M. Gardemal, A. C. Neumann, B. D'Argenio, and E. Marsella, 1991, Geology of the strata exposed on the Florida Escarpment: Marine Geology, v. 91, p. 177–194.

Pomar, L., 1993, High-resolution sequence stratigraphy in prograding carbonates: Application to seismic interpretation, *in* R. Loucks and I. F. Sarg, eds., Recent advances and applications of carbonate sequence stratigraphy: AAPG Memoir 57, p. 389–407.

Pomar, L., W. C. Ward, and D. G. Green, 1996, Upper Miocene reef complex of the Llucmajor area, Mallorca, Spain, *in* E. K. Franseen, M. Esteban, W. C. Ward, and J. -M. Rouchy, eds., Models for carbonate stratigraphy from Miocene reef complexes of the Mediterranean regions: SEPM Concepts in Sedimentology and Paleontology 5, p. 191–225.

Ramsayer, G. R., 1979, Seismic stratigraphy, a fundamental exploration tool: Offshore Technology Conference Proceedings, v. 3, p. 1859–1867.

Read, J. F., 1985, Carbonate platform facies models: AAPG Bulletin, v. 69, p. 1–21.

Rudolph, K. W., W. Schlager, and K. T. Biddle, 1989, Seismic models of a carbonate foreslope-to-basin transition, Picco di Vallandro, Dolomite Alps, northern Italy: Geology, v. 17, p. 453–456.

Sanders, D., 1994, Carbonate platform growth and erosion: The Cretaceous to Tertiary of Montagna della Maiella, Italy: Ph.D. dissertation, Swiss Federal Institute of Technology (ETH), Zurich, Switzerland, 122 p.

Sarg, J. F., 1988, Carbonate sequence stratigraphy, *in* C. K. Wilgus, B. S. Hastings, C. G. St. C. Kendall, H. W. Posamentier, C. A. Ross, J. C. Van Wagoner, eds., Sea-level changes: An integrated approach: SEPM Special Publication 42, p. 155–181.

Sarg, J. F., 1989, Middle–Late Permian depositional sequences, Permian Basin, west Texas and New Mexico, *in* A. W. Bally, ed., Atlas of seismic stratigraphy, v. 2: AAPG Studies in Geology No. 27, p. 140–155.

Schlager, W., 1992, Sedimentology and sequence stratigraphy of reefs and carbonate platforms: AAPG Continuing Education Course Note Series No. 34, 71 p.

Schlager, W., and R. N. Ginsburg, 1981, Bahama carbonate platforms— The deep and the past: Marine Geology, v. 44, p. 1–24.

Schlager, W., F. Bourgeois, G. MacKenzie, and J. Smit, 1988, Boreholes at Great Isaac and Site 626 and the history of the Florida Straits, *in* J. A. Austin Jr., W. Schlager, et al., Proceedings of the Ocean Drilling Program, Scientific Results: Ocean Drilling Program, Texas A&M University, College Station, Texas, v. 101, p. 425–437.

Schwab, A. M., and G. P. Eberli, 2000, Synthetic seismic model of the Miette buildup margin and a comparison to subsurface seismic data of the Redwater Reef margin, *in* P. W. Homewood and G. P. Eberli, eds., Genetic stratigraphy on the exploration and production scales: Bulletin du Centre de Recherches Elf Exploration-Production Mémoire 24, p. 203–222.

Sheridan, R. E., J. T. Crosby, G. M. Bryan, and P. L. Stoffa, 1981, Stratigraphy and structure of southern Blake plateau, Northern Florida Straits, and Northern Bahama Platform from multi-channel seismic reflection data: AAPG Bulletin, v. 65, p. 2571–2593.

Stafleu, J., and W. Schlager, 1993, Pseudo-toplap in seismic models of the Schlern-Raibl contact, Sella Platform, northern Italy: Basin Research, v. 5, p. 55–65.

Stafleu, J., and M. D. Sonnenfeld, 1994, Seismic models of a shelf-margin depositional sequence: Upper San Andres Formation, Last Chance Canyon, New Mexico: Journal of Sedimentary Research, v. B64, no. 4, p. 481–499.

Stössel, I., 1999, Rudists and carbonate platform evolution: The Late Cretaceous Maiella carbonate platform margin, Abruzzi, Italy: Memorie di Scienze Geologiche, v. 51, no. 2, p. 333–413.

Strasser, A., B. Pittet, H. Hillgärtner, and J.-B. Pasquier, 1999, Depositional sequences in shallow-dominated sedimentary systems: Concepts for a high-resolution analysis: Sedimentary Geology, v. 128, p. 201–221.

Surlyk, F., and J. R. Ineson, 1992, Carbonate gravity flow deposition along a platform margin scarp (Silurian, North Greenland): Journal of Sedimentary Petrology, v. 62, p. 400–410.

Tyrell, W. W., and R. G. Davis, 1989, Miocene carbonate shelf margin, Bali-Flores Sea, Indonesia, *in* A. W. Bally, ed., Atlas of seismic stratigraphy, v. 2: AAPG Studies in Geology No. 27, p. 174–179.

Van Konijnenburg, J.-H., 1997, Sedimentology and stratigraphic architecture of a Cretaceous to lower Tertiary carbonate base-of-slope succession, Gran Sasso d'Italia, central Apennines, Italy: Ph.D. thesis, Swiss Federal Institute of Technology (ETH), Zurich, Switzerland, 174 p.

Van Konijnenburg, J.-H., D. Bernoulli, and M. Mutti, 1999, Stratigraphic architecture of a Lower Cretaceous–lower Tertiary carbonate base-of-slope succession: Gran Sasso d'Italia (central Apennines, Italy, *in* P. M. Harris, A. H. Saller, and J. A. Simo, eds., Advances in carbonate sequence stratigraphy: Application to reservoirs, outcrops, and models: SEPM Special Publication 63, p. 291–315.

Vecsei, A., 1998, Bioclastic sediment lobes on a supply dominated Upper Cretaceous carbonate platform, Montagna della Maiella: Sedimentology, v. 45, p. 473–487.

Vecsei, A., 1991, Aggradation und Progradation eines Karbonatplattformrandes: Kreide bis mittleres Tertiär der Montagna della Maiella, Abruzzen: Ph.D. thesis, Mitteilungen des Geologischen Institutes der Eidgenössischen Technischen Hochschule und der Universität, Zürich, 170 p.

Vecsei, A., D. Sanders, D. Bernoulli, G. P. Eberli, and J. S. Pignatti, 1998, Sequence stratigraphy and evolution of the Maiella carbonate platform margin, Cretaceous to Miocene, Italy, *in* P. C. De Graciansky, J. Hardenbol, T. Jacquin, and P. R. Vail, eds., The Mesozoic and Cenozoic sequence stratigraphy of European basins: SEPM Special Publication 60, p. 53–74.

Wagner, P. D., 1997, Seismic signature of carbonate diagenesis, *in* F. J. Marfurt, and A. Palaz, eds., Carbonate seismology: Society of Exploration Geophysicists Geophysical Developments Series 6, p. 307–320.

Walles, F. E., 1993, Tectonic and diagenetically induced seal failure within the south-western Great Bahamas Bank: Marine Petroleum Geology, v. 10, p. 14–28.

Weimer, P., and T. L. Davis, eds., 1996, Applications of 3-D seismic data to exploration and production: AAPG Studies in Geology 42/SEG Geophysical Development Series 5, 270 p.

Wilber, R. J., J. D. Milliman, and R. B. Halley, 1990, Accumulation of bank-top sediment on the western slope of Great Bahama Bank: Rapid progradation of a carbonate megabank: Geology, v. 18, p. 970–974.

Winterer, E. L., and A. Bosellini, 1981, Subsidence and sedimentation on a Jurassic passive continental margin (Southern Alps, Italy). AAPG Bulletin, v. 65, p. 394–421.

Zachos, J. C., L. D. Stott, and K. C. Lohmann, 1994, Evolution of the early Cenozoic marine temperatures: Paleoceanography, v. 9, p. 353–387.

11

Borgomano, J. R. F., and J. M. Peters, 2004, Outcrop and seismic expressions of coral reefs, carbonate platforms, and adjacent deposits in the tertiary of the Salalah Basin, South Oman, *in* Seismic imaging of carbonate reservoirs and systems: AAPG Memoir 81, p. 251–266.

Outcrop and Seismic Expressions of Coral Reefs, Carbonate Platforms, and Adjacent Deposits in the Tertiary of the Salalah Basin, South Oman

Jean R. F. Borgomano

Shell Research and Technical Services, Rijswijk, The Netherlands; University of Provence, Marseille, France

Jeroen M. Peters

Petroleum Development Oman, Muscat, Sultanate of Oman

ABSTRACT

The integration of outcrop, seismic, and well data in the Salalah area, south Oman, allows a sequence-stratigraphic interpretation of the Salalah carbonate margin that is related to the opening of the Gulf of Aden during the Tertiary. The outcrop expression of platform, coral reefs, faulted shelf edge, and base-of-slope carbonates is briefly described to support the seismostratigraphic interpretation of an analog interval in the subsurface. The rapidly increasing rate of subsidence, combined with the differentiation of faulted highs and troughs, resulted in the stepwise drowning of shallow-marine carbonates and the upward increase in basinal marls and turbidites. The growth of coral reefs on the edge of fault blocks preceded the end of carbonate deposition.

INTRODUCTION

In 1994, Petroleum Development Oman drilled the first onshore hydrocarbon exploration well in the Salalah Basin situated in Dhofar, the southernmost province of the Sultanate of Oman (Figure 1). The objective of this well (Salalah Plain-1) was identified on seismic as a moundlike structure and was interpreted, by analogy to nearby Tertiary outcrops, as a Tertiary reef (Figure 2). The well was dry but confirmed the stratigraphic model and structural interpretations. It yielded information relevant to the regional geology and especially to the evolution of the carbonate sedimentation in relation to the opening of the Gulf of Aden during the Oligocene–Miocene (for regional setting and geology, see Platel and Roger, 1989; Roger et al., 1989, 1997).

This chapter focuses on the outcrop and seismic expression of the carbonate platform, reef, and adjacent deposits that characterize the Eocene–Miocene interval in the Salalah Basin. Outcrop and subsurface stratigraphy are interpreted within the tectonic context of the Gulf of Aden rift (Figures 2B, 3). The Salalah Basin offers a unique opportunity to study, from combined outcrop and subsurface data, the thick (500–2500 m) Cenozoic carbonate sequence deposited on the southeastern margin of the Arabian Peninsula (Figure 2A).

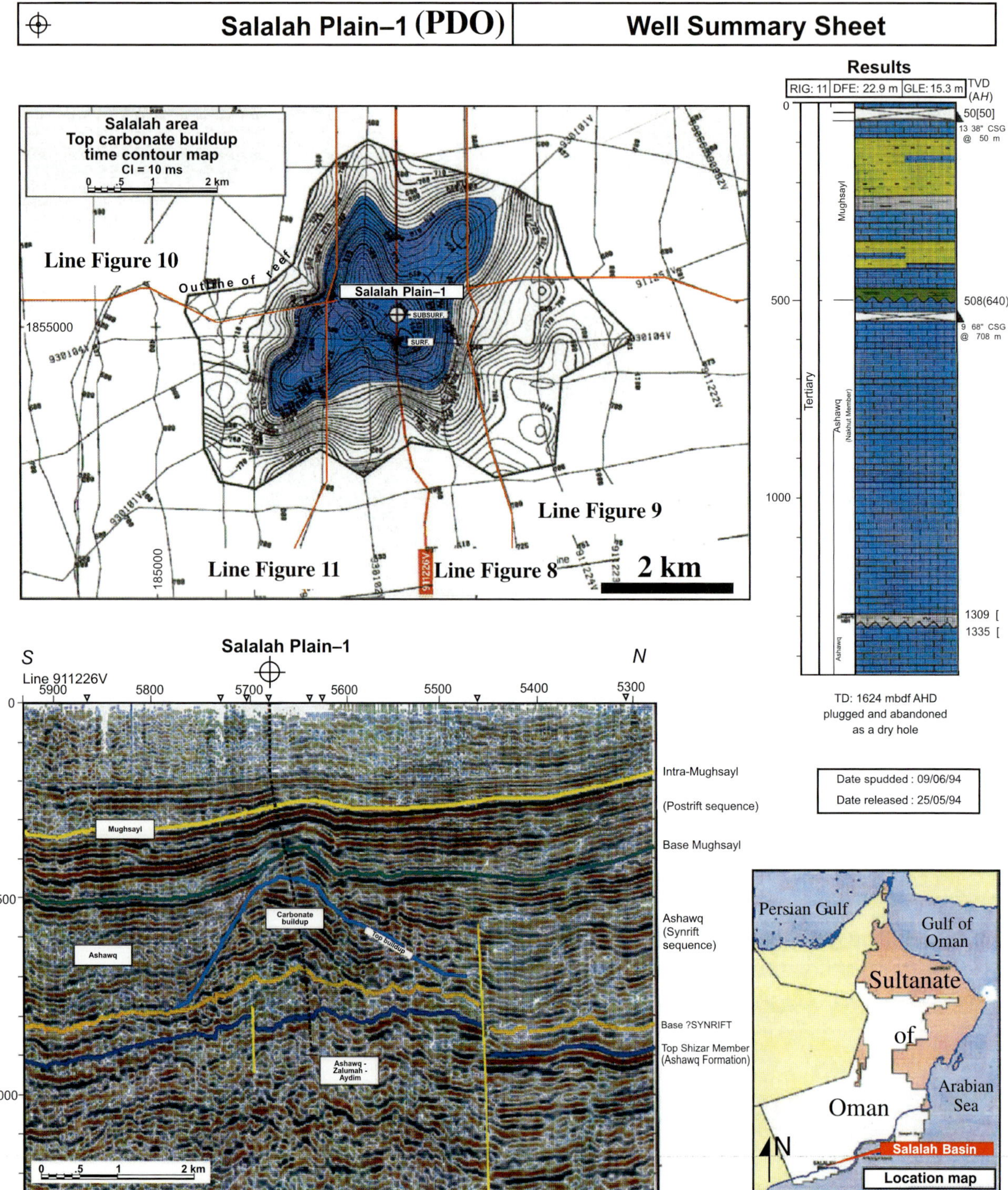

FIGURE 1. Location of the study area and summary sheet of the Petroleum Development Oman (PDO) well Salalah Plain-1.

SUBSURFACE DATA

Figure 1 summarizes the subsurface data acquired in relation with the Salalah Plain-1 well. The seismic survey included several two-dimensional (2-D) lines with 25-m shooting and recording spacing, 5–60-Hz sweep in 10 s, a maximum offset of 3200 m, and 72-bin fold (Figure 1). A 300-ms two-way traveltime (TWT) stratigraphic closure was interpreted between 460 and 800 ms TWT. The blue horizon (Figure 1) that materializes this structural interpretation is in fact not a chronostratigraphic surface but the interpreted envelope of the reefal facies in relation to the mound anomaly. Following the integration of the well results, an alternative to this trap interpretation will also be proposed in the next paragraphs.

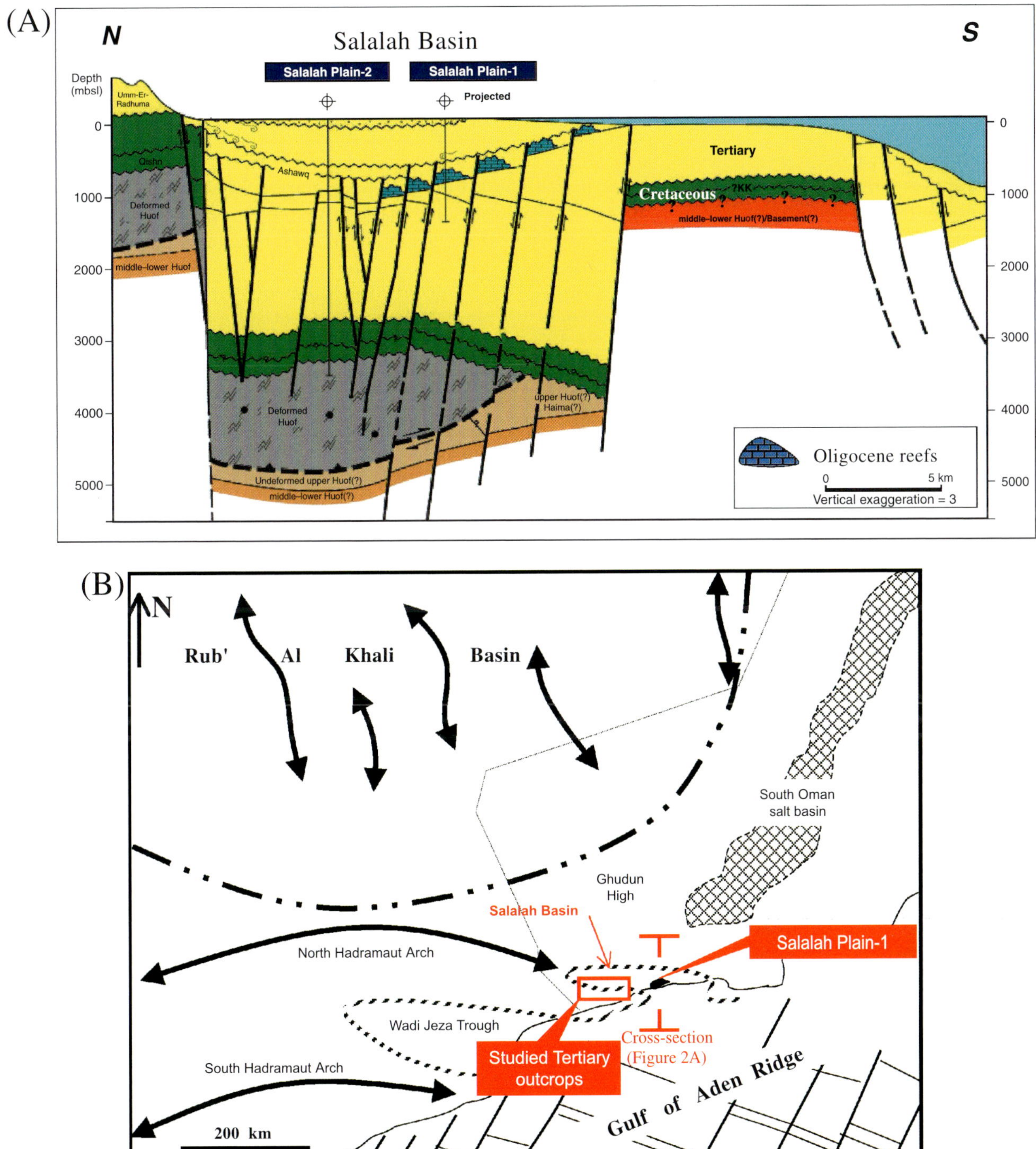

FIGURE 2. General geologic framework of the southern Arabian Peninsula. (A) Geologic cross section in northeast Yemen. (B) Simplified map of the main structural elements in the southern Arabian Peninsula.

The well was deviated in response to surface constraints and penetrated the crest of the mapped structure at 560 m true vertical depth (TVD). The total depth of the well was 1450 m TVD. A standard log suite was acquired that confirmed the good reservoir quality of the objective; the structure was fully water bearing, presumably because of lack of charge. Lithologic and biostratigraphic data were obtained from cuttings and sidewall samples and integrated in further interpretation of the log and seismic data. No cores were taken.

The stratigraphic interpretation of the seismic, based on the stratigraphic nomenclature defined in the surrounding outcrops (Figure 3, Platel and Roger, 1989), was confirmed by the well results and included:

- a prerift platform sequence (Eocene–early Oligocene) represented by the restricted and shallow-marine marls and limestones of the basal Ashawq Formation (Shizar Member),
- a synrift reefal sequence (early Oligocene) represented by the coral-rich shallow-marine limestones of the upper Ashawq Formation (Nakhlit Member), and
- a mainly postrift basinal sequence (late Oligocene–Miocene) represented by the pelagic and turbiditic limestones and marls of the Mughsayl Formation.

One of the alternative interpretations discussed in this chapter is the possible synchronism between the uppermost Ashawq Formation and the lowermost Mughsayl Formation and its geodynamic implications.

OUTCROP EXPRESSION OF TERTIARY REEF, PLATFORM, AND ADJACENT DEPOSITS

Most of the outcrop data discussed in this section were collected during several surveys organized by Petroleum Development Oman between 1990 and 1994 in the Salalah area. The 1:100,000 scale geologic maps (Platel et al., 1987) published by the Ministry of Petroleum and Minerals of Oman (MPM) and the French Geological Survey (BRGM) provided the basis for our stratigraphic, sedimentologic, and structural analyses. The present chapter focuses on the outcrop area of Mughsayl that is dominated by the Eocene–Miocene carbonates of the Mughsayl and Ashawq Formations (Figure 4A). The outcrop area, located approximately 40 km to the west of the Salalah Plain-1 well, is characterized by 1000-m-high coastal cliffs and corresponds to the western edge of the Salalah Basin that was exhumed during the upper Miocene uplift of the Gulf of Aden rift shoulder (Platel and Roger, 1989). A faulted paleoshelf edge, mapped by Platel et al. (1987), separates the shallow-marine carbonates (Ashawq Formation) from the slope-to-basin carbonates (Mughsayl Formation) (Figure 4B). A spectacular steep road with numerous hairpin turns climbs the cliff formed by this paleoshelf edge and offers unique insights into this complex depositional system (Figure 5A).

Reefs

The most relevant outcrop features consist of several moundlike structures (Figure 5) that can be observed along the paleoshelf edge west of Mughsayl (Figure 4). The present-day topography and bioclastic grainstone layers draping the flank of the structures reveal pyramidal shapes. The apparent elevation of the mounds is approximately 100 m, which is two times less than their estimated horizontal dimensions. The uppermost mounds are separated by paleogullies dipping toward the basin (Figure 5). The mounds are located at the top of the Ashawq Formation and form an apparent backstepping set of carbonate buildups (Figure 4A). This overall "transgressive trend" is consistent with the parasequence set observed on the flank of one the mounds showing an "opening-upward" succession (e.g., Borgomano, 2000) from restricted miliolid wackestones to coral boundstones and fining-upward bioclastic rudstone-grainstones (Figure 5B).

The mounded lithofacies consists of coral boundstones, including a significant volume fraction (30–40% bulk volume [BV]) of inter- and intraskeletal calcite cements and geopetal mud sediments. Coral fragments, micritic grains, and benthonic foraminifera dominate the overlying bioclastic grainstones. These grainstones contain a moderate volume of interparticle cements (10–30% BV) and have moderate porosity (10–20%). At the small scale of the outcrop, the bioclastic sediments drape the topography created by the coral boundstone. Both biologic (coral growth) and diagenetic (carbonate cementation) processes contributed to the buildup of the mound. These shelf-edge coral mounds can be considered as ecological reefs (Longman, 1981) that strongly influenced depositional processes in adjacent zones and were deposited on a topographic high.

Platform

The broadest part of the plateau to the west of the paleoshelf edge (Figure 4A) consists of a 200- to 300-m thick succession of horizontal limestone and dolomite beds belonging to the Ashawq Formation (Figure 6A). According to Roger et al. (1997), this succession of horizontal beds corresponds to a carbonate platform system (Wilson, 1975), which was dominated by mudstone-wackestone deposited in restricted shallow-marine

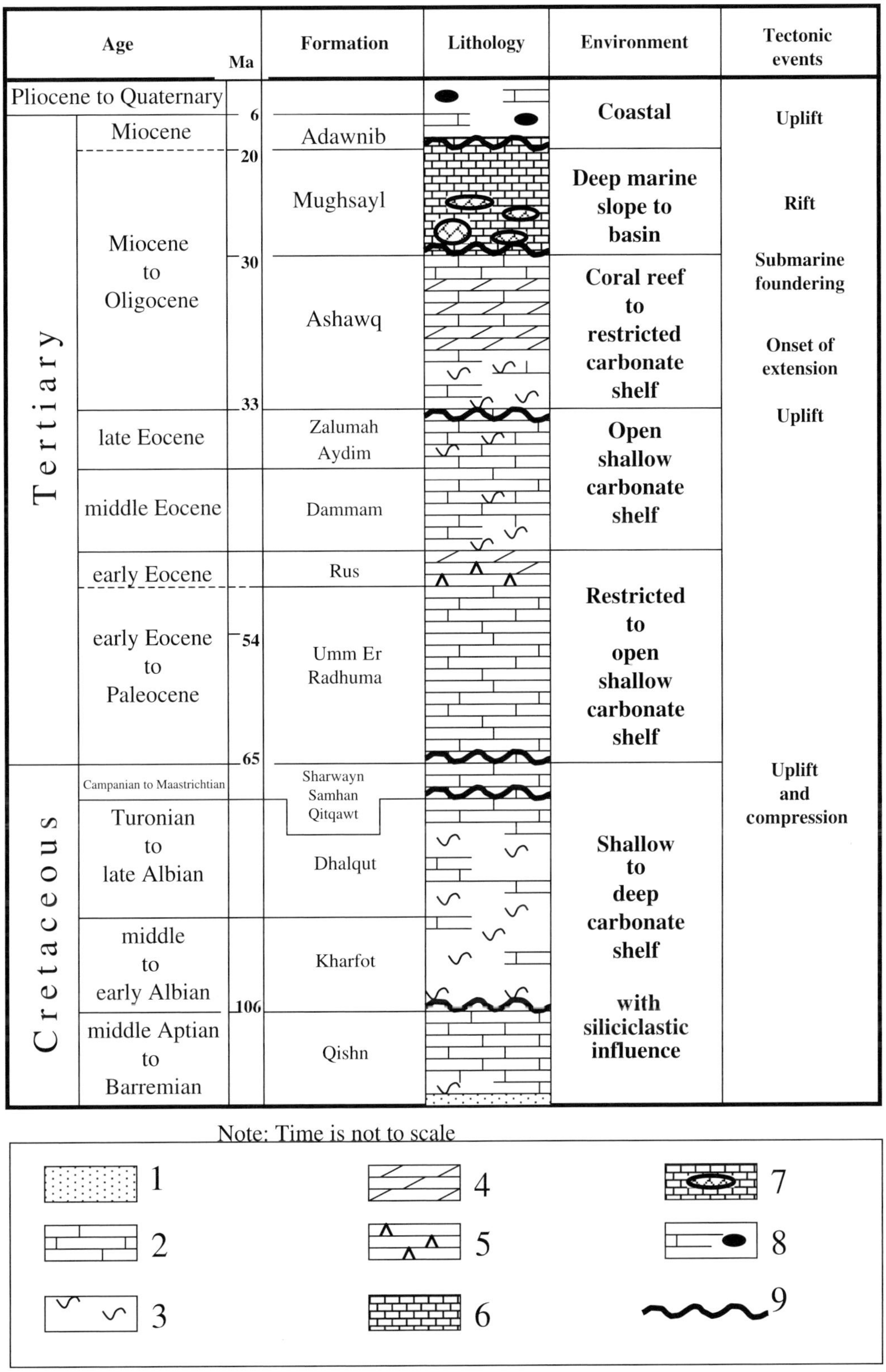

Age		Ma	Formation	Lithology	Environment	Tectonic events
Pliocene to Quaternary		6			Coastal	Uplift
Tertiary	Miocene	20	Adawnib			
	Miocene to Oligocene	30	Mughsayl		Deep marine slope to basin	Rift
		33	Ashawq		Coral reef to restricted carbonate shelf	Submarine foundering Onset of extension
	late Eocene		Zalumah Aydim		Open shallow carbonate shelf	Uplift
	middle Eocene		Dammam			
	early Eocene		Rus		Restricted to open shallow carbonate shelf	
	early Eocene to Paleocene	54 65	Umm Er Radhuma			
Cretaceous	Campanian to Maastrichtian		Sharwayn Samhan Qitqawt		Shallow to deep carbonate shelf with siliciclastic influence	Uplift and compression
	Turonian to late Albian		Dhalqut			
	middle to early Albian	106	Kharfot			
	middle Aptian to Barremian		Qishn			

Note: Time is not to scale

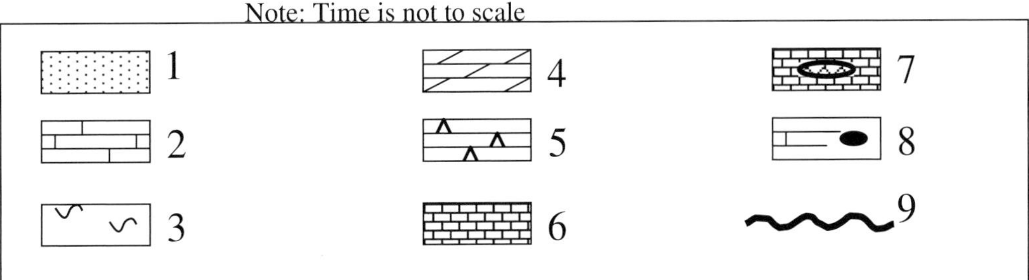

FIGURE 3. Stratigraphic chart of the Cretaceous–Tertiary in Dhofar (south Oman), modified from Roger et al. (1989). 1 = argillaceous sandstone; 2 = neritic limestone; 3 = marls; 4 = dolomite; 5 = sulfates; 6 = turbiditic limestones; 7 = limestone olistostrome; 8 = conglomeratic limestone; 9 = regional unconformity.

conditions similar to modern lagoons. The spatial relationships between the reefs and the platform succession have not been mapped in detail, but our investigations suggest that the upper part of the platform sequence is laterally equivalent to the marginal reefs and can be considered as a "back-reef" succession.

Base of Slope to Basin

On the basinward side of the paleoshelf edge (Figure 4B), the Mughsayl Formation consists of 400–500 m of resedimented carbonates dominated by grainstones, rudstones, breccias, and olistostromes (Figures 6B, 7).

According to Roger et al. (1997) and our investigation, this formation fills a fault-controlled submarine canyonlike incision within the shallow-marine carbonates of the Ashawq Formation. It is also important to observe that the Mughsayl Formation locally overlies the Ashawq Formation on both sides of the paleoshelf

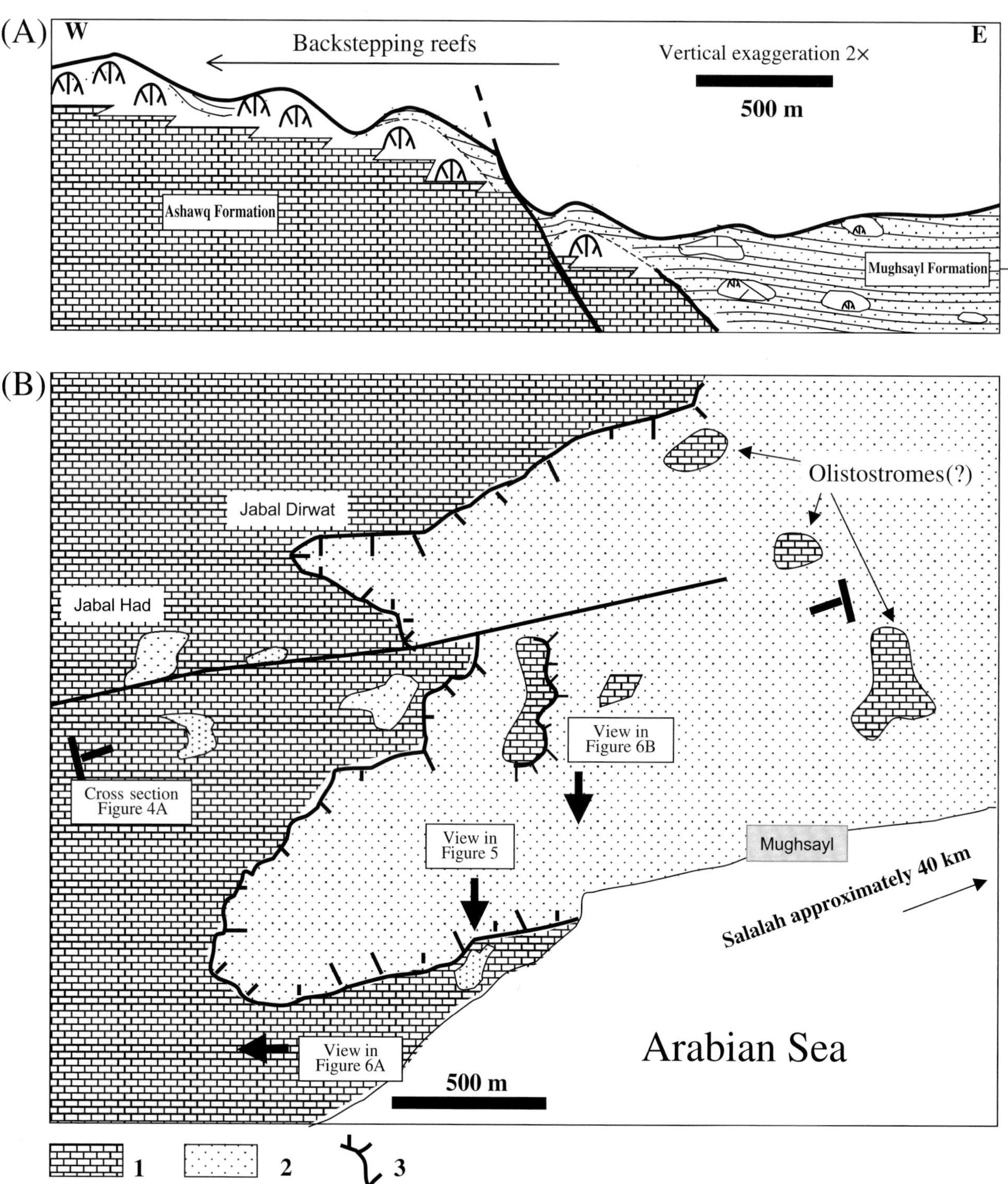

FIGURE 4. Geologic framework of the studied Tertiary outcrops in the Mughsayl area. (A) Cross section and (B) map showing the spatial relationships between the shallow-marine carbonates of the Ashawq Formation (Eocene–early Oligocene) and the deeper-marine carbonates of the Mughsayl Formation (early–late Oligocene) (adapted from Platel et al., 1987). 1 = Shallow-marine carbonate of the Ashawq Formation (shelf and coral reefs); 2 = deeper marine carbonates of the Mughsayl Formation (low- and high-density turbidites, breccias, and olistostromes); 3 = paleoshelf edge marked by a faulted sharp contact between the two formations.

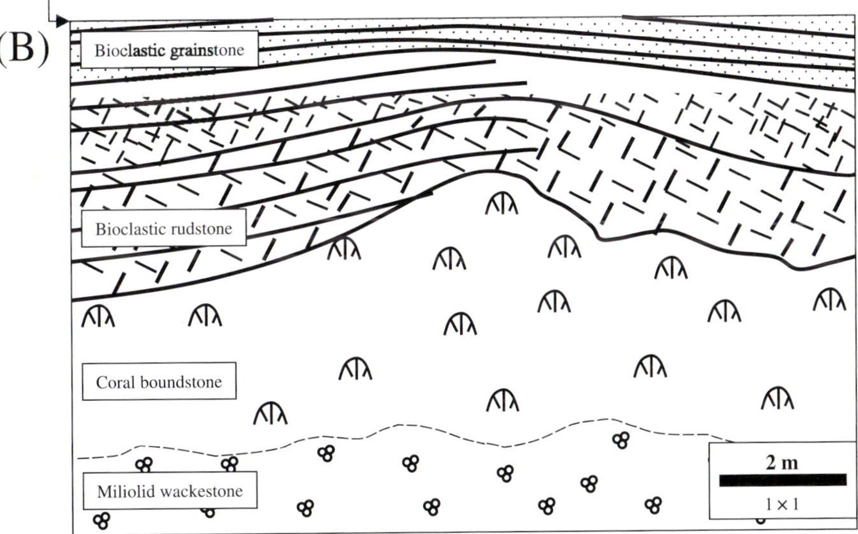

FIGURE 5. (A) Overview of the coral reefs (Ashawq Formation) and adjacent deposits in the Mughsayl area (for location, see Figure 4). Notice the paleogully that separates the two reefs. The arrows indicate the dipping directions of the layers that flank the reefs. The dashed lines follow the crests of the reefs. The dips of the base-of-slope layers (Mughsayl Formation) are represented by the plain lines. A near-vertical fault, parallel to the plane of the photo, controls the contact between the two formations. (B) Detailed cross section at the crest of the Ashawq reef, showing the interpreted onlap and drape of the coral boundstone by bioclastic grainstones. The vertical succession, from restricted inner shelf miliolid wackestone to high energy rudstone-grainstone, matches with the overall backstepping of the shallow-marine carbonates.

edge (Figure 4). The Mughsayl Formation is in sharp contact with the Ashawq Formation: It onlaps and drapes the topography formed by the Ashawq mounds. Gradual transition or interfingering between the two units was not observed in the outcrop.

The Mughsayl Formation was deposited in deep-water environments at the foot of a fault-related scarp. This interpretation is supported by the presence of interbedded deep-water pelagic mudstones and the products of gravity-flow processes, including debris flows (Figure 6B), grain flows, turbidity currents (Figure 7A), and slumps (Figure 7B). These gravity deposits are commonly formed by sheetlike meter-thick beds dominated by bioclastic and lithoclastic gainstone-rudstone with high interparticle porosity (20–35% BV). This detrital carbonate material is of neritic origin (coral

(A)

(B)

Figure 6. (A) General view of the shelf carbonates belonging to the Ashawq Formation (for location, see Figure 4B). This succession of horizontal beds, dominated by mudstone-wackestone, could correspond to the back-reef setting. (B) View of typical olistostromes (megablocks) and slumped layers in the Mughsayl Formation at the base of the paleoslope (for location, see Figure 4B).

and pelecypod skeletal material, foraminifera, and peloids) and includes a high proportion of resedimented "marine cement" in lithoclasts. The megablocks have also a neritic origin (coral reef) and contain marine carbonate cements. Although the generation of carbonate grain flows and turbidites is not necessarily controlled by steep slopes or shelf break (Cook, 1982), carbonate debris flows, with such a high proportion of megablocks (Figure 6B), suggest the existence of an erosive shelf edge and a significant fault scarp during the Oligocene. Overall, the Mughsayl Formation has the characteristics of a proximal base-of-slope complex with high proportion of debris flows, lithoclast gravels, and megablocks (Borgomano, 2000). The transition to a more distal (dominated by turbidite or bioclast sand) or basinal system has not been observed in the outcrops.

SEISMIC EXPRESSIONS OF REEF, PLATFORM, AND ADJACENT DEPOSITS

In view of the lack of well data in the area, the surrounding outcrops were used for the interpretation of the 2-D seismic survey in the Salalah Basin. The Oligocene mound structure drilled by the Salalah Plain-1 well (Figure 1) is developed on an east-west–trending fault block parallel to the axis of the Salalah Basin and the Wadi Jeza Trough (Figure 2). This buried faulted block in the Salalah Basin may correspond to the eastern extension of the uplifted structures outcropping in the Mughsayl area (Figure 4). The interpretations of three dip lines and one strike line are displayed in the Figures 8–11. The consistent record of terminations and seismic facies (Vail et al., 1977; Vail, 1987; Schlager, 1992) integrated with the well data and a qualitative extrapolation of the outcrop observations forms the basis of this seismic-stratigraphic interpretation.

Only two genetic types, observed in the outcrops, have been differentiated in relation to the seismic expression (Figures 8–10): the shallow-marine carbonates that build up, grow, and create topographies and the base-of-slope carbonates that fill and drape existing topographies. Rock-property analyses from the outcrop indicate significant difference between the two types: shallow-marine carbonates are very consolidated and have moderate porosity (10–20% BV),

(A)

(B)

FIGURE 7. (A) View of undisturbed layers in the Mughsayl Formation formed by alternating calciturbidites (low and high density), grain flows, and pelagic mud. (B) View of slumped and faulted layers in the Mughsayl Formation. The lithology of the sedimentary layers in A and B is similar. The visible deformations suggest the slope instability and synsedimentary tectonic activities. Neptunian dikes are visible above the geologist.

(A)

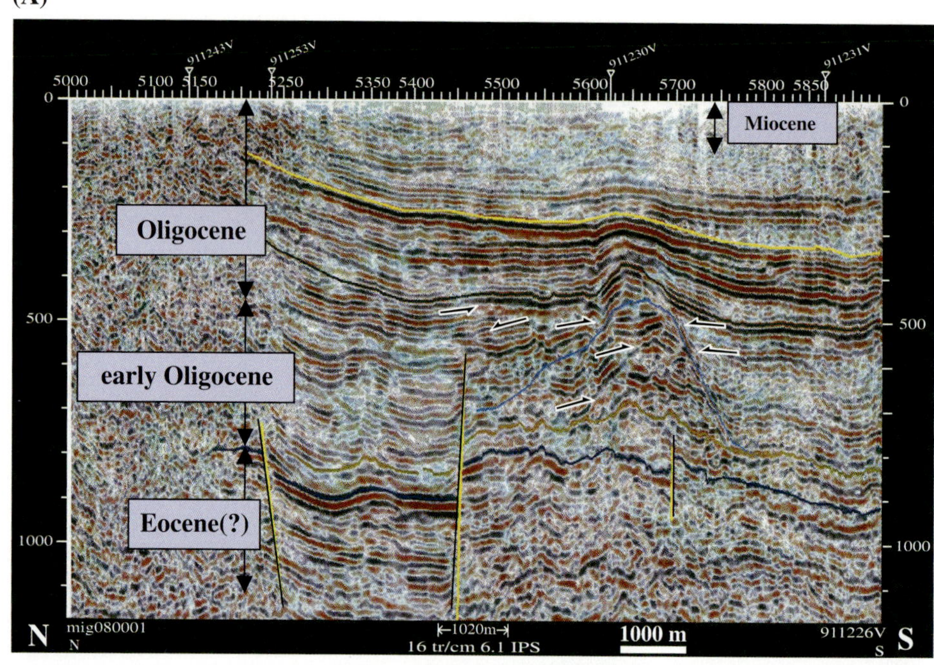

(B)

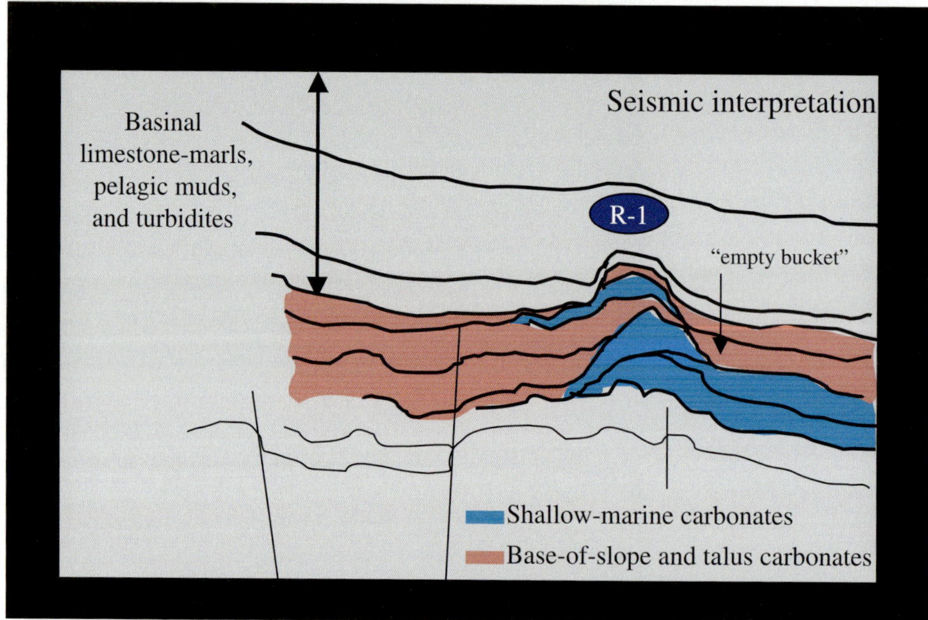

(C)

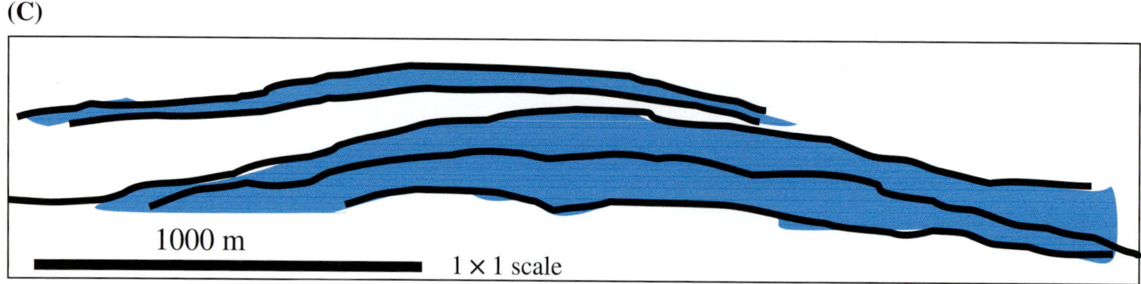

Figure 8. (A–B) Seismic-stratigraphic interpretation of 2-D seismic line (for location, see Figure 1). The growth of shallow-marine carbonates is initially limited to the south, then to the platform rim, leaving an "empty bucket" at the place of the early platform. (C) One-to-one scale representations of the reef geometry showing the steep slope angle. This dip line is oriented toward the basin. R-1 = reef 1.

(A)

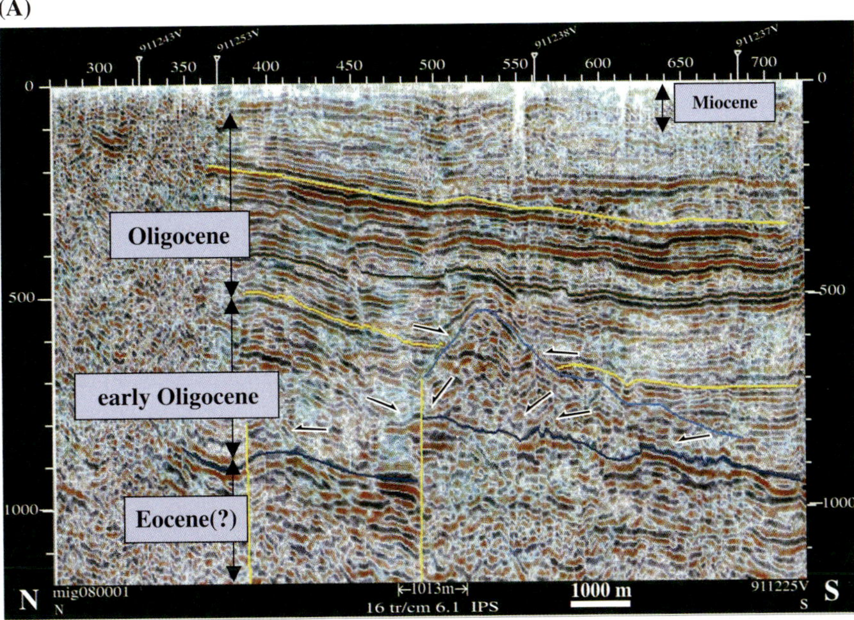

(B)

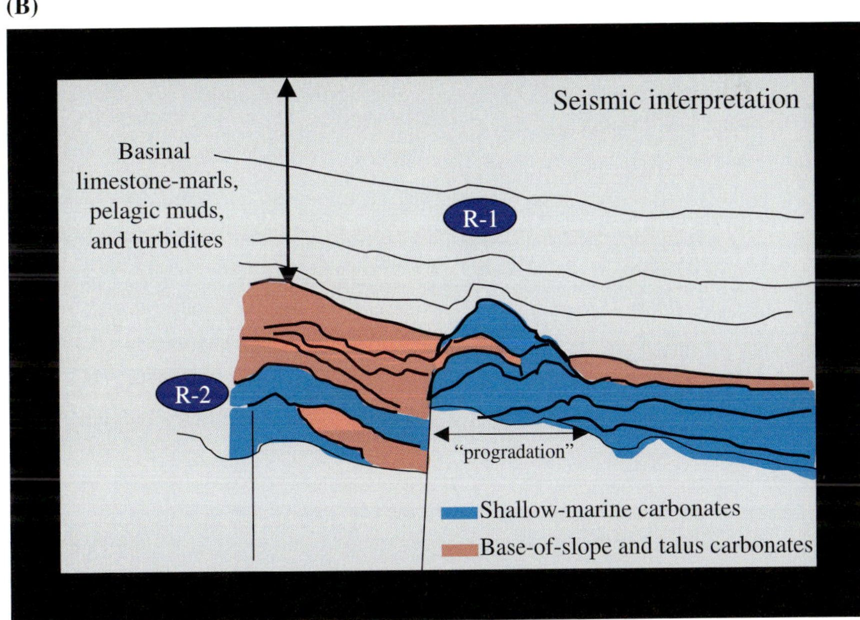

FIGURE 9. (A–B) Seismostratigraphic interpretation of 2-D seismic line (for location, see Figure 1). The line is oriented toward the basin (dip line). Platform progradation toward the north is interpreted based on the downlaps onto the basal horizon. R-1 = reef 1; R-2 = reef 2.

whereas base-of-slope carbonates are less consolidated and have higher porosity (20–35% BV). In this context, seismic facies characterized by parallel reflections ("tram lines"), offlaps (prograding), and mound shape (aggrading) associated with a strong reflectivity, are interpreted as shallow-marine carbonates. Seismic facies characterized by onlap (infill), draping onto the shallow-marine units, and wedge shape associated to a weaker refectivity are interpreted as base-of-slope units. A possible gully incising the shallow-marine unit is also observed on the strike line, suggesting active submarine erosion (Figure 10B). Two reefs (R-1 and R-2) are identified and several periods of reef growth are interpreted (three for R-1 and two for R-2). Downlapping terminations, which are interpreted to be related to a prograding shelf, predate the growth of the reefs (Figure 9B). This is significantly different than the predrill interpretation, which shows only a single symmetric build-up (Figure 1). Our present interpretation fits better with the asymmetric system observed in outcrops (Figures 4, 5), where the reefs sensu stricto occupy the margin of a broader platform that can be defined as a "rimmed platform" (Wilson, 1975). This seismic-stratigraphic interpretation also suggests that during the early Oligocene, the upper part of the Ashawq Formation (shallow-marine unit) and the lower part of the Mughsayl Formation (base-of-slope unit) were synchronous and interfingering at the seismic scale.

A true-scale representation of the reef R-1 (Figure 8C) shows dimensions that are three to four times greater than the individual outcrop reef dimensions but similar to the paleoslope profile displayed in Figure 4A. The low frequency of the seismic precludes the identification of small individual buildups that are most likely present on the margin of the shelf. The slope profile (10–15°) interpreted from both outcrop and seismic falls within the range of "Gaussian slope profiles" recorded in a majority of carbonate platform margins (Adams and Schlager, 2000). The steeper slope (45°) related to the flanks of the outcropping reefs could correspond to the edge of the "gullies" identified on seismic. Beyond seismic resolution, it is difficult to illustrate transition between the shallow-marine and base-of-slope carbonates. Whether this transition takes place as sharp contacts or gradient deserves additional seismic interpretation. The fault-related edge between the two formations in the outcrop suggests a sharp, onlap-like transition. Overall, the seismic

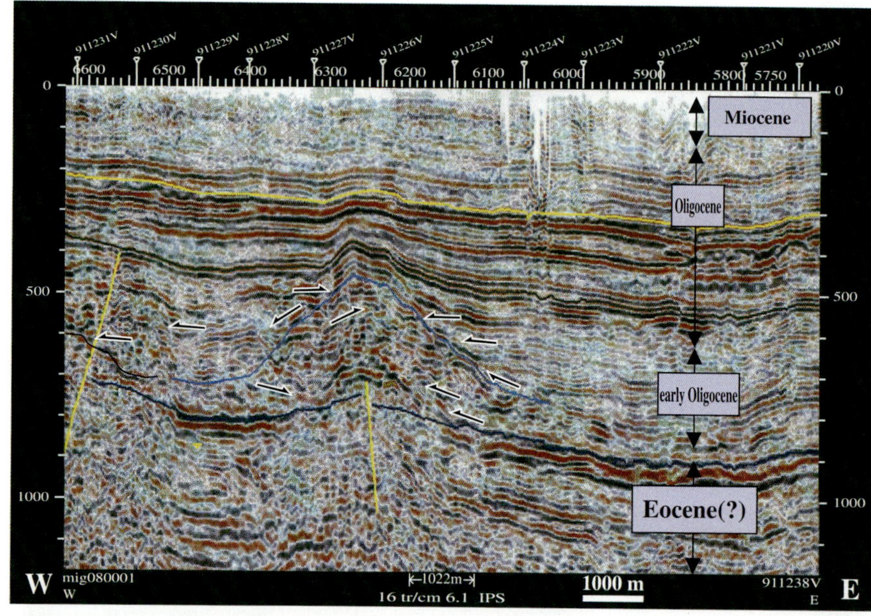

(B)

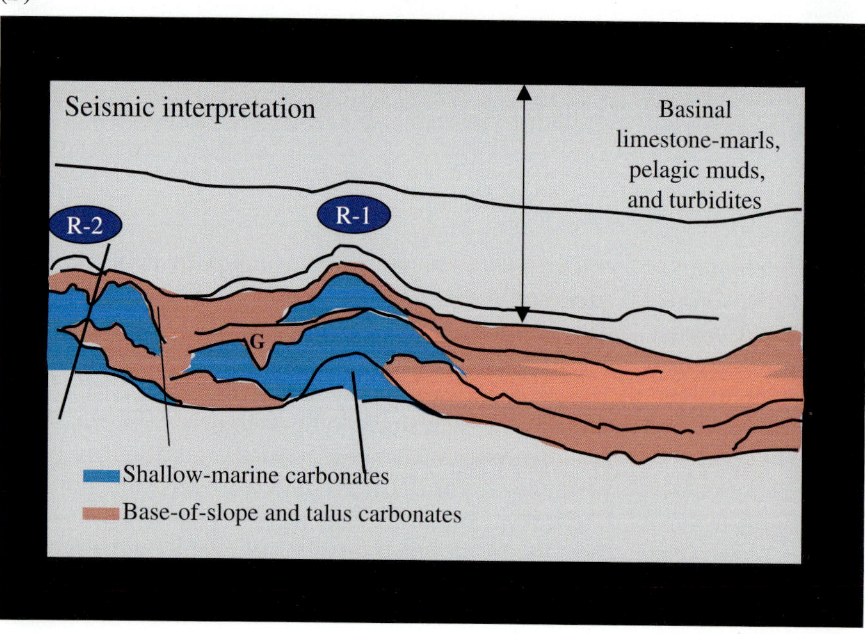

FIGURE 10. (A–B) Seismostratigraphic interpretation of 2-D seismic line (for location, see Figure 1). The line is oriented parallel to the basin axis (strike). This line shows the main difference in seismic expression between the shallow-marine and the base-of-slope carbonates: the shallow-marine carbonates buildup, grow, and create topographies (mound) and have a strong reflectivity and the base-of-slope carbonates fill and drape existing topographies (onlap) and have a lower reflectivity. R-1 = reef 1; R-2 = reef 2; G = gully.

FIGURE 11. (A–B) Seismostratigraphic interpretation of 2-D seismic line (for location, see Figure 1). The line is oriented toward the basin (dip line). R-1 = reef 1; R-2 = reef 2.

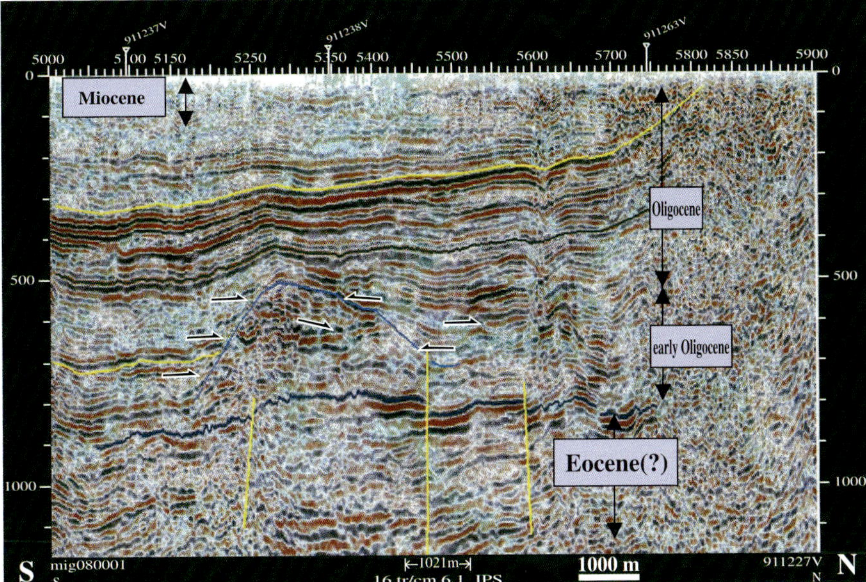

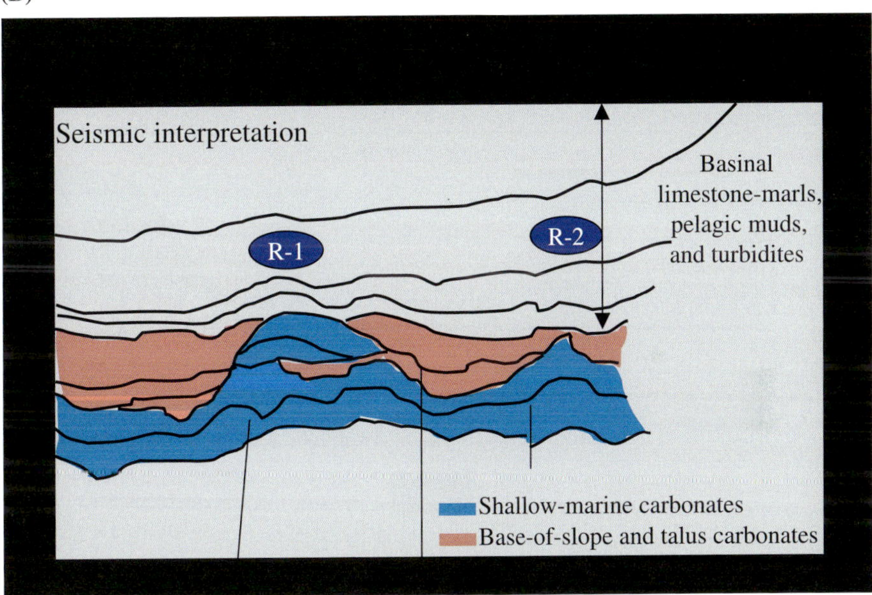

responses reflect structures observed on the nearby outcrops: draped mounds, onlaps, horizontal strata, incised gullies, and steep slopes related to a faulted shelf edge.

EVOLUTION OF THE SALALAH CARBONATE MARGIN DURING THE TERTIARY

The evolution of the Salalah carbonate margin has been interpreted according to the models of accumulation geometries of carbonate rocks summarized by Schlager (1999). The balance of two rates can explain this evolution: the rate of creation of accommodation space and the rate of carbonate production and growth. The difference between the growth rate of the platform rim and the inner shelf is also an important parameter as is the spatial variability of the rate of accommodation creation as a function of the tectonic subsidence (Borgomano, 2000). Four major phases in this evolution have been identified (Figure 12):

1) Aggrading platform (prerift, Figure 12A): During the Eocene, stable conditions and a uniform

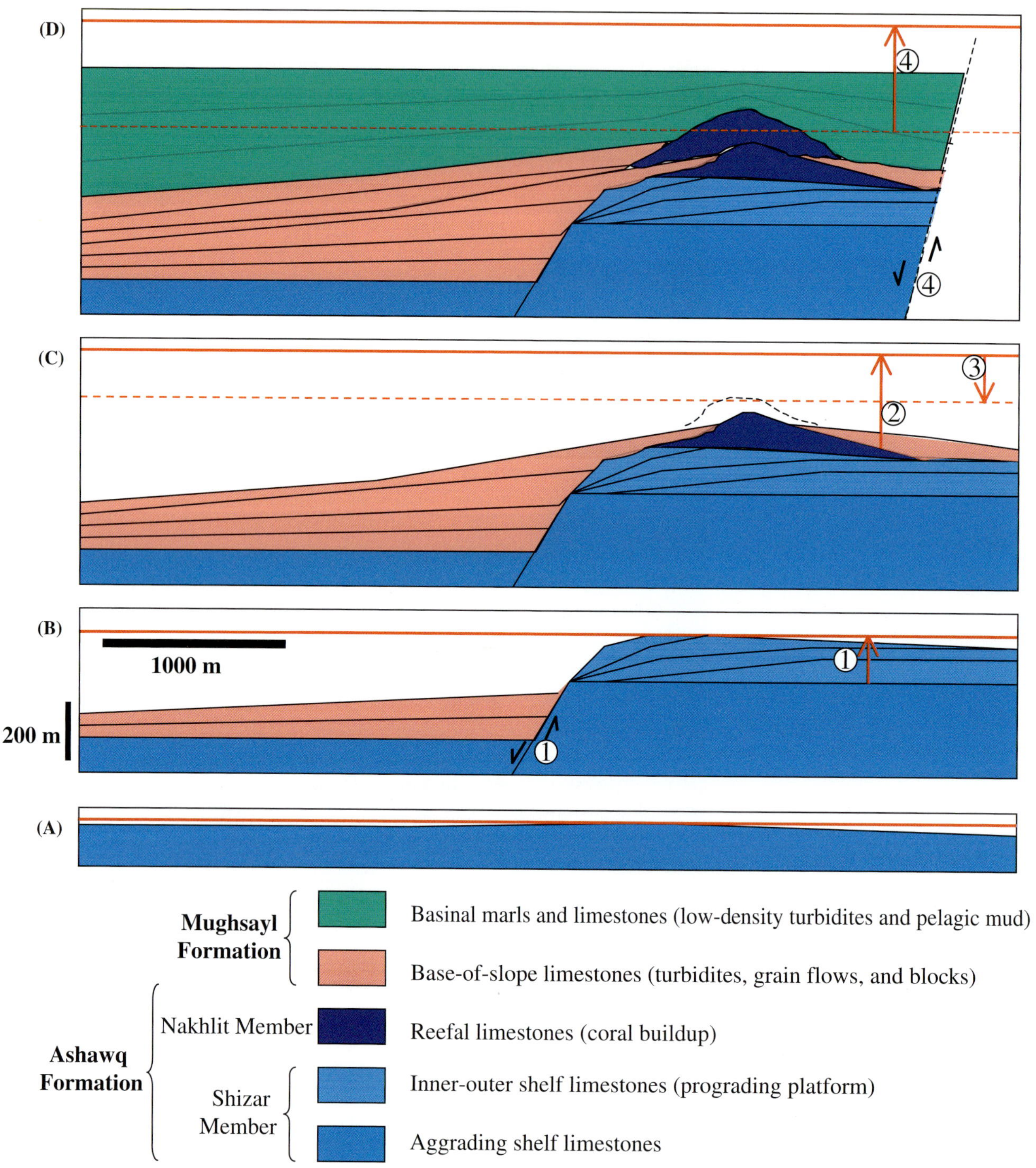

FIGURE 12. Schematic evolution of the western Salalah carbonate margin during the Eocene–Oligocene. (A) Aggrading platform during the Eocene (prerift). (B) Prograding platform developed at the beginning of the Oligocene (early rift). An increase in accommodation and formation of a faulted edge occur (1). The shelf cannot build out across the fault, and surplus carbonate production is transported into the trough. (C) Reef growth during the early Oligocene (middle rift). The rate of accommodation increase is controlled by tectonic subsidence and exceeds the growth rate of the inner shelf. Only the shelf rim is able to keep up with this increasing rate of subsidence and the buildup of coral reefs is promoted (2). A second phase of coral buildup is identified, suggesting an intervening regression (3) controlled by tectonic uplift or a sea level drop. (D) Reef drowning at the end of the early Oligocene (late postrift). The acceleration of the tectonic subsidence (4) results in the final drowning of the reef. Low-density turbidites and pelagic mud directly overlie the reef.

subsidence and/or sea level rise results in the accumulation of a thick aggrading carbonate platform at a rate of 120 m/m.y. (Platel and Roger, 1989).

2) Prograding platform (early synrift, Figure 12B): At the beginning of the Oligocene, the formation of grabens and the overall increase in tectonic subsidence result in the differentiation of two domains: an incised shallow-marine shelf and deeper-marine troughs. The accumulation rate of the shallow-marine carbonates exceeds the rate of accommodation increase on the structural highs and results in progradation of the shelf (Figure 9B). The steep slope and water depth below the photic zone in the troughs inhibit the outbuilding of the shelf beyond the faulted edge. The surplus of carbonate production from the platform is shed into the troughs. This is typical of highstand systems tracts on faulted platform margins (Borgomano, 2000). The accumulation rate on the platform probably ranges from 100 to 150 m/m.y.
3) Reef growth (synrift, Figure 12C): During the early Oligocene, the rate of accommodation increase is controlled by strong tectonic subsidence that exceeds the growth rate of the inner shelf (Figure 9B). Only the shelf rim is able to keep up with this increasing rate, through the buildup of coral reefs. This stage corresponds to the "empty-bucket" geometry within a platform transgressive system tract (Schlager, 1999). Two phases of reef growth can be identified and suggest that intermediate minor regression controlled by tectonics or sea level drop punctuated the reef growth. Carbonate shelves on the surrounding highs supply the bulk of carbonate material shed into the preserved troughs. The growth rate of the coral reefs is estimated to be between 200 and 300 m/m.y.
4) Reef drowning (late postrift, Figure 12D): At the end of the early Oligocene, tectonic subsidence results in the final drowning of the reefs and carbonate platform in the Salalah Basin. Low-density turbidites, supplied from the edges of the basin, and pelagic mud directly overlie the last reef (R-1). According to Platel and Roger (1989), the rate of subsidence— exceeded 300 m/m.y. in the Mughsayl area, but our estimate from the Salalah Plain-1 well, in a more basinal position, is two to three times greater. This rate of subsidence exceeded the growth potential of the reef, allowing its preservation and subsequent burial. During the late Oligocene and early Miocene, deeper-marine limestones and marls progressively filled the Salalah Basin.

The upper Miocene uplift of the Gulf of Aden rift shoulders marks the end of the marine evolution of the Salalah carbonate margin.

CONCLUSION

The evolution of the Salalah carbonate margin was strongly controlled by the opening of the Gulf of Aden during the Eocene–Miocene interval. The rapidly increasing rate of subsidence combined with the differentiation of faulted highs and troughs result in the disappearance of the carbonate systems in four major steps: from aggrading shelf to prograding shelf and reef growths and finally reef drowning. The rapid tectonic subsidence was only balanced during the initial stage of the rift by the carbonate production resulting in the formation of reefs on the highs and accumulation of thick wedge of detrital material in the troughs.

The uppermost part of the Ashawq Formation (coral reefs) and the lowermost part of the Mughsayl Formation (base-of-slope carbonates) are synchronous and apparently interfinger at the seismic scale. Both seismic and outcrops revealed similar stratigraphic architectures, sedimentary topography, and frequency of spatial variability. An additional important conclusion is that the single "buildup" initially interpreted from the seismic consists in reality of a prograding shelf rim at the base and an aggrading shelf rim at the top (coral reef sensu stricto).

ACKNOWLEDGMENTS

The authors thank the Ministry of Oil and Gas in Oman who authorized the publication of this chapter. We would also like to express our gratitude to our Petroleum Development Oman colleagues who contribute to the field surveys and drilling campaigns, especially Wytse Sikkema and Salim Al Maskiry.

REFERENCES CITED

Adams, E. W., and W. Schlager, 2000, Basic types of submarine slope curvature: Journal of Sedimentary Research, v. 70, no. 4, p. 814–828.

Borgomano, J. R. F., 2000, The Upper Cretaceous carbonates of the Gargano-Murge region, southern Italy: A model of platform-to-basin transition: AAPG Bulletin, v. 84, p. 1561–1588.

Cook, H. E., 1982, Carbonate submarine fan versus carbonate debris aprons: Facies patterns, depositional processes and models: Geological Society of America Bulletin, v. 14, p. 466–467.

Longman, M. W., 1981, A process approach to recognizing facies of reef complexes, *in* D. F. Toomey, ed., European fossil reef models: SEPM Special Publication 30, p. 4–40.

Platel, J.-P., and J. Roger, 1989, Evolution dynamique du Dhofar pendant le Cretace et le Tertaire en relation avec l'ouverture du Golfe d'Aden: Bulletin de la Société Géologique de France, v. V, no. 2, p. 253–263.

Platel, J. P., A. Berthiaux, J. Roger, A. Ferrand, and C. Robelin, 1987, Cartes geologiques a 1/100000 Hawf, Marbat, Sadh: Ministry of Petroleum and Minerals, Muscat and Salalah, Sultanate of Oman, 3 sheets.

Roger, J., J.-P. Platel, C. Cavelier, and C. Bourdillon de Grissac, 1989, Donnees nouvelles sur la stratigraphie et l'histoire geologique du Dhofar (Sultanat d'Oman): Bulletin de la Société Géologique de France, v. V, no. 2, p. 265–277.

Roger, J., J.-P. Platel, C. Cavelier, and C. Bourdillon de Grissac, 1997, Geology of Dhofar. Geology and geodynamic evolution during the Mesozoic and the Cenozoic: MPM geological documents, Muscat, Sultanate of Oman, 259 p.

Schlager, W., 1992, Sedimentology and sequence stratigraphy of reefs and carbonate platforms, A short course: AAPG Continuing Education Course Note Series No. 34, 76 p.

Schlager, W., 1999, Sequence stratigraphy of carbonate rocks: Leading Edge (August), p. 901–907.

Vail, P. R., 1987, Seismic stratigraphy interpretation using sequence stratigraphy. Part 1: Seismic stratigraphy interpretation procedure, *in* A. W. Bally, ed., Atlas of seismic stratigraphy, v. 1: AAPG Studies in Geology No. 27, p. 1–10.

Vail, P. R., R. M. Mitchum, and S. Thompson, 1977, Seismic stratigraphy and global changes of sea level, Part 3, *in* C. E. Payton, ed., Seismic stratigraphy—Applications to hydrocarbon exploration: AAPG Memoir 26, p. 61–83.

Wilson, J. L., 1975, Carbonate facies in geologic history: Berlin, Springer-Verlag, 471 p.

12

Belopolsky, A. V., and A. W. Droxler, 2004, Seismic expressions of prograding carbonate bank margins: Middle Miocene, Maldives, Indian Ocean, *in* Seismic imaging of carbonate reservoirs and systems: AAPG Memoir 81, p. 267–290.

Seismic Expressions of Prograding Carbonate Bank Margins: Middle Miocene, Maldives, Indian Ocean

Andrei V. Belopolsky[1]
Rice University, Houston, Texas, U.S.A.

André W. Droxler
Rice University, Houston, Texas, U.S.A.

ABSTRACT

Carbonate bank margins in the Maldives, a large isolated carbonate platform in the equatorial Indian Ocean, prograded significantly in the middle Miocene. The interpretation of a dense grid of two-dimensional seismic data located prograding margins and imaged the internal architecture of the prograding sequences in detail. Three individual prograding bank margins were positioned tens of kilometers away from each other. The margins prograded from both east and west toward the central seaway. Each prograding complex consists of five depositional sequences bounded by unconformities and correlative conformities. Seismic correlation shows that the deposition of sequences in individual prograding complexes was synchronous and driven by a mechanism of regional scale.

One prograding complex, complex I, is examined in detail in three dimensions on a set of seismic lines. Each prograding sequence is subdivided into two packages: strong- and weak-amplitude reflection packages. Strong-amplitude reflection packages consistently display a basinward shift in onlap and/or downstepping and are interpreted as having formed during falling relative sea level. Weak-amplitude reflection packages are interpreted as having formed during the rise and highstand of relative sea level. Each sequence thus represents a complete sea level cycle. The terminal middle Miocene sequence is characterized by voluminous downslope deposition of sediments interpreted as gravity-flow deposits. This regional event is interpreted as a response to a significant sea level fall at the end of the middle Miocene.

INTRODUCTION

Prograding Margins

Progradation is lateral outbuilding of sedimentary bodies and is common in various depositional settings (siliciclastic shelves, deltas, carbonate margins and banks, mixed silicilastic and carbonate environments). Prograding strata are easily recognized on both seismic and outcrop sections because of their characteristic clinoform shape in dip profile. Progradation of carbonate platforms

[1]*Present address:* BP, Houston, Texas, U.S.A.

and bank margins has been described at different scales in the geologic record of basins worldwide (e.g., Bosellini, 1984; Eberli and Ginsburg, 1989; Pomar, 1993; Sonnenfeld and Cross, 1993; Tinker, 1998). Prograding margins typically consist of vertically and laterally stacked unconformity-bounded sedimentary units or sequences. Each sequence represents a pulse in sedimentation controlled by interplay between sediment supply and changing accommodation space.

In carbonate depositional systems, progradation has been documented for carbonate shelves (e.g., Tyrrell and Davis, 1989; Pomar, 1993), ramps (e.g., Sarg, 1988; Handford, 1995) and large isolated platforms (e.g., Eberli and Ginsburg, 1989). Progradation does not typically develop on small isolated carbonate platforms and atolls because they commonly have steep slopes surrounded by deep waters. In this case, progradation of the bank margins is prevented because the excess carbonate material produced on the platform tops is transported downslope by catastrophic events in the form of slumps and turbidite flows. As a result, the material shed from a small isolated platform or bank typically bypasses the slope and is deposited at its toe as sheets or aprons (e.g., Sarg, 1988; Handford and Loucks, 1993).

Prograding margins may rapidly change the geographic position and shape of a carbonate system. Progradation is commonly responsible for the coalescence of isolated carbonate banks as in the Great Bahama Bank ("the Strait of Andros," Eberli and Ginsburg, 1989), Saya de Malha Bank in the Indian Ocean (Purdy and Bertram, 1993, their figure 29), and Segitiga Platform in the East Natuna Sea, Indonesia (Bachtel et al., 2004).

Many spectacular examples of prograding carbonate margins have been studied in outcrop and on seismic profiles. To this day, however, almost all of the progradation models, with a few exceptions (e.g., Osleger and Tinker, 1999), are based on two-dimensional (2-D) cross sections. Because the deposition of sediments occurs in three-dimensional (3-D) space, sediment distribution and the resulting stratal geometries may vary significantly along strike. A single 2-D cross section is not likely to adequately represent the anatomy of the prograding margin, and a comprehensive 3-D model is desired for the documentation and full understanding of the process.

Maldives Physiography

The Maldives Archipelago is located in the equatorial Indian Ocean and occupies the central part of the Chagos-Laccadives ridge. The origin of the ridge is attributed to the Réunion hot-spot activity (Duncan and Hargraves, 1990). The atolls form an 800-km-long chain extending in a north-south direction (Figure 1). The physiography of the Maldives is unique and makes the archipelago remarkably different from other isolated

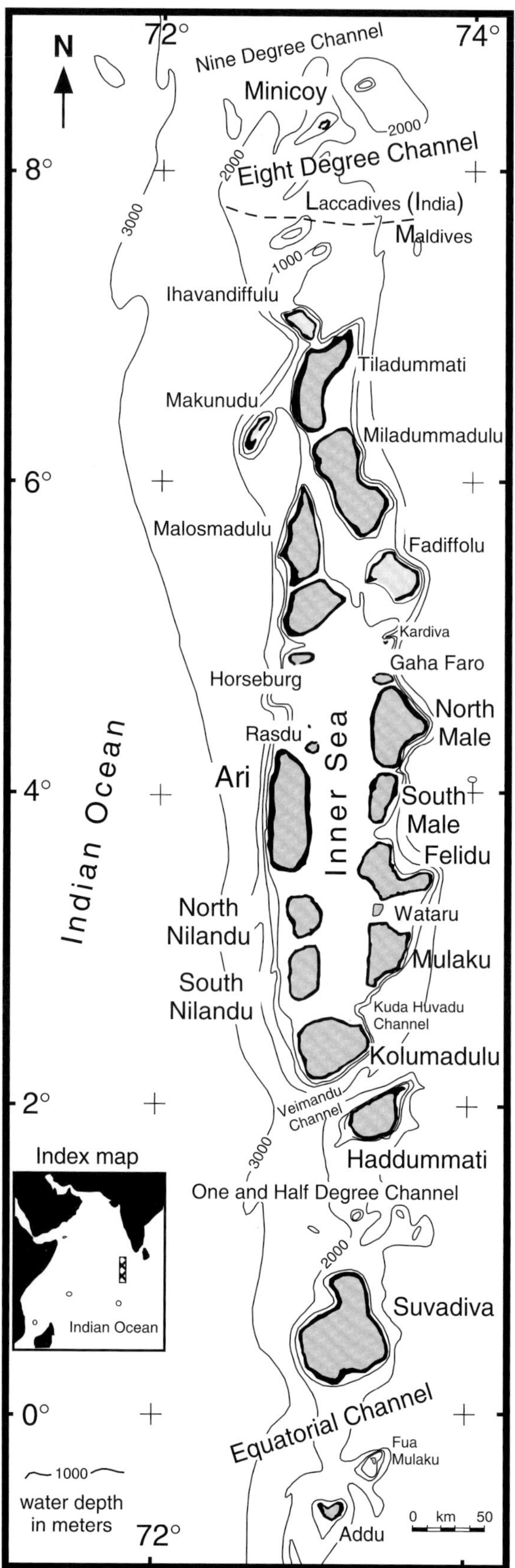

FIGURE 1. Atolls of the Maldives Archipelago, central equatorial Indian Ocean. Black outlines represent atolls; shaded areas indicate atoll lagoons.

oceanic carbonate platforms and guyots. The first main difference is the shear size of the Maldives Platform (820 × ~120 km). The second differentiating feature is two parallel chains of atolls in the central part of the archipelago separated by a 50-km-wide relatively shallow (water depth 200–600 m) Inner Sea (Figure 2). The oceanward slopes of the atolls, in contrast, are steep and rapidly reach water depths more than 2500 m.

The middle Miocene prograding carbonate bank margins were first identified by Aubert and Droxler (1992, 1996) and Purdy and Bertram (1993). This study is based on the interpretation of more recent seismic

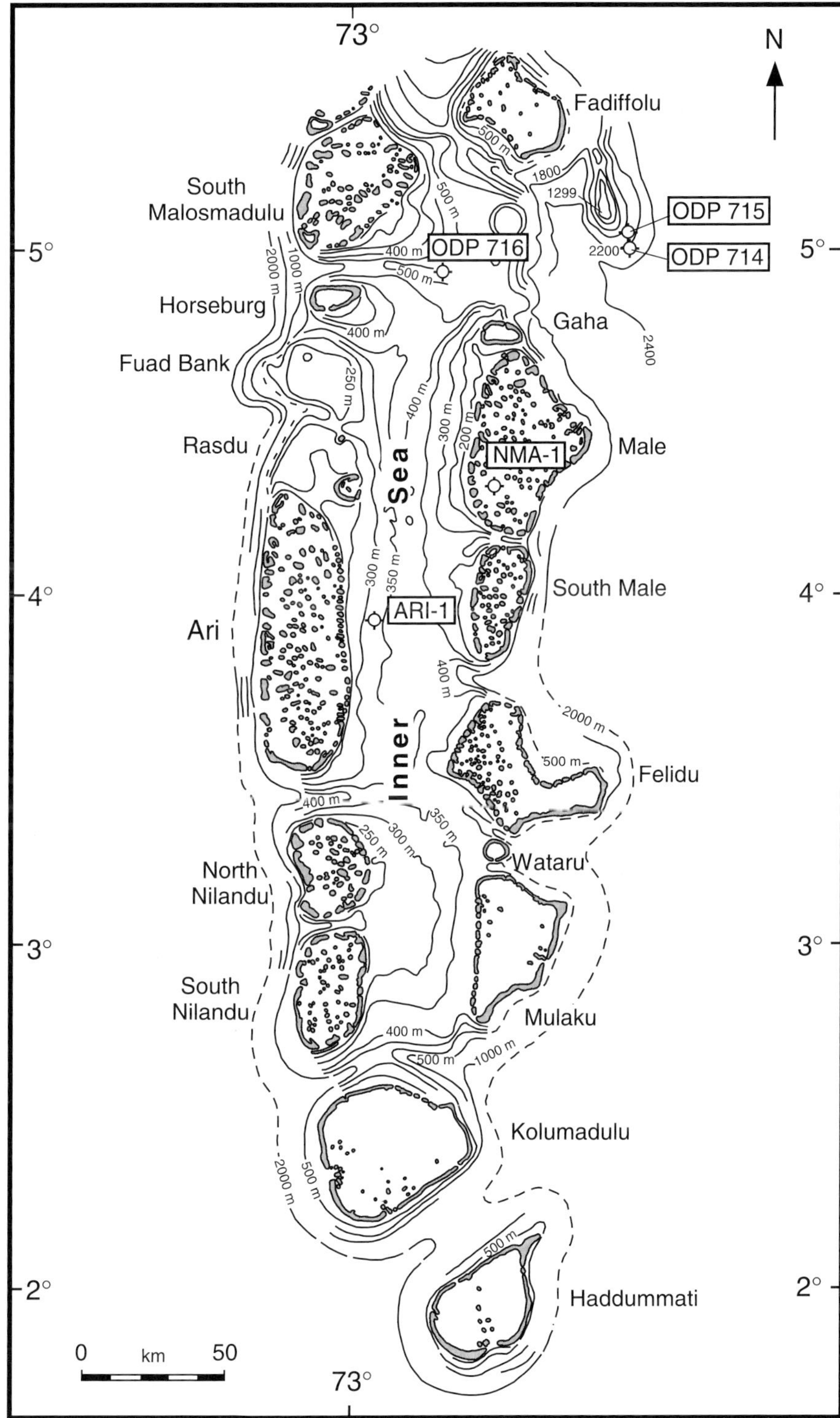

FIGURE 2. Central Maldives atolls and bathymetry map. Location of Ocean Drilling Program (ODP) Sites 714, 715, and 716 and industry wells ARI-1 and NMA-1 are shown.

data that allows the detailed documentation of the internal architecture of the individual prograding bank margins. The along-strike variation of stratal geometries can be studied on seismic profiles of various spatial orientations. A series of isopach maps of individual prograding sequences captures the evolution and spatial relationship of the clinoforms through time. The timing of sequence formation is based on correlation to two deep industry wells (NMA-1 and ARI-1) and Ocean Drilling Program (ODP) Site 716 (Figure 2).

In this chapter, we will (1) document the morphology and internal architecture of prograding sequences in 3-D on a set of strike and dip lines, (2) display the evolution of prograding margins through time, and (3) establish the timing of progradation in the Inner Sea. Subsequently, we will discuss the scale (local vs. regional) and the driving force of progradation in the Maldives.

DATA AND METHODS

Seismic Data

Previous studies in the Maldives were based on the early 1970s Elf Aquitaine seismic data (Aubert and Droxler, 1992, 1996; Purdy and Bertram, 1993; Aubert, 1994). The Elf data were collected both in the Inner Sea and within the atoll lagoons, providing a good regional coverage. Seismic processing did not include time migration, and diffractions from dipping events, such as bank margins, significantly distorted seismic displays and hampered interpretation.

The seismic data set used in this study comprise 6000 km of 2-D multichannel seismic profiles acquired by Royal Dutch Shell Oil Company in 1989 (Figure 3). The seismic grid covers the entire Inner Sea area but does not extend beyond its limits into the atoll lagoons. The average line spacing is 2 km between east-west lines and 4 km between north-south lines. The 60-fold multichannel data were collected with four air gun arrays with an individual 4804-in.^{3} (78,723-cm^3) gun capacity, towed at a depth of 6 m, and recorded using a 3000-m-long 240-channel streamer with 12.5-m group spacing. Seismic processing included spherical divergence and geometric spreading compensation, filtering, predictive deconvolution, velocity analysis at every 2 km, normal moveout correction, phase compensation, and time migration. The quality of the zero-phase seismic data is good to excellent down to 2 s, whereas it varies at deeper levels because of the raypath problems under the carbonate bank margins. A few seismic sections were sampled for frequency content and the measured dominant frequency was 50 Hz at 1 s, 35 Hz at 1.5 s, and 25 Hz at 2 s. The vertical resolution for a zero-phase wavelet equals one-fourth of the wavelength (Kallweit and Wood, 1982). Using the velocities from a vertical seismic profile (VSP)

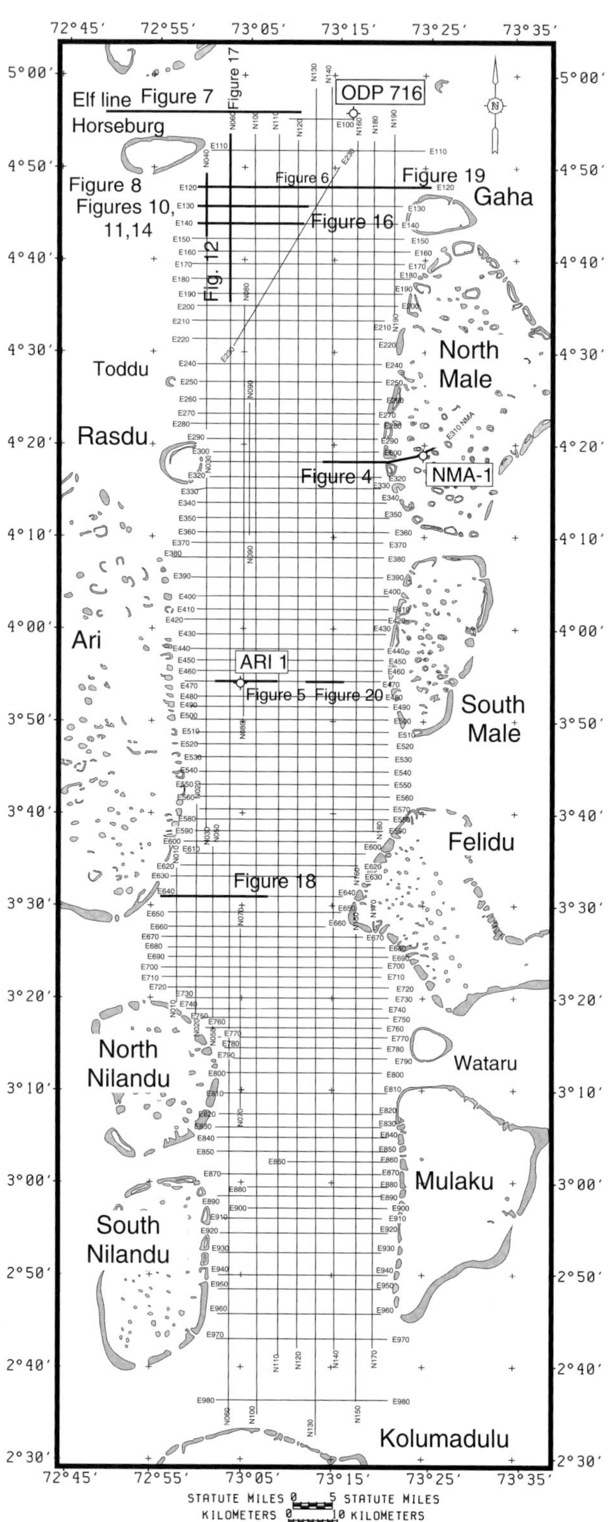

FIGURE 3. Shell 2-D multichannel seismic grid and locations of Elf NMA-1 and Shell ARI-1 wells and Ocean Drilling Program (ODP) Site 716. Thick lines indicate seismic lines discussed in text. The location of the Elf seismic line reproduced from Aubert and Droxler (1996) is also shown.

survey in the Shell ARI-1 well, the average vertical resolution in the window of interest would be 10 m at 1 s two-way traveltime (TWT) and 15 m at 1.5 s TWT. This resolution should be sufficient to image the true stratal geometries of the prograding sequences.

Well Data

Well data are crucial for the correlation and ground truthing of seismic interpretations. No wells were drilled through the prograding bank margins themselves. The nature and age of sediments in the Maldives, however, are known from industry wells, NMA-1 and ARI-1, and ODP Site 716. Correlation of seismic reflections to these wells is used to determine the age of the prograding sequences.

NMA-1 well was drilled in 1976 by the Elf Aquitaine–operated consortium in the North Male Atoll lagoon in 45 m of water (Figures 2, 3). The well penetrated 2106 m of Eocene to Pleistocene carbonate sediments and bottomed in 116 m of weathered basalts (Figure 4). The well recovered late Oligocene–middle Miocene sediments (Purdy and Bertram, 1993; Aubert and Droxler, 1996). The well's lithologic and stratigraphic record is incomplete because of significant gaps in sediment recovery caused by frequent losses of circulation while drilling and caving.

The ARI-1 well was drilled in 1991 by Shell in the central part of the Inner Sea in 348 m of water (Figures 2, 3). The well encountered 3315 m of late Eocene to modern carbonates and recovered 50 m of weathered basalts (Figure 5). Despite the fact that the upper 450 m of sediments (uppermost late Miocene and Pliocene–Pleistocene) were not recovered, and a few sample gaps exist in the lower part of the well record, the ARI-1 well provides reliable information on the composition, biostratigraphy, and paleobathymetry of the sediments in the basin. Biostratigraphic analyses of sidewall core and cutting samples included identification of foraminifera, nannoplankton, and palynomorphs. The well was logged using a standard suite of logging tools (caliper, gamma ray, resistivity, sonic, spontaneous potential, and density). A VSP provided seismic velocities essential for depth conversion and correlation between the seismic and well data.

Ocean Drilling Program Site 716 (Figures 2, 3) was drilled in 544 m of water during Leg 115 in 1987 and recovered 264 m of periplatform carbonate sediments of late Miocene to Holocene age (Backman et al., 1988; Droxler et al., 1990; Malone et al., 1990). The sediments are represented by foraminifera-rich calcareous oozes with significant amount of bank-derived aragonite needles (Droxler et al., 1990; Malone et al., 1990). The ODP Site 716 record complements the information from the

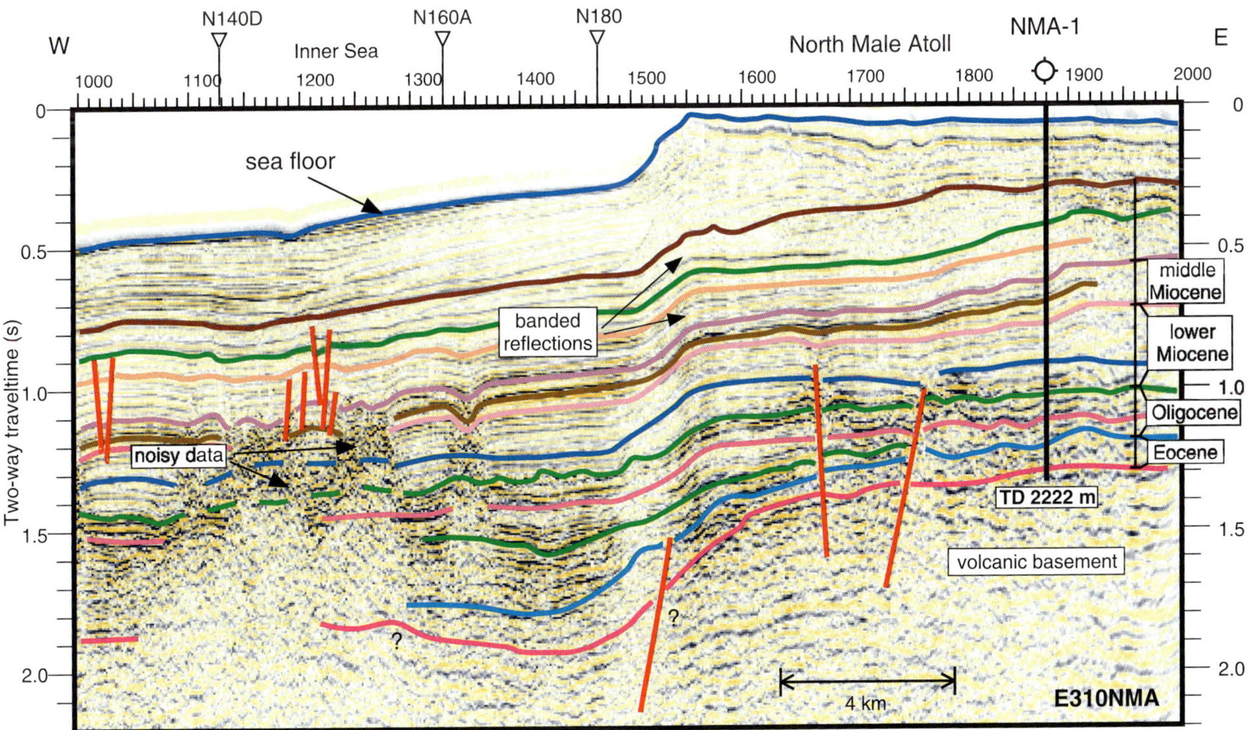

FIGURE 4. Elf NMA-1 well drilled in the lagoon of North Male Atoll and Shell seismic line E310NMA. Uneven sea floor bathymetry causes bending of the reflections under the atoll margin.

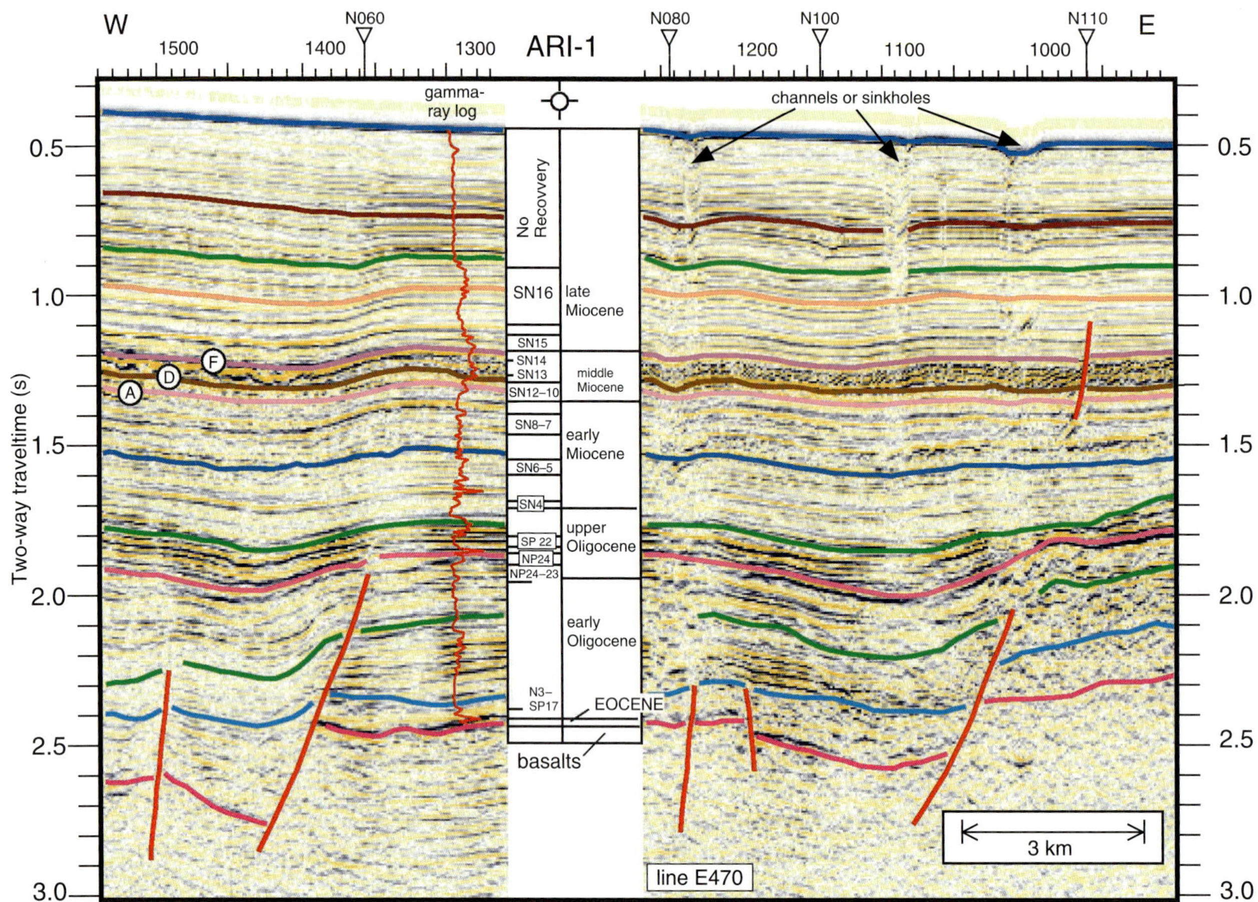

FIGURE 5. Shell ARI-1 well tie to the interpreted segment of seismic line E470. The well's biostratigraphy and gamma-ray log are shown. Letters A, D, and F mark sequence boundaries of the middle Miocene sequences that will be discussed subsequently in the text. The seismic section was split into two panels to show the location of the well.

industry wells where the uppermost sections were not recovered.

GENERAL GEOLOGIC SETTING OF THE MALDIVES

Figure 6 is a west-east–oriented seismic line across the Inner Sea, displaying the basement and entire sedimentary package within the Maldives Platform. Neritic carbonate production in the Maldives was established in the early Eocene (Purdy and Bertram, 1993; Aubert and Droxler, 1996; Belopolsky, 2000). Carbonate banks became established on the topographic volcanic basement highs that were separated by deep seaways (Belopolsky, 2000). Carbonate banks continued to accrete in the Eocene and early Oligocene. In the late Oligocene, the banks began developing elevated rims along their margins that separated bank interiors from open waters. In the early Miocene, relative sea level rise led to the drowning of bank interiors (Belopolsky, 2000). Where possible, the banks backstepped onto higher ground and migrated upslope. Ultimately, by the end of the early Miocene time, a series of flat-topped banks were established around the periphery of the ancestral Inner Sea (Belopolsky, 2000), the wide trough in the interior of the Maldives Platform that still persists today. During the middle Miocene, the bank margins prograded toward the central axis of the paleo-Inner Sea. During Miocene and Pliocene time, the central part of the Inner Sea was partially filled with periplatform sediments whereas the flat-top carbonate banks were aggraded.

MIDDLE MIOCENE PROGRADATION IN THE MALDIVES

Previous Studies Based on the Elf Data

Aubert and Droxler (1992) described the middle Miocene progradation from both east and west toward the central part of the Inner Sea using seismic data in the northern part of the Inner Sea. They related the onset of the bidirectional progradation to the seasonal switch in

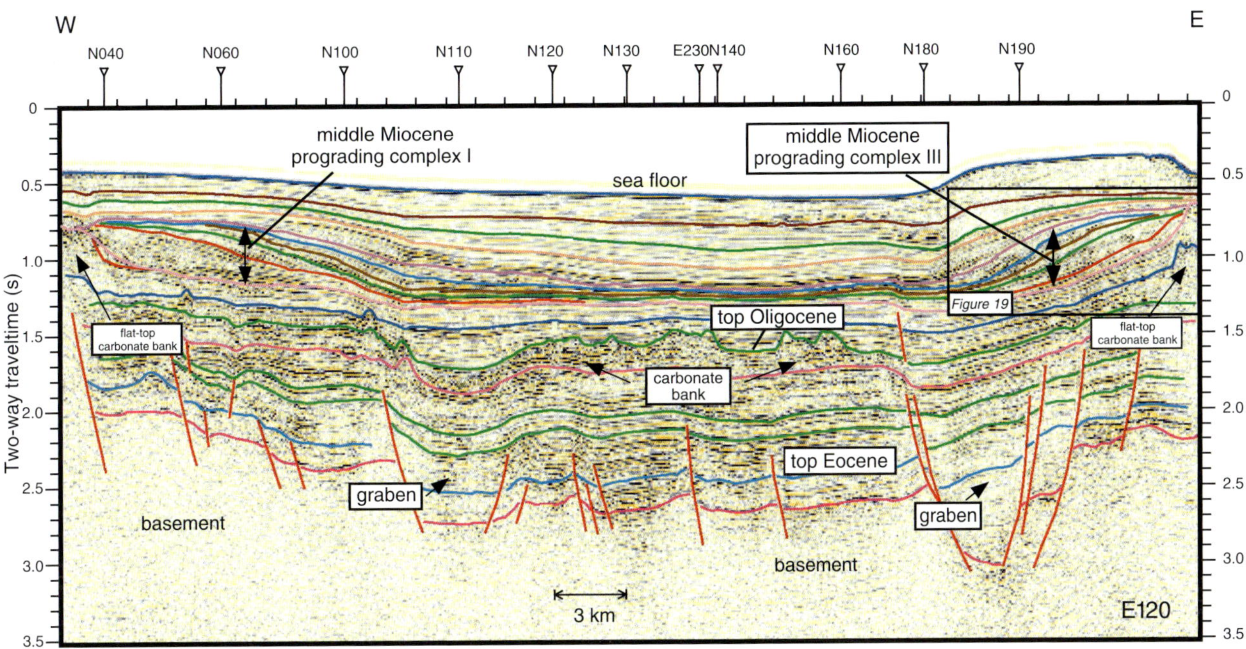

FIGURE 6. Interpreted Shell seismic line E120 showing volcanic basement dissected by grabens and overlying Tertiary carbonate sequences.

the wind patterns caused by the establishment of the Indian monsoons. Purdy and Bertram (1993) further documented the regional bidirectional progradation in the Maldives. They reported different styles of progradation for the eastern and western rows of atolls. Aubert (1994) and Aubert and Droxler (1996) described the bidirectional progradation in the Inner Sea as having started in the late middle Miocene and continuing until the late Pliocene. Maximum progradation occurred in the late middle Miocene–early late Miocene. The progradation was described as a series of flat-topped sigmoidal sequences attached to early Miocene "nucleating mounds." On some seismic profiles, individual sequences separated by downward shifts in onlap where defined but "no coherent regional sea level signal could be obtained" because of correlation difficulties (Aubert and Droxler, 1996, p. 518). On one seismic profile, nine high-order sequences were defined (Figure 7; Aubert and Droxler, 1996). The sequences showed an evolution from low-angle aggrading ramps to increasingly steeper prograding margins. The interpretation of stratal geometries suggested a long-term regression punctuated by short-term transgressions and regressions (Aubert and Droxler, 1996).

Interpretation of Shell Data

The Shell seismic grid provides dense coverage of the Inner Sea but does not extend into the modern atolls lagoons or channels between the atolls. The interpretation of the Shell seismic grid shows that a series of flat-top carbonate banks ("nucleating mounds" of Aubert and Droxler, 1996) were established in the peripheral parts of the paleo-Inner Sea at the end of the early Miocene (Belopolsky, 2000). The location of these banks results from significant backstepping of platform margins in the late Oligocene and the early Miocene (Figure 8) in response to a rapidly rising relative sea level (Belopolsky, 2000). In the latest early Miocene, flat-top carbonate banks grew vertically by as much as 300 m. In the middle Miocene, the bank margins prograded significantly on the both sides of the paleo-Inner Sea toward the center of the platform.

The seismic lines used in this study have a few minor imaging problems. V- and U-shaped features that are either channels or dissolution sinkholes in the shallower section create a velocity pull-down effect and diffractions that propagate deeper into the section. This effect distorts the shape of some of the reflectors forming the middle Miocene clinoforms. Solution collapses and gas chimneys caused by CO_2 and H_2S gases are not uncommon in carbonate sediments (Story et al., 2000; Ianello and Dorobek, 2001). These imaging problems were taken into account during the interpretation of stratal geometries.

Age Control on Sequence Development

The timing of progradation in the Maldives was established by correlation to wells where the age of sediments was determined from micropaleontologic analyses of sidewall cores and cuttings. The biostratigraphic record from the Shell ARI-1 well is more complete than that from the Elf NMA-1 well, and correlation of seismic

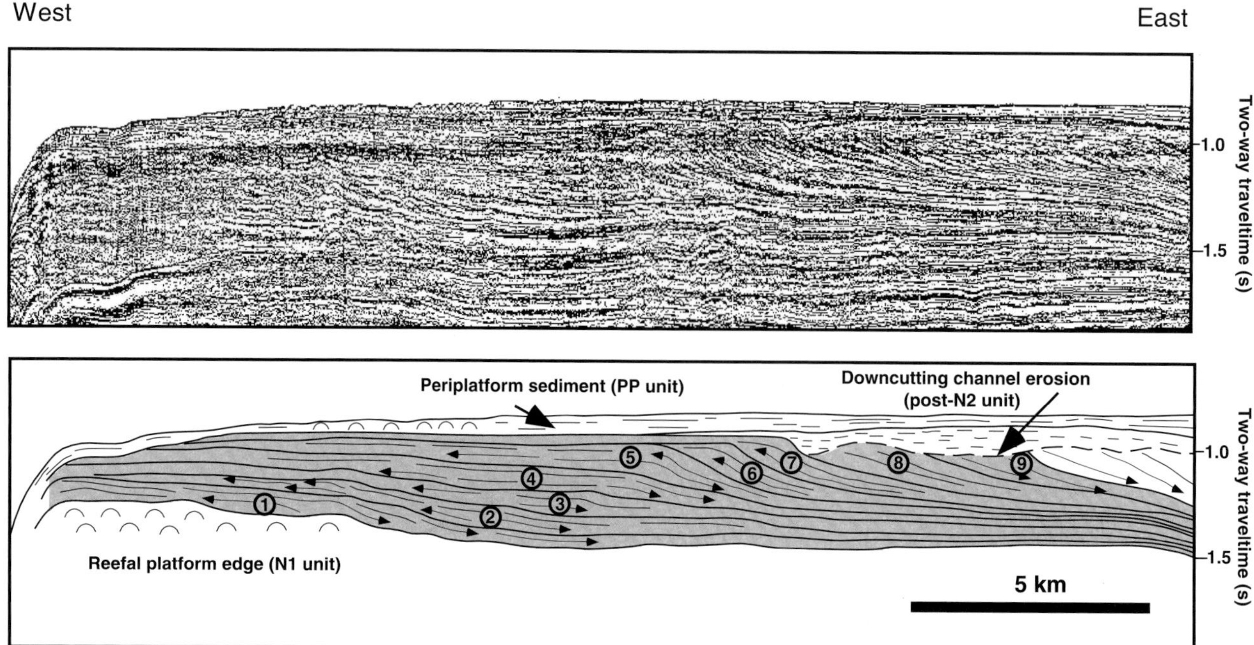

FIGURE 7. Progradation geometries within the middle to late Miocene carbonates in the western margin of the Maldives, based on the interpretation of Elf seismic data (figure 13 of Aubert and Droxler, 1996). Prograding sequences are attached to the "reefal platform edge" or a "nucleating mound." Nine prograding sequences are defined (marked by numbers 1–9). Their formation is related to high-frequency sea level fluctuation by Aubert and Droxler (1996). The location of this seismic line is shown in Figure 3.

reflections to ARI-1 is straightforward. The ARI-1 well, however, was drilled in the central part of the Inner Sea and recovered only the thin, basinal equivalents of the prograding sequences (Figure 5). This overall "condensed" stratigraphy in the ARI-1 well results in a limited age control of the individual sequences. Nevertheless, the timing of the initiation and development of progradation is fairly well constrained. The Shell/Worldwide foraminifer zones (Van Morkhoven and Schroeder, 1986; Styzen, 1996) were used to define the biostratigraphy of the section. These zones are based on the last appearance datums (LADs) instead of first appearances (Blow, 1979). *LADs* are defined as the first encounter of the species downhole, which is a common practice when working with well cuttings. The base of the prograding sequences was dated as early middle Miocene (Shell/Worldwide Zone SN10-12). The top of the last prograding sequence built away from the slopes of the early Miocene flat-top banks is constrained by Zone SN14 and the base of Zone SN15. This establishes the last sequence as latest middle Miocene or earliest late Miocene in age. Progradation of bank margins continued locally until the end of the early Pliocene.

Prograding Complexes

The progradation of the bank margins in the Maldives began in the middle Miocene. The progradation occurred from both the east and west sides of the Maldives toward the central seaway (paleo-Inner Sea). The areas of progradation are easily identified on the isochron map of the middle Miocene unit (Figure 9) as areas of local thickness increase. Thick accumulations represent areas where prograding clinoforms are stacked. The term *prograding complex* is used here to describe an area where laterally extensive and vertically significant sets of two or more clinoforms were constructed. This term should not be confused with the *prograding complex* of Mitchum et al. (1994) who restricted this term to describe a lowstand prograding wedge. The area of thick middle Miocene sediments directly northwest off Felidu Atoll is not composed of prograding clinoforms but by the time-equivalent gravity-flow deposits and slumps that will be discussed subsequently in the text.

On the isochron map, three prograding complexes were identified within the limits of the Shell seismic grid (Figure 9). None of the complexes were covered by the Shell seismic grid in their entirety because complexes extend under the modern atolls. Complex I is located in the northwestern corner of the grid near Horseburg Atoll. In map view, the imaged part of the complex has an oval shape with dimensions of 20 × 30 km. Fifteen Shell seismic lines offer different cross-sectional views through the clinoforms of complex I. Prograding complex II is located in the west-central part of the Inner Sea between Ari and North Nilandu Atolls. The complex covers an area of 20 × 30 km, and it is imaged by 20 Shell

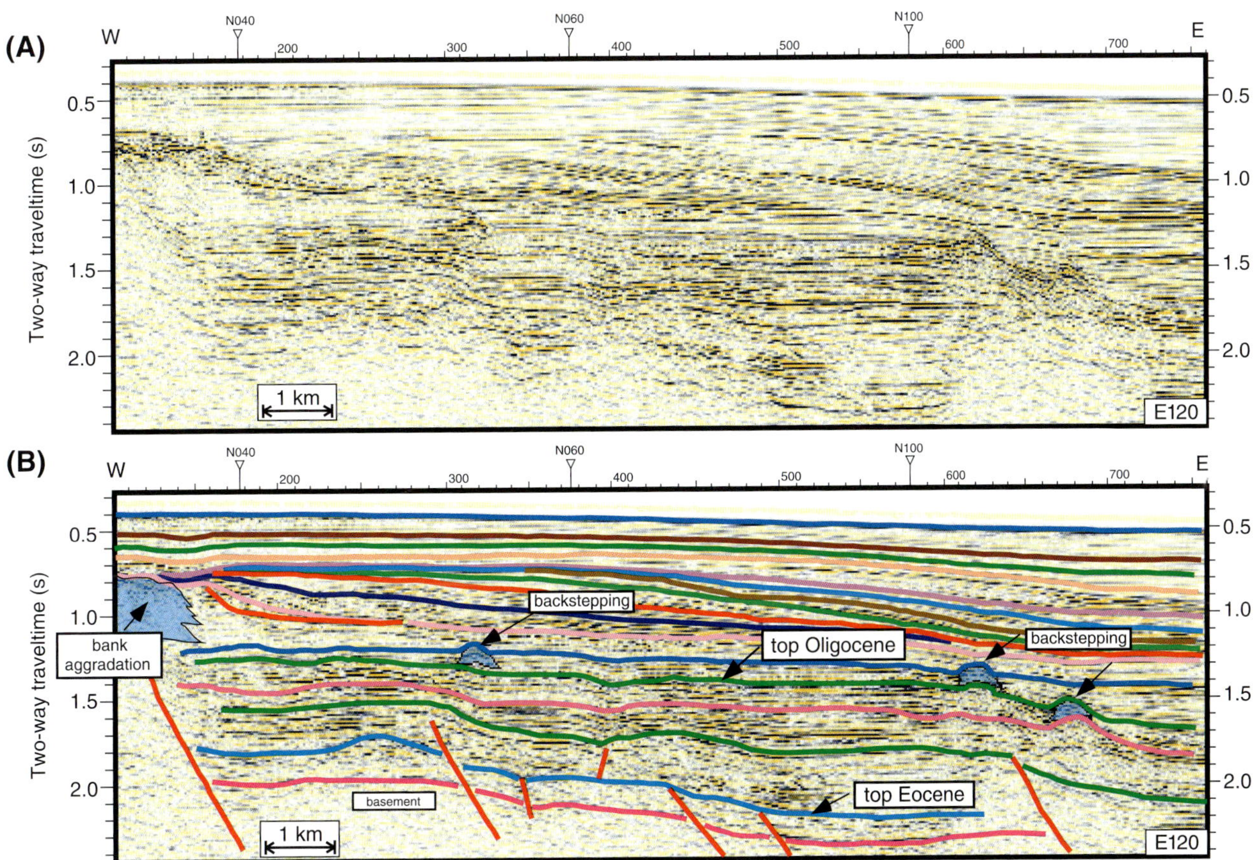

FIGURE 8. Uninterpreted (A) and interpreted (B) segments of seismic line E120. The line shows late Oligocene–early Miocene backstepping of the carbonate bank margin and the establishment of the early Miocene flat-top bank.

seismic lines. Prograding complex III is located in the northeastern part of the Shell seismic grid north of Gaha Atoll and is imaged by only three Shell seismic lines.

Progradation was also reported outside the area covered by Shell seismic lines on the Elf lines acquired in the atoll lagoons and in the northern part of the Inner Sea (Aubert and Droxler, 1992, 1996; Purdy and Bertram, 1993; Aubert, 1994). The shaded areas on Figure 9 indicate locations where the middle Miocene progradation was documented by Purdy and Bertram (1993) and Aubert and Droxler (1996). Aubert and Droxler (1996) stated that the progradation from the western side appears to have been more substantial than that from the eastern side, in both the thickness of sediments and the lateral migration of the margins. Absence of middle Miocene progradation under the Felidu Atoll was reported by Purdy and Bertram (1993).

Seismic Expression of Prograding Sequences

Five depositional sequences are defined within each prograding complex based on the stratigraphic relationships, variation in seismic facies, and stratal geometries of the seismic units. The sequences have clinoform shape in dip view and thin out in the basinward direction where they are commonly represented by a single reflector (Figure 10). The sequence boundaries are named A–E starting from the oldest mapped sequence (Figure 10).

Each middle Miocene sequence can be divided into two seismic packages: a strong-amplitude reflection package (SARP) that forms the lower part of the sequence and a weak-amplitude reflection package (WARP) that comprises the upper part of the sequence (Figure 11). Strong-amplitude reflection packages consistently display basinward shift in onlap relative to the underlying reflection. In dip view, the internal reflections comprising SAPRs are typically oblique or sigmoidal oblique. The lowest SARP reflections onlap the sequence boundary, and the shallower reflections that form the topsets of SARPs display either toplap or downstepping (onlap onto the underlying reflection at a lower level). The clinoform toes downlap onto the sequence boundary. On seismic dip profiles, SARPs commonly have a convex-upward shape (Figures 10, 11).

The topsets of the WARPs in the upper part of the prograding sequences extend in a shoreward direction and progressively onlap the underlying SARPs. The

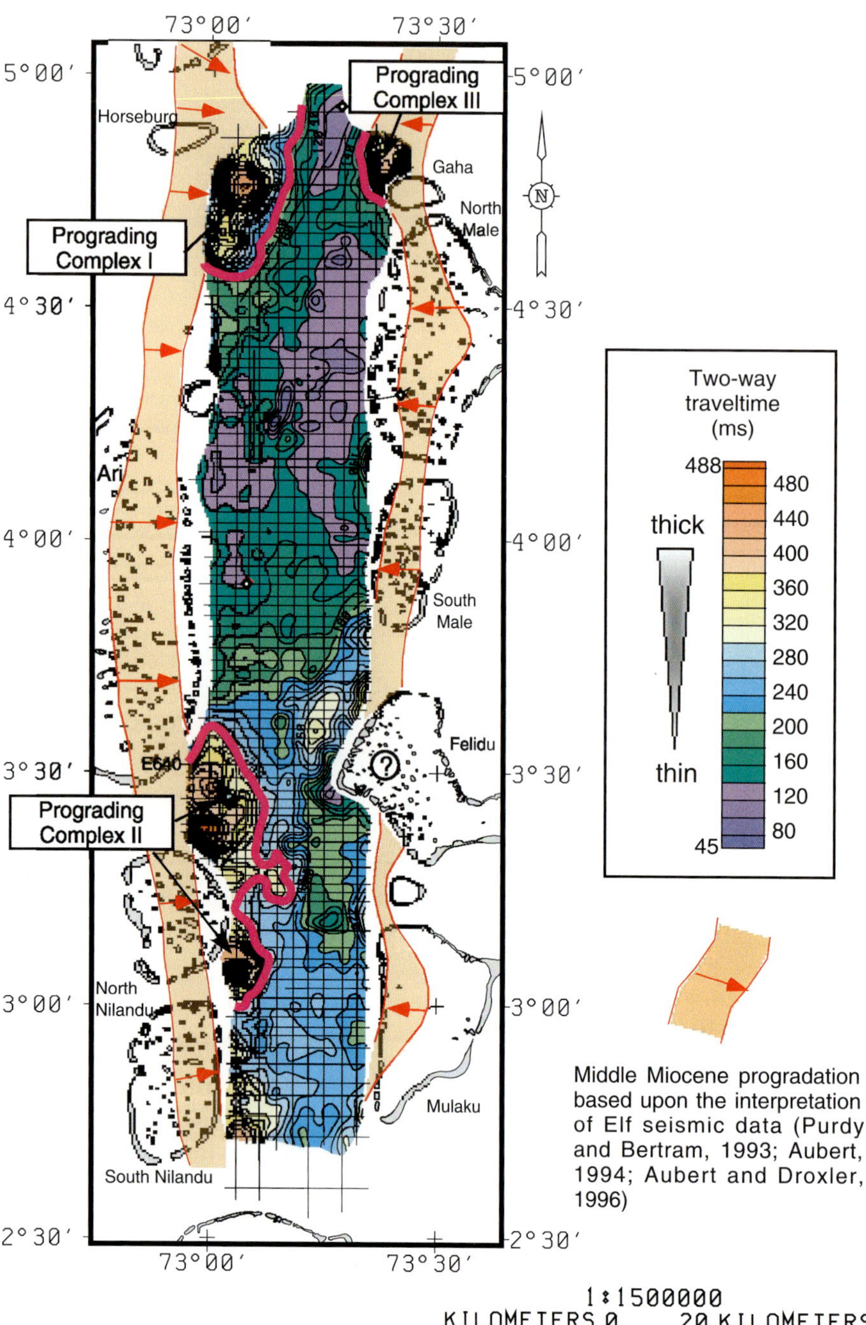

FIGURE 9. Isochron map of the middle Miocene prograding sequences. Thick areas indicate prograding bank margins and are labeled as prograding complexes I–III. Shaded areas under modern atolls and in channels between them mark areas where Miocene progradation was identified on the Elf seismic data by Purdy and Bertram (1993) and Aubert and Droxler (1996). Evidence of progradation is absent under the Felidu Atoll (Purdy and Bertram, 1993).

topsets of the uppermost WARP reflectors display toplap, and the bottomsets have downlap terminations.

ANATOMY OF PROGRADING COMPLEX I

Prograding complex I, located in the northwest corner of the Inner Sea (Figure 9), is imaged on the dense grid of 2-D seismic lines showing its internal architecture in detail. The prograding clinoforms are attached to the edge of an early Miocene bank (Figures 6, 11). Unfortunately, Shell lines do not extend far enough west to image the bank in its entirety. This bank, however, is imaged on the Elf seismic profiles (Figure 7, Aubert and Droxler, 1996). The carbonate bank has a flat top. The average gradient of the bank's eastern margin is about 7°. Sequence boundary "A" was picked as the base of the prograding sequences (Figure 11) and was correlated around the basin.

A detailed description of stratal geometries of prograding complex I is possible from the dense seismic grid. Table 1 summarizes the main parameters of prograding sequences of complex I. The parameters include maximum sequence thickness, lateral extent and orientation

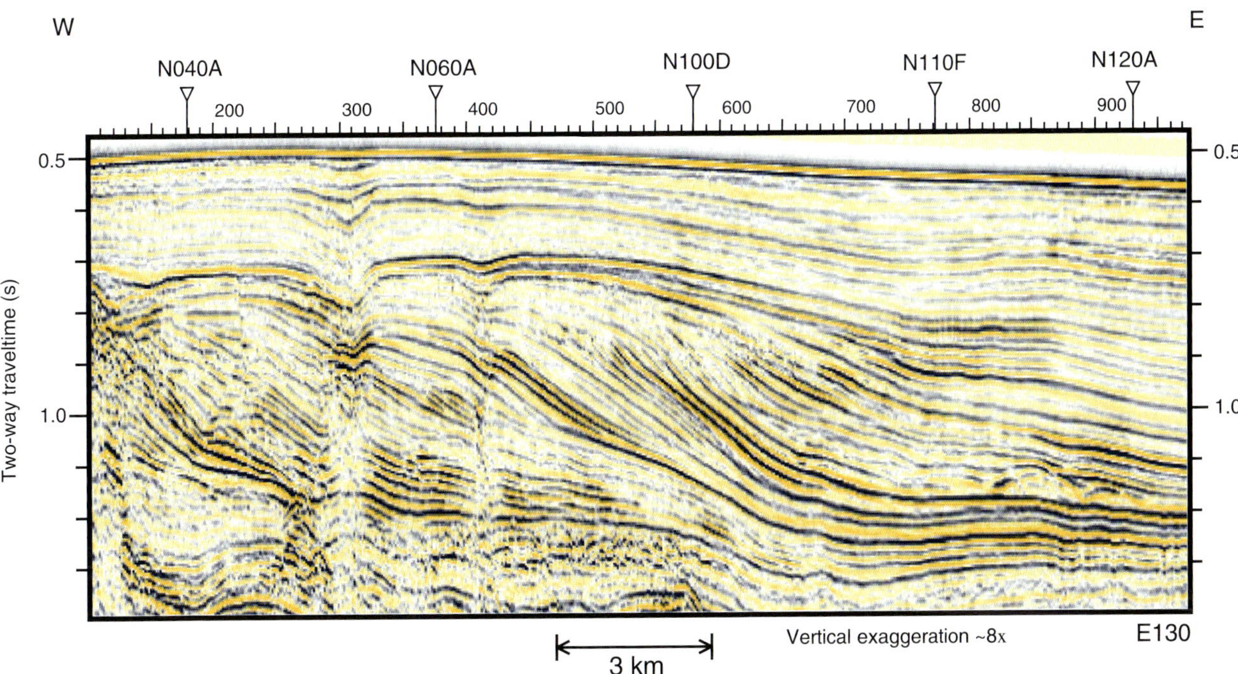

FIGURE 10. Uninterpreted segment of line E130 showing middle Miocene clinoforms.

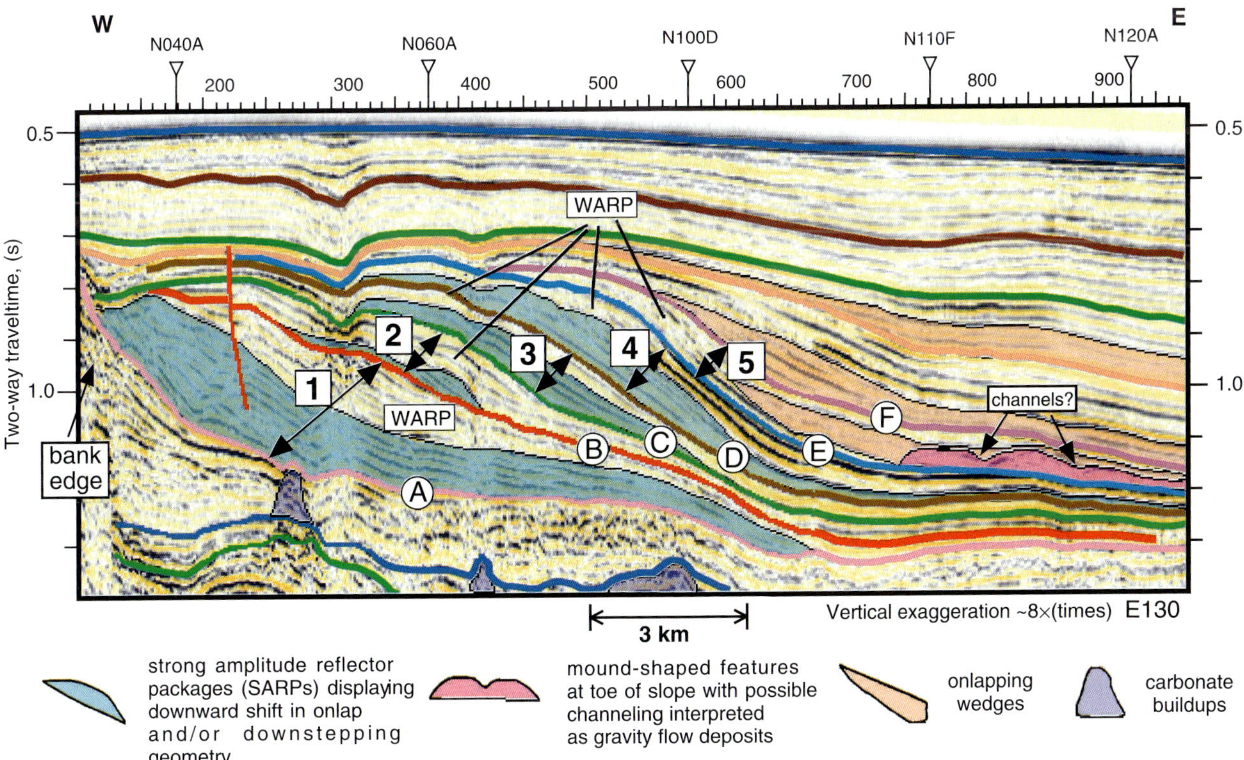

FIGURE 11. Interpreted segment of line E130 showing five middle Miocene prograding sequences (numbered 1–5). Letters indicate sequence boundaries. Each sequence is divided into strong-amplitude reflection package (SARP, shaded) and weak-amplitude reflection package (WARP, no color fill).

Table 1. Main parameters of the middle Miocene prograding sequences of complex I.

Sequence	*Maximum thickness (ms TWT)*	*Maximum Thickness (m)*	*Area (km^2)*	*Maximum dip angle (°)*	*Orientation of prograding margin*	*Lateral migration of prograding margin (km)*
SARP 1	220	275	110	7	North-south	
WARP 1	128	160	86	4	North-south	3
2	125	156	107	6	North-northeast–south-southwest	1.5
3	107	134	128	6	North-northeast–south-southwest	2
4	154	192.5	93	7.5	North-northeast–south-southwest	2
5	166	208	126	9	Northeast-southwest	3

Two-way traveltime (TWT) values were converted to depth using 2500 m/s sonic velocity.

of the prograding margin, and maximum measured angle of clinoform dip. An average sonic velocity of 2500 m/s was used to convert the seismic data to depth. This value was determined by analysis of stacking velocities and velocities from the VSP survey in the ARI-1 well.

Sequence 1

Sequence boundary A forms the base sequence (sequence 1) (Figure 11). The lower part of sequence 1 SARP shows undulating and mounded reflectors on some dip lines (Figures 10, 11). A U-shaped seismic unconformity, present on the north-south-strike–oriented seismic line N040 (Figure 12), is interpreted as a slide scar created by slope failure. The slide created a 6-km-wide depression that was later filled with sediments of sequence 1. The sediments above Sequence A appear as chevron-shaped seismic reflections interpreted as rotated blocks (Figure 12). The detachment surface is listric and concave-upward in cross-sectional view (Figure 11) and likely has an amphitheater or spoonlike shape in map view. Gravitational collapse structures of similar geometry have been reported from different settings and occur at various scales (e.g., Hesthammer and Fossen, 1999). The causes of such gravitational collapses include seismic shocks, oversteepening of slopes, rapid rates of sedimentation, and changes in pore fluid pressure (Hesthammer and Fossen, 1999).

The area of the collapse feature is coincident with a limited area of thickening on the sequence 1 SARP isochron map (Figure 13A). This example illustrates creation of new accommodation space by the local removal of sediment by sliding. Sequence 1, however, is not limited to the filling of the space created by the slide and has a

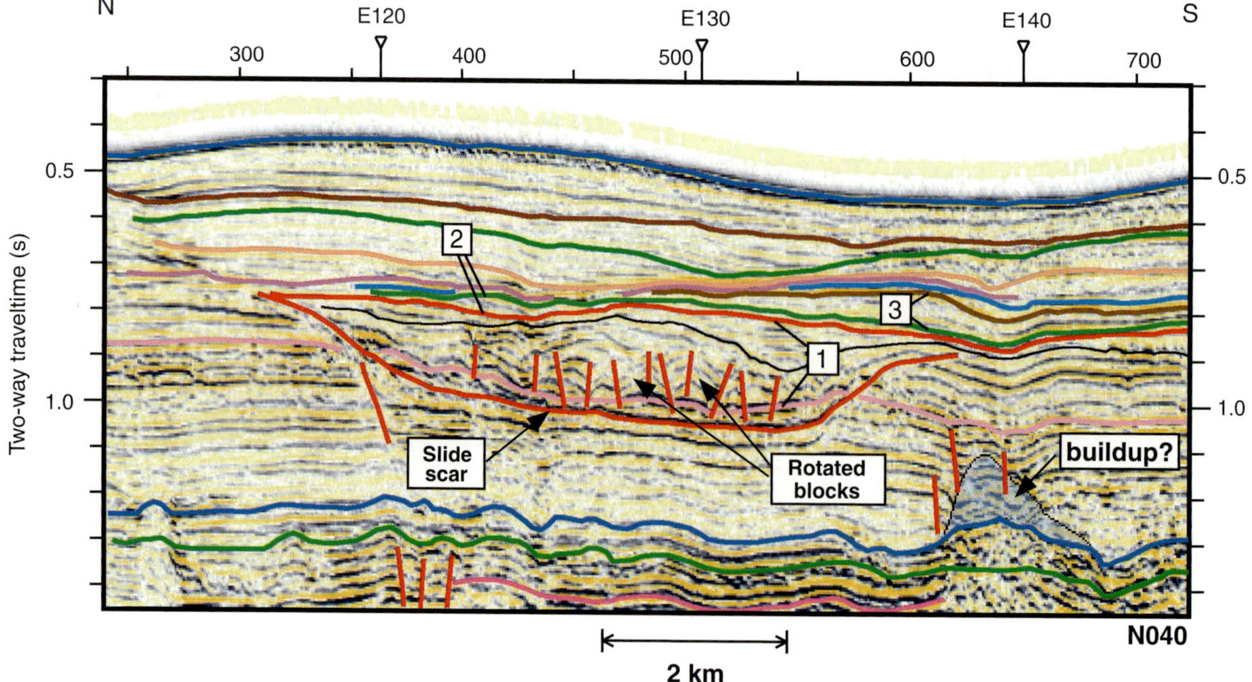

Figure 12. Interpreted segment of seismic line N040 showing gravitational slide at the base of sequence 1 in strike view.

more regional character. Figure 13A shows that SARP 1 is as much as 275 m thick locally (Table 1) in the area where the sediments filled the space created by the gravitational sliding. The sediments of sequence 1 WARP form a wedge with a maximum thickness of 160 m (Figure 13B). Comparison of the SARP 1 and WARP 1 maps shows that the locus of deposition shifted 3 km to the east from its previous position (Figure 13A and 13B).

Sequence 2

Sequence boundary B separates sequences 1 and 2 (Figure 11). On line E130 (Figures 10, 11), sequence 2 SARP appears as a convex-upward wedge with oblique internal reflections. The clinoform of sequence B is shifted significantly (~4 km) in the basinward direction relative to the topset of WARP 1. The vertical component of this downward shift is 87.5 m. The magnitude of this shift, however, appears to be smaller on other seismic lines. The magnitude of the downward shift on line E130 is overestimated as a result of the line cutting at an angle oblique to the true dip. The overlying reflectors of sequence 2 WARP onlap onto the sequence 2 SARP and sequence boundary B. The maximum angle of the clinoform foresets observed in sequence 2 is 6°, and the maximum thickness of sequence 2 is 156 m (Table 1). The locus of sedimentation of sequence 2 shifted 1.5 km in the southeast direction relative to the locus of sequence 1 WARP (Figure 13C). The orientation of the prograding wedge in map view changed from north-south to a north-northeast–south-southwest orientation.

Sequence 3

Sequence boundary C forms the base of sequence 3 and was dated by correlation to the ARI-1 well (Figure 5), where a sample collected just above sequence boundary C was placed in the Shell/Worldwide foraminifer Zone SN13. Zone SN13 represents late Serravallian age (11.75–11.5 Ma). Sequence 3 SARP is thicker and laterally more extensive than the SARP of sequence 2 (Table 1). The topset reflections of SARP 3 are shifted about 2 km basinward relative to the topset of WARP 2. The overlying WARP 3 is 70 ms TWT (87.5 m) in its thickest part. The isochron map of sequence 3 shows further migration of the prograding margin toward a southeast direction (Figure 13D).

Sequence 4

Sequence boundary D separates sequences 3 and 4. The internal reflections of sequence 4 SARP on line E130 show systematic downstepping of the bank margin (Figure 14). A series of unconformities are arranged in a steplike manner at progressively lower topographic levels. This geometry is commonly referred to as *erosive regression* (Curray, 1964) or *forced regression* (Posamentier et al., 1990; Hunt and Tucker, 1992, 1995; Posamentier et al., 1992). Sequence 4 imaged on line E120 (Figure 15) does not display the downstepping of the shelf margins, possibly because of the oblique orientation of the section. Within the sequence, however, a reflection with a steep break is interpreted as an erosional cliff (Figure 15). On line 140 (Figure 16), the lower part of sequence 4 is expressed as a series of downstepping SARPs and possibly a slumped SARP. The reflectors of sequence 4 WARP onlap onto the previous sequence. The slope of the prograding bank margin in sequence 4 is steeper than in previous sequences and reaches 7.5°. Sequence 4 is also characterized by a well-defined slope break that contrasts with the smoother slopes in preceding sequences. The height of the slope, or vertical distance between the clinoform breakpoint and the basin floor, is 400 m (320 ms TWT).

Sequence 5

Sequence boundary E marks the boundary between sequences 4 and 5 (Figure 11). The stratal geometries of this sequence are unique. First, convex-upward reflections with bidirectional downlap onto sequence boundary E are observed at the distal part of the clinoform (Figures 10, 11, 16). These mounded seismic facies bear a striking resemblance to basin-floor fans, as described in siliciclastic deep-water settings (e.g., Van Wagoner et al., 1990; Mitchum et al., 1994). Individual mounded features have a positive relief as much as 50 m and are as much as 2 km long (Figures 11, 16). Their base typically rests on the flat surface of sequence boundary E. On line E130 (Figure 11), the mounds are located 6 km west of the sequence 4 clinoform break. Some individual moundlike features appear to be joined together or overlap one another, whereas others are separated by U-shaped features that may be channels (Figure 16). It is important that these basin-floor mounded features are unique to sequence 5 and are present across the whole basin.

The moundlike bodies of sequence 5 are covered by an apron of sediments onlapping onto the sequence boundary E (Figure 11). Strong-amplitude reflection packages are absent on lines E120 and E130 but a well-defined SARP is present on line E140 (Figure 16). The upper part of sequence 5 shows oblique reflections that form a prograding wedge. The isochron map for sequence 5 (Figure 13F) clearly shows that thicker sediments were deposited at the toe of the bank slope.

Strike View

Prograding strata are easy to recognize in dip-oriented seismic and outcrop sections but their appearance in depositional strike view has not been well

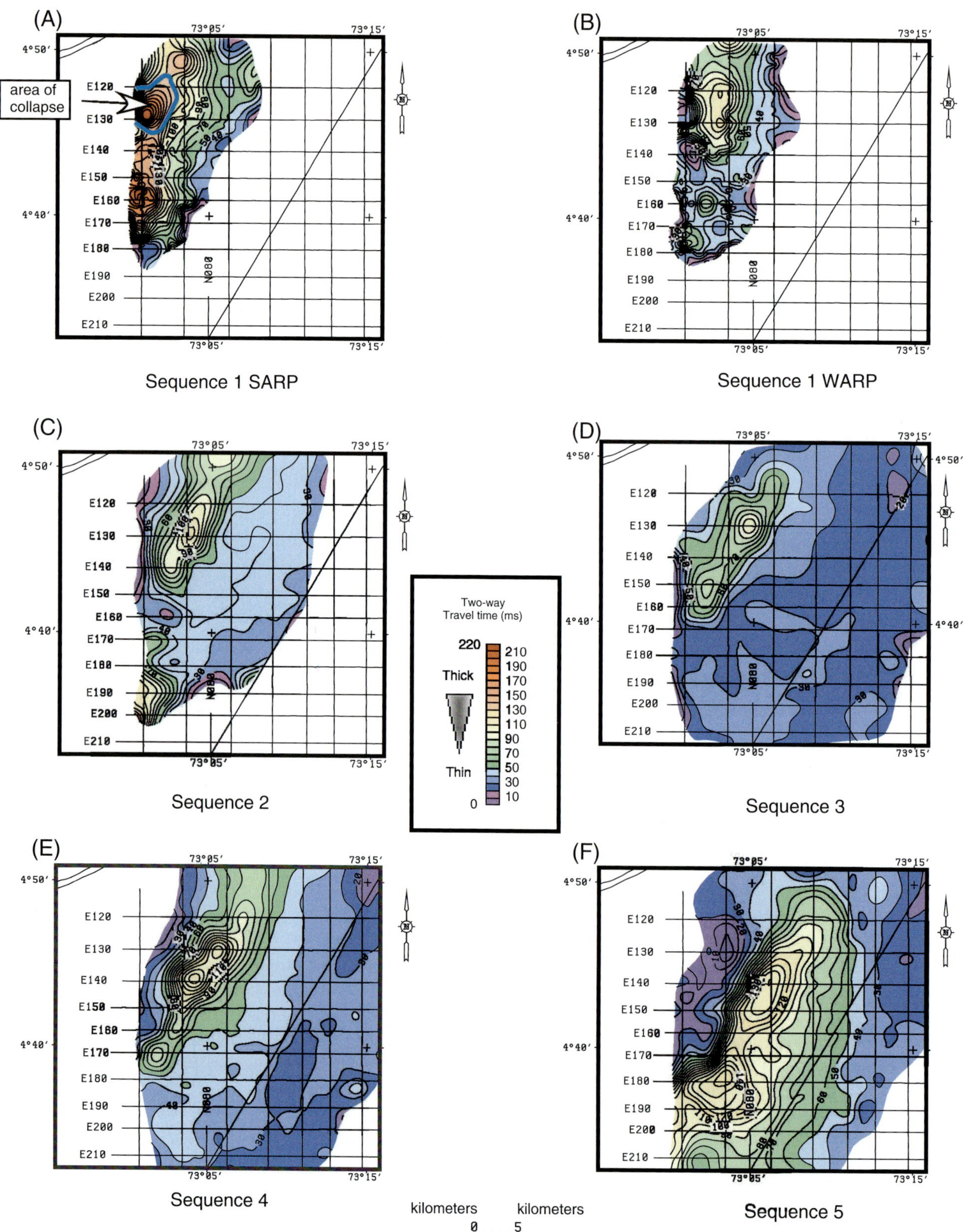

FIGURE 13. (A–F) Isochron maps of the middle Miocene sequences of prograding complex I. The maps show thicknesses of individual prograding sequences and lateral migration of the prograding margin.

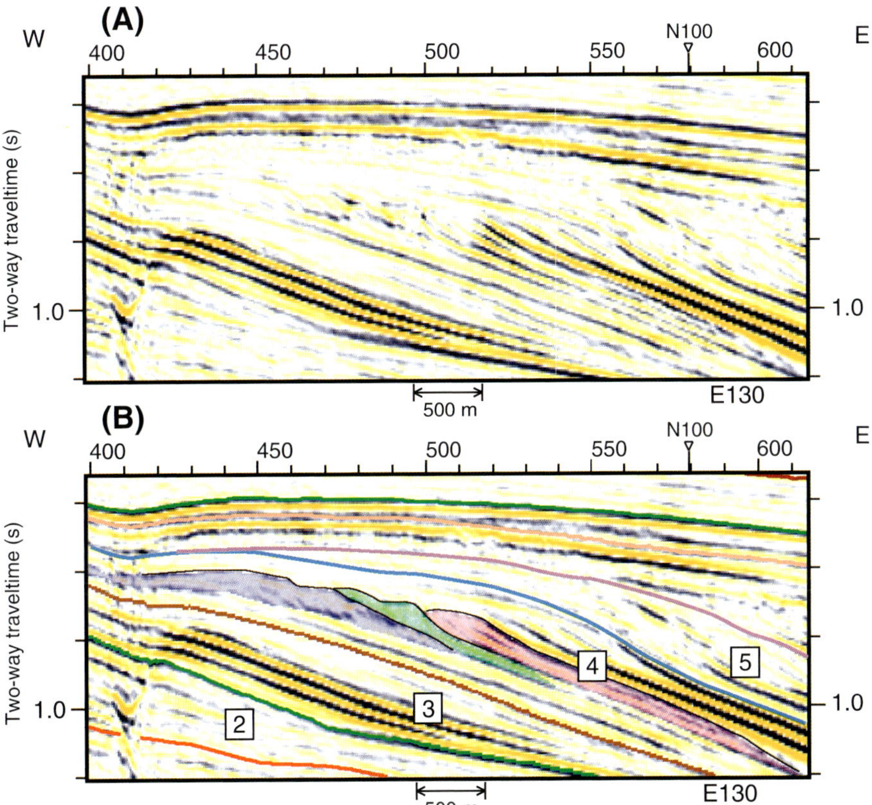

FIGURE 14. Uninterpreted (A) and interpreted (B) segments of line E130 showing downstepping margin ("forced regressive geometry") of the middle Miocene sequence 4 of prograding complex I.

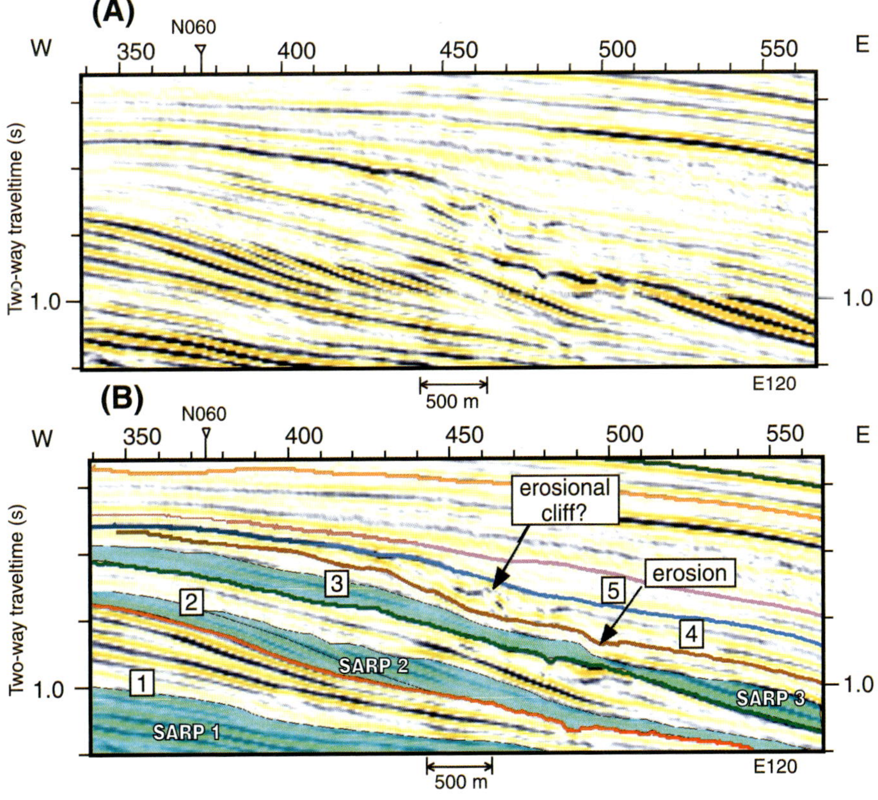

FIGURE 15. Uninterpreted (A) and interpreted (B) segment of line E120 showing internal geometry of middle Miocene sequences of prograding complex I.

documented. Shell seismic line N060 cuts prograding complex I at an angle close to depositional strike (Figure 17). In this cross section, the prograding sequences have broad mounding geometries with numerous small channels oriented perpendicular to the margin. The channels are present at both the sequence boundaries

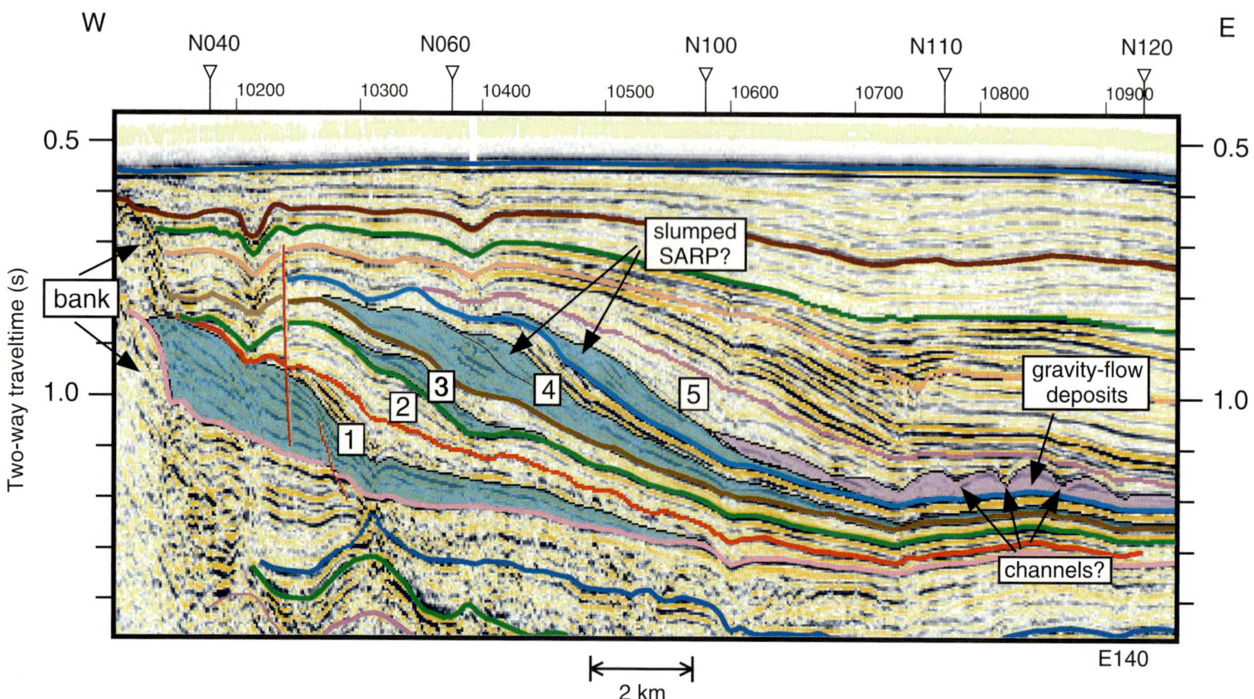

FIGURE 16. Interpreted segment of line E140 showing the middle Miocene prograding sequences of prograding complex I. Shaded areas represent SARPs and areas with no color fill represent WARPs. Mounded bodies at the toe of the clinoform are unique to sequence 5 and are interpreted as gravity-flow deposits, possibly reworked by currents. Many individual mounds are separated by U-shaped features that may be channels.

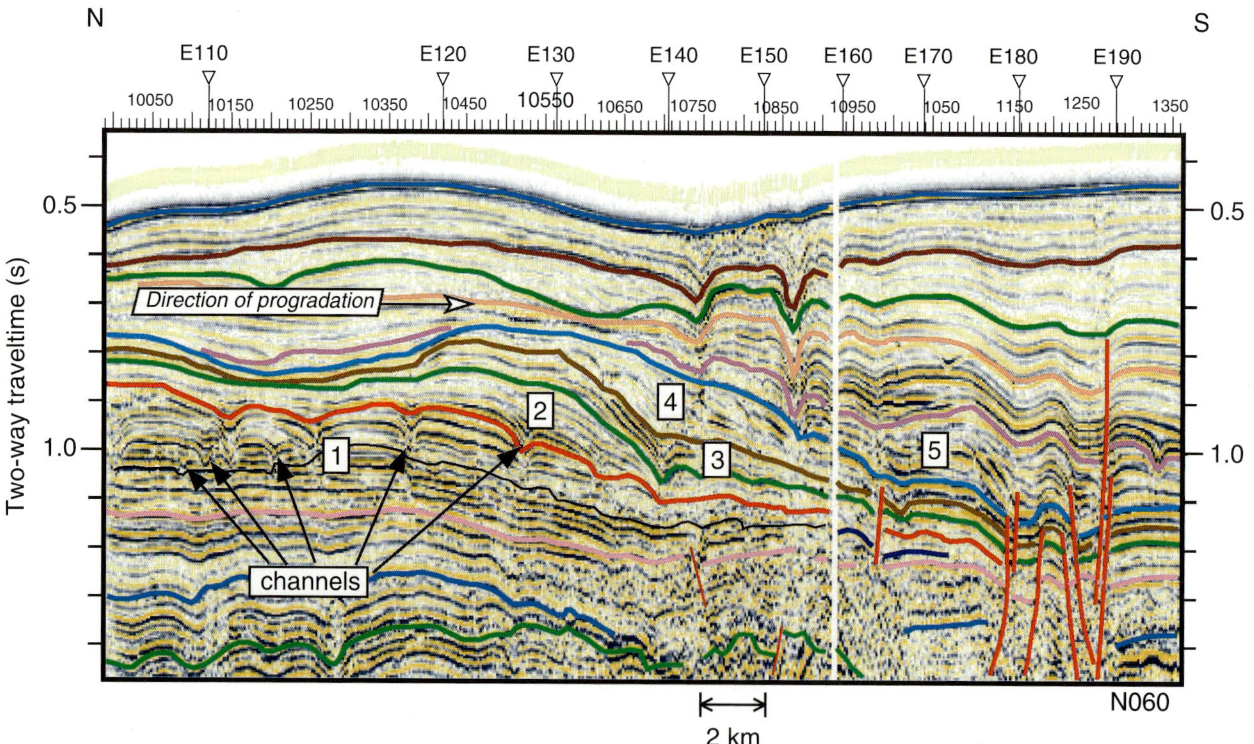

FIGURE 17. Segment of line N060 showing strike view of the middle Miocene prograding complex I. In this cross section, prograding sequences have broad mounded shape. Numerous channels oriented normal to the margin likely served as conduits of carbonate material derived from the bank top. The southern component of the southeast migration of prograding margin is evident on this profile.

and within the sequences at different levels (Figure 17). The widths of the channels range from 150 to 700 m (with an average width of 300 m), and their depths range from to 25 to 75 m. Many channels display multiple cut-and-fill events. These geometries are similar to the submarine canyons of the prograding Great Bahama Bank margin in the Straits of Florida (Anselmetti et al., 2000). The canyons in the Bahamas, however, are significantly deeper than channels in the Maldives.

The channels within the Maldives' prograding margin likely served as the conduits that brought the carbonate material from the bank top to the outbuilding clinoform fronts. This configuration demonstrates that during progradation, the sediments were delivered by a line source rather than supplied by a single point source. The strike view also shows the migration of the sedimentation locus through time. On line N060 (Figure 17), the thickest part of the margin migrated from north to south from the time of sequence 1 deposition for approximately 20 km.

Overlying Sequences

Vigorous progradation was halted at the end of the middle Miocene and was followed by a transgressive filling of the basin with sediments. The overlying upper Miocene sequences display local progradation of bank margins and filling of the central basin. The shallower reflections progressively onlap sequence 5 and completely cover the middle Miocene prograding complexes.

INTERNAL ARCHITECTURE OF PROGRADING COMPLEXES II AND III

Prograding Complex II

Prograding complex II occupies the area between Ari and North Nilandu Atolls (Figure 9). Five sequences were identified in complex II. The reflections that mark the base of sequences 1 and 2 and the top of sequence 3 in complex I were correlated across the basin and form corresponding sequence boundaries in complex II (Belopolsky, 2000). The lower three sequences can be divided into SARPs and WARPs (Figure 18), just like the sequences in complex I. Sequence 4 is significantly thinner than sequence 4 from prograding complex I, where forced regressive geometry was observed on seismic line E130 (Figure 14). Moundlike features, similar to those described in prograding complex I and interpreted as gravity-flow deposits, are found at the base of

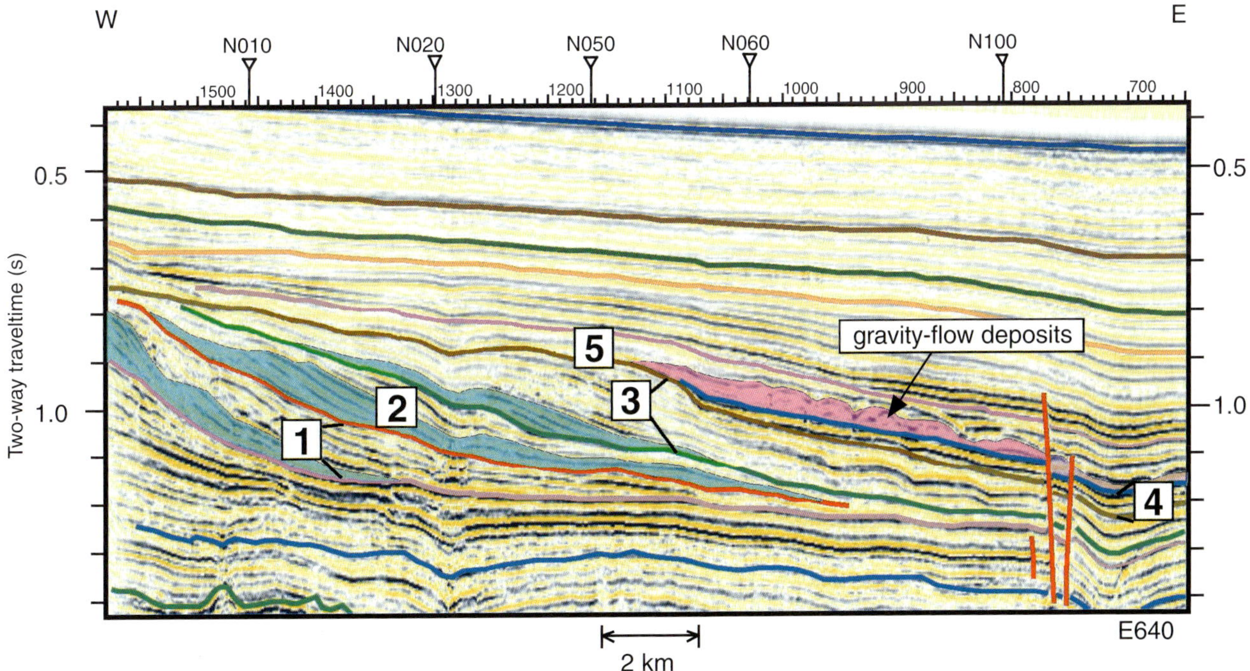

FIGURE 18. Interpreted segment of line E640 through the middle Miocene prograding complex II. Five middle Miocene prograding sequences are identified by numbers 1–5. Shaded areas indicate SARPs; areas with no color fill represent WARPs. Correlation of seismic reflections between prograding complexes I and II demonstrates that the deposition of prograding sequences was synchronous on the margins situated 150 km apart. The internal geometries of prograding sequences in complexes I and II are similar. Sequence 4 of complex II is thin on seismic line E640, partially because of the orientation of this particular seismic section.

sequence 5 in prograding complex II (Figure 18). Deposition of these deposits must have been synchronous in prograding complexes I and II.

The number, internal architecture, and seismic character of prograding sequences of complex II matches, with minor differences, the sequences of complex I. Seismic correlation suggests that the deposition of prograding sequences in complex II was synchronous with deposition of sequences in complex I. Thus, margins that were separated approximately 150 km from each other simultaneously produced similar stratal geometries. This implies that deposition of the middle Miocene prograding sequences was controlled by an external factor of regional scale.

Prograding Complex III

Prograding complex III is located north of Gaha Atoll and is imaged on only three Shell seismic lines (Figure 9). Line E130 imaged the edge of the flat-top bank to which complex III is attached (Figure 19).

Prograding complex III is located approximately 30 km to the east of prograding complex I on the eastern side of the Maldives Platform. Five prograding sequences were identified within complex III (Figure 19). Sequence 1 is characterized by a basinward shift in onlap. The lower part of sequence 2 has forced regressive geometry with four downstepping wedges (Figure 19). Sequences 3, 4, and 5 in complex III are thinner than in complex I. This may be explained by an oblique crosscutting of the sequences by seismic lines in complex III. Differentiation into SARPs and WARPs in the sequences of complex III is sometimes not obvious, which again may be attributed to the limited coverage by seismic profiles and crosscutting at an angle to the dip by seismic lines.

Correlation of sequences between prograding complexes I and III is straightforward by simply tracing

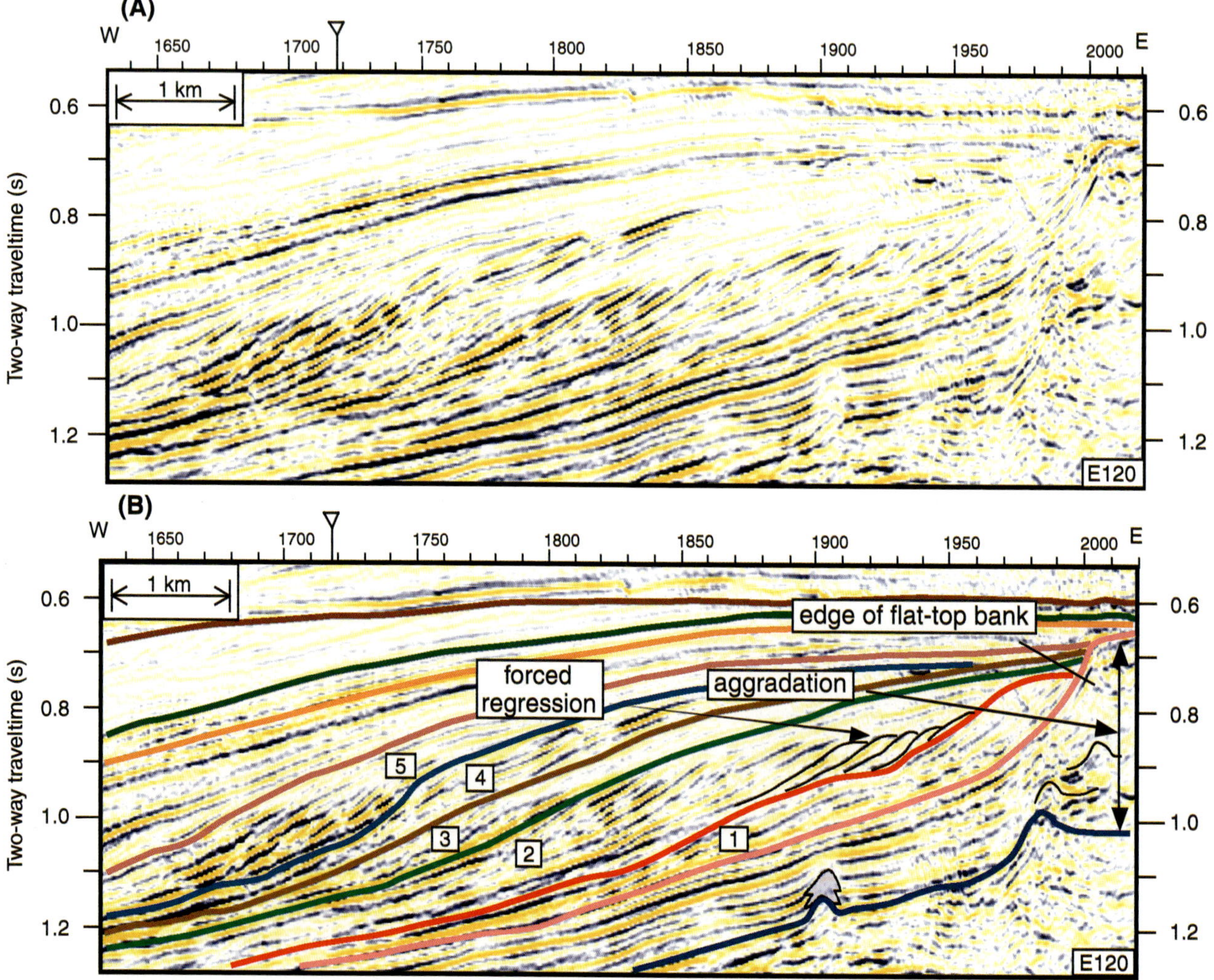

Figure 19. Uninterpreted (A) and interpreted (B) segments of line E120 showing the middle Miocene sequences (1–5) of prograding complex III. The sequences are attached to the slope of a late early Miocene flat-top carbonate bank that aggraded significantly in the early Miocene. Note the downstepping geometry (forced regression) in sequence 2.

sequence-bounding reflections on a single seismic line (Figure 6). The correlation shows that deposition of sequences in these two complexes was synchronous. This simultaneous progradation of bank margins in the Maldives in the middle Miocene was therefore driven by an external forcing factor.

INTERPRETATION OF SEISMIC FACIES AND STRATAL GEOMETRIES

SARPs and WARPs

Based on the interpretation of stratal patterns, stratigraphic relationships, and seismic facies, SARPs are interpreted as sediments deposited during periods of falling relative sea level. We assume that all carbonate sediments are produced locally on the flooded bank top at sea level. The basinward shift in onlap represented by SARPs is equivalent to the downward shift in coastal onlap. Falling base level resulted in the exposure of the preceding sequence topsets and downward (basinward) migration of the bank margin. This downstepping, or forced regression, is clearly expressed on some of the seismic profiles. This characteristic geometry is a solid indicator of falling relative sea level because neither variation in sediment supply nor environmental changes may be responsible for the downward migration of the bank margin (Schlager, 1992).

The difference in the amplitude of internal reflections between SARPs and WARPs within prograding sequences may be explained by differences in lithology and/or diagenetic overprint. The actual composition of the prograding sequences is unknown because wells did not penetrate the prograding clinoforms. The material forming the sequences must be mostly, if not entirely, carbonate and of local origin, given the remoteness of the Maldives from any landmass. Seismic properties of carbonate sediments and rocks depend on a wide range of parameters including porosity, pore type and shape, pore fluid, and saturation (Wang, 1997), with pore type being most important. The reflectivity of sediments is caused by changes in acoustic impedance, which is a product of sediment sonic velocity and density. Diagenetic alteration is one of the most important factors influencing the sonic velocity of carbonate sediments. Coarse-grained sediments are more susceptible to diagenetic alteration and reach higher sonic velocities after shallow burial faster than fine-grained sediments (Anselmetti, 1994).

Strong-amplitude reflection packages are likely to be composed of beds with relatively coarse grains, different grain-to-mud ratio, and higher degree of diagenetic alteration. The sediments that form SARPs were probably winnowed by currents and have a high grain-to-mud ratio resulting from the removal of fine-grained material. Because of their high primary porosity, these sediments are also likely to be more quickly and more extensively cemented and consequently have higher sonic velocity values than the overlying WARPs.

Weak-amplitude reflection packages are interpreted as sediments deposited during relative sea level rise and subsequent highstand. The onlapping strata of the lower part of WARPs represent the initial flooding after the preceding base level drop. Flooding results in a greater area available for sediment production and an increase in the amount of muddier sediment produced by the carbonate factory (Droxler and Schlager, 1985; Schlager et al., 1994). The prograding uppermost part of a WARP represents deposition during the sea-level highstand when excess sediment is shed from the bank top. Weak-amplitude reflection packages are likely to be composed of more homogeneous sediments and are probably more mud-rich, based on their overall poor reflectivity.

The Relationship of the Middle Miocene Prograding Sequences to Sea Level Changes

Each prograding sequence consists of two packages: a basal SARP, which formed during the lowering of relative sea level, and an upper WARP, which formed during flooding and sea-level highstand. Each sequence therefore represents a complete sea level cycle. The cycle begins with a relative sea level fall during which an unconformity is developed and carbonate production is reduced and shifted in a basinward direction. The subsequent flooding causes the backstepping of the bank margin and an increase in the amount of carbonate sediment produced. During the last part of the cycle, a relative sea level highstand, the bank margin progrades and the bank top sheds the excess material.

The slopes of the prograding margins has become progressively steeper and reached its maximum declivity at the time of sequence 5 formation (Table 1). Steepening of prograding carbonate margins through time has been reported from different localities around the world (e.g., Read, 1985; Eberli and Ginsburg, 1989; Pomar, 1993; Tinker, 1998). In the case of the middle Miocene prograding margins in the Maldives, the steepening of the clinoform slope was possibly related to a significant sea-level fall during the deposition of sequence 4 and resulting forced regressive geometry. Erosion resulting from exposure and the formation of narrow shelves during falling sea level led to the creation of a steeper margin. The slope of the overlying sequence 5 was developed on the antecedent steep slope and continued to steepen until it failed.

The deep-water mounded features at the bottomsets of the clinoforms in sequence 5 are interpreted as gravity-flow deposits. Their deposition is connected to the steepening of platform margins and their failure. The channels on top of the mounds or lobes likely served as the conduits for downslope dispersal of carbonate material. It does not seem possible, however, to conclude whether the channels originated updip at top of the clinoforms. These downslope deposits were also possibly reworked by currents that are common along bank margins. The mapping of individual lobes and channels is not possible at this time because their size is typically smaller than half-spacing of seismic lines. It is also difficult to determine whether the formation of the mounded features occurred as a result of a single or multiple events. It is important to note that the lobes of sequence 5 are found throughout on the paleo-Inner Sea seafloor but are unique to sequence 5. This implies a regional and unique set of conditions for their formation and possibly a common trigger mechanism that facilitated their deposition.

The deep-water mounded deposits of sequence 5 may be similar to the Late Cretaceous coarse-grained lobes described by Eberli et al. (1993) in the Maiella Platform in Italy. The lobes described in Maiella had a positive relief of as much as 70 m and extended laterally for as much as 2 km. Internally, the lobes consisted of stacked channel complexes. These coarse-grained lobes were interpreted as lowstand slope fan deposits (Eberli et al., 1993).

In the Maldives, the mounded seismic facies on the basin floor above sequence boundary E are also similar to the description of the Pleistocene turbidites and debris-flow deposits in the Tongue of the Ocean (Schlager and Chermak, 1979) and in the Exuma Sound (Crevello and Schlager, 1980) in the Bahamas. These deposits formed coalescing mounded lobes at the toe of slope and basin floor. The turbidites in the Exuma Sound were mainly composed of the clean graded skeletal sands and lithoclasts, and the gravity-flow deposits were composed of pebbly mud with rubble and graded carbonate sands (Crevello and Schlager, 1980). Biostratigraphic dating showed that the debris-flow deposits formed 75,000–80,000 yr B.P. during the falling of sea level (Droxler, 1984).

Betzler et al. (1999) also described the mounded morphology of Miocene turbidite lobes on the western side of the Great Bahama Bank. The turbidite lobes contained flat to convex-down reflections in their central part, which are interpreted as turbidite feeder channels.

In siliciclastic systems, the deposition of gravity-flow deposits on the basin floor is associated with relative sea level falls (e.g., Van Wagoner et al., 1988). In carbonate systems, however, the deposition of turbidites and debris-flow deposits in the basin is not considered to be diagnostic for establishing relative sea level position. It has been demonstrated that the export of shallow-water carbonate material in the form of turbidites is more common during sea level high stands (Droxler and Schlager, 1985; Schlager et al., 1994). Under the "highstand shedding" scenario, the carbonate factory produces and exports a greater amount of material when the bank top is flooded. Late Quaternary highstand turbidites typically are composed of aragonite and high-Mg calcite nonskeletal bank–derived material (Haak and Schlager, 1989). During sea level lowstands, a large part of the bank is exposed, which results in a complete or partial shutdown of carbonate production. This exposure commonly results in sediment erosion and redistribution of sediments downslope in the form of coarse-grained debris flows, slumps, and megabreccias that are largely derived from platform margin and upper slope sediments (Grammer et al., 1993; Handford and Loucks, 1993).

A study by Spence and Tucker (1997) showed that the probability of carbonate megabreccia occurrence is higher during sea level falls because of overpressuring in confined layers as pore fluid drains from sediments when the platform top becomes exposed. The link between a relative sea level fall and the development of overpressure becomes important when we consider the basinal counterparts of sequences 4 and 5 in the Maldives. In the central and deepest part of the basin, sequences 4 and 5 are represented by a seismic unit showing a lateral change in character from parallel internal reflectors to disrupted reflections resembling imbrication (Figure 20). Aubert and Droxler (1996) first noted the seismic character of this unit. This basinal equivalent of prograding sequences 4 and 5 is as much as 137 m (110 ms TWT) thick. Three possible explanations for this layer-bounded disturbance are considered. The first explanation relates the deformation to gravity-driven failure and compressional faulting above a specific detachment surface within the basin-floor strata. Alternatively, the deformation may be related to early compaction of fine-grained sediments caused by dewatering. Regionally extensive polygonal fault systems related to volumetric contraction during early dewatering have been documented in basins worldwide (Cartwright, 1996; Cartwright and Dewhurst, 1998; Dewhurst et al., 1999), especially within marine deposits composed of ultrafine-grained claystones or carbonate chalks. The third explanation is that these sediments are large-scale sediment waves created by bottom currents. Resolution of standard industry seismic data typically is not sufficient to resolve true geometries of these sediments. But high-resolution seismic data from the west margin of the Great Bahama Bank imaged "thin, discontinuous seismic facies" within the drift deposits that were interpreted as "migrating sediment waves produced by currents that were strong enough to cause bedload traction" (Anselmetti et al., 2000, and references therein). The increased activity of such currents is commonly related to the sea level falls (Anselmetti et al., 2000). It is possible that the

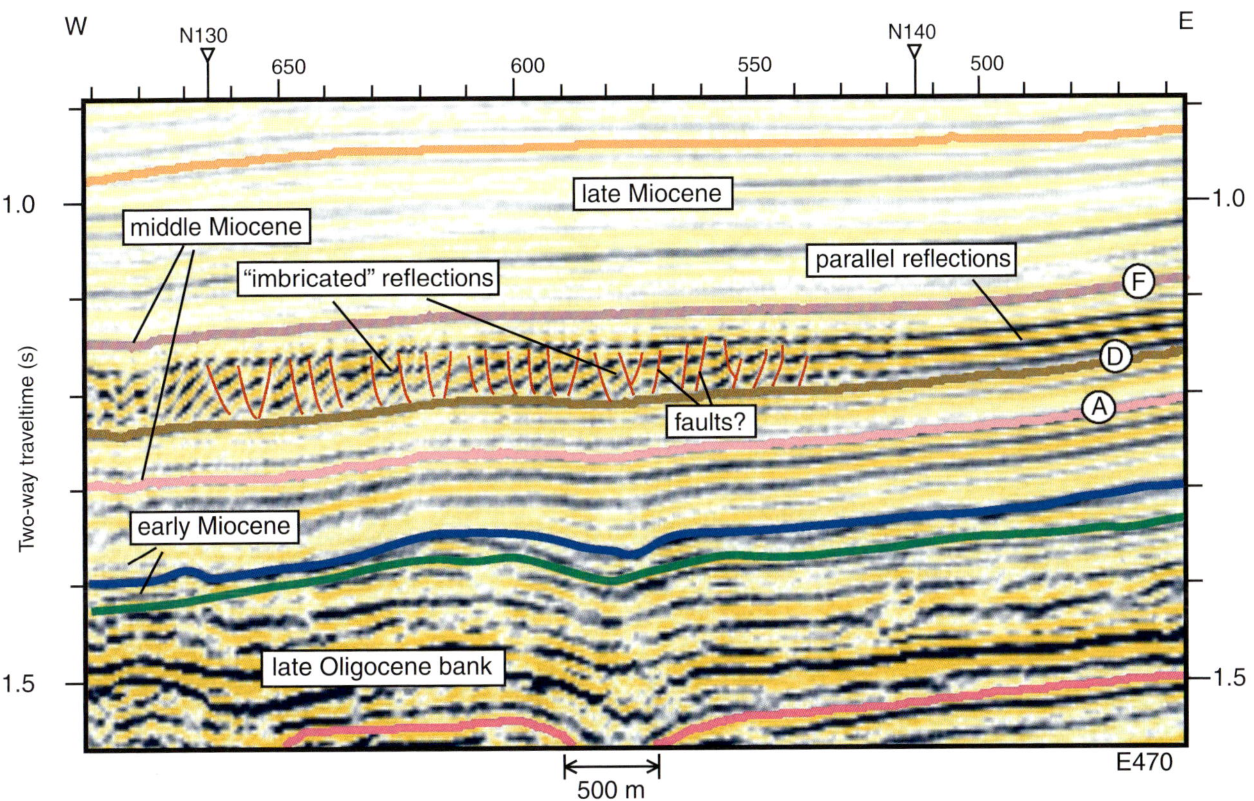

FIGURE 20. Interpreted segment of line E470 showing basinal counterparts of the prograding middle Miocene sequences (bounded by sequence boundaries A and F). Layer-bounded disrupted deposits of the 4 and 5 are bounded by sequence boundaries D and F. Parallel continuous reflectors on the east grade into disrupted, "imbricated" reflectors to the west. Possible interpretations of these seismic facies include regionally extensive polygonal (in map view) fault systems related to volumetric contraction resulting dewatering, or large-scale sediment waves that are not fully resolved in this seismic data set.

resolution of the seismic data in the Maldives is not sufficient to image the true geometries for these sediments. There is not, however, any evidence of drift deposits produced by strong current in the paleo-Inner Sea in the middle Miocene, which was much shallower than the Straits of Florida.

The second and third explanations are favored because the gravity-driven compressional faulting should not be unique to one horizon and would be expected in strata both above and below the described unit. The mounded gravity-flow deposits of sequence 5 are thus interpreted as lowstand deposits related to a significant relative sea level fall. The sea level fall possibly caused the overpressuring of the low-permeability fine-grained basin sediments that resulted in the geologically instantaneous fluid expulsion.

The differences in individual prograding margins within Maldives exist but are not significant— the number and general geometries of sequences remain the same whereas their thickness and certain elements vary. It is not easy to adequately document the differences because the seismic grid, albeit dense, is 2-D and may introduce some aliasing because of the way the profiles go through a margin. However, there is little doubt, that local factors, such as currents, antecedent geometry, local tectonic movements, and possibly some variation in the carbonate-producing biota, are important in shaping the character of individual prograding margins. It is difficult to access these factors from the seismic data alone. The strong general similarity of the margins located as far as 150 km apart is far more important because it implies a regional driving force that caused the progradation. This force is relative sea level changes that very likely had a strong eustatic component. The middle Miocene was the time of significant changes in the Earth's climatic and oceanographic conditions. Evidence suggests that the West Antarctica Ice Sheet was formed in the early–middle Miocene and its early evolution included repeated episodes of advance and retreat across the continental shelves (Abreu and Anderson, 1998). Temperature-corrected oxygen isotope record indicates that 85% of the benthic $\delta^{18}O$ increase in the late middle Miocene can be attributed to the increase of continental ice volume (Lear et al., 2000). This record indicates a lowering of sea level in the middle Miocene results from the removal and storage of the part of the oceanic water masses in the form of continental ice. These eustatic sea level changes must have had a primary effect on the evolution of the prograding margins.

SUMMARY

Interpretation of seismic and well data in the Maldives shows that significant progradation of carbonate bank margins occurred in the middle Miocene on a regional scale. Three prograding carbonate bank margins located tens of kilometers from each other recorded the same number of prograding sequences with similar internal geometries. Each sequence typically contained a lower SARP and an upper WARP. Based on the reflection geometry and relationship, SARPs were interpreted as having formed during falling relative sea level. Weak-amplitude reflection packages were interpreted as having formed during the subsequent relative sea level rise and highstand. In strike view, prograding strata appear as broad mounds with numerous channels oriented normal to the strike. These channels might have acted as conduits of carbonate material delivered from the bank top to the outbuilding fronts.

Five prograding sequences recorded carbonate bank response to relative sea level fluctuations. Their prograding geometries are interpreted as representing five complete sea level cycles. The last sequence recorded a sea level fall of significant magnitude and is associated with the accumulation of massive gravity-flow deposits.

ACKNOWLEDGMENTS

We thank the government of the Republic of Maldives and Royal Dutch Shell for providing seismic and well data and Shell E&P Technology Laboratory in Bellaire, Texas, for usage of a seismic workstation. We also thank Mitch Harris, Albert Bally, Ed Purdy, John Karlo, and Peter Vail for fruitful discussions of the presented material. Constructive reviews by Gregor Eberli, Evan Franseen, and Steven Dorobek improved the quality of the manuscript. Funding for this research was provided by an NSF Grant OCE-9730954 to A.W.D. and a Rice University Mills Bennett Fellowship and an SEPM Grant-in-Aid to A.V.B.

REFERENCES CITED

Abreu, V. S., and J. B. Anderson, 1998, Glacial eustasy during the Cenozoic: Sequence stratigraphic implications: AAPG Bulletin, v. 82, no. 7, p. 1385–1400.

Anselmetti, F. S., 1994, Physical properties and seismic response of carbonate sediments and rocks: Ph.D. dissertation, Geological Institute, Swiss Federal Institute of Technology, Zurich, 157 p.

Anselmetti, F. S., G. P. Eberli, and Z.-D. Ding, 2000, From the Great Bahama Bank into the Straits of Florida: A margin architecture controlled by sea-level fluctuations and ocean currents: Geological Society of America Bulletin, v. 112, no. 6, p. 829–844.

Aubert, O., 1994, Origin and stratigraphic evolution of the Maldives (central Indian Ocean). Ph.D. dissertation, Rice University, Houston, Texas, 258 p.

Aubert, O., and A. W. Droxler, 1992, General Cenozoic evolution of the Maldives carbonate system (equatorial Indian Ocean): Bulletin du Centre de Recherches Elf Exploration Production Elf Aquitaine, v. 16, no. 1, p. 113–136.

Aubert, O., and A. W. Droxler, 1996, Seismic stratigraphy and depositional signatures of the Maldive carbonate system (Indian Ocean): Marine and Petroleum Geology, v. 13, no. 5, p. 503–536.

Bachtel, S. L., R. D. Kissling, D. Martono, S. Rahardjanto, P. A. Dunn, B. A. MacDonald, 2004, Seismic stratigraphic evolution of the Miocene–Pliocene Segitiga Platform, East Natuna Sea, Indonesia: The origin, growth, and demise of an isolated carbonate platform, *in* G. P. Eberli, J. L. Masaferro, and J. F. Sarg, eds., Seismic imaging of carbonate reservoirs and systems: AAPG Memoir 81, p. 309–328.

Backman, J., R. A. Duncan, et al., 1988, Proceedings of the Ocean Drilling Program, Initial Reports: Ocean Drilling Program, Texas A&M University, College Station, Texas, v. 115, p. 1073.

Belopolsky, A. V., 2000, Tectonic and eustatic controls on the evolution of the Maldive carbonate platform: Ph.D. dissertation, Rice University, Houston, Texas, 267 p.

Betzler, C., J. J. G. Reijmer, K. Bernet, G. P. Eberli, and F. Anselmetti, 1999, Sedimentary patterns and geometries of the Bahamian outer carbonate ramp (Miocene–Lower Pliocene, Great Bahama Bank): Sedimentology, v. 46, no. 6, p. 1127–1143.

Blow, W. H., 1979, The Cainozoic Globigerinidea: Leiden, E. J. Brill, 1413 p.

Bosellini, A., 1984, Progradation geometries of carbonate platforms: Examples from the Triassic of the Dolomites, northern Italy: Sedimentology, v. 31, p. 1–24.

Cartwright, J. A., 1996, Polygonal fault systems: A new type of fault structure revealed by 3-D seismic data from the North Sea Basin, *in* P. Weimer and T. L. Davis, eds., Applications of 3-D seismic data to exploration and production: AAPG Studies in Geology 42/SEG Geophysical Developments Series 5, p. 225–230.

Cartwright, J. A., and D. N. Dewhurst, 1998, Layer-bound compaction faults in fine-grained sediments: Geological Society of America Bulletin, v. 110, no. 10, p. 1242–1257.

Crevello, P. D., and W. Schlager, 1980, Carbonate debris sheets and turbidites, Exuma Sound, Bahamas: Journal Sedimentary Petrology, v. 50, p. 1121–1148.

Curray, J. R., 1964, Transgressions and regressions, *in* R. L. Miller, ed., Papers in marine geology: New York, Macmillan, p. 175–203.

Dewhurst, D. N., J. A. Cartwright, L. Lonergan, 1999, The development of polygonal fault systems by syneresis of colloidal sediments: Marine and Petroleum Geology, v. 16, p. 793–810.

Droxler, A. W., 1984, Late Quaternary glacial cycles in the Bahamian deep basins and in the adjacent Atlantic Ocean: Ph.D. dissertation, University of Miami, Florida, Coral Gables, 165 p.

Droxler, A. W., and W. Schlager, 1985, Glacial versus interglacial sedimentation rates and turbidite frequency in the Bahamas: Geology, v. 13, p. 799–802.

Droxler, A. W., G. A. Haddad, D. A. Mucciarone, and J. L. Cullen, 1990, Pliocene–Pleistocene aragonitic cyclic variations in Ocean Drilling Program holes 714A and 716B (the Maldives) compared to hole 633A (Bahamas): Records of climate-induced $CaCO_3$ preservation at intermediate water depths, *in* R. A. Duncan et al., eds., Proceedings of the Ocean Drilling Program, Scientific Results: Ocean Drilling Program, Texas A&M University, College Station, Texas, v. 115, p. 539–577.

Duncan, R. A. and R. B. Hargraves, 1990, $^{40}Ar/^{39}Ar$ geochronology of basement from the Mascarene Plateau, the Chagos Bank and the Maldives Ridge, *in* R. A. Duncan et al., eds., Proceedings of the Ocean Drilling Program, Scientific Results: Ocean Drilling Program, Texas A&M University, College Station, Texas, v. 115, p. 43–51.

Eberli, G., and R. N. Ginsburg, 1989, Cenozoic progradations of northwestern Great Bahama Bank, a record of lateral platform growth and sea-level fluctuations, *in* P. D. Crevello, J. L. Wilson, J. F. Sarg, and J. F. Read, eds., Controls of carbonate platform and basin development: SEPM Special Publication 44, p. 339–351.

Eberli, G. P., D. Bernoulli, D. Sanders, A. Vecsei, 1993, From aggradation to progradation: The Maiella Platform, Abruzzi, Italy, *in* J. A. Simo, R. W. Scott, and J.-P. Masse, eds., Cretaceous carbonate platforms: AAPG Memoir 56, p. 213 232.

Grammer, G. M., R. N. Ginsburg, and P. M. Harris, 1993, Timing of deposition, diagenesis, and failure of steep carbonate slopes in response to a high-amplitude/high-frequency fluctuation in sea level, Tongue of the Ocean, Bahamas, *in* R. G. Loucks and J. F. Sarg, eds., Carbonate sequence stratigraphy— Recent development and applications: AAPG Memoir 57, p. 107–131.

Haak, A. B., and W. Schlager, 1989, Compositional variations in calciturbidites due to sea-level fluctuations, late Quaternary, Bahamas: Geologische Rundschau, v. 78, p. 477–486.

Handford, C. R., 1995, Baselap patterns and the recognition of lowstand exposure and drowning— A Mississippian-ramp example and its seismic signature: Journal of Sedimentary Research, v. B65, no. 3, p. 323–337.

Handford, C. R., and R. G. Loucks, 1993, Carbonate depositional sequences and system tracts— Responses of carbonate platforms to relative sea-level changes, *in* R. G. Loucks and J. F. Sarg, eds., Carbonate sequence stratigraphy— Recent development and applications: AAPG Memoir 57, p. 3–41.

Hesthammer, J., and H. Fossen, 1999, Evolution and geometries of gravitational collapse structures with examples from the Statjord field, northern North Sea: Marine and Petroleum Geology, v. 16, p. 259–281.

Hunt, D., and M. Tucker, 1992, Stranded parasequences and the forced regressive wedge system tract: Deposition during base-level fall: Sedimentary Geology, v. 81, p. 1–9.

Hunt, D., and M. Tucker, 1995, Stranded parasequences and the forced regressive wedge system tract: Deposition during base-level fall— Reply: Sedimentary Geology, v. 95, p. 145–160.

Iannello, C., and S. Dorobek, 2001, Regional characteristics, timing, and significance of dissolution-collapse features in Lower Cretaceous carbonate platform strata, DeSoto Canyon area, offshore Alabama-Florida (abs.): AAPG Bulletin, v. 85.

Kallweit, R. S., and L. C. Wood, 1982, The limits of resolution of zero-phase wavelets: Geophysics, v. 47, no. 7, p. 1035–1046.

Lear, C. H., H. Elderfield, and P. A. Wilson, 2000, Cenozoic deep-sea temperatures and global ice volumes from Mg/Ca in benthic foraminiferal calcite: Science, v. 287, no. 5451, p. 269–272.

Malone, M. J., P. A. Baker, S. J. Burns, and P. K. Swart, 1990, Geochemistry of periplatform carbonate sediments, Leg 115, Site 716 (Maldives Archipelago, Indian Ocean), *in* R. A. Duncan et al., eds., Proceedings of the Ocean Drilling Program, Scientific Results: Ocean Drilling Program, Texas A&M University, College Station, Texas, v. 115, p. 647–659.

Mitchum, R. M., J. B. Sangree, P. R. Vail, and W. W. Wornardt, 1994, Recognizing sequences and system tracts from well logs, seismic data, and biostratigraphy: Examples from the Late Cenozoic of the Gulf of Mexico, *in* P. Weimer and H. W. Posamentier, eds., Siliciclastic sequence stratigraphy: Recent developments and applications: AAPG Memoir 58, p. 163 197.

Osleger, D. A., and S. W. Tinker, 1999, Three-dimensional architecture of Upper Permian high-frequency sequences, Yates-Capitan shelf margin, Permian Basin, U.S.A., *in* J. A. Howell and J. F. Aitken, eds., High resolution sequence stratigraphy: Innovations and applications: Geological Society Special Publication 104, p. 169–185.

Pomar, L., 1993, High-resolution sequence stratigraphy in prograding Miocene carbonates: Application to seismic interpretation, *in* R. G. Loucks and J. F. Sarg, eds., Carbonate sequence stratigraphy— Recent development and applications: AAPG Memoir 57, p. 389–407.

Posamentier, H. W., G. P. Allen, and D. P. James, 1990, Aspects of sequence stratigraphy: Recent and ancient examples of forced regression (abs.): AAPG Annual Meeting Program, v. 74, p. 742.

Posamentier, H. W., G. P. Allen, D. P. James, and M. Tesson, 1992, Forced regression in a sequence stratigraphic framework: Concepts, examples, and exploration significance: AAPG Bulletin, v. 76, p. 1687–1709.

Purdy, E. G., and G. T. Bertram, 1993, Carbonate concepts from the Maldives, Indian Ocean: AAPG Studies in Geology 34, 56 p.

Read, J. F., 1985, Carbonate platform facies models: AAPG Bulletin, v. 69, p. 1–21.

Sarg, J. F., 1988, Carbonate sequence stratigraphy, *in* C. K. Wilgus, H. Posamentier, J. Van Wagoner, C. A. Ross, and C. G. St. C. Kendall, eds., Sea-level changes: An integrated approach: SEPM Special Publication 42, p. 155–181.

Schlager, W., 1992, Sedimentology and sequence stratigraphy of reefs and carbonate platforms. AAPG Continuing Education Course Note Series No. 34, 71 p.

Schlager, W., and A. Chermak, 1979, Sediment facies of platform-basin transition, Tongue of the Ocean, Bahamas, *in* L. J. Doyle and O. H. Pilkey, eds., Geology of continental slopes: SEPM Special Publication 27, p. 193–208.

Schlager, W., J. J. G. Reijmer, and A. W. Droxler, 1994, Highstand shedding of carbonate platforms: Journal of Sedimentary Research, v. B64, p. 270–281.

Sonnenfeld, M. D., and T. A. Cross, 1993, Volumetric partitioning and facies differentiation within the Permian upper San Andres Formation of Last Chance Canyon, Guadalupe Mountains, New Mexico, *in* R. G. Loucks and J. F. Sarg, eds., Carbonate sequence stratigraphy— Recent development and applications: AAPG Memoir 57, p. 435–474.

Spence, G. H., and M. E. Tucker, 1997, Genesis of limestone megabreccias and their significance in carbonate sequence stratigraphic models: A review: Sedimentary Geology, v. 112, p. 163–193.

Story, C., P. Peng, C. Heubeck, C. Sullivan, and J. D. Lin, 2000, Liuhua 11-1 field, South China Sea: A shallow carbonate reservoir developed using ultrahigh-resolution 3-D seismic, inversion, and attribute-based reservoir modeling: Leading Edge (August), v. 19, no. 8, p. 834–844.

Styzen, M. J., 1996, Late Cenozoic chronostratigraphy of the Gulf of Mexico: Gulf Coast Section, SEPM Foundation, chart in two sheets.

Tinker, S. W., 1998, Shelf-to-basin facies distribution and sequence stratigraphy of a steep-rimmed carbonate margin: Capitan depositional system, McKittrick Canyon, New Mexico and Texas: Journal of Sedimentary Research, v. 68, no. 6, p. 1146–1174.

Tyrrell, W. W., and R. G. Davis, 1987, Miocene carbonate shelf margin, Bali-Flores Sea, Indonesia, *in* A. W. Bally, ed., Atlas of seismic stratigraphy: AAPG Studies in Geology No. 27, v. III, p. 174–179.

Van Morkhoven, F. P. C. M., and A. G. Schroeder, 1986, Atlas of planktonic foraminifera: Internal publication, Shell Oil Company, Houston, Texas, Shell International Petroleum Maatschappij, The Hague.

Van Wagoner, J. C., H. W. Posamentier, R. M. Mitchum, P. R. Vail, J. F. Sarg, T. S. Loutit, and J. Hardenbol, 1988, An overview of the fundamentals of sequence stratigraphy and key definitions, *in* C. K. Wilgus, H. Posamentier, J. Van Wagoner, C. A. Ross, and C. G. St. C. Kendall, eds., Sea-level changes: An integrated approach: SEPM Special Publication 42, p. 39–45.

Van Wagoner, J. C., R. M. Mitchum, K. M. Campion, and V. D. Rahmanian, 1990, Siliciclastic sequence stratigraphy in well logs, cores, and outcrops: AAPG Methods in Exploration Series 7, 55 p.

Wang, Z., 1997, Seismic properties of carbonate rocks, *in* I. Palaz and K. Marfurt, eds., Carbonate seismology: Society of Exploration Geophysicists Geophysical Development Series 6, p. 29–52.

13

Isern, A. R., F. S. Anselmetti, and P. Blum, 2004, A neogene carbonate platform, slope, and shelf edifice shaped by sea level and ocean currents, Marion Plateau (northeast Australia), *in* Seismic imaging of carbonate reservoirs and systems: AAPG Memoir 81, p. 291–307.

A Neogene Carbonate Platform, Slope, and Shelf Edifice Shaped by Sea Level and Ocean Currents, Marion Plateau (Northeast Australia)

Alexandra R. Isern
National Science Foundation, Arlington, Virginia, U.S.A.

Flavio S. Anselmetti
Swiss Federal Institute of Technology (ETH), Zurich, Switzerland

Peter Blum
Ocean Drilling Program, College Station, Texas, U.S.A.

ABSTRACT

More than 1700 km of high-resolution seismic data were collected over the Marion Plateau, northeast Australia, to investigate the influence of sea level and oceanography on subtropical carbonate platforms growing on the plateau surface. Seismic data, interpreted in combination with sediments recovered during Ocean Drilling Program Leg 194 and modern oceanographic data, have enabled characterization of the parameters controlling platform growth and development in this region.

Most modern carbonate platforms, such as the Bahamas Platform, have sedimentation patterns that reflect the prevailing wind direction where sediments are forced off the platform on the leeward side, leaving the windward side relatively sediment-starved. This results in platform asymmetry with steep windward and gentler leeward slopes. The seismic data presented here indicate that the carbonate platforms off northeast Australia, although similar in morphology to the Bahamas Platform in many respects, are dominated by oceanographic currents as the primary energy source creating a similar asymmetrical platform geometry where the upcurrent side of the platform is relatively sediment starved and most sediment is deposited on the downcurrent slope. Currents in the study area are dominated by the southward-flowing East Australian Current that generally flows opposite to the prevailing Southeast Trade Winds. This current likely determines not only the morphology, but also the growth potential of the platforms, as well as the volume and final location of sediment transported from the platform top.

Despite the massive, tablelike structures exhibited in the seismic data, Leg 194 drilling demonstrated that the platforms are almost entirely composed of the remains of cool, subtropical organisms, such as red algae, bryozoans, and larger benthic foraminifera. Coralline algae were notably absent from most sequences. These calcite-dominated organic remains have a low diagenetic potential, resulting in uncemented and friable slope successions. Nevertheless, the platform tops are well cemented. The fact that the cool subtropical faunal assemblages produce platform geometries that are similar to tropical carbonates suggests that physical parameters, such as current flow and sea level change, may be more important than biofacies in establishing platform architectures.

INTRODUCTION

The geometries and evolution of carbonate platforms and their slopes are generally controlled by sea level change and wind direction, with sea level defining the long-term aggradation and winds defining the major direction of platform progradation (Eberli and Ginsburg, 1987). In some cases, ocean currents can influence downslope carbonate shedding and produce unique carbonate platform and slope geometries (Glaser and Droxler, 1993; Anselmetti et al., 2000).

The Marion Plateau, basinward of the central Great Barrier Reef, northeast Australia, is a unique carbonate-dominated province, where ocean currents flow opposite the general wind direction (Tomczak and Godfrey, 1994) (Figure 1). Two major isolated carbonate platform complexes and numerous smaller platforms exist on the surface of the plateau (Figures 1, 2). Previous investigations of the seismic stratigraphic architecture of the Marion Plateau produced a model for the evolution of these carbonate platforms (Davies et al., 1989; Davies et al., 1991; Pigram et al., 1992) that has subsequently been tested using the seismic data described here as well as cores from Ocean Drilling Program (ODP) Leg 194. These investigations have shown that currents flowing over the Marion Plateau are the dominant transport mechanism for moving sediments off the carbonate platforms to the platform slopes and the plateau surface. This sediment transport is modulated by the influence of sea level variability, which determines the morphology and development of the Marion Plateau carbonate platforms and adjacent sediment sequences.

This chapter presents the seismic stratigraphic analysis of a high-resolution two-dimensional seismic grid covering a large part of the Marion Plateau surface. This seismic data set, collected during a 1-month cruise in April 1999, images the Miocene carbonate platforms, adjacent slopes, and open-shelf areas on the plateau and was part of the site survey for ODP Leg 194 drilled in January 2001 (Isern et al., 2002). Lithologies recovered during Leg 194 drilling are used to calibrate the seismic stratigraphic observations and interpretations. Together, seismic data and core analyses provide a unique database to document the evolution and factors controlling this carbonate-dominated continental shelf.

METHODS AND DATA

A grid of 1700 km of high-resolution multichannel seismic profiles was acquired from the R/V *Franklin* (FR 03/99; Australian Geological Survey Organisation [AGSO] 209) operated by the Commonwealth Scientific and Industrial Research Organisation. Seismic data were acquired using two 45/105-in.^{3} GI airguns, a 24-channel Innovative Transducers Inc. nonfluid, stealth array cable consisting of six hydrophones per channel, 12.5 m group spacing, 30 m deck leader, 105 m tow leader, and 2×150 m active sections. Shot spacing was 25 m. The streamer was held at a water depth of 5 ± 1 m using three Digicourse 5010 active birds. Navigation and positioning were achieved using a DGPS system with reference to the Optus Townsville base station. Recording length was 3 s with a sampling rate of 1 ms. The profiles shown are the result of a standard multichannel processing suite, including band-pass filtering between 20–25 and 400–450 Hz. The sixfold stack is displayed unmigrated with an automatic gain control of 250 ms.

To correlate core and log information accurately with the seismic data, a time-depth curve was constructed for each drill site from Leg 194 (Figure 1; Isern et al., 2002). These curves were obtained by integrating velocity data from sonic logs and shipboard velocity measurements, calibrated with results from check-shot surveys when available. For sites where no logging data were available, tie points were defined that link prominent high-amplitude reflections to unique horizons in the cores, commonly hardgrounds or exposure surfaces at megasequence boundaries. Near the drill-hole terminus, the basement reflection was used as a tie point to fix the critical lower part of the time-depth

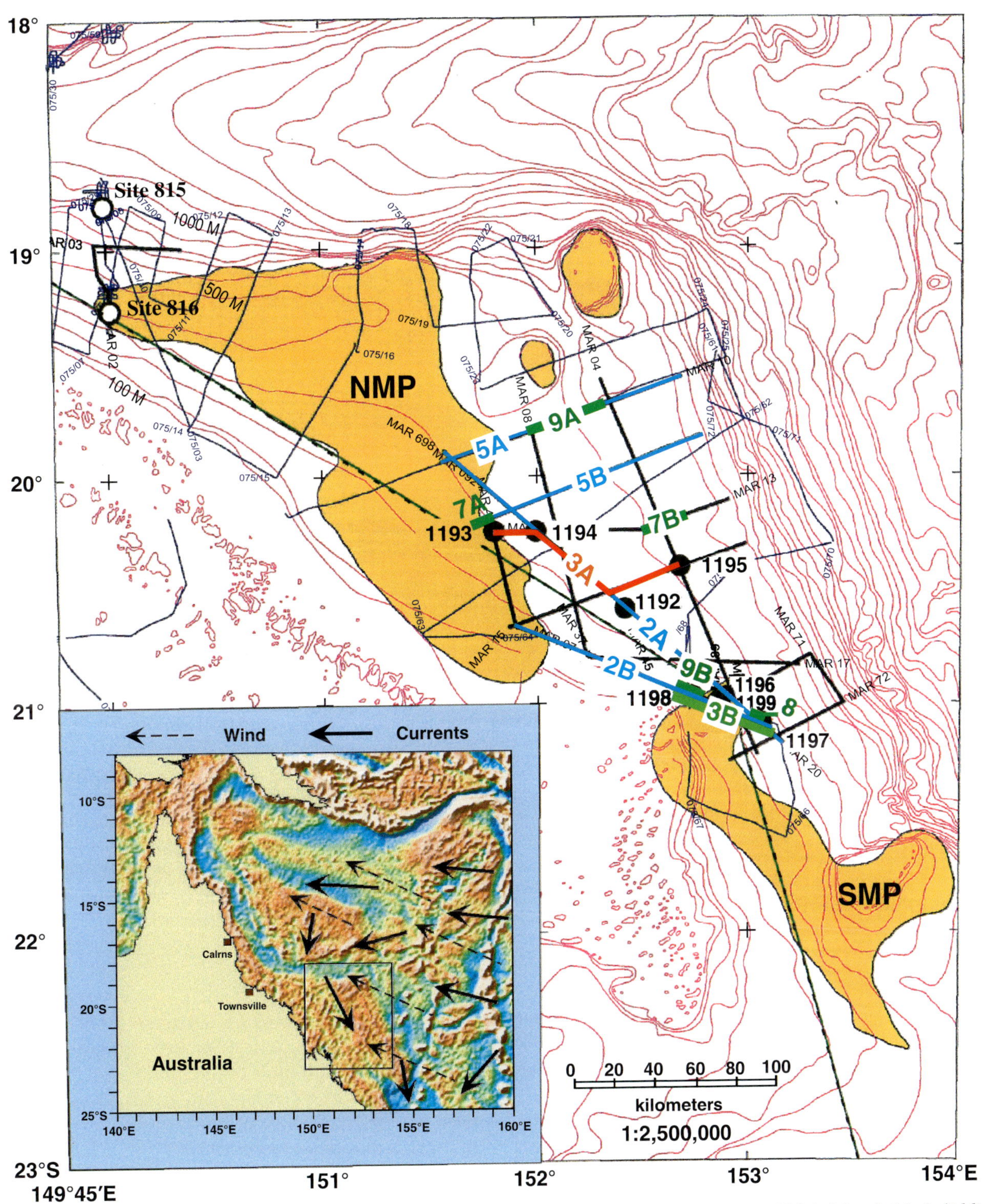

FIGURE 1. Site map showing Leg 194 drill sites (black dots) and Sites 815 and 816 from Leg 133 (white dots). Solid lines show the locations of multichannel seismic lines from the AGSO (thinner lines; Survey 75) and from the Leg 194 site survey (heavier lines). Heavy lines overlying the Leg 194 site survey track lines indicate the locations of figures in this manuscript: blue represents figures where whole seismic lines are shown, red indicates composite lines, and green indicates partial lines. Filled area (yellow) shows the estimated extents of the Northern Marion Platform (NMP; early to middle Miocene) and Southern Marion Platform (MP3; early to late Miocene). Dashed line shows the boundary of the Great Barrier Reef Marine Park. Inset map shows the directions of winds (dashed arrows) and currents (solid arrows) off northeast Australia and the outline of the area shown in the larger map. Winds are dominated by the southeast trades, whereas the dominant current direction is the result of the southward deflection of an arm of the South Equatorial Current, which develops into the East Australia Current immediately south of the study area.

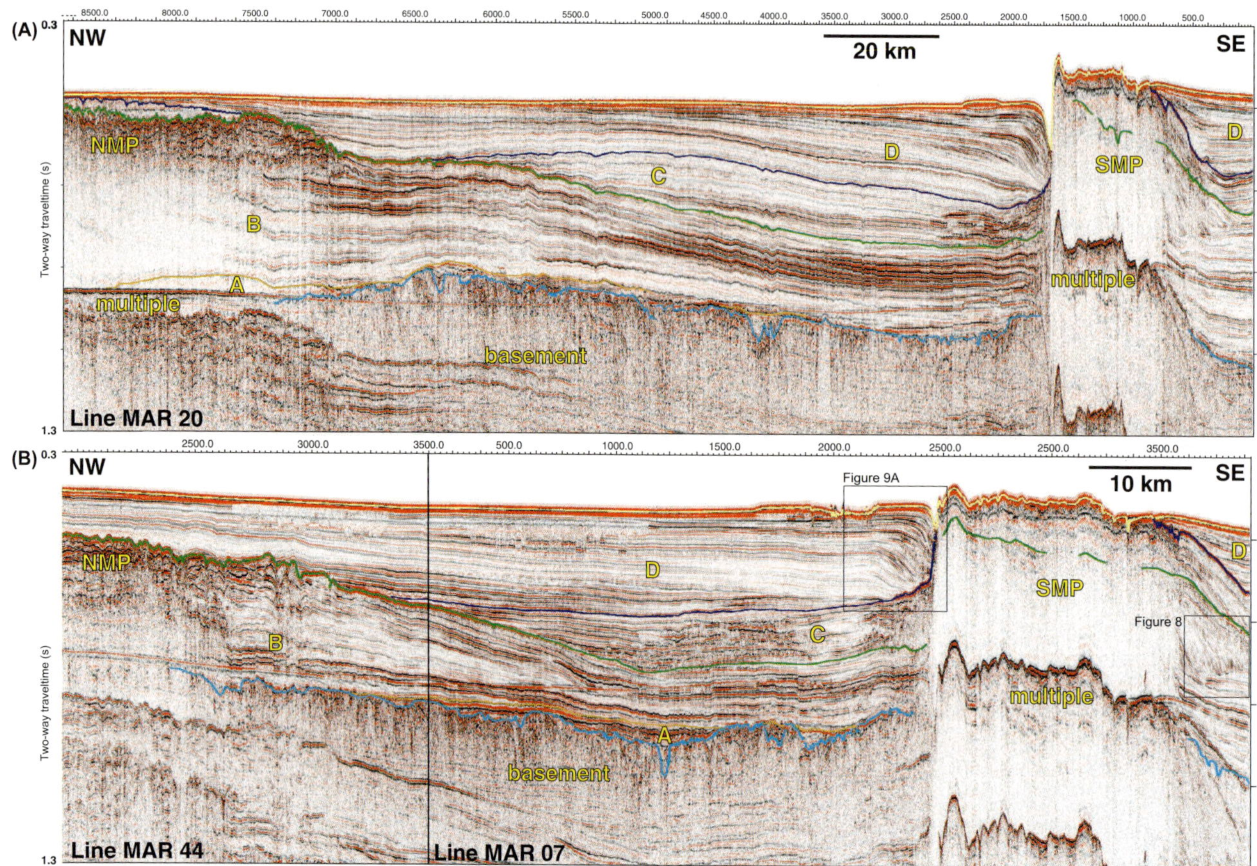

FIGURE 2. Both A and B show the nature of periplatform sediment deposition resulting from sediments shed by the Northern Marion Platform (NMP; megasequences A and B) and Southern Marion Platform (SMP; megasequences A, B, and C). They also show the removal of sediments in the up-current direction (left) adjacent to the SMP and deposition in the downcurrent direction (right). Sediment drifts resulting from strong sea-floor currents occur in megasequences C and D. (A) Seismic line MAR 20 displaying characteristics of seismic megasequences and the carbonate platforms observed in the study area. The location of the sea-floor multiple is indicated on the left side of the figure as well as in the SMP on the right. (B) Seismic overview linking seismic lines MAR 44 and MAR 07. This section displays characteristics of seismic megasequences along a line somewhat parallel-trending to that shown in (A). Boxes indicate the location of enlarged sections shown in Figures 8 and 9A. For the location of these profiles, see Figure 1.

curve. Variations in the slopes of the time-depth curves reflect the high sonic velocities in the platform sediments (Sites 1193, 1196, and 1199), the medium velocities at the proximal slope sections (Sites 1194, 1197, and 1198), and the low velocities in the most distal locations (Sites 1192 and 1195).

SEISMIC AND DEPOSITIONAL FACIES OF MARION PLATEAU MEGASEQUENCES

Seismic data collected for this study provided excellent images of late Oligocene–Holocene sedimentation on the Marion Plateau. Following the nomenclature of Pigram (1993), four unconformity-bounded seismic megasequences (A–D) were identified overlying acoustic basement. These five seismic units can be mapped through most of the survey area. Time-depth curves, combined with shipboard age-depth curves, were used to assign chronostratigraphic datums to the seismic sequence boundaries at each site. These correlations could then be compared across the entire seismic grid (Isern et al., 2002). The match between the ages of the sequence boundaries at the different sites was good within the limits of seismic and shipboard biostratigraphic resolution. This finding confirms the chronostratigraphic significance of these seismic sequence boundaries (Figure 3). Seismic sequence boundaries separating megasequences are discussed using uppercase letters representing the names of the overlying and underlying megasequences (A–B, B–C, etc.). An exception to the nomenclature of Pigram is made when

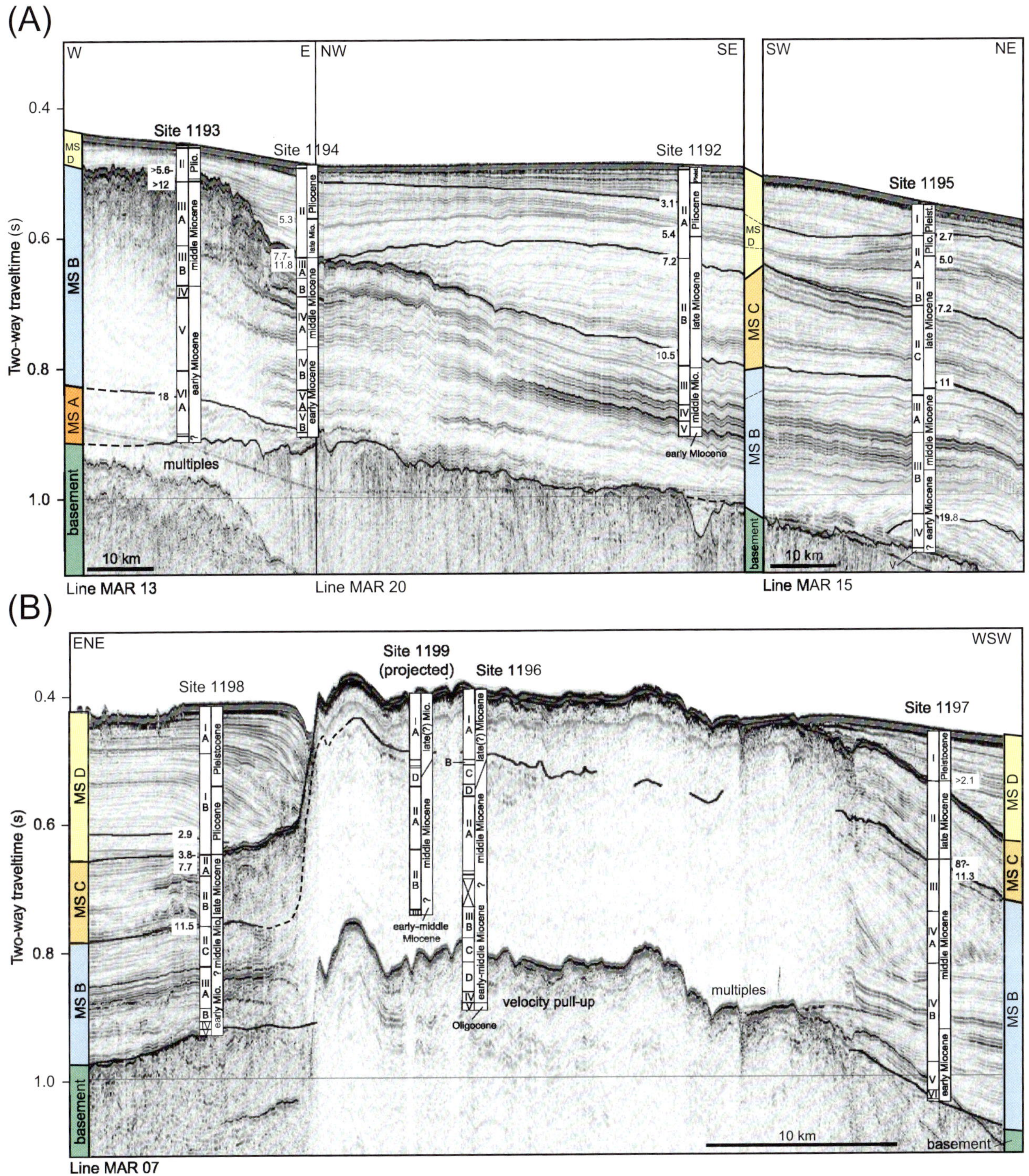

FIGURE 3. At ODP sites, Roman numerals in the left columns at each site indicate lithologic units. These do not correlate from site to site because unit definition is based on site-independent shipboard sedimentologic descriptions. Right-hand columns indicate epoch boundaries defined using shipboard age-depth relationships. Numbers adjacent to the sites indicate the absolute ages of seismic-sequence boundaries derived from time-depth conversion and shipboard ages (modified from Isern et al., 2002). (A) Seismic transect linking lines MAR 13, MAR 20, and MAR 15 (locations of these lines can be found in Figure 1). ODP Leg 194 sites 1192, 1193, 1194, and 1195 and megasequence boundaries are indicated. (B) Seismic line MAR 07 (location can be found on Figure 1). ODP Leg 194 Sites 1196, 1197, 1198, and 1199 are indicated. Site 1199 is approximately 5 km off-line and was projected margin-parallel onto line MAR 07.

discussing the carbonate platform sequences because their ages, which were calibrated by drilling, are substantially different from those of Pigram (1993) (Figures 3, 4). Instead, we name the two imaged carbonate platforms the Northern Marion Platform (NMP) and the Southern Marion Platform (SMP) (Isern et al., 2002). All seismic-stratigraphic elements discussed can be recognized on Figures 2 and 5, which show four large-scale seismic transects. Figure 3 summarizes these findings. For more details on ages and depths of seismic megasequence boundaries as well as lithologic unit boundaries at each site, see Isern et al. (2002).

Leg 194 drilling penetrated and sampled a Neogene sedimentary record used to calibrate the sequence-stratigraphic architecture of the Marion Plateau megasequences by identifying their lithologic signatures and providing a chronostratigraphic framework (Figure 3). This information constrains carbonate platform histories and the magnitude and timing of sea level changes on the Marion Plateau. The following sections discuss the major findings for each of the megasequences.

Acoustic Basement

Seismic Facies

Acoustic basement is characterized by a high-amplitude reflection at the interface with overlying sediments along with numerous diffractions caused by the irregular bedrock surface. In some places, distinct morphologic structures, such as narrow highs and depressions, can be recognized clearly, even on unmigrated seismic profiles. Over most of the plateau, the basement surface occurs at a fairly uniform two-way traveltime (0.9–1.15 s) with a slight northeastward dip toward the edge of the plateau where it is downfaulted to the Cato Trough (Figures 5, 6).

Sedimentary Facies and Ages

The nature of acoustic basement is highly variable, as shown by the range of seismic signatures observed in the study area and by samples recovered at four drill sites (1193, 1194, 1197, and 1198). Generally, recovered samples of acoustic basement were dominated by altered basalt flows and volcaniclastic breccias-conglomerates.

Seismic Megasequence A (Paleogene(?)–Early Miocene)

Seismic Facies

Megasequence A, the oldest depositional megasequence over basement, is generally restricted to the eastern part of the plateau, which has only limited seismic coverage. This thin sequence is characterized by highly continuous reflections prograding westward over basement. Because megasequence A overlies and infills basement irregularities, it has variable thickness.

Sedimentary Facies and Ages

Megasequence A is not easily traced seismically or lithologically between Leg 194 drill sites. In the northern part of the survey area, this megasequence consists of a siliciclastic substrate directly overlying acoustic basement (Sites 1193 and 1196). These siliciclastic sediments are comprised of bioclastic, glauconite-rich, poorly sorted quartz sandstone with varying amounts

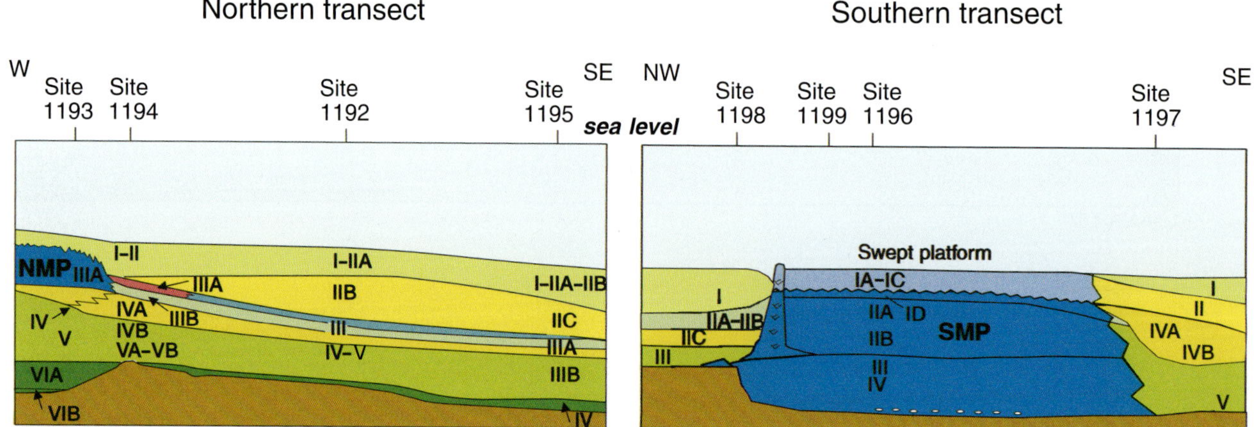

FIGURE 4. Schematic diagram reconstructing the depositional history of the northern (Sites 1193, 1194, 1192, and 1195) and southern (Sites 1198, 1199, 1196, and 1197) drilling transects based on recovered lithologies. This figure shows the sequence relationships between the Northern Marion Platform (NMP) and the Southern Marion Platform (SMP). Blue areas represent carbonate platform growth and the red area represents a lowstand carbonate ramp. Numbers on the sequences correspond to shipboard lithologic units (modified from Isern et al., 2002).

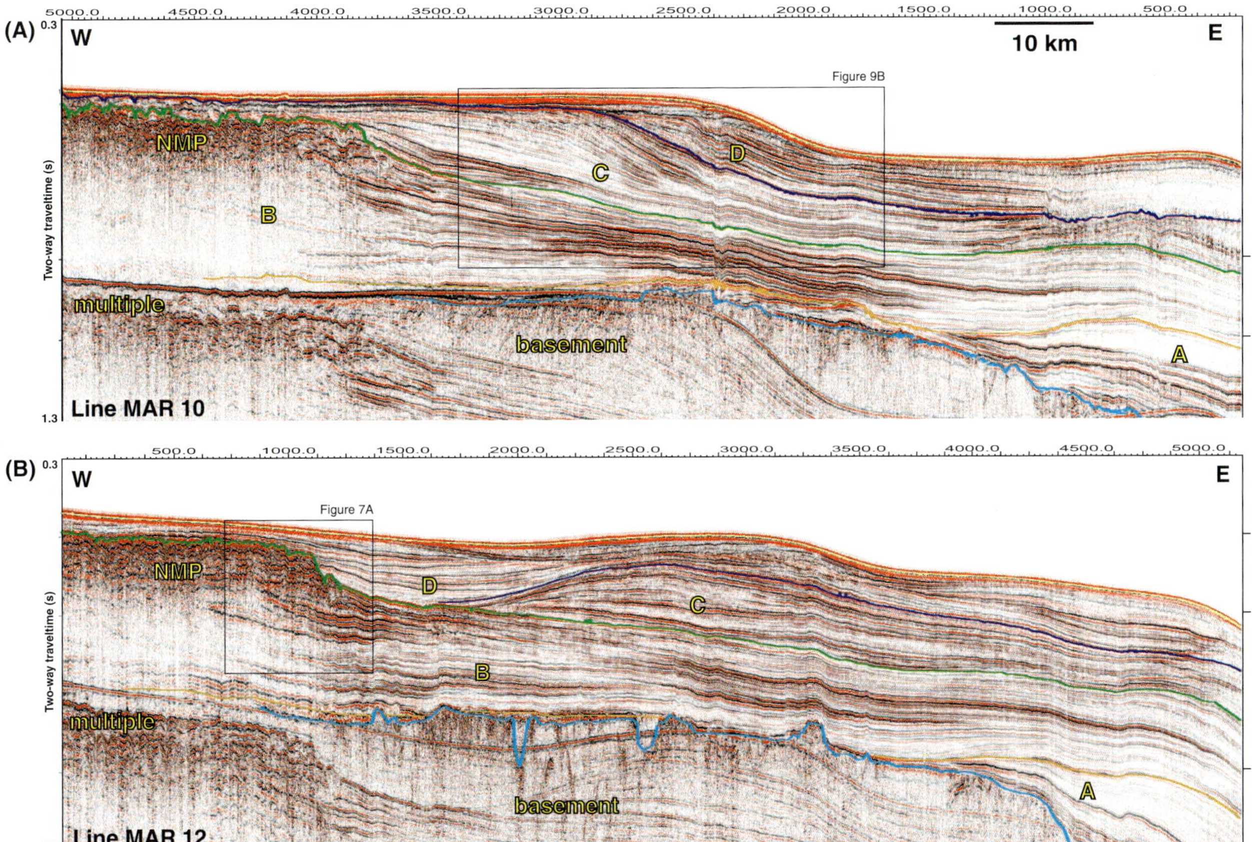

FIGURE 5. Seismic lines MAR 10 (A) and MAR 12 (B), both of which are dip sections perpendicular to the middle Miocene Northern Marion Platform (NMP) (see Figure 1 for locations). These sections show the nature of sediment drifts that make up seismic megasequences C and D. Also shown are the carbonate platform-derived sediments of megasequence B. In these sections, megasequence A could not be distinguished from B. The irregular nature of seismic basement is evident on both sections. Boxes indicate the location of Figures 7A and 9B.

of pebble-sized carbonate clasts. At the southern sites (Sites 1197 and 1198), a thin veneer of coarse carbonate deposits of possibly Eocene age overlies basement. These carbonates are characterized by the presence of small oysters, pectenids, and large bioclasts of massive celleporiform bryozoans in a matrix of larger benthic foraminifera, bryozoans, echinoderms, and coralline algae.

Seismic Megasequence B (Early Miocene–Middle Miocene)

Seismic Facies

The NMP, which is part of megasequence B, is characterized by chaotic, moderately hummocky, internal platform reflections (Figure 7A). This platform sequence overlies dipping slope reflections (Figures 5, 7A). Rather than a gradual progradation of the platform over its slope, as seen in other examples, such as the Great Bahama Bank (Eberli and Ginsburg, 1987), the NMP has a sharp lower base indicating a rapid nucleation followed by vertical aggradation. The NMP top shows prominent karstification with channels and gullies cut in the platform surface (Figure 7A). This karstification resulted from subaerial exposure during the late middle Miocene sea level lowstand. This lowstand led to the nucleation of a small lowstand ramp adjacent to the NMP (Isern et al., 2002) (Figure 3A; Site 1194 Unit IIIA).

Megasequence B is distinct from overlying megasequences C and D because it does not show strong seismic evidence of sediment-reworking by currents. Platform slope sediments adjacent to the NMP are characterized by a seismic facies with high-amplitude and nearly continuous seismic reflections (Figures 2, 3, 5). Because much of megasequence B sediment is derived from the NMP, this megasequence thickens to the west. This pattern is well documented by isopach maps of megasequences A and B combined (Figure 6C), which show both units thinning toward the east before they thicken again at the plateau edge (Figure 5).

Unlike the NMP, the SMP has few internal seismic reflections (Figures 2, 3). Below the weak reflection separating the megasequences B and C platform phases (Figure 3), the SMP is nearly transparent, offering no

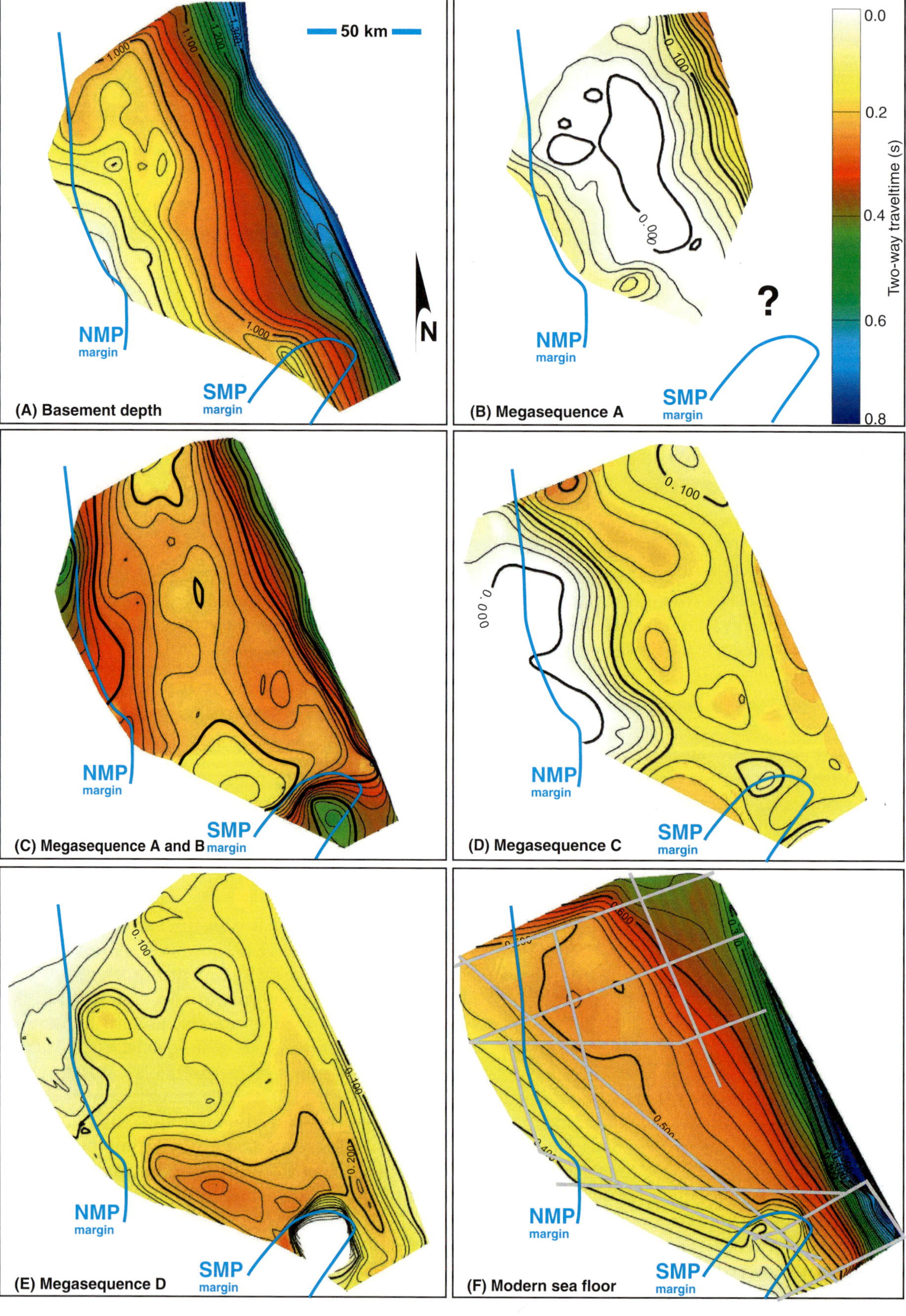
50 km
N
NMP margin
SMP margin
(A) Basement depth
(B) Megasequence A
(C) Megasequence A and B
(D) Megasequence C
(E) Megasequence D
(F) Modern sea floor
?
0.0
0.2
0.4
0.6
0.8
Two-way traveltime (s)

seismic insights into its depositional architecture. The prograding SMP slope sediments, however, are characterized by spectacular sedimentary geometries displaying sharp transitions from relatively steeply dipping slope to nearly horizontal sediments beyond the toe of slope (Figures 2, 8). This transition coincides with a rapid decrease in thickness at the toe of slope so that slope reflections terminate and downlap sharply (Figure 8). The alignment of all these reflection terminations creates a pseudo-unconformity that crosses time lines, as described by Schlager (1981). This pseudo-unconformity is seismically imaged because the flat-lying basinward continuation of slope reflections (indicated by red arrows on Figure 8B) is too thin and has insufficiently high impedance contrasts for it to be imaged seismically.

Sedimentary Facies and Ages

At Site 1193 on the NMP, megasequence B consists of early Miocene inclined slope deposits underneath the platform cap (Figure 7). At Sites 1193 and 1194, these slope deposits were mostly fine silt-sized carbonate debris, which became mixed with a pelagic fraction in a periplatform environment. The top of these inclined sediments, immediately underneath the carbonate platform at Site 1193, was dated at ~16 Ma. No age-diagnostic markers were found within the overlying platform deposits, although the platform top (megasequence B surface) was shown to have an age of ~11.0 Ma through seismic correlation along the megasequence B–C boundary. Thus, NMP growth spans at least ~5 m.y. in the middle Miocene (Isern et al., 2002).

Both the NMP and SMP are characterized by a massive, tablelike morphology, normally reflecting construction by tropical carbonates. These tropical carbonates have a dominantly metastable mineralogy that undergoes early diagenetic cementation, enabling the construction of such massive, steep-sided structures. However, despite their geometries, both NMP and SMP are dominated by the remains of cool, subtropical organisms (red algae, bryozoans, and larger foraminifera with coralline algae being minor to absent). This result is important for seismic interpretation because it shows that platform morphology does not have a simple relationship to the constituent organisms.

In the late middle Miocene, sediments within megasequence B record a significant sea level fall, possibly related to a major ice-building phase in Antarctica. This sea level fall exposed the Marion Plateau carbonate platforms and resulted in karstification of platform surfaces (Figures 2, 3A). In addition, this regression caused a shift in carbonate production from platform surfaces to their adjacent slopes. At Site 1194, this slope deposition is a 30-m-thick lowstand carbonate ramp composed of skeletal packstone-floatstone dominated by bryozoans (Subunit IIIA) deposited between circa 13 and 11 Ma (Figure 3A). This shallow-water ramp sequence overlies neritic upper-slope and hemipelagic sediments, indicating a shallowing-upward trend and thus a major 86 ± 50 m sea level lowering in the latest middle Miocene (Isern et al., 2002).

Megasequence B on the open plateau is dominated by a mixture of distal periplatform and pelagic components. In the site farthest from the Miocene platforms (1195), the top of megasequence B coincides with a 20- to 30-m-thick interval rich in glauconite overlying distal periplatform sediments. The absence of shallow-water components indicates a reduction of neritic carbonate production at the end of the middle Miocene; a likely result of lowered sea level.

The shape of the SMP paleoplatform, formed by the top of megasequence B (equivalent to the top of the NMP), is represented by a weak reflection that occurs at a subsurface depth of approximately 100–130 m below sea floor (mbsf) (Figure 3). The SMP (drilled at Sites 1196 and 1199), which is more than 500 m thick, cannot be further subdivided seismically below this presumed B–C boundary surface because of the transparent seismic facies within the platform core. Recovered cores, however, document that after initiation of SMP on a substrate of latest Oligocene age, this platform complex developed as a product of several growth phases during the early and middle Miocene.

FIGURE 6. Sediment isopach maps in two-way traveltime. The margins of the Northern (NMP) and Southern Marion Platforms (SMP) are indicated by blue lines. Fine contours are spaced every 20 ms, which is equal to approximately 20 m. The color bar on the right is applicable to all thickness maps. (A) Topography of acoustic basement showing its gently dipping nature. Basement cannot be mapped below the NMP or SMP because scattering and attenuation from the platform top hinder penetration of the seismic signal. (B) Megasequence A thickness showing the depocenter off the edge of the Marion Plateau. Megasequence A cannot be mapped in the southeast part of the study area. (C) Thickness of megasequences A and B combined showing the depocenters of the NMP and SMP and the thick sequence of sediments shed off the eastern edge of the Marion Plateau. (D) Megasequence C thickness showing its lobe-shaped, current-controlled sediment distribution in the center of the study area. (E) The thickness of megasequence D was influenced by current-controlled sedimentation that infilled existing relief on the surface of megasequence C, resulting in a depocenter north of the SMP. (F) Modern sea-floor topography, with the sediment showing its smooth surface gradually dipping to the northeast.

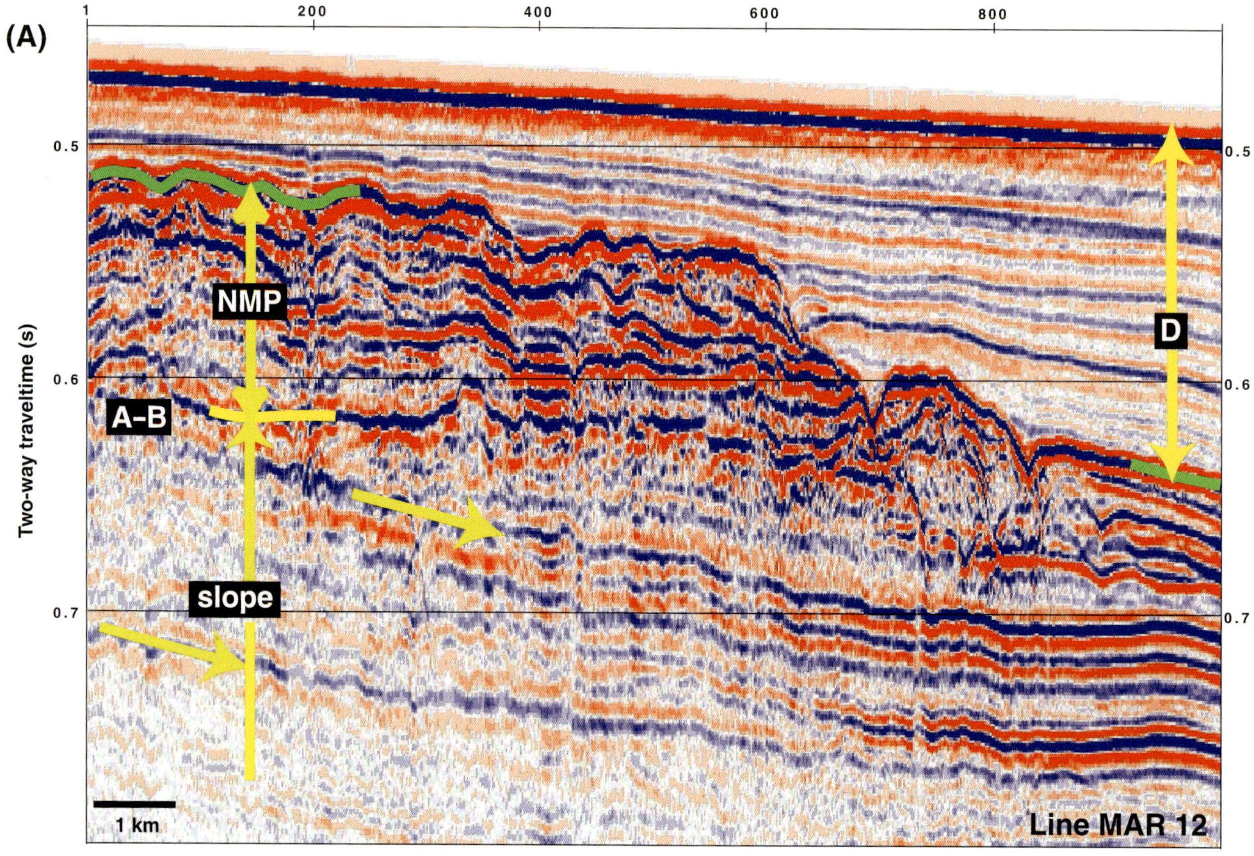

(A)
1
200
400
600
800
0.5
0.6
0.7
Two-way traveltime (s)
NMP
A–B
slope
D
1 km
Line MAR 12

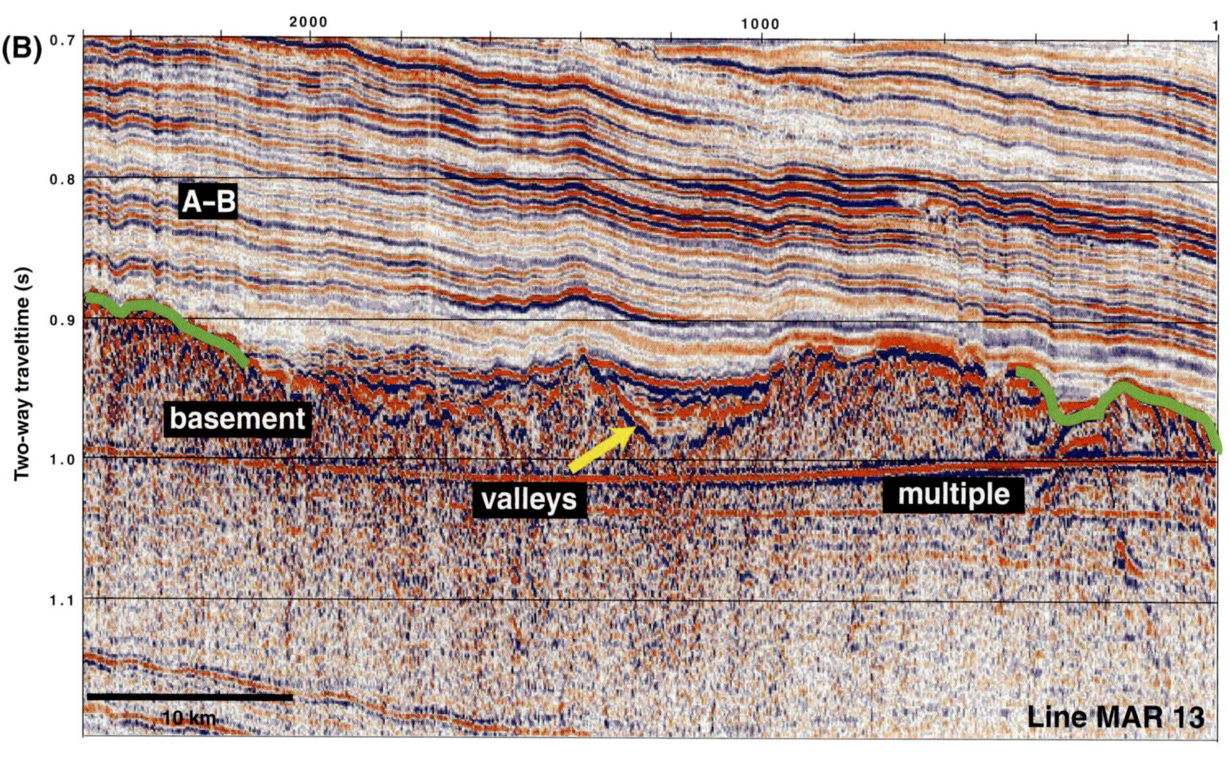

(B)
2000
1000
1
0.7
0.8
0.9
1.0
1.1
Two-way traveltime (s)
A–B
basement
valleys
multiple
10 km
Line MAR 13
SW
NE

At Site 1198, in an upcurrent position adjacent to the SMP escarpment, the top 70 m of megasequence B appears to thicken toward the platform, documenting the shedding of neritic material throughout the upper middle Miocene (Figure 3B). These periplatform sediments overlie hemipelagic deposits near the base of megasequence B. Interestingly, at both sites on the SMP slopes (Sites 1197 and 1198), no early Miocene platform shedding is indicated, either by debris in the periplatform sediments or geometrically by thickening toward the platform. This observation may indicate the importance of a strong southerly current that helped shape the geometry of the platform by transporting much of the periplatform sediment to a more distal location or to a lack of significant platform growth during this time. At Site 1197, the B–C boundary caps the coarsest sediment interval at this site, which forms a thick prograding slope unit that undergoes a downcore transition from a steep-dipping slope to nearly flat hemipelagic deposits.

Seismic Megasequence C (Late Miocene)

Seismic Facies

In the northern and eastern parts of the survey area, megasequence C sediments form a series of drift lobes displaying patterns and seismic facies similar to overlying megasequence D (Figures 2A, 5). In the southern part of the study area, the upper boundary of megasequence C is the top of the seismically opaque SMP edifice (Figures 2, 3). Scattering of the seismic signal from the well-indurated SMP top makes it difficult to determine the geometry and architecture of megasequence C within this platform.

Prior to drilling, it was believed that the entire SMP was of late Miocene age and thus contained within megasequence C. As a result, the SMP would have been initiated on distal sediments of NMP during the middle to late Miocene sea level lowstand (Pigram et al., 1992). Ocean Drilling Program Leg 194 drilling showed, however, that only the top 100–150 m of the SMP consists of a platform phase that is time-equivalent to open-shelf megasequence C sediments. As the adjacent sides of the platform could not be linked seismically, no accurate stratigraphic prediction could be made before drilling.

On seismic data, the southeastern margin of the SMP displays a completely different geometry than the northwestern margin (Figures 2, 3). The upcurrent northwestern margin consists of a steep escarpment, partially onlapped by talus debris at the base, and a nondepositional upper part. Megasequence C sediments thicken toward the platform, providing evidence that these sediments accumulated from platform-derived carbonates in a talus at the base of the escarpment. This talus is recognized as a discontinuous to chaotic seismic facies that interfingers basinward with the continuous reflections of the open plateau sediments. A significant break in seismic facies and geometry across the C–D boundary in this area is present (Figure 9A), indicating the sudden halt of platform-derived sedimentation caused by the drowning of the SMP. In contrast, the southeastern margin shows a thick prograding package dominated by megasequence C sediments (Figure 2).

The deeper-water facies of megasequence C is restricted toward the west by the topographic high of the NMP on which megasequence C onlaps and wedges out. In the east, megasequence C thickens slightly toward the plateau edge (Figure 5). In the North, a thin blanket of megasequence C sediments overlies the NMP in a stratigraphic arrangement that, farther south, is occupied by megasequence D sediments (Figures 2, 5, 9B).

The significant mixed-carbonate-siliciclastic sediment drifts within megasequence C are characterized by large-scale, convex-upward, rounded geometries with continuous reflectors typical of sediment-drift deposition. Interestingly, these drifts are also characterized by prograding geometries as has also been observed in the Bahamas (Eberli et al., 1997). In the northern part of the study area, these drifts indicate active carbonate production where the resulting sediments are transported by bottom currents, which created the large sediment drifts observed (Figures 5A, 9B; also note the depocenter within megasequence C in the northeast part of Figure 6D). The age of these sediments makes it unlikely that they are derived from the NMP because both Legs 133 and 194 drillings—correlated to seismic data—have shown that this platform was drowned in the

FIGURE 7. (A) Detail of seismic line MAR 12 showing the Northern Marion Platform (NMP) margin. The location of this section is shown on Figure 5B. The NMP and slope sediments that underlie the platform carbonates are indicated on the left side of the figure along with the direction of sediment progradation (yellow arrows). The location of major boundaries are indicated in green and the horizontal scale of the image can be seen on the lower left of the figure. (B) Detail of seismic line MAR 13 showing the nature of acoustic basement here overlain by sediments of megasequence A that transgressed over basement. These sediments are overlain by periplatform sediments of megasequence B that were shed from the adjacent NMP. Green lines indicate the location of the basement reflector. The seafloor multiple is also indicated.

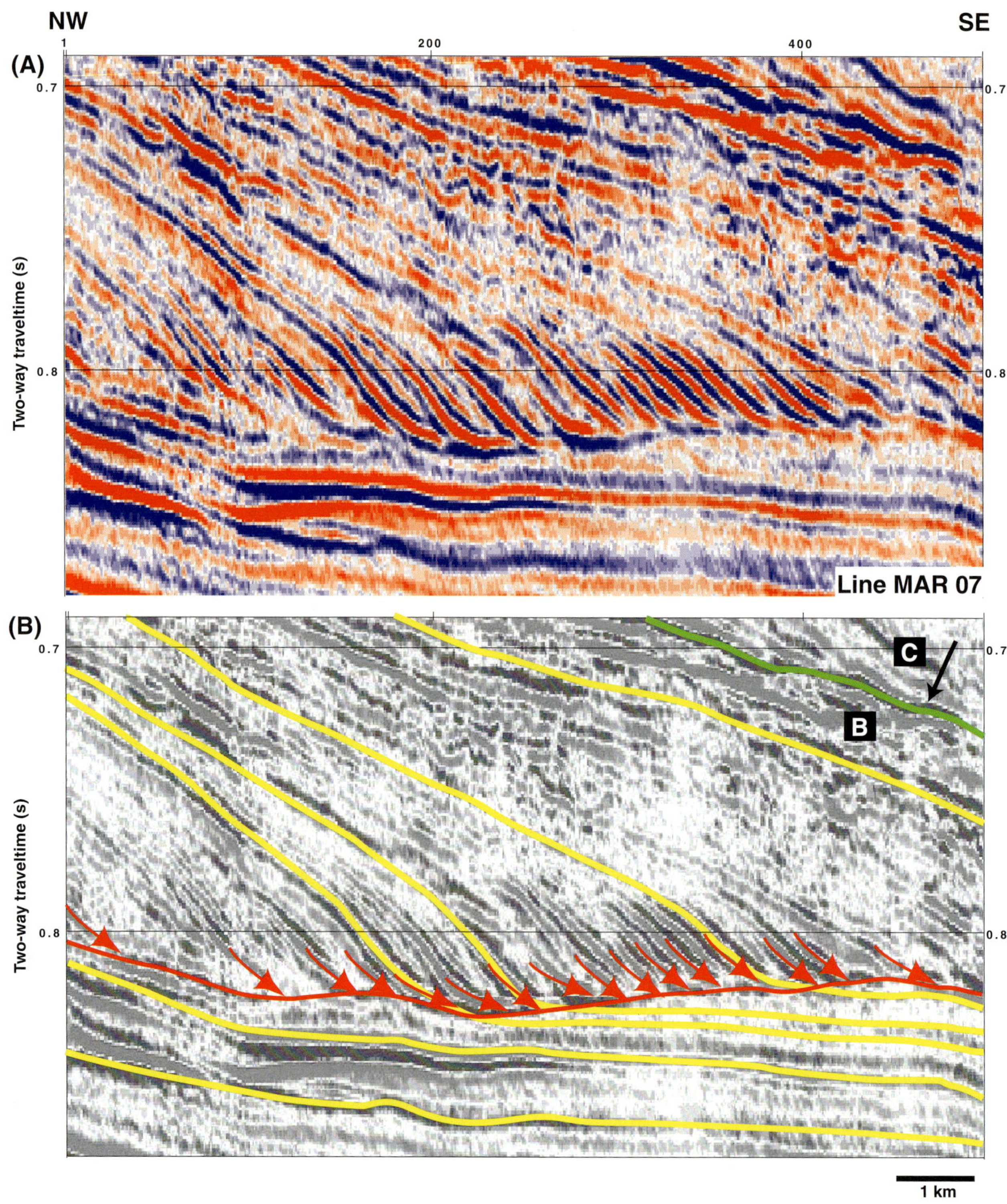

FIGURE 8. Detail of seismic line MAR 07. The location of this section is shown on Figure 2B. (A) Uninterpreted image showing sediments from the prograding carbonate platform slope of the Southern Marion Platform (SMP). (B) Interpreted image showing an apparent seismic unconformity (red horizon) that is formed by reflection terminations of slope beds (red arrows) that thin basinward to below seismic resolution. The resulting horizon is a pseudo-unconformity that crosses time lines. The location of the megasequence B–C boundary is indicated in green. Yellow lines trace other representative horizons.

middle Miocene prior to the onset of megasequence C. Sediments within megasequence C exhibit form a spectacular clinoform complex that prograded toward the southeast (Figures 5, 9B). These sediments interfinger with current deposits at the base of the slope in a similar manner to that described from the western edge of the Great Bahama Bank (Anselmetti et al., 2000).

The facies change to current-dominated depositional environments observed in megasequences C and D is marked by a major regional unconformity represented by the megasequence B–C boundary. This horizon is onlapped from the east over a 50-km-long surface until the sediments reach the NMP top (Figure 5). Strong currents that swept upslope areas clean while gradually infilling the surface relief created this unconformity.

Sedimentary Facies and Ages

Megasequence C is best developed near the SMP where it records the platform-derived sedimentation from the youngest phase of platform growth. Platform-proximal sediments were drilled at Sites 1197 and 1198 and, despite very low recovery, their age and constituents indicate the existence of a late Miocene carbonate platform growth phase. At Site 1197, megasequence C consists of relatively fine to medium grainstones, whereas at Site 1198 it consists of coarse rudstone and floatstone deposited at the base of an escarpment. Farther from the platform, these periplatform sediments interfinger with time-equivalent hemipelagic drift deposits. This transition is clearly imaged on the seismic data (Figure 2).

Previous studies interpreted the downlap of megasequence C drift deposits onto slope deposits of megasequence B northwest of Site 1198 as indicating that the SMP was entirely of late Miocene age (Figure 2B). Drilling through the SMP (Sites 1196 and 1199) revealed however, that only the uppermost 100–180 m potentially consist of late Miocene carbonates. Biostratigraphic information below this depth gave a middle Miocene age. A low-amplitude, low-frequency reflection beneath the SMP top is a candidate for the boundary between these two carbonate platform-growth phases (Figure 3). The interval between the sea floor and approximately 110–130 mbsf dips gently to the southeast and coincides approximately with megasequence C sediments on the downcurrent slope at Site 1197. Age models for the B–C boundary indicate ages of 10.5 (Site 1192), 11.0 (Site 1195), 11.5 (Site 1198), less than 11.8 (Site 1194), and less than 11.3 Ma (Site 1197) (Isern et al., 2002). Giving more weight to the open-plateau section, which has the best seismic coherency and age control, an age of circa 11.0 Ma is postulated for the B–C boundary. This boundary correlates with the karstic top of the northern platform, placing an upper age limit on the middle Miocene platform growth of the NMP.

Seismic Megasequence D (Pliocene–Holocene)

Seismic Facies

Despite the preexisting high relief created by the Miocene carbonate platforms and their adjacent slopes, the Marion Plateau today has little relief (Figure 6F). This pattern results from the infilling of preexisting topography by seismic megasequence D (Figures 2, 5). Megasequence D is characterized by generally continuous, strong reflectors with numerous downlapped and onlapped unconformities and lobe-shaped sediment packages that resulted from current-induced redistribution of sediments over the plateau. In contrast, topographic highs, such as the NMP, were mostly current-swept so that only a thin veneer of sediment accumulated.

Adjacent to the SMP, which outcrops on the modern sea floor with little or no sediment on its surface, an upcurrent moat lies immediately adjacent to the exposed paleoplatform escarpment (Figures 2, 9A). This moat, an excellent example of a contourite, existed throughout deposition of megasequence D. Time-equivalent features were observed in seismic data collected for Leg 133 on the northern margin of the Marion Plateau (Davies et al., 1991). Some evidence of sediment slumping to the southeast into the moat can be also observed (Figure 9A).

Sequence geometry indicates a predominant progradation of drift units within megasequence D resulting from a current flowing in a generally north-south direction. Major hiatuses at the C–D and B–C boundaries and erosional unconformities within drift packages are likely to be the result of strong currents preventing sediment deposition (Figure 9). Isopach maps clearly show the depocenter of megasequence D between the two former carbonate platforms (Figure 6E). They also show a second depocenter in the northwest where a depression created by a drift lobe of the underlying megasequence C in the east and the NMP high in the West was filled (Figures 5B, 6D).

The influence of strong bottom currents on sedimentation is an important characteristic of Marion Plateau megasequences C and D. This influence continues today as evidenced by extensive sediment winnowing on the Plateau surface. Although progradational units are not commonly interpreted as drifts, there are several lines of evidence to support this interpretation. Most significant is the fact that these sequences were deposited in the downcurrent direction. This argues against turbidity currents as the cause for these features because many of the sequences are

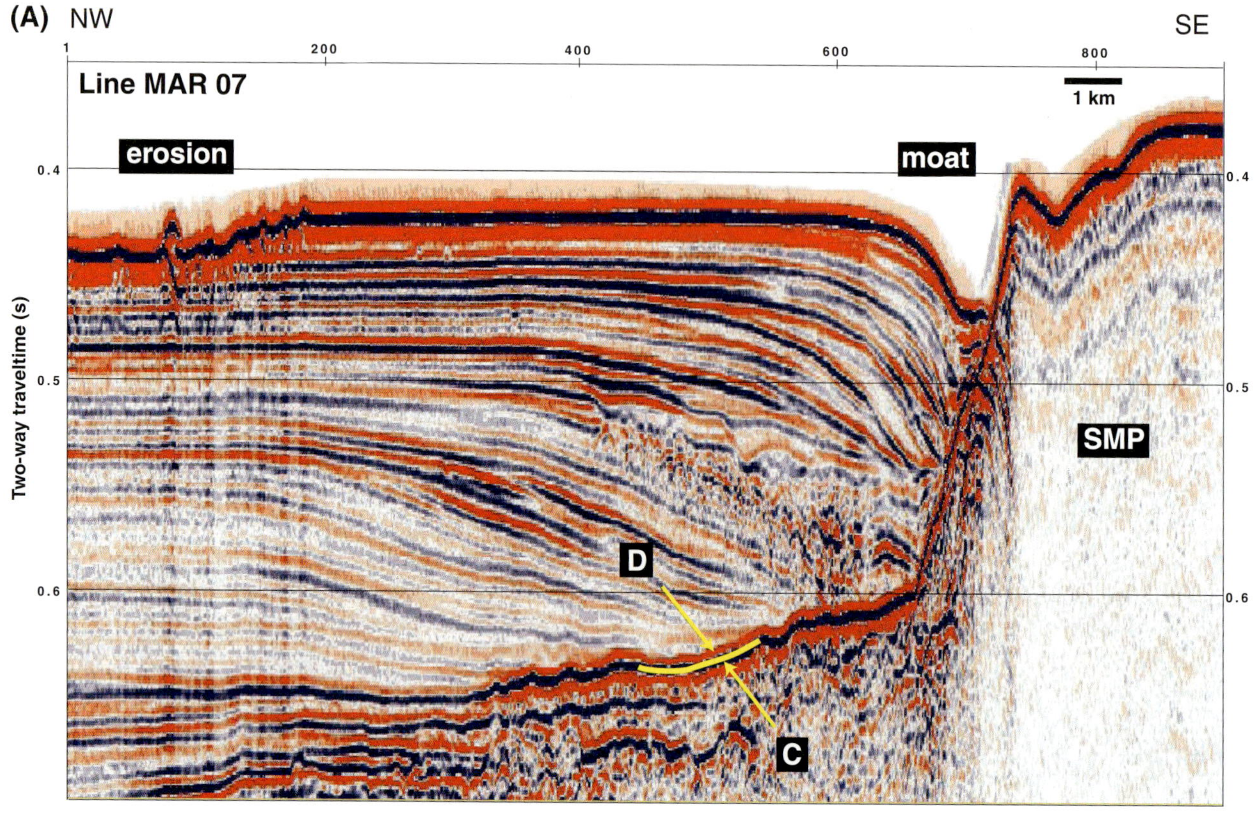
(A) NW
SE
1
200
400
600
800
Line MAR 07
1 km
erosion
moat
Two-way traveltime (s)
0.4
0.5
0.6
SMP
D
C

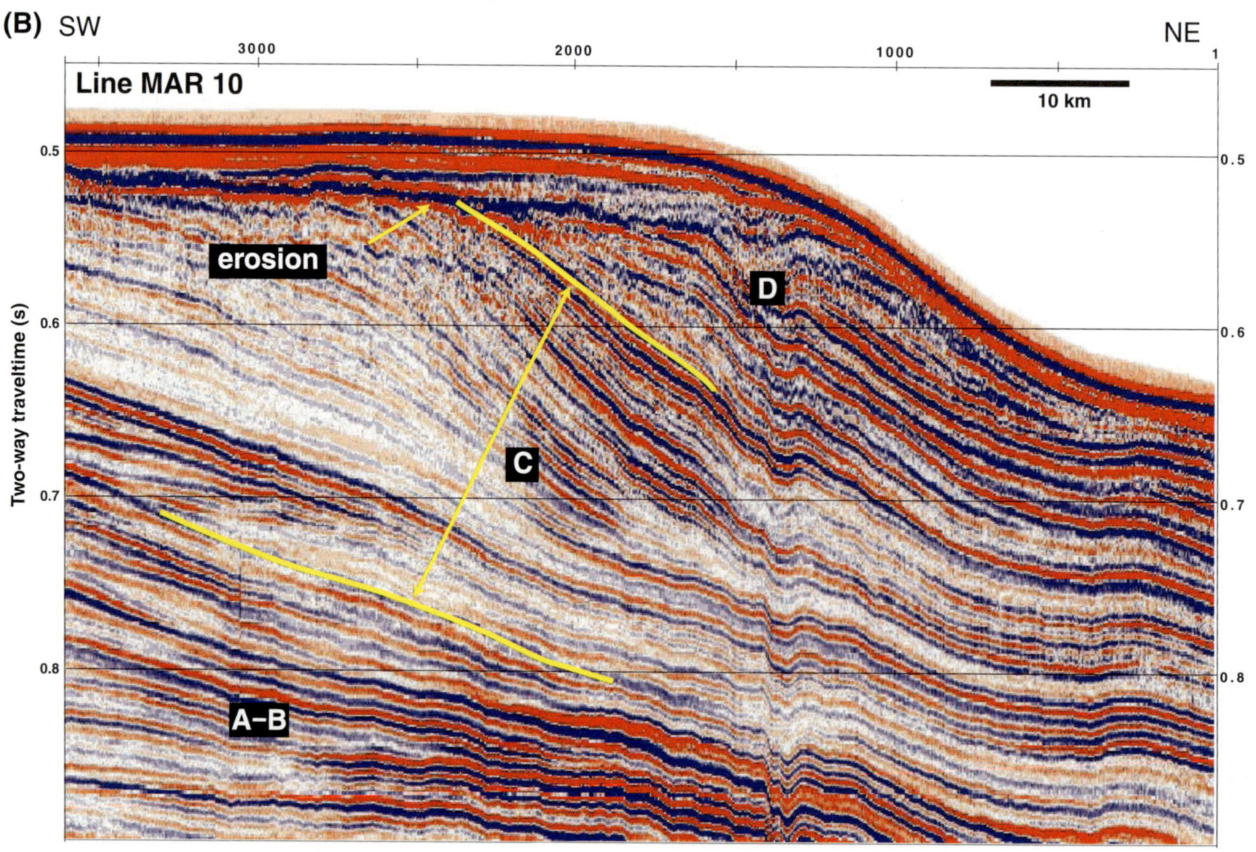
(B) SW
NE
3000
2000
1000
1
Line MAR 10
10 km
erosion
D
C
A–B
Two-way traveltime (s)
0.5
0.6
0.7
0.8

aligned parallel, not perpendicular, to topographic highs (Mountain and Tucholke, 1989). In addition, as documented on the Bermuda Rise, longitudinal furrows are commonly seen in association with sediment waves in areas influenced by strong bottom currents, as seen on the Marion Plateau (Figure 5) (Driscoll and Laine, 1996). Also important is that the sediment source for these drifts is likely to have been some distance from the study area because, during the deposition of megasequences C and D, there were no major sediment sources active in this area. Thus, to explain the high sedimentation rates observed, current-influenced movement of sediment is the most likely explanation. This interpretation is also well supported by seismic and drilling data from the northern edge of the Marion Plateau (Davies et al., 1991).

Sedimentary Facies and Ages

The top of the SMP is characterized by an iron-stained, bored, and encrusted (both hard and soft organisms) hardground surface as shown by dredged sediments. Sediment samples and sea-floor photographs taken during the site survey cruise show prominent current ripples and document the strong influence of currents on the modern sea floor of the Marion Plateau (Isern et al., 2002). The study area is characterized by only subtle bathymetric changes because drift sedimentation in the Pliocene–Pleistocene infilled much of the preexisting relief.

Strong currents at the sea floor have favored early submarine diagenesis resulting in a cemented, iron-stained, submarine hardground encrusted with serpulids and bryozoans on the SMP surface (Isern et al., 2002). This hardground has been swept free of sediment with the exception of small patches infilling surface depressions. Bryozoans are common in both encrusting and branching forms. Many organisms on the surface of the crusts were still living at the time of recovery. The hardground appears to form an approximately 2-cm-thick layer over the sea floor as several "plates" of similar thickness were recovered in the dredges collected. The matrix of the hardground is a cemented sandstone although finer-grained cemented intervals were also recovered. Some recovered pieces of the hardground are massive, whereas others are more vesicular. In one of the dredges, highly phosphatized(?) shark teeth and gastropods were found, which showed some rounding resulting from reworking. Sediments sampled from the remainder of the plateau surface consist of wackestone to packstone containing abundant planktonic foraminifera. Minor skeletal grains include bryozoans, scaphopods, solitary corals, sponge spicules, and pteropods.

At all Leg 194 sites, megasequence D, the youngest of the Marion Plateau megasequences, is composed of hemipelagic drift deposits. The sediments in this megasequence mainly consist of unlithified, weakly to strongly bioturbated alternating sequences of mudstones and skeletal wackestones and packstones deposited in depths greater than 200 m (Isern et al., 2002). Coarser intervals are often well sorted. Megasequence D occurs over the entire study area with the exception of parts of the SMP surface (Sites 1196 and 1199) where megasequence C outcrops at the sea floor. The base of megasequence D is an unconformity observed at all proximal platform sites characterized by submarine hardgrounds. This surface provides an excellent link among seismic unconformities, biostratigraphic hiatuses, and nondepositional processes on the sea floor.

At all proximal sites, the age of the onset of megasequence D sedimentation is dependent on the location of the site relative to the depocenters of drift deposition. These ages are always younger than those in areas distal to platform sedimentation where the sequence boundary is conformable. Sites 1192, 1194, and 1195 were drilled into this conformable succession at the base of megasequence D, recovering the oldest sediments within this megasequence. At these sites, the base of megasequence D was dated at 7.2 (Site 1192), 7.7 (Site 1194), and 7.2 Ma (Site 1195) (Figure 3) (Isern et al., 2002). Sites 1192 and 1195 have the best constrained age model because of their distance from platform sedimentation, thus megasequence D was likely to have initiated circa 7.2 m.y. ago. Two smaller-scale seismic unconformities, representing current-controlled hiatuses in sedimentation and lateral shifts of depocenters, can be traced throughout the seismic grid. Their age ranges along the transect are 2.8–3.1 and 5.0–5.4 Ma, respectively (Figure 3).

The C–D boundary at either side of the SMP coincides with the end of platform-derived shedding of

Figure 9. (A) Detail of seismic line MAR 07 showing the moat formed by current activity on the northern edge of the Southern Marion Platform (SMP). The removal of sediment by currents exposed the upcurrent edge of the SMP escarpment and kept the SMP surface nearly free of sediment. The location of this image is shown on Figure 2B. The megasequence C–D boundary is indicated in yellow. (B) Detail of seismic line MAR 10 showing the prograding sediments of megasequences C and D. Megasequence boundaries are indicated in yellow. These sediments are greatly influenced by current activity, which results in the formation of thick drift deposits. The location of the image is shown on Figure 5A.

neritic constituents onto the platform slopes and may date the drowning of the southern platform at least 1 m.y. earlier than originally predicted (Pigram, 1993).

SUMMARY AND CONCLUSIONS

Seismic data provide high-resolution images of late Oligocene to Holocene carbonate platform and slope sedimentation on the Marion Plateau. These data, in combination with ODP cores and modern oceanographic data, have demonstrated the importance of deep currents in modulating the influence of sea level variation on the morphology and evolution of Marion Plateau carbonate platforms and adjacent sedimentary sequences. The morphology of the Marion platforms is controlled by oceanographic currents and to a lesser extent by sea level. Sea-floor currents create an asymmetrical platform geometry; the upcurrent side is relatively sediment starved and most sediments are deposited in the downcurrent direction. These currents likely determine not only the morphology, but also the growth potential of the platforms, as well as the amount and ultimate location of sediment transported from the platform top, whereas sea level variability dominantly influenced platform aggradation. The recognition that currents control Marion Plateau carbonate platform sedimentation has important implications for the interpretation of platforms imaged on seismic data. For example, although the platforms of the Marion Plateau are similar in morphology to the Bahamas Platform, the controlling factors differ greatly. For the Great Bahama Bank, wind direction and sea level are the primary control, whereas for the platforms of the Marion Plateau, the forcing is from strong sea-floor currents and sea level.

Another important finding concerns interpretation of the biofacies that constructed the Marion Plateau carbonate platforms as a function of their morphologies. The platforms drilled are massive, tablelike structures of the type normally interpreted to be the result of tropical-subtropical carbonate-secreting organisms. Despite this, the sediments recovered from Leg 194 are almost entirely composed of the remains of cool, subtropical organisms, such as red algae, bryozoans, and larger benthic foraminifera, with coralline algae being minor to absent. These calcite-dominated organic remains have a lower diagenetic potential than their aragonite-dominated counterparts in the tropical realm. The fact that the cool subtropical faunal assemblages produce platform geometries that are similar to those built by tropical carbonates suggests that diagenesis, as well as physical parameters such as current flow and sea level change, may be more important in controlling platform architecture than the dominant biofacies.

ACKNOWLEDGMENTS

We acknowledge the excellent support provided by the Australian Geological Survey Organisation (AGSO) seismic technicians who faced many challenges resulting from the use of a new system on a vessel that had not previously performed such a survey as on FR 03/99. Mike Sexton, shore-based at AGSO, was essential for solving processing bugs and faxing software updates to the ship. We also greatly appreciate the hard work of the R/V *Franklin* crew. Their efforts helped make the cruise a great success. Above all, we acknowledge the funding for this cruise provided by AGSO and the Australian Research Council. We also thank the scientists, technical staff, and the crew of the JOIDES Resolution for their hard work during ODP Leg 194. This manuscript was greatly improved by reviews from Christian Betzler, Craig Fulthorpe, and Gregor Eberli.

REFERENCES CITED

Anselmetti, F. S., G. P. Eberli, and Z.-D. Ding, 2000, From the Great Bahama Bank into the Straits of Florida: A margin architecture controlled by sea level fluctuations and ocean currents: Geological Society of America Bulletin, v. 112, p. 829–844.

Davies, P. J., P. A. Symonds, D. A. Feary, and C. J. Pigram, 1989, The evolution of the carbonate platforms of northeast Australia, *in* P. D. Crevello, J. L. Wilson, J. F. Sarg, and J. F. Read, eds., Controls on carbonate platform and basin development: SEPM Special Publication 44, p. 233–258.

Davies, P. J., J. A. McKenzie, A. Palmer-Julson, et al., 1991, Proceedings of the Ocean Drilling Program, Initial Reports: Ocean Drilling Program, Texas A&M University, College Station, Texas, v. 133, 810 p.

Driscoll, N. W., and E. P. Laine, 1996, Abyssal current influence on the southwest Bermuda Rise and surrounding region: Marine Geology, v. 130, p. 231–263.

Eberli, G. P., and R. Ginsburg, 1987, Segmentation and coalescence of platforms, Tertiary, NW Great Bahama Bank: Geology, v. 15, p. 75–79.

Eberli, G. P, P. K. Swart, M. Malone, et al., 1997, Proceedings of the Ocean Drilling Program, Initial Reports: Ocean Drilling Program, Texas A&M University, College Station, Texas, v. 166, 850 p.

Glaser, K. S., and A. W. Droxler, 1993, Controls and development of late Quaternary periplatform carbonate stratigraphy in Walton Basin (northeastern Nicaragua Rise, Caribbean Sea): Paleoceanography, v. 8, p. 243–274.

Isern, A. R., F. S. Anselmetti, P. Blum, et al., 2002, Proceedings of the Ocean Drilling Program, Initial Report: Ocean Drilling Program, Texas A&M University, College Station, Texas, v. 194, p. 1–88.

Mountain, G. S., and B. E. Tucholke, 1989, Abyssal sediment waves, *in* A. W. Bally, ed., Atlas of seismic stratigraphy, v. 3: AAPG Studies in Geology No. 27, p. 233–236.

Pigram, C. J., 1993, Carbonate platform growth, demise and sea level record: Marion Plateau, northeast Australia: Ph.D. thesis, Australian National University, Canberra, Australia, 316 p.

Pigram, C. J., P. J. Davies, D. A. Feary, and P. A. Symonds, 1992, Absolute magnitude of the second-order middle to late Miocene sea-level fall, Marion Plateau, northeast Australia: Geology, v. 20, p. 858–862.

Schlager, W., 1981, The paradox of drowned reefs and carbonate platforms: Geological Society of America Bulletin, v. 92, p. 197–211.

Tomczak, M., and J. S. Godfrey, 1994, Regional oceanography: An introduction: Holland, Pergamon Press, 421 p.

14

Bachtel, S. L., R. D. Kissling, D. Martono, S. P. Rahardjanto, P. A. Dunn, and B. A. MacDonald, 2004, Seismic stratigraphic evolution of the Miocene–Pliocene Segitiga Platform, East Natuna Sea, Indonesia: The origin, growth, and demise of an isolated carbonate platform, *in* Seismic imaging of carbonate reservoirs and systems: AAPG Memoir 81, p. 309–328.

Seismic Stratigraphic Evolution of the Miocene–Pliocene Segitiga Platform, East Natuna Sea, Indonesia: The Origin, Growth, and Demise of an Isolated Carbonate Platform

Steven L. Bachtel
ExxonMobil Upstream Research Company, Houston, Texas

Randal D. Kissling
ExxonMobil Development Company, Houston, Texas

Dwi Martono
Pertamina, Inc., Jakarta, Indonesia

Setya P. Rahardjanto
ExxonMobil Oil Indonesia, Jakarta, Indonesia

Paul A. Dunn
ExxonMobil Upstream Research Company, Houston, Texas

Bruce A. MacDonald
ExxonMobil Production Company, Houston, Texas

ABSTRACT

A high-resolution, two-dimensional seismic survey covering 7500 km^2 provides an unprecedented view of the evolution of a Miocene–Pliocene carbonate platform in the East Natuna–Sarawak Sea, Indonesia. The Segitiga Platform (1400 km^2) contains Terumbu Formation carbonate strata as much as 1800 m thick that were deposited in platform interior, reef and shoal margin, and slope to basin environments.

The Segitiga Platform was subdivided into 12 seismic sequences that demonstrate a history of (1) initial isolation, (2) progradation and coalescence, (3) backstepping and shrinkage, and (4) terminal drowning. Interpretations of seismic facies maps for each sequence were used to help illustrate platform history. These seismic facies maps indicate that the Segitiga Platform originated as three smaller platforms on extensional fault-block highs. Deep intraplatform seaways separated these smaller platforms. Progradation of shallow-water carbonates filled the seaways during a phase of coalescence and the three platforms were amalgamated to form a merged composite platform (1400 km^2; middle–upper Miocene). A rapid relative rise in sea level at the end of Miocene time caused a major backstepping of the carbonate margins (and a concomitant drowning of the adjacent Natuna field carbonate platform to the east) resulting in a platform of greatly reduced size (600 km^2) during the lower Pliocene. Rapid subsidence, combined with an eustatic rise at the end of the early Pliocene, caused terminal drowning of the Segitiga Platform. The platform was buried by younger siliciclastics of the Muda Formation.

Eustatic sea level change controlled the timing of sequence-boundary formation, but structural movements modified internal sequence character and facies distribution. Faulting created topography that acted as templates for the initiation of carbonate platform deposition and provided pedestals for the localization of backstepped platforms. Cessation of faulting may have instigated progradation of the platform resulting from the deceleration of accommodation-space production. Regional subsidence may have controlled the location and extent of platform backstepping. Geographic variability in sequence stacking of coeval platform margins is observed over relatively short distances. Progradation is most strongly developed on the leeward side of the platform, but increased accommodation resulting from the rapid local subsidence or changing oceanographic currents also influenced the direction and magnitude of progradation.

INTRODUCTION

During the late Oligocene and early–middle Miocene time, offshore southeast Asia was characterized by development of shallow-water carbonates and reefs strongly influenced by structural patterns provided by a complex plate tectonic history (Fulthorpe and Schlanger, 1989). Isolated carbonate platforms developed throughout region and form hydrocarbon reservoirs in Indonesia (Rudolph and Lehmann, 1989; Jordan and Abdullah, 1992; Saller et al., 1993), Malaysia (Epting, 1989), Vietnam (Mayall et al., 1997), and the Philippines (Longman, 1980; Grötsch and Mercadier, 1999).

A regional two-dimensional (2-D) seismic grid was acquired west of the Natuna "L" structure approximately 200 km northeast of Natuna Island in the East Natuna–Sarawak Sea (Figure 1). The high-resolution 2-D seismic data spectacularly image the seismic facies and stratal geometries throughout the entire life cycle of an isolated carbonate platform. The Segitiga Platform lies structurally updip and westward of the giant Natuna gas field on the "Terumbu carbonate shelf" of May and Eyles (1985). The "Terumbu carbonate shelf" is actually a collage of at least six separate isolated carbonate platforms (Figure 2), the Segitiga Platform being one of these.

The Segitiga Platform is one of the best seismically imaged isolated carbonate platforms in the world and illustrates the entire platform life cycle from origination, to growth and expansion, to eventual backstepping and drowning. Dense, high-quality, regional 2-D seismic data allow precise mapping of seismic facies within a sequence-stratigraphic framework. These data allow for the study of coeval carbonate margins sequentially through time and illustrate the relative effects of structural subsidence and environmental factors (e.g., prevailing winds and oceanographic currents) on the internal sequence architecture. Dense 2-D seismic coverage of the platform interior also allows for definition of intraplatform seaways and patterns of subsequent progradational fill. Good seismic quality at depth permits the imaging of basement structure and shows the influence of structure on platform development.

Bosellini (1984) recognized prograding carbonate margins on isolated carbonate platforms from the Triassic of northern Italy. Eberli and Ginsburg (1989) first described progradation as an important mechanism in the lateral expansion and coalescence of isolated platforms in the Great Bahama Bank. At the Great Bahama Bank, initially separate platforms were observed to laterally coalesce as progradation filled the deeper intraplatform seaways between the platforms. Progradation in the Bahamas is strongly indicative of windward to leeward sediment flux (Eberli and Ginsburg, 1989). Seismic stratigraphic interpretation suggests that the evolution of the Segitiga Platform was similar to that of the Great Bahama Bank, but the Segitiga Platform displays a subsequent phase of retrogradation ("backstepping") and terminal drowning that is not present in the Bahamas. Platform evolution of the Segitiga Platform is highly dynamic and controlled by several mechanisms, including eustatic changes in sea level, local and regional structural subsidence, and oceanographic or environmental factors.

The objectives of this chapter are to (1) describe and illustrate seismic facies of the platform within a sequence-stratigraphic framework, (2) describe the carbonate platform evolution of the Segitiga Platform based on changes in the distribution of these facies and geometries, and (3) discuss and illustrate controls on the stratal architecture of this carbonate platform. The usefulness of this platform as an analog to other

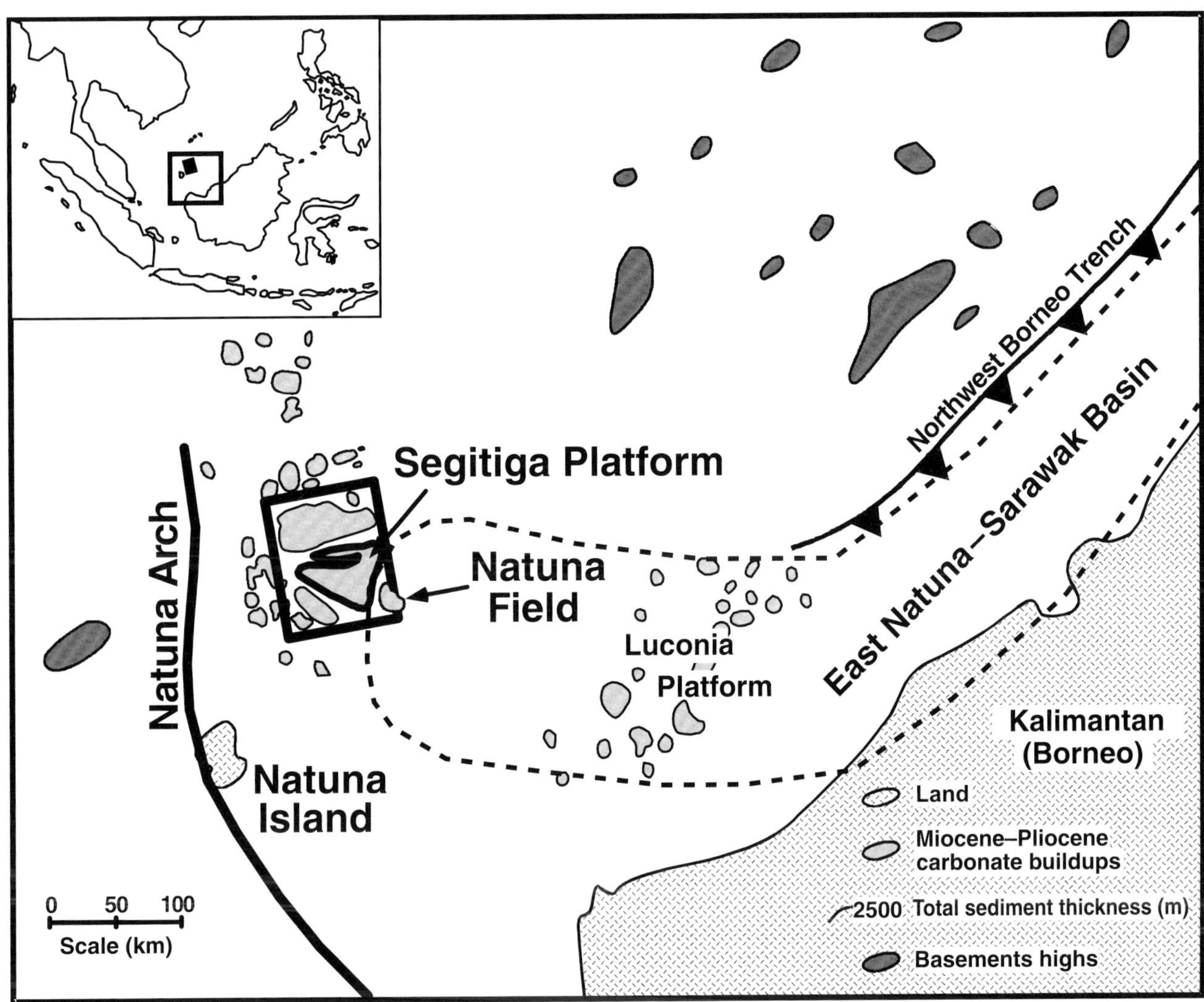

FIGURE 1. The Segitiga Platform is one of several isolated carbonate platforms located west of the Natuna gas field on the eastern flank of the Natuna arch. Reactivation of extensional faulting created during the opening of the South China Sea provided a structural template for the origin and growth of isolated carbonate platforms. Dashed line represents outline of the greater East Natuna–Sarawak Basin (modified from Dunn et al., 1993).

isolated carbonate platforms is a function of the quality and quantity of seismic data and of the completeness of the existing stratal evolution.

METHODS

Regional 2-D seismic data were acquired over the study area in the fall and winter of 1997. The data cover an area of 80 × 90 km and consist of 7500 line-km of seismic (Figure 2A). The seismic grid spacing is 1½ km in the north-south direction and 3 km in the east-west direction. Over the eastern half of the Segitiga Platform, the density of the grid spacing was increased to 1½ × 1½ km (Figure 2A).

The seismic facies method offers a technique to illustrate seismic character in map view and interpret changes in depositional environment on a sequence-by sequence-basis. The seismic facies method is summarized in Mitchum and Vail (1977) and Ramsayer (1979). The seismic data were interpreted by picking seismic sequences using IESX geophysical software. The Terumbu Limestone was subdivided into 12 seismic sequences (Figure 3) based on reflector termination patterns of toplap and downlap at sequence boundaries. The sequences were then mapped throughout the 2-D data. Interpretation concentrated on the platform top and was less rigorous in basinal and slope areas. The internal geometry and character of each sequence were mapped noting highstand reflector geometry, platform-margin position, and toe-of-slope position. Progressive changes in interpreted seismic facies for each sequence can then be used to (1) predict variations in depositional trends, (2) describe the long-term platform evolution, and (3) predict

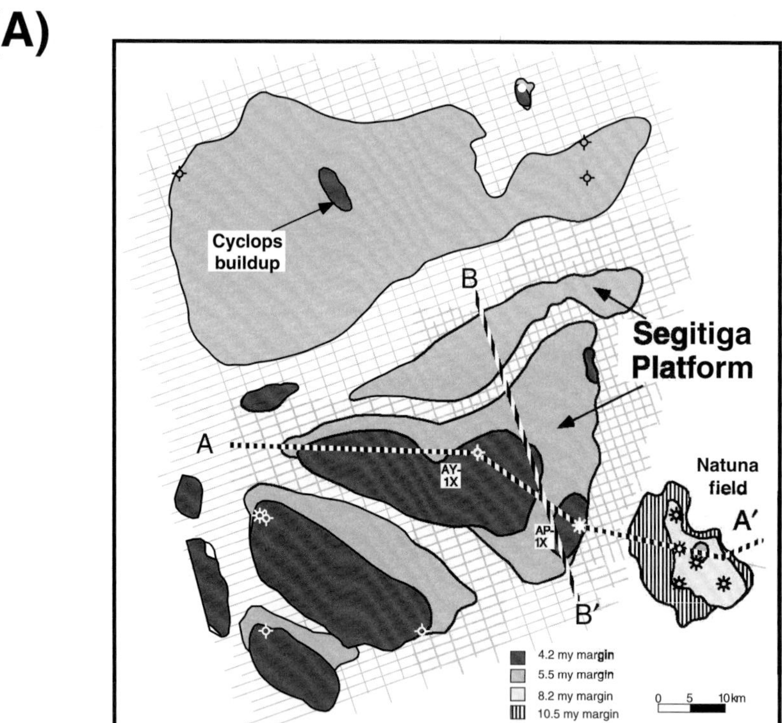

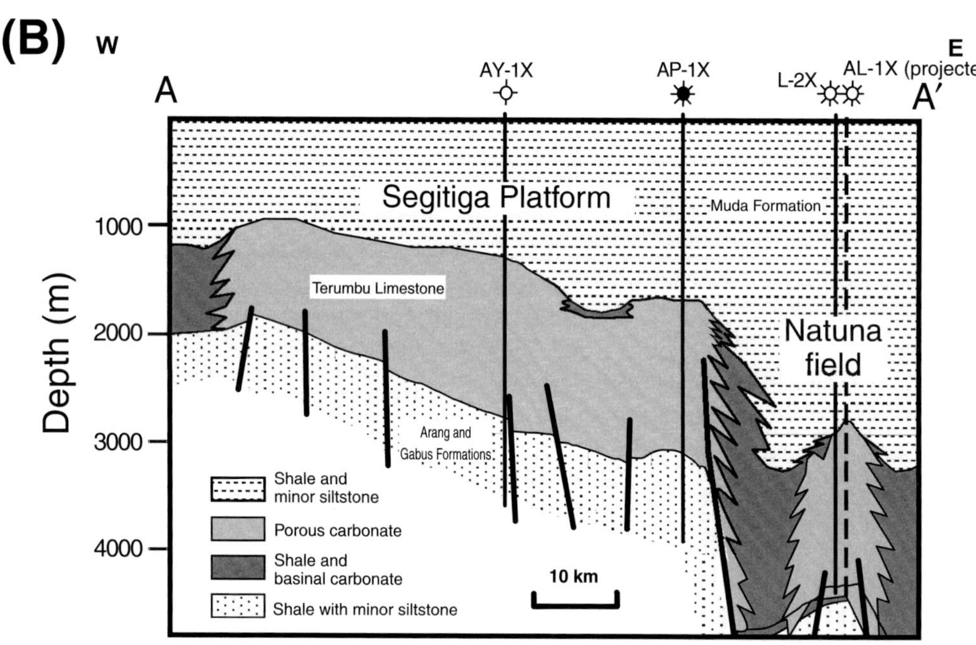

Figure 2. (A) High-resolution 2-D seismic covers an area of 7200 km^2 and blankets at least five major isolated carbonate platforms west of the Natuna field. Note the backstepping and shrinkage of platforms through time (compare the smaller 4.2-m.y. platform outlines with the larger 5.5-m.y. outlines). The Natuna field drowned earlier than the Segitiga Platform (based on biostratigraphy and correlation of time lines from seismic) at a time when the Segitiga Platform was at maximum expansion. (B) Cross section AA′ shows the regional eastward tilt of the platform and relation between the Natuna field and the Segitiga Platform; cross section BB′ refers to Figure 8.

reservoir properties (ϕ, k) when combined with rock-property data derived from core and well logs. Detailed interpretations of seismic facies are critical to the prediction of reservoir facies in exploration plays of limited well control, but can also be used in production scenarios when calibrated to rock-property data from well data.

GEOLOGIC SETTING

Tectonic Setting

The East Natuna–Sarawak Basin is an arcuate basin bounded on the south by accreted terranes of Kalimantan and to the west by the Natuna arch (Figure 1).

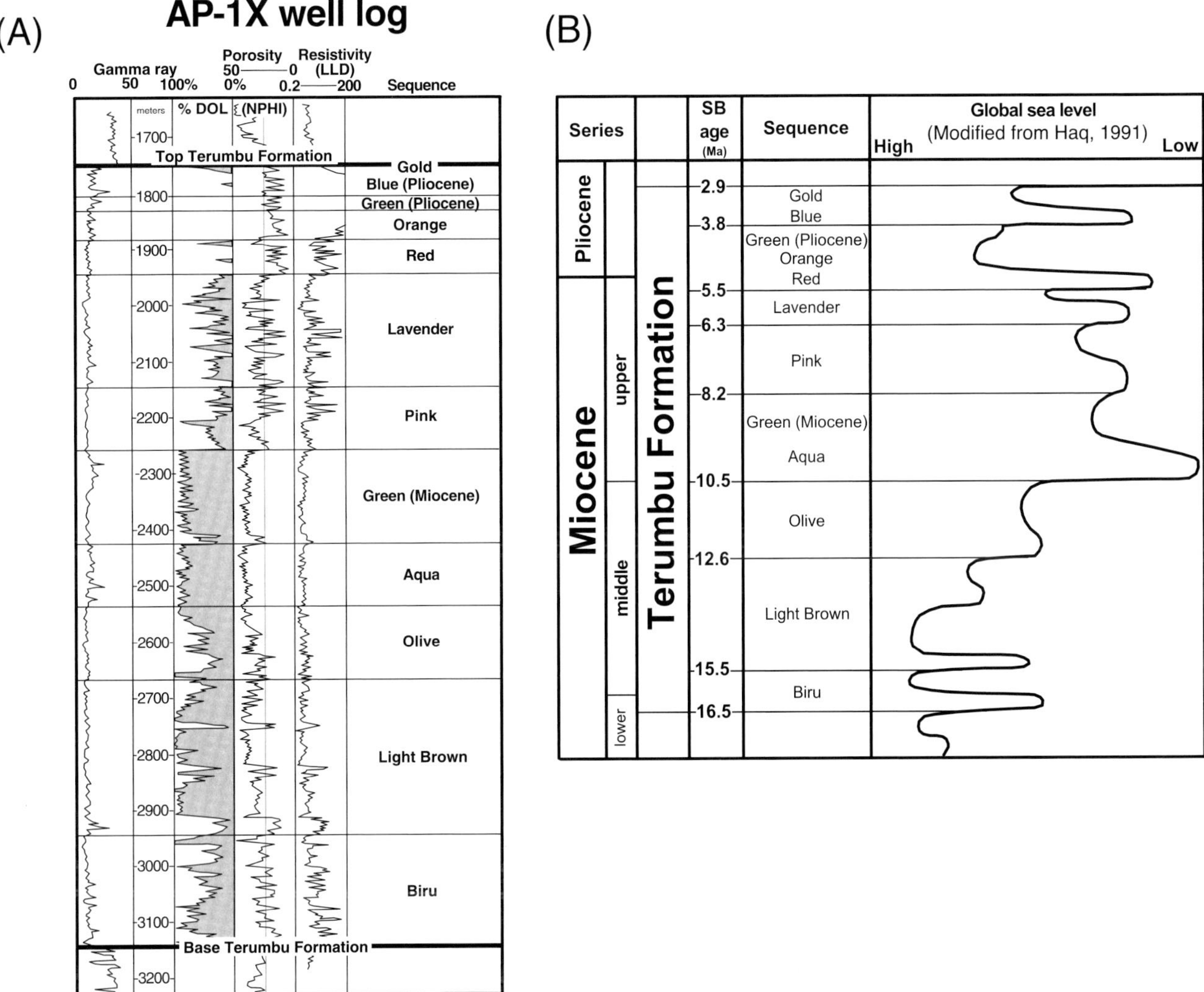

FIGURE 3. (A) Electric log from the AP-1X well. The Terumbu Formation consists of as much as 1800 m of limestone and dolomite. Terumbu Formation porosity from two wells on the platform averages 24%. Sequence names are shown in the right column. (B) Correlation of sequences to sea level curve (modified from Haq, 1991) based on biostratigraphic picks between middle and upper Miocene and platform drowning at the end of the Miocene.

To the north, the basin opens into relatively shallow waters of the southern South China Sea, which is floored by attenuated continental crust. Formation of the East Natuna–Sarawak Basin was related to subduction-related subsidence in the Northwest Borneo Trough during early Eocene to late Cenozoic time (Hamilton, 1979).

The Segitiga Platform is centered on a series of basement-involved extensional fault blocks at the western end of the East Natuna–Sarawak Basin (Figure 1). The fault blocks are most likely comprised of foundered continental crust and formed during extension and opening of the South China Sea during middle Mesozoic time (Hamilton, 1979). Ru and Pigott (1986) recognize three stages of rifting from early Cretaceous to late early Miocene time. Geohistory analysis from four wells in the Natuna field by Rudolph and Lehmann (1989) indicates a period of rifting during late Oligocene to early Miocene time. Postrift sag is evident on the geohistory curve during the Miocene when most carbonate deposition took place. Phases of increased subsidence at the end of the Miocene caused platforms to backstep and finally drown. This increased rapid subsidence may have been a result of subduction along the Northwest Borneo Trench (Figure 1).

Basement rocks in the area are comprised of Cretaceous igneous rocks, encountered at the base of the AP-1X and AY-1X wells (Figure 2A) and in outcrops on Natuna Island (Rudolph and Lehmann, 1989). Sedimentary rocks in the study area can be subdivided into three major packages (Figure 2B): (1) a deepening-upward package of coastal plain to marginal marine siliciclastics (Gabus and Arang Formations), (2) an overall deepening-upward package of marine carbonates (Terumbu Formation), and (3) a shallowing-upward package of deep-marine to coastal-plain siliciclastics (Muda Formation).

Lithologic Composition

The Segitiga Platform is comprised of thick middle Miocene–lower Pliocene carbonates of the Terumbu Formation. The Terumbu Formation is between 2400 and 5000 ft (731 to 1524 m) thick and consists of mixtures of limestone and dolomite. The shallow-water carbonate platforms are separated by relatively deep intraplatform seaways (as much as 300 m deep) and probably filled by muddy carbonates and grainy carbonate sediment gravity flows. Longman et al. (1987) reported similar intraplatform seaways in the Ramba field in the South Sumatra Basin. Core control there indicates that the seaway at Ramba field is filled with tight carbonate mudstone and thin, intercalated carbonate conglomerate and foraminifera grainstone.

The lack of core on the Segitiga Platform limits rock-property information, but cores from the Natuna field are used to predict lithologies on the Segitiga Platform. Lithofacies recognized in Natuna field core include (1) coral-red algal boundstone, (2) coral-red algal-echinoderm packstone-grainstone, (3) red algal-mollusk packstone, and (4) planktonic foraminifer-ostracode wackestone (Rudolph and Lehmann, 1989; Kozar and Oswald, 1993). These lithofacies are interpreted to represent reef and shoal margin, inner platform, and slope-to-basin environments. Porosity in the Terumbu Formation of the Segitiga Platform averages 24% (average based on only two wells, see Figure 3A). The porosity is mostly moldic, caused by the dissolution of aragonitic skeletal debris—most commonly, corals and red algae. A well log from the AP-1X well is provided in Figure 3A and sequences are loosely correlated to the sea level curve of Haq (1991). Correlations are based on in-house biostratigraphy reports that provide good correlation to the middle–upper Miocene boundary and moderate correlation of the Miocene–Pliocene boundary.

SEQUENCE DEFINITION AND SEISMIC FACIES DESCRIPTIONS

The Segitiga Platform is subdivided into 12 seismic sequences (Figure 3) and interpreted based on seismic facies character. Sequence boundaries were picked based on seismic reflector terminations, typically defined in areas of increased accommodation where clear geometric relations (onlap, toplap, and downlap) could be seen between internal reflectors and sequence boundaries. Terminations of the intrasequence reflectors against the upper and lower sequence boundaries were described, as well as the reflector geometry within each sequence. The seismic facies method is summarized by Ramsayer (1979). Six seismic facies were identified (Figure 4) in the Segitiga Platform (in descending order of predicted reservoir quality): (1) mounded, (2) progradational, (3) chaotic, (4) parallel (platform), (5) inclined (slope), and (6) parallel (basinal). Parallel (platform) seismic facies were further subdivided into subfacies based on topographic expression (e.g., low, high, or undifferentiated topographic position). Examples of seismic facies are shown in Figure 5.

Mounded Seismic Facies

Mounded seismic facies (Figures 4, 5A) are characterized by bidirectional downlap of internal reflectors. The reflectors are commonly parallel to an upper sequence boundary and the internal geometry is convex up and typically described as "mounded." A thickening of the sequence may occur locally where mounded facies occur. The mounded facies is an end-member facies expression and is transitional with the chaotic seismic facies.

Interpretation

Mounded seismic facies are interpreted to represent shelf-margin or shelf-interior reefs and associated grainstone shoals. High porosity and permeability would be predicted in these coral-algal boundstone and grainstone sediments because of secondary dissolution of aragonitic skeletal debris and the possibility of dolomitization at the shelf margin.

Progradational Seismic Facies

Progradational seismic facies (Figures 4, 5A) are identified by reflectors that toplap against an upper sequence boundary and downlap onto a thin maximum flooding surface (supra-adjacent to the lower sequence boundary). Internally, the reflectors exhibit either a low-relief sigmoidal geometry or a steeper oblique geometry. These facies are typically inclined in the direction of sediment transport.

Interpretation

Progradational facies are interpreted to be grain-dominated lithofacies (including boundstone) associated with the platform margin, reef flat, and platform interior facies that prograde away from the shelf margin or localized banks. Progradational geometries are observed at two different scales. First, progradational

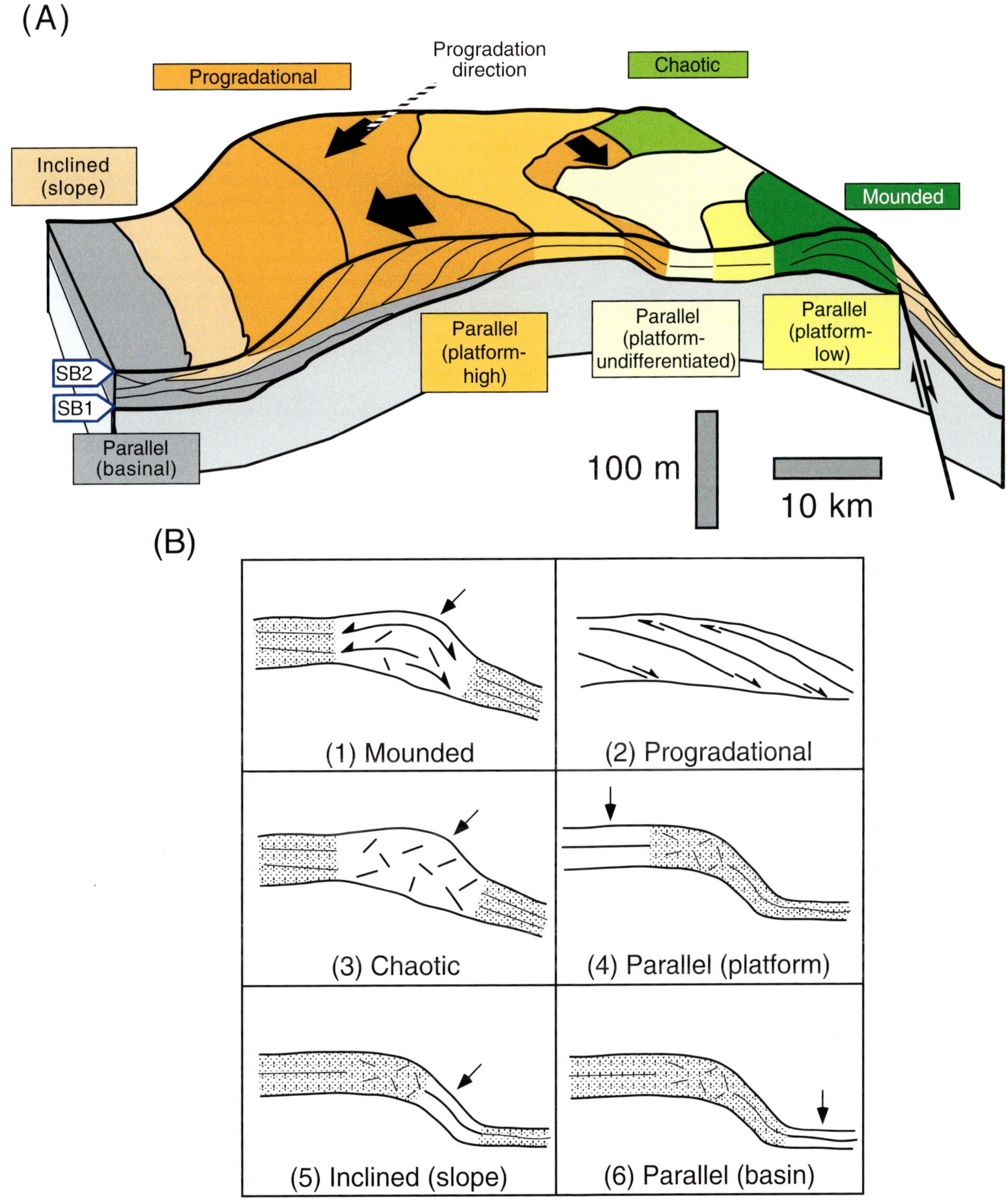

FIGURE 4. (A) Eight seismic facies are recognized on the Segitiga Platform. Distribution of seismic facies can be used in low-density data settings to interpret and predict reservoir properties across the platform (modified from Dunn et al., 1993). (B) Line-drawing examples of seismic facies illustrating differences between geometric and platform-location criteria for the definition of seismic facies. Figure 5 shows seismic examples of the facies illustrated here.

seismic facies are recognized as well developed, large-scale geometries prograding basinward from a well-developed shelf margin. Secondly, very subtle small-scale prograding geometries are observed in the platform interior where it is interpreted that grainy sediments prograded away from platform interior skeletal banks and reefs. Progradational geometries are interpreted to be developed during relative highstand of sea level.

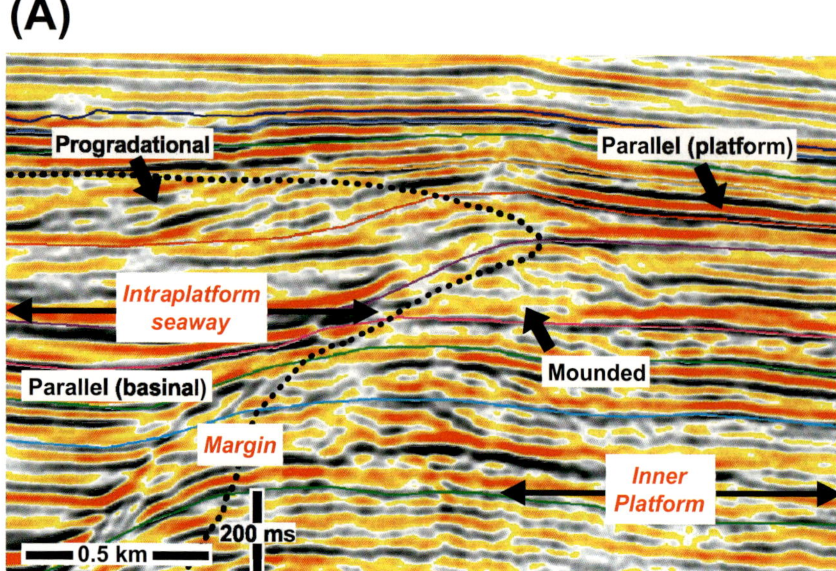

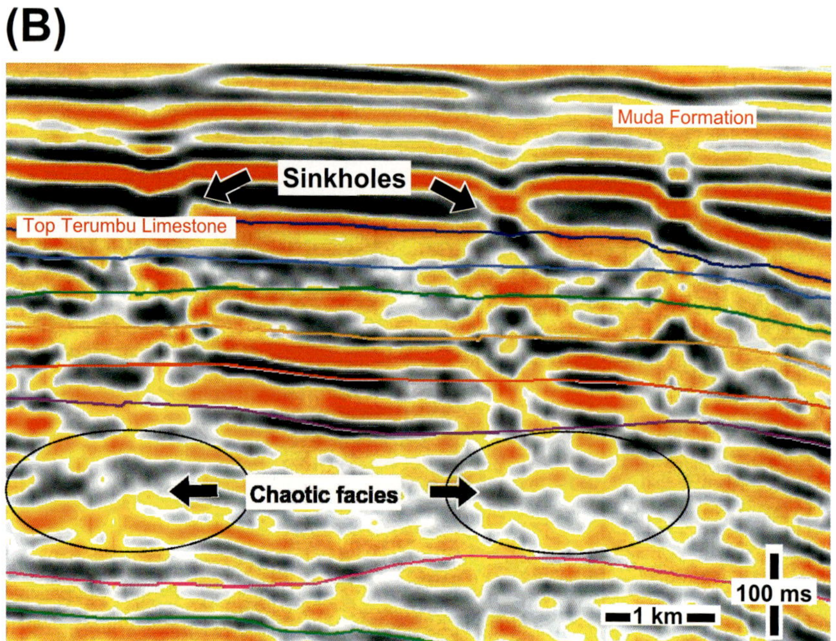

FIGURE 5. (A) Seismic facies examples from the southwestern margin of the Segitiga Platform. Note the aggradational to backstepping margin, followed by prograding deposits as the intraplatform seaway is filled (dotted line traces position of platform margin through time). (B) Chaotic seismic facies examples from the central part of the Segitiga Platform. Note the inverted cone of data disruption below karst (sinkholes). Muda Formation shales show drape or collapse into the sinkholes. Ellipses outline areas of chaotic seismic facies.

This facies is predicted to have excellent reservoir quality with high amounts of moldic porosity in grainstones caused by the secondary dissolution of aragonitic skeletal debris.

Chaotic Seismic Facies

Chaotic seismic facies (Figures 4, 5B) are identified by reflectors with internally disrupted reflectors that have a "chaotic" character. Chaotic facies can occur as linear bodies at the platform margin or as isolated ellipses in the platform interior. These facies are transitional with mounded seismic facies.

Interpretation

The interpretation of chaotic facies is difficult because several geologic processes can be represented by chaotic seismic facies. First, chaotic seismic facies can be interpreted as shelf-margin or shelf-interior patch reefs, analogous to the mounded seismic facies, although with more disruption of the seismic signal. Chaotic facies can also be recognized as data disruption below a significant karst surface (Figure 5B) or in an intensely faulted area. Reservoir properties of chaotic facies are predicted to be variable because of multiple origins, but would be similar to mounded facies or display enhanced permeability because of karsting and fracturing. Chaotic facies in areas of data disruption (e.g., intensely faulted areas) should be considered areas of "no data."

Parallel (Platform) Seismic Facies

Parallel (platform) seismic facies (Figures 4, 5A) are identified by concordant and parallel relationships between intrasequence seismic reflectors and the sequence bounding surfaces. Internal reflectors between sequence boundaries are parallel, straight to slightly wavy, and continuous to semicontinuous. Parallel (platform) seismic facies are further divided into subfacies based on the topographic expression of where the facies is located: areas of high topography, parallel (platform-high), areas of undifferentiated topography, parallel (platform-undifferentiated), or areas of low topography, parallel (platform-low).

Interpretation

Parallel (platform) seismic facies occur in platform interior positions and could represent a wide range of rock types from grain- to mud-dominated lithofacies. Parallel (platform-high) seismic facies are interpreted to have good reservoir properties because of enhanced secondary dissolution on topographic highs during relative lowstands of sea level. Parallel (platform-low) seismic facies are interpreted to have less favorable reservoir properties because topographically lower areas of carbonate platform interiors are prone to muddier lagoonal facies. Parallel (platform-undifferentiated) seismic facies have a poorly defined topographic expression and could display reservoir properties of either high or low position. Areas of parallel (platform) facies that are characterized by high amplitudes may be areas with layered lithology or porosity relationship (perhaps mixed muddy and grainy platform interior facies).

Inclined (Slope) Seismic Facies

Inclined (slope) facies are present basinward of the shelf margin and landward of the toe of slope (Figure 4). These seismic facies are typically gently inclined reflectors that decrease in gradient toward the toe of slope, but reflectors show parallel geometric relations with sequence boundaries. The lack of toplap and downlap geometric relationships differentiates this facies from progradational seismic facies. Locally, toplap and downlap geometries are observed in this slope position and are captured as progradational seismic facies.

Interpretation

Reservoir properties of the slope seismic facies are predicted to be gradational from relatively good near the platform margin to relatively poor toward the toe of slope. This basinward degradation in reservoir properties is caused by a decrease in grain size and abundance of skeletal debris as well as an increase in muddy sediment fabrics. A basinward decrease in skeletal debris and a basinward increase in muddier fabrics would be predicted to produce less favorable reservoir properties in more distal slope positions.

Parallel (Basinal) Seismic Facies

Basinal seismic facies are present basinward of the toe of slope and are generally comprised of high-amplitude parallel seismic facies (Figures 4, 5).

Interpretation

Basinal seismic facies are interpreted to have relatively poor porosity and permeability because of muddy fabrics and fine grain size of skeletal debris (Longman et al., 1987). Interbedded grainy sediment gravity flows, sourced from the shelf margin, are possible and may provide thin flow units within the otherwise tight, muddy basinal facies. Thin interbeds of argillaceous material may be present between mudstone beds.

SEISMIC FACIES MAPS AND PLATFORM EVOLUTION

A series of 12 seismic facies maps illustrates the distribution of seismic facies and progressive platform evolution of the Segitiga Platform. Seismic facies can then be used to predict lateral variation in reservoir properties based on correlation of core data with well logs tied to interpreted seismic facies. The stratigraphic evolution of the Terumbu Formation of the Segitiga Platform can be divided into three distinct stages: (1) platform initiation and structurally controlled isolation (Figure 6A), (2) platform coalescence and expansion (Figure 6B), and (3) platform backstepping and shrinkage (Figure 6C). The platform was terminally drowned and buried by younger siliciclastics (Figure 6D).

Stage 1: Initiation and Structurally Controlled Isolation

The Segitiga Platform originated in early middle Miocene time as three smaller platforms isolated on

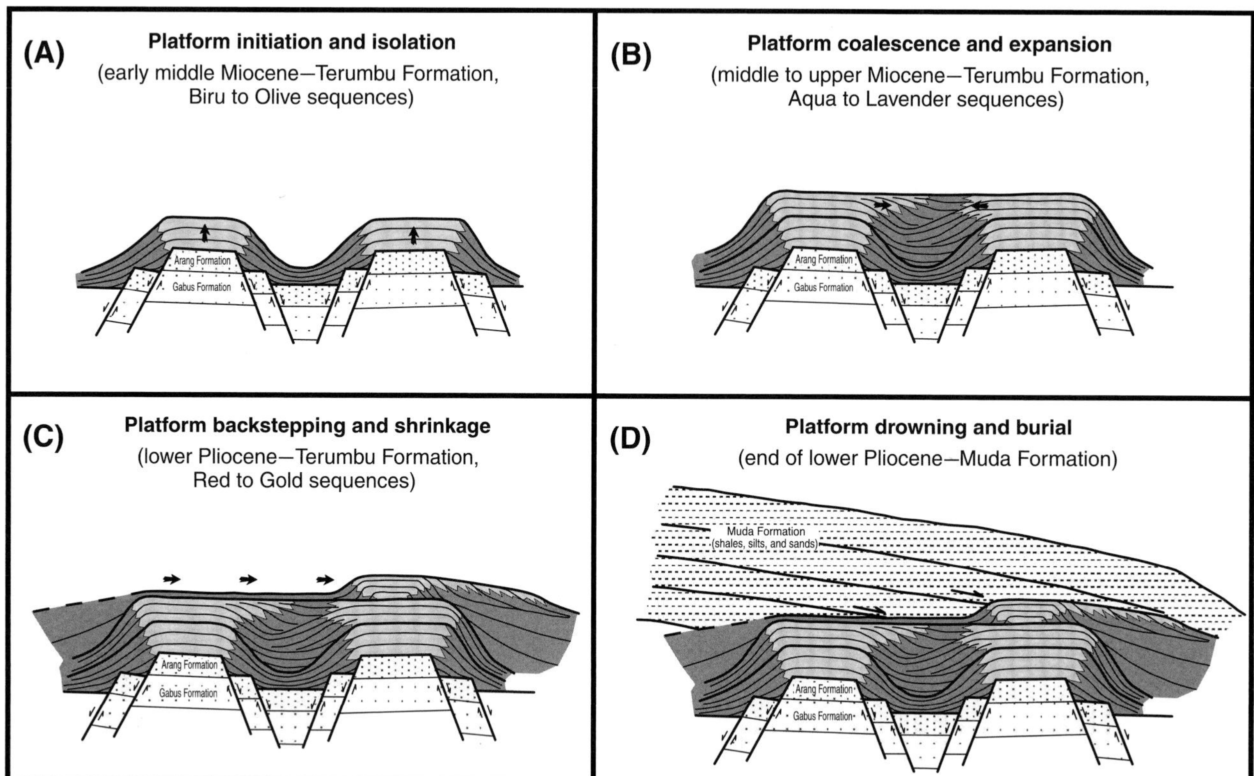

FIGURE 6. Platform evolution of the Segitiga Platform. (A) Platform initiation as three individual platforms, localized on faulted highs and separated by deep intraplatform seaways. (B) The platforms coalesced as prograding deposits filled the seaways and formed a large composite platform. (C) Rapid subsidence, combined with eustatic rise at the end of the Miocene, caused the platforms to backstep. (D) The platform was terminally drowned during the lower Pliocene and younger siliciclastics of the Muda Formation buried the platform.

structural highs (Figure 7A). Deep intraplatform seaways separated the individual platforms and varied in depth from 100 to 200 m. This initiation-isolation phase is evident during deposition of Biru, Light Brown, and Olive sequences (Figure 7A). Platform margins are dominated by mounded and chaotic seismic facies, interpreted to be reef-rimmed platform margins. These platform margins mostly aggraded through time. Margins characterized by progradational seismic facies are less abundant and are interpreted as grain-dominated platform margins. Note the northeast-southwest–trending area of relatively thin, parallel-high seismic facies (Biru to Olive sequences; Figure 7A). This area marks the position of the horst block that controlled initial establishment of the platform (Figures 6A, 8). Thicker, parallel-low seismic facies were deposited off the flanks of this horst block as the block was actively subsiding during deposition of Biru to Olive sequences (Figure 8). Progradational seismic facies show transport of sediment away from this structurally high area (Biru to Olive sequences; Figure 7A). The position of the northeastern margin of the Segitiga Platform during Biru to Olive sequence deposition is 6 km south of the subsequent Aqua to Lavender Platform margin (compare Figure 7A margin with Figure 7B margin).

Stage 2: Platform Coalescence and Expansion

Platform coalescence and expansion took place from Aqua to Lavender sequences as the Segitiga Platform expanded to its maximum size of 1400 km^2. Mounded seismic facies dominate most platform margins and are interpreted as reef-rimmed margins. Coalescence of the Segitiga Platform was initiated by strong northward progradation of the northeast margin during deposition of the Aqua sequence (Figure 7B). The margin, interpreted as a grain-dominated lithofacies, migrated approximately 6 km to the north. As a result, progradational seismic facies nearly filled the intraplatform channel to the southwest. Intraplatform seaways were nearly filled on the eastern part of the platform, but remained open to the sea toward the west, where seaways retain depths as much as 120 m. The south fork of the intraplatform seaway was filled by deposition of the Green (Miocene) sequence. Note that progradation in the intraplatform seaway switched directions during the Green (Miocene) sequence, with sediments now being transported from north to south (Figure 9A). Areas that were topographically high during the initiation-isolation stage gradually migrated

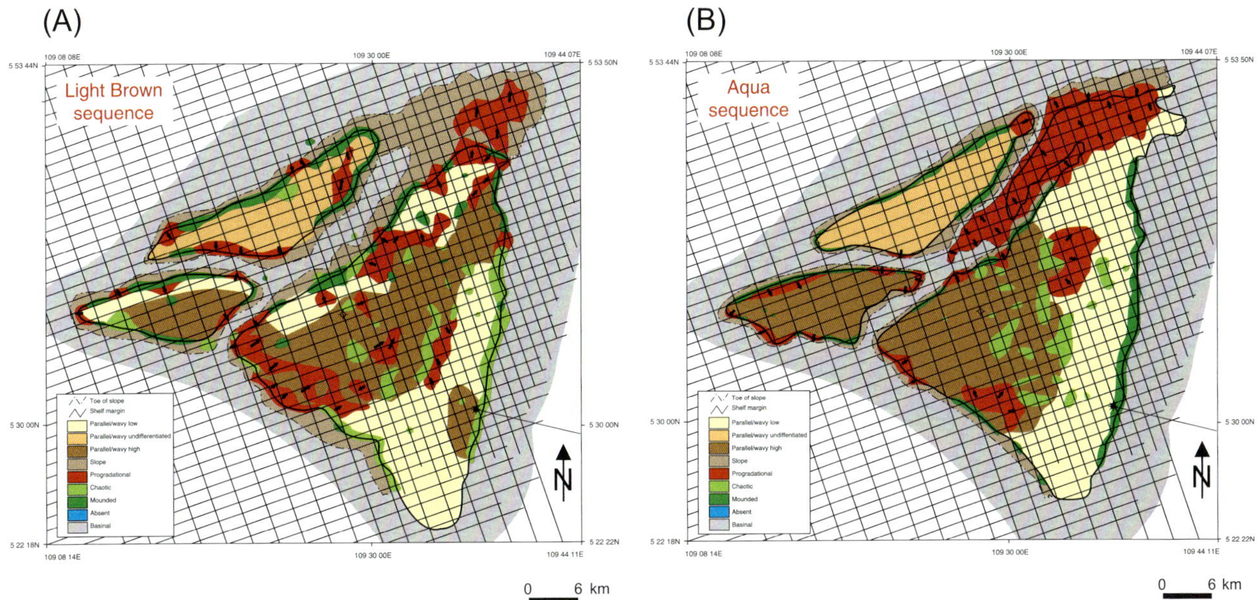

FIGURE 7. Seismic facies maps for the Light Brown and Aqua sequences. The Light Brown sequence (middle Miocene) illustrates the three separate platforms that comprised the Segitiga Platform at inception. The Aqua sequence (upper Miocene) shows an area of strong northward progradation as the intraplatform seaways began to fill and the separate platforms began to coalesce.

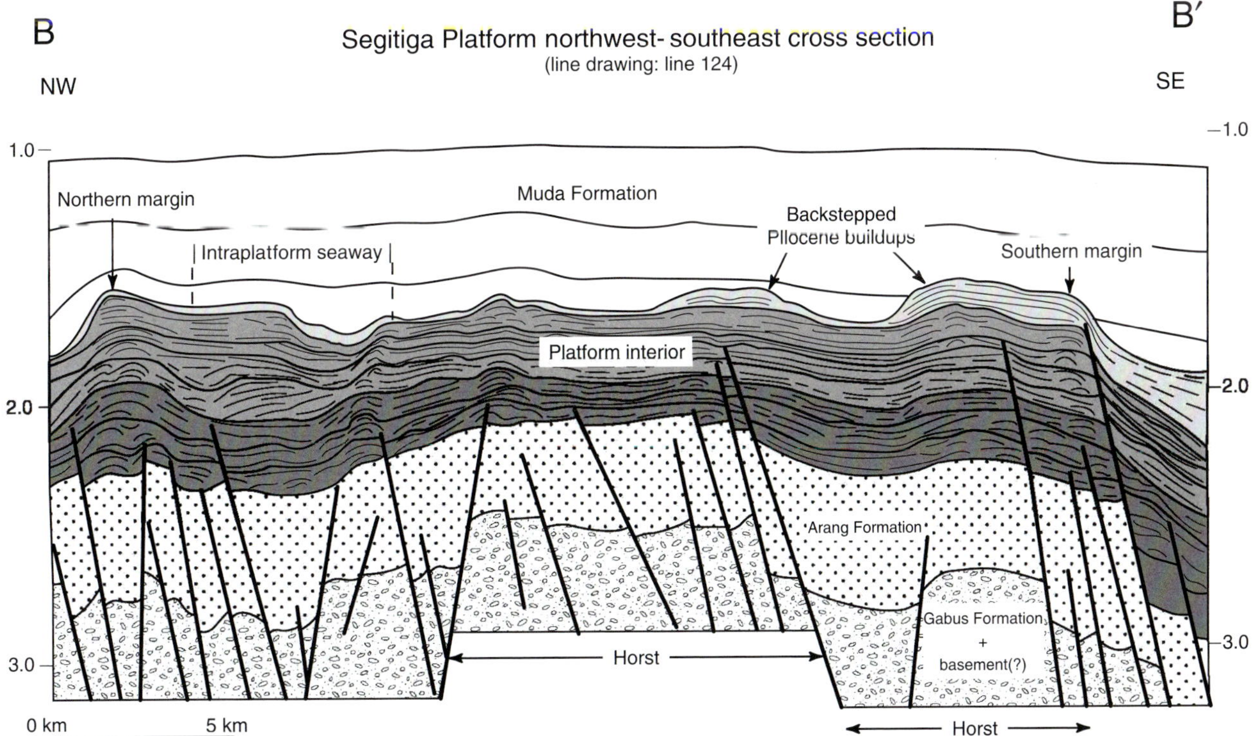

FIGURE 8. Line drawing of seismic line 124 illustrates features of the Segitiga Platform. The platform initiated over the horst block in the central part of the platform. Note the thinner strata over the horst indicative of slower subsidence. Also note the progradational fill of the intraplatform seaway. Younger faults may control the location of the back-stepped Pliocene buildups. Location of cross section on Figure 2A.

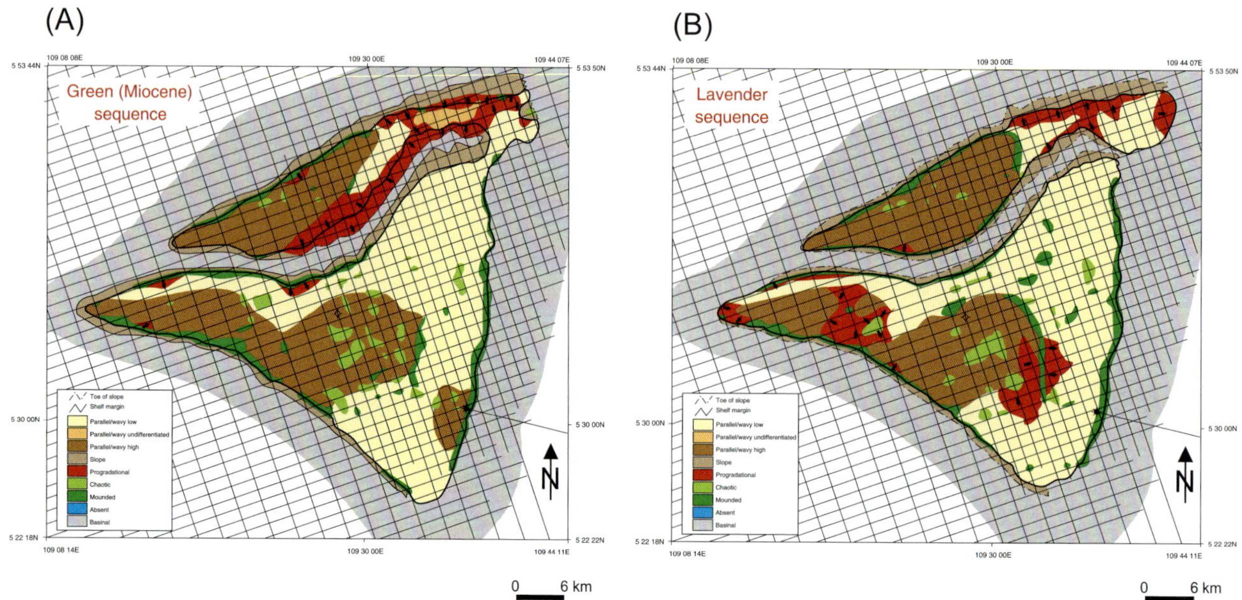

FIGURE 9. Seismic facies maps for the Green (Miocene) and Lavender sequences (upper Miocene). These sequences show the Segitiga Platform at maximum coalescence and expansion. The south fork of the intraplatform seaway has been filled and the remaining seaway is very shallow to filled in its eastern reaches. Note the reversal of progradation direction in the Green (Miocene) sequence when compared to the Aqua sequence (Figure 7B).

to the southwest and formed a "dumbbell-shaped" area of parallel (platform-high) seismic facies during deposition of Green (Miocene) to Lavender sequences (Figure 9). Local faulting may have controlled these younger topographically high areas, because the main horst-bounding structures are inactive by deposition of the Aqua sequence (Figure 7B).

Stage 3: Platform Backstepping and Shrinkage

The Segitiga Platform backstepped after Lavender sequence deposition and was reduced in area to approximately 600 km^2 (Figure 10). During deposition of the Red and Orange sequences, the Segitiga Platform was reduced to four relatively small platforms. The major buildups are to the south and they have been penetrated by the AY-1X and AP-1X wells, respectively. The platforms backstepped to the topographically high areas of Stage 2 deposition (previously defined "dumbbell-shaped" area). The two outlier platforms to the north were terminally drowned prior to deposition of the Green (Pliocene) sequence. Although backstepping dominated during this phase, the southern margin of the platform prograded as much as 3 km to the south from Red to Gold sequences (Figures 5A, 10). Chaotic facies mapped within the Gold sequence are interpreted to be areas of karst collapse on top of the Terumbu Formation as evidenced by draping or collapse of younger Muda Formation siliciclastics into karst caverns (Figure 5B). The Segitiga Platform was terminally drowned after Gold sequence deposition and buried by younger deep-water shale and siltstone of the Muda Formation.

CONTROLS ON PLATFORM INITIATION AND GROWTH

The Segitiga Platform provides insight into the controls on the growth and evolution of isolated platforms as well as the mechanisms that influenced platform architecture. Differential structural subsidence, eustatic change in sea level, and environmental influences (prevailing winds and ocean currents) were the primary controls on platform evolution and stratal architecture. Eustasy was the primary control on timing of sequence formation, but structural subsidence greatly modified the internal facies, stacking, and thickness trends of the sequences. The following sections summarize the structural and environmental controls on the stratal architecture of the Segitiga Platform.

Structural Controls on Platform Initiation and Development

Structural movements influenced the stratal architecture of the Segitiga Platform in several ways: (1) horst blocks provided a structural template for platform nucleation, (2) cessation of normal faulting may have instigated platform progradation and coalescence, (3) faulting may have provided partial control on facies distribution as well as providing a structural template

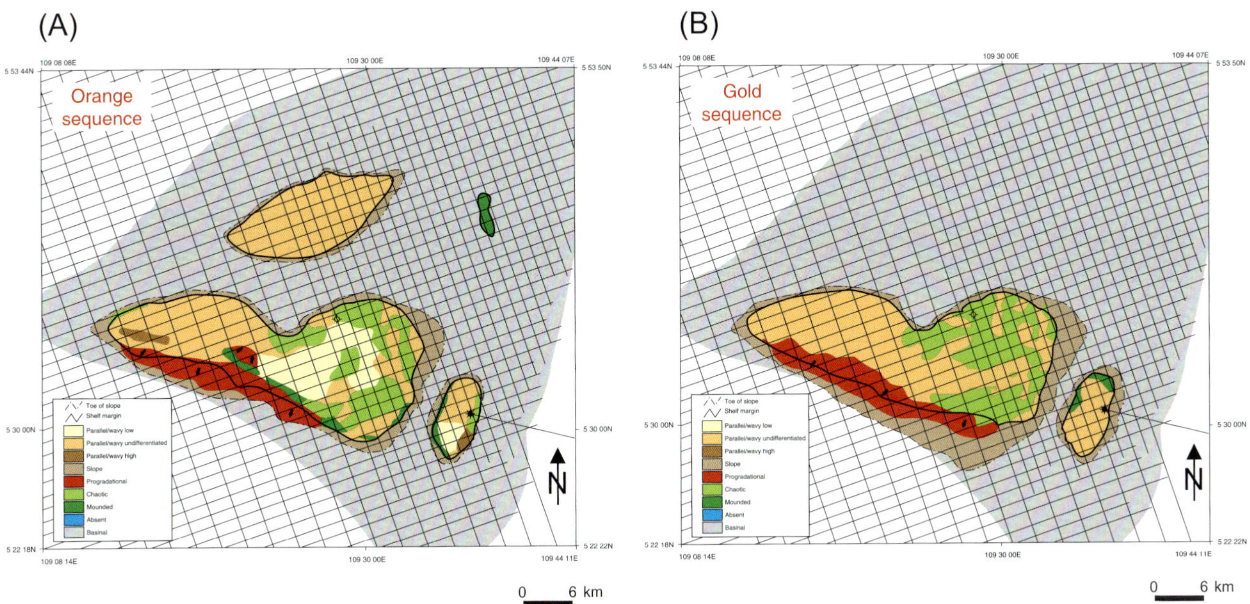

FIGURE 10. Seismic facies maps of the Orange and Gold sequences (Pliocene). These maps illustrate the progressive backstepping of the lower Pliocene sequences. The Gold sequence represents the last deposition before terminal drowning of the Segitiga Platform. Note that progradation continues on the southern margin of the platform.

for backstepping platforms, (4) local rapid subsidence may have caused variations in sequence stacking, and (5) increased regional subsidence may have controlled where and when terminal drowning occured.

Horst Blocks as Structural Template for Platform Initiation

Initial growth of the platforms centered on topographically high horst blocks (Figures 6A, 8), which were probably a result of extension during the opening of the South China Sea. The horst blocks remained topographically high enough to provide environments conducive for shallow-water carbonate production. Intervening grabens were much deeper and collected deeper-water muddy carbonates and allochthonous carbonate sediment transported from the platform tops. Relief across the horst-bounding normal faults must have been sufficiently high or movement sufficiently rapid to keep the platform from laterally prograding and filling the interplatform seaways, especially during Stage 1 (Biru to Olive sequences) deposition where most margins are observed to aggrade. Strata deposited during Stage 1 are noticeably thinner on top of horst structures and thicken into adjacent lows, suggesting that faulting may have remained active during this phase of deposition (Figure 8).

Cessation of Faulting: Control on Platform Progradation and Coalescence

Most fault movement is interpreted to have stopped by the end of Olive sequence deposition because very few faults offset the top of the Olive sequence (Figure 11). This time is also coincident with the initiation of the expansion and coalescence of the Segitiga Platform caused by progradation during Stage 2 deposition. This progradation is strongly evident in the northeast corner of the platform where the margin migrated northward approximately 6 km (Figures 7B, 12). Progradation also filled the intraplatform seaways in the north-central part of the platform (Figure 13), allowing originally separate platforms to coalesce into a single composite platform (Figures 7B, 8). It is suggested that the apparent coincidence in timing between the cessation of faulting and the progradation and coalescence of initially separate platforms is in part a result of the deceleration of accommodation production caused by decreased fault-induced subsidence.

The initiation of progradation may also have been controlled by other factors, such as the filling of intraplatform seaways with enough sediment to reduce shelf-to-basin relief (Eberli and Ginsburg, 1989) and an increase in carbonate productivity.

Faulting: Control on Distribution of Facies and Location of Backstepping Platforms

Horst blocks were structurally high areas on the platform that localized parallel-high (platform) seismic facies (Figure 10A). These facies are predicted to have more favorable reservoir characteristics because of a topographically higher position that would be prone to secondary dissolution during relative falls in sea level. With time, this topographically high area (and associated facies) migrated to the southwest (compare

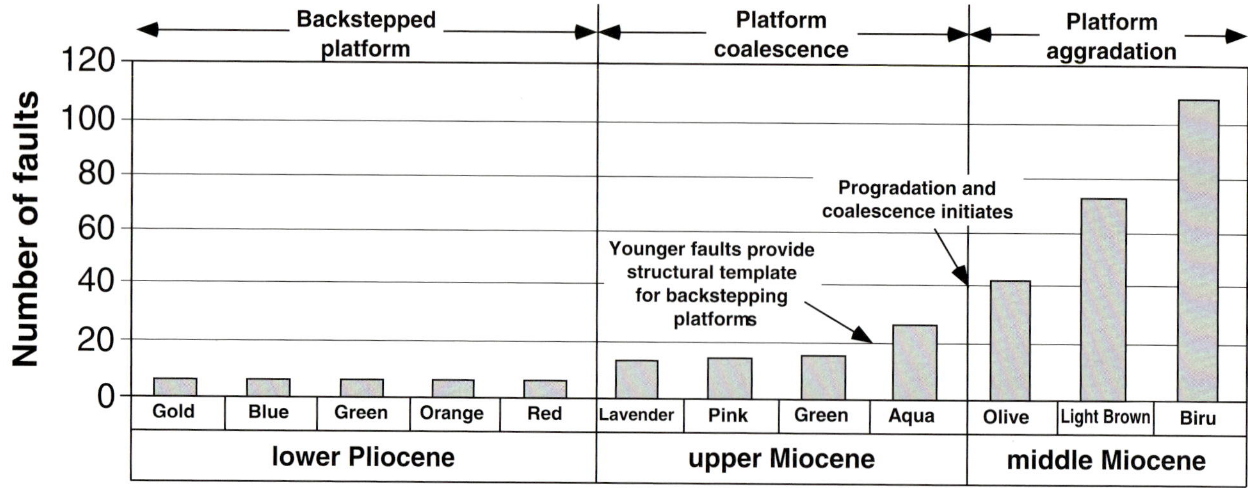

FIGURE 11. Histogram illustrating the age and abundance of faults mapped throughout the 2-D study area. The cessation of faulting and associated subsidence after Olive time may have decreased the rate of accommodation increase on the platform, thus allowing platform sediments to prograde. Younger faults, those that cut sequences above Olive, created topographic highs that functioned as a structural foundation for backstepping Pliocene platforms.

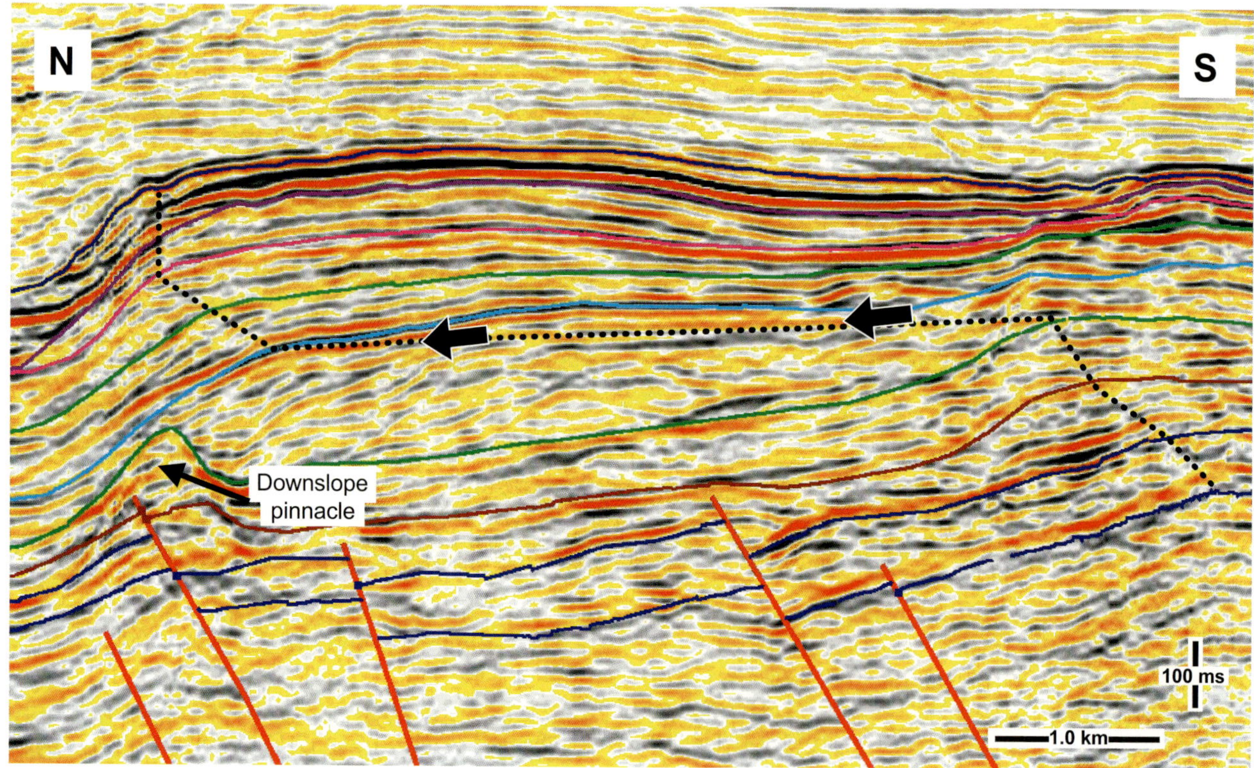

FIGURE 12. South-to-north progradation was common on the northeast corner of the Segitiga Platform and may have been a result of local subsidence of the platform in this area. Northward prograding deposits represent the initial fill of the intraplatform seaway (see Figure 13). The dotted line marks the platform margin through time. Arrows denote the major interval of progradation during deposition of the Aqua sequence.

dark brown areas of Figures 7A, 9A). The areas that remained topographically high may have been controlled by younger faults that continued movement. The remaining high (Figure 9) is a remnant of the horst block developed in the central part of the platform during Stage 1 (Figure 7A). The remnant topographic

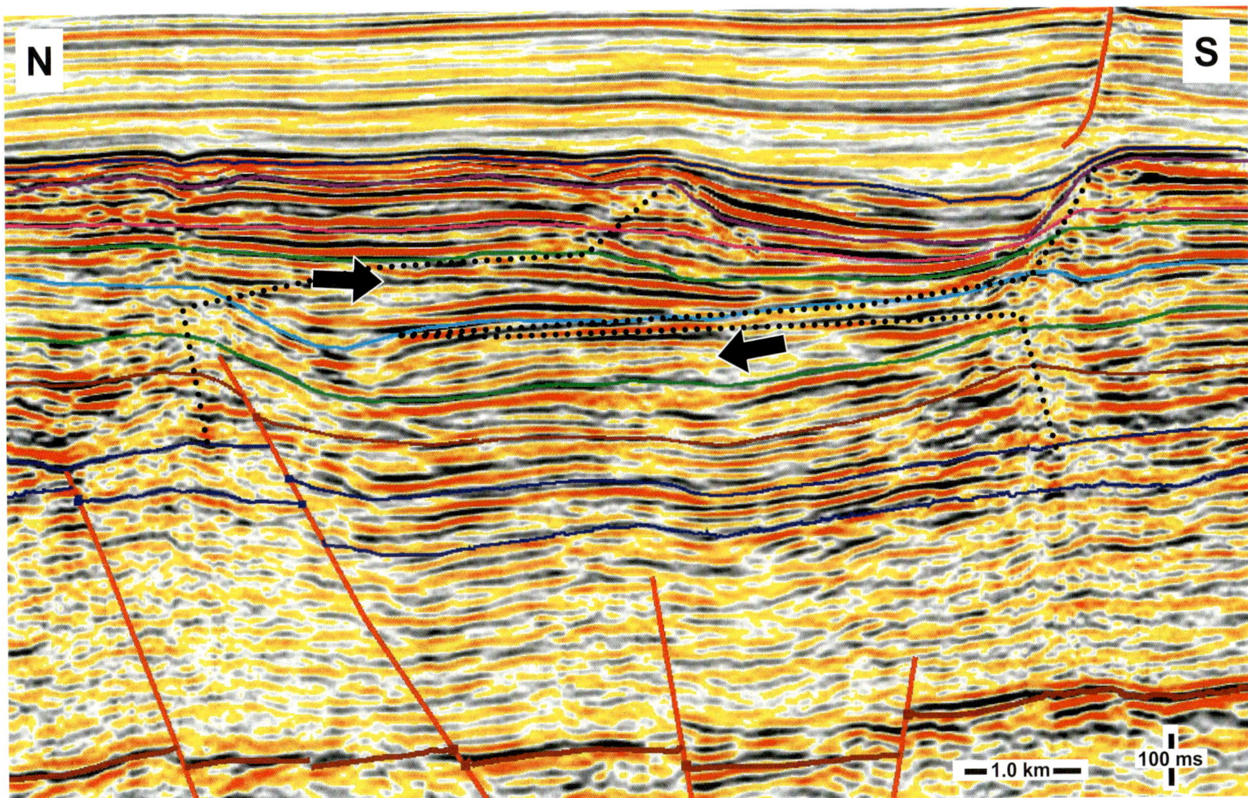

FIGURE 13. Filling of the intraplatform seaway was initially from the south to north and most likely controlled by structural subsidence. Later, progradation from north to south began, probably in response to windward-leeward sediment flux or changing ocean currents. The dotted line marks the position of the two platform margins through time. Arrows denote direction of progradation. Note the compaction-induced fault above the carbonate platform on the right.

highs provided a foundation for backstepping platforms of Stage 3 (Figure 10). A comparison of highs in the Green and Lavender sequences with the areal distribution of backstepped platforms in the Orange and Gold sequences shows a strong correlation. Thus, as relative sea level rose, these topographic highs were the only remaining parts of the platform with the ability to support a robust carbonate factory. Figure 14 shows that the backstepped lower Pliocene platform (located on the platform north of Segitiga; Cyclops buildup of Figure 2) seated on the topographically highest part of the platform, was controlled by a fault that cuts as high as the Olive sequence. Then, the platform prograded southward in the leeward direction (Figure 14).

Local Subsidence Effects: Variation in Sequence Stacking

Strong variations in sequence stacking (e.g., progradational, retrogradational, or aggradational sequence stacking) are observed locally when coeval margins are compared at different geographic localities (Figures 15, 16). Differences in sequence stacking are caused by differences between rates of sediment production and platform accommodation. Accommodation increase may be supplied via tectonic subsidence, eustatic sea level rise, or possibly compaction. Progradation occurs if the rate of sediment production is greater than the rate of accommodation increase. If the rate of sediment production is less than the rate of accommodation increase, then retrogradation (or backstepping) of sequences will occur, or drowning if the rate of accommodation greatly exceeds the rate of production. If the rate of sediment production is approximately equal to the rate of accommodation gain, then aggradation, or vertical building of sequences, will result. Figure 15 illustrates a small part of the northwest part of the Segitiga Platform that shows these geographic variations can take place over relatively small distances. Figure 16 shows seismic examples of these margins. The northern side of the platform is strongly aggradational (Figures 15, 16A), the western corner is retrogradational (Figures 15, 16C), and the southern side, facing the intraplatform seaway, is strongly progradational (Figures 15, 16B). The aggradational and progradational margins are interpreted as windward (north) and leeward (south) margins, respectively, using the direction of progradation as evidence of leeward sediment flux. The retrogradational

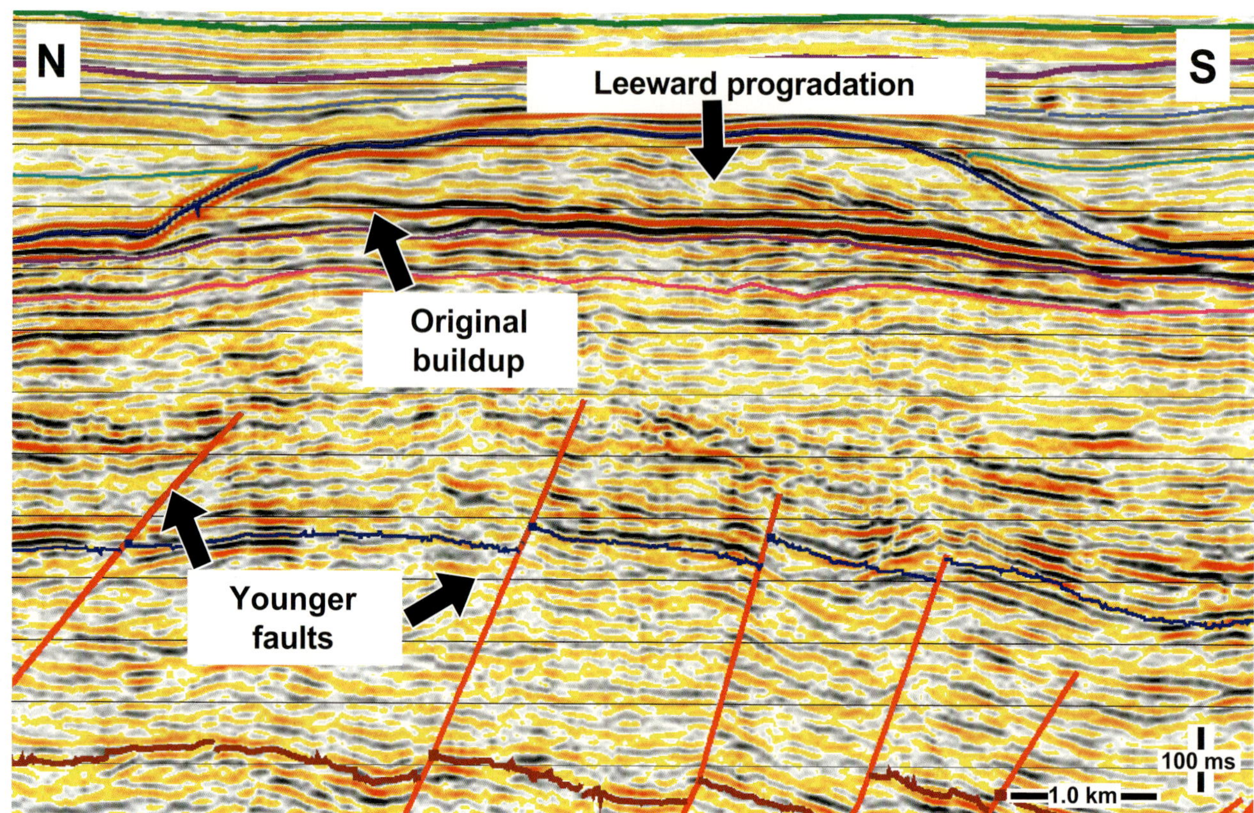

FIGURE 14. Lower Pliocene Cyclops buildup on the platform north of the Segitiga Platform illustrates the influence of faulting on backstepped platforms. The original buildup initiated on top of a topographic high caused by younger faults, which cut above the Olive sequence. The buildup then prograded southward in the leeward direction. See Figure 2A for location of the Cyclops buildup.

margin is controlled by rapid increased subsidence caused by local structural instability of the western "corner" of this platform.

Regional Subsidence and Distribution of Backstepping Platforms

The areal distribution of the backstepped platforms (Figure 2A) suggests that regional subsidence may have influenced which platforms drowned and which thrived. Figure 2A shows that backstepping was much more pronounced in the northern and eastern parts of the 2-D survey. The Natuna field (Figure 2) was terminally drowned at the same time the major backstepping took place on the Segitiga Platform to the west (end of Miocene time). The large platform to the north of the Segitiga Platform (Figure 2A) shows a greater than 90% reduction in platform size between the Lavender sequence and the Gold sequence. The Segitiga Platform shows a 57% reduction in size over the same interval. The two smaller platforms to the southwest show relatively minor decreases in platform size, and in fact, the Pliocene platforms actually prograde to the southwest (Figure 2A). It is interpreted here that increased regional subsidence to the north and east may have caused significant drowning of platforms to the northeast whereas platforms to the southwest were relatively unaffected.

Sea Level Change and Environmental Controls on Stratal Architecture

Relative changes in sea level and environmental changes, such as prevailing wind direction and paleo-oceanographic currents, are important factors in the stratal architecture of the Segitiga Platform.

Relative Changes in Sea Level and Karst Development

Relative changes in sea level are responsible for the timing of sequence-boundary formation, the stacking of successive sequences, and the secondary enhancement of reservoir properties resulting from the dissolution of primarily aragonitic skeletal material during lowstands of sea level. Major drops in sea level exposed the entire platform top and karst was formed locally on

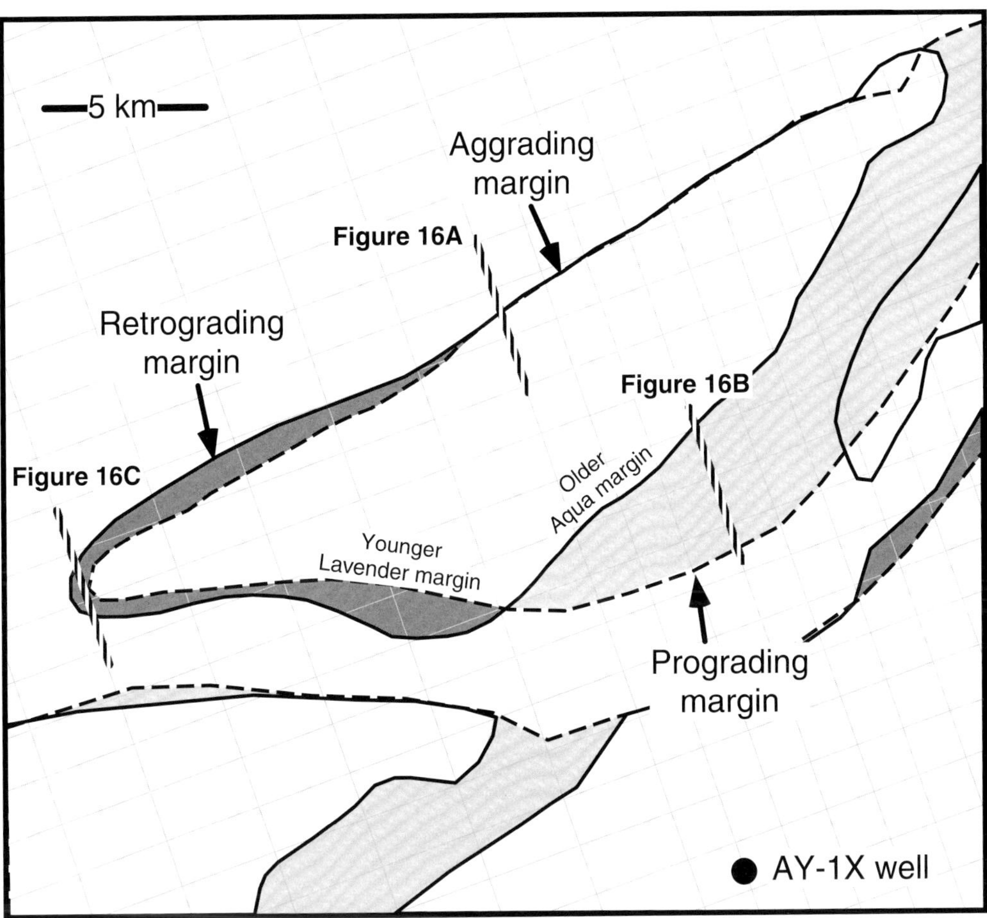

Figure 15. The northwest part of the Segitiga Platform illustrates the geologic variation in sequence stacking of coeval platform margins. Aggrading margins are interpreted to be windward margins, whereas prograding margins (light gray) represent sediment flux to the leeward side of the platform. Retrograding margins (dark gray) are most likely controlled by local subsidence at this "corner" of the platform. Note location of seismic examples (hatchured lines) shown in Figure 16.

some sequence boundaries. Karst features have been recognized only on top of the Miocene and on top of the lower Pliocene. Collapse features create a cone of data disruption (chaotic seismic facies) beneath the sinkhole because of the attenuation of the seismic signal. Figure 17 shows that the relict platform margins may have had an influence on the distribution of karst in younger sequences. Note the rugose or rough character of the Lavender sequence boundary above the relict platform area. This relationship suggests that the relict platform provided topographically high regions in younger sequences that experienced diagenetic enhancement (e.g., karsting).

Prevailing Wind Direction

Past studies of platform evolution (Eberli and Ginsburg, 1989) have stressed the importance of windward-leeward sediment transport on the development of carbonate platforms. Windward-leeward sediment transport is apparent on the Segitiga Platform as north-to-south progradation, especially evident on the southern margin of the Gold and Orange sequences (Figure 10), filling the intraplatform seaway on the Green sequence (Figures 9, 13), and in the Cyclops buildup (Figure 14).

South-to-north progradation, opposite sediment flux from the apparent windward-leeward trend, was strong on the northeast margin and filled the intraplatform seaways during Aqua-sequence deposition (Figures 7B, 12, 13). The cause of this progradation in an opposite direction from windward-leeward sediment flux could be from several sources: (1) changing oceanographic currents as the intraplatform seaway filled, (2) tectonic subsidence of the northeast part of the platform, (3) a change in prevailing wind direction, and (4) the possibility of seasonal changes in the direction of monsoon approach. Subtle inner platform progradation is most commonly oriented away from topographic highs on the platform interior (Figure 7A). No apparent control other than topography can be offered for the explanation of inner platform sediment flux.

CONCLUSIONS

The Terumbu Formation is subdivided into 12 seismic sequences. Eustatic sea level change controlled the timing of sequence development, but structural movements, both locally and regionally, had a significant

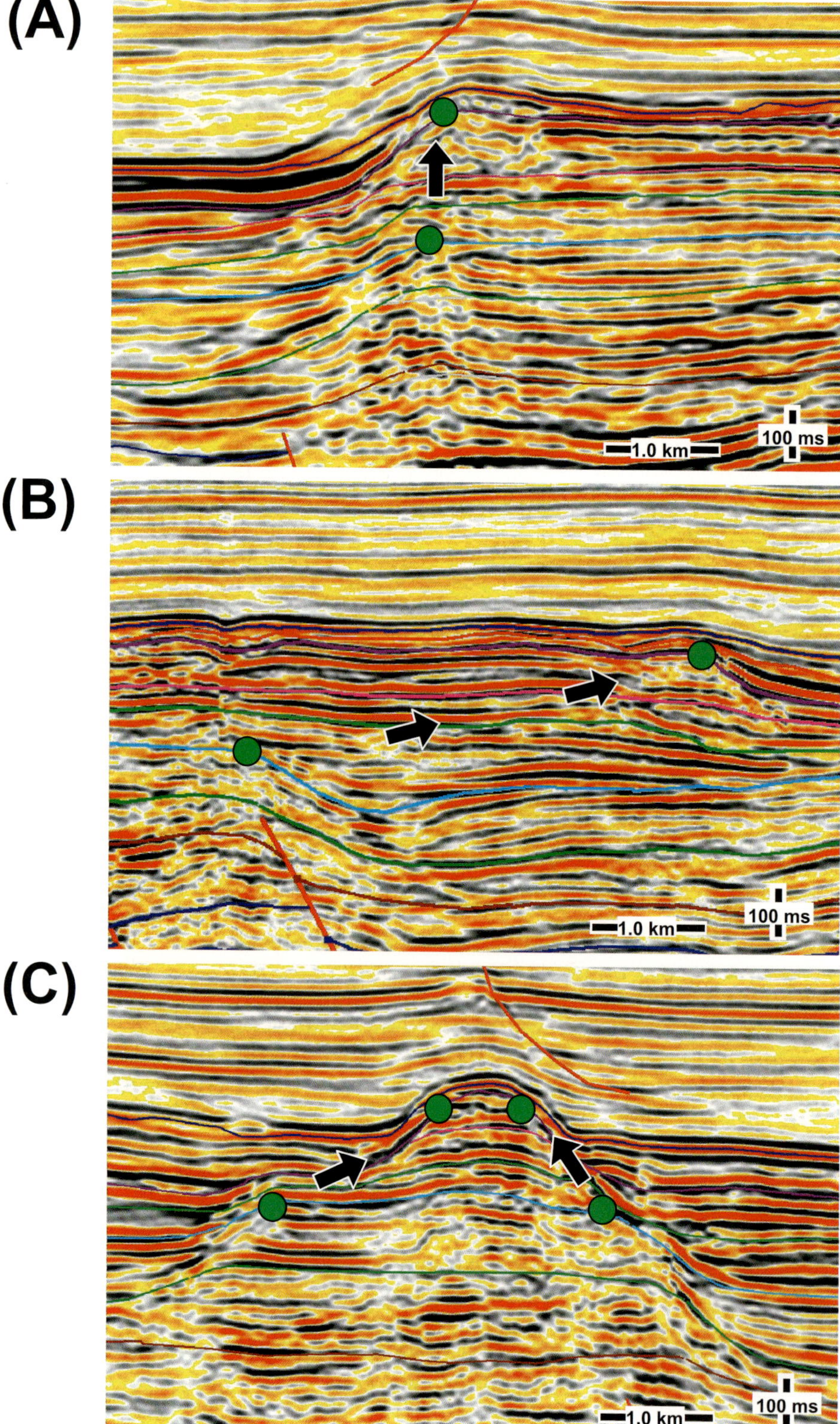

FIGURE 16. Seismic examples of geographic variability in sequence stacking over short distances (locations on Figure 15). Sequences stack (A) aggradationally on the northern margin, (B) progradationally on the southern margin, and (C) retrogradationally on the western "corner" of the platform. Green dots are interpreted shelf-margin positions for Aqua and Lavender sequences.

effect on the modification and growth history of the platform architecture.

The Segitiga Platform initiated in middle Miocene time as a series of three individual carbonate platforms separated by relatively deep (300 m) intraplatform seaways. The seaways were filled and platforms coalesced with time forming a large composite platform. A period of rapid subsidence combined with eustatic sea level fall caused the platforms to backstep and finally drown during lower Pliocene time.

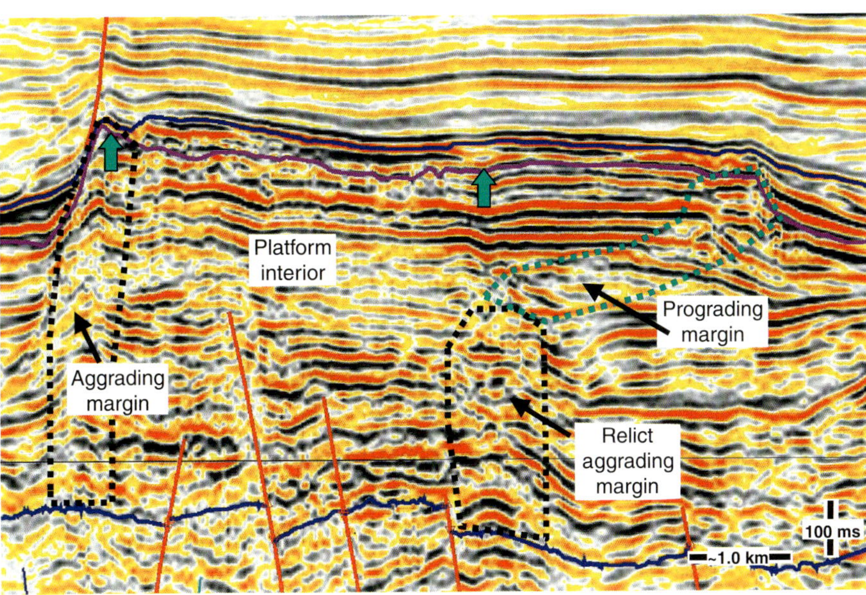

FIGURE 17. The location of relict platform margins may have had partial control on the areal distribution of karst. Note the irregular and rugose character of the Lavender sequence boundary (between green arrows), which may represent karsting. The area of interpreted karst is located over an older and smaller platform bounded by a relict aggrading margin. As younger strata prograded across this relict margin, the strata compacted over the margin, leaving the area above the older platform topographically higher. This area may have experienced enhanced meteoric dissolution during a relative fall in sea level and the subsequent formation of a karst plain.

Geographic variability in platform-margin stacking is observed locally over relatively short distances. Coeval margins are observed to prograde, retrograde, and aggrade over a 10-km^2 area. This variability is thought to be a function of localized structural subsidence and environmental controls, such as prevailing winds and ocean currents.

Faulting provided topographically high areas for platform initiation as well as a structural foundation for backstepping platforms. Topographically high areas of the platforms may have been diagenetically enhanced and karsted more strongly. Cessation of faulting, the filling of intraplatform seaways, or increased carbonate productivity may have instigated progradation and platform coalescence.

The new 2-D seismic data have allowed a more dynamic view of carbonate platform evolution and facies distribution in the Segitiga Platform that can serve as a seismic analog for other ancient isolated platforms worldwide.

ACKNOWLEDGMENTS

Thanks to various divisions of ExxonMobil Corporation and Pertamina, Inc., for permission to publish the results of this study. Special thanks goes to numerous past ExxonMobil geologists who first studied these platforms and the Natuna field based on more widely spaced 2-D data. Thanks to Lyndon Yose and Jim Markello who edited early versions of this manuscript. Volker Vahrenkamp, Andrei Beloposky, and Jose Luis Massaferro provided excellent reviews and technical edits for the manuscript. Thanks to Steve Dorobek for providing the impetus to finish this chapter.

REFERENCES CITED

Bosellini, A., 1984, Progradation geometries of carbonate platforms: Examples from the Triassic of the Dolomites, northern Italy: Sedimentology, v. 31, p. 1–24.

Dunn, P. A., M. H. Feeley, and R. D. Kissling, 1993, Sequence stratigraphic study of the Natuna aquifer area, offshore Indonesia: Exxon Production Research Geoscience Research Application Report EPR.34ES. 93, 72 p.

Eberli, G. P., and R. N., Ginsburg, 1989, Cenozoic progradation of northwestern Great Bahama Bank, a record of lateral platform growth and sea-level fluctuations, *in* P. D. Crevello, J. L. Wilson, J. F. Sarg, and J. F. Read, eds., Carbonate platform and basin development: SEPM Special Publication 44, p. 339–351.

Epting, M., 1989, The Miocene carbonate buildups of Central Luconia, offshore Sarawak, *in* A. W. Bally, ed., Atlas of seismic stratigraphy: AAPG Studies in Geology No. 27, p. 168–173.

Fulthorpe, C. S., and S. O. Schlanger, 1989, Paleo-oceanographic and tectonic settings of early Miocene reefs and associated carbonates of offshore southeast Asia: AAPG Bulletin, v. 73, p. 729–756.

Grötsch, J., and C. Mercadier, 1999, Integrated 3-D reservoir modeling based on 3-D seismic: The Tertiary Malampaya and Camago buildups, offshore Palawan, Philippines: AAPG Bulletin, v. 83, p. 1703–1728.

Hamilton, W., 1979, Tectonics of the Indonesian region: U.S. Geological Survey Professional Paper 1078, 345 p.

Haq, B. U., 1991, Sequence stratigraphy, sea-level change, and significance for the deep sea, *in* D. I. M. MacDonald, ed., Sedimentation, tectonics, and eustasy: Sea-level changes at active margins: International Association Sedimentologists Special Publication 12, p. 3–39.

Jordan, C. F., Jr., and M. Abdullah, 1992, Arun field—Indonesia, North Sumatra Basin, Sumatra, *in* N. F. Foster and E. A. Beaumond, comps., Stratigraphic traps III: AAPG Treatise of Petroleum Geology, Atlas of Oil and Gas Fields, p. 1–39.

Kozar, M. G., and E. J. Oswald, 1993, Integrated structural and stratigraphic study of the D-Alpha block, offshore Indonesia: Exxon Production Research Geoscience Research Application Report EPR.35ES. 93, 65 p.

Longman, M. W., 1980, Carbonate petrology of the Nido B-3A core, offshore Palawan, Philippines, *in* R. B. Hally and R. G. Loucks, eds., Carbonate reservoir rocks: SEPM Core Workshop Notes No. 1, p. 161–183.

Longman, M. W., R. J. Maxwell, A. D. M. Mason, and L. R. Beddoes Jr., 1987, Characteristics of a Miocene intrabank channel in Batu Raja Limestone, Ramba field, South Sumatra, Indonesia: AAPG Bulletin, v. 71, p. 1261–1273.

May, J. A., and D. R. Eyles, 1985, Well log and seismic character of Tertiary Terumbu carbonate, South China Sea, Indonesia: AAPG Bulletin, v. 69, p. 1339–1358.

Mayall, M. J., A. Bent, and D. M. Roberts, 1997, Miocene carbonate buildups offshore Socialist Republic of Vietnam, *in* A. J. Fraser, S. J. Matthews, and R. W. Murphy, eds., Petroleum geology of southeast Asia: Geological Society Special Publication 126, p. 117–120.

Mitchum, R. M., and P. R. Vail, 1977, Seismic stratigraphy and global change of sea level, part 7: Seismic stratigraphic interpretation procedure, *in* C. E. Payton, ed., Seismic stratigraphy—Application to hydrocarbon exploration: AAPG Memoir 26, p. 135–143.

Ramsayer, G. R., 1979, Seismic stratigraphy, a fundamental exploration tool: 11th Annual Offshore Technology Conference Proceedings, p. 1859–1862.

Ru, K., and J. D. Pigott, 1986, Episodic rifting and subsidence in the South China Sea: AAPG Bulletin, v. 70, p. 1136–1155.

Rudolph, K. W., and P. J. Lehmann, 1989, Platform evolution and sequence stratigraphy of the Natuna Platform, South China Sea, *in* P. D. Crevello, J. L. Wilson, J. F. Sarg, and J. F. Read, eds., Carbonate platform and basin development: SEPM Special Publication 44, p. 353–361.

Saller, A., R. Armin, L. O. Ichram, and C. Glenn-Sullivan, 1993, Sequence stratigraphy of aggrading and backstepping carbonate shelves, Oligocene, central Kalimantan, Indonesia, *in* R. G. Loucks and J. F. Sarg, eds., Carbonate sequence stratigraphy: Recent developments and applications: AAPG Memoir 57, p. 267–290.

15

Vahrenkamp, V. C., F. David, P. Duijndam, M. Newall, and P. Crevello, 2004, Growth architecture, faulting, and karstification of a middle Miocene carbonate platform, Luconia Province, offshore Sarawak, Malaysia, *in* Seismic imaging of carbonate reservoirs and systems: AAPG Memoir 81, p. 329–350.

Growth Architecture, Faulting, and Karstification of a Middle Miocene Carbonate Platform, Luconia Province, Offshore Sarawak, Malaysia

Volker C. Vahrenkamp[1]
Sarawak Shell Berhad, Miri, Malaysia

Frank David
Sarawak Shell Berhad, Miri, Malaysia

Peter Duijndam
Sarawak Shell Berhad, Miri, Malaysia

Mark Newall[1]
Sarawak Shell Berhad, Miri, Malaysia

Paul Crevello
PetrexAsia Consultants, Kuala Lumpur, Malaysia

ABSTRACT

The Mega Platform is a 30- × 50-km-large and 1.2-km-thick middle Miocene carbonate platform located in the Luconia Province, offshore Sarawak, Borneo. The platform originated in the late early to early middle Miocene on a regional fault-bounded structural high, first aggraded and then backstepped during a series of third-order sea level fluctuations during the middle Miocene (TB2.3–2.6).

The Jintan Platform termination with an area of 8 × 12 km is one of the prominent backsteps toward the top of the Mega Platform. Three-dimensional (3-D) seismic indicates that growth on Jintan ceased relatively early with continued carbonate aggradation in adjacent smaller terminations (M1, M1-East). Spectacular reservoir architecture and diagenesis are revealed by the seismic. Several transgressive, aggradational, and progradational cycles are overprinted by repeated karst events. Dissolution features and

[1]*Present address:* Petroleum Development Oman, Muscat, Sultanate of Oman.

bank-margin collapse are aligned to a deep-seated regional fault system, which periodically became reactivated during carbonate growth. A large triangular-shaped graben formed during one of the faulting periods but subsequently healed by a prograding reef-margin sequences.

Two alternative scenarios are presented to explain the ultimate demise of the platform. The first proposes drowning resulting from a combination of subsidence and eustatic sea level rise. The second evokes a much-later drowning, which was preceded by a long period of exposure resulting from a second-order sea level fall and an initial decrease in subsidence caused by the onset of tectonism in Borneo during the late Miocene. In any case, following a hiatus of about 5 m.y., the platform was finally buried by deep-marine siliciclastics that prograded into the basin from the large delta systems of northwest Borneo.

Recognition of growth architecture, faulting, and karstification is a key to exploiting the hydrocarbon reservoirs of the Mega Platform. A 30-m-thick low-porosity and -permeability layer shields the gas trapped in Jintan from the underlying aquifer. Penetrated by only one well, the extent of the layer and areas of breaching caused by faulting and karstification are identified on seismic. Interpretation of the seismic is critical to assessing whether and how the underlying aquifer is felt during reservoir depletion and whether there is pressure communication between adjacent reservoirs connected via the aquifer. Cores and logs from three wells provide ground truthing of reservoir architecture, karst features, and faulting derived from the interpretation of reflection and inversion seismic. The interpretation is then imported into static and dynamic 3-D models to constrain reservoir properties, predict dynamic behavior, and guide optimum field development.

INTRODUCTION

The late Oligocene to Miocene was a period of widespread carbonate deposition in southeast Asia (Epting, 1980; Fulthorpe and Schlanger, 1989; Ehrlich et al., 1993; Saller et al., 1993; Gucci and Clark, 1993; Sun and Esteban, 1994). Many of these late Tertiary carbonate sequences have been targets of hydrocarbon exploration and numerous oil and gas reservoirs being discovered [i.e., Malaysia: Central Luconia (Ho, 1978; Epting, 1980; Epting, 1989; Sulaiman, 1995); Philippines: Nido (Withjack, 1985) and Malampaya (Grötsch and Mercadier, 1999); Indonesia: Arun (Abdullah and Jordan, 1987; Jordan and Abdullah, 1992), northwest Java Sea (Yaman et al., 1991); Ramba (Longman et al., 1987); South Lho Sukon (Maliki and Soenarawi, 1991), Natuna (May and Eyles, 1985; Rudolph and Lehmann, 1989); China: Liuhua, Pearl River (Moldovanje et al., 1995)].

The Luconia province, offshore Sarawak, Malaysia, is one of the largest of these southeast Asian carbonate provinces (Figure 1). More than 70 Miocene carbonate platforms have been mapped in an area of 240 × 240 km and range in size from a few to more than 200 km^2 (Figure 1). To date, more than 40 of them have been drilled and more than 20 were gas bearing with reserves estimated to exceed 40 tscf.

The Luconia Province is believed to be a microcontinent that moved southward toward Sundaland during the Late Cretaceous to Paleocene, resulting in collision and development of the Rajang Group accretionary prism on Borneo (e.g., Ru and Pigott, 1986). Prior to carbonate growth, during the late Oligocene to early Miocene, the southern, near-Borneo part of the Luconia Province appears to have been situated in a coastal plain to inner neritic setting (Ho, 1978). In the north, significant synsedimentary faulting segmented the province into a series of south-southwest– to north-northeast–trending horsts and grabens with the depositional environments ranging from inner to outer neritic. Dextral strike-slip movement complemented the dominantly extensional component of fault motion and controlled the horst and graben development essential for carbonate platform development.

Widespread carbonate deposition commenced in the late early Miocene and lasted to the end of the middle Miocene, as revealed by detailed biostratigraphy of the pre- and post-carbonate clastics and Sr-isotope dating of the carbonates from several Luconia Platforms (Vahrenkamp et al., 1996; Vahrenkamp, 1998). Growth and architecture of the platforms were dictated by a relative constant subsidence rate over the province (based on relatively uniform platform thicknesses of about 1200 m),

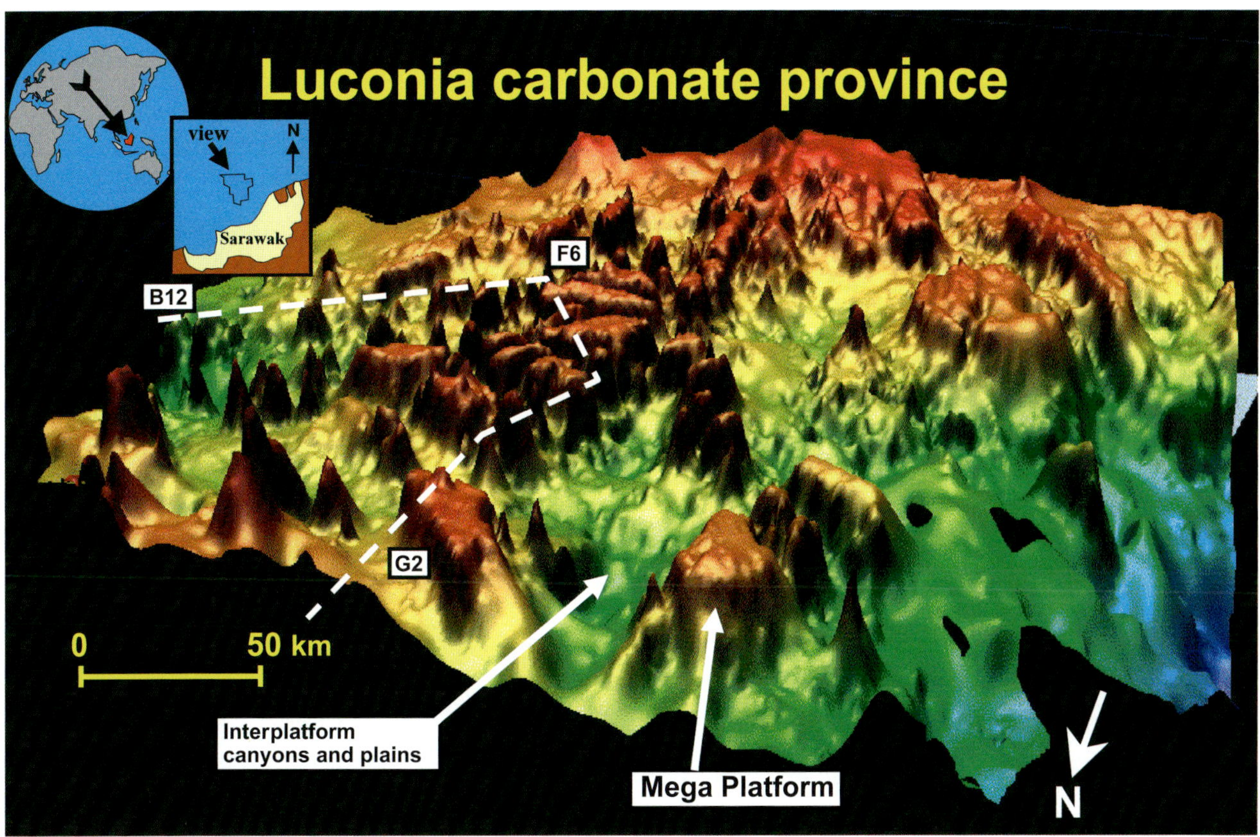

FIGURE 1. The Luconia carbonate province, offshore Sarawak, Borneo, in the South China Sea. The interplatform bathymetry reaches 1000 m toward the end of the main phase of platform growth at the end of the middle Miocene (ER Mapper image of top carbonate from 16,500 km; 2-D seismic merged with several 3-D seismic surveys; courtesy of N. Casson). White dashed line is the section of Enclosure 1 (see seismic section fold out, inside back cover, this Memoir).

and the interplay with eustatic sea level fluctuations and the monsoonal wind system of the middle Miocene affecting the area (Figures 2, 3). Compared with other carbonate depositional systems (Schlager, 1999), average sedimentation rates were quite high (150–200 m/m.y.).

Toward the end of the middle Miocene, deltaic progradation from the south and southeast split the province into a southern area that contains low-relief banks with carbonate and siliciclastic interfingering and a central and northern area where high-relief buildups occur (Vahrenkamp, 1998). Subaerial exposure and burial below shallow-marine clastics terminated carbonate growth in the low-relief areas of the south. In contrast, starting in the earliest late Miocene, the high-relief carbonate buildups of the north experienced a prolonged period of nondeposition and diagenesis, probably as a result of exposure and karstification. Finally, they drowned in front of successive sequences of late Miocene to Pliocene prograding marine deltaics originating from the Baram Delta of Borneo to the east-southeast (Enclosure 1 in Figure 1; Ho, 1978; Epting, 1980; Aigner et al., 1989; Epting, 1989). Eventually, most platforms were buried by these clastics. Only the most basinward platforms in the extreme northeast of the province outboard of the present-day shelf break continued to grow or were restarted (when is not known, Pliocene is speculated). Carbonate deposition is still ongoing in the present-day north Luconia shoals.

The Mega Platform, which is located close to the farthest extent of the siliciclastic wedge, originated on one of the structural highs formed by regional extension and faulting during the Paleocene to early Miocene (Figures 2, 4; Enclosure 1 in Figure 1). It is a large platform covering an area of 1500 km^2 (30 × 50 km), reaching an estimated thickness of 1200 m, and is topped by seven prominent culminations, each one of which forms a gas reservoir (Figure 5). This chapter covers the growth and demise of the Mega Platform and especially the architecture of one of its culminations, the Jintan Reservoir (Figure 5), as revealed by three-dimensional (3-D) seismic interpretation. As such, it focuses on the final third of the Mega Platform growth.

DATA AND METHODS

Seismic Data

The 335.6 km^2 M1/Jintan 3-D data set was acquired by Geco-Prakla between October 23, 1992, and January 6,

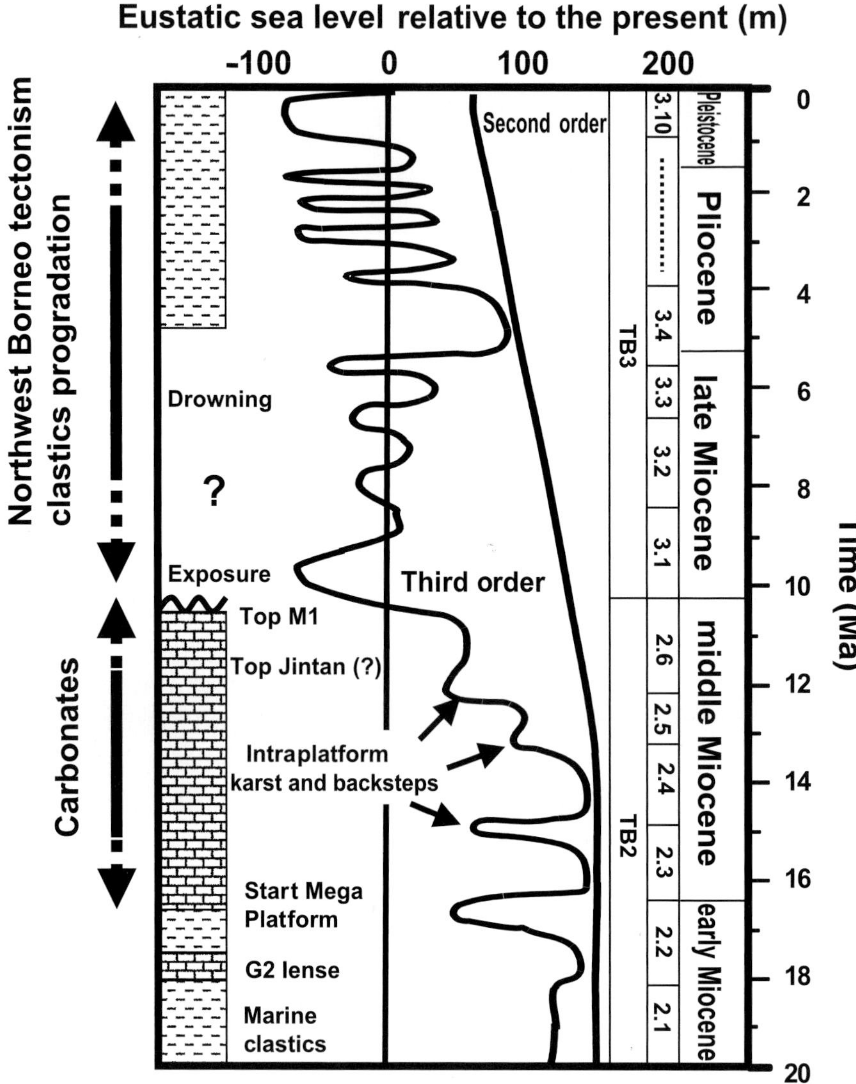

FIGURE 2. Based on Sr isotope stratigraphy, Luconia main platform growth correlates with a period of sea level highstand during the middle Miocene on the global sea level curve of Haq et al. (1988) (supercycle TB2 and cycles 2.3–2.6). A deeper carbonate interval has been dated using Sr isotopes as about 18 m.y. The demise of the Luconia Platforms coincides with a major eustatic sea level drop at the end of the middle Miocene (boundary TB2–TB3) and the onset of Borneo tectonism. Siliciclastics directly above the Mega Platform have been dated with nannoplankton as early Pliocene, indicating a 5-m.y. gap between carbonate and clastic deposition in this area. Despite evidence for karstification of the top of the M1 termination, it is unclear how much of this hiatus the platform spent subaerially exposed. Ultimately, platform drowning is related to increased subsidence caused by tectonic events in northwest Borneo and sediment loading from the Baram Delta (Enclosure 1). We speculate that the major karst horizon described in this chapter (top of lower carbonate sequence) coincides with the third-order sea level drop at about 12 m.y. The top of the Jintan termination would then relate to a fourth- or fifth-order event of the subsequent third cycle.

1993, using a dual-source, triple-streamer configuration and a cable length of 3000 m. The M1/Jintan survey consists of 660 subsurface lines shot in the 158/338° direction with an average line spacing of 25 m. The data were acquired at 2 ms to 5.0 s, with a 240-trace, 40-fold configuration and stacking bin size of 6.25 × 25 m. A total of 22,320 km of data was processed. There are 14,035 prime line-km, with infill making up another 8285 km, or 37% of the total kilometers. The data were processed with a Kirchhoff 3-D DMO, a two-pass velocity analysis, K-F filtering, and a one-pass finite-difference migration. The data are zero phased. Apart from the gas-chimney areas above the M1 and M1-East fields (Figure 4), the quality of the data is good and little acquisition imprint is recognizable in the amplitude maps at objective levels.

The top of the reservoir is typically characterized by a relatively high positive amplitude (e.g., Figures 6, 7). The seismic energy is concentrated at around 25 Hz, and there is little seismic energy below about 6 Hz or above 50 Hz. No obvious indications of processing artifacts have been observed.

The reflection seismic was further analyzed using Shell's proprietary attribute extraction techniques and Landmark software (dip and azimuth maps, coherency filters, etc.). The seismic has been inverted using Jason Workbench software with the data limited to a vertical gate of 0.5–1.5 s two-way traveltime (TWT).

Horizon and Fault Data

In addition to the regionally available top- and base-carbonate horizons interpreted on a coarse two-dimensional (2-D) seismic grid, nine horizons were interpreted on the 3-D survey to constrain the reservoir architecture of the Jintan termination (e.g., the upper part of the Mega Platform in the study area). Horizons were interpreted on true-amplitude reflectivity data, until

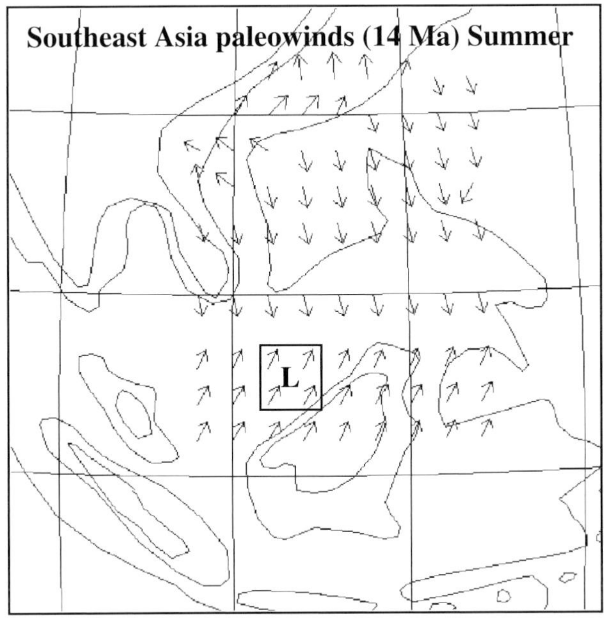

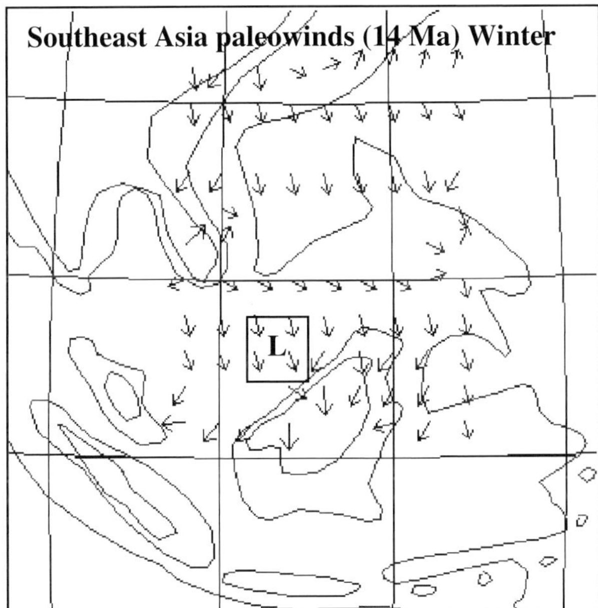

FIGURE 3. A simulation of wind circulation for the middle Miocene indicates summer and winter monsoon winds at an angle of about 100° for the Luconia Province (square). This wind pattern had a profound influence on geometries and internal architecture of Luconia carbonate platforms (Vahrenkamp et al., 1996).

a first-pass seed grid of interpreted in- and cross-lines was established. The final line-grid interpretation is on a square line grid of every 125th in-line (line) and 25th cross-line (trace). Autotracking was used for the higher-quality areas and handpicking and interpolating were used in poorer-data areas. The interpretation was refined by using acoustic-impedance data and information from cores and well logs (Figure 7). The central platform area of Jintan is characterized by parallel, subhorizontal reflections and zones, which are fairly variable in reflectivity and impedance, suggesting underlying variations in porosity (Figures 6, 7).

Fault mapping was done iteratively using both reflection and impedance seismic, as well as semblance and coherence cubes.

Seismically Derived Property Data

Porosity estimates of the platform sediments were derived from a porosity/impedance correlation and sparse spike and stochastic inversion using Jason Workbench software. Porosity cubes were depth converted and transferred into a Shell proprietary 3-D reservoir modeling package (GEOCAP) for visualization and static and dynamic reservoir modeling.

Wells and Well Logs

Approximately 28 exploration, appraisal, and production wells penetrate the Mega Platform, all with a standard modern suite of log data (Figure 4). The Jintan termination has been penetrated by three wells (Jin-1 to -3; Figures 5, 7), two of which have been extensively cored (Jin-2: 55.2 m; Jin-3: 181.7 m) with nearly 100% recovery. Jintan-3 drilled the longest section and reached the deeper water-wet section to provide calibration for a Mega Platform wide positive reflection. Jintan-1 and -2 drilled a shorter section (with Jintan-2 the shortest) just reaching the water-leg. All three wells have reasonable to good seismic-to-well matches at the main intrareservoir interfaces (Figure 7). The well-to-seismic match is poorest for Jintan-1, which is caused by a lower-quality sonic log. Detailed core and thin-section descriptions and their geologic interpretation provide the calibration and ground truth for the seismic interpretation. In addition, information was integrated from cores available from three of the other six platform terminations (M1, M3, and M4) and from numerous wells in similar platforms of the Luconia province.

RESULTS

Platform Thickness

The Mega Platform reaches a maximum thickness of about 1200 m. In lieu of a complete well penetration, this estimate is based on a 600-ms-TWT carbonate interval derived from a top and base horizon interpretation on seismic reflection data (Figure 4) and a velocity of 4000 m/s (limestone with an average porosity of 25%). The estimate is corroborated by the thickness of the carbonate section of two nearby platforms with complete

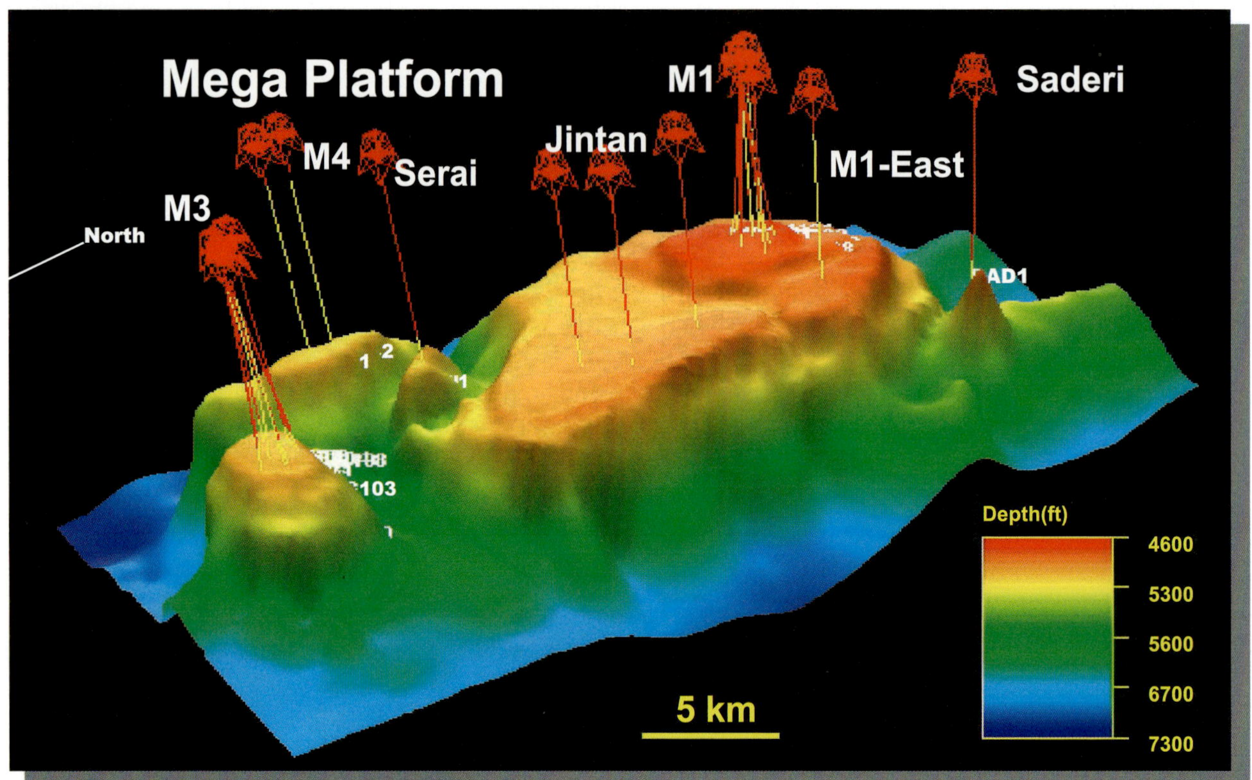

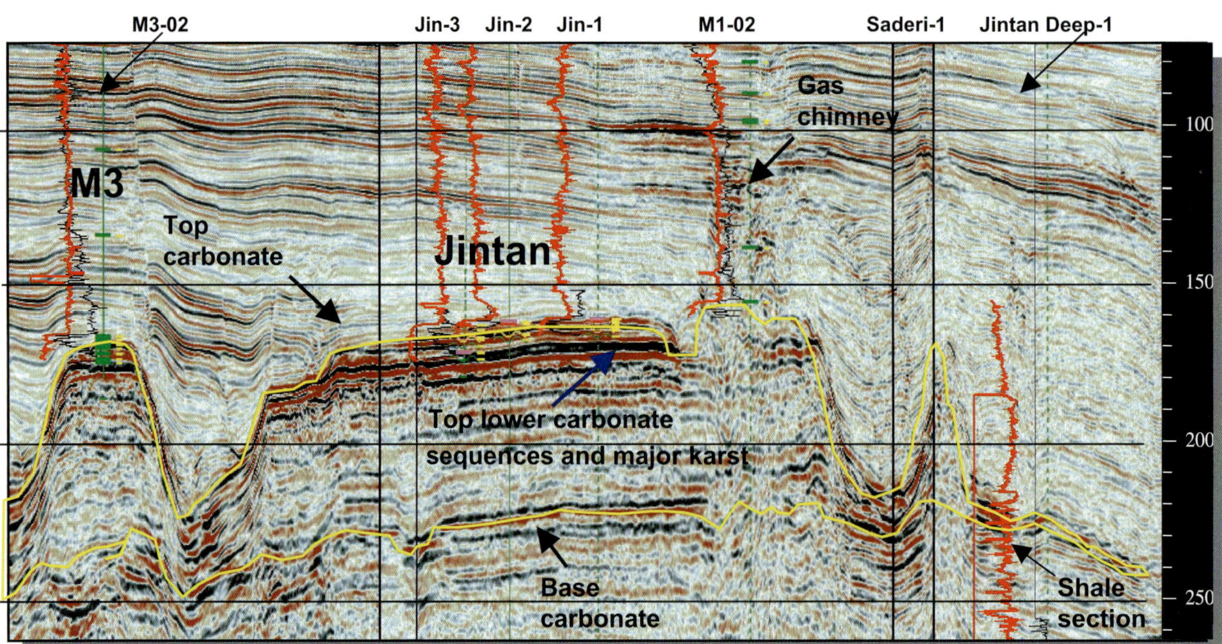

FIGURE 4. The Mega Platform. Top carbonate from merged 3- and 2-D seismic. Approximately 28 wells in seven terminations penetrate the platform. The seismic section shows the top-carbonate and base-carbonate horizons. Based on fossils, the shale section below base carbonate has been dated as early Miocene. It contains a carbonate lense, which has been dated in the nearby G2 Platform with Sr isotopes as 18 m.y. The blue arrow marks the top of the heavily karstified lower carbonate sequence (red reflector) and subsequent flooding (black reflector). This boundary can be traced clearly between terminations across the whole Mega Platform. Note the gas chimney over part of the M1 termination.

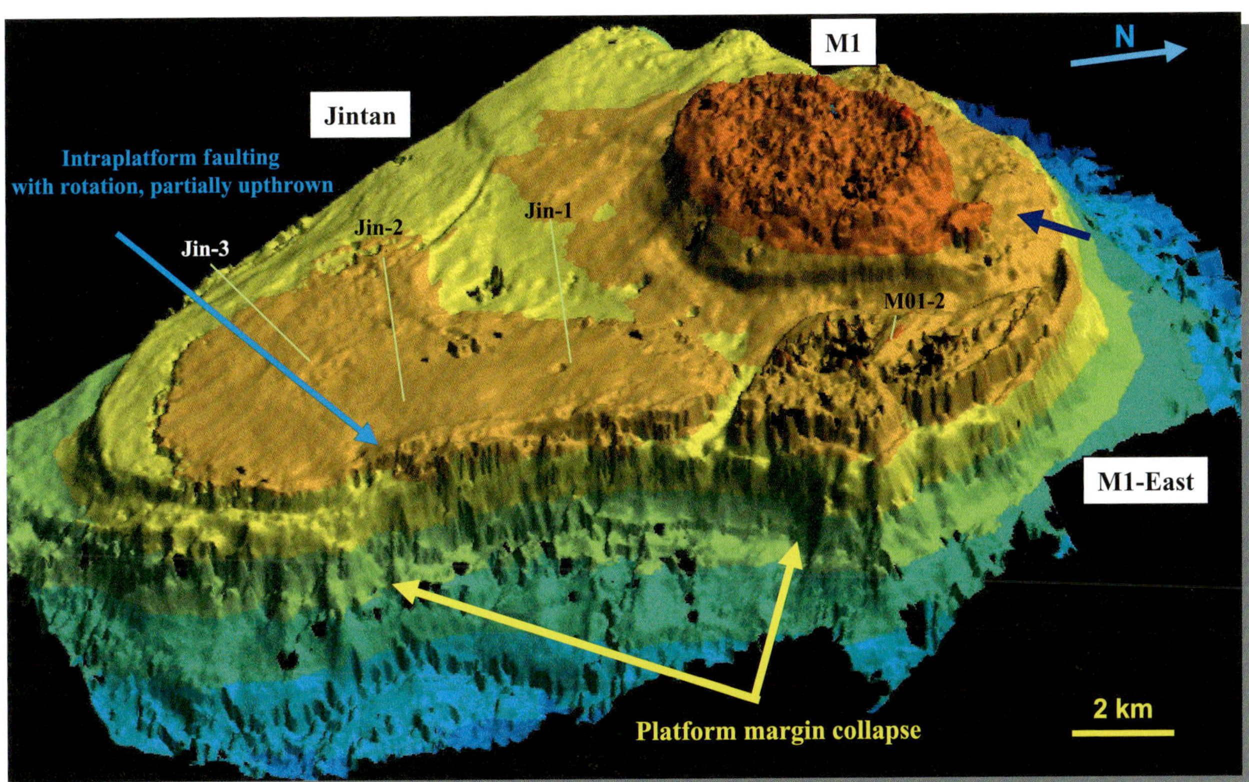

FIGURE 5. Top carbonate of the M1, M1-East, and Jintan terminations of the Mega Platform (M1 has 11 well penetrations not shown here). The color shades highlight elevation and hence the prominent backsteps. Whereas Jintan and M1-East stopped at the end of the orange period, M1 continued to grow albeit leaving a backstep terrace behind (blue arrow). The convex scar at the platform margin along Jintan–M1-East is thought to be the result of a large-scale margin collapse. Ridges in the platform (cyan arrow) delineate faults with some of the fault blocks upthrown, possibly caused by rotation or compression during overburden compaction.

well penetration (one inboard and one outboard of the Mega Platform, respectively, F6 and G2).

Large-Scale Depositional Cyclicity

Core evaluations have shown that overall three principal facies exist:

a. platform margin and interior chaotic reflection patterns are interpreted to be associated with reefal facies,
b. platform interior high-impedance layers are associated with occasionally slightly argillaceous wackestone/packstone/grainstone units with relatively low and predominantly moldic porosity, interpreted to have been deposited during a transgressive flooding of the platforms after intermittent exposure (Figures 6, 8, 10),
c. platform interior low-impedance layers are associated with high-porosity packstone/grainstone sediments assumed to have been deposited during sea level highstands (Figures 6, 9, 10).

The repetitive occurrence of thick layers of (b) and (c) over the whole carbonate interval (Figure 4) suggests that overall the sequence was deposited during many cycles of relative sea level fluctuations. Based on age constraints derived from Sr isotope analyses, this cyclic deposition took place during the middle Miocene (Vahrenkamp et al., 1996; Vahrenkamp, 1998) (Figure 2).

Large-Scale Platform Architecture (Backstepping)

Overall, the Mega Platform had at least four major growth intervals (Figure 4): an initial platform, which initiated on several platforms and coalesced to cover probably the entire Mega Platform extent, a first disintegration phase during which the platform split into five subplatforms (M3, M4, Serai, Jintan/M1/M1-East, and Saderi). During the next stage, M1, M1-East, and Jintan split into individual smaller platforms. During the last phase, Jintan, M3, M4, and Serai drowned, whereas the more northern platforms of M1 and Saderi continued growing. Finally, growth stopped on these platforms as

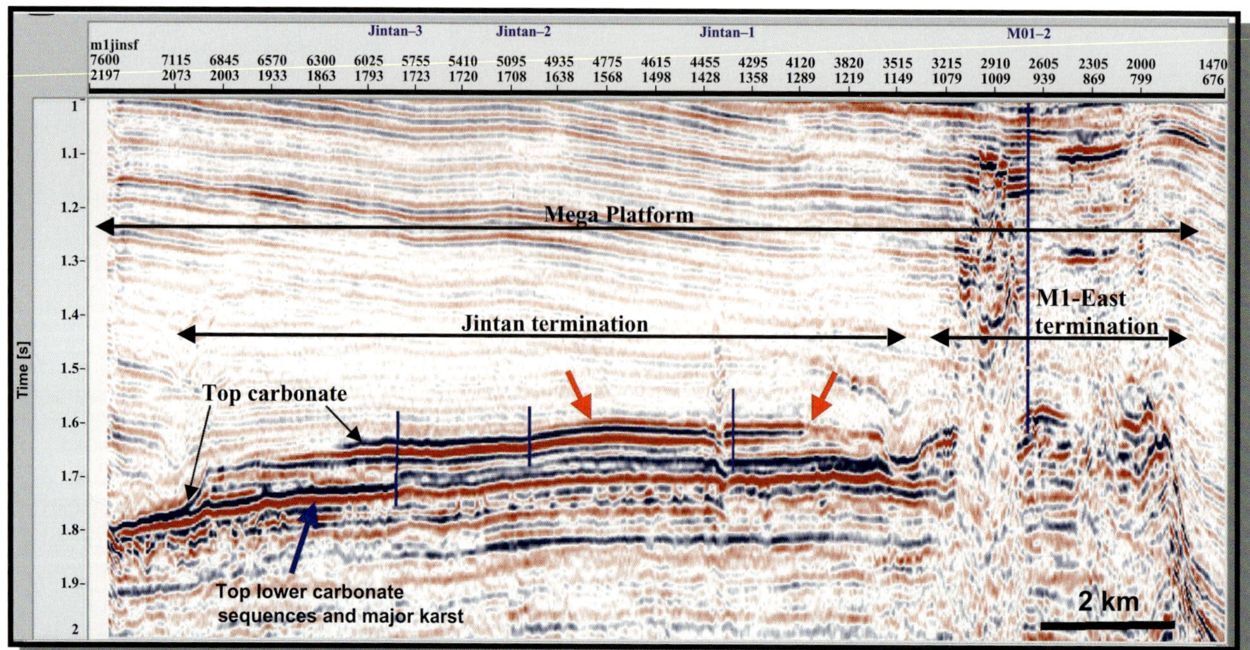

FIGURE 6. Section through the 3-D seismic volume that passes through all Jintan well locations. The logged intervals in the wells are indicated by the blue lines. The problems caused by shallow gas can be seen around well M1–2 in the M1-East termination. The central platform area is characterized by parallel, subhorizontal reflections. The blue arrow marks the top of the heavily karstified lower carbonate sequence (red reflector) and subsequent flooding of the middle carbonate sequence (black reflector). This boundary can be traced clearly between terminations across the whole Mega Platform. Top carbonate is generally characterized by a large positive reflection (black). However, the anomaly seen above the reflector (red arrows) delineates a final growth phase (Z3b, see also Figure 13).

well. An illumination of the top carbonate horizon of the Mega Platform clearly reveals the final major backsteps (Figure 5), whereas the reflection seismic allows the correlation of the individual growth segments (Figure 4). At least some of the lines of platform separations seem to be aligned with the deep-seated north-northeast–south-southwest– and northwest-southeast–trending regional fault systems (Figures 4, 16). A correlation of the large-scale growth and disintegration episodes with the middle Miocene third-order sequences as indicated on the eustatic sea level curve (Figure 2) is a possibility. However, a detailed age constraint to that level is not available.

Detailed Depositional Architecture of the Upper Mega Platform at the Jintan Termination

Based on wire-line log interpretations (Figure 11), core studies, and seismic interpretation (Figure 7), the upper section of the Mega Platform at Jintan has been subdivided into three larger depositional sequences containing multiple high-frequency sequences and nine reservoir zones (Table 1).

Lithofacies descriptions, faunal content, and key bounding surfaces were used to define the depositional sequences. These were tied to the well logs and seismic, using the synthetic seismic traces of the wells. The reservoir zones were also used to constrain seismic inversion and 3-D geologic, porosity, and permeability models.

Lower Carbonate Sequence

Reservoir Zones 1a/1b (Shallow Lagoon)

This sequence was only penetrated in Jintan-3 and only Zone 1b was cored. Although bioclast grainstone-dominated lithologies were found at the base of the core, it is characterized in its upper section by corals in a mud-dominated lithofacies deposited in a low-energy shallow lagoonal setting. This lagoonal lithofacies is capped by a pronounced subaerial unconformity, which is interpreted as a sequence boundary separating the lower from the middle carbonate sequences. On seismic data, the zone appears as a moderate to low acoustic-impedance unit (Figure 7B), consistent with the fairly high porosities. Zone 1a, which has not been cored, shows a similar character on acoustic-impedance data as cored Zone 1b. Karst structures are abundant (see separate karst section below). Although plug-derived porosity and permeability values are high (Figure 10), well-test permeabilities in Jintan-2 and -3 are even higher (200 md), suggesting possible

(A)

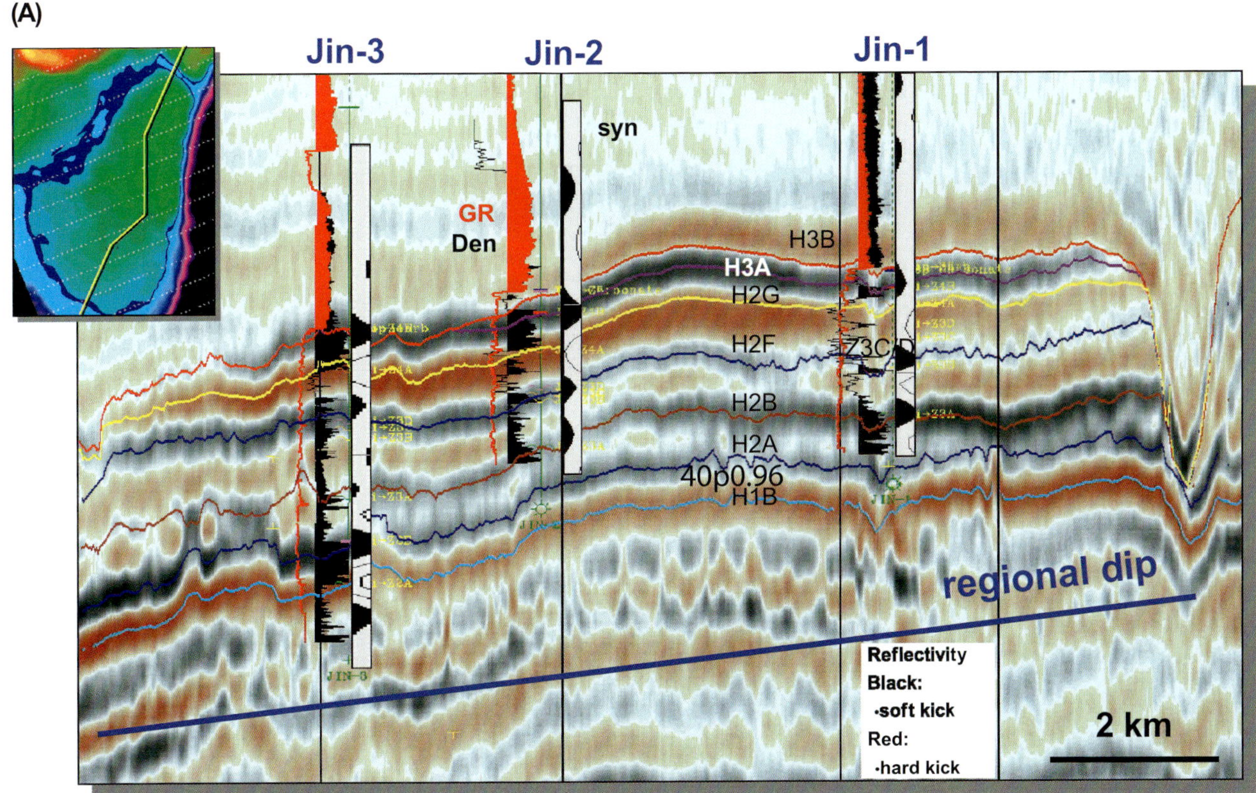

(B)

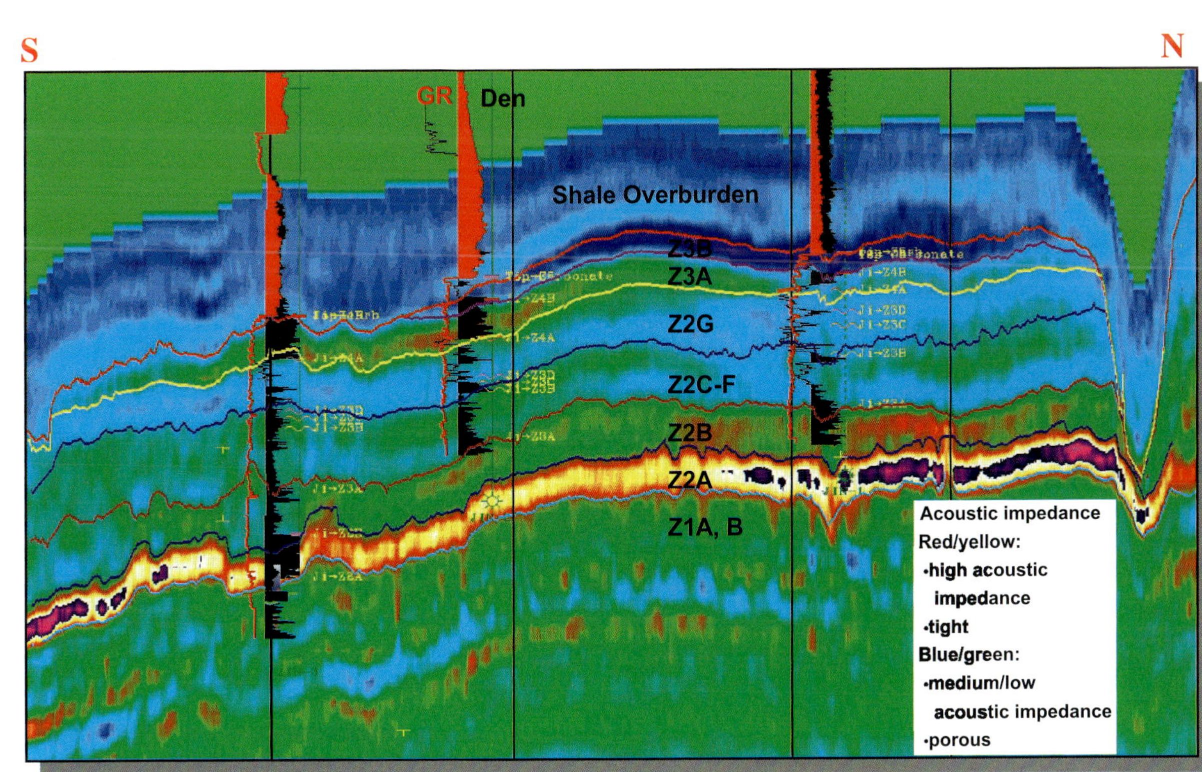

FIGURE 7. (A) Reflection and (B) acoustic-impedance section through the Jintan termination at well locations. Synthetic seismic logs have a good match with the reflection seismic. Seven of the 10 mapped horizons are shown (H1b, H2a, etc.). Horizons define reservoir zones shown on the acoustic impedance (Z1a, Z2a, etc.). Varied reflectivity and impedance suggest variations in porosity.

FIGURE 8. Rock types associated with transgressive systems tracts are characterized by a deep-marine fauna, relatively high argillaceous content, and low porosity. A mixture of wackestone and packstone/grainstone associations and rhodolites indicate that energy conditions were variable, i.e., deposition probably took place near the storm-wave base.

permeability enhancement through fracturing and/or karsting. On seismic, the unit is capped by one of the most robust and continuous markers of the Mega Platform (Zone 2a).

Middle Carbonate Sequence Set

This is the thickest sequence of the Jintan reservoir (>150 m), which was completely cored in Jintan-3 and partly cored in Jintan-2. It is interpreted to represent a complete third-order depositional sequence and is a composite of stacked, higher-order (fourth to fifth) depositional sequences with their bounding unconformities (Figure 12). Its base is the prominent top Z1b unconformity described above, which is interpreted to represent a lowstand phase during which the whole Mega Platform emerged with widespread erosion and karstification (see karst section below). Apart from the flooding unit Z2a, most units of this sequence represent highstand bank-top deposits with moderate to good reservoir properties. The coalescence of the Zones Z2d, f, and g over the crest of the platform, the lowstand (or not-quite-as-high highstand) onlap unit Z2e, as well as the associated weathering and karstification recognized in core, indicate significant periods of emergence between these high-frequency sequences. Although accommodation

FIGURE 9. Rock types associated with highstand systems tracts are characterized by a shallow-marine fauna, low argillaceous content, and high porosity. Larger pores are biomoldic with vugs connected via a network of small-sized interparticle pores.

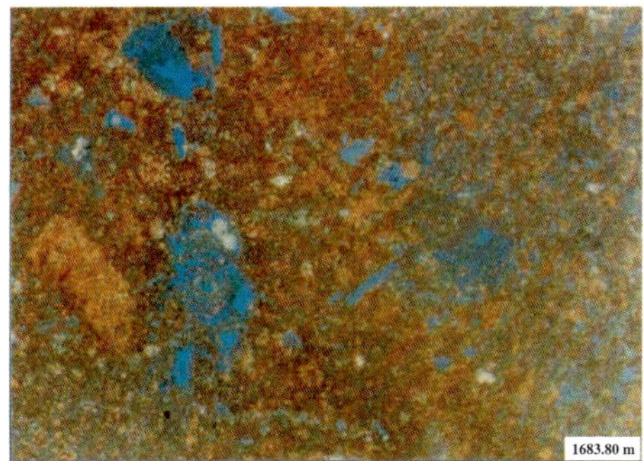

Mottled fine- to medium-grained bioclast packstone with separate vug and micron-size interparticle matrix pores

Mollusk bioclast-peloid packstone with separate vug pores

FIGURE 10. Crossplot of porosity vs. permeability plug data from Jintan-3 divided into nine rock fabrics. Note the log/log axis and the three grain-size-dependent poroperm fields for interparticle porosities of Lucia (1995). The overall low permeability of these rocks with grain sizes predominantly more than 100 μm is related to the microporosity of the matrix, which connects the larger biomoldic vugs. The variety of facies and fabrics recovered from Jintan-3 cores is representative for Jintan. RF-1 and -2 form the high-impedance and low-porosity and -permeability flooding layers (transgressive systems tracts). RF-3 to -9 are characteristic of highstand systems tracts.

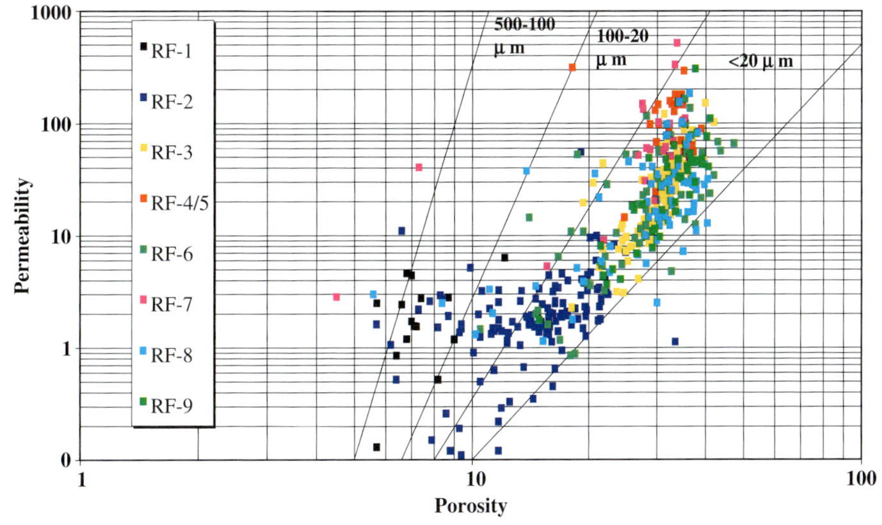

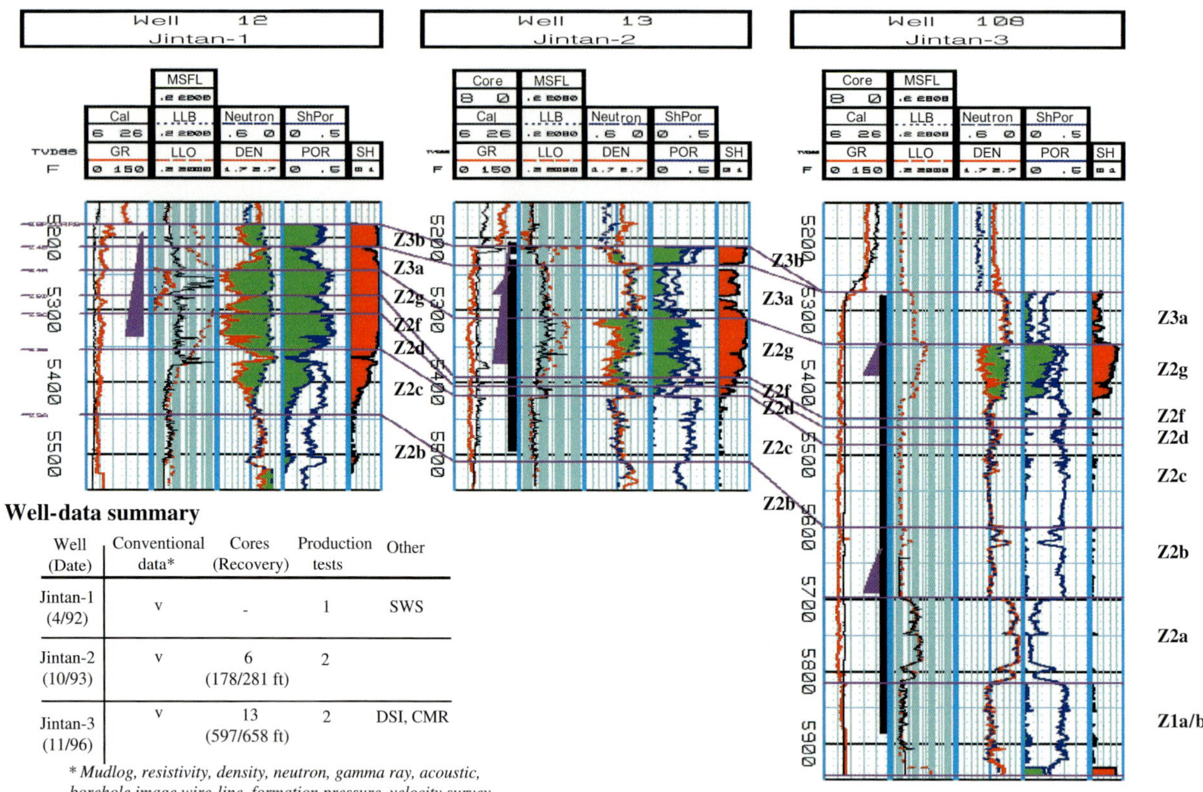

Well-data summary

Well (Date)	Conventional data*	Cores (Recovery)	Production tests	Other
Jintan-1 (4/92)	v	-	1	SWS
Jintan-2 (10/93)	v	6 (178/281 ft)	2	
Jintan-3 (11/96)	v	13 (597/658 ft)	2	DSI, CMR

** Mudlog, resistivity, density, neutron, gamma ray, acoustic, borehole image,wire-line formation pressure, velocity survey*

FIGURE 11. Data overview of Jintan wells 1, 2, and 3. In addition to the suite of logs, cores are available from Jin-2 and -3. Horizons mapped on seismic are correlated between wells. Note dense flooding Zone Z2a drilled and cored in Jin-3, the change in thickness and porosity of the prominent flooding Zone Z3a, and the absence of the high-porosity shallow-water carbonates of Z3b in well Jin-3. SWS = sidewall sample; DSI = dipole shear image log; CMR = combinable magnetic resonance log.

space was sparse on the crest, off-platform progradation (Figure 12A) in a direction oblique or opposite to the prevailing paleowinds (Figure 2) healed a structural and karst-modified graben to the northwest of the platform (Figure 12; see also fault section below).

Reservoir Zone Z2a (Tight Zone/Drowning Unit)

Based on log character and lithofacies, the unit can be divided into lower and upper sections. The lower section consists of a platy coral floatstone/wackestone, with minor shale and an argillaceous, foram-rhodolite, bioclast packstone lithofacies representing a condensed, transgressive, and deepening-upward shoreface platform (Figure 8). The upper section predominantly consists of coarse, coral-lithoclast packstones and breccias. It is interpreted as a lowstand unit of an overlying sequence with platform-derived reef talus and abruptly overlies the lower, deeper water unit. The base-level fall it represents occurs very rapidly (representing a higher-order sequence boundary). Porosity and permeability are modest to low for both units (Table 1). Zone 2a is a pronounced regional marker on seismic and shows high acoustic-impedance values and corresponding low porosity and permeability (Figure 10), as well as an intense overprinting by karstification (see karst section below).

Reservoir Zone Z2b (Open-Marine Platform/Leeward Talus)

This zone was cored only in Jintan-3. The bottom part has a consistent log character and is a fine- to medium-sized bioclastic grainstone with pervasive sand-sized vugs and interparticle porosities. Two tight zones near the top of the unit are tight bioclastic grainstones. The fossil content, lithofacies, and seismic data suggest that this unit represents at the Jintan-3 location a leeward prograding, off-reef sand and talus apron with low-relief clinoforms. It is interpreted to represent a highstand systems tract.

Reservoir Zone Z2c (Open-Marine Platform/Lagoon/ Margin Reef Sand Belt)

This is another highstand systems tract, which was partially cored in Jintan-2 and -3. The cored interval in Jintan-2 consists of bioclastic grainstones and packstones with coral-rich intervals. The correlative interval in Jintan-3 consists of coral floatstone and mud-dominated bioclastic packstones/wackestone. This

Table 1. The subdivision of the reservoir section into depositional sequences and reservoir zones.

Carbonate sequence	*Reservoir zone*	*Description/ Depositional setting*	*Systems tracts*	*Penetrated*	*Cored*
Upper carbonate sequence	3b	Open-marine platform	Highstand	Jin-1 and -2	Jin-2 (partially)
	3a	Deep platform/Drowning or transgressive unit	Transgressive	Jin-1 to -3	Jin-2 and -3
Middle carbonate sequence	2g	Shallow platform and reef sand belt	Highstand	Jin-1 to -3	Jin-2 and -3
	2e		Lowstand or not-so-high highstand	seismic onlap onto crest	
	2d and f	Reef rubble and karsts	Highstand	Jin-1 to -3	Jin-2 and -3
		Stacked unconformities. No space creation on platform crest			
	2c	Open-marine platform lagoon/Reef-margin sand belt	Highstand	Jin-1 to -3	Jin-2 and -3
	2b	Open-marine platform lagoon/Reef-margin sand belt	Highstand	Jin-1 to -3	Jin-3
	2a	Lowstand unit and reef talus	Lowstand	Jin-3	Jin-3
		Deep platform/Drowning or transgressive unit	Transgressive	Jin-3	Jin-3
Lower carbonate sequence	1b	Shallow lagoon	Highstand	Jin-3	Jin-3
	1a	Shallow lagoon(?)	Highstand	Jin-3	

Systems tract interpretation, well penetration, and core availability are also listed.

implies an aggrading platform margin, reef sand belt in Jintan-2, and a mud-dominated fill of a slightly deeper open-marine platform, or restricted shallow lagoon in Jintan-3, which lies more toward the leeward interior of the platform. The core-plug measurements (Figure 10) show a similar distribution of porosity and permeability for both the grain- and mud-dominated fabric.

Reservoir Zones Z2d and Z2f (Reef Rubble and Karsts/Lagoon)

These zones were fully cored in both wells and are relatively thin and limited to a few lithofacies types with high permeability: coral rubble rudstones and grainstones, coral breccia in micrite matrix, and burrowed bioclast packstones, which are interpreted as shallow, near-reef rubble and lagoonal patch-reef debris. The lithofacies appear to have been strongly diagenetically altered and weathered. Deposition occurred probably in a highstand carbonate platform setting with compound subaerial exposure unconformities. The ϕk data show a wide scatter of porosities and permeabilities (RF3 of Figure 10).

Reservoir Zone Z2e (Lowstand or Not-So-High Highstand Onlap)

This unit has not been penetrated in any of the wells and can only be seen on seismic, onlapping the prograding sequences of Z2b–d. It is interpreted to represent shallow-water carbonates deposited during a minor (high-frequency) relative lowstand (or an intermittent minor sea level highstand, which did not reflood the entire platform).

Reservoir Zone Z2g (Shallow Platform/Reef Sand Belt)

This zone was cored in both wells, fully in Jintan-3 and partially in Jintan-2. In Jintan-3, coral floatstone is typical, as well as mottled and burrowed bioclastic packstone. These more mud-dominated lithofacies were probably deposited in a protected, low-energy lagoon setting. The top section is grain-dominated and interpreted to represent a phase of base-level fall and increase in energy, prior to the exposure of the platform. In Jintan-2, most rocks are grain-dominated and were probably deposited near a rimmed reef with sand shoals to the east. Although reservoir quality is overall similar, Jintan-3 shows more scatter in both permeability and porosity, because of the larger variety in lithofacies types compared to the more uniform grainstones of Jintan-2.

Upper Carbonate Sequence

This sequence is bounded at its base by the subaerial unconformity, which caps the middle carbonate sequence at reservoir Zone Z2g. It is overlain by marine

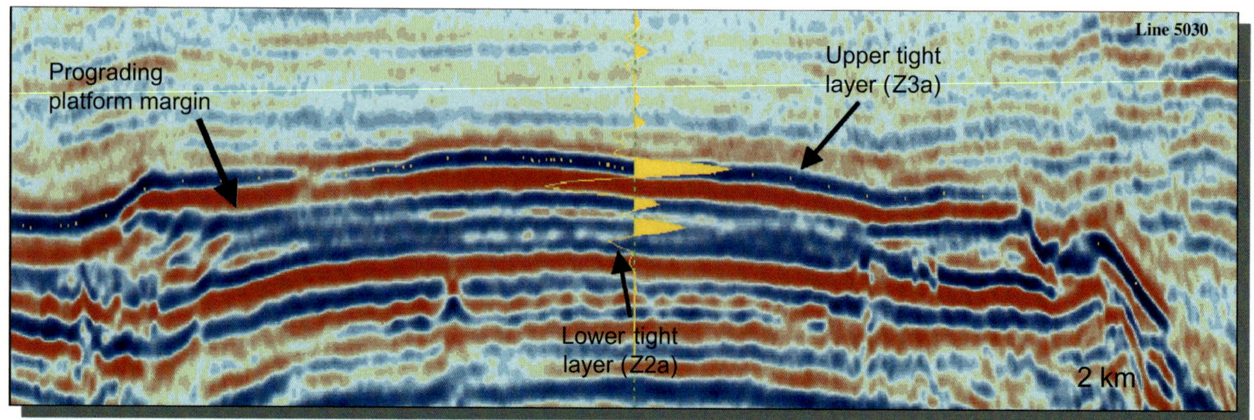

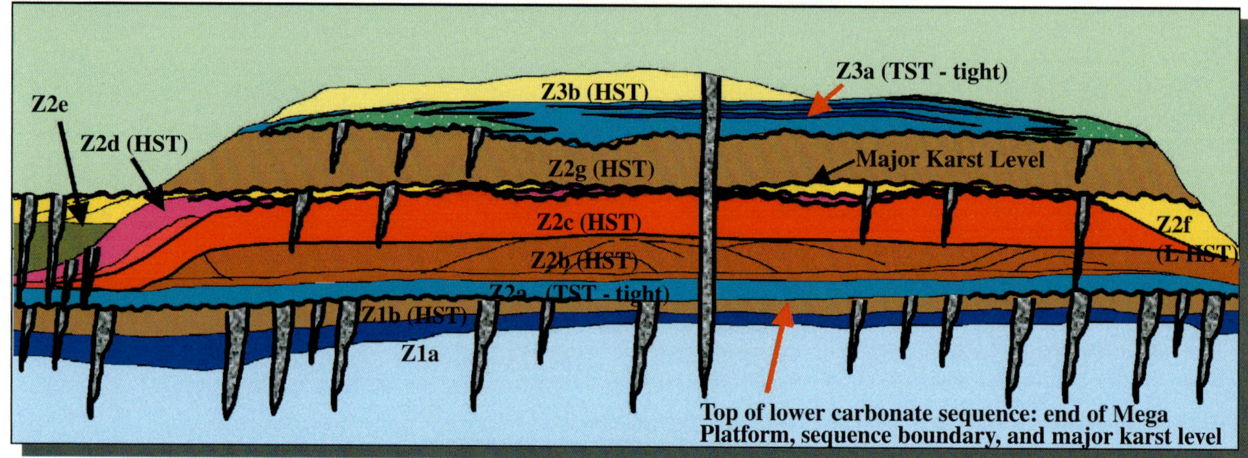

FIGURE 12. Reflection seismic section of the Jintan termination and a line-drawing interpretation based on this and other lines and core and log data. Three major sequences are distinguished by prominent exposure horizons and major flooding zones. A complex pattern of clinoforms in sequence 2 is interpreted as a prograding reef-margin, healing an intraplatform depression created by syndepositional graben faulting. Karst is prominent at many levels with younger karst features often reactivating older ones. HST = highstand systems tracts; TST = transgressive systems tract.

shales, which drape the platform deposits. The thickness of this unit ranges between 65 and 100 ft (30 m). Reservoir Zones Z3a and Z3b represent a complete depositional sequence from a basal unconformity, a transgressive grainstone, and early highstand low-energy open platform (Zone 3a) to porous, high-energy grainstones, deposited in a highstand, open-marine platform environment (Zone 3b).

Reservoir Zone Z3a (Deep Platform/Drowning Unit)

Reservoir Zone 3a was penetrated in all three wells and displays a marked thickening from 32 ft (10 m) in Jintan-1 to 72 ft (22 m) in Jintan-2 and -3 (Figures 11, 12). It consists of mixed lithofacies, which are tight, argillaceous, and mud-dominated in Jintan-3, moderately porous in Jintan-2, and porous and grain-dominated in Jintan-1. Acoustic-impedance data show that this zone is tight in the area around Jintan-3 and a porous reservoir zone to the northeast of Jintan-1 (Figure 7B). This indicates a shallow-platform grain-shoal setting around Jintan-1 and a deep, subtidal depositional environment in the Jintan-3 area. Such a differentiation is further supported by the depositional topography derived from seismic (Figure 13A).

In Jintan-3, the logs indicate a gradual deepening-upward and increased siliciclastic content. Core-plug data of Jintan-3 show the low ϕk character of this zone (Figure 10). The ϕk plot for Jintan-2 shows more scattering and better porosity and permeability because of the presence of a mixture of rock types.

Reservoir Zone Z3b (Open-Marine Platform)

This porous zone, which is characterized on wireline logs by low density and velocity, is not directly expressed on seismic. It is present in Jintan-1 and -2 only, but was not deposited in Jintan-3 because of the final backstepping of the carbonate platform to the northeast. In Jintan-2, where 1 m of core was retrieved, the diverse bioclastic grainstone/packstone composition is interpreted to be indicative of a moderately current-swept open-marine platform. Because it is only about 10 m thick, this unit is not marked by a separate reflector

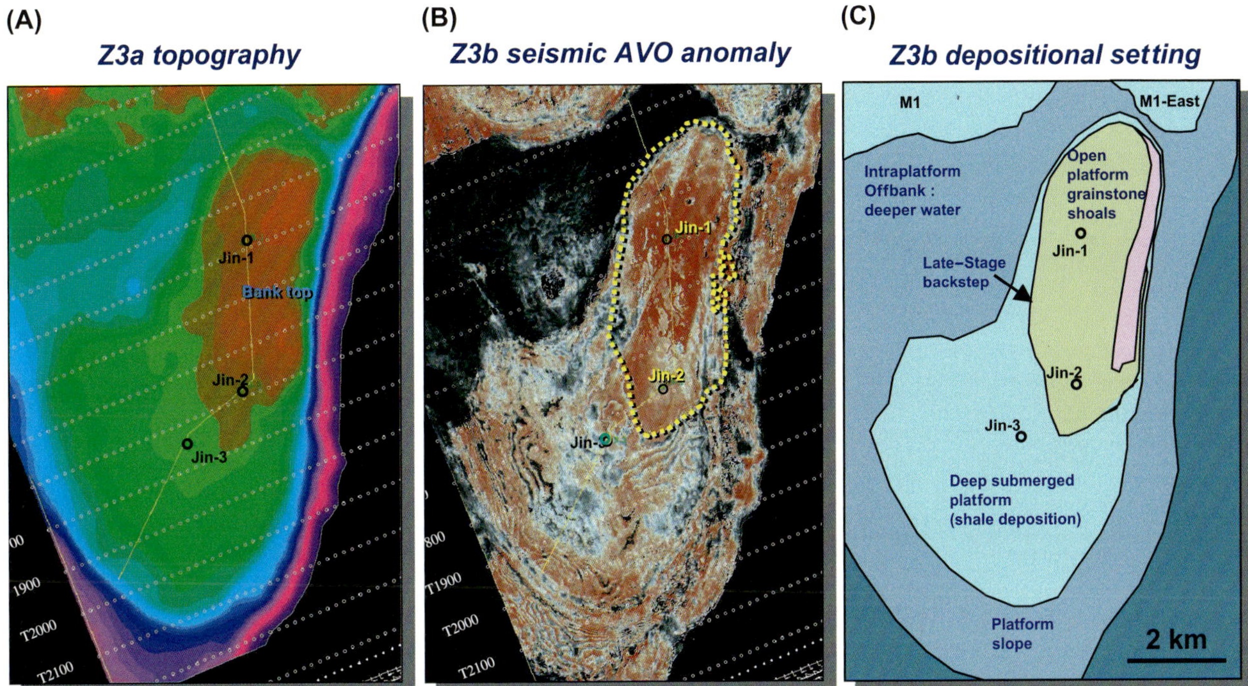

FIGURE 13. The determination of the areal extent of the last Jintan growth phase (Z3b) is guided by the depositional geometry of the underlying flooding Zone Z3a. The Z3a depth map is a reasonable representation of the substratum. It shows a structural high over the Jintan-1 and -2 area with an open platform environment (from Jintan-2 core). Jintan-3 is located in a deeper submerged platform setting. The Z3b amplitude-vs.-offset (AVO) anomaly restricts the lateral extent of the shallow-water grainstone shoals to the Z3a topographic high. Determination of the lateral extent of this zone is important both for reservoir volumetrics (high porosity and saturation in the crestal area of the reservoir) and development options (high permeability above the tight flooding Zone Z3a).

on seismic. Instead, it can be detected by an amplitude anomaly, which produces a brightening over the crest of the field, and amplitude-vs.-offset (AVO) modeling indicates gas-filled reservoir (Figure 13). Its extent corresponds to the high-lying areas defined by the preexisting topography of Z3a and is constrained by the extent of the amplitude anomaly.

Syndepostional Karstification, Faulting, and Depositional Response

Core, Formation MicroImager log, and seismic data all indicate that Jintan and the adjacent M1 and M3 Platforms have experienced several periods of subaerial exposure and karstification. A closer look at the karst features reveals a complicated pattern of karst creation and reactivation, which is, in addition, influenced by syndepositional tectonic deformation of the platform. The visualization of the karst systems is best on seismic, especially horizon-slice semblance maps (Figures 14, 15). These maps are generated by first-processing reflectivity data to vertical semblance data and then resampling it parallel to a flattened reference horizon.

The most extensive case of karstification is found at and below the top of the lower depositional sequence. It covers the whole Mega Platform but has been drilled only in the Jintan-3 well. Seismic reflections near and at the top of the lower depositional sequence (top Z1b) seem to have many disruptions over a thick interval, which on first view or on 2-D seismic could be misinterpreted as "noisy" seismic (Figure 14A). On a semblance slice, however, these features line up, revealing an extensive karst network, which covers most of the Mega Platform and extends deeply into the underlying sequences (Figure 15A). In core, the top of this zone, which is associated with evidence for emergence and extensive karstification, is interpreted as a major sequence boundary. A circular feature several hundred meters in diameter is found at the convergence of the two limbs of the heart-shaped Jintan Platform (Figure 15). On seismic sections, the feature is a deep depression with chaotic reflection patterns. It is interpreted as a large cave that collapsed, possibly during a period of renewed faulting (see below).

Subsequent to the exposure of the lower depositional sequence the platform was reflooded and covered by a 35-m-thick unit of transgressive carbonates of low porosity and permeability (Z2a). However, the hard reflection associated with this transgression again shows lateral discontinuities. A complex dendritic network can be observed on a semblance slice along the transgressive horizon in the off-bank areas west of Jintan with two orthogonal orientations roughly aligned to the

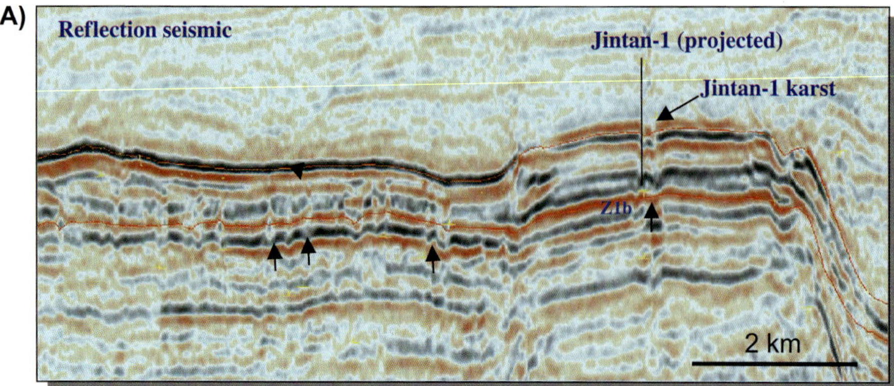

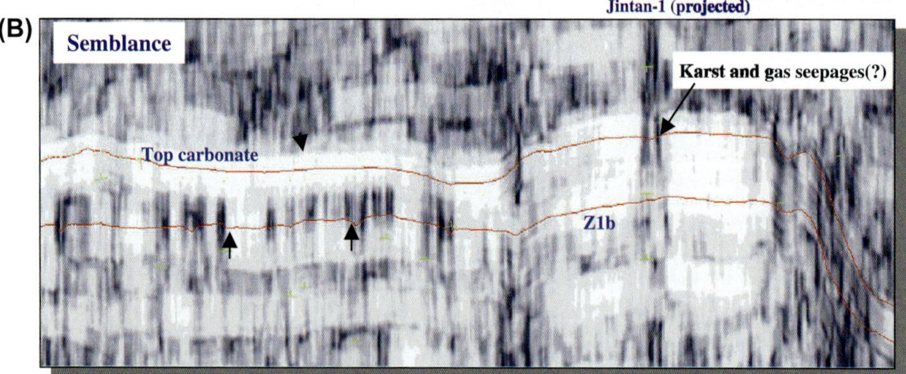

FIGURE 14. (A) Seismic reflection and (B) semblance section of the Jintan termination (see Figure 15 for location). Interruptions of the reflection lines are interpreted as a combination of faulting/fracturing and karsting. The dark streaks on the semblance slice show maximum seismic discontinuity vertically reaching up to the top of Zone 2. Most karst is associated with the top Zones Z1b and Z2a in the area between Jintan and M1. However, some features reach top carbonate in the center of the Jintan termination.

regional fault pattern. On closer analysis, the seismic disruptions extend upward into younger off-bank deposits of the middle depositional sequence (Figure 14). We interpret this as evidence that karstification of the transgressive unit occurred during repeated exposure in the later stages of the deposition of the middle sequence. Interestingly, the horizon is seismically less disturbed over the crest of Jintan despite evidence from core of extensive karstification associated with exposure of units Z2c–g in the crestal parts of the platform. The western extent of the Jintan Platform appears to have acted as a boundary separating different karst domains.

After deposition of the transgressive unit Z2a, a severely faulted wedge-shaped graben formed along the north-northeast–south-southwest and northwest-southeast strike of the old regional fault pattern to the northwest of the Jintan (Figure 16). The timing of this syndepostional structural deformation can be constrained by the constant thickness of Zone Z2a and the progradation of the overlying units into the newly created accommodation space. We also relate the collapse of the large cave in unit Z1b to this structural deformation episode. The additional accommodation space created by the extensional faulting was filled by the prograding sequence of the middle depositional sequence (Z2B–Z2G).

A large north-northeast–south-southwest–trending karst system is located directly at and to the northeast of well Jintan-1, where it can be traced off-bank into the spillpoint area between Jintan and M1 (Figure 15). This system extends upward to the top of the platform. Seismic reveals gas-chimney effects above this location indicating that the karst influenced the seal integrity possibly caused by the collapse after seal emplacement. In Jintan-1, permeabilities from well tests were estimated at or even higher than 1000 md. These high permeabilities are likely a consequence of the proximity to the karst system.

Bank-Margin Collapse

The eastern margin of the Jintan termination is a steep constructional platform edge with topography of several hundred meters. The serrate outline of the margin indicates repeated and spectacular bank-margin collapse (Figure 15). It is likely that the platform margin failed and that large chunks collapsed into the abyss. Seismic reveals a complex pattern of collapse with some fault blocks being wedged and thrown up, now forming elevated rim pieces (Figure 5). This reversal is likely the result of later compression during compaction of the overlying siliciclastics.

INTERPRETATION

Platform Growth and Demise

The balance among a rapidly subsiding depositional system, eustatic sea level fluctuations, and significant

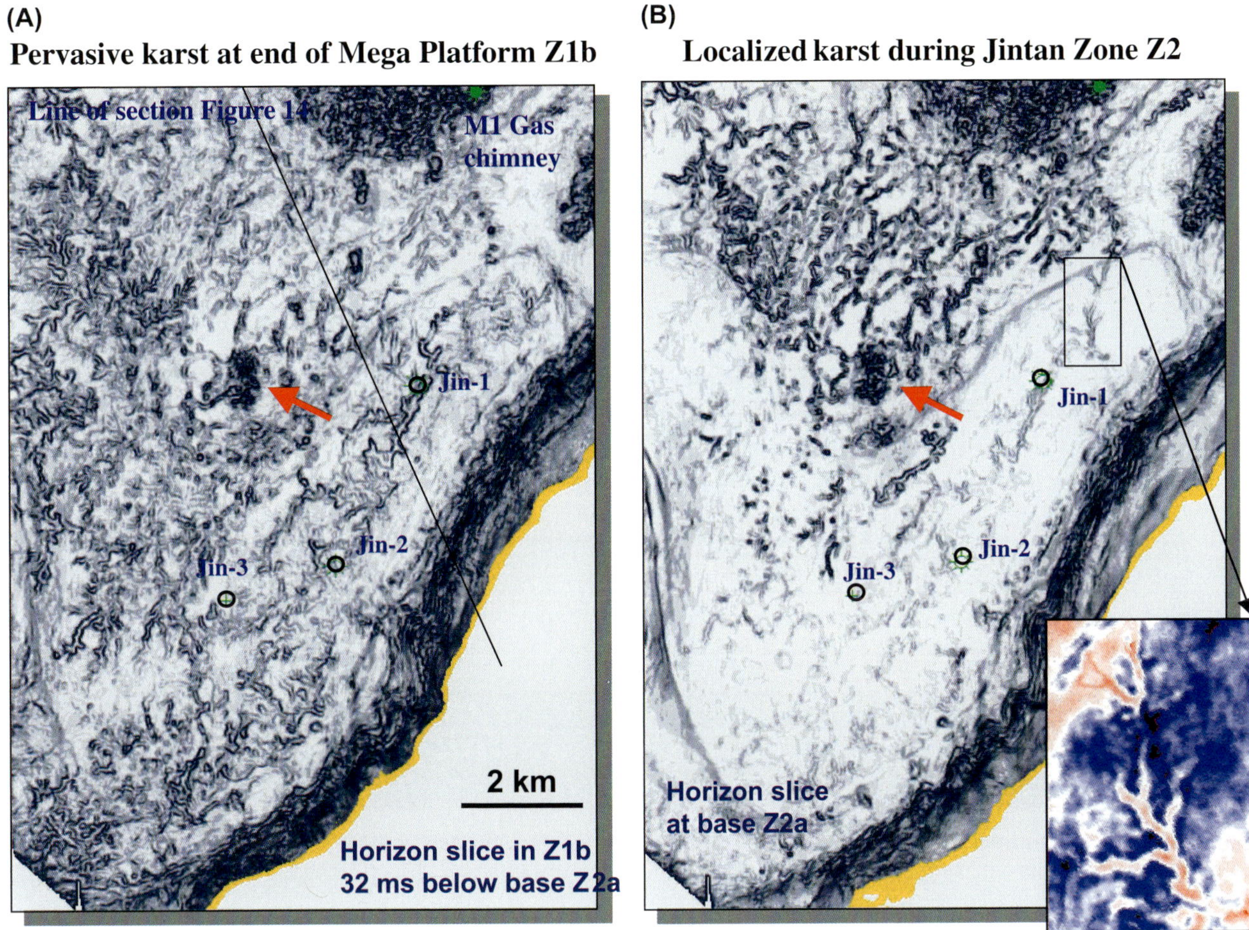

FIGURE 15. Horizon semblance maps of Z1b and Z2a showing two episodes of karsting. (A) Karstification of Zone Z1b is Mega Platform wide and pervasive. (B) Karstification of Z2a is more localized in areas that have been affected by syndepositional faulting during platform stage Z2b. A dark blotch (red arrow) at the convergence of the two Jintan platform branches is interpreted to represent a collapsed cave in Zone 1a/b. Karst patterns are dendritic in places, in others resemble drainage patterns (close-up). Black line in left image shows the section of Figure 14.

carbonate growth has created a 1200-m-thick carbonate sequence over a comparatively short time period of 6–8 m.y. This plots at the upper limit of growth potential for comparative sedimentary systems (Schlager, 1999), indicating that the demise of the Mega Platform may have been a story of failed keep-up of the carbonate system with rapid creation of accommodation space. This would place the prominent backsteps seen on the Mega Platform and elsewhere in the Luconia Province into the context of insufficient time during periods of shallow-water deposition to fill accommodation place over the platform. Periods of exposure, and more importantly, flooding below a depth of rapid sediment accumulation increased until the platforms finally drowned. The hiatus between the top carbonate at the end of the middle Miocene and burial under prodeltaic clastics in the early Pliocene was spent essentially under water—certainly below the depth of carbonate deposition, but more likely at several hundred meters depth based on foreset geometries in the overlying siliciclastics (see Enclosure 1 in Figure 1). The absence of pelagic sediments covering the platforms is enigmatic. However, ocean currents are capable of preventing deposition on or eroding deep-marine deposits from the tops of drowned platforms (Grötsch et al., 1993).

An alternative possibility is to put the backstepping of the platforms into the context of an overall decreasing accommodation space toward the end of the middle Miocene. In this scenario, deposition does not have the opportunity to recapture the full area of the platforms because the time from exposure to reflooding to renewed exposure is too short to start the carbonate factory and fill all of the accommodation space. Thus, overall decreasing accommodation space and extended periods of exposure led to backstepping and platform demise. In this case, a significant duration of the hiatus between the top carbonate and the overlying siliciclastics is spent exposed prior to a much-later drowning event. This scenario is supported by the multiple events of extensive karstification documented below and at

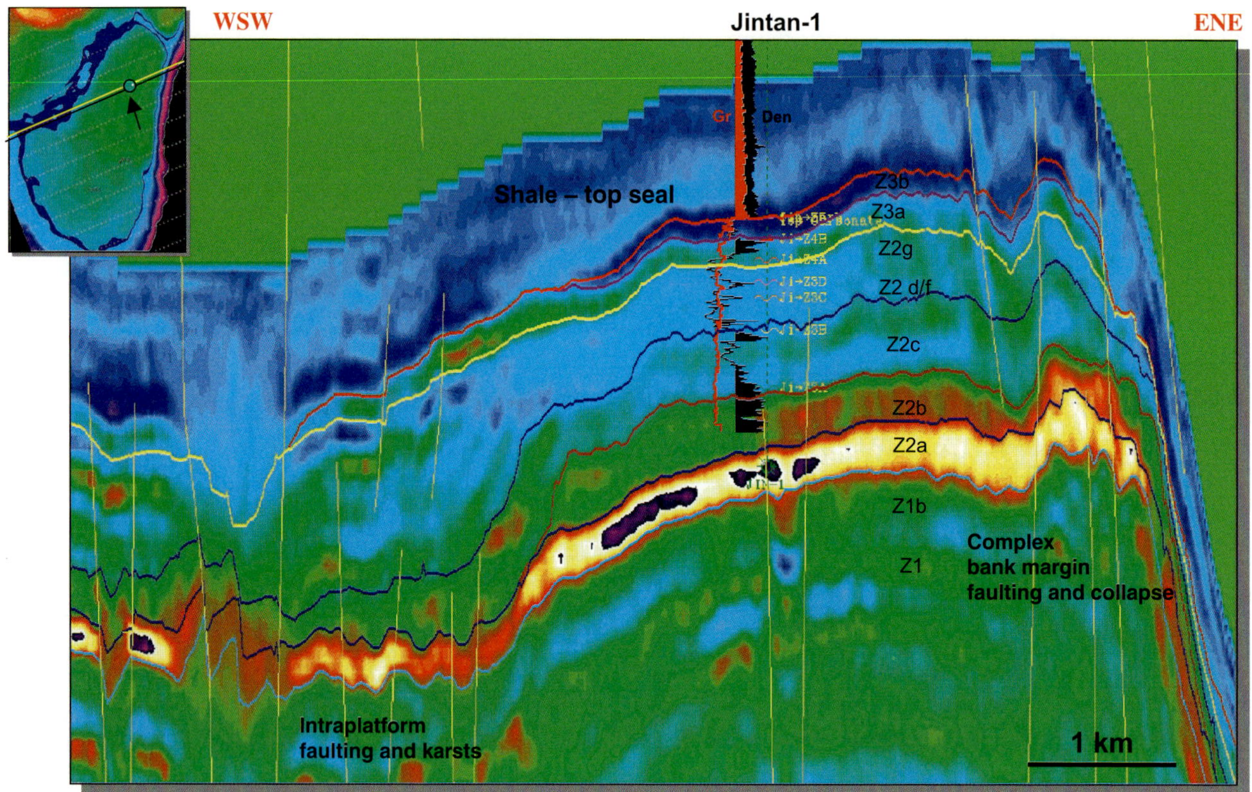

FIGURE 16. Acoustic-impedance section through the Jintan showing intense faulting both in the intraplatform area and at the platform margin. Yellow/red = relatively high acoustic impedance = low porosity; green/blue = relatively low acoustic impedance = high porosity.

the top of the Mega Platform and other carbonate platforms of the province (Figure 2; Vahrenkamp, 1998). Marine shales of Pliocene age have been found washed into vugs 100 m below top carbonate in the center of the M1 termination. The present-day pore water composition of the platforms is brackish (21,000 ppm dissolved solids) indicating that at least some flushing of the platforms from original seawater has occurred. Diagenetic features of the limestone and particularly porosity creation have been associated with meteoric diagenesis (Epting, 1980). However, in view of similar solution and precipitation features described from marine environments (Schlager and James, 1978; Melim et al., 1995), these arguments may be equivocal.

Resolving this issue will require a better database, a higher time resolution of the Sr isotope stratigraphy or any other dating technique, and a more detailed investigation of the subsidence history of the Luconia province.

Diagenesis

Spectacular karst features can be observed in the 3-D seismic of the Mega Platform, even evidence consistent with the collapse of a large cave several hundred meters in diameter (Figures 14, 15). However, the most prominent feature is the dendritic pattern recognized in the flooding layer Z2a (Figure 15). Its formation is likely closely related to faulting and associated fracturing of the rocks, explaining why it is best developed in the tight and brittle transgressive unit. During platform exposure, meteoric waters were channeled into the fractures and fissures of the otherwise tight layer. Eventually, the fracture network was widened to a scale that is recognizable on the resolution of seismic data. Orientation of dissolution along structural trends is a common feature documented in many cave maps of the karst literature (Palmer, 1995). However, the displays from the Mega Platform show that it can also be observed on 3-D seismic. Similar patterns have been observed on 3-D seismic in other ancient carbonate platforms as well (e.g., Droste and Van Steenwinkel, 2004, their figure 15 of the Cretaceous Natih Platform of Oman).

Another related feature is the reactivation of already existing karst during repeated exposure events. The pervasive karst associated with the top of the lower depositional sequence seems to have grown upward during karstification associated with exposure periods of the middle sequence (Zone 2c–f). Similar upward growth (or reactivation of existing karst features during renewed exposure) has been described regarding the shallow subsurface of Florida (Evans et al., 1994). In Jintan, however, the upward growth is localized to those areas also disturbed during the fracturing event described above. Over the platform proper, where Zone Z2a is apparently not pervasively fractured, the karst could not grow upward except in a few localized fault zones (Figure 15). We speculate that this may also be related to the platform area being covered by thick sequences of grainy deposits that were (a) less brittle and thus did not provide a connected fracture pattern for karst channeling and (b) had a more even pore distribution to start out with. As meteoric waters percolated through the grainy sediments unchanneled, they may have created a more regular dissolution pattern with a homogeneous porosity and permeability enhancement.

Bank-margin collapse is a common feature in carbonate platforms, with scars having been described as scalloped margins in the Bahamas (Mullins and Hine, 1989). However, the complicated pattern of the Jintan margin shows that upward thrusting of smaller fault slices could also be a feature of such a collapse (if it is not related to later compactional faulting of the overburden). This introduces fault planes and layer discontinuities into reservoir sequences, which may be of significance for hydrocarbon production (e.g., water breakthrough along fault planes) and offset of low permeability layers.

Application of Results for Reservoir Management

The assessment of reserves volumes in the gas business is a prerequisite for securing long-term sales contracts and economic planning of reserves development. In an expensive offshore environment with sparse well data, reliable and quantitative information of the interwell geology and reservoir properties is essential for economic success. For the reservoirs of the Mega Platform in general and Jintan in particular, the results of the seismic, core, and log evaluation can be used to quantify reservoir properties, predict reservoir behavior, and optimize development. The export of porosity cubes from impedance data into 3-D reservoir modeling and simulation programs at a resolution that preserves detailed geologic features is currently a routine workflow. In addition to the determination of hydrocarbon volumes of the seven Mega Platform reservoirs, it is of particular interest to assess the shared pressure regime via the common aquifer and the potential influx of water into the reservoirs during depletion. To show the impact and value of the 3-D seismic data on assessing these issues, they are discussed briefly below:

1) In the Jintan reservoir, the predicted influx of water from the underlying aquifer is greatly reduced by the thick low-permeability transgressive layer Z2a, which shields much of the reservoir area from the underlying aquifer. The continuity of this zone, penetrated only in one well (Jintan-3), can be deduced from the depositional model (platform-wide major transgression) and is clearly visible on seismic (Figures 6, 7, 17). However, faulting and karstification of this layer in certain areas below the reservoir and in the off-platform area to the northwest of Jintan suggest that connectivity exists to the large aquifer across this zone. High-porosity and -permeability clinoforms dip from the platform into the structural graben and the karst field to the northwest of Jintan (Figures 7, 17), providing pathways for the advance of water from the faulted and karsted flank onto the crest of the reservoir. Recognition of these features greatly aides in controlling uncertainty during the simulation of fluid flow through the reservoir and helps in deciding how to develop the reservoir, where to place the offshore platform, and where (and where not) to drill development wells.
2) The interpolation of porosity and hence volume data over a large area, such as the Jintan reservoir, based on sparse well data is risky. The incorporation of quantitative seismic data significantly reduces volumetric uncertainty. A perfect example is the porosity prediction for the transgressive layer Z3a. This layer is penetrated by all three wells and contains rocks of low porosity and permeability (Figure 11). However, although lateral variations in porosity and layer thickness are already significant based on the well data, the full impact of the lateral variation only becomes quantifiable using the seismic data (Figure 18). In this case, the combination of seismically derived high porosity in the crestal area of the field and porosity-dependent saturation-height functions has led to a significant increase in the hydrocarbon volume estimates.
3) Seven gas reservoirs are found on the Mega Platform sharing the same aquifer and hence the same pressure regime. Because not all reservoirs will be developed at the same time, the early production of gas from one reservoir will reduce the system pressure and hence change the reservoir condition of the other, undeveloped reservoirs. The magnitude of this influence depends on the connectivity in the water leg between the platforms. For example, decrease of reservoir pressure will lead to the expansion of gas and, in the case of reservoirs filled to spillpoint, this gas may escape the structure and be lost for recovery. Recognition of the large-scale architecture of the platform and the reservoir properties associated with the individual layers allows construction of large 3-D models to quantify these effects. In the case of the Mega Platform, the platform-wide occurrence of low-porosity

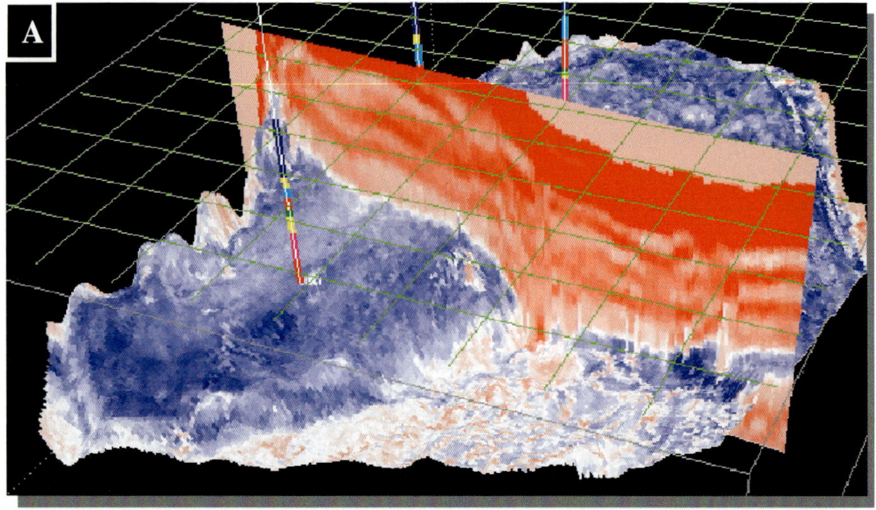

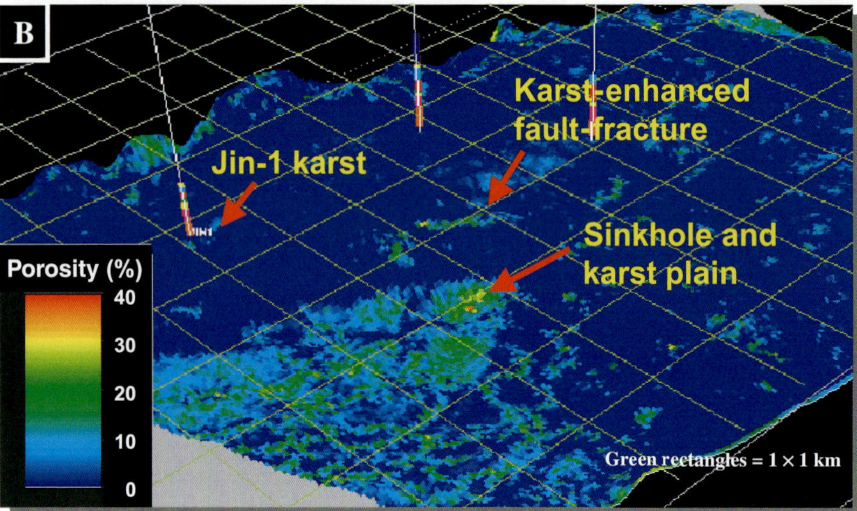

FIGURE 17. (A) Horizon and section view of a 3-D impedance cube imported into a 3-D static reservoir model (GEOCAP software) showing the tight flooding Zone Z2a over the crest of the field (blue) and in the downfaulted and karsted off-platform area (red and blue colors). The three Jintan wells are also shown with the color coding corresponding to the overlying reservoir zones. Note the off-platform clinoforms on the section image dipping toward the karsted lowlands. (B) Equivalent layer of a 3-D porosity model derived from seismic inversion. The porosity model shows exceptional conservation of the karst and fault features, which breach the tight flooding layer with porosities above 20%. This is likely to significantly impact water encroachment across the tight layer from the underlying aquifer.

and -permeability transgressive layers significantly reduces communication between the reservoirs. However, in the case of Jintan, which is filled to spillpoint, pressure depletion has been measured in a well only a few years after production commenced in the nearby M1 reservoir. Using the architecture derived from 3-D seismic captured in a 3-D static reservoir model, it was shown in dynamic simulations that the volumes escaping the structures are insignificant to warrant an acceleration of Jintan reservoir development.

CONCLUSIONS

1) The middle Miocene Mega Platform was initiated together with many other platforms of the Luconia Province on regionally oriented structural highs during the late early Miocene. It grew to a thickness of 1200 m and its growth terminated in several backsteps at the end of the middle Miocene.
2) Principle architectural elements are tight argillaceous limestones, which formed during flooding and deep submergence, and porous clean limestones deposited in several environments during sea-level highstands.
3) The Jintan termination, one of the backsteps, preserved at least two (third-order?) late middle Miocene depositional sequences after the breakup of the Mega Platform. After periods of transgression and the deposition of flooding units, the platform first aggraded and then prograded in several higher-order sequences interrupted by several periods of subaerial exposure.
4) Repeated exposure, syndepositional faulting, and fracturing provided the prerequisites for a spectacular multistory karst system with potential significant impact on hydrocarbon recovery during reservoir development.

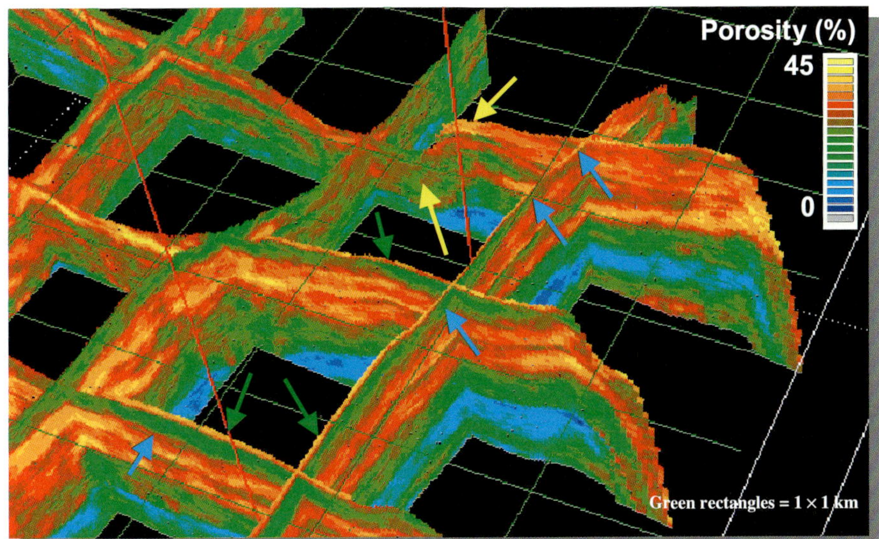

FIGURE 18. Fence diagram of a 3-D porosity model derived from a layer-based stochastic seismic inversion using the Jason Workbench. The porosity cube is imported into a 3-D static modeling program (GEOCAP) for further manipulation (e.g., facies-dependent porosity/permeability transfer) before a seamless export to dynamic simulation. This model is superior to a porosity model based on well data only (kriged or interpolated porosity cubes). The seismically derived model faithfully preserves (a) the pinch-out of the tight flooding Zone Z3a (blue arrows) in the crest of the field, with an impact on volumetrics of an additional 0.5 tscf gas initially in place; (b) the clinoforms dipping into the off-platform area where the continuity of tight flooding Zone Z2a is interrupted by severe karsting (yellow arrows, see also Figure 17); dynamic simulation shows a potential significant impact on the encroachment of water during future production; (c) the correct placement of the thin high-porosity final growth layer Zone Z3b (green arrows). This layer could be an ideal target for optimum placement of high-capacity horizontal producers.

5) Reflection and impedance 3-D seismic play a key role in recognizing platform architecture, structural disturbances, and diagenesis. The seismic data allow quantification of these features in terms of pore volume and pore-fluid flow in static and dynamic reservoir models and have potentially significant economic impact.

ACKNOWLEDGMENTS

We wish to express our sincere thanks to Sarawak Shell Berhad, Nippon Oil Exploration Limited, Petronas Carigali Berhad, and Petroliam Nasional Berhad Malaysia for permission to publish this chapter. This chapter benefited substantially from the knowledge and experience of many Shell geologists and petroleum engineers that have worked over the last 30 years on Luconia geology. We would like to particularly acknowledge Christian Hoecker, who made the first images of the Jintan Karst, Liew Shiew Ling, Ceri Powell, Anyi Ngau, Gordon Taylor, Ante Frens, and Simon Roddy. Mark Sams of Jason Geosystems did the seismic inversion. His enthusiasm and in-depth technical expertise was a most valuable contribution to our work. We especially thank AAPG reviewers Rick Sarg and John Snedden for their insight and constructive criticism.

REFERENCES CITED

Abdullah, M., and C. F. Jordan, 1987, The geology of Arun field, Miocene reef complex: Proceedings of the Indonesian Petroleum Association, 16th Annual Convention, p. 65–96.

Aigner, T., M. Doyle, D. Lawrence, M. Epting, and A. Van Vliet, 1989, Quantitative modeling of carbonate platforms: Some examples, *in* P. D. Crevello, J. L. Wilson, J. F. Sarg, and J. F. Read, eds., Controls on carbonate platform and basin development: SEPM Special Publication 44, p. 27–37.

Droste, H., and M. Van Steenwinkel, 2004, Stratal geometries and patterns of platform carbonates: The Cretaceous of Oman, *in* G. P. Eberli, J. L. Masaferro, and J. F. Sarg, eds., Seismic imaging of carbonate reservoirs and systems: AAPG Memoir 81, p. 185–206.

Epting, M., 1980, Sedimentology of Miocene carbonate build-ups, Central Luconia, offshore Sarawak: Geological Society of Malaysia Bulletin, v. 12, p. 17–30.

Epting, M., 1989, Miocene carbonate build-ups of central Luconia, offshore Sarawak, *in* A. W. Bally, ed., Atlas of seismic stratigraphy: AAPG Studies in Geology No. 27, p. 168–173.

Ehrlich, R. N., A. P. Longo Jr., and S. Hyare, 1993, Response of carbonate platform margins to drowning: Evidence of environmental collapse, *in* R. G. Loucks and J. F. Sarg, eds., Carbonate sequence stratigraphy: AAPG Memoir 57, p. 241–266.

Evans, M. W., S. W. Snyder, and A. C. Hine, 1994, High-resolution seismic expression of karst evolution within the Upper Aquifer system: Crooked Lake, Polk County, Florida: Journal of Sedimentary Research, v. B64, p. 232–244.

Fulthorpe, C. S., and Seymor O. Schlanger, 1989, Paleo-oceanographic and tectonic settings of early Miocene reefs and associated carbonates of offshore southeast Asia: AAPG Bulletin, v. 73, p. 729–756.

Grötsch, J., and C. Mercadier, 1999, Integrated 3-D reservoir modeling based on 3-D Seismic: The Tertiary Malampaya and Camago buildups, offshore Palawan, Philippines: AAPG Bulletin, v. 83, p. 1703–1728.

Grötsch, J., R. Schroeder, S. Noé, and E. Flügel, 1993, Carbonate platforms as recorders of high amplitude eustatic sea level fluctuations: The late Albian appeninica event: Basin Research, v. 5, p. 197–212.

Gucci, M. A., and M. H. Clark, 1993, Sequence stratigraphy of a Miocene carbonate buildup, Java Sea, *in* R. G. Loucks and J. F. Sarg, eds., Carbonate sequence stratigraphy: AAPG Memoir 57, p. 291–304.

Haq, B. U., J. Hardenbol, and P. R. Vail, 1988, Mesozoic and Cenozoic chronostratigraphy and eustatic cycles, *in* C. Wilgus, B. Hastings, C. Kendall, H. Posamentier, C. Ross, and J. van Wagoner, eds., Sea level changes: An integrated approach: SEPM Special Publications 42, p. 71–108.

Ho, K. F., 1978, Stratigraphic framework for oil exploration in Sarawak: Geological Society of Malaysia Bulletin, v. 10, p. 1–13.

Jordan, C. F., and M. Abdullah, 1992, The Arun field—Indonesia, North Sumatra Basin, Sumatra, *in* A. F. Beaumont and N. H. Foster, comps., Stratigraphic traps III: AAPG Treatise of Petroleum Geology, Atlas of Oil and Gas Fields, p. 1–39.

Longman, M. W., R. J. Maxwell, A. D. M. Mason, and L. R. Beddoes Jr., 1987, Characteristics of a Miocene intrabank channel in Batu Raja Limestone, Ramba field, south Sumatra, Indonesia: AAPG Bulletin, v. 71, p. 1261–1273.

Lucia, J. F., 1995, Rock fabric/petrophysical classification of carbonate pore space for reservoir characterization: AAPG Bulletin, v. 79, p. 1275–1300.

Maliki, M. A., and S. Soenarawi, 1991, South Lho Sukon-D1 discovery, north Sumatra: Proceedings of the Indonesian Petroleum Association, 20th Annual Convention, p. 235–254.

May, J. A., and D. R. Eyles, 1985, Well log and seismic character of Tertiary Terumbu carbonate, South China Sea, Indonesia: AAPG Bulletin, v. 69, p. 1339–1358.

Melim, L. A., P. K. Swart, and R. G. Maliva, 1995, Meteoric-like fabrics forming in marine waters: Implications for the use of petrography to identify diagenetic environments: Geology, v. 23, p. 755–758.

Moldovanje, E. P., H. C. Tanner, and J. Y. Zhang, 1995, Regional exposure events and platform evolution of Zhujiang Formation carbonates, Pearl River Mouth Basin: Evidence from primary and diagenetic seismic facies, *in* D. A. Budd, A. H. Saller, and P. M. Harris, eds., Unconformities and porosity in carbonate strata: AAPG Memoir 63, p. 125–140.

Mullins, H. T., and A. C. Hine, 1989, Scalloped bank margins: The beginning of the end of a carbonate platform?: Geology, v. 17, no. 1, p. 30–33.

Palmer, A. N., 1995, Geochemical models for the origin of macroscopic solution porosity in carbonate rocks, *in* D. A. Budd, A. H. Saller, and P. M. Harris, eds., Unconformities and porosity in carbonate strata: AAPG Memoir 63, p. 77–102.

Ru, K., and J. D. Pigott, 1986, Episodic rifting and subsidence in the South China Sea: AAPG Bulletin, v. 70, p. 1136–1155.

Rudolph, K. W., and P. J. Lehmann, 1989, Platform evolution and sequence stratigraphy of the Natuna Platform, South China; *in* P. D. Crevello, J. L. Wilson, J. F. Sarg, and J. F. Read, eds., Controls on carbonate platform and basin development: SEPM Special Publication 44, p. 353–361.

Saller, A., R. Armin, L. O. Ichram, and C. Glenn-Sullivan, 1993, Sequence stratigraphy of aggrading and backstepping carbonate shelves, Oligocene, central Kalimantan, Indonesia, *in* R. G. Loucks and J. F. Sarg, eds., Carbonate sequence stratigraphy: AAPG Memoir 57, p. 267–290.

Schlager, W., 1999, Scaling of sedimentation rates and drowning of reefs and carbonate platforms: Geology, v. 27, no. 2, p. 183–186.

Schlager, W., and N. P. James, 1978, Low-Mg calcite limestones forming at the deep-sea floor, Tongue of the Ocean, Bahamas: Sedimentology, v. 25, p. 675–702.

Sulaiman, M., 1995, Cyclic carbonate deposition, facies succession, and diagenesis of a Central Luconia build-up, offshore Sarawak: M.S. thesis, University of Brunei Darussalam, 73 p.

Sun, S. Q., and M. Esteban, 1994, Paleoclimatic controls on sedimentation, diagenesis, and reservoir quality: Lessons from Miocene carbonates: AAPG Bulletin, v. 78, p. 519–543.

Vahrenkamp, 1998, Sr-isotope stratigraphy of Miocene carbonates, Luconia Province, Sarawak, Malaysia: Implications for platform growth and demise and regional reservoir behavior (abs.): AAPG Annual Meeting Program.

Vahrenkamp, V. C., Y. Kamari, and S. A. Rahman, 1996, Three dimensional geological model and multiple scenario volumetrics of the F23 Miocene carbonate build-up, Luconia Province, offshore Sarawak: Proceedings of the Petronas Research and Technology Forum, Genting Highlands, Malaysia.

Withjack, E. M., 1985, Analysis of naturally fractured reservoirs with bottomwater drive: Nido A and B fields, offshore northwest Palawan, Philippines: Journal of Petroleum Technology: SPE Paper 12019, p. 1481–1490.

Yaman, F., T. Ambismir, and T. Bukhari, 1991, Gas exploration in Parigi and Pre-Parigi carbonate buildups, NW Java Sea: Proceedings Indonesian Petroleum Association, 20th Annual Convention, IPA 91-11.20, p. 319–346.

16

Bracco Gartner, G. L., W. Schlager, and E. W. Adams, 2004, Seismic expression of the boundaries of a Miocene carbonate platform, Sarawak, Malaysia, *in* Seismic imaging of carbonate reservoirs and systems: AAPG Memoir 81, p. 351–365.

Seismic Expression of the Boundaries of a Miocene Carbonate Platform, Sarawak, Malaysia

Guido L. Bracco Gartner[1]
Vrije Universiteit, Amsterdam, The Netherlands

Wolfgang Schlager
Vrije Universiteit, Amsterdam, The Netherlands

Erwin W. Adams[2]
Vrije Universiteit, Amsterdam, The Netherlands

ABSTRACT

The origin of seismic reflections in slope deposits of a Miocene carbonate platform, offshore Sarawak, was studied using cores, well-log data, and two-dimensional seismic. This isolated carbonate platform has slope angles ranging from 2 to 25°. Our interpretation of the seismic data is that the asymmetric and high-rising platform (250–300 m relief) has different stratigraphic character for the southern and northern flanks. The southern slope was characterized by bypass or erosion throughout the aggrading phase of platform development. It was subsequently buried by shale with downbending, onlapping beds that indicate terrigenous sediment transport from the south. An alternative is folding during tectonic deformation. On the northern flank, the shale already started to pile up during platform aggradation. Phases of erosional or bypass conditions were short and alternated with two phases formed when platform debris interfingered with surrounding shale. Shale intercalations can be recognized seismically by negative reflections that quickly lose amplitude away from the platform. Although the overall shape of the platform is probably related to an older structural pattern of the Luconia Province, the asymmetry of the platform architecture and the distribution of sediments are most likely the results of paleowinds.

[1]*Present address:* Shell International E & P, Rijswijk, The Netherlands.
[2]*Present address:* Massachusetts Institute of Technology, Cambridge, Massachusetts, U.S.A.

INTRODUCTION

Delimiting the precise nature of carbonate platforms in seismic data is not an easy task because of limited resolution and interference patterns. For example, it can be hard to discriminate between onlap and lithologic interfingering. Facies transitions and genuine unconformities on the flanks of carbonate platforms encased in shale have produced particularly distinct pseudo-unconformities in seismic models of outcrops (Rudolph et al., 1989; Schlager et al., 1991; Biddle et al., 1992; Stafleu and Schlager, 1993). These models demonstrated how geometric unconformities could be generated in the process of transforming stratal relationships into the seismic image. From a reservoir point of view, misinterpreting a pseudo-unconformity may have considerable practical implications. Most important in this respect is the question whether the observed seismic structure has sealing properties. Onlap of shale on carbonates is likely to produce a better reservoir seal than lateral interfingering of carbonates and shales, with numerous tongues of potentially porous carbonates extending into the shale. Second, a false lapout pattern may severely influence volume predictions of hydrocarbon reserves. With regard to sedimentology and stratigraphy, pseudo-unconformities mask true stratal relations and genesis.

This study focuses on the geometry of carbonate platform flanks and their seismic architecture. The key question is: Where do the carbonate flank deposits interfinger with the surrounding shale, and where are they simply onlapped by it? The concept of accretional, bypass, and erosional slopes is highly relevant in this context. Carbonate platform slopes tend to steepen as they grow higher. This increase in slope angle leads to a change in the depositional regime from accretion to bypass and finally to erosion (Mullins and Neumann, 1979; Schlager and Chermak, 1979; Schlager and Camber, 1986). An important difference between accretional slopes and bypass or erosional slopes is that on accretion-type slopes, porous debris tongues, shed from the platform, are physically connected with the platform. On bypass and erosional slopes, they are physically separated from the main platform body and form lenses at the toe of slope. Criteria used to distinguish bypass or erosional slopes from accretional slope regimes are the continuity of the reflections within the platform edifice, but also the geometry of the surrounding shale.

The contrasting models of interfingering vs. onlap boundaries and accretion vs. erosion are tested at a Miocene carbonate platform in the Central Luconia Province, offshore Sarawak (herein referred to as the *EX Platform*). The studied platform is an isolated carbonate platform, i.e., detached from the island shelf, with slope angles ranging from 2° to 25°, and is entirely encased by shale. Epting (1980) first described this platform and interpreted the flanks as constructed by numerous tongues of carbonate debris ("carbonate stringers") embedded in basinal shales.

In the context of this study, we worked on the lithostratigraphy of two boreholes and tied seismic data to log borehole stratigraphy, petrophysics, and synthetic seismic traces. Finally, we analyzed the patterns of seismic reflections and complex-trace attributes at the carbonate-shale facies transition on the flanks of the platform.

Geologic Setting and Platform Development

The Central Luconia Province is situated in the South China Sea. It is bounded by the extensional South China Basin to the north and the compressional Balingian Province to the south (Epting, 1980). Seafloor spreading in the South China Basin during the Oligocene to middle Miocene affected the continental crust to the south, which resulted in the formation of a southwest-northeast–trending horst-graben system (Epting, 1989). These elevated blocks initiated the growth of numerous carbonate platforms (Figure 1). Carbonate deposition started in the early Miocene, but was most prolific during the middle to late Miocene. Successive stages of progradation, aggradation, and retrogradation governed the overall architecture of the platforms. According to Epting (1989), platform growth ended with a rapid rise in relative sea level accompanied by clastic input from the hinterland of Borneo.

The EX Platform has been discussed in Epting (1980, 1989) and Aigner et al. (1989). Epting (1980)

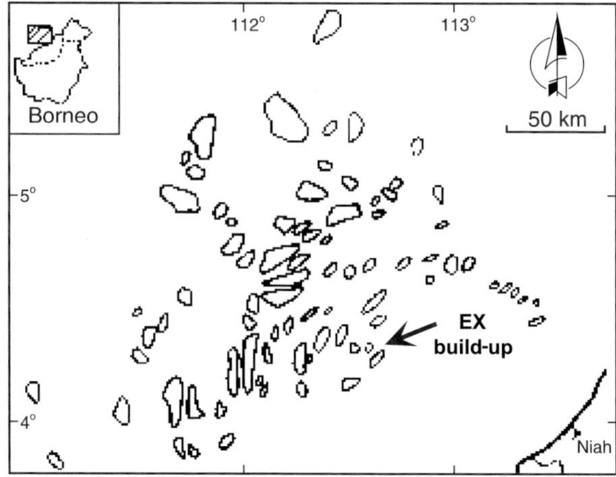

FIGURE 1. Map of the major carbonate platforms in Central Luconia (modified after Epting, 1980).

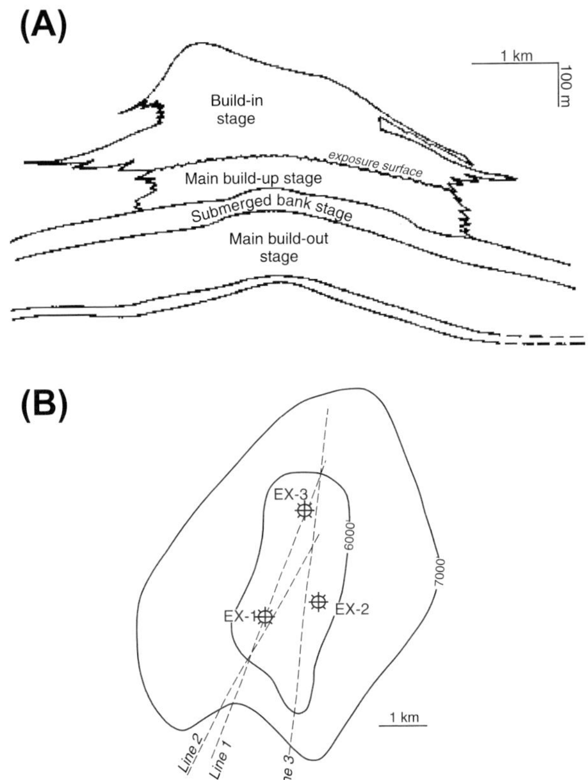

Figure 2. (A) Interpretation of platform development based on the geometry seen on seismic line 1 (see B) (Epting, 1980, 1989). (B) Subsea depth contour lines to top carbonate of the studied platform with the positions of three wells and three seismic lines (modified after Epting, 1989).

provides a detailed model of the platform's growth history, which was controlled by carbonate production, subsidence, sea level fluctuations, and clastic input. In the first of four carbonate growth stages, the "main build-out stage," carbonate production exceeded the relative rise in sea level resulting in platform expansion (Figure 2A). During the next stage (submerged bank stage), the platform was drowned and carbonate production was halted. Production resumed in the third stage and was in equilibrium with the relative rise in sea level. This "main build-up stage" is characterized by near-vertical aggrading margins of the platform. Finally, during the "build-in stage," relative rise in sea level slowly exceeded the rate of carbonate production resulting in retrogradation, and ultimately drowning of the platform (Figure 2A).

DATA SET AND METHODS

The data set included three 2-dimensional (2-D) poststack-migrated seismic sections (lines 1–3), and wire-line logs, cores, and thin sections from two wells, EX-2 and EX-3 (Figure 2B). Well EX-1 was never cored and we did not have access to log data. The only difference with Epting's data set is that the seismic data used herein were reprocessed in 1993 and 1999. Main enhancements were a better dip preservation (e.g., by dip moveout and velocity analysis, normal moveout correction, and Kirchoff migration) and an increase in resolution by zero phasing. Additionally, true amplitude recovery improved the use of seismic reflection-strength displays. We calibrated core data to the wire-line response of gamma-ray, sonic-velocity, density, neutron-porosity, and caliper logs to predict zones of platform-derived debris in intervals without core control. Based on these petrophysical characteristics, the seismic characteristics of the breccia intervals were defined. Additionally, we generated a synthetic seismogram from the EX-3 wire-line-log data. With the velocity-log data, the depth data were converted into traveltime data. The impedance profile was then resampled at an interval of 6 ms (the sample rate of the seismic acquisition) and transformed into a reflection-coefficient log. To create the vertical synthetic seismic profile, the reflection-coefficient log was convolved with the source signature of the subsurface seismic data (35-Hz wavelet). Attenuation processes, including energy loss with depth through transmission, spherical divergence, and absorption, have all been omitted from the modeling process. These factors almost certainly reduce the quality of the deeper parts of seismic profiles. The model also assumes normal incident waves and no significant multiple-reflection generation.

RESULTS

Lithology

Figure 3 shows lithologic sections of wells EX-2 and EX-3 subdivided into the four platform development stages of Epting (1980). The first carbonates at the base of section EX-2 consist of lime mudstone overlain by a 10-m-thick sulfurous shale interval. The overlying "main build-out stage" reaches a thickness of 180 m in EX-2. The interval starts with meter-thick wackestones with platy corals and platform-derived breccia intervals followed by heavily dolomitized grainstones with a common occurrence of boundstone. Dolomitization decreases toward the top of the interval with an increase of the planktonic foraminifera content. The 20-m-thick "submerged bank stage" has an erosional base. It is overlain by a bioturbated argillaceous lime mudstone with abundant red algae (rhodolites) fragments. The "main build-up stage" (130 m) consists of boundstones with numerous colonies of massive and branching corals that alternate

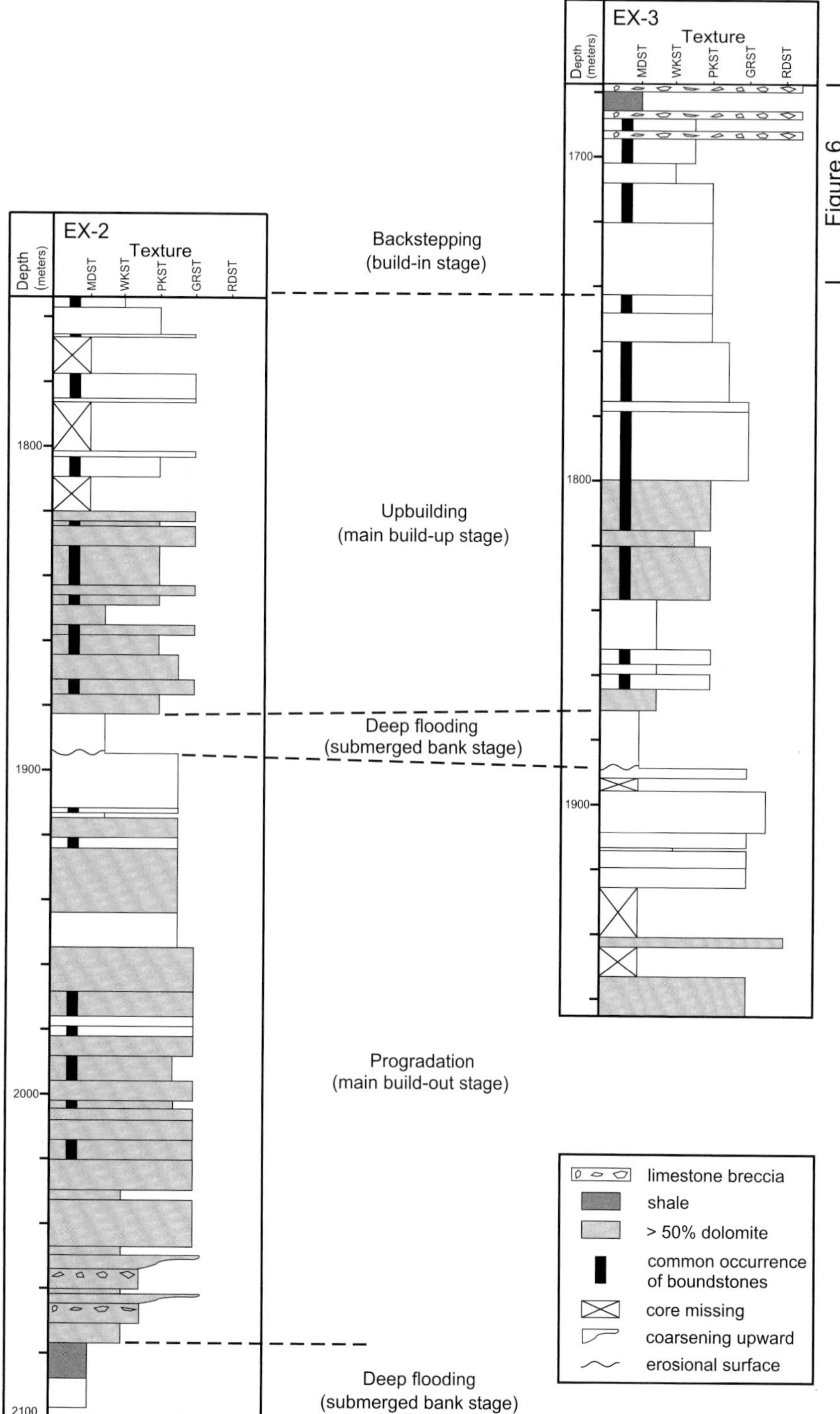

FIGURE 3. Sedimentary logs of wells EX-2 and EX-3 showing lithology, texture, and main stages of platform growth. Stage names between brackets are compared to those of Epting (1980). Details on the uppermost limestone breccia interval in EX-3 are shown in Figure 6. MDST = mudstone; WKST = wackestone; PKST = packstone; GRST = grainstone; RDST = rudstone.

with foraminifera-rich packstone and grainstone layers. The "build-in stage" was cored for 60 m in EX-3, but is estimated to have a maximum thickness of 250 m at the highest point of the platform, based on seismic. Foraminifera-rich packstones alternate with in-situ boundstones.

Distinct intervals of slope debris can be found in the top of EX-3. Lithoclasts of grainstone, packstone, and wackestone with diameters ranging from 0.5 to 5 cm are components of these breccia intervals (Figure 4). The lithoclasts of these breccias are derived from the platform margin. They contain coral fragments, bivalves, bryozoans, echinoderms, red algae, and benthic foraminifera, such as *Amphistegina*, *Rotalia*, and *Elphidium* (Figure 5). Fine calcareous shale separates these breccia intervals.

Log Petrophysics and Synthetic Seismogram

From the gamma-ray log, a clear differentiation can be made between the carbonates and their siliciclastic cover (Figure 6). The gamma-ray response immediately drops below 50 API units on entering the carbonate interval below 1670 m. A few small intervals containing shale or fine carbonate mud cause the gamma-ray response to increase above 50 API units again. Below 1693 m, the lithology is mainly carbonate. Three limestone breccia intervals with a minimum thickness of 1 m were recognized at the top of core EX-3. Figure 6 shows that these intervals have the following log characteristics: low natural gamma ray (<45 API units), high velocity (>3500 m/s), high density (>2.3 g/cm^3), low neutron porosity (<20%), low caliper, and small thin spikes on the sonic and density logs. Just above the cored interval, we found two zones with equal petrophysical characteristics, which we consequently interpreted as limestone breccias.

FIGURE 4. Examples of limestone breccia in the upper part of core EX-3. Clasts (c) consist of grainstone, packstone, and wackestone from the platform. Depth is "driller's depth" in meters.

Figure 7 shows the synthetic seismic section plus the impedance log in time of well EX-3. The marked breccia-shale interval at 1670 m core depth, which has a total thickness of 24 m, corresponds to one single positive reflection at 1465 ms. Reflections inside the platform are an indication of impedance differences caused mainly by variations in porosity. The positive reflection at 1575 ms marks an impedance change caused by a lithologic transition from argillaceous lime mudstone to boundstone (Figure 7).

SEISMIC INTERPRETATION AND DISCUSSION

We have used amplitude, spacing of reflections, reflection strength, as well as lapout patterns to determine the outlines of the EX Platform (Figure 8). The platform is seismically characterized by high-amplitude reflections with a spacing of 40–60 ms. Reflections in the covering shale are of low amplitude and spaced 15–20 ms apart with intermediate amplitude events at 60–90 ms. Differences in amplitude are clearly visible in the reflection-strength display (Figure 8). Reflection strength filters the amplitude from the complex trace and provides a good indication of the distribution of the two major lithologies (Taner and Sheriff, 1977; Taner et al., 1979). The boundaries between the two reflection domains are mostly sharp. Reflections tend to be discontinuous at this boundary, with abrupt changes in dip (amplitude display, Figure 8A) and amplitude (reflection-strength display, Figure 8B). We interpret this boundary as the outline of the platform. Seismic models of outcrops in similar settings support the validity of this approach (Bracco Gartner and Schlager, 1999).

(A)

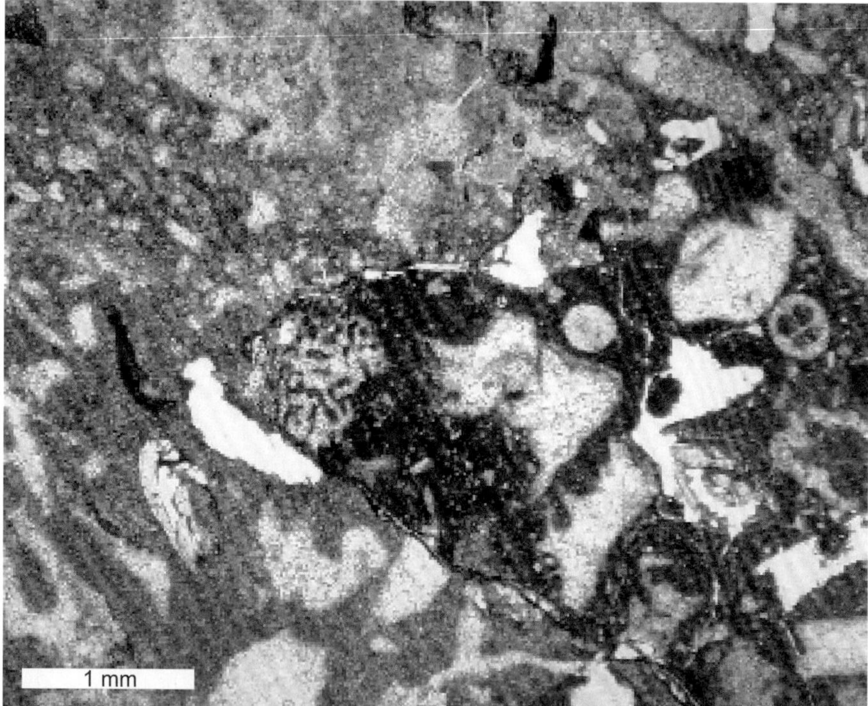

(B)

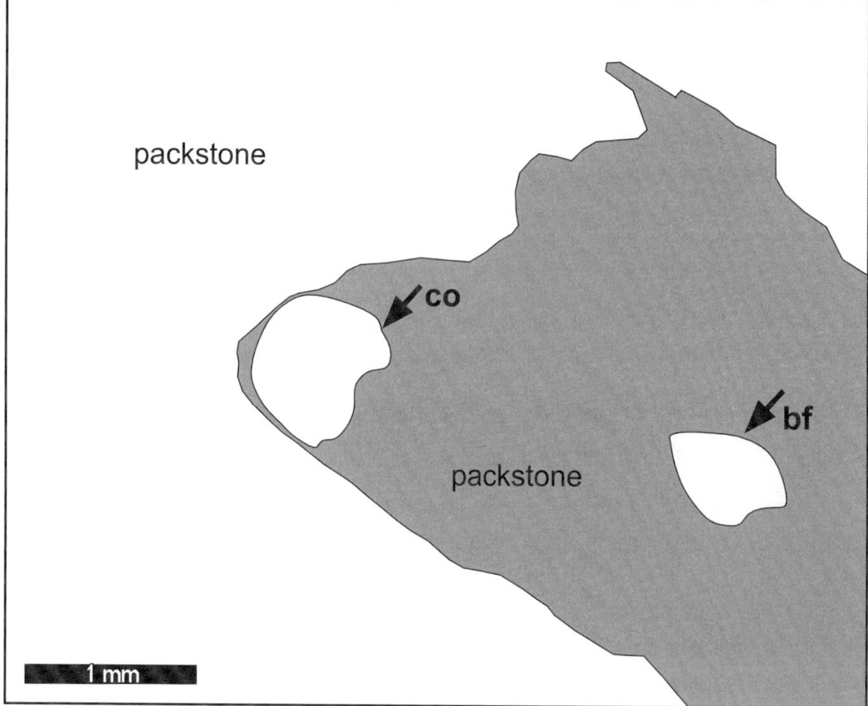

FIGURE 5. (A) Thin section from a breccia interval at a depth of 1681 m. (B) Line drawing of A showing a packstone lithoclast (right) with coral fragments (co), benthic foraminifera (bf), and other bioclasts in a dark micrite matrix surrounded by a packstone with a light-colored micrite matrix.

The outline of the platform in cross section shows significant variation in slope angle (Figure 8A). Estimates of slope angles are fairly reliable because the covering shale is rather homogeneous such that pull-up effects are minor. However, the sparse 2-D nature of the seismic coverage still causes some uncertainties in the true slope angles because of possible oblique slope intersection as well as "side-swipe" effects. The southern side has a uniform angle of 13°. The northern side of the platform starts at the top with a slope of 1–2°, becomes horizontal for about 350 m, increases to 11° for 650 m, and finally reaches a maximum of 25°. We will discuss the southern and northern flanks separately.

Figure 9 shows two lines across the southern flank of the platform. Line 1 shows a vertical zone with noise, between traces 195 and 370, most likely caused by gas. The result is that the reflections are significantly disturbed. The reflection geometries are better visible in the second line (line 2) as a series of downbending reflections that onlap the platform. Similar reflections on a well-documented Miocene platform on the Marion Plateau are demonstrably younger hemipelagic sediments that prograded toward the platform and buried it after scouring by ocean currents had ceased (Figure 10; Davies et al., 1991; Pigram et al., 1993; Isern et al., 2002; Isern et al., 2004). The geometric similarity between the EX Platform and the one on the Marion Plateau is striking. We believe that the interpretation of a steep platform slope that was first swept clean by bottom currents and then buried by prograding clastics is highly probable also for the EX Platform. An alternative interpretation could be that slumping causes these reflections. However, this is unlikely

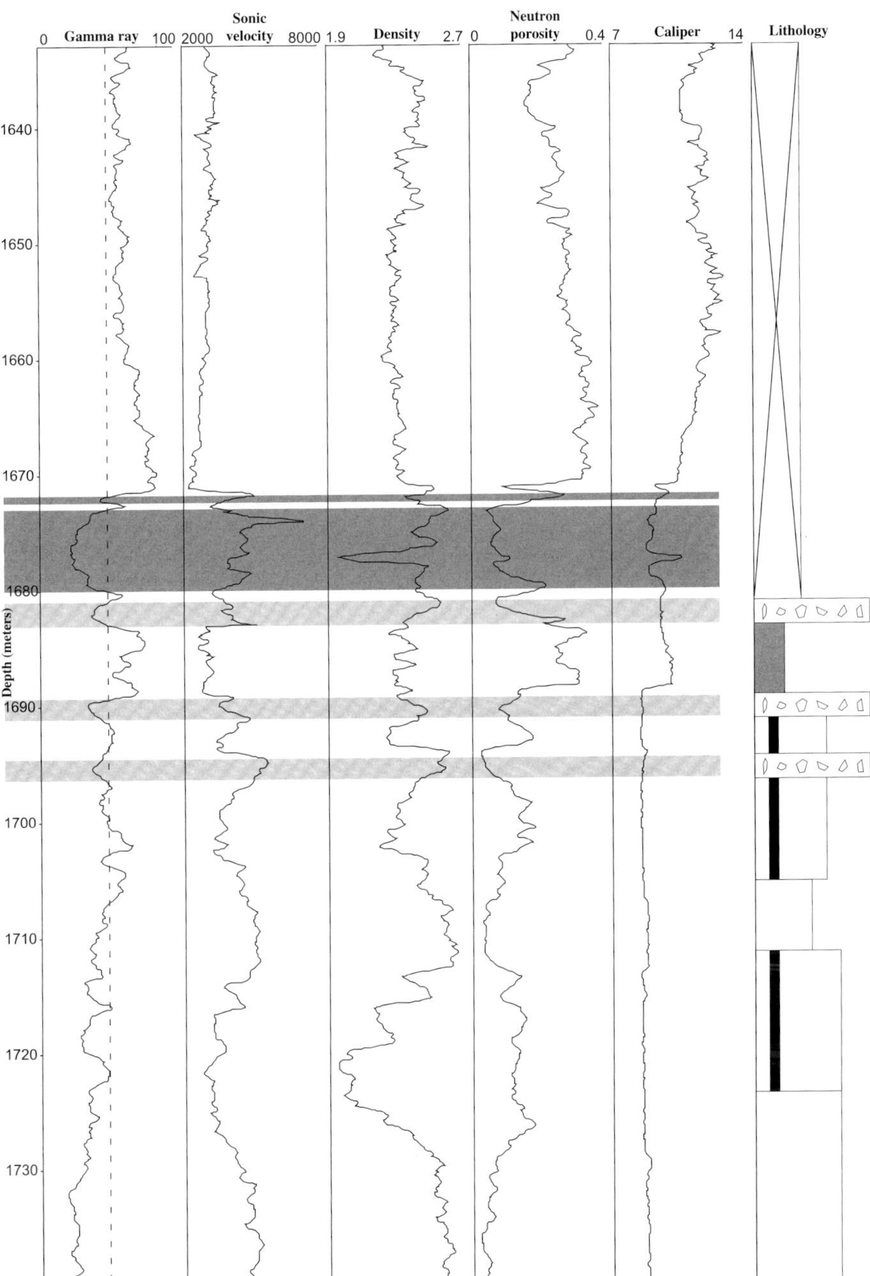

FIGURE 6. Five wire-line logs from well EX-3 and the top of the sedimentary log of Figure 3. In the gamma-ray log, the "shale-line" at 50 API units (dashed) is used to separate the carbonate interval from the siliciclastic cover. The characteristics of the breccia intervals with a minimum thickness of 1 m are displayed in five different petrophysical logs. Light-gray intervals indicate the three breccia intervals recognized in the cores. The dark gray zones have not been cored, but are interpreted as breccias because of their petrophysical characteristics. See Figure 3 for the key to the sedimentary log.

because the layering of the presumed slump is smooth and undisturbed. Second, there are many successive layers between 1.40 and 1.65 s that have the same gradual thickening trend toward the platform, thus filling the trough that had formed at about 1.60 s (Figure 9).

The northern flank shows a backstepping system that is less steep overall than the southern one but includes steps that are steeper than the southern flank. Two seismic lines across the northern flank, lines 1 and 3 (Figure 2B), show horizontal platform interior reflections truncated by a steeply inclined reflection (Figure 11). Below 1.5 s, the angle of the reflection reaches 25° (black arrow in Figure 11A). We interpret this situation as a bypass or an erosional slope analogous to slopes in recent Bahamian platforms (Schlager and Camber, 1986; Grammer et al., 1993). If slope angles exceed 12–15°, sediment gravity flows containing platform material in a muddy matrix may become so vigorous that they start to erode the slope. These sediment gravity flows commonly come to rest at the toe of slope and form an apron (Mullins and Neumann, 1979; Schlager and Chermak, 1979). Only sand- and gravel-rich depositional systems can build steeper slopes (Kenter, 1990). Surveys of modern and ancient examples show that most carbonate systems are mud-dominated below the fair-weather wave base, where nannoplankton and aragonitic mud are the fine fraction. Noncohesive slopes are uncommon. An enlargement of the part of the northern-slope seismics clearly shows onlap geometry and truncated platform reflections (Figure 12A). Together with the presence of lens-shaped reflections to the right of the slope reflection, this suggests the presence of an erosional or bypass slope with an adjacent debris apron. The most prominent lens-shaped reflections are negative reflections at 1.48 s/trace 459 and 1.56 s/trace 487. They are visible on the reflection-strength display (arrows in Figure 8B). The associated positive events, interpreted as the breccia intervals, also rapidly

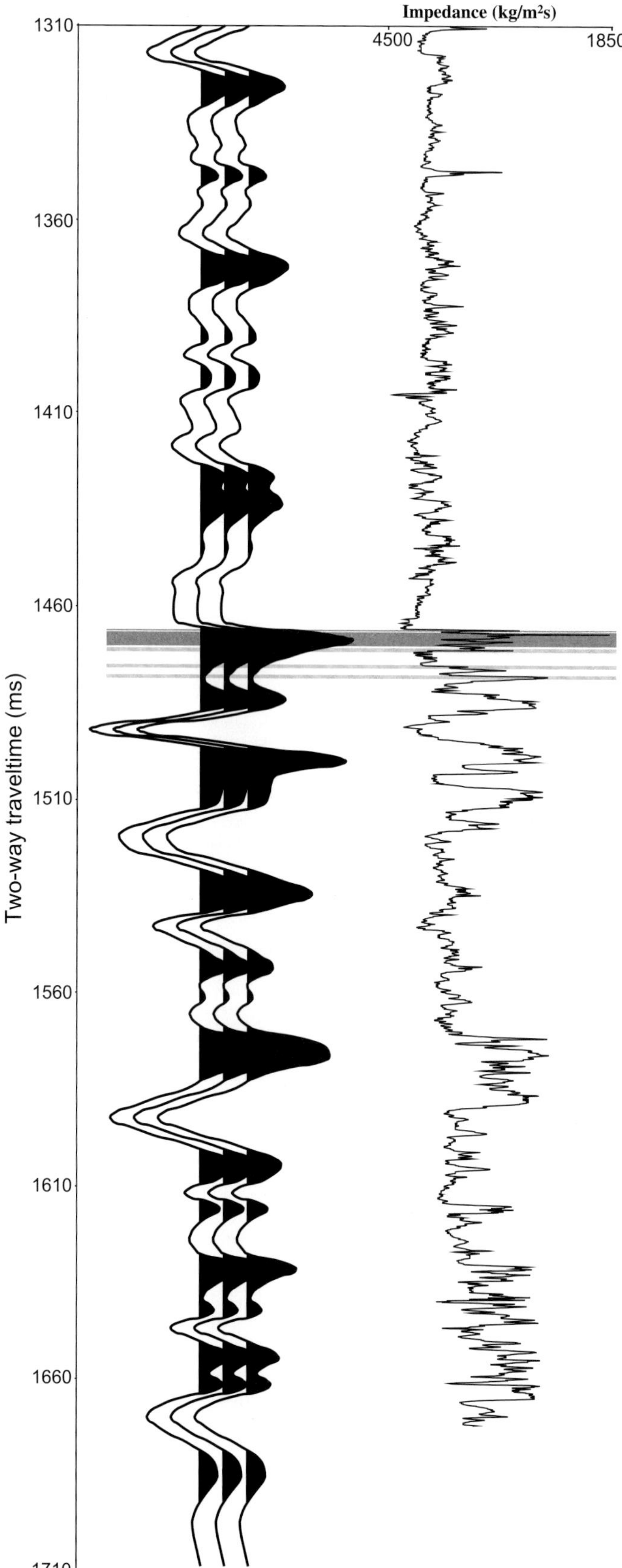

FIGURE 7. Synthetic seismic section (35 Hz) and impedance log. Horizontal gray bands over the logs indicate the breccia intervals from Figure 6. This breccia zone is visible on the synthetics by one single positive reflection. Vertical axis is scaled in time for both logs, which corresponds to a depth of 1433–2089 m in units of time (driller's depth).

decrease in amplitude away from the platform. We interpret this amplitude and reflection pattern as two stacked erosional or bypass slopes with debris tongues at their toe (Figure 13C). The high-amplitude reflections disappear in the platform direction because the shales wedge out. In the off-platform direction, the amplitudes decrease because the breccias thin out. The result is that there is a physical separation of the toe-of-slope debris from the source on the platform, which is characteristic for bypass-erosional slopes. Alternatively, bypass conditions could have been created because the slope was for a while an oversteepened slump scar.

It is possible that the high-angle slope reflection at 1.56 ms is accentuated by a normal fault (between white arrows in Figure 11). This fault would have an approximate throw of 20 m and could be the result of the compaction of shales intercalated between breccia tongues.

Two debris tongues can be interpreted by the terminating reflection at 1.43 ms and the "bridge" reflection at 1.47 ms (Figures 11, 12A). This second reflection corresponds to the single positive reflection of the synthetic seismic trace that correlates to the breccia interval located in the core (arrow in Figure 12B). This reflection is mainly caused by the youngest shale-to-carbonate transition.

Viewed overall, the platform geometry shows a persistent asymmetry with alternations of erosion and interfingering in the north and continuous bypass or erosion with subsequent onlap in the south. The slope geometry of the southern flank of the platform shows characteristics similar to the Miocene Liuhua Platform in the Pearl River Mouth Basin (Erlich et al., 1990, 1993). On both platforms, flank reflections steepen and are truncated in a basinward direction, suggesting an erosional character of the slope before burial. This could have occurred under bypass conditions or by severe erosion of the slope (Figure 13A, B). Prevailing winds or currents may have caused north-south asymmetry of the platform. The elevated rim and stationary growth on the south side suggests that this side was facing windward. Numerous studies have revealed the influence of wind on platform architecture and the distribution of

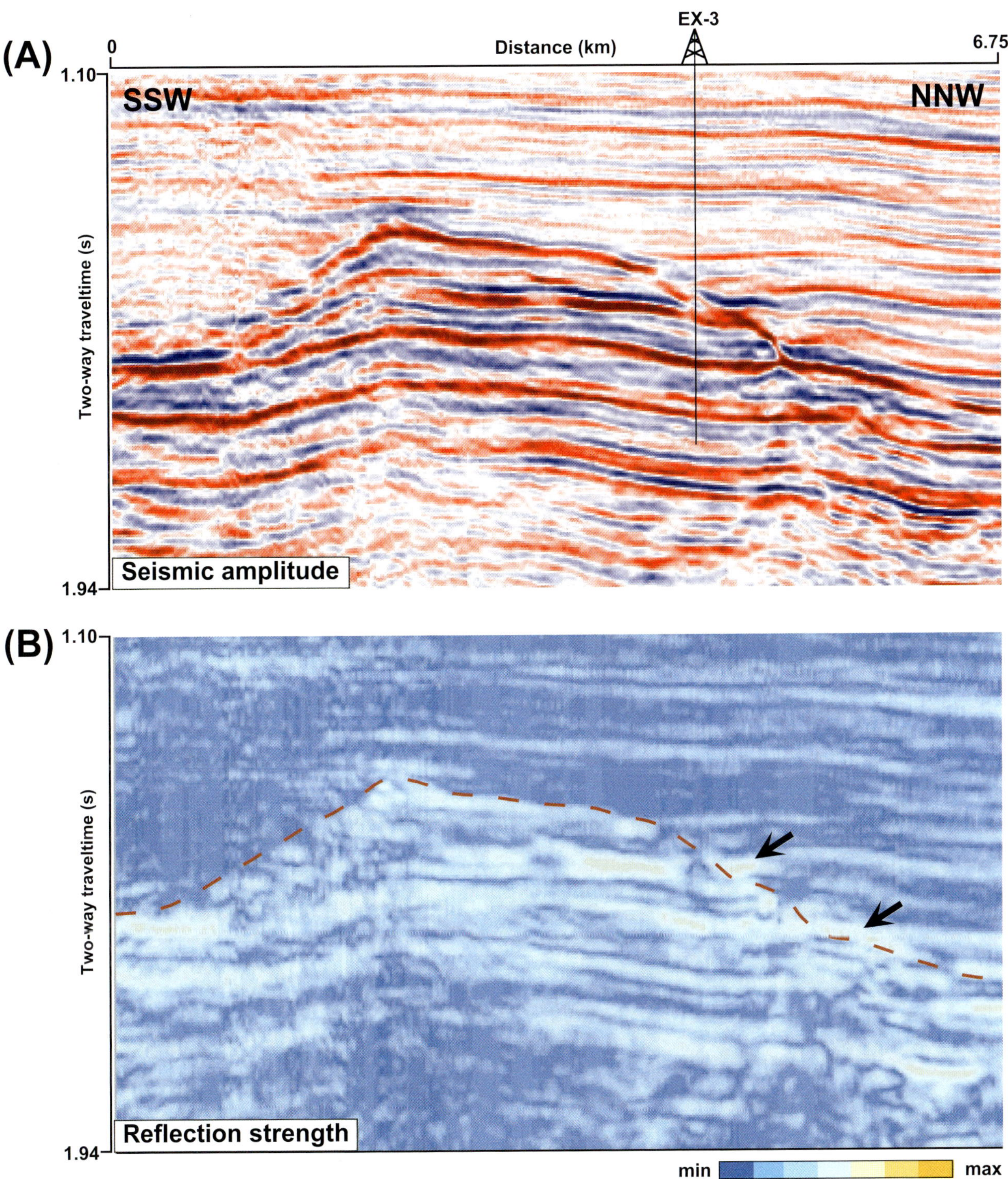

Figure 8. (A) Seismic-amplitude and (B) reflection-strength displays of line 1 and the position of well EX-3. The dashed line marks the outline of the platform. Except for two high-amplitude wedges (arrows) caused by a high shale-breccia impedance contrast, the boundary between high and low reflection magnitude is clearly visible.

sediments within platforms (e.g., Purser, 1973; Hine and Neumann, 1977; Eberli and Ginsburg, 1989). During the middle Miocene, the Luconia Province had a monsoonal wind pattern similar to today, with southwesterly winds during summer and northwesterly winds during the winter (Vahrenkamp, 1996, 1998). In our working hypothesis, this wind pattern would have been dominated by the summer monsoon. This

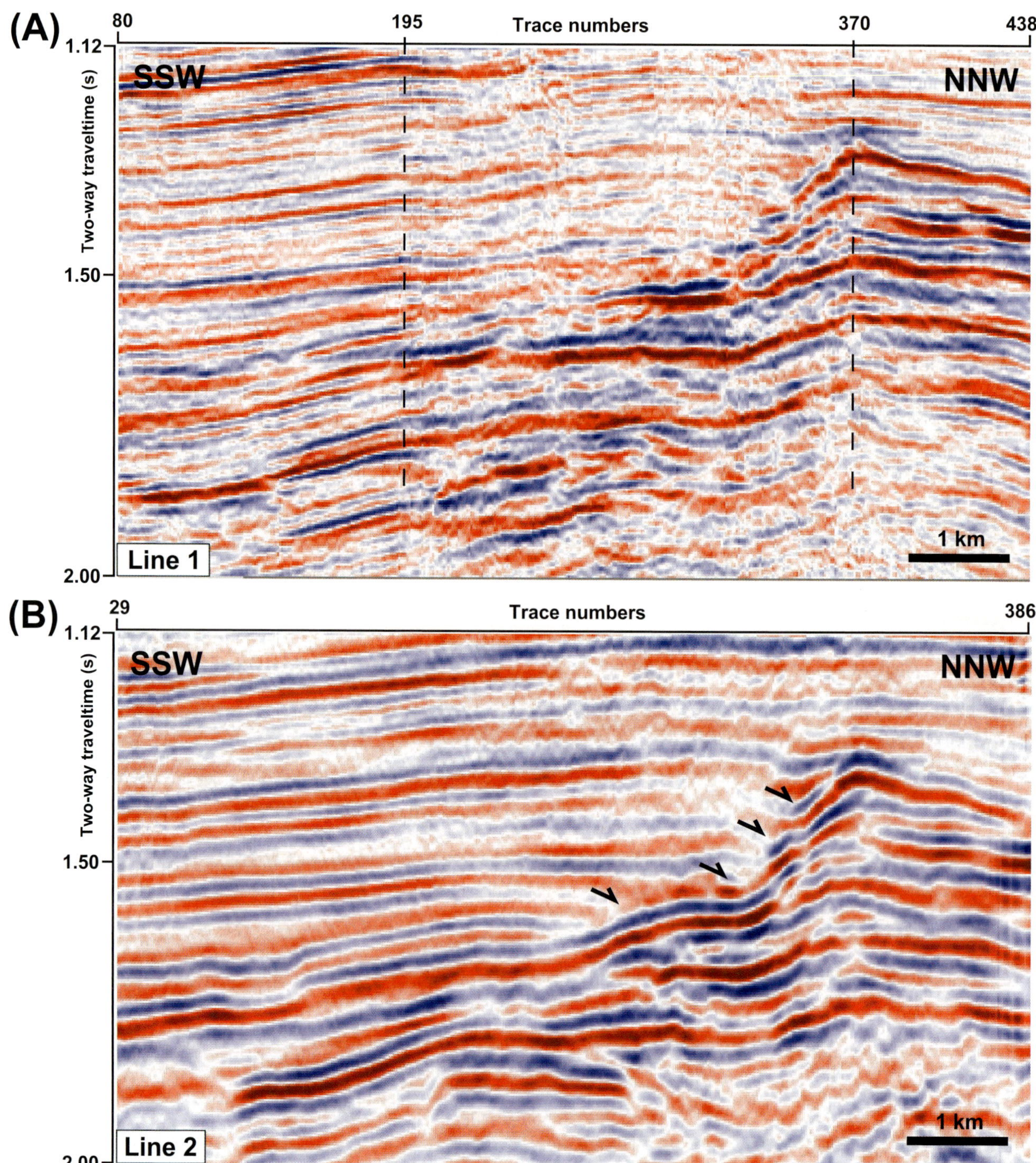

FIGURE 9. The southern flank of the platform in lines 1 and 2. (A) Line 1 shows a broad band of noise between traces 195 and 370 (dashed lines), which disturbs the basinal reflections. This band of noise is interpreted as "gas effect." (B) In line 2, the basinal reflections are seen as a series of downbending reflections that onlap against the platform (arrows).

situation is similar to the Liuhua Platform (Erlich et al., 1993). Both platforms show the same asymmetrical growth history in such that the platform shrinkage occurred in several stages. An alternative (or addition) to wind are ocean currents related to the water exchange between the Pacific and Indian Ocean. For the Miocene, these currents are poorly known, but it is probable that the Pacific Ocean was donating a significant volume of surface water to the Indian Ocean then as now.

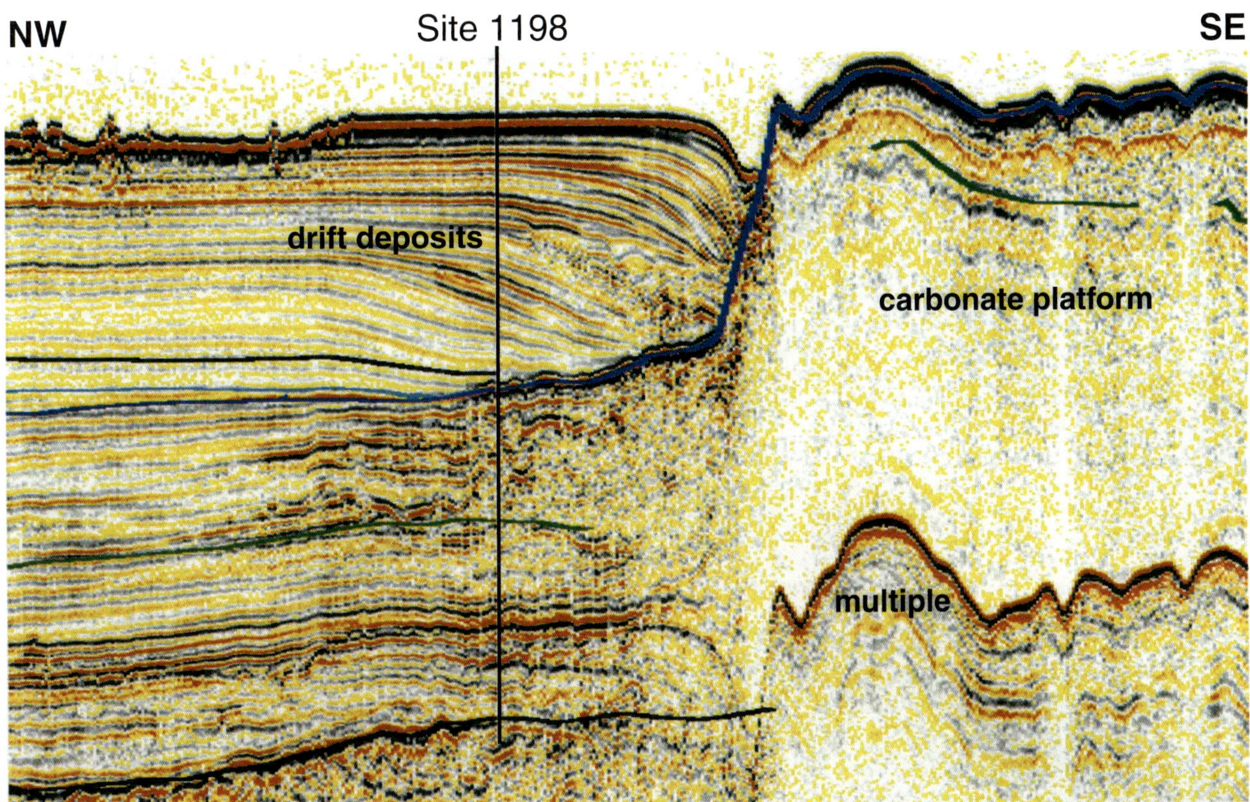

FIGURE 10. Seismic line across the Miocene platform on the Marion Plateau off eastern Australia with Ocean Drilling Project Site 1198. The northwestern part of the platform shows downbending onlap similar to the southern flank of the studied platform. Onlap interpretation on the Marion Plateau was proved by core dating of the burying sediments on sites north of the platform. Here, a strong ocean current might have transported both planktonic carbonates and fine-grained terrigenous detritus. Sedimentation and scouring occurred where additional turbulence was created by the interaction of the ocean currents with the sharp platform topography (modified after Isern et al., 2004).

Comparison with Epting's Interpretation

Figure 14 shows a comparison of our interpretation (Figure 14B) and that of Epting (1980, 1989) (Figure 14C). There are three key differences. The first difference is the continuous southern slope in our interpretation vs. the presence of long debris tongues in the Epting model. This is mainly because of the fact that Epting based his interpretation almost solely on a single seismic line (line 1 on Figure 9). On that line, some reflections on the southern flank could indeed be interpreted as "carbonate stringers." However, the "gas effect" (line 1), the clear downbending onlap pattern (line 2), and the similarity with the coeval platform on the Marion Plateau strongly support our interpretation. The second difference is the "main build-up stage" that is truncated in the north by partial (bypass) or complete erosion of the slope in our interpretation. The third difference is the initial retreat of the "build-in stage" in our interpretation vs. a prograding stringer in the Epting model.

These differences are partly based on better reflection definitions, more recently processed seismic data, and the use of color displays. Furthermore, we used two sedimentologic concepts about the EX Platform that were published after Epting's initial paper. First is the concept of bypass and erosional slopes, where breccia tongues do not need to be laterally continuous with the main part of the platform. The second concept is that of the downbending onlap reflections. The high-amplitude positive reflection at 1.5 s marks the top of the "main build-up stage," but is more likely to be caused by lithologic transition from boundstone to packstone than by an exposure surface. Finally, from the available data, we do not see a reason for assuming an inclined shape for the "submerged bank stage." In both wells, we found this horizon at exactly the same depth. The limited data that we had available

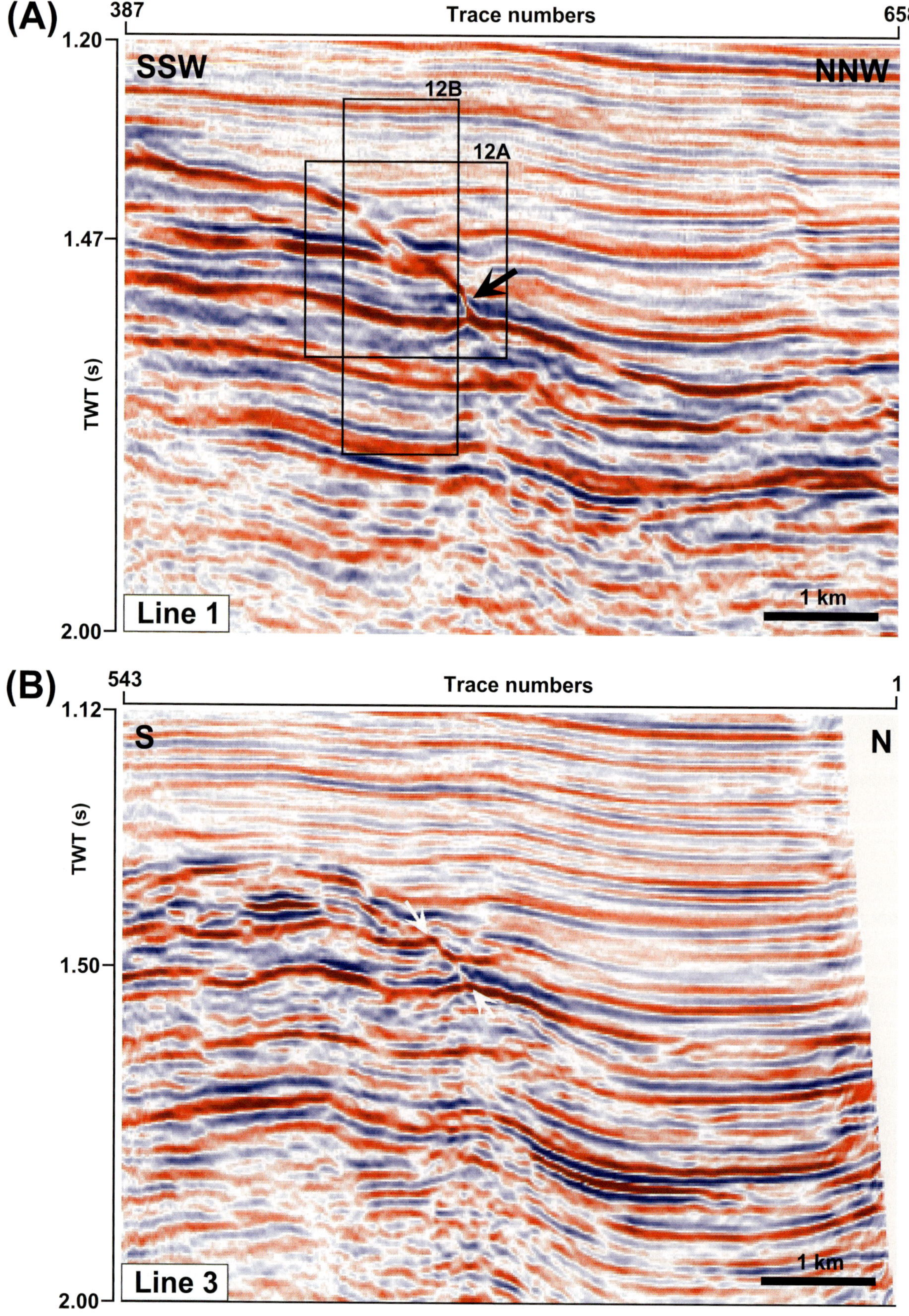

FIGURE 11. The northern flank of the platform seen in lines (A) 1 and (B) 3 on the same scale. The steep positive reflection (black arrow on line 1) is interpreted as an erosional slope (see Figure 13C), although a small normal fault is a plausible alternative (between white arrows on line 3). Boxes indicate the positions of the blow-up displays in Figure 12.

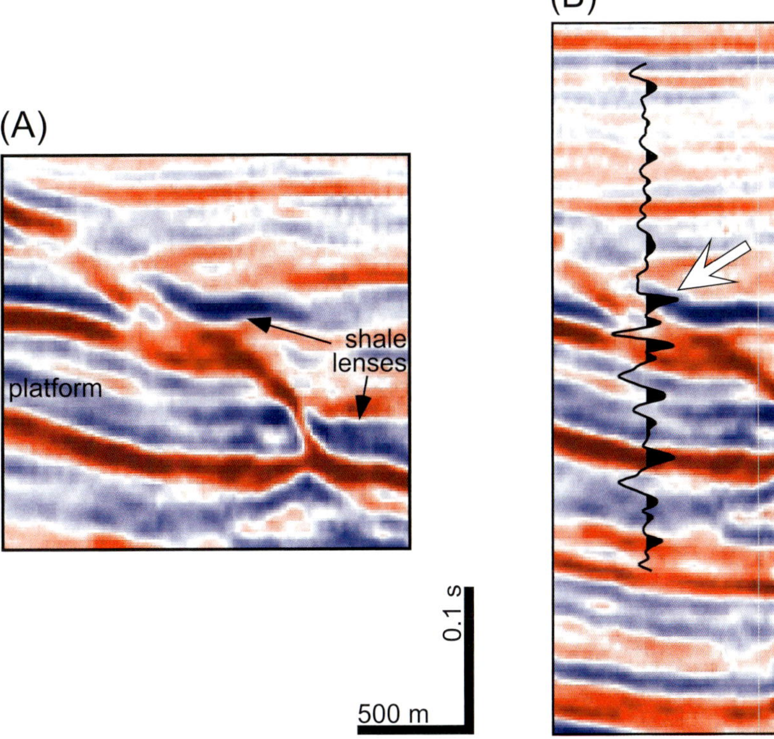

FIGURE 12. Enlargements of the seismic data outlined by the boxes in Figure 11. (A) Discontinuity of the reflections. Platform reflectors are clearly separated from basin reflectors. (B) The position of the synthetic trace and the reflection interpreted as breccia (white arrow).

makes a pull-up on the time section of the "submerged bank stage" the most plausible explanation.

CONCLUSIONS

Lithologic interfingering and onlapping geometries were found on slopes of an asymmetric carbonate platform. The southern slope of the EX Platform was characterized by bypass or erosion throughout the aggrading phase of platform development. The drift deposits that subsequently buried the slope show downbending, onlapping reflectors on seismic. On the northern flank, the shale started to pile up during platform aggradation. Phases of erosional or bypass conditions are short and alternate with two phases in which platform debris interfingered with the surrounding shale. Shale intercalations can be recognized seismically on both seismic-amplitude and reflection-strength displays by negative reflections that quickly lose amplitude away from the platform. Bypassing might also have occurred because the slope was for a while an oversteepened slump scar.

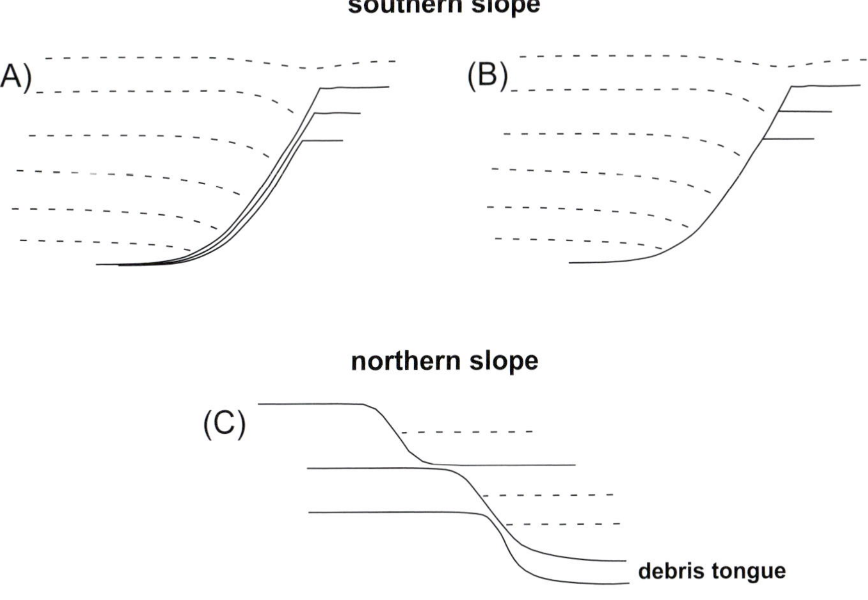

FIGURE 13. Genetic depositional models of the slopes of the studied platform. (A, B) Two interpretive line drawings for the southern flank show two consecutive stages of slope development and basin infill. (A) This assumes aggradation of the platform margin and bypass conditions, followed by a series of downbending onlap of basinal sediments. (B) This assumes severe erosion on the slope. (C) The northern flank of the platform with slope angles between 11° and 25° is either a bypass slope with sedimentary controlled onlap or is truncated by erosion. Both hypotheses will cause the high-angle reflections as seen on the northern flank seismic (black arrow in Figure 11).

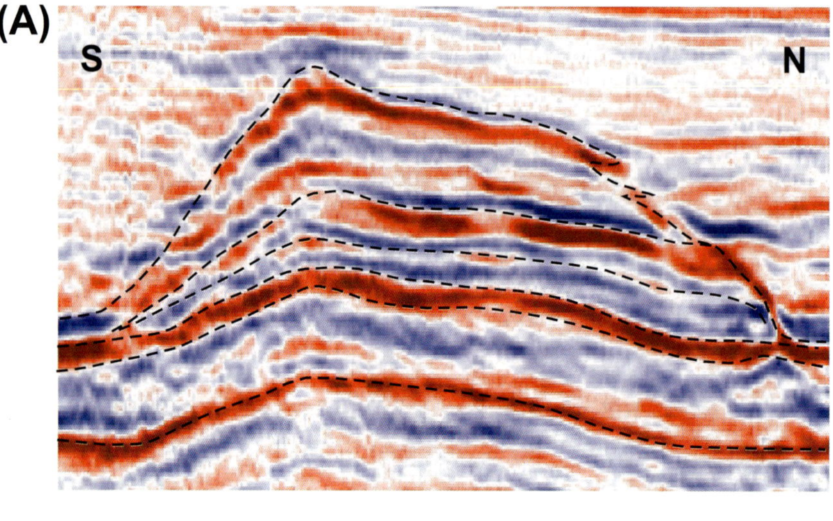

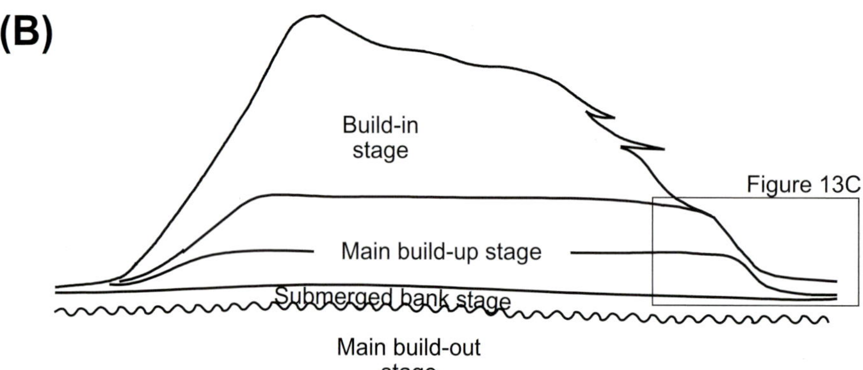

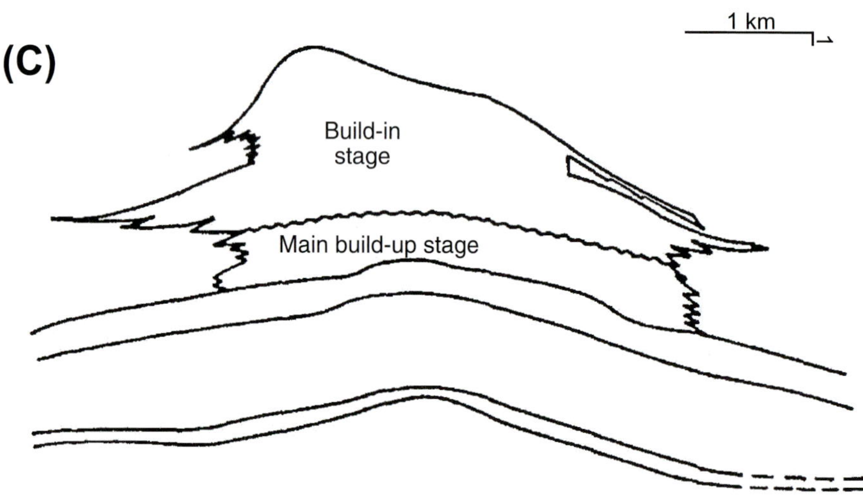

Figure 14. Interpretations of line 1 in (A) time and (B, C) depth. (A, B) Result of this study. (C) Model by Epting (1989) on the same scale, slightly modified. Epting's model assumes significantly more interfingering than the interpretation proposed here, especially on the southern flank.

ACKNOWLEDGMENTS

We thank Petronas Nasional Berhad, especially Mr. Chua Beng Yap and Mr. Muhammad Adib, for the release of the data and the permission to publish this study. Mr. Tan E. Kim and his crew from the core shed are also gratefully acknowledged. Arnout-Jan Everts, Volker Vahrenkamp, and Guy Mueller (Sarawak Shell Berhad) gave important advice on studying both the cores and the seismics. Reviewers Xavier Janson and Chris Kendall are thanked for helpful comments. This manuscript also benefited from earlier reviews by Steve Bachtel, Moyra Wilson, Art Saller, Rick Sarg, and David Budd.

REFERENCES CITED

Aigner, T., M. Doyle, D. Lawrence, M. Epting, and A. Van Vliet, 1989, Quantitative modeling of carbonate platforms: Some examples, *in* P. D. Crevello, J. L. Wilson, J. F. Sarg, and J. F. Read, eds., Controls on carbonate platform and basin development: SEPM Special Publication 44, p. 27–37.

Biddle, K. T., W. Schlager, K. W. Rudolph, and T. L. Bush, 1992, Seismic model of a progradational carbonate platform, Picco di Vallandro, the Dolomites, northern Italy: AAPG Bulletin, v. 76, p. 14–30.

Bracco Gartner, G. L., and W. Schlager, 1999, Discrimination between onlap and lithologic interfingering in seismic models of outcrops: AAPG Bulletin, v. 83, p. 952–971.

Davies, P. J., J. A. McKenzie, A. Palmer-Julson, et al., 1991, Proceedings of the Ocean Drilling Program, Initial Report 133: Ocean Drilling Program, Texas A&M University, College Station, Texas, 810 p.

Eberli, G. P., and R. N. Ginsburg, 1989, Cenozoic progradation of Northwestern Great Bahama Bank, a record of lateral platform progradation and sea-level fluctuations, *in* P. D. Crevello, J. L. Wilson, J. F. Sarg, and J. F. Read, eds., Controls on carbonate platform and basin development: SEPM Special Publication 44, p. 339–352.

Epting, M., 1980, Sedimentology of Miocene carbonate buildups, Central Luconia, offshore Sarawak: Geological Society of Malaysia Bulletin, v. 12, p. 17–30.

Epting, M., 1989, Miocene carbonate buildups of central Luconia, offshore Sarawak, *in* A. W. Bally, ed., Atlas of seismic stratigraphy, v. 3: AAPG Studies in Geology No. 27, p. 168–173.

Erlich, R. N., S. F. Barrett, and G. B. Ju, 1990, Seismic and geologic characteristics of drowning events on carbonate platforms: AAPG Bulletin, v. 74, p. 1523–1537.

Erlich, R. N., A. P. Longo Jr., and S. Hyare, 1993, Response of carbonate platform margins to drowning: Evidence of environmental collapse, *in* R. G. Loucks and J. F. Sarg, eds., Carbonate sequence stratigraphy: Recent developments and applications: AAPG Memoir 56, p. 241–266.

Grammer, G. M., R. N. Ginsburg, and P. M. Harris, 1993, Timing of deposition, diagenesis, and failure of steep carbonate slopes in response to a high-amplitude/high-frequency fluctuation in sea level, Tongue of the Ocean, Bahamas, *in* R. G. Loucks and J. F. Sarg, eds., Carbonate sequence stratigraphy: Recent developments and applications: AAPG Memoir 56, p. 107–131.

Hine, A. C., and A. C. Neumann, 1977, Shallow carbonate-bank margin growth and structure, Little Bahama Bank, Bahamas: AAPG Bulletin, v. 61, p. 376–406.

Isern, A. R., F. S. Anselmetti, P. Blum, et al., 2002, Proceedings of the Ocean Drilling Program, Initial Report 194: Ocean Drilling Program, Texas A&M University, College Station, Texas, CD-ROM.

Isern, A. R., F. S. Anselmetti, and P. Blum, 2004, A Neogene carbonate platform, slope, and shelf edifice shaped by sea level and ocean currents, Marion Plateau (Northeast Australia), *in* G. P. Eberli, J. L. Masaferro, and J. F. Sarg, eds., Seismic imaging of carbonate reservoirs and systems: AAPG Memoir 81, p. 291–308.

Kenter, J. A. M., 1990, Carbonate platform flanks: Slope angle and sediment fabric: Sedimentology, v. 37, p. 777–794.

Mullins, H. T., and A. C. Neumann, 1979, Deep carbonate bank margin structure and sedimentation in the northern Bahamas, *in* L. J. Doyle and O. H. Pilkey, eds., Geology of continental slopes: SEPM Special Publication 27, p. 165–192.

Pigram, C. J., P. J. Davies, and G. C. H. Chaproniere, 1993, Cement stratigraphy and the demise of the early-middle Miocene carbonate platform on the Marion Plateau, *in* J. A. McKenzie, P. J. Davies, and A. Palmer-Julson, eds., Proceedings of the Ocean Drilling Program, Scientific Result 133: Ocean Drilling Program, Texas A&M University, College Station, Texas, p. 499–512.

Purser, B. H., 1973, The Persian Gulf. Holocene carbonate sedimentation and diagenesis in a shallow epicontinental sea: New York, Springer-Verlag, 471 p.

Rudolph, K. W., W. Schlager, and K. T. Biddle, 1989, Seismic models of carbonate foreslope-to-basin transition, Picco di Vallandro, Dolomite Alps, northern Italy: Geology, v. 17, p. 453–456.

Schlager, W., and O. Camber, 1986, Submarine slope angles, drowning unconformities and self-erosion of limestone escarpments: Geology, v. 14, p. 762–765.

Schlager, W., and A. Chermak, 1979, Sediment facies of platform-basin transition, Tongue of the Ocean, Bahamas, *in* L. J. Doyle and O. H. Pilkey, eds., Geology of continental slopes: SEPM Special Publication 27, p. 193–208.

Schlager, W., K. T. Biddle, and J. Stafleu, 1991, Picco di Vallandro (Dürrenstein)— A platform-basin transition in outcrop and seismic model: Ortisei, Italy, Dolomieu Conference, Guidebook Excursion D, 22 p.

Stafleu, J., and W. Schlager, 1993, Pseudo-toplap in seismic models of the Schlern-Raibl contact (Sella Platform, northern Italy): Basin Research, v. 5, p. 55–65.

Taner, M. T., and R. E. Sheriff, 1977, Application of amplitude, frequency, and other attributes to stratigraphic and hydrocarbon determination, *in* C. E. Payton, ed., Seismic stratigraphy— Applications to hydrocarbon exploration: AAPG Memoir 26, p. 301–328.

Taner, M. T., F. Koehler, and R. E. Sheriff, 1979, Complex seismic trace analysis: Geophysics, v. 44, p. 1041–1063.

Vahrenkamp, V. C., 1996, Miocene carbonates of the Luconia Province, offshore Sarawak: Implications for regional geology and reservoir properties from Strontium isotope stratigraphy: Petronas Technology Forum Proceedings, 10 p.

Vahrenkamp, V. C., 1998, Sr-isotope stratigraphy of Miocene carbonates, Luconia Province, Sarawak, Malaysia: Implications for platform growth and demise and regional reservoir behavior (abs.): AAPG Annual Meeting Program, p. 4.

Index

D

E

F

G

H

I

J

K

L

M

N

O

P

Q

R

S

T

U

V

W

X

Y

Z